STATISTICA™

System Reference

D1721254

StatSoft®

www.statsoft.com

STATSOFT SOFTWARE LICENSE AGREEMENT

The following constitutes the terms of the License Agreement between a single user (User) of this software system, and the producer of the system, StatSoft, Inc. (called StatSoft hereafter). By opening the system, you (the User) are agreeing to become bound by the terms of this License Agreement. If you do not agree to the terms of this agreement do not open the system, and contact the StatSoft Customer Service Department at a local StatSoft office (or an authorized StatSoft reseller) in order to obtain an authorization number for the return of the system. This License Agreement pertains also to all third party software included in or distributed with StatSoft products.

License

Unless explicitly stated on the packaging of the program media (CD or disks) and/or the product registration card, the enclosed software system is sold to be used on one computer system by one user at a time. This License Agreement explicitly excludes renting or loaning the software system or any of its components. Unless explicitly stated on the program media, this License Agreement explicitly excludes the use of this system on multi-user systems, networks, any time sharing systems, or via the Internet. (Contact StatSoft concerning Multi-user License Programs and Web Server licensing.) The user is allowed to install the software system on one hard disk drive. However, the software will never be installed on more than one hard disk drive at a time. The documentation accompanying this software system (or any of its parts) shall not be copied or reproduced in any form.

Disclaimer of Warranty

Although producing error free software is obviously a goal of every software manufacturer, it can never be guaranteed that a software program is actually free of errors. Business and scientific application software is inherently complex (and it can be used with virtually unlimited numbers of data and command settings, producing idiosyncratic operational environments for the software); therefore, the User is cautioned to verify the results of his or her work. This software system is provided "as is" without warranty of any kind. StatSoft and distributors of StatSoft software products make no representation or warranties with respect to the contents of this software system and specifically disclaim any implied warranties or merchantability or fitness for any particular purpose. In no event shall StatSoft be liable for any damages whatsoever arising out of the use of, inability to use, or malfunctioning of this software system. StatSoft does not warrant that this software system will meet the User's requirements or that the operation of the software system will be uninterrupted or error free.

Limited Warranty

If within 30 days from the date when the software system was purchased (i.e., invoice date), the program media (CD or disks) are found to be defective (i.e., they are found to be unreadable by the properly aligned media drive of the computer system on which the system is intended to run), StatSoft will replace the media free of charge. After 30 days, the User will be charged for the replacement a nominal media replacement fee. If within 30 days from the date when the software system was purchased (i.e., invoice date), the software system was found by the User not capable of performing any of its main (i.e., basic) functions described explicitly in promotional materials published by StatSoft, StatSoft will provide the User with replacement media free of defects (or a replacement component downloadable from the StatSoft Web site), or if the replacement cannot be provided within 90 days from the date when StatSoft has acknowledged receiving notification from the User about the defect, the User will receive a refund of the purchasing price of the software system.

Updates, Corrections, Improvements

The User has a right to purchase all subsequent updates, new releases, new versions, and modifications of the software system introduced by StatSoft for an update fee or for a reduced price (depending on the scope of the modification). However, purchasing an update or upgrade (for a reduced price) constitutes a replacement of an existing license and not acquisition of a new license. StatSoft is not obligated to inform the User about new updates, improvements, modifications, and/or corrections of errors introduced to its software systems. In no event shall StatSoft be liable for any damages whatsoever arising out of the failure to notify the User about a known defect of the software system.

STATISTICA SYSTEM REFERENCE

Table of Contents

STATISTICA SYSTEM REFERENCE

Frequently Asked Questions (FAQ)
Detailed Table of Contents

CHAPTER

STATISTICA –
A GENERAL OVERVIEW

continued ➜

CHAPTER

STATISTICA – A GENERAL OVERVIEW

INTRODUCTION

STATISTICA is a comprehensive, integrated data analysis, graphics, database management, and custom application development system featuring a wide selection of basic and advanced analytic procedures for business, data mining, science, and engineering applications.

Analytic Facilities

STATISTICA includes not only general purpose statistical, graphical, and analytic data management procedures, but also comprehensive implementations of specialized methods for data analysis (e.g., data mining, business, social sciences, biomedical research, or engineering applications).

All analytic tools offered in the *STATISTICA* line of software are available to you as part of an integrated package. These tools can be controlled through a selection of alternative user interfaces and a comprehensive, industry standard (Visual Basic-based) programming language. Interactive user interfaces can be easily customized and the language (*STATISTICA* Visual Basic) can be used to automate tasks of any complexity, from simple macros recorded to automate a routine operation, to advanced, large scale development projects (e.g., custom analytic extensions that integrate *STATISTICA* with other applications or with large, enterprise-wide, Internet or intranet-wide computing environments).

Unique Features

Some of the unique features of the *STATISTICA* line of software include:

- the breadth of selection and comprehensiveness of implementation of analytical procedures,

- the unparalleled selection, quality, and customizability of graphics integrated seamlessly with every computational procedure,

- the efficient and user-friendly user interface,

- the fully integrated, industry standard *STATISTICA* Visual Basic that adds more than 10,000 new functions to the comprehensive syntax of Microsoft Visual Basic, thus comprising one of the most extensive development environments available, and

- a wide selection of advanced software technologies (see *Software Technology*, below) that are responsible for *STATISTICA*'s practically unlimited capacity, performance (speed, responsiveness), and application customization options.

One of the most unique and important features of the *STATISTICA* family of applications is that these technologies allow even inexperienced users to tailor *STATISTICA* to their specific preferences. You can customize practically every aspect of *STATISTICA,* including even the low-level procedures of its user interface. The same version of *STATISTICA* can be used:

- By novices to perform routine tasks using the default (e.g., **Quick**) analysis startup dialogs (containing just a few, self-explanatory buttons), or even by accessing *STATISTICA* with their Web browsers (and a highly simplified "front end"), and

- By experienced analysts, professional statisticians, and advanced application developers who can integrate any of *STATISTICA*'s highly optimized procedures (more than 10,000 functions) into custom applications or computing environments, using any of the cutting edge, object-oriented, and/or Web-embedded software technologies.

The General "Philosophy" of the *STATISTICA* Approach

STATISTICA's default configuration (its general user interface and system options) is a result of years of listening carefully to our users. We have received feedback from tens of thousands of our users, representing hundreds of thousands of users from all continents and, practically speaking, all walks of life. One of the most important facts that we learned from these users is

how different their needs and preferences are (both across individuals and projects or applications). In order to meet those differentiated needs, *STATISTICA* is designed to offer perhaps one of the most flexible and easily customizable user interfaces of any contemporary application.

Although *STATISTICA* provides access to a powerful arsenal of advanced software technologies (see *Software Technology*, below), you do not even need to know about them because they are designed to work automatically and intuitively. A novice user may never see more than a few self-explanatory buttons. Advanced options, however, are only one tab or mouse click away. Practically every aspect of *STATISTICA* (from the startup configuration, to the way the output is generated and managed by the system, to how *STATISTICA* prompts you to choose your next step) can be changed with a mouse click. Moreover, *STATISTICA* remembers your changes until you change your mind. Practically every dialog used to select an analysis or perform a routine operation can be easily replaced (e.g., simplified, enhanced, or combined with custom, user-designed procedures). *STATISTICA* will always look and work the way you want.

Software Technology
(A Technical Note)

The performance, flexible customizability, and the wide selection of options that can be tailored to your needs mentioned in the previous section would not be possible if *STATISTICA* did not feature the advanced technologies that drive all functions of the application. *STATISTICA* uses and/or supports virtually all the relevant leading edge software technologies available today. Every one of the thousands of *STATISTICA* features is accessible from the object model, and each can be accessed transparently for selective or global integration with other applications (e.g., using VB, C++, or Java). *STATISTICA* incorporates advanced implementations of the OLE/ActiveX technology in client and server mode for all document types; practically no limitations are imposed in terms of either the amount or complexity of data that can be stored. *STATISTICA* also is optimized for Web and multimedia applications. Computational and graphics procedures are driven by countless proprietary optimizations (such as, for example, the "quadruple precision" computational technology that allows us to overcome the limitations of the IEEE floating point storage standards). As a result, *STATISTICA* offers unmatched speed, numerical precision, and responsiveness, which is aided by multithreading. Data access is based on a flexible streaming technology that allows *STATISTICA* to work effortlessly with both the simple input datafiles stored on the local drive and queries of multidimensional databases containing terabytes of data and stored in remote

data warehouses, and processed in-place (i.e., without having to import them to a local storage; this feature is available in enterprise versions of *STATISTICA*).

For example, you can simultaneously run multiple instances of *STATISTICA*, each running multiple analyses of data from multiple and simultaneously open input datafiles and queries, and the results can be organized into separate projects. *STATISTICA*'s input and output datafiles and graphs can be of practically unlimited size, comprising hierarchies of documents of various types. The output can be directed to a multitude of output channels such as high performance workbooks, reports, the Internet, etc.

Web Enablement

Because *STATISTICA* is fully Web enabled, both "output from" and "input into" *STATISTICA* can be handled via the Internet.

In addition to sending output from analyses directly into a Web server, *STATISTICA* provides methods to set up very sophisticated, automated systems that run scripts that can, for example, periodically draw data from external sources, perform specific sets of analyses on them, and update the HTML output on specified Web servers.

STATISTICA can accept Internet "input" via two types of functionality (note that these features are offered in the versions properly licensed for Web server installations):

- Being able to access "Internet data," which means supporting XML data input (the technology allowing users to read data stored in relatively "proprietary" formats, where the instructions as to how to read and encode the data are stored with the data and thus can be followed by XML compatible applications like *STATISTICA*).

- Being able to "input instructions" into *STATISTICA* using an Internet browser as a front end for *STATISTICA* (a "thin client"), and thus being able to run *STATISTICA* from any computer in the world as long as it is connected to the Internet. *Note that accessing STATISTICA this way requires appropriate licensing and cannot be used with a stand-alone version of STATISTICA.*

STATISTICA Web Server (a technical note). This latter functionality is well supported in *STATISTICA* because its comprehensive automation interfaces (based on the object model architecture) can be executed via a scripting language (e.g., VBScript, Jscript, or C++). If *STATISTICA* is installed on the server, then you can access the automation model to process data and get back results that you can send down to the client in the form of HTML tables and graphics (JPG/PNG) files. Moreover, if *STATISTICA* is on the client, you could also use client-side scripting and take advantage of the fact that your Internet browser is an ActiveX

document host, allowing you to open *STATISTICA* Spreadsheets and Graphs within the browser and still have the *STATISTICA* toolbars and menus available.

Note also that this Web browser-based user interface (which is offered from StatSoft in the form of a selection of simplified "dialog templates" invoked from the Internet browser, allowing you to easily perform common types of analyses and graphs) can be flexibly customized by modifying the scripts of analyses that are executed on the server side. Therefore, this browser-based user interface is not only fully customizable, but it also allows you to easily incorporate in it other applications that can interface with *STATISTICA*.

Record of Recognition

We are pleased to report that, as of this release, *STATISTICA* has received the highest rating in every published, independent comparative review in which it was featured. In the history of the software industry, very few products have ever achieved such a record. For more information about StatSoft and *STATISTICA*'s record of recognition, please visit our Web site at http://www.statsoft.com.

EXAMPLE 1: CORRELATIONS (INTRODUCTORY OVERVIEW)

Starting *STATISTICA*. After installing *STATISTICA*, you can start the program by selecting **STATISTICA** from the **Programs** menu in the Windows Start menu (you can also double-click on either **STATIST.exe** in Windows *Explorer* or the icon of any *STATISTICA* file, e.g., a spreadsheet). When you start *STATISTICA*, the last used datafile opens. If you are using *STATISTICA* for the first time, a blank spreadsheet opens.

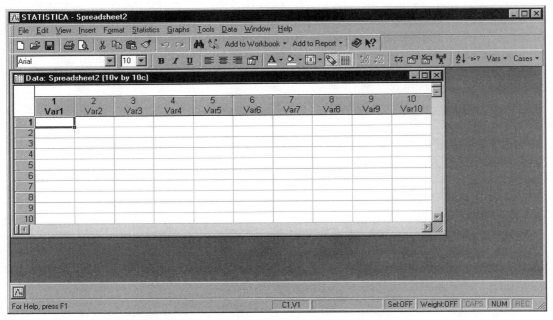

Customization of *STATISTICA*. Note that practically all aspects of the behavior and appearance of *STATISTICA* (even many elementary features illustrated in this example, such as where all output is directed) can be permanently customized to match your preferences. For example, even the first step (opening *STATISTICA*) can be customized; you can change the default full-screen opening mode, the appearance of the data spreadsheet, toolbars, etc.

Selecting a datafile. For this example, use ***Adstudy.sta*** (located in the Examples\Datasets subfolder of your *STATISTICA* installation folder). You can open it using the ***File - Open*** menu, one of the ⬚ Open Data buttons on any startup panel, or the 🗁 toolbar button.

Data spreadsheets (multimedia tables). *STATISTICA* datafiles are always displayed in a spreadsheet (i.e., one spreadsheet is one datafile). All *STATISTICA* Spreadsheets are displayed using StatSoft's powerful multimedia table technology, which will be illustrated later, and they can contain not only practically unlimited amounts of data but also sound, video, embedded documents, automation scripts, and custom user interfaces.

It is possible to have more than one data spreadsheet open at a time (with each spreadsheet connected to a different analysis); thus, most output produced by *STATISTICA* is displayed in spreadsheets (multimedia tables). Note that data management facilities are available from the ***Data*** menu whenever any spreadsheet is open.

The ***Spreadsheet*** toolbar contains the Vars ▾ (Variables) and Cases ▾ buttons, which display menus that contain options to restructure the datafile (e.g., ***Add***, ***Move***, ***Shift*** variables).

Vars ▾ button menu: Cases ▾ button menu:

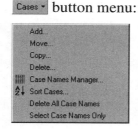

All the above options are described in Chapter 4 - *STATISTICA Spreadsheets* and the *Electronic Manual*.

Variable specifications. The variable (column) headers in the spreadsheet contain variable names. Double-click on a variable header to display its ***Variable*** specifications dialog.

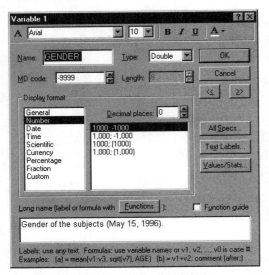

Spreadsheet formulas. In this dialog, you can change the variable name and/or format, enter a formula to recalculate the values of the variable, etc. If the **Long name** box starts with an equal sign (**=**), *STATISTICA* will interpret it as a formula [a comment can follow after a semicolon(;)]. For example, if you enter into the **Long name** box (of variable one) **=(v2+v3+v4)/3** or **=mean(v2:v4)**, the current values of that variable will be replaced by the average of variables two through four, separately for each case (row) of the spreadsheet.

Specifications of all variables can also be reviewed and edited together in a "combined" **Variable Specifications Editor** dialog, accessed by clicking the **All Specs** button on the **Variable** specifications dialog.

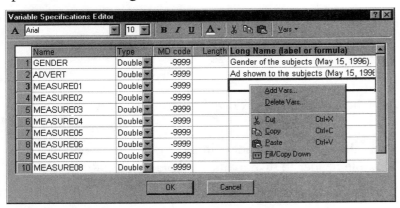

Shortcut menus in the spreadsheet. A useful feature of the spreadsheet is the list of commands available from its shortcut menus. Shortcut menus are dynamic menus that are

displayed by right-clicking on an item (e.g., a cell in the spreadsheet). The spreadsheet shortcut menu includes a selection of specific data management operations and other options related to the current variable (column), case (row), and/or block of cells.

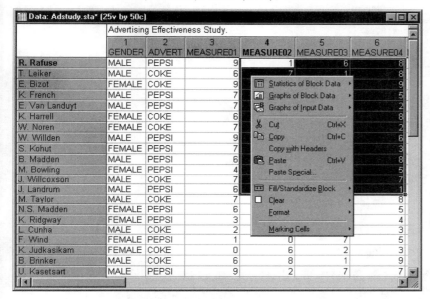

Three ways of handling output. You can customize the way output is managed in *STATISTICA* (see Chapter 2 – *STATISITCA Output Management*). When you perform an analysis, *STATISTICA* generates output in multimedia tables (spreadsheets) and graphs. You can direct all output to three basic channels:

- workbooks (see page Chapter 3 – *STATISITCA Workbooks*),

- reports (see page Chapter 5 – *STATISITCA Reports*), and

- stand-alone windows (see Chapter 2 – *STATISITCA Output Management*).

These three output channels, controlled by the options on the **Output Manager** tab of the **Options** dialog (accessible from the **File – Output Manager** menu or the **Tools - Options** menu), can be used in many combinations (e.g., a workbook and report simultaneously), and each output channel can be customized in a variety of ways. Also, all output objects (spreadsheets and graphs) can contain other embedded and linked objects and documents, so *STATISTICA* output can be hierarchically organized in a variety of ways.

Calculating a correlation matrix. Now, compute a correlation matrix for the variables in the datafile. To display the *Basic Statistics and Tables* (Startup Panel), select *Basic Statistics/Tables* from the *Statistics* menu,

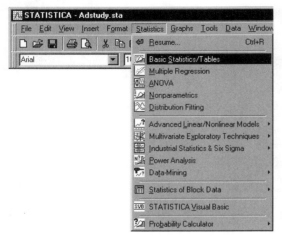

or from the *STATISTICA* Start button menu in the lower-left corner of the screen.

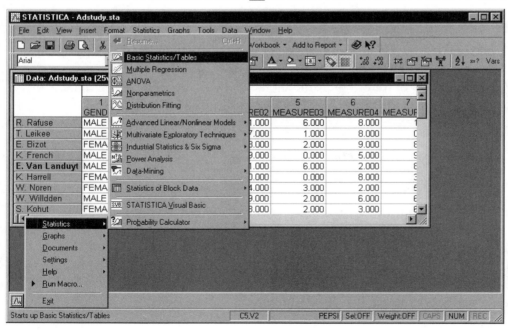

At this point, make sure that a block (a group of selected cells) is not selected in the spreadsheet (to deselect a block, click the cursor in any cell in the spreadsheet). If a block is

selected, *STATISTICA* assumes that the variables corresponding to the block are intentionally preselected for the analysis, and when you later click the **OK** or **Summary** button, instead of prompting you to select variables, *STATISTICA* will automatically produce the correlations for the selected block variables.

In the Startup Panel shown below,

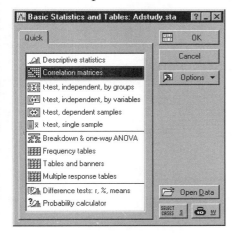

select **Correlation matrices** by double-clicking it (or selecting it and clicking the **OK** button).

After you have selected **Correlation matrices** on the Startup Panel, the **Product-Moment and Partial Correlations** dialog is displayed.

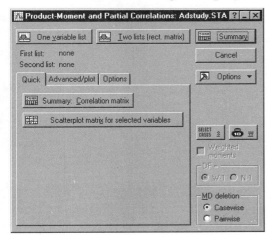

Quick vs. advanced analyses. As with most analyses dialogs (and several other types of *STATISTICA* dialogs), the **Product-Moment and Partial Correlations** dialog is organized by the type of options available. Typically, at least two types of analyses are available.

The **Quick** tab of a dialog contains the most commonly used options, enabling you to quickly specify a basic analysis without having to search through a variety of options.

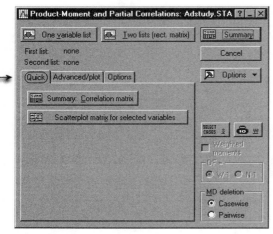

The **Advanced** tab contains the options available on the **Quick** tab as well as a variety of less commonly used options (e.g., in this case, options to save matrices, produce less commonly requested statistics, or produce a variety of plots). Additional tabs are often available as well, depending on the type of analysis being specified.

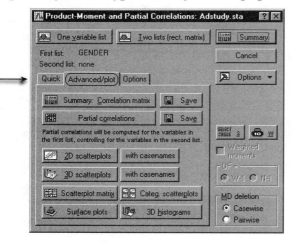

Note that in some cases, only a **Quick** tab is available. As with all dialogs in *STATISTICA*, pressing F1 or clicking the ![?] button in the upper-right corner displays a help topic with information on the options available on the currently selected tab.

The "self-prompting" nature of all dialogs. All dialogs in *STATISTICA* follow the "self-prompting" dialog convention, which means that whenever you are not sure what to select next, simply click the **OK** button or the **Summary** button and *STATISTICA* will proceed to the next logical step, prompting you for the specific input needed (e.g., variables to be analyzed).

Variables button. The default tab in the **Product-Moment and Partial Correlations** dialog is a typical quick analysis definition dialog that contains options to specify the variables to be analyzed and to review summary statistics. Every analysis definition dialog in *STATISTICA* contains at least one **Variables** button, which is used to specify variables to be analyzed. For this example, click the **One variable list** button (or press ALT+V).

Variable selection dialog. Click the **One variable list** button (or the **Summary** button if no variables in your spreadsheet have been selected) to display a variable selection dialog. (Note that, as mentioned earlier, if you have "played" with the data spreadsheet before and selected a block, the variables selected in the block will be automatically selected, and when you click **Summary**, the default correlation matrix for the variables selected in the block will be produced.)

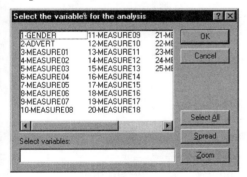

The variable selection dialog supports various ways of selecting variables (including the standard Windows SHIFT+click and CTRL+click conventions to select ranges and discontinuous lists of variables, respectively).

The dialog also offers various shortcuts and options to review the contents of the datafile. For example, you can spread the variable list to review their long names or formulas (by clicking the **Spread** button); or you can zoom in on a variable (by clicking the **Zoom** button) to review a sorted list of all values and descriptive statistics for the variable, as shown below.

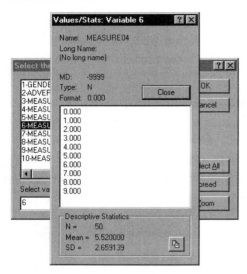

For this example, click the *Select All* button, and then click the *OK* button to return to the *Product-Moment and Partial Correlations* dialog. Next, click the *Summary* button to generate a default correlation matrix for the selected variables.

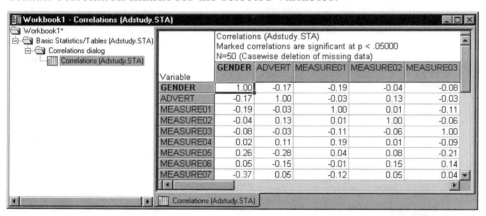

Note that instead of clicking the *Summary* button, you could have clicked the *Summary: Correlation matrix* button. Also, depending on the defaults you have specified for handling output, the Correlations spreadsheet can be displayed in a report or a stand-alone window rather than in a workbook as shown above.

Results spreadsheets (multimedia tables). In addition to storing data, spreadsheets are used in *STATISTICA* to display most of the numeric output. Note that spreadsheets offer many display features and options, and in this example, significant correlations are marked with a different format to help distinguish them; by default the color is red). Spreadsheets can

hold anywhere from a short line to gigabytes of output, and they offer a variety of options to facilitate reviewing the results and visualizing them in predefined and custom-defined graphs, as will be seen later in this example. Also, as mentioned earlier, *STATISTICA* Spreadsheets are managed using StatSoft's powerful multimedia table technology. They can handle not only virtually unlimited amounts of data, but also video, sound, custom user interfaces, and auto-executing scripts, as well as offer virtually unlimited customization options (see Chapter 4 – *STATISITCA Spreadsheets* for further details on spreadsheets).

Spreadsheet options. As mentioned earlier, most spreadsheet facilities are accessible via buttons on the **Spreadsheet** toolbar and the shortcut menus (shortcut menus are displayed by right-clicking in any cell). You can try these options to see how they work, or you can review their descriptions by pressing the help key (F1) or clicking the 🕮 toolbar button and then clicking on the respective toolbar button. For example, you can change all aspects of the display formats for each column, edit the output, or append blank cases and variables to make room for notes or output pasted from other sources. Spreadsheets can be printed in a variety of ways (by default, in presentation-quality tables with gridlines). Also, since spreadsheets are used for input, you can easily specify an analysis using the results from a previous analysis (for example, you could use this correlation matrix to specify a multidimensional scaling analysis). To use a results spreadsheet as an input spreadsheet, select **Input Spreadsheet** from the **Data** menu when that spreadsheet is active.

Analysis workbooks and other output options. All results can be displayed (and stored) in stand-alone windows, reports, or workbooks, which represent the default (and perhaps the most versatile) way of handling output from analyses (see Chapter 3 – *STATISTICA Workbooks* for further details on workbooks). Depending on your selections in the **Output Manager** (accessible via the **File – Output Manager** menu, see the next paragraph), results can be put in a single workbook that holds the results from all analyses, a separate analysis workbook that holds the results (spreadsheets and graphs) from a single analysis, the workbook that contains the original datafile, or a preexisting workbook. Additionally, you can choose to have the results sent to a workbook automatically, or you can send them to the workbook yourself by clicking the `Add to Workbook ▾` toolbar button to send selected stand-alone spreadsheets or graph windows to a workbook.

Output Manager. Which type of workbook you choose, or whether or not you even choose to use a workbook, depends entirely on how you prefer to store your data and results. To change the output destination for results of a particular analysis only, click the `🗔 Options ▾` button on any analysis or graph specification dialog and select **Output** to display the **Analysis/Graph Output Manager** dialog.

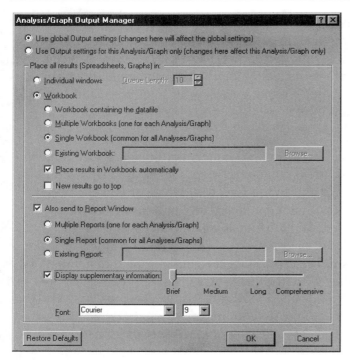

To change output options for all analyses, use the (global) **Output Manager** (the **Output Manager** tab of the **Options** dialog, accessible from the **File – Output Manager** menu or the **Tools - Options** menu), or select the **Use global Output settings (changes here will affect the global settings)** option button on the **Analysis/Graph Output Manager** dialog.

As with all workbooks, individual documents (e.g., spreadsheets, graphs) or groups of documents can be printed, extracted, copied, and deleted from an analysis workbook. See Chapter 3 – *STATISTICA Workbooks* for more details; see also the *Electronic Manual*.

Copy vs. Copy with Headers. Contents of spreadsheets can be copied to the Clipboard via either the default **Copy** by pressing CTRL+C (which copies only the contents of the selected block) or selecting **Copy with Headers** from the **Edit** menu (which copies the block along with its respective variable and case names). If pasted into a word processor document, spreadsheets will appear as active (in-place editable) *STATISTICA* objects, standard RTF-formatted tables, or tab-delimited text (depending on your choice in the **Paste Special** dialog of the word processor).

Printing spreadsheets. To produce a hard copy of the output spreadsheets, select **Print** from the **File** menu (or press CTRL+P) to display the **Print** dialog, in which you specify printing options. You can also use the shortcut method by clicking the printer toolbar button . This

shortcut method does not display the **Print** dialog, but prints the entire current document. If you want to print a document from within a workbook, make sure the document is selected in the workbook and select the **Selection** option button on the **Print** dialog. You can also extract a copy of the document from the workbook (by dragging it from the tree pane or using the **Workbook** menu) and then print it.

Optional reports of all output. Workbooks offer perhaps the most flexible options to manage your output (see Chapter 3 – *STATISTICA Workbooks*). In some circumstances, however, it may be useful to automatically produce a log of all results (contents of all spreadsheets and/or graphs) in a traditional word processor style report format where comments and annotations can be inserted in arbitrary locations, objects can be placed side by side, etc. (see Chapter 5 – *STATISTICA Reports* for further details on reports).

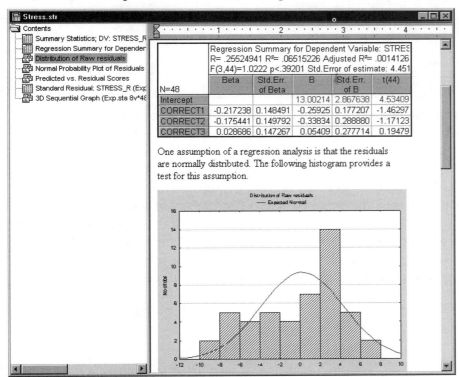

In order to create such a report, select the **Also send to Report Window** check box and either the **Single Report (common for all Analyses/graphs)**, the **Multiple Reports (one for each Analysis/graph),** or the **Existing Report** option button on the **Output Manager** dialog. As mentioned above, to display this dialog, select **Output Manager** from the **File** menu or **Options** from the **Tools** menu (for global changes), or click the  button on any analysis or

graph specification dialog and select **Output** (for local changes). Using the **Output Manager**, you can also specify the amount of supplementary information to be included with the spreadsheet results.

Interpretation of the results (Electronic Statistics Textbook). Now return to the example and the correlation matrix that has been produced.

	Correlations (Adstudy.STA) Marked correlations are significant at p < .05000 N=50 (Casewise deletion of missing data)				
Variable	GENDER	ADVERT	MEASURE01	MEASURE02	MEASURE03
GENDER	1.00	-0.17	-0.19	-0.04	-0.08
ADVERT	-0.17	1.00	-0.03	0.13	-0.03
MEASURE01	-0.19	-0.03	1.00	0.01	-0.11
MEASURE02	-0.04	0.13	0.01	1.00	-0.06
MEASURE03	-0.08	-0.03	-0.11	-0.06	1.00
MEASURE04	0.02	0.11	0.19	0.01	-0.09
MEASURE05	0.26	-0.28	0.04	0.08	-0.21
MEASURE06	0.05	-0.15	-0.01	0.15	0.14
MEASURE07	-0.37	0.05	-0.12	0.05	0.04

Workbook1 - Correlations (Adstudy.STA) — Workbook1* / Basic Statistics/Tables (Adstudy.STA) / Correlations dialog / Correlations (Adstudy.STA)

Each of the cells of the correlation matrix represents a value (in the range of –1.00 to +1.00) that reflects the relation between the variables (see the respective variable and case headers). The higher the absolute value of the correlation coefficient, the closer the relation; if the value is positive, the relation is "positive" (high values of one variable correspond to high values of the other variable; likewise, low values of one variable correspond to low values of the other variable). If the value is negative, then the opposite is true (low values of one variable correspond to high values of the other variable). To learn more about how to interpret values of correlations, please review a comprehensive, illustrated discussion of the topic in the *Electronic Manual*, which features the complete contents of the StatSoft *Electronic Statistics Textbook* (an award-wining general resource on statistics that has been recommended by *Encyclopedia Britannica*

Britannica
Internet Guide Award

StatSoft web site praised for Quality, Accuracy, Presentation and Usability.

for its "Quality, Accuracy, Presentation, and Usability.") and the **Statistical Advisor** (see page 26). To display the *Electronic Manual*, select **Contents and Index** from the **Help** menu. Then enter the respective term (e.g., **Correlations**) into the **Type in the word(s) to search for** box on the **Search** tab of the *Electronic Manual*, click the **List Topics** button, and then select the desired topic in the **Select topic** box (in this case **Correlations – Introductory Overview**):

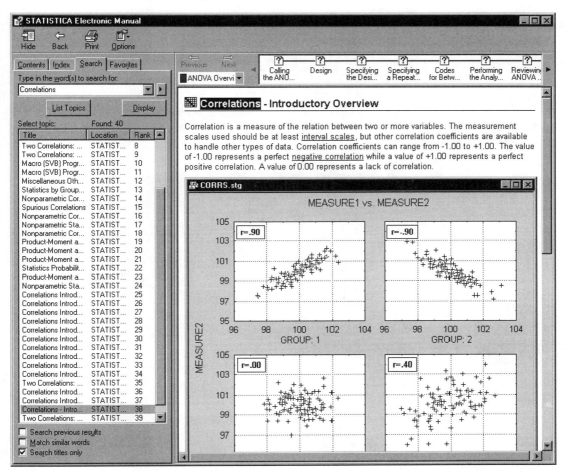

One of the important (and often overlooked) issues discussed in the *Electronic Manual* is the importance of scatterplots in examining correlations. For example, even very large and highly statistically significant correlation coefficients can be entirely due to one unusual data point ("outlier"), and if that is the case, then the correlation coefficient (even if statistically significant) would have no value to us (e.g., it would have no "predictive validity"). Following this concern, and the advice of the *Electronic Statistics Textbook*, let us examine a scatterplot that will visualize a relation between the variables and thus visualize a particular correlation coefficient from the table.

Producing graphs from spreadsheets. While examining the spreadsheet, you may view the correlations graphically, for example, to visualize the correlation between variables *Measure09* and *Measure05*. To produce a scatterplot for these two variables, right-click on the

respective correlation coefficient (*-.47*). In the resulting shortcut menu, select *Graphs of Input Data*, and then select one of the graphs in the submenu, shown below.

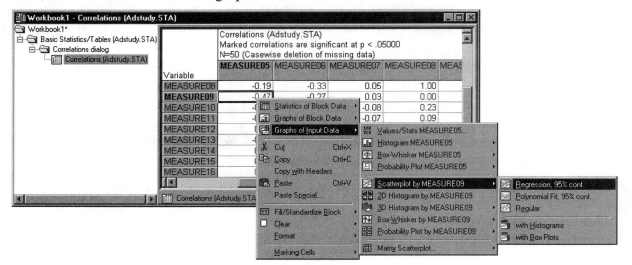

The requested graph will be displayed on the screen.

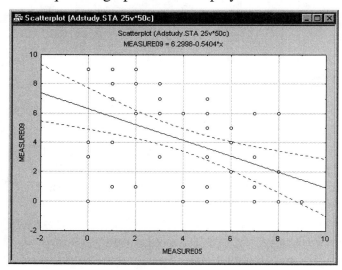

As we can learn from the graph, there are no unusual patterns of data, thus there is no reason for being concerned about outliers (see the section above, see also the topic on outliers in the *Electronic Manual*).

Graph customization. Note that now, when the focus is on the graph window, the toolbar has changed. The ***Graph Tools*** toolbar (which accompanies all graph windows) looks different from the toolbar for the spreadsheets.

It contains a variety of graph customization and drawing tools. All of these options are also available from menus, and most of them are available from the shortcut menus by right-clicking on specific parts of the graph. Note that the options on shortcut menus are hierarchical, meaning that the first one or two options apply specifically to the graph element you have selected, while lower options will display dialogs that offer more options on a greater variety of graph elements related to the element you have selected. If you right-click anywhere on the empty space outside the graph axes, a menu of global options is displayed (as shown below).

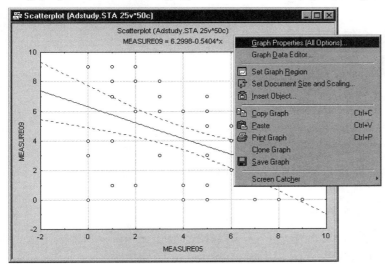

For more information on graph customization, see Chapter 7 – *STATISTICA Graphs: Creation and Customization* or the *Electronic Manual*.

Now let us return to the spreadsheet.

Split scrolling in spreadsheets. Spreadsheets can be split into up to four sections (panes) by dragging the split box (the small rectangle at the top of the vertical scrollbar or to the left of the horizontal scrollbar). This is useful if you have a large amount of information and you want to review results from different parts of the spreadsheet. When you move the

mouse pointer to the split box, the mouse pointer changes to ⫞ or ⫟. Now, to position the split, drag it to the desired position.

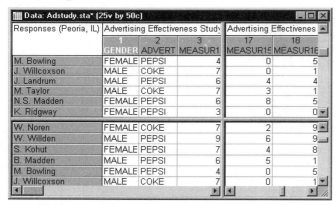

You can change the position of the split by dragging the split box (now located between panes) to a new position.

Note that vertically split panes scroll together when you scroll horizontally; horizontally split panes scroll together when you scroll vertically. For information about highlighting blocks of data across split panes and about variable-speed highlighting of blocks of data, see *How Can I Expand a Block in the Spreadsheet Outside the Current Screen?* in Chapter 4 – *STATISTICA Spreadsheets.*

Drag-and-drop. *STATISTICA* supports the complete set of standard spreadsheet (Excel-style) drag-and-drop facilities. For example, in order to move a block, point to the border of the selection (the mouse pointer changes to an arrow) and drag it to the new location.

To copy a block of data point to the border of the selection (the mouse pointer changes to an arrow) then drag the selection to a new location while pressing the CTRL key. Note that when you are dragging the selection, a plus sign (+) is displayed next to the mouse pointer to indicate you are copying the text rather than moving it (see the illustration, below).

To insert a block between columns or rows, point to the border of the selection (the mouse pointer changes to an arrow) and then drag the selection while pressing the SHIFT key.

If you point between rows, an insertion bar is displayed between the rows, and when you release the mouse button, the block is inserted between those two rows [creating new case(s)]. If you point between columns, an insertion bar is displayed between the columns, and when you release the mouse button, the block is inserted between those two columns [creating new variable(s)].

Note that if you also press the CTRL key while you are dragging the selection, the block will be copied and inserted instead of moved and inserted; a plus will be displayed next to the mouse pointer (as shown in the illustration below).

Additionally, a series of values within a block can be extrapolated (AutoFilled) by dragging the Fill Handle (a small, solid square located on the lower-right corner of the block border).

Electronic Manual. For more information on any of the menu commands, press the help key (F1) when the command is selected. *STATISTICA* provides a comprehensive *Electronic Manual* for all program procedures and all options available in a context-sensitive manner. It is accessed by pressing the F1 key or clicking the help button **?** on the caption bar of all dialogs (there is a total of over 100 megabytes of compressed documentation included).

Due to its dynamic hypertext organization, organizational tabs (e.g., **Contents**, **Index**, **Search**, and **Favorites**), and various facilities used to customize the help system, it is faster to use the *Electronic Manual* than to look for information in the traditional manuals.

Note also that the status bar on the bottom of the *STATISTICA* window also displays short explanations of the menu commands or toolbar buttons when an item is selected or a button is clicked.

Statistical Advisor. A **Statistical Advisor** facility is built into the *Electronic Manual*. When you select **Statistical Advisor** from the **Help** menu, *STATISTICA* asks you a set of simple questions about the nature of the research problem and the type of your data. Then the advisor suggests the statistical procedures that appear most relevant and tells you where to look for them in the *STATISTICA* system.

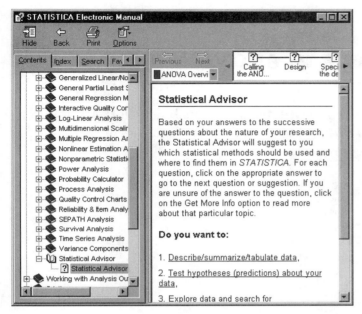

Direct jumps (hypertext links) are available from the **Statistical Advisor** topics to the corresponding *Introductory Overviews* (StatSoft's *Electronic Statistics Textbook*) discussing in detail the respective statistical methods and procedures.

EXAMPLE 2: ANOVA

Calling the ANOVA module. To start an ***ANOVA/MANOVA*** analysis, select ***ANOVA*** from the ***Statistics*** menu to display the ***General ANOVA/MANOVA*** (Startup Panel).

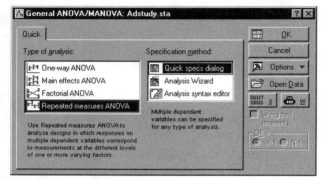

This dialog is used to specify very simple analyses (e.g., via ***One-way ANOVA*** - designs with only one between-group factor) and more complex analyses (e.g., via ***Repeated measures ANOVA*** – designs with between-group factors and a within-subject factor).

Design. For this example of a 2 x 2 (between) x 3 (repeated measures) design, open the datafile ***Adstudy.sta***. Select ***Repeated measures ANOVA*** as the ***Type of analysis*** and ***Quick specs dialog*** as the ***Specification method*** and then click the ***OK*** button on the ***General ANOVA/MANOVA*** (Startup Panel) to display the ***ANOVA/MANOVA Repeated Measures ANOVA*** dialog.

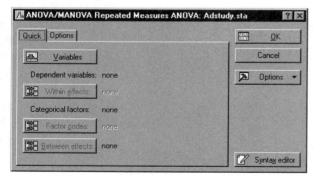

Specifying the design (variables). The first (between-group) factor is **Gender** (with 2 levels: **Male** and **Female**). The second (between-group) factor is **Advert** (with 2 levels: **Pepsi** and **Coke**). The two factors are crossed, which means that there are both **Male** and **Female** subjects in the **Pepsi** and **Coke** groups. Each of those subjects responded to 3 questions (this repeated measure factor will be called **Response**: it has 3 levels represented by variables **Measure01**, **Measure02**, and **Measure03**).

Click the **Variables** button (on the **ANOVA/MANOVA Repeated Measures ANOVA** dialog) to display the variable selection dialog. Select **Measure01** through **Measure03** as dependent variables (in the **Dependent variable list** field) and **Gender** and **Advert** as factors [in the **Categorical predictors (factors)** field].

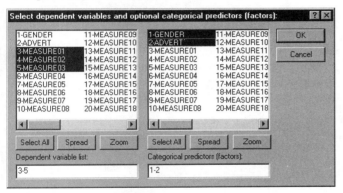

Then click the **OK** button to return to the previous dialog.

The repeated measures design. Note that the design of the experiment that we are about to analyze can be summarized as follows:

	Between-Group Factor #1: *Gender*	Between-Group Factor #2: *Advert*	Repeated Measure Factor: *Response*		
			Level #1: *Measure01*	Level #2: *Measure02*	Level #3: *Measure03*
Subject 1	Male	Pepsi	9	1	6
Subject 2	Male	Coke	6	7	1
Subject 3	Female	Coke	9	8	2
⋮	⋮	⋮	⋮	⋮	⋮

Specifying a repeated measures factor. The minimum necessary selection is now completed and, if you did not care about selecting the repeated measures factor, you would be ready to click the **OK** button and see the results of the analysis. However, for our example, you need to specify that the three dependent variables you have selected be interpreted as three

levels of a repeated measures (within-subject) factor. Unless you do so, *STATISTICA* assumes that those are three "different" dependent variables and runs a MANOVA (i.e., multivariate ANOVA).

In order to define the desired repeated measures factor, click the **Within effects** button to display the **Specify within-subjects factors** dialog.

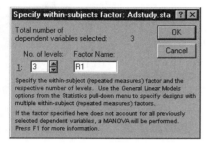

Note that *STATISTICA* has suggested the selection of one repeated measures factor with **3** levels (default name **R1**). You can only specify one within-subject (repeated measures) factor via this dialog. To specify multiple within-subject factors, use the **General Linear Models** module (available in the optional **Advanced Linear/Nonlinear Models** package). Press the F1 key (or click **?**) in this dialog to review a comprehensive discussion of repeated measures and examples of designs in the *Electronic Manual*. Edit the name for the factor (e.g., change the default **R1** into **RESPONSE**), and click the **OK** button to exit the dialog.

Codes (defining the levels) for between-group factors. You do not need to manually specify codes for between-group factors [e.g., instruct *STATISTICA* that variable **Gender** has two levels: **1** and **2** (or **Male** and **Female**)] unless you want to prevent *STATISTICA* from using, by default, all codes encountered in the selected grouping variables in the datafile. To enter such custom code selection, click the **Factor codes** button to access the **Select codes for indep. vars (factors)** dialog.

This dialog contains various options. For example, you can review values of individual variables before you make your selections by clicking the **Zoom** button, scan the file and fill the codes fields (e.g., **Gender** and **Advert**) for some individual or all variables, etc. For now, click the **OK** button; *STATISTICA* automatically fills in the codes fields with all distinctive values encountered in the selected variables,

and closes the dialog.

Performing the analysis. When you click the *OK* button upon returning to the *ANOVA/MANOVA Repeated Measures ANOVA* dialog, the analysis is performed, and the *ANOVA Results* dialog is displayed. Various kinds of output spreadsheets and graphs are now available.

Note that this dialog is tabbed, which allows you to quickly locate results options. For example, if you want to perform planned comparisons, click the *Comps* tab. To view residual statistics, click the *Resids* tab. For this simple overview example, we will only use the results available on the *Quick* tab.

Reviewing ANOVA results. Start by looking at the ANOVA summary of all effects table by clicking the *All effects* button (the one with a *SUMM*-ary icon ▦).

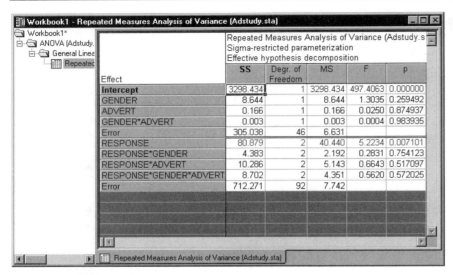

The only effect (ignoring the *Intercept*) in this analysis that is statistically significant (*p = .007*) is the *RESPONSE* effect. This result may be caused by many possible patterns of means of the *RESPONSE* effect (for more information, consult the *ANOVA – Introductory Overview* in the *Electronic Manual*). We will now look graphically at the marginal means for this effect to see what it means.

To bring back the *ANOVA Results* dialog (that is, "resume" the analysis), press CTRL+R, select *Resume* from the *Statistics* menu, or click the *ANOVA Results* button on the *Analysis* bar. When the *ANOVA Results* dialog is displayed, click the *All effects/Graphs* button to review the means for individual effects.

This dialog contains a summary *Table of all effects* (with most of the information you have seen in the *All effects* spreadsheet) and is used to review individual effects from that table in

the form of the plots of the respective means (or, optionally, spreadsheets of the respective mean values).

Plot of means for a main effect. Double-click on the significant main effect **RESPONSE** (the one marked with an asterisk in the **p** column) to see the respective plot.

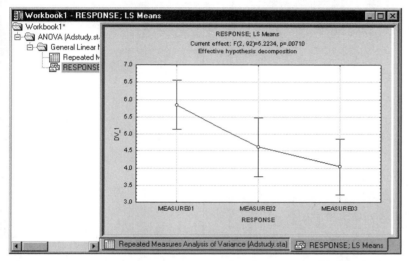

The graph indicates that there is a clear decreasing trend; the means for the consecutive three questions are gradually lower. Even though there are no significant interactions in this design (see the discussion of the *Table of all effects*, above), we will look at the highest-order interaction to examine the consistency of this strong decreasing trend across the between-group factors.

Plot of means for a three-way interaction. To see the plot of the highest-order interaction, double-click on the row marked **RESPONSE*GENDER*ADVERT**, representing the interaction between factors 1 (**Gender**), 2 (**Advert**), and 3 (**Response**), on the **Table of All Effects** dialog. An intermediate dialog, **Specify the arrangement of the factors in the plot**, is displayed, which is used to customize the default arrangement of factors in the graph.

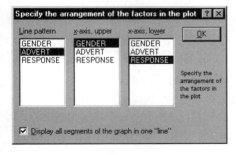

Note that unlike the previous plot of a simple factor, the current effect can be visualized in a variety of ways. Click the **OK** button to accept the default arrangement and produce the plot of means.

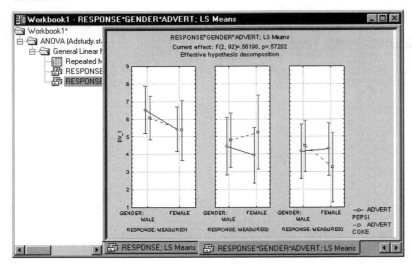

As you can see, this pattern of means (split by the levels of the between-group factors) does not indicate any salient deviations from the overall pattern revealed in the first plot (for the main effect, **RESPONSE**). Now you can continue to interactively examine other effects; run post-hoc comparisons, planned comparisons, and extended diagnostics; etc., to further explore the results.

Interactive data analysis in *STATISTICA*. This simple example illustrates the way in which *STATISTICA* supports interactive data analysis. You are not forced to specify all output to be generated before seeing any results. Even simple analysis designs can, obviously, produce large amounts of output and countless graphs, but usually you cannot know what will be of interest until you have a chance to review the basic output. With *STATISTICA*, you can select specific types of output, interactively conduct follow-up tests, and run supplementary "what-if" analyses after the data are processed and basic output reviewed. *STATISTICA*'s flexible computational procedures and wide selection of options used to visualize any combination of values from numerical output offer countless methods to explore your data and verify hypotheses.

Automating analyses (macros and *STATISTICA* Visual Basic). Any selections that you make in the course of the interactive data analysis (including both specifying the designs and choosing the output options) are automatically recorded in the industry standard Visual Basic code. You can save such macros for repeated use (you can also

assign them to toolbar buttons, modify or edit them, combine with other programs, etc.). For more information, see Chapter 11 - *STATISTICA Visual Basic* or the *STATISTICA Visual Basic Primer*.

USER INTERFACE IN *STATISTICA*

General Features

Customized Operation

The *STATISTICA* system can be controlled in several ways. The following sections summarize the features of the three main alternative user interfaces of *STATISTICA*:

1. interactive interface (see page 39),

2. *STATISTICA* Visual Basic (see page 48), and

3. Web browser-based interfaces (see page 49).

However, note that:

- Many aspects of these user interfaces do not exclude each other; thus, depending on your specific applications and preferences, you can combine them;

- The customizable menus and toolbars can be used to integrate the alternative user interfaces and, for example, to provide quick access to macro (Visual Basic) programs or commonly used files; and

- Almost all features of these alternative user interfaces can be customized (leading to different appearance and behavior of *STATISTICA*); it is usually recommended that you customize your system in order to take full advantage of *STATISTICA*'s potential to meet your preferences and optimal requirements of the tasks that you need to accomplish (see *Customization of the Interactive User Interface* on page 51).

Alternative Access to the Same Facilities; Custom Styles of Work

Even without any customization, the default settings of *STATISTICA* offer alternative user interface means and solutions to achieve the same results. This "alternative access" principle present in every aspect of its user interface allows *STATISTICA* to support different styles of work. For example, most of the commonly used tools can be accessed alternatively:

- From traditional menus;

- Via keyboard shortcuts;

- By using the toolbars and the clickable fields on the status bar;

- Via custom toolbars (user-defined toolbars with buttons and special controls, which can include macros and commands); and

- From the shortcut menus associated with specific objects (cells, workbook icons, parts of graphs), which are displayed by right-clicking on the item (shortcut menus).

It is suggested that you explore the alternative user interface facilities of *STATISTICA* before becoming attached to one style or another.

Multiple Analysis Support

As mentioned before, you can have several copies of *STATISTICA* open at the same time. Each of them can run the same or different types of analyses (traditionally called modules), such as **Basic Statistics**, **Multiple Regression**, **ANOVA**, etc. Moreover, in one *STATISTICA* application, multiple analyses can be open simultaneously. They can be of the same or a different kind (e.g., five **Multiple Regressions** and two **ANOVAs**), and each of them can be performed on the same or a different input datafile (multiple input datafiles can be opened simultaneously).

Individual "analyses" – functional units of your work. In order to facilitate taking advantage of this "multitasking" functionality, your work with *STATISTICA* is organized into functional units called "analyses" that are represented with buttons on the **Analysis** bar at the bottom of the application window (above the status bar, see the illustration below, where **Basic Statistics**, **Cluster Analysis**, and **Canonical Analysis** are running simultaneously). Normally, at least one analysis button is created, and consecutive buttons are added as you start new analyses. A variety of options is provided to control (and/or permanently configure) this aspect of *STATISTICA*.

By default, when you select specific output from a results dialog, the output (a table or a graph) is displayed and the dialog is automatically minimized into its respective analysis button on the bottom of the screen. Click that button (or press CTRL+R) to display the dialog again and resume the analysis.

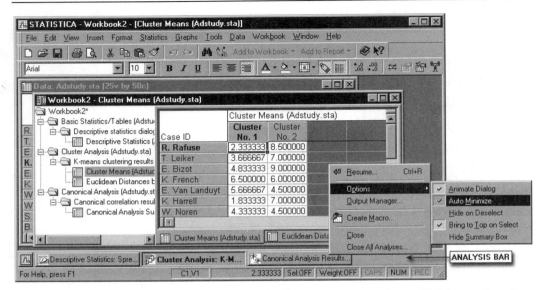

A selection of options pertaining to analysis management is available on the shortcut menu (accessed by right-clicking on an analysis button on the **Analysis** bar) related to the respective analysis buttons (as shown above).

A useful hint for users with large screens. Users with large screens can turn off the default minimization of the analysis dialogs and take advantage of the fact that most of these dialogs are small, and thus can be kept permanently on the screen and used as "semi-toolbars" from which consecutive output objects are selected. You can adjust this option either for a particular analysis (clear the **Auto Minimize** command on the analysis button shortcut menu, shown above), or globally for the entire program (use the **Analyses/Graphs** tab of the **Options** dialog, accessible via the **Tools - Options** menu).

When you run multiple analyses and the *STATISTICA* workspace becomes cluttered, you can hide all windows related to specific analyses (or close them altogether via the analysis button shortcut menu command **Close All Analyses**); you can also open new *STATISTICA* applications, which offers another simple way to organize and manage your work.

1. Interactive User Interface

Overview

Main components of the interactive user interface of *STATISTICA*.

Although the interactive user interface of *STATISTICA* is not the only one available (see *Customizing the Operation and Appearance of STATISTICA*, page 51, and Chapter 11 – *STATISTICA Visual Basic*), it is in most cases the easiest and most commonly used.

Many components of this user interface can be seen on the *STATISTICA* screen.

First, similar to most software programs, menu bars and various toolbars are displayed at the top of the screen. These are customizable and displayed in the most appropriate manner for your tasks.

At the bottom of the screen, the **Analysis** bar (containing minimized analysis/graph dialogs) and the status bar are displayed. Additionally, shortcut menus are available when you right-click in appropriate places.

Datafiles can be displayed in spreadsheets, workbooks, reports, or individual windows. Results spreadsheets or graphs can be displayed in workbooks, reports, or individual windows. Note that additional documents (such as Microsoft Word or Bitmap images) can also be displayed in spreadsheets, workbooks, or reports. Finally, *STATISTICA* Visual Basic code is displayed in macro windows.

Normally you would not simultaneously see all of these facilities and tools at one time. You always have the ability to make the user interface of *STATISTICA* as simple or complex as your particular needs and comfort level demand (see page 51). These various tools and facilities are described in detail in the *Electronic Manual*.

Modules. While *STATISTICA* offers a variety of statistical and graphical procedures, each procedure can be performed in the same application of *STATISTICA*. This means that, for example, it is possible to calculate residual statistics using options in the **Multiple Regression** module, then immediately use that output in the **Factor Analysis** or another exploratory module without first starting another application of *STATISTICA*. For more information on using results as input data, see *Can I Use the Results of One Analysis to Perform Another Analysis?* in Chapter 4 – *STATISTICA Spreadsheets* or the *Electronic Manual*.

The Flow of Interactive Analysis

Startup Panel. When a particular statistical procedure is selected from the *Statistics* menu, a respective Startup Panel is displayed (as shown below and also in the context of *Example 1: Correlations* on page 8).

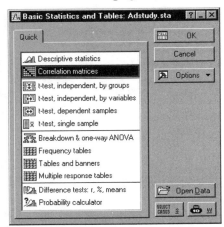

Each Startup Panel contains a list of the types of analyses available in that particular module. Clicking anywhere outside the panel automatically minimizes it as a button on the *Analysis* bar. Users of systems with high-resolution screens can change this default and keep the consecutive dialogs (in each analysis sequence) on the screen. This way they can be used as convenient "toolbars" from which options (e.g., output) can be requested (see page 38).

Toolbars. If you prefer to use buttons rather than menus to select statistical analyses, you can activate the *Statistics* toolbar (which has buttons for every module) by right-clicking on any toolbar and selecting *Statistics* from the shortcut menu that lists available toolbars. Alternatively, you can select *Statistics* from the *View – Toolbars* menu. You can also create your own toolbar that holds the toolbar buttons for analyses you use most frequently (as illustrated on page 47). For more on toolbars, see the *Toolbar Overview*, page 56.

"Analysis definition" and "output selection" dialogs. When the desired analysis and (when requested) a new datafile are selected on the Startup Panel, the analysis definition dialog is displayed, in which you select the variables to be analyzed and other options and features of the task to be performed. Often, these dialogs have several tabs that group the options, analyses, and/or results in logical categories to make it easier to locate specific features.

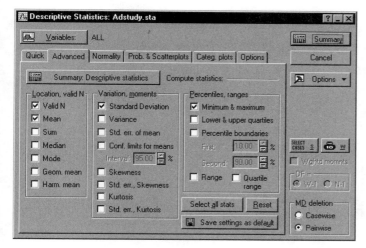

In some simple analyses (such as descriptive statistics, as shown on the sample screen above), the analysis definition dialog also serves as an output selection dialog where you can request the type and format of the output (e.g., some specific spreadsheets or graphs).

Output. As described in more detail in Chapter 2 – *STATISTICA Output Management* (and as illustrated in *Example 1: Correlations* (page 8) and *Example 2: ANOVA* (page 28), the consecutive output spreadsheets and graphs are displayed by default in workbooks. These workbooks can be saved and later reopened, making it easy to return to specific results as needed.

Additionally, you can send all output to an analysis report (see Chapter 5 – *STATISTICA Reports*), which produces an easily organized (via the report tree), easily formatted, and easily printed report of a specific analysis. You can also choose to send all results, regardless of what analysis it comes from, to a single report. Alternatively, the output can be directed to separate windows.

In either case, output options for a single analysis or session can be set by clicking the ▣ Options ▾ button on the analysis or graph specification dialog and selecting **Output**. Global output options are available by selecting **Options** from the **Tools** menu and accessing the **Output Manager** tab of the **Options** dialog or by selecting **Output Manager** from the **File** menu. For more information, see Chapter 2 – *STATISTICA Output Management* or the *Electronic Manual*.

Features of Analysis

STATISTICA provides direct access to all statistical and graphical analyses dialogs via the
Statistics

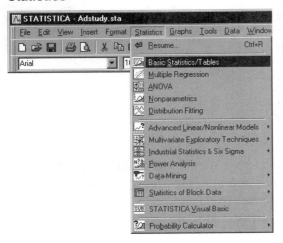

and **Graphs** menus

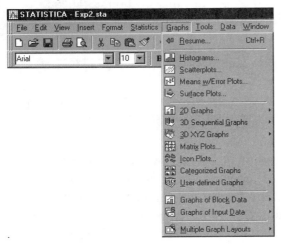

(as well as from the **Statistics** toolbar and the **Graphs** toolbar). These menus are never disabled, i.e., they are available whenever any input data document is open. The **Statistics** menu provides access to all available analysis types within *STATISTICA*. The **Graphs** menu provides direct access to a variety of commonly used graph types (e.g., scatterplots, histograms, means/error plots, etc.) as well as hierarchical access to all graph types in *STATISTICA*

including **2D Graphs**, **3D Sequential** and **XYZ Graphs**, **Matrix Plots**, **Icon Plots**, **Categorized Graphs**, **User-defined Graphs**, **Graphs of Block Data**, and **Graphs of Input Data**. It also provides access to **Multiple Graph Layouts**. Comprehensive discussions of all various types of statistics and graphs offered by *STATISTICA* are available in the glossary of the *Electronic Manual*. See also *Appendix C: STATISTICA Family of Products* in *STATISTICA: The Small Book* for more information on all members of the comprehensive selection of data analysis applications from the *STATISTICA* family of products.

Using the Analysis bar. To take advantage of *STATISTICA*'s "multitasking" functionality (see *Multiple Analysis Support*, page 37), *STATISTICA*'s analyses are organized as functional units that are represented with buttons on the **Analysis** bar at the bottom of the application window (above the status bar, see the illustration below, where **Basic Statistics**, **Cluster Analysis**, and **Canonical Analysis** are running simultaneously). Normally, at least one analysis button is created, and consecutive buttons are added as you start new analyses.

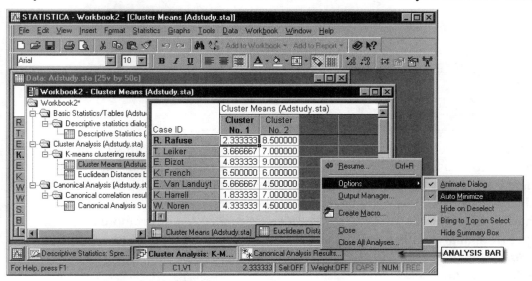

Minimizing dialogs (and a hint for large screen users). Depending on your preferences, you can choose to minimize all analysis dialogs when you select another window in *STATISTICA* or another application. By default the **Auto Minimize** command is checked; however, when your screen is large enough to accommodate several windows, it is recommended that you clear this option. This keeps the analysis dialogs on screen while the respective output created from these dialogs is produced, thus allowing you to use the dialogs as "toolbars" from which output can be selected. See page 38 for information on how to adjust this command.

Continuing analyses/graphs. It is easy to continue the current analysis or graph (i.e., to change the focus to the current dialog for a particular analysis). Select **Resume Analysis/Graph** from the **Tools - Analysis Bar** menu, press CTRL+R, or click the analysis/graph button on the **Analysis** bar. When multiple analyses are running, you can also select the specific analysis from the **Tools - Analysis Bar - Select Analysis/Graph** menu (as shown below).

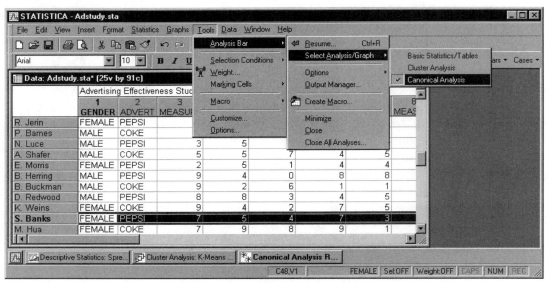

Hiding windows. To further facilitate the organization of windows from various analyses, you can hide all windows associated with a particular analysis when that analysis is deselected by checking **Hide on Select** from the **Tools - Analysis Bar - Options** menu. By default, this command is cleared. Note that this command only applies when the results are sent to individual windows; see Chapter 2 – *STATISTICA Output Management* for more details on managing output from analyses. In addition, there is a command to close all document windows, **Window – Close All** (or CTRL+L), and a command to close all analyses, **Tools – Analysis Bar – Close All Analyses**.

Bringing windows to the top. Check **Bring to Top on Select** from the **Tools - Analysis Bar - Options** menu to activate (bring to the top of *STATISTICA*) all windows associated with a particular analysis when that analysis is selected, replacing whatever dialogs were on top. This command also facilitates the organization of individual windows from various analyses. By default, this command is checked. Note that this command only applies when the results are sent to individual windows; see Chapter 2 – *STATISTICA Output Management* for more details on managing output from analyses.

Hiding the summary box. By default, a summary box is located at the top of certain results dialogs (such as *Multiple Regression - Results*) and contains basic summary information about the analysis. You can hide an individual summary box by clicking the ⬦ button in the lower-right corner of the summary box. You can also suppress the display of all summary boxes globally by checking *Hide Summary Box* from the *Tools - Analysis Bar - Options* menu.

Document Types

STATISTICA uses five principal document types:

- Workbooks (see Chapter 3 – *STATISTICA Workbooks*)
- Spreadsheets (multimedia tables) (see Chapter 4 – *STATISTICA Spreadsheets*)
- Reports (see Chapter 5 – *STATISTICA Reports*)
- Graphs (see Chapter 6 – *STATISTICA Graphs: General Features*)
- Macros (*STATISTICA* Visual Basic programs) (see Chapter 11 – *STATISTICA Visual Basic*)

Using these five document types, you can manage data of various types, perform data entry and analysis, generate graphs of the highest quality, develop custom applications of any degree of complexity, and create custom-formatted reports.

You can quickly access the most recently used documents. Click the *STATISTICA* Start button 📊 (in the lower-left corner of the screen, see illustration below) and select *Documents*. On the *General* tab of the *Options* dialog (accessed via the *Tools – Options* menu), you can specify how many recently used documents (by default *9*) to display.

For more detailed information about each document type, see the respective chapters for workbooks, spreadsheets, reports, graphs, and macros (*STATISTICA* Visual Basic programs) in the *System Reference*; for further information, see the *Electronic Manual*.

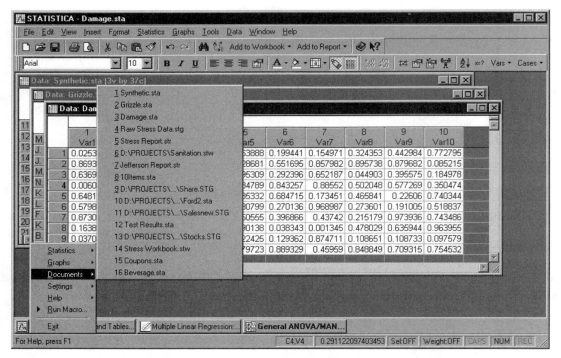

Toolbars related to types of active document windows. Each of the main types of *STATISTICA* document windows (see above) manages data in a different way, and thus offers different customization and management options. These differences are reflected in the toolbars that accompany each type of window. Menu commands and toolbar buttons for each of the main types of documents are discussed in detail in the respective chapters in the *System Reference*.

Note that workbooks do not have a specialized toolbar (although the **Standard** toolbar is always available) because the toolbars that are offered by workbooks depend on the currently displayed document. Therefore, when you are editing a spreadsheet, graph, report, macro, or foreign document (e.g., a Microsoft Excel spreadsheet) within a workbook, the toolbars and menus relevant for that document type are available.

When you select an "empty node" in the workbook tree pane, by default, the **Statistics** toolbar is displayed (in place of the specific document type toolbar) in order to maintain the same size and proportions of the application workspace.

User-defined toolbars. In addition to the variety of toolbars provided in *STATISTICA*, you can also create user-defined toolbars. These toolbars can include any command available in *STATISTICA*, as well as special controls (i.e., font name, font size, graph styles, etc.). The

toolbars can be given any name and can be designated to open depending on the active document type.

Also, you can customize all toolbars (including existing toolbars) by adding commands and special controls.

To create a toolbar (or edit an existing one) use the **Toolbars** tab of the **Customize** dialog accessible from the **Tools - Customize** menu. Customizing a toolbar is as easy as dragging commands from the dialog to the toolbars, as shown in the illustration below.

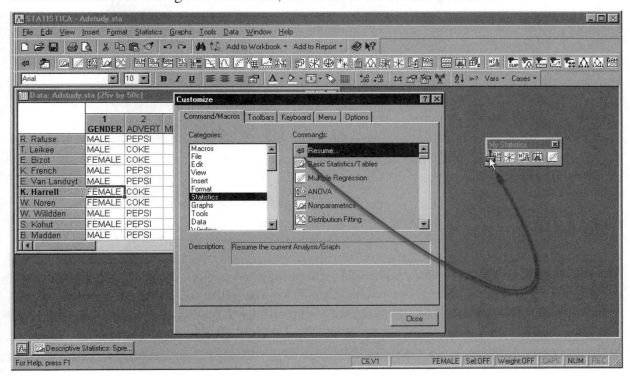

Shapes and locations of toolbars can be easily adjusted (e.g., all toolbars can be docked or free floating). All of these options make it possible for you to create unique toolbars that provide you with a very specialized user interface. The *Electronic Manual* includes simple to follow, step-by-step instructions on how to make those customizations. See *Create New Toolbar*, page 57, for more details.

User-defined menus. Customizing the menus is equally easy and can be performed using the **Menu** tab of the **Customize** dialog shown above (see *Customize Menu*, page 59, or the *Electronic Manual* for details).

2. *STATISTICA* Visual Basic and Controlling *STATISTICA* from Other Applications

The industry standard *STATISTICA* Visual Basic language (integrated into *STATISTICA*) provides another (alternative) user interface to the entire functionality of *STATISTICA*, and it offers incomparably more than just a "supplementary application programming language" that can be used to write custom extensions. *STATISTICA* Visual Basic takes full advantage of the object model architecture of *STATISTICA* and can be used to access programmatically every aspect and virtually every detail of the functionality of *STATISTICA*. Even the most complex analyses and graphs can be recorded into Visual Basic macros and later be run repeatedly or edited and used as building blocks of other applications. *STATISTICA* Visual Basic adds an arsenal of more than 10,000 new functions to the standard comprehensive syntax of Microsoft Visual Basic, thus comprising one of the largest and richest development environments available. For more information, see Chapter 11 - *STATISTICA Visual Basic*.

Controlling *STATISTICA* from other applications. One of the features that make the *STATISTICA* Visual Basic environment so powerful is the ability to integrate and manipulate various applications and their environments into a single program. For example, you can record or write a *STATISTICA* Visual Basic program that computes predictions via the *STATISTICA* **Time Series** module and execute that program from within an Excel spreadsheet or a Microsoft Word document. The exchange of information between different applications is accomplished by exposing those applications to the Visual Basic programs as Objects. So, for example, you can run statistical analyses in the *STATISTICA* **Basic Statistics** module from a Visual Basic program in Excel by declaring inside the program an object of type **Statistica.Application**.

Once an object has been created, the Visual Basic program then has access to the properties and methods contained in that object. Properties can be mostly thought of as variables, methods can be mostly thought of as subroutines or functions that perform certain operations or computations inside the respective application object. You can call *STATISTICA* procedures directly from many other applications and programming languages (e.g., C++, Java, and others).

3. Web Browser-Based
User Interface

Because *STATISTICA* is fully Web enabled, both "output from" and "input into" *STATISTICA* can be handled via the Internet. In addition to sending output from analyses directly into a Web server, *STATISTICA* provides methods to set up very sophisticated, automated systems that run scripts that can, for example, periodically draw data from external sources, perform specific sets of analyses on them, and update the HTML output on specified Web servers.

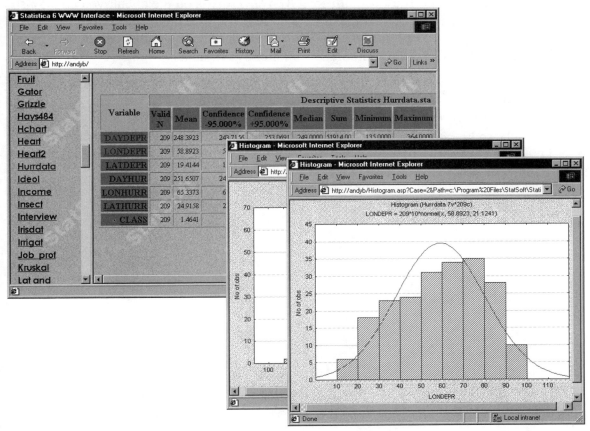

Note that these features are offered only in the versions properly licensed for Web server installation.

STATISTICA can also be run via custom-tailored Web browser-based user interfaces, where an Internet browser is serving as a front end for *STATISTICA* (a "thin client"). Thus, the analytic power of *STATISTICA* can be accessed from any computer in the world as long as it is connected to the Internet. *Note that accessing STATISTICA this way requires appropriate licensing and cannot be used with a stand-alone version of STATISTICA.*

STATISTICA Web Server (a technical note). This latter functionality is well supported in *STATISTICA* because its comprehensive automation interfaces (based on the object model architecture) can be executed via a scripting language (e.g., VBScript, Jscript, or C++). If *STATISTICA* is installed on the server, then you can access the automation model to process data and get back results that you can send down to the client in the form of HTML tables and graphics (JPG/PNG) files. Moreover, if *STATISTICA* is on the client, you could also use client-side scripting and take advantage of the fact that your Internet browser is an ActiveX document host, allowing you to open *STATISTICA* Spreadsheets and Graphs within the browser and still have the *STATISTICA* toolbars and menus available.

Note also that this Web browser-based user interface (which is offered from StatSoft in form of a selection of simplified "dialog templates" invoked from the Internet browser, allowing you to easily perform common types of analyses and graphs) can be flexibly customized by modifying the scripts of analyses that are executed on the server side. Therefore, this browser-based user interface is not only fully customizable, but it also allows you to easily incorporate in it other applications that can interface with *STATISTICA*.

CUSTOMIZING THE OPERATION AND APPEARANCE OF *STATISTICA*

Overview

STATISTICA offers the flexibility of fully customizable user interfaces. The program recognizes the necessity of adjusting the standard user interface to better suit your specific needs. In fact, *STATISTICA* "anticipates" your needs in that it remembers various choices as you make them, essentially learning as you go. For example, if you launch an analysis from the **Advanced** tab on an analysis specification dialog, then the **Advanced** tab will be selected for you (instead of the **Quick** tab) the next time you display that dialog.

Practically all aspects of the user interface can be customized starting with such elementary controls as the menus, toolbars, and the keyboard. The process for customizing these screen components is quick and straightforward (for example, see the illustration of customizing the toolbar on page 47). You can set both global and local customizations for graphs, spreadsheets, workbooks, reports, etc., and maintain different configurations of *STATISTICA* (for a single user as well as for network users). You can also define entirely new user interfaces (see *Toolbar Overview*, page 56, *Menu Overview*, page 59, and *Keyboard Overview*, page 61).

Customization of the Interactive User Interface

As mentioned before, *STATISTICA* offers facilities to define entirely new user interfaces including the Internet browser-based user interfaces (see page 49). However, also practically all aspects of the default, interactive user interface can easily be adjusted in a variety of ways. For example, you can either add to the default options, simplify them, or keep changing them as your needs change. Depending on the requirements of the tasks to be performed as well as your personal preferences for particular "modes" of work (and aesthetic choices), you can suppress all icons, toolbars, status bars, long menus, workbook facilities, drag-and-drop facilities, dynamic (automatic) links between graphs and data, 3D effects in tables, and 3D effects in dialog boxes; request "bare-bones" sequential output with simple, paper-white

spreadsheets and monochrome graphs; and set the system to automatically maintain no more than one simple report at a time (see the left panel on the illustration, below);

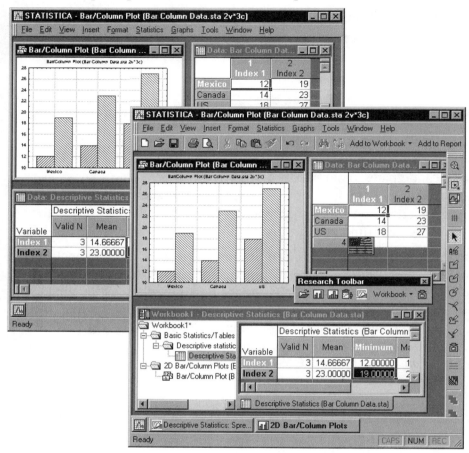

or alternatively, you could define elaborate local and global toolbars; take full advantage of all special tools and controls, icons, toolbars, macros (e.g., assign particular tasks to specific new menu options, toolbars, or keys), elaborate multimedia tables, workbook facilities, and drag-and-drop facilities; establish multiple dynamic (automatic) links between graphs and data and internal links between graphical objects; customize the output windows with colors, special fonts, and highlights; adjust the default graph styles and their display modes; and send the results to separate hierarchically organized workbooks to create an elaborate, "multi-layered" data analysis environment that facilitates the exploration of complex datafiles and allows you to compare different aspects of the output (see the right panel on the illustration, above).

Customization
of Documents

There are a variety of comprehensive, specialized tools to customize the layout and operation of *STATISTICA* (and its documents). For example, *STATISTICA* has a comprehensive system of managing defaults of every aspect of graphs and combining customizations into hierarchically organized "styles." Similarly, you can create custom layouts and formats for spreadsheets (multimedia tables) and even customize events (e.g., what happens when you double-click on a table). See the *Electronic Manual* for further details.

Local vs. Permanent
Customizations

Many aspects of the appearance of *STATISTICA* can be adjusted in both the **View** and **Tools** menus. Each of these two methods, however, has a different function. Specifically:

View menu. The changes requested in the **View** menu affect the current appearance of *STATISTICA* (e.g., hides the toolbar) or the current document window (e.g., changes font in the spreadsheet).

Tools menu. The options available via the **Tools – Options** menu (discussed in more detail in the next section) are used to adjust the permanent program defaults. Note, however, that the global options that are applicable to documents of a particular type (e.g., a graph or a spreadsheet) will not change the current document. Instead, they will only be stored as program defaults that will affect the creation of the next (i.e., new) document of the respective type.

For example, if you change the **Default Spreadsheet Layout** on the **Spreadsheets** tab of the **Options** dialog (available from the **Tools – Options** menu), you will see the new Spreadsheet Layout applied only when you create a new spreadsheet. These defaults will not affect any files opened from the disk, however, because those spreadsheets are displayed with the specific appearance with which they were previously saved (use the **View** menu to customize the existing objects).

General Defaults

Customization of the general system defaults. The general default settings can be adjusted at any point in *STATISTICA*. They can be adjusted on the respective tabs of the **Options** dialog (accessible via the **Tools - Options** menu).

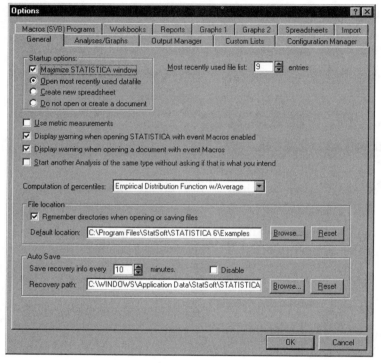

They control:

- The general aspects of the behavior of *STATISTICA* (such as maximizing *STATISTICA* on startup, workbook and report facilities, file locations, or custom lists, etc.),

- The way in which the output is produced (e.g., analysis workbooks, analysis reports, etc.),

- The general appearance of the application window (icons, toolbars, etc.), and

- The appearance of document windows.

All these and other general settings are accessible regardless of the type of the document window that is currently active (e.g., a spreadsheet or a graph). For more information about a specific tab, see the *Electronic Manual*.

Switching between alternative sets of defaults (configurations). Options are provided on the **Configuration Manager** tab of the **Options** dialog to maintain "libraries" of settings and switch between them for different projects (or users). For further details, see *Maintaining Different Configurations of STATISTICA*, page 56, and the **Configuration Manager** tab in the *Electronic Manual*.

Graph Customization

Interactive graph customization. The customization options in *STATISTICA* graphics include hundreds of features and tools that can be used to adjust every detail of the display and associated data processing. However, these options are arranged in a hierarchical manner, so those used most often are accessible directly via shortcuts by double-clicking or right-clicking on the respective element of the graph.

Permanent settings and automation options. The initial (default) settings of all of these features can be easily adjusted so that even the default appearance and behavior of *STATISTICA* graphs will match your specific needs and/or will require very little intervention on your part. Various aspects of *STATISTICA* Graphs can be permanently adjusted by using:

1. the **Options** dialog (accessible from the **Tools - Options** menu),

2. the comprehensive system of graph styles,

3. user-defined graphs, and

4. *STATISTICA* Visual Basic.

They are briefly reviewed in Chapter 6 – *STATISTICA Graphs: General Features*. For more information, please refer to the *Electronic Manual*.

There are no limits to how "deeply customized" your *STATISTICA* custom graphs can be, because *STATISTICA* Visual Basic (with all its powerful custom drawing tools as well as the *STATISTICA*-based library of graphics procedures) can be used to produce virtually any graphics or multimedia output supported by the contemporary computer hardware. Those custom developed displays or multimedia output can be assigned to *STATISTICA* toolbars, menus, or dialogs and become a permanent part of "your" *STATISTICA* application.

See Chapter 7 - *STATISTICA Graphs: Creation and Customization* or the *Electronic Manual* for further details on these graph customization methods.

Maintaining Different Configurations of *STATISTICA*

STATISTICA stores all program settings when you exit *STATISTICA*, and restores them the next time you start the application. You can create different configurations of these settings by using the **Configuration Manager** tab of the **Options** dialog (available via the **Tools – Options** menu). With the configuration manager, you can save the current program state into a new or existing configuration, or you can restart *STATISTICA* using a different configuration. Other options include the ability to import or export configuration to a separate file so they can be shared among *STATISTICA* installations.

Customized Configurations for Individual Users on a Network

The same principle described in the previous paragraph applies to network installations of *STATISTICA*. On a network, *STATISTICA* is installed in only one location (on a server), but each user can still configure *STATISTICA* differently because the setting configuration information is stored locally. Note that you need to choose **Network Installation** in the **STATISTICA Setup** program in order to install it properly on a non-local drive (network server). Note that a network version of *STATISTICA* is necessary to assure its reliable operation when used by more than one user at a time or even one user if *STATISTICA* is not installed on the local system.

Toolbar Overview

STATISTICA includes a set of toolbars that contain the most needed buttons or special controls for a given task. By default, the menu bar, **Standard** toolbar, and **Spreadsheet** toolbar are visible when *STATISTICA* is initially opened. Other toolbars are visible when they are needed and the toolbar(s) associated with the respective document type(s) replace the **Spreadsheet**

toolbar. For example, when a graph is the active document, the **Graph Tools** toolbar is visible. Toolbar visibility can be toggled from the **View – Toolbars** menu as well as from the **Toolbars** tab of the **Customize** dialog (available via the **Tools – Customize** menu).

Note that *STATISTICA* toolbars are hidden during in-place activation of non-*STATISTICA* objects. Therefore, you should not expect to see *STATISTICA* toolbars when editing a Microsoft Word (or similar document) that is embedded in a *STATISTICA* document.

In addition to the default set of toolbars, you can create your own toolbars using any combination of buttons or special controls (e.g., font name and size controls, font color controls, or graph styles). As with the *STATISTICA* toolbars, visibility can be dependent on the activity for which the toolbar is needed.

Create New Toolbar

You can create a new toolbar using the **Customize** dialog. To display that dialog, select **Customize** from the **Tools** menu (or the **View - Toolbars** menu).

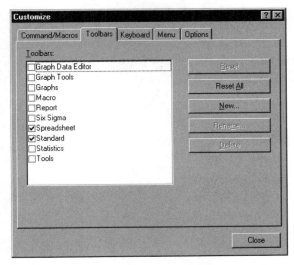

To create a new toolbar, click the **New** button on the **Toolbars** tab of the **Customize** dialog. This displays the **Toolbar Name** dialog. In the **Toolbar Name** box, type the name of the new toolbar (e.g., My Statistics) and click the **OK** button. A small block (empty) toolbar is displayed in the middle of the screen.

Once you have created the toolbar, you can build the toolbar using various tabs of the *Customize* dialog (see below).

Customize Toolbar

Adding a toolbar button. When you create a new toolbar (see above), you will notice the new toolbar (which will remain blank until you add buttons to it) floating in the middle of your screen; drag this floating toolbar down until it is located to the side of the dialog. With the toolbar in this position, you can easily add toolbar buttons using drag-and-drop features. To do this, select the toolbar button from an existing toolbar to be added to the new toolbar, and then drag it onto the new toolbar. Note that this will remove the toolbar button from its previous toolbar and add it to the new toolbar. To copy the toolbar button (and hence it will remain on the previous toolbar and be added to the new one), press the CTRL key while you are dragging it (a plus sign will be added to the mouse pointer ⬚).

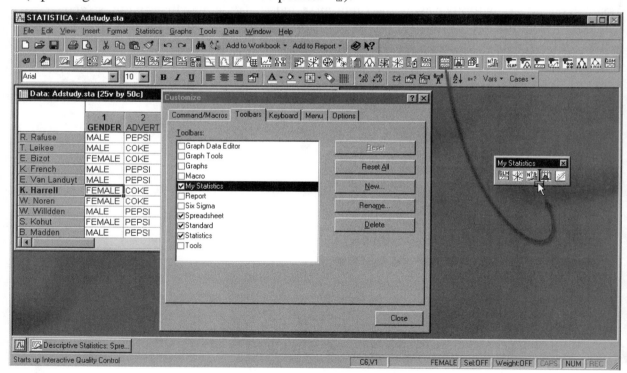

Note that you can add toolbar buttons to other toolbars without the *Customize* dialog displayed. To do this, press and hold the ALT key, select the toolbar button, and then drag it onto a different toolbar. You can copy the button by pressing both the ALT and the CTRL keys.

Menu Overview

Like toolbars (see *Toolbar Overview*, page 56), menus in *STATISTICA* are customizable. Although they cannot house all the special controls that are available for toolbars (see *Toolbars and Menus in STATISTICA*, page 61, for a comparison of the two types of command bars), it is possible to customize existing menus by adding simple options. Essentially, you can drag any command or macro onto a menu (see below). Additionally, you can customize shortcut menus (the menus accessed by right-clicking your mouse).

The current view affects what menu bars are displayed. For example, if you are viewing a graph, the menu bar for graphs will be visible with options for layouts, drawing objects, and inserting objects, as well as other options. Graph specific menus (e.g., **Layouts**, **Draw**, and **Object**) would not be available from the menu bar when a spreadsheet or macro is active.

Unlike toolbars, which can be toggled via the **Toolbars** tab of the **Customize** dialog (available via the **Tools – Customize** menu) or the **View – Toolbars** menu command, menu bars cannot be turned off manually. However, they are replaced with real menus for in-place editing of non-*STATISTICA* objects. For example, if you are editing a Microsoft Word document within a *STATISTICA* Workbook, a Microsoft Word menu bar is available.

There are six main menu bars in *STATISTICA*: graph, macro, non-document, report, spreadsheet, and workbook. Fourteen additional menu bars exist and are available when one type of *STATISTICA* document is being edited within another *STATISTICA* document (e.g., the report graph menu bar is available when a graph embedded within a report is edited).

Customize Menu

Adding a command. Commands (menu options) and global macros can be added to toolbars using the **Commands** tab of the **Customize** dialog. Select a category of functions from the **Categories** list to display the available commands in the **Commands** list. Once you have located the appropriate command, drag the command from the **Commands** list to the toolbar.

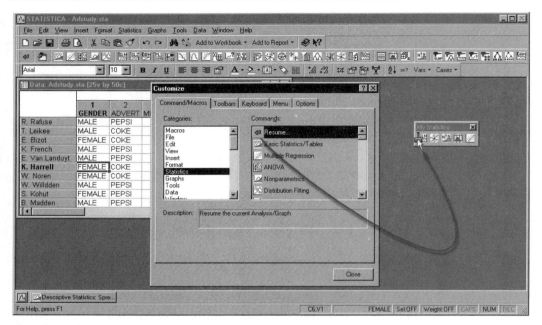

Continue this process for each command you would like to add to the toolbar. When you are finished, click the **Close** button. Note that commands can also be dragged from the menus themselves in a similar manner (when the **Customize** dialog is displayed), as illustrated below.

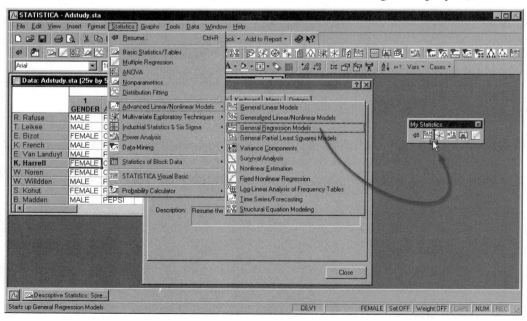

Removing a toolbar button or command. To remove toolbar buttons or commands from any toolbar, first select *Customize* from the *Tools* menu to display the *Customize* dialog, then right-click on the item to be removed and select *Delete* from the resulting shortcut menu.

Toolbars and Menus in *STATISTICA*

Although both toolbars and menus can be customized in *STATISTICA*, there are a few differences between the two items. Primarily, toolbars can be customized with a greater variety of controls than menus, and user-defined toolbars are easier to create than user-defined menus.

Special controls vs. simple buttons. As mentioned in the *Toolbar Overview*, in addition to housing standard commands and macros, toolbars can also be customized with a variety of special controls. For example, it is possible to add *Font Size* and *Name* controls, *Font Colors*, *Graph Styles*, and other specialized controls that would normally appear on a toolbar as a drop-down list box. Menus, on the other hand, can only be customized with simple macros and commands (or buttons).

Turning off toolbars. Another difference between toolbars and menu bars is that toolbars can easily be turned off (via the *Customize* dialog or the *View - Toolbars* menu), while menu bars cannot be turned off without some programming effort. Essentially, this ensures that menus (that are indispensable for normal operation of *STATISTICA)* will always be available regardless of whether a *STATISTICA* document is open or not. For a list of the menus available in different situations (i.e., when a report is open, when a spreadsheet is open, etc.), refer to the *Menu Overview* (page 59).

User-defined toolbars. An additional difference between toolbars and menus is that you can create up to sixteen user-defined toolbars using the simple facilities illustrated above, whereas some programming effort is necessary to create user-defined menus. Note however, that you can add or remove items from existing menus when the *Customize* dialog is displayed in a similar manner as toolbar buttons, see above.

Keyboard Overview

Keyboard shortcuts (e.g., accelerator keys or shortcut keys) can be added for any command. Using the *Customize* dialog available from the *Tools* menu, you can specify the specific command and when it will work. More specifically, you can specify that the created hot key will work when any *STATISTICA* document is open (e.g., a global shortcut) or only when specific document type (e.g., a report or workbook) is open. For example, you could assign

CTRL+SHIFT+/ (i.e., CTRL+SHIFT+?) to the **Help** menu's **Contents and Index** command. Pressing CTRL+SHIFT+/ would then launch the *STATISTICA Electronic Manual*.

Keyboard Interface

Keyboard combinations can be used as shortcut keys in order to access various *STATISTICA* procedures. Shortcut keys can be classified as global (available in every window) or local (for specific windows). You can assign a shortcut key to a macro via the **Customize** dialog (available via **Tools – Customize**).

Global Shortcut Keys

The following shortcut keys are available in all document windows:

File I/O:	Open file of the current type	CTRL+O
	Save	CTRL+S
	Save As	F12
	Create New Document	CTRL+N
Printing:	Print	CTRL+P
Edit:	Select All	CTRL+A
	Clear Selection	DEL
Clipboard:	Cut	CTRL+X
		SHIFT+DELETE
	Copy	CTRL+C
		CTRL+INSERT
	Paste	CTRL+V
		SHIFT+INSERT
Screen Catcher:	Capture Rectangle	ALT+F3
Statistics:	Resume	CTRL+R
Graphs:	Resume	CTRL+R
STATISTICA Visual Basic:	SVB Editor	ALT+F11
Windows:	Close All	CTRL+L

Cascade	SHIFT+F6
Tile Horizontally	ALT+F6
Tile Vertically	ALT+SHIFT+F6
Help	F1
Exit	ALT+F4
Switch to	CTRL+ESC
Context Sensitive Help	SHIFT+F1

Spreadsheet Window Shortcut Keys

In addition to the common (global) keys available in every window, the following keyboard combinations are specific to the spreadsheet window:

Recalculate:		
	Recalculate All Formulas	F9
	Displays **Recalculate** dialog	SHIFT+F9
Edit:		
	Bold	CTRL+B
	Italic	CTRL+I
	Underline	CTRL+U
	Undo	CTRL+Z
	Redo	CTRL+Y
	Expanding Blocks	SHIFT+cursor
	Edit Cell	F2
	Enter New Line	CTRL+ENTER or ALT+ENTER
	Find	CTRL+F
	Replace	CTRL+H
	Repeat Find/Replace	F3
	Go To	CTRL+G

Expanding blocks. You can also expand blocks by positioning the cursor in one corner of the block to be selected, scrolling to the location of the opposite (diagonal) corner, and clicking in that corner location while holding down the SHIFT key.

Report Window Shortcut Keys

The following keyboard combinations (local shortcut keys) are specific to the *STATISTICA* Report window:

Edit:	Undo	CTRL+Z
	Redo	CTRL+Y
	Find	CTRL+F
	Replace	CTRL+H
	Repeat Find/Replace	F3
	Page Break	CTRL+ENTER
Format:	Font	CTRL+F9
Characters:	Regular	CTRL+0 (zero)
	Bold	CTRL+B
	Italic	CTRL+I
	Underline	CTRL+U
	Switch Focus	CTRL+F2

Graph Window Shortcut Keys

The following keyboard combinations (local shortcut keys) are specific to the graph window (note that the Object Alignment and Format shortcut keys require that an object be selected):

Object Alignment:	Alignment Grid	CTRL+G
	Snap to Grid	hold TAB
Format:	Move to Front	+
	Move to Back	CTRL+- (- on number pad)
	Move Forward	=
	Move Back	-
Edit:	Undo	CTRL+Z*
	Redo	CTRL+Y

* This is a multilevel undo with a separate queue maintained for each graph.

Rotating text using cursor keys. Select the added text and use the PAGE DOWN and PAGE UP shortcut keys to rotate text objects selected in the graph clockwise or counter-clockwise, respectively, in 5° increments. To rotate in 1° increments, hold down the CTRL key while pressing PAGE DOWN and PAGE UP.

Moving and resizing objects using cursor keys. Select an object and position the cursor on the object (to resize or move it), and then use the keyboard cursor keys to drag it to a new position.

Navigating through graph objects. Select an object and then press the TAB key to move from object to object in your graph.

Macro Window Shortcut Keys

The following keyboard combinations (local shortcut keys) are specific to the macro window:

Edit:	Undo	CTRL+Z
	Redo	CTRL+Y
	Find	CTRL+F
	Replace	CTRL+H
	Repeat Find/Replace	F3
View:	Object Browser	F2
Debug:	Step Into	F8
	Step Over	SHIFT+F8
	Step Out	CTRL+F8
	Run to Cursor	F7
	Add Watch	CTRL+F9
	Quick Watch	SHIFT+F9
	Toggle Breakpoint	F9
Run:	Run Macro	F5
	Break	CTRL+BREAK
	Pause	CTRL+ESC

Workbook Window Shortcut Keys

The following keyboard combinations (local shortcut keys) are specific to the workbook window:

Workbook:	Rename Item in the Tree Pane	F2
	Delete	DEL
	Cut	SHIFT+DEL
	Previous Page	PAGE UP
	Next Page	PAGE DOWN
	Previous Document	UP ARROW
	Next Document	DOWN ARROW
	Top of Workbook	HOME
	Bottom of Workbook	END
	Switch Focus	CTRL+F2
	Insert Workbook Item dialog	INSERT

STATISTICA Query Shortcut Keys

The following keyboard combinations (local shortcut keys) are specific to *STATISTICA* Query:

Query:	Return Data to *STATISTICA*	F5
	Refresh Query	F5
	Cancel Query	SHIFT+F5

GENERAL MENUS AND TOOLBARS

Statistics Menu

The following commands are accessible from the **Statistics** menu, which is available whenever any *STATISTICA* document is open. The following sections include brief overviews of the statistical procedures available with each command; for more information, see the *Electronic Manual*.

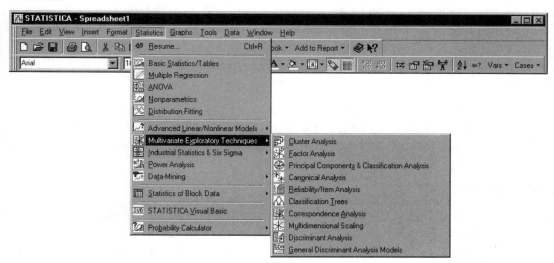

The top portion of the **Statistics** menu includes several most commonly used statistical procedures (or modules) and they are included in the **STATISTICA Base** package. As mentioned before, it is very easy in *STATISTICA* to customize the menus and, for example, add to this list additional modules that you use most often.

The middle portion of the **Statistics** menu includes a set of hierarchical options representing all add-on packages available for *STATISTICA*. For more information on the various add-on packages for *STATISTICA*, refer to the *Electronic Manual* and the StatSoft Web page (www.statsoft.com).

The remainder of the **Statistics** menu provides access to other commonly used tools for analyzing data.

 Resume

Select **Resume** from the **Statistics** menu (or press CTRL+R) to continue the most recently displayed analysis or graph. This option is particularly useful if you want to display the previous dialog after you have created a results spreadsheet or graph.

Basic Statistics/Tables

Select **Basic Statistics/Tables** from the **Statistics** menu to display the **Basic Statistics and Tables** (Startup Panel). The **Basic Statistics and Tables** module provides options for calculating descriptive statistics, correlation analyses, *t*-tests for dependent, independent, and single samples, breakdowns and one-way ANOVAs, frequency tables, crosstabulation and stub-and-banner tables, and multiple response/dichotomies tables. You can also access the probability calculator via this module as well as perform other significance tests using only summary statistics (e.g., tests for differences between proportions, correlations, and means when all you know are the hypothesized values and sample statistics).

See also the **Nonparametric Statistics** module for alternative tests and statistics similar to those included in the **Basic Statistics and Tables** module; those tests do often not rely on any particular assumptions (e.g., regarding underlying distributions) and tend to be more robust.

Multiple Regression

Select **Multiple Regression** from the **Statistics** menu to display the **Multiple Linear Regression** (Startup Panel). The general purpose of multiple regression is to analyze the relationship between several independent or predictor variables and a dependent or criterion variable.

This module will perform least-squares multiple linear regression and compute detailed residual statistics. Methods for forward and backwards stepwise selection of predictor variables are also provided. Both intercept and non-intercept models can be evaluated, and different methods can be chosen for computing the overall ANOVA table and R-square statistic (for non-intercept models). *STATISTICA* will compute all standard multiple regression results statistics, and perform extensive residual analyses. Predicted values can also be computed for user-defined values of the predictors.

See also the **General Regression Models (GRM)** module for complex linear models, best-subset selection, and other advanced techniques; see the **General Linear Models (GLM)** module to fit complex linear models; see **Generalized Linear/Nonlinear Models (GLZ)** and **Nonlinear**

Estimation to fit nonlinear models; see also *Generalized Additive Models (GAM)*, *Partial Least Squares Models (PLS)*, and *General Discriminant Analysis Models (GDA)* for more specialized procedures.

ANOVA

Select *ANOVA* from the *Statistics* menu to display the *General ANOVA/MANOVA* (Startup Panel). The purpose of analysis of variance (*ANOVA*) is to test for significant differences between means in different groups or variables (measurements), usually arranged by an experimenter in order to evaluate the effects of different treatments or experimental conditions, or combinations of treatments or conditions, on one (*ANOVA*) or more (*MANOVA*) outcome measures (dependent variables).

The *ANOVA/MANOVA* module is a subset of the *General Linear Models (GLM)* module, and can perform univariate (ANOVA) and multivariate (MANOVA) analysis of variance of factorial designs with or without a repeated measure. *STATISTICA* will use, by default, the sigma restricted parameterization for factorial designs, and apply the effective hypothesis approach (see Hocking, 1985) when the design is unbalanced or incomplete. Type I, II, III, and IV hypotheses can also be computed, as can Type V and Type VI hypotheses that will perform tests consistent with the typical analyses of fractional factorial designs in industrial and quality improvement applications (see also the *Experimental Design* module). Results include summary ANOVA tables, univariate and multivariate results for repeated measures factors with more than 2 levels, the Greenhouse-Geisser and Huynh-Feldt adjustments, plots of interactions, detailed descriptive statistics, detailed residual statistics, planned and post-hoc comparisons, testing of custom hypotheses and custom error terms, detailed diagnostic statistics and plots (e.g., histogram of within-cell residuals, homogeneity of variance tests, plots of means versus standard deviations, etc.)

See also the *General Linear Models (GLM)* module to analyze any kind of linear model with categorical and/or continuous predictor variables, random effects, and multiple repeated measures factors; stepwise and best-subset selection of predictor effects is available in the *General Regression Models (GRM)* module; to fit nonlinear models see the *Generalized Linear/Nonlinear Models (GLZ)* and *Nonlinear Estimation*; see also *Generalized Additive Models (GAM)*, *Partial Least Squares Models (PLS)*, and *General Discriminant Analysis Models (GDA)* for more specialized procedures.

 Nonparametrics

Select **Nonparametrics** from the **Statistics** menu to display the **Nonparametric Statistics** (Startup Panel). Nonparametric statistics were developed for use in cases when the researcher does not know the parameters of the distribution of the variable of interest in the population (hence the name nonparametric). In more technical terms, nonparametric methods do not rely on the estimation of parameters (such as the mean or the standard deviation) describing the distribution of the variable of interest in the population. Therefore, these methods are also sometimes (and more appropriately) called parameter-free methods or distribution-free methods.

The **Nonparametric Statistics** module features a comprehensive selection of inferential and descriptive statistics including all common tests and some special application procedures. Available statistical procedures include the Wald-Wolfowitz runs test, Mann-Whitney U test (with exact probabilities for small samples), Kolmogorov-Smirnov tests, Wilcoxon matched pairs test, Kruskal-Wallis ANOVA by ranks, Median test, Sign test, Friedman ANOVA by ranks, Cochran Q test, McNemar test, Kendall coefficient of concordance, Kendall *Tau* (b, c), Spearman rank order R, Fisher's exact test, *Chi*-square tests, V-square statistic, *Phi*, *Gamma*, Sommer's d, contingency coefficients, and others.

Specialized nonparametric tests and statistics are also part of many modules, e.g., **Survival Analysis**, *STATISTICA* **Process Analysis**, and others.

 Distribution Fitting

Select **Distribution Fitting** from the **Statistics** menu to display the **Distribution Fitting** (Startup Panel). The **Distribution Fitting** module is used to fit a variety of continuous and discrete distributions to the data.

The fit can be evaluated via the *Chi*-square test or the Kolmogorov-Smirnov one-sample test (the fitting parameters can be controlled); the Lilliefors and Shapiro-Wilk's tests are also supported. In addition, the fit of a particular hypothesized distribution to the empirical distribution can be evaluated in customized histograms (standard or cumulative) with overlaid selected functions; line and bar graphs of expected and observed frequencies, discrepancies and other results can be produced from the output spreadsheets.

Other distribution fitting options are available in *STATISTICA* **Process Analysis**, where you can compute maximum-likelihood parameter estimates for the *Beta*, Exponential, Extreme Value (Type I, Gumbel), *Gamma*, Log-Normal, Rayleigh, and Weibull distributions. Also

included in that module are options for automatically selecting and fitting the best distribution for the data, as well as options for general distribution fitting by moments (via Johnson and Pearson curves). User-defined 2- and 3-dimensional functions can also be plotted and overlaid on the graphs. The functions can reference a wide variety of distributions such as the *Beta*, Binomial, Cauchy, *Chi*-square, Exponential, Extreme value, F, *Gamma*, Geometric, Laplace, Logistic, Normal, Log-Normal, Pareto, Poisson, Rayleigh, *t* (Student), or Weibull distribution, as well as their integrals and inverses.

Advanced Linear/Nonlinear Models

The options on the *Advanced Linear/Nonlinear Models* menu offer a wide array of the most advanced modeling and forecasting tools on the market, including automatic model selection facilities and extensive interactive visualization tools. Note that these modules are available as part of the optional add-on package *STATISTICA Advanced Linear/Non-Linear Models*.

General Linear Models. Select *General Linear Models* from the *Statistics - Advanced Linear/Nonlinear Models* menu to display the *General Linear Models (GLM)* (Startup Panel). The *General Linear Models (GLM)* module provides a generalization of the linear regression model, such that effects can be tested (1) for categorical predictor variables, as well as for effects for continuous predictor variables and (2) in designs with multiple dependent variables as well as in designs with a single dependent variable.

GLM is a complete implementation of the *General Linear Model*. You can choose simple or highly customized one-way, main-effect, factorial, or nested ANOVA or MANOVA designs, repeated measures designs, simple, multiple and polynomial regression designs, response surface designs (with or without blocking), mixture surface designs, simple or complex analysis of covariance designs (e.g., with separate slopes), or general multivariate MANCOVA designs. Factors can be fixed or random (in which case synthesized error terms will be computed). *GLM* offers both the overparameterized and *Sigma*-restricted parameterization for categorical factor effects. *STATISTICA* will compute the customary Type I through IV sums of squares for unbalanced and incomplete designs; *GLM* also offers two additional methods for analyzing missing cell designs: Hockings (1985) "effective hypothesis decomposition," and a method that will automatically drop effects that cannot be fully estimated (e.g., when the least squares means do not exist for all levels of the respective main effect or interaction effect). The latter method is the one commonly applied to the analysis of highly fractionalized designs in industrial experimentation (see also Experimental Design). Results statistics computed by *GLM* include ANOVA tables with univariate and multivariate tests, descriptive statistics, a comprehensive selection of different types of plots of means (observed, least squares, weighted) for higher-order interactions, with error bars (standard errors) for effects involving

between-group factors as well as repeated measures factors; extensive residual analyses and plots (for the "training" or computation sample, for a cross-validation or "verification" sample, or for a prediction sample), desirability profiling, specifications of custom error terms and effects; comprehensive post-hoc comparison methods for between-group effects as well as repeated measures effects, and the interactions between repeated measures and between effects including: Fisher LSD, Bonferroni, Scheffé, Tukey HSD, Unequal N HSD, Newman Keuls, Duncan, and Dunnett's test (with flexible options for estimating the appropriate error terms for those tests), tests of assumptions (e.g., Levene's test, plots of means vs. standard deviations, etc.).

The *General Regression Models (GRM)* module offers methods for stepwise and best-subset selection of effects in a general linear model; see also the *Generalized Linear/Nonlinear Model (GLZ)* module for non-linear alternatives to *GLM*.

Generalized Linear/Nonlinear Models. Select *Generalized Linear/Nonlinear Models* from the *Statistics - Advanced Linear/Nonlinear Models* menu to display the *Generalized Linear/Nonlinear Models* (Startup Panel). The *Generalized Linear/Nonlinear Models (GLZ)* module provides a generalization of the linear regression model such that (1) nonlinear, as well as linear, effects can be tested (2) for categorical predictor variables, as well as for continuous predictor variables, using (3) any dependent variable whose distribution follows several special members of the exponential family of distributions, as well as for any normally distributed dependent variable.

A wide range of distributions (from the exponential family) can be specified for the response variable: Normal, Poisson, *Gamma*, Binomial, Multinomial, Ordinal Multinomial, and Inverse Gaussian. Available link functions include: Log, Power, Identity, Logit, Probit, Complimentary Log-Log, and Log-Log links. In addition to the standard model fitting techniques, *GLZ* also provides unique options for exploratory analyses, including model building facilities like forward- or backward-only selection of effects (effects can only be selected for inclusion or removal once during the selection process), standard forward or backward stepwise selection of effects (effects can be entered or removed at each step, using a p to enter or remove criterion), and best-subset regression methods (using the likelihood score statistic, model likelihood, or Akaike information criterion). These methods can be applied to categorical predictors (ANOVA-like designs; effects will be moved in or out of the model as multiple-parameter blocks) as well as continuous predictors. The module will compute all standard results statistics, including likelihood ratio tests, and Wald and score tests for significant effects, parameter estimates and their standard errors and confidence intervals, etc. In addition, for ANOVA-like designs, tables and plots of predicted means (the equivalent of least squares means computed in the general linear model) with their standard errors could be computed to

aid in the interpretation of results. **GLZ** also includes a comprehensive selection of model checking tools such as spreadsheets and graphs for various residuals and outlier detection statistics, including raw residuals, Pearson residuals, deviance residuals, studentized Pearson residuals, studentized deviance residuals, likelihood residuals, differential *Chi*-square statistics, differential deviance, and generalized Cook distances, etc. Predicted and residual statistics can be requested for observations that were used for fitting the model and those that were not (i.e., for the cross-validation sample).

STATISTICA also includes the **Nonlinear Estimation** module for fitting arbitrary regression functions using least-squares or user-defined loss functions; see also the *Data–Mining* modules (page 94) for extensions to the **Generalized Linear/Nonlinear Model**.

General Regression Models. Select **General Regression Models** from the **Statistics - Advanced Linear/Nonlinear Models** menu to display the **General Regression Models** (Startup Panel). The **General Regression Models (GRM)** module is called a "general" regression models program because it applies the methods of the general linear model, allowing it to build models for designs with multiple-degrees-of-freedom effects for categorical predictor variables, as well as for designs with single-degree-of-freedom effects for continuous predictor variables. **GRM** implements stepwise and best-subset model-building techniques for univariate and multivariate analysis of variance (ANOVA/MANOVA), regression, and analysis of covariance (ANCOVA/MANCOVA) designs.

The **General Regression Models (GRM)** module offers all standard and unique results options described in the context of the **GLM** module, including desirability profiling, predicted and residual statistics for the computation or training sample, cross-validation or verification sample, and prediction sample; tests of assumptions, means plots, etc. In addition, unique regression-specific results options are also available, including Pareto charts of parameter estimates, whole model summaries (tests) with various methods for evaluating no-intercept models, partial and semi-partial correlations, etc.

To analyze simple or complex ANOVA/ANCOVA or MANOVA/MANCOVA designs, see also the **General Linear Models (GLM)** module; *STATISTICA* also includes an implementation of **Generalized Linear/Nonlinear Models (GLZ)** and **Generalized Additive Models (GAM)**.

Partial Least Squares Models. Select **Partial Least Squares** from the **Statistics - Advanced Linear/Nonlinear Models** menu to display the **Partial Least Squares Models** (Startup Panel). Partial least squares is a linear regression method that forms components (factors, or latent variables) as new independent variables (explanatory variables, or predictors) in a regression model. The components in partial least squares are determined by both the response variable(s) and the predictor variables. A regression model from partial least squares can be

expected to have a smaller number of components without an appreciably smaller R-square value.

STATISTICA **PLS** implements the two most general algorithms for partial least squares analysis: **SIMPLS** and **NIPALS**. Like **GLM** and **GLZ**, **PLS** offers both the overparameterized and sigma restricted parameterization methods for categorical predictors in ANOVA/ANCOVA-like models. **PLS** will compute all the standard results for a partial least squares analysis, and also offers a large number of results options and in particular graphics options that are usually not available in other implementations; for example, graphs of parameter values as a function of the number of components, two-dimensional plots for all output statistics (parameters, factor loadings, etc.), two-dimensional plots for all residual statistics, etc. Also, like **GLM**, **GRM**, **GDA**, and **GLZ**, the **Partial Least Squares** module offers extensive residual analysis options, and predicted and residual statistics can be requested for observations that were used for fitting the model (the "training" sample), those that were not (i.e., the cross-validation or verification sample), and for cases without observed data on the dependent (response) variables (the prediction sample).

STATISTICA includes a large number of modules for fitting linear or nonlinear models, and for reducing the dimensionality of a variable space, e.g., **General Linear Models (GLM)**, **General Regression Models (GRM)**, **Generalized Linear/Nonlinear Models (GLZ)**, **Generalized Additive Models (GAM)**, and **Factor Analysis**, to name only a few.

⊞ Variance Components. Select **Variance Components** from the **Statistics - Advanced Linear/Nonlinear Models** menu to display the **Variance Components and Mixed Model ANOVA/ANCOVA** (Startup Panel). Variance components are used in the context of experimental designs with random effects, to denote the estimate of the (amount of) variance that can be attributed to those effects. For example, if you are interested in the effect that the quality of different schools has on academic proficiency, you could select a sample of schools to estimate the amount of variance in academic proficiency (component of variance) that is attributable to differences between schools.

The **Variance Components** module will allow you to analyze designs with any combination of fixed effects, random effects, and covariates. *STATISTICA* will analyze standard factorial (crossed) designs as well as hierarchically nested designs, and compute the standard Type I, II, and III analysis of variance sums of squares and mean squares for the effects in the model. In addition, you can compute the table of expected mean squares for the effects in the design, the variance components for the random effects in the model, the coefficients for the denominator synthesis, and the complete **ANOVA** table with tests based on synthesized error sums of squares and degrees of freedom (using Satterthwaite's method). Other methods for estimating variance components are also supported (e.g., MIVQUE0, Maximum Likelihood [ML], Restricted

Maximum Likelihood [REML]). For maximum likelihood estimation, both the Newton-Raphson and Fisher scoring algorithms are used, and the model will not be arbitrarily changed (reduced) during estimation to handle situations where most components are at or near zero. Several options for reviewing the weighted and unweighted marginal means, and their confidence intervals, are also available. Extensive graphics options can be used to visualize the results.

The ***Generalized Linear/Nonlinear Models (GLM)*** module will also compute variance components (based on denominator synthesis) of complex ANOVA/ANCOVA designs.

Survival Analysis. Select ***Survival Analysis*** from the ***Statistics - Advanced Linear/Nonlinear Models*** menu to display the ***Survival and Failure Time Analysis*** (Startup Panel). Survival analysis (exploratory and hypothesis testing) techniques include descriptive methods for estimating the distribution of survival times from a sample, methods for comparing survival in two or more groups, and techniques for fitting linear or nonlinear regression models to survival data. A defining characteristic of survival time data is that they usually include "censored observations," e.g., observations that "survived" to a certain point in time, and then dropped out from the study (patients who are discharged from a hospital). Instead of discarding such observations from the data analysis altogether (i.e., unnecessarily lose potentially useful information) survival analysis techniques can accommodate censored observations, and "use" them in statistical significance testing and model fitting.

This module features a comprehensive implementation of a variety of techniques for analyzing censored data. In addition to computing life tables with various descriptive statistics and Kaplan-Meier product limit estimates, you can compare the survivorship functions in different groups using a large selection of methods (including the Gehan test, Cox *F*-test, Cox-Mantel test, Log-rank test, and Peto & Peto generalized Wilcoxon test). Also, Kaplan-Meier plots can be computed for groups (uncensored observations are identified in graphs with different point markers). *STATISTICA* also features a selection of survival function fitting procedures (including the Exponential, Linear Hazard, Gompertz, and Weibull functions) based on either unweighted and weighted least squares methods. *STATISTICA* also offers full implementations of four general explanatory models (Cox's proportional hazard model, exponential regression model, log-normal and normal regression models) with extended diagnostics, including stratified analyses and graphs of survival for user-specified values of predictors. For Cox proportional hazard regression, you can choose to stratify the sample to permit different baseline hazards in different strata (but a constant coefficient vector), or you can allow for different baseline hazards as well as coefficient vectors. In addition, general facilities are provided to define one or more time-dependent covariates.

For engineering applications, see also Weibull Analysis options in the **Process Analysis** module.

Nonlinear Estimation. Select **Nonlinear Estimation** from the **Statistics - Advanced Linear/Nonlinear Models** menu to display the **Nonlinear Estimation** (Startup Panel). Nonlinear estimation involves finding the best fitting relationship between the values of a dependent variable and the values of a set of one or more independent variables (it is used as either a hypothesis testing or exploratory method).

You can specify any type of model by typing in the respective equation into an equation editor. The equations can include logical operators; thus, discontinuous (piecewise) regression models and models including indicator variables can also be estimated. The equations can also include a wide selection of distribution functions and cumulative distribution functions. The models can be fit using least squares or maximum-likelihood estimation, or any user-specified loss function. When using the least-squares criterion, the very efficient Levenberg-Marquardt and Gauss-Newton algorithms can be used to estimate the parameters for arbitrary linear and nonlinear regression problems. When using arbitrary loss functions, you can choose from among four very different procedures (quasi-Newton, Simplex, Hooke-Jeeves pattern moves, and Rosenbrock pattern search method of rotating coordinates). You have full control over all aspects of the estimation procedure (e.g., starting values, step sizes, convergence criteria, etc.). The most common nonlinear regression models are predefined in the **Nonlinear Estimation** module, and can be chosen simply as menu commands. Those regression models include stepwise Probit and Logit regression, the exponential regression model, and linear piecewise (break point) regression. Standard results nonlinear include the parameter estimates and their standard errors, the variance/covariance matrix of parameter estimates, the predicted values, residuals, and appropriate measures of goodness of fit (e.g., log-likelihood of estimated/null models and *Chi*-square test of difference, proportion of variance accounted for, classification of cases and odds-ratios for Logit and Probit models, etc.). Predicted and residual values can be appended to the datafile for further analyses.

STATISTICA also includes implementations of powerful algorithms for fitting **Generalized Linear/Nonlinear Models (GLZ)**, including probit and multinomial logit models, and **Generalized Additive Models (GAM)**; see the respective descriptions for additional details.

Fixed Nonlinear Regression. Select **Fixed Nonlinear Regression** from the **Statistics - Advanced Linear/Nonlinear Models** menu to display the **Fixed Nonlinear Regression** (Startup Panel). The **Fixed Nonlinear Regression** module is used to specify nonlinear transformations of your variables. These transformed variables are then used in your regression analysis. The available transformations are: X to the second, third, fourth, or fifth power, the square root of

X, natural log of X, log base 10 of X, Euler (e) = 2.71… to the power of X, 10 to the power of X, and the inverse of X.

The **Fixed Nonlinear Regression** module includes the same options as the **Multiple Regression** module for specifying multiple regression models, and/or to request stepwise forward or backward selection of predictors. Like the **Multiple Regression** module, the **Fixed Nonlinear Regression** facilities will calculate a comprehensive set of statistics and extended diagnostics including the complete regression table (with standard errors for B, *Beta* and intercept, R-square and adjusted R-square for intercept and non-intercept models, and **ANOVA** table for the regression), part and partial correlation matrices, correlations and covariances for regression weights, the sweep matrix (matrix inverse), the Durbin-Watson d statistic, Mahalanobis and Cook's distances, deleted residuals, confidence intervals for predicted values, and many others. The extensive residual and outlier analysis options include a large selection of plots, including a variety of scatterplots, histograms, normal and half-normal probability plots, detrended plots, partial correlation plots, different casewise residual and outlier plots and diagrams, and others.

See also the **General Regression Models (GRM)** and **General Linear Models (GLM)** modules for methods for fitting response surface and polynomial regression models; nonlinear regression models can also be fit in the **General Nonlinear Estimation** and **Generalized Linear/Nonlinear Models (GLZ)** modules.

▦ Log-Linear Analysis of Frequency of Tables.
Select **Log-Linear Analysis of Frequency Tables** from the **Statistics - Advanced Linear/Nonlinear Models** menu to display the **Log-Linear Analysis** (Startup Panel). Log-linear analysis provides a "sophisticated" way of looking at crosstabulation tables (to explore the data or verify specific hypotheses), and it is sometimes considered an equivalent of ANOVA for frequency data. Specifically, it is used to test the different factors that are used in the crosstabulation (e.g., gender, region, etc.) and their interactions for statistical significance.

This module offers a complete implementation of log-linear modeling procedures for multi-way frequency tables. Frequency tables can be computed from raw data, or can be entered directly into *STATISTICA*. You can, at all times, review the complete observed table as well as marginal tables, and fitted (expected) values, and can evaluate the fit of all partial and marginal association models or select specific models (marginal tables) to be fitted to the observed data. *STATISTICA* also offers an intelligent automatic model selection procedure that first determines the necessary order of interaction terms required for a model to fit the data, and then, through backwards elimination, determines the best sufficient model to satisfactorily fit the data (using criteria determined by you). The standard output includes G-square (Maximum-Likelihood *Chi*-square), the standard Pearson *Chi*-square with the appropriate degrees of freedom and significance levels, the observed and expected tables, marginal tables, and other

statistics. Graphics options available in the **Log-Linear** module include a variety of 2D and 3D graphs designed to visualize 2-way and multi-way frequency tables (including interactive, user-controlled cascades of categorized histograms and 3D histograms revealing "slices" of multi-way tables), plots of observed and fitted frequencies, plots of various residuals (standardized, components of Maximum-Likelihood *Chi*-square, Freeman-Tukey deviates, etc.), and many others.

See also the **Generalized Linear/Nonlinear Models (GLZ)** module, which provides options for analyzing binomial and multinomial logit models with coded ANOVA/ANCOVA-like designs.

Time Series/Forecasting. Select **Time Series/Forecasting** from the **Statistics - Advanced Linear/Nonlinear Models** menu to display the **Time Series Analysis** (Startup Panel). A time series is a sequence of measurements, typically taken at successive points in time. Time series analysis includes a broad spectrum of exploratory and hypothesis testing methods that have two main goals: (a) identifying the nature of the phenomenon represented by the sequence of observations, and (b) forecasting (predicting future values of the time series variable). Both of these goals require that the pattern of observed time series data is identified and more or less formally described. Once the pattern is established, you can interpret and integrate it with other data (i.e., use it in our theory of the investigated phenomenon, e.g., seasonal commodity prices).

The **Time Series** module contains a wide range of descriptive, modeling, decomposition, and forecasting methods for both time and frequency domain models. These procedures are integrated, that is, the results of one analysis (e.g., ARIMA residuals) can be used directly in subsequent analysis (e.g., to compute the autocorrelation of the residuals). Various methods for transforming and smoothing the time series (prior to an analysis) are supported, including: de-trending, removal of autocorrelation, moving average smoothing (unweighted and weighted, with user-defined or Daniell, Tukey, Hamming, Parzen, or Bartlett weights), moving median smoothing, simple exponential smoothing (see also the description of all exponential smoothing options below), differencing, integrating, residualizing, shifting, 4253H smoothing, tapering, Fourier (and inverse) transformations, and others. Autocorrelation, partial autocorrelation, and crosscorrelation analyses can also be computed. Specialized time-series fitting and modeling procedures include ARIMA and interrupted time series (intervention) analysis, a complete implementation of all 12 common exponential smoothing models (with or without trend and seasonal components), classical seasonal decomposition (Census Method I), X-11 Monthly and Quarterly Seasonal Decomposition and Seasonal Adjustment (Census Method II), Polynomial Distributed Lag Models, Spectrum (Fourier) and Cross-Spectrum Analysis, etc. *STATISTICA* also includes numerous options for plotting the data, autocorrelations and cross-correlations, components, etc.

Structural Equation Modeling. Select *Structural Equation Modeling* from the *Statistics - Advanced Linear/Nonlinear Models* menu to display the *Structural Equation Modeling* (Startup Panel). Structural equation modeling is a very general, very powerful multivariate analysis technique that includes specialized versions of a number of other analysis methods as special cases. Major applications of structural equation modeling include causal modeling or path analysis, confirmatory factor analysis, second order factor analysis, regression models, covariance structure models, and correlation structure models.

SEPATH is a complete implementation that includes numerous advanced features: *STATISTICA* can analyze correlation, covariance, and moment matrices (structured means, models with intercepts). Simple or complex factor or path models can be specified via dialogs, a simple path-language, or via step-by-step wizards. The *SEPATH* module will compute, using constrained optimization techniques, the appropriate standard errors for standardized models, and for models fitted to correlation matrices. The results options include a comprehensive set of diagnostic statistics including the standard fit indices as well as noncentrality-based indices of fit, reflecting the most recent developments in the area of structural equation modeling. You can fit models to multiple samples (groups) and specify for each group fixed, free, or constrained (to be equal across groups) parameters. When analyzing moment matrices, these facilities allow you to test complex hypotheses for structured means in different groups. The module includes powerful Monte Carlo simulation options: you can generate (and save) datafiles for predefined models, based on normal or skewed distributions. Bootstrap estimates can be computed, as well as distributions for various diagnostic statistics, parameter estimates, etc. over the Monte Carlo trials. Numerous flexible graphing options are available to visualize the results (e.g., distributions of parameters) from Monte Carlo runs.

See also the *Factor Analysis* module for maximum likelihood factor analysis; the *General Linear Models (GLM)* module includes various facilities for testing custom-hypotheses, for example, to test for the equality of parameter estimates in a linear model.

Multivariate Exploratory Techniques

The options available from the *Multivariate Exploratory Techniques* menu provide a broad selection of exploratory techniques for various types of data, with extensive, interactive visualization tools. Note that these modules are available as part of the optional add-on package *STATISTICA Multivariate Exploratory Techniques*.

Cluster Analysis. Select *Cluster Analysis* from the *Statistics - Multivariate Exploratory Techniques* menu to display the *Clustering Method* (Startup Panel). Cluster analysis

encompasses a number of different classification algorithms that can be used to develop taxonomies (typically as part of exploratory data analysis).

This module includes a comprehensive implementation of clustering methods (k-means, hierarchical clustering, 2-way joining). *STATISTICA* can process data from either raw datafiles or matrices of distance measures (e.g., correlation matrices), and can cluster cases, variables, or both based on a wide variety of distance measures (including Euclidean, squared Euclidean, City-block (Manhattan), Chebychev, Power distances, Percent disagreement, and 1-r) and amalgamation/linkage rules (including single, complete, weighted and unweighted group average or centroid, Ward's method, and others). Matrices of distances can be saved for further analysis with other modules of the *STATISTICA* system. In k-means clustering, you have full control over the initial cluster centers.

Alternative methods for detecting clusters (structure) in observations and/or variables are available in the **Factor Analysis**, the **Principal Components and Classification Analysis**, **Correspondence Analysis**, and **Neural Networks**.

❄ **Factor Analysis.** Select **Factor Analysis** from the **Statistics - Multivariate Exploratory Techniques** menu to display the **Factor Analysis** (Startup Panel). Factor analysis is used (1) to reduce the number of variables and (2) to detect structure in the relationships between variables.

The **Factor Analysis** module contains a wide range of statistics and options, and provides a comprehensive implementation of factor (and hierarchical factor) analytic techniques with extended diagnostics and a wide variety of analytic and exploratory graphs. It will perform principal components, common, and hierarchical (oblique) factor analysis. The output includes eigenvalues (regular, cumulative, relative), factor loadings, factor scores (which can be appended to the input datafile, reviewed graphically as icons, and interactively recoded), and a number of more technical statistics and diagnostics. Available rotations include Varimax, Equimax, Quartimax, Biquartimax (either normalized or raw), and Oblique rotations. The factorial space can be plotted and reviewed "slice by slice" in either 2D or 3D scatterplots with labeled variable-points; other integrated graphs include Scree plots, various scatterplots, bar and line graphs, and others. Both raw datafiles and matrices of correlations can be used as input.

Confirmatory factor analysis and other related analyses can be performed with the **Structural Equation Modeling and Path Analysis (SEPATH)** module, where a designated **Confirmatory Factor Analysis Wizard** will guide you step-by-step through the process of specifying the model. See also the **Principal Components and Classification Analysis** module for advanced methods for classifying and mapping cases and variables.

⊕ **Principal Components and Classification Analysis.** Select *Principal Components & Classification Analysis* from the *Statistics - Multivariate Exploratory Techniques* menu to display the *Principal Components and Classification Analysis* (Startup Panel). The *PCCA* module is useful for mapping variables and cases (observations) into the factor space (dimensions) computed from a set of analysis variables and cases; for example, you could compute a factor space (dimensions) from a set of key variables and cases that represent well the area or phenomenon of interest, and then apply the dimensions thus derived to a new set of cases and variables.

The *PCCA* module computes principal components for large numbers of variables, and a wide range of associated statistics. You can specify supplementary variables and supplementary cases (observations), which will not be used for the extraction of principal components, but can be mapped into the coordinate system (factor structure) determined from the variables and cases selected for the analysis. Supplementary variables and cases will be included in all results tables and graphs, for example in the scatterplot of factor loadings, where supplementary variables and cases are labeled and identified by different point markers. The *PCCA* module computes all standard results statistics, including factor coordinates of variables and cases (and supplementary variables and cases), contributions of variables and cases to the variance, eigenvalues and eigenvectors, factor scores, factor score coefficients, cosine-squares, etc. A large number of 2D and 3D plots are available including plots of eigenvalues, simple scatterplots, and, of course, plots of the factor coordinates for variables and cases, including supplementary variables and cases.

Factor analysis and similar analyses can also be performed with various other *STATISTICA* modules, such as *Factor Analysis*, *Structural Equations Modeling and Path Analysis (SEPATH)*, *Correspondence Analysis*, and *Multidimensional Scaling*.

✳✳ **Canonical Analysis.** Select *Canonical Analysis* from the *Statistics - Multivariate Exploratory Techniques* menu to display the *Canonical Analysis* (Startup Panel). Canonical analysis is used to investigate the relationship between two sets of variables (it is used as either a hypothesis testing or exploratory method).

This module offers a comprehensive implementation of canonical analysis procedures; it can process raw datafiles or correlation matrices and it computes all of the standard canonical correlation statistics (including eigenvectors, eigenvalues, redundancy coefficients, canonical weights, loadings, extracted variances, significance tests for each root, etc.) and a number of extended diagnostics. The scores of canonical variates can be computed for each case and visualized via integrated icon plots (they can also be appended to the datafile). The *Canonical Analysis* module also includes a variety of integrated graphs (including plots of eigenvalues, canonical correlations, scatterplots of canonical variates, and many others).

Confirmatory analyses of structural relationships between latent variables can also be performed via the **SEPATH (Structural Equation Modeling and Path Analysis)** module; advanced stepwise and best-subset selection of predictor variables for **MANOVA/MANCOVA** designs (with multiple dependent variables) are available in the **General Regression Models (GRM)** module.

Reliability/Item Analysis. Select **Reliability/Item Analysis** from the **Statistics - Multivariate Exploratory Techniques** menu to display the **Reliability and Item Analysis** (Startup Panel). Reliability and item analysis can be used to construct reliable measurement scales, to improve existing scales, and to evaluate the reliability of scales already in use. Specifically, reliability and item analysis techniques aid in the design and evaluation of sum scales, that is, scales that are made up of multiple individual measurements (e.g., different items, repeated measurements, different measurement devices, etc.).

The **Reliability/Item Analysis** module includes a comprehensive selection of procedures for the development and evaluation of surveys and questionnaires. You can calculate reliability statistics for all items in a scale, interactively select subsets, or obtain comparisons between subsets of items via the "split-half" (or split-part) method. In a single run, you can evaluate the reliability of a sum-scale as well as subscales. When interactively deleting items, the new reliability is computed instantly without processing the datafile again. The output includes correlation matrices and descriptive statistics for items, Cronbach *Alpha*, the standardized *Alpha*, the average inter-item correlation, the complete ANOVA table for the scale, the complete set of item-total statistics (including multiple item-total R's), the split-half reliability, and the correlation between the two halves corrected for attenuation.

Classification Trees. Select **Classification Trees** from the **Statistics - Multivariate Exploratory Techniques** menu to display the **Classification Trees** (Startup Panel). The **Classification Trees** module is used to predict membership of cases or objects (i.e., classify cases) in the classes of a categorical dependent variable from their measurements on one or more predictor variables.

The **Classification Trees** module provides a comprehensive implementation of the most recently developed algorithms for efficiently producing and testing the robustness of classification trees. Classification trees can be produced using categorical predictor variables, ordered predictor variables, or both, and using univariate splits or linear combination splits. *STATISTICA* includes options for performing exhaustive splits (as in THAID and C&RT™) or discriminant-based splits, unbiased variable selection (as in QUEST), direct stopping rules (as in FACT) or bottom-up pruning (as in C&RT), pruning based on misclassification rates or on the deviance function, generalized *Chi*-square, G-square, or Gini-index goodness-of-fit measures. Priors and misclassification costs can be specified as equal, estimated from the data, or user specified. You can also specify the v value for v-fold cross-validation during tree

building, v value for v-fold cross-validation for error estimation, size of the SE rule, minimum node size before pruning, seeds for random number generation, and *Alpha* value for variable selection. Integrated graphics options are provided to explore the input and output data.

Advanced methods for tree classifications, including flexible options for model building and interactive tools to explore the trees are also available in the **General Classification/Regression Tree Models (GTrees)** and **General CHAID (Chi-square Automatic Interaction Detection) Models** facilities.

Correspondence Analysis. Select **Correspondence Analysis** from the **Statistics - Multivariate Exploratory Techniques** menu to display the **Correspondence Analysis (CA): Table Specifications** (Startup Panel). Correspondence analysis provides a descriptive/exploratory technique designed to analyze simple two-way and multi-way tables containing some measure of correspondence between the rows and columns. The results provide information that is similar in nature to those produced by factor analysis techniques, and they allow you to explore the structure of categorical variables included in the table.

This module features a full implementation of simple and multiple correspondence analysis techniques. *STATISTICA* will accept input datafiles with grouping (coding) variables that are to be used to compute the crosstabulation table, datafiles that contain frequencies (or some other measure of correspondence, association, similarity, confusion, etc.) and coding variables that identify (enumerate) the cells in the input table, or datafiles with frequencies (or other measure of correspondence) only (e.g., you can directly type in and analyze a frequency table). For multiple correspondence analysis, you can also directly specify a Burt table as input for the analysis. *STATISTICA* will compute various tables, including the table of row percentages, column percentages, total percentages, expected values, observed minus expected values, standardized deviates, and contributions to the *Chi*-square values. Results include the generalized eigenvalues and eigenvectors, all standard diagnostics including the singular values, proportions of inertia for each dimension, standard coordinate values for column and row points, etc. In addition to the 3D histograms that can be computed for all tables, you can produce a line plot for the eigenvalues, and 1D, 2D, and 3D plots for the row or column points. Row and column points can also be combined in a single graph, along with any supplementary points (each type of point will use a different color and point marker, so the different types of points can easily be identified in the plots).

To analyze the structure (dimensions) of variables in a correlation or covariance matrix, and to apply that structure to supplementary variables and observations, you can also use the **Principal Components and Classification Analysis (PCCA)** module.

✳ **Multidimensional Scaling.** Select *Multidimensional Scaling* from the *Statistics - Multivariate Exploratory Techniques* menu to display the *Multidimensional Scaling* (Startup Panel). Multidimensional scaling (MDS) can be considered to be an alternative to factor analysis, and it is typically used as an exploratory method. In general, the goal of the analysis is to detect meaningful underlying dimensions that allow the researcher to explain observed similarities or dissimilarities (distances) between the investigated objects. In factor analysis, the similarities between objects (e.g., variables) are expressed in the correlation matrix. With MDS you can analyze not only correlation matrices but also any kind of similarity or dissimilarity matrix.

The *Multidimensional Scaling* module includes a full implementation of (nonmetric) multidimensional scaling. Matrices of similarities, dissimilarities, or correlations between variables (i.e., "objects" or cases) can be analyzed. The starting configuration can be computed by *STATISTICA* (via principal components analysis) or specified by you. *STATISTICA* employs an iterative procedure to minimize the stress value and the coefficient of alienation. You can monitor the iterations and inspect the changes in these values. The final configurations can be reviewed via spreadsheets and via 2D and 3D scatterplots of the dimensional space with labeled item points. The output includes the values for the raw stress (raw F), Kruskal stress coefficient S, and the coefficient of alienation. The goodness of fit can be evaluated via Shepard diagrams (with d-hats and d-stars).

To analyze correlation matrices or covariance matrices, you can use the *Factor Analysis* module, or the *Principal Components and Classification Analysis* module; to analyze the underlying dimensions (structure) for frequency data, you can also use the *Correspondence Analysis* module.

▦ **Discriminant Analysis.** Select *Discriminant Analysis* from the *Statistics - Multivariate Exploratory Techniques* menu to display the *Discriminant Function Analysis* (Startup Panel). Discriminant function analysis is used to determine which variables discriminate between two or more naturally occurring groups (it is used as either a hypothesis testing or exploratory method).

The *Discriminant Analysis* module is a full implementation of multiple stepwise discriminant function analysis. *STATISTICA* will perform forward or backward stepwise analyses, or enter user-specified blocks of variables into the model. In addition to the numerous graphics and diagnostics describing the discriminant functions, *STATISTICA* also provides a wide range of options and statistics for the classification of old or new cases (for validation of the model). The output includes the respective Wilk's *Lambdas*, partial *Lambdas*, F to enter (or remove), the *p*-levels, the tolerance values, and the R-square. *STATISTICA* will perform a full canonical analysis and report the raw and cumulative eigenvalues for all roots, and their *p*-levels, the raw

and standardized discriminant (canonical) function coefficients, the structure coefficient matrix (of factor loadings), the means for the discriminant functions, and the discriminant scores for each case (which can also be automatically appended to the datafile). Integrated graphs include histograms of the canonical scores within each group (and all groups combined), special scatterplots for pairs of canonical variables (where group membership of individual cases is visibly marked), a comprehensive selection of categorized (multiple) graphs allowing you to explore the distribution and relations between dependent variables across the groups (including multiple box-and-whisker plots, histograms, scatterplots, and probability plots), and many others. The *Discriminant Analysis* module will also compute the standard classification functions for each group. The classification of cases can be reviewed in terms of Mahalanobis distances, posterior probabilities, or actual classifications.

STATISTICA also includes the *General Discriminant Analysis Models* module for fitting ANOVA/ANCOVA-like designs to categorical dependent variables, and to perform various advanced types of analyses (e.g., best-subset selection of predictors, based on misclassification rates in an independent validation sample; profiling of posterior probabilities, etc.).

General Discriminant Analysis Models. Select *General Discriminant Analysis Models* from the *Statistics - Multivariate Exploratory Techniques* menu to display the *General Discriminant Analysis (GDA)* (Startup Panel). The *General Discriminant Analysis (GDA)* module is called a "general" discriminant analysis program because it applies the methods of the general linear model to the discriminant function analysis problem. In short, the discriminant function analysis problem is "recast" as a general multivariate linear model, where the dependent variables of interest are (dummy) coded vectors that reflect the group membership of each case.

In addition to traditional (standard) stepwise discriminant analysis (as available in the *Discriminant Function Analysis* module), the *GDA* module provides support for continuous and categorical predictors. You can specify simple and complex ANOVA/ANCOVA-like designs, e.g., mixtures of continuous and categorical predictors, polynomial (response surface) designs, factorial designs, nested designs, etc. *GDA* also supports multiple degree of freedom effects in stepwise selection, as well as best-subset selection of predictor effects. *GDA* allows you to perform model building (selection of predictor effects) not only based on traditional criteria (e.g., p-to-enter/remove; Wilk's *Lambda*), but also based on misclassification rates; in other words *STATISTICA* will select those predictor effects that maximize the accuracy of classification, either for those cases from which the parameter estimates were computed, or for a cross-validation sample (to guard against over fitting). *STATISTICA* computes detailed results and diagnostic statistics and plots; *GDA* provides a large number of auxiliary information to help you judge the adequacy of the chosen discriminant analysis model (descriptive statistics

and graphs, Mahalanobis distances, Cook distances, and leverages for predictors, etc.). **GDA** also includes an adaptation of the general **GLM** (**GRM**) response profiler; these options allow you to quickly determine the values (or levels) of the predictor variables that maximize the posterior classification probability for a single group, or for a set of groups in the analyses; in a sense, you can quickly determine the typical profiles of values of the predictors (or levels of categorical predictors) that identify a group (or set of groups) in the analysis.

For traditional and stepwise discriminant analysis, see also the **Discriminant Analysis** module. Categorical dependent (criterion) variables can also be analyzed via the **Log-Linear Analysis** module, as well as the **Generalized Linear/Nonlinear Models (GLZ)** module.

Industrial Statistics & Six Sigma

The commands on the **Industrial Statistics & Six Sigma** menu provide access to fully customizable (e.g., callable from other environments), easy and quick to use, versatile quality control charts with a selection of automation options, and user-interface shortcuts to simplify routine work. Additionally access to process capability, Gage R&R, and other quality control/improvement applications and the largest selection of DOE and related visualization techniques including interactive desirability profilers are also available.

The Six Sigma shortcuts provide fast access to the options and statistical analysis procedures that are commonly applied in Six Sigma quality control programs. Six Sigma methodology and management strategies have recently become very popular to organize the quality control effort at every level of the organization (see, for example, Harry and Schroeder, 2000; Pyzdek, 2001). *STATISTICA* contains all necessary analytic methods to implement Six Sigma strategies, and various advanced techniques to extend this approach.

Quality Control Charts. Select **Quality Control Charts** from the **Statistics - Industrial Statistics & Six Sigma** menu to display the **Quality Control Charts** (Startup Panel). In all production processes, the extent to which products meet quality specifications must be monitored. In the most general terms, there are two "enemies" of product quality: (1) deviations from target specifications, and (2) excessive variability around target specifications. During the earlier stages of developing the production process, designed experiments are often used to optimize these two quality characteristics (see **Design of Experiments**); the methods provided in **Quality Control Charts** are on-line or in-process quality control procedures to monitor an on-going production process.

The **Quality Control Charts** module will compute all standard Shewhart control charts for variables (e.g., X-bar, X, R, S) or attributes (P, Np, C, U), and includes a large number of options to handle short-runs, sets, causes, actions, user-defined alarms, etc. These options are

---

highly customizable, and designed to support complex, real-time, quality control monitoring applications. Note that the **Quality Control Charts** facilities in *STATISTICA* are specifically designed to support large-scale quality control charting setups, and can take advantage of distributed data processing on large networks.

▦ Process Analysis. Select *Process Analysis* from the **Statistics - Industrial Statistics & Six Sigma** menu to display the **Process Analysis Procedures** (Startup Panel). In industrial settings, process analysis refers to a collection of analytic methods that can be used to ensure adherence of a product to quality specifications. These methods include cause-and-effects diagrams (Ishikawa charts), process (machine) capability analysis, fitting measurements to non-normal distributions, analysis of Gage repeatability and reproducibility, Weibull and reliability/failure time analysis, and options for generating sampling plans.

The **Process Analysis** module provides an option to compute cause-and-effect (Ishikawa, or fishbone) charts from data; all general graphics facilities for labeling, including drawings, bitmaps, symbols, etc. are supported in those charts, to provide the tools for producing highly customized charts specifically tailored to the process of interest. This module also includes a comprehensive selection of options for computing process capability indices for grouped and ungrouped data (e.g., Cp, Cr, Cpk, Cpl, Cpu, K, Cpm, Pp, Pr, Ppk, Ppl, Ppu), normal/distribution-free tolerance limits, and corresponding process capability plots; in addition, you choose estimates based on general non-normal distributions (Johnson and Pearson curve fitting by moments), as well as all other common continuous distributions. Repeatability/reproducibility experiments with single or multiple trials can be generated and analyzed; results include estimates of the components of variance (repeatability or equipment variation, operator or appraiser variation, part variation, operator-by-part variation, operators-by-trials, parts-by-trials, operators-by-parts-by-trials). Results can be computed based on the range method or the ANOVA table. Additional statistics for the variance components can include the percent of tolerance, process variation, and total variation. The Weibull analysis options provide powerful graphical techniques for exploiting the power and the ability of the Weibull distribution to be generalized; you can produce Weibull probability plots and estimate the parameters of the distribution, along with confidence intervals for reliability. Probability plots can be computed for complete, single-censored, and multiple-censored data, and parameters can be estimated from hazard plots of failure orders. Estimation methods include maximum likelihood (for complete and censored data), weighting factors based on linear estimation techniques for complete and single-censored data, and modified moment estimators. *STATISTICA* includes graphical goodness-of-fit tests, and the Hollander-Proschan, Mann-Scheuer-Fertig, and Anderson-Darling tests of goodness of fit. The options for generating sampling plans include fixed and sequential sampling plans for normal and binomial means, or

Poisson frequencies; results include the sample sizes, operating characteristic (OC) curves, plots of the sequential plans with or without data, expected (H_0/H_1) run lengths, etc.

Comprehensive selections of methods for estimating variance components for random effects are also available in the designated *STATISTICA* **Variance Components** module, and the **General Linear Models (GLM)** module. The **Power Analysis** module also provides options for computing required sample sizes and power estimates for a large number of research designs (e.g, ANOVA) and data types (e.g., for binary counts, censored failure time data, etc.).

Experimental Design (DOE). Select **Experimental Design (DOE)** from the **Statistics - Industrial Statistics & Six Sigma** menu to display the **Design & Analysis of Experiments** (Startup Panel). Experimental design techniques apply analysis of variance principles to product development. The primary goal is usually to extract the maximum amount of unbiased information regarding the factors affecting a production process from as few (costly) observations as possible.

The **Design of Experiments** module offers an extremely comprehensive selection of procedures to design and analyze the experimental designs used in industrial (quality) research: $2^{(k-p)}$ factorial designs with blocking (for more than 100 factors, including unique, highly efficient search algorithms for finding minimum aberration and maximum unconfounding designs, where you can specify the interaction effects of interest that are to be unconfounded), screening designs (for over 100 factors, including Plackett-Burman designs), $3^{(k-p)}$ factorial designs with blocking (including Box-Behnken designs), mixed-level designs, central composite (or response surface) designs (including small central composite designs), Latin square designs, Taguchi robust design experiments via orthogonal arrays, mixture designs and triangular surfaces designs, vertices and centroids for constrained surfaces and mixtures, and D- and A-optimal designs for factorial designs, surfaces, and mixtures.

STATISTICA includes an extremely large number of other computational methods for analyzing data collected in experiments, and for fitting ANOVA/ANCOVA-like designs to continuous or categorical outcome variables. Specifically, the *STATISTICA* family of products includes complete implementations of **General Linear Models (GLM)** and **General Regression Models (GRM)** with sophisticated model-building procedures (stepwise and best-subset selection of predictor effects), **Generalized Linear/Nonlinear Models (GLZ)**, **General Discriminant Analysis Models (GDA)**, **General Classification/Regression Trees Models (GTrees)** and **General CHAID Models**.

Six Sigma (DMAIC) Shortcuts

Six Sigma methodology and management strategies provide an overall framework for organizing company wide quality control efforts. These methods have recently become very popular, due to numerous success stories from major US-based as well as international corporations. For reviews of Six Sigma strategies, refer to Harry and Schroeder (2000), or Pyzdek (2001).

The submenu commands that are displayed after selecting **Six Sigma (DMAIC) Shortcuts** from the **Statistics - Industrial Statistics & Six Sigma** menu provide shortcuts to the modules and specific analytic facilities in those modules that are commonly part of Six Sigma management strategies. These are organized into the categories of activities that make up the Six Sigma effort: **Define** (*D*), **Measure** (*M*), **Analyze** (*A*), **Improve** (*I*), **Control** (*C*); or *DMAIC* for short. The following is a brief description of the Six Sigma shortcuts. Note that not all of the shortcuts may be available to you, depending on your configuration of *STATISTICA*.

Define. The **Define** phase is concerned with the definition of project goals and boundaries, and the identification of issues that need to be addressed to achieve the higher sigma level. Shortcuts are provided to the following modules and analytic facilities:

>> **Cause and Effect Diagrams.** Select **Cause and Effect Diagrams** from the **Statistics - Industrial Statistics & Six Sigma - Six Sigma (DMAIC) Shortcuts - Define** menu to produce a cause-and-effect (Ishikawa, or fishbone) diagram; links to the **Process Analysis** module.

Blank Graph. Select **Blank Graph** from the **Statistics - Industrial Statistics & Six Sigma - Six Sigma (DMAIC) Shortcuts - Define** menu to produce a blank graph; use drawing tools to create highly customized charts, maps, etc.

Six Sigma Calculator. Select **Six Sigma Calculator** from the **Statistics - Industrial Statistics & Six Sigma - Six Sigma (DMAIC) Shortcuts - Define** menu to display the **Six Sigma Calculator** dialog. The **Six Sigma Calculator** is an extension or special case of the **Probability Calculator** (see page 99) for the normal distribution; instead of the normal distribution z and p, the **Six Sigma Calculator** expects as input either **Sigma** or **DPMO** (defects per million opportunities), and will compute each quantity from the other.

Measure. The goal of the **Measure** phase of the Six Sigma strategy is to gather information about the current situation, to obtain baseline data on current process performance, and to identify problem areas. Shortcuts are provided to the following modules and analytic facilities:

Descriptive Statistics. Select *Descriptive Statistics* from the *Statistics - Industrial Statistics & Six Sigma - Six Sigma (DMAIC) Shortcuts - Measure* menu to display the *Descriptive Statistics* dialog of the *Basic Statistics/Tables* module.

SixGraph Summary. Select *SixGraph Summary* from the *Statistics - Industrial Statistics & Six Sigma - Six Sigma (DMAIC) Shortcuts - Measure* menu to display the *Defining Variables for X-bar and R Chart* dialog of the *Quality Control* module; after specifying the respective variables and chart parameters, a *SixGraph Summary Graph* is produced.

Missing and Out of Range Data. Select *Missing and Out of Range Data* from the *Statistics - Industrial Statistics & Six Sigma - Six Sigma (DMAIC) Shortcuts - Measure* menu to display the *Missing and 'Out of Range' Data Plots* dialog.

Box and Whisker Plot. Select *Box and Whisker Plot* from the *Statistics - Industrial Statistics & Six Sigma - Six Sigma (DMAIC) Shortcuts - Measure* menu to display the *2D Box Plots* dialog.

Gage R & R. Select *Gage R & R* from the *Statistics - Industrial Statistics & Six Sigma - Six Sigma (DMAIC) Shortcuts - Measure* menu to display the *Repeatability & Reproducibility Analysis* dialog of the *Process Analysis* module.

Sampling. Select *Sampling* from the *Statistics - Industrial Statistics & Six Sigma - Six Sigma (DMAIC) Shortcuts - Measure* menu to display the *Sampling Plans* dialog of the *Process Analysis* module.

Power Analysis. Select *Power Analysis* from the *Statistics - Industrial Statistics & Six Sigma - Six Sigma (DMAIC) Shortcuts - Measure* menu to display the startup panel of the *Power Analysis* module.

Process Capability. Select *Process Capability* from the *Statistics - Industrial Statistics & Six Sigma - Six Sigma (DMAIC) Shortcuts - Measure* menu to display the *Process Capability Analysis Setup -- Raw Data* dialog in the *Process Analysis* module.

Distribution Fitting. Select *Distribution Fitting* from the *Statistics - Industrial Statistics & Six Sigma - Six Sigma (DMAIC) Shortcuts - Measure* menu to display the *Process Capability Analysis Setup -- Raw Data* dialog in the *Process Analysis* module; use the options on the *Distribution* tab to fit various distributions, and to identify the distribution and parameter values that provide the best fit.

Frequency Tables. Select *Frequency Tables* from the *Statistics - Industrial Statistics & Six Sigma - Six Sigma (DMAIC) Shortcuts* - *Measure* menu to display the *Frequency Tables* dialog in the *Basic Statistics/Tables* module.

Time Series Plots. Select *Time Series Plots* from the *Statistics - Industrial Statistics & Six Sigma - Six Sigma (DMAIC) Shortcuts* - *Measure* menu to display the startup panel of the *Time Series* module.

Analyze. The goal of the *Analyze* phase of the Six Sigma quality effort is to identify the root cause(s) of quality problems, and to confirm those causes using the appropriate data analysis tools. Shortcuts are provided to the following modules and analytic facilities:

Cause and Effect Diagrams. Select *Cause and Effect Diagrams* from the *Statistics - Industrial Statistics & Six Sigma - Six Sigma (DMAIC) Shortcuts - Analyze* menu to produce a cause-and-effect (Ishikawa, or fishbone) diagram; links to the *Process Analysis* module.

t-test. Select *t-test* from the *Statistics - Industrial Statistics & Six Sigma - Six Sigma (DMAIC) Shortcuts - Analyze* menu to display the *t-test for Independent Samples by Groups* dialog of the *Basic Statistics/Tables* module.

Paired t-test. Select *Paired t-test* from the *Statistics - Industrial Statistics & Six Sigma - Six Sigma (DMAIC) Shortcuts - Analyze* menu to display the *t-test for Dependent Samples* dialog of the *Basic Statistics/Tables* module.

Nonparametric Tests. Select *Nonparametric Tests* from the *Statistics - Industrial Statistics & Six Sigma - Six Sigma (DMAIC) Shortcuts - Analyze* menu to display the startup panel of the *Nonparametric Statistics* module.

Stratification Analysis. Select *Stratification Analysis* from the *Statistics - Industrial Statistics & Six Sigma - Six Sigma (DMAIC) Shortcuts - Analyze* menu to display the *Statistics by Groups (Breakdown)* dialog of the *Basic Statistics/Tables* module.

Correlations. Select *Correlations* from the *Statistics - Industrial Statistics & Six Sigma - Six Sigma (DMAIC) Shortcuts - Analyze* menu to display the *Product-Moment and Partial Correlations* dialog of the *Basic Statistics/Tables* module.

2D Scatterplots. Select *2D Scatterplots* from the *Statistics - Industrial Statistics & Six Sigma - Six Sigma (DMAIC) Shortcuts - Analyze* menu to display the *2D Scatterplots* dialog.

3D Scatterplots. Select *3D Scatterplots* from the *Statistics - Industrial Statistics & Six Sigma - Six Sigma (DMAIC) Shortcuts - Analyze* menu to display the *3D Scatterplots* dialog.

Regression. Select *Regression* from the *Statistics - Industrial Statistics & Six Sigma - Six Sigma (DMAIC) Shortcuts - Analyze* menu to display the startup panel of the *Multiple Regression* module.

Experimental Design. Select *Experimental Design* from the *Statistics - Industrial Statistics & Six Sigma - Six Sigma (DMAIC) Shortcuts - Analyze* menu to display the startup panel of the *Experimental Design* module.

GLM. Select *GLM* from the *Statistics - Industrial Statistics & Six Sigma - Six Sigma (DMAIC) Shortcuts - Analyze* menu to display the startup panel of the *General Linear Models (GLM)* module.

GLZ. Select *GLZ* from the *Statistics - Industrial Statistics & Six Sigma - Six Sigma (DMAIC) Shortcuts - Analyze* menu to display the startup panel of the *Generalized Linear Models (GLZ)* module.

Improve. The goal of the *Improve* phase is to implement solutions that address the problems (root causes) identified during the previous (*Analyze*) phase. Shortcuts are provided to the following modules and analytic facilities:

Quality Control Charts. Select *Quality Control Charts* from the *Statistics - Industrial Statistics & Six Sigma - Six Sigma (DMAIC) Shortcuts - Improve* menu to display the startup panel of the *Quality Control Charts* module.

SixGraph Summary. Select *SixGraph Summary* from the *Statistics - Industrial Statistics & Six Sigma - Six Sigma (DMAIC) Shortcuts - Improve* menu to display the *Defining Variables for X-bar and R Chart* dialog of the *Quality Control* module; after specifying the respective variables and chart parameters, a *SixGraph Summary Graph* is produced.

Frequency Tables. Select *Frequency Tables* from the *Statistics - Industrial Statistics & Six Sigma - Six Sigma (DMAIC) Shortcuts - Improve* menu to display the *Frequency Tables* dialog in the *Basic Statistics/Tables* module.

Histogram. Select *Histogram* from the *Statistics - Industrial Statistics & Six Sigma - Six Sigma (DMAIC) Shortcuts - Improve* menu to display the *2D Histograms* dialog.

Pareto Chart. Select *Pareto Chart* from the *Statistics - Industrial Statistics & Six Sigma - Six Sigma (DMAIC) Shortcuts - Improve* menu to display the variable specifications dialog for the Pareto chart of the *Quality Control* module.

Process Capability. Select *Process Capability* from the *Statistics - Industrial Statistics & Six Sigma - Six Sigma (DMAIC) Shortcuts - Improve* menu to display the *Process Capability Analysis Setup -- Raw Data* dialog in the *Process Analysis* module.

Experimental Design. Select *Experimental Design* from the *Statistics - Industrial Statistics & Six Sigma - Six Sigma (DMAIC) Shortcuts - Improve* menu to display the startup panel of the *Experimental Design* module.

Six Sigma Calculator. Select *Six Sigma Calculator* from the *Statistics - Industrial Statistics & Six Sigma - Six Sigma (DMAIC) Shortcuts - Improve* menu to display the *Six Sigma Calculator* dialog. The *Six Sigma Calculator* is an extension or special case of the *Probability Calculator* (see page 99) for the normal distribution; instead of the normal distribution z and p, the *Six Sigma Calculator* expects as input either *Sigma* or *DPMO* (defects per million opportunities), and will compute each quantity from the other.

Stratification Analysis. Select *Stratification Analysis* from the *Statistics - Industrial Statistics & Six Sigma - Six Sigma (DMAIC) Shortcuts - Improve* menu to display the *Statistics by Groups (Breakdown)* dialog of the *Basic Statistics/Tables* module.

Stratification Charts. Select *Stratification Charts* from the *Statistics - Industrial Statistics & Six Sigma - Six Sigma (DMAIC) Shortcuts - Improve* menu to display the *2D Categorized Histograms* dialog.

Control. The goal of the *Control* phase is to evaluate and monitor the results of the previous phase (*Improve*). Shortcuts are provided to the following modules and analytic facility:

Quality Control Charts. Select *Quality Control Charts* from the *Statistics - Industrial Statistics & Six Sigma - Six Sigma (DMAIC) Shortcuts - Control* menu to display the startup panel of the *Quality Control Charts* module.

Customize. The commands in the *Statistics - Industrial Statistics & Six Sigma - Six Sigma (DMAIC) Shortcuts - Customize* menu are used to add the Six Sigma shortcut options described above to the menu bar, or to add a toolbar to *STATISTICA* with these DMAIC shortcuts.

Display Six Sigma Menu. Select *Display Six Sigma Menu* from the *Statistics - Industrial Statistics & Six Sigma - Six Sigma (DMAIC) Shortcuts - Customize* menu to add the Six Sigma commands to the menu bar: ‖ File Edit View Insert Format Statistics Graphs Six Sigma Tools Data Window Help

Display Six Sigma Toolbar. Select *Display Six Sigma Toolbar* from the *Statistics - Industrial Statistics & Six Sigma - Six Sigma (DMAIC) Shortcuts - Customize* menu to display the *Six Sigma* toolbar. The toolbar will contain a button for each group of options (*Define*, *Measure*, *Analyze*, *Improve*, and *Control*), and a shortcut to the *Six Sigma Calculator*, clicking on the toolbar buttons will display the respective menu of all shortcuts as described above.

Power Analysis

Select *Power Analysis* from the *Statistics* menu to display the *Power Analysis and Interval Estimation* (Startup Panel). The *Power Analysis* module (available as an optional add-on package) implements the techniques of statistical power analysis, sample size estimation, and advanced techniques for confidence interval estimation. The main goal of the first two techniques is to help you decide, while in the process of designing an experiment, (a) how large a sample is needed to allow statistical judgments that are accurate and reliable, and (b) how likely your statistical test will be to detect effects of a given size in a particular situation. The third technique is useful in implementing objectives (a) and (b) above, and in evaluating the size of experimental effects in practice.

The *Power Analysis* module is used to perform power, sample size, and related computations for a large number of different tests, including 1-sample *t*-tests, 2-sample independent sample *t*-tests, 2-sample dependent sample *t*-tests, planned contrasts, 1-way ANOVA (fixed and random effects), 2-way ANOVA, *Chi*-square test on a single variance, *F*-test on 2 variances, *Z*-test (or *Chi*-square test) on a single proportion, *Z*-test on 2 independent proportions, McNemar's test on 2 dependent proportions, *F*-test of significance in multiple regression, *t*-test for significance of a single correlation, *Z*-test for comparing 2 independent correlations, log-rank test in survival analysis, test of equal exponential survival, with accrual period, test of equal exponential survival with accrual period and dropouts, *Chi*-square test of significance in structural equation modeling, tests of "close fit" in structural equation modeling confirmatory factor analysis, etc.

Data-Mining

The commands available from the *Data-Mining* menu offer a comprehensive selection of data mining solutions, with an icon-based, extremely easy-to-use user interface. The modules here feature a selection of completely integrated, and automated, ready to deploy "as is" (but also easily customizable) systems of specific data mining solutions for a wide variety of business

applications. The modules can be used interactively, and/or used to build, test, and deploy new solutions.

General Slicer/Dicer Explorer with OLAP. Select *General Slicer/Dicer Explorer with OLAP* from the *Statistics - Data-Mining* menu to explore your data via OLAP (On Line Analytic Processing) multi-dimensional tables using a flexible, intelligent data mining user interface. A wide variety of options is included to "drill down" the data using various methods to create views and subsets including crosstabulations, breakdowns, and virtually countless data visualization tools. Data from remote servers can be directly queried and processed in place, without having to import them to the local storage. Note that this is a module of the *STATISTICA Enterprise-Wide Data Mining System*.

General Classifier (Trees and Clusters). Select *General Classifier (Trees and Clusters)* from the *Statistics - Data-Mining* menu to explore a large suite of data mining methods for determining the optimal way to classify observations into meaningful clusters, nodes, or market segments. Data from remote servers can be directly queried and processed in place, without having to import them to the local storage. Successful models can be deployed for repeated or routine use. Note that this is a module of the *STATISTICA Enterprise-Wide Data Mining System*.

General Modeler and Multivariate Explorer. Select *General Modeler and Multivariate Explorer* from the *Statistics - Data-Mining* menu to explore a large suite of linear and nonlinear modeling methods for predicting one or more dependent criterion variables, from continuous and/or categorical predictors (and their interactions). A flexible, intelligent, data mining user interface will aid in the process of selecting and setting up the appropriate model. A variety of very efficient algorithms are provided for fitting models based on least squares, maximum likelihood, and arbitrary loss functions. Data from remote servers can be directly queried and processed in place, without having to import them to the local storage. Successful models can be deployed for repeated or routine use. Note that this is a module of the *STATISTICA Enterprise-Wide Data Mining System*.

General Forecaster. Select *General Forecaster* from the *Statistics - Data-Mining* menu to explore a large suite of methods for predicting new values of variables based on the patterns of historical data. A flexible, intelligent data mining user interface is provided to aid in the process of setting up a model for the forecasting. Data from remote servers can be directly queried and processed in place, without having to import them to the local storage. Successful models can be deployed for repeated or routine use. Note that this is a module of the *STATISTICA Enterprise-Wide Data Mining System*.

System Reference – 95

General Neural Network Explorer. Select *General Neural Network Explorer* from the *Statistics - Data-Mining* menu to apply neural network methodology to classification and prediction (forecasting) problems. A large number of different network architectures are supported, and automatic methods are provided to compare the effectiveness of those architectures. Data from remote servers can be directly queried and processed in place, without having to import them to the local storage. Successful models can be deployed for repeated or routine use. Note that this is a module of the *STATISTICA Enterprise-Wide Data Mining System*.

Specialized Data Mining Modules

A large portion of analytic functionality used by *STATISTICA* Data Miner are driven by the computational engines of modules that are included in various other *STATISTICA* products (please refer to the respective parts of *Appendix A - STATISTICA Family of Products* for detailed information about these modules). However, there are three modules that include selections of highly specialized data mining and data mining modeling techniques that are offered only as part of *STATISTICA* Data Miner. The following three paragraphs include technical information about these three modules.

General Classification/Regression Tree Models. Select *General Classification/ Regression Tree Models* from the *Statistics - Data-Mining* menu to display the *General Classification and Regression Trees (General Trees)* (Startup Panel). Classification and regression trees are used to classify (divide) cases based on a set of predictor variables. Unlike linear or nonlinear regression-like algorithms, this module will find hierarchical decision rules to provide optimal separation between observations with regard to a categorical or continuous criterion variable, based on splits on one or more continuous and/or categorical predictor variables. This module is a comprehensive implementation of the methods described as CART® by Breiman, Friedman, Olshen, and Stone (1984). However, the *General Trees* module contains various extensions and options that are typically not found in implementations of this algorithm, and that are particularly useful for data mining applications. In addition to standard analyses, the implementation of these methods in *STATISTICA* allows you to specify ANOVA/ANCOVA-like designs with continuous and/or categorical predictor variables, and their interactions.

General CHAID Models. Select *General CHAID Models* from the *Statistics - Data-Mining* menu to display the *General CHAID Models* (Startup Panel). The CHAID (*Chi*-square Automatic Interaction Detection) method is an alternative to the *General Trees* options for classifying (dividing) cases based on a set of categorical or continuous predictor variables. CHAID uses a different algorithm (from *General Trees*) to determine a final hierarchical classification tree where each split (node) can produce multiple branches (unlike *General Trees*,

where all splits are binary). The **CHAID** analysis can be performed for both continuous and categorical dependent (criterion) variables. In addition to the traditional **CHAID** analysis, you can combine categorical and continuous predictor variables into ANOVA/ANCOVA-like designs and perform the analysis using a design matrix for the predictors. You can also request exhaustive searches for the best solution (**Exhaustive CHAID**).

Generalized Additive Models. Select **General Additive Models** from the **Statistics - Data-Mining** menu to display the **Generalized Additive Models** (Startup Panel). Like generalized linear models, generalized additive models can be used to predict a dependent variable from various distributions (e.g., Binomial, Normal, *Gamma*, Poisson), from one or more continuous or categorical predictor variables, using various link functions (e.g., Logit, Log, Inverse, Identity); in addition, in generalized additive models each predictor variable is smoothed via a cubic spline smoother, to yield the best fit of the model, and best prediction of the dependent variable. The *STATISTICA* **Generalized Additive Models** facilities are an implementation of methods developed and popularized by Hastie and Tibshirani (1990). The module will handle continuous and categorical predictor variables and allow you to choose from a wide variety of distributions for the dependent variable and link functions for the effects of the predictor variables on the dependent variable.

Statistics of Block Data

The **Statistics of Block Data** commands provide easy access to statistical procedures for each row or column in a selected block in the spreadsheet. The selected statistics will be computed and added to (i.e., appended to the end of) the spreadsheet as additional rows or columns. Selecting a statistic from the **Statistics of Block Data -** **Block Columns** menu will append the resulting statistic in a new row and the case name will include the name of the selected statistic and the range description. Selecting a statistic from the **Statistics of Block Data -** **Block Rows** will append the resulting statistic in a new column and the variable name will contain the name of the selected statistic and the range description. Note that the **Statistics of Block Data** commands are also available from the shortcut menu obtained by right-clicking on a selected block in spreadsheet.

Means. Select **Means** from either the **Statistics - Statistics of Block Data - Block Columns** or **Statistics - Statistics of Block Data - Block Rows** menu to compute the mean for each column or row in a selected block in the spreadsheet. The means will be computed and added to (i.e., appended to the end of) the spreadsheet as additional rows or columns (as described above), as shown in the illustration below.

Data: Spreadsheet1* (3v by 7c)			
	1 Var1	**2** Var2	**3** Var3
1	19	7	66
2	17	6	62
3	15	3	58
4	16	4	64
5	11	9	67
6	18	5	60
MEAN case 1-6	16	5.666667	62.83333

Medians. Select *Medians* from either the *Statistics - Statistics of Block Data - Block Columns* or *Statistics - Statistics of Block Data - Block Rows* menu to compute the median for each column or row in a selected block in the spreadsheet. The medians will be computed and added to (i.e., appended to the end of) the spreadsheet as additional rows or columns (as described above).

SD's. Select *SD's* from either the *Statistics - Statistics of Block Data - Block Columns* or *Statistics - Statistics of Block Data - Block Rows* menu compute the standard deviation for each column or row in a selected block in the spreadsheet. The standard deviations will be computed and added to (i.e., appended to the end of) the spreadsheet as additional rows or columns (as described above).

Valid N's. Select *Valid N's* from either the *Statistics - Statistics of Block Data - Block Columns* or *Statistics - Statistics of Block Data - Block Rows* menu to compute the valid sample size for each column or row in a selected block in the spreadsheet. The sample sizes will be computed and added to (i.e., appended to the end of) the spreadsheet as additional rows or columns (as described above).

Sums. Select *Sums* from either the *Statistics - Statistics of Block Data - Block Columns* or *Statistics - Statistics of Block Data - Block Rows* menu to compute the sum for each column or row in a selected block in the spreadsheet. The sums will be computed and added to (i.e., appended to the end of) the spreadsheet as additional rows or columns (as described above).

Min's. Select *Min's* from either the *Statistics - Statistics of Block Data - Block Columns* or *Statistics - Statistics of Block Data - Block Rows* menu to determine the minimum value for each column or row in a selected block in the spreadsheet. The minimums will be found and added to (i.e., appended to the end of) the spreadsheet as additional rows or columns (as described above).

Max's. Select *Max's* from either the *Statistics - Statistics of Block Data - Block Columns* or *Statistics - Statistics of Block Data - Block Rows* menu to determine the maximum value for each column or row in a selected block in the spreadsheet. The maximums will be found and

added to (i.e., appended to the end of) the spreadsheet as additional rows or columns (as described above).

25%'s. Select *25%'s* from either the *Statistics - Statistics of Block Data - Block Columns* or *Statistics - Statistics of Block Data - Block Rows* menu to compute the 25th percentile for each column or row in a selected block in the spreadsheet. The 25th percentile will be computed and added to (i.e., appended to the end of) the spreadsheet as additional rows or columns (as described above).

75%'s. Select *75%'s* from either the *Statistics - Statistics of Block Data - Block Columns* or *Statistics - Statistics of Block Data - Block Rows* menu to compute the 75th percentile for each column or row in a selected block in the spreadsheet. The 75th percentile will be computed and added to (i.e., appended to the end of) the spreadsheet as additional rows or columns (as described above).

All. Select *All* from either the *Statistics - Statistics of Block Data - Block Columns* or *Statistics - Statistics of Block Data - Block Rows* menu to compute all the statistics mentioned above for each column or row in a selected block in the spreadsheet. All the statistics will be computed and added to (i.e., appended to the end of) the spreadsheet as additional rows or columns (as described above).

STATISTICA Visual Basic

Select *STATISTICA Visual Basic* from the *Statistics* menu to display the *Macros* dialog. You can use this dialog to edit, delete, or run existing macro (SVB) programs as well as to create new macros. For more information on using this dialog or *STATISTICA* Visual Basic, refer to Chapter 11 - *STATISTICA Visual Basic*, the *STATISTICA Visual Basic Primer*, or the *Electronic Manual*.

Probability Calculator

This option includes a general application for calculating probabilities (the *Probability Distribution Calculator*) and a utility for calculations based on the Pearson Product Moment Correlation Distribution. Note that the *STATISTICA Power Analysis* program is designed to allow you to compute statistical power and estimate required sample size while planning experiments, and to evaluate experimental effects in your existing data. You will find many features in this module designed to allow you to perform these calculations quickly and effectively in a wide variety of data analysis situations. It also includes a variety of probability distribution calculators.

Distributions. Select *Distributions* from the *Statistics - Probability Calculator* menu to display the *Probability Distribution Calculator*.

This facility allows you to compute, for various theoretical distributions and parameters (e.g., F, df_1, df_2), either (1) the upper or lower tail areas, given a user-specified variate value (two sided probability values can be computed for symmetrical distributions; e.g., the normal distribution), or (2) the critical values, given a user specified probability value.

A variety of distributions is available as shown in the illustration below. For each distribution, you can specify whether the density function and distribution function graph icons will be drawn using fixed or non-fixed scaling.

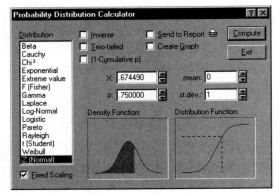

For more information on the *Probability Distribution Calculator*, refer to the *Electronic Manual*.

Correlations. Select the *Correlations* from the *Statistics - Probability Calculator* menu to display the *Pearson Product Moment Correlation Distribution* dialog. Use this dialog to compute the *p* value for a given correlation coefficient or you can enter a desired *p* value and determine the correlation coefficient (based on a specified sample size).

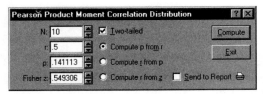

The parameters used in this dialog are *N* (the sample size), *r* (the correlation coefficient), *p* (the significance level), and *Fisher z* (the critical value of the distribution). These parameters can be specified or computed. To use this dialog, enter values in two of these fields (leaving the other two blank) and then click the *Compute* button to compute the remaining two values (see below).

Using either one- or two-tailed tests, you can compute the *p*-level and *Fisher z* from a given *r* and sample size (*N*), compute *r* and its *Fisher z* value from a given *p* and sample size (*N*), or compute *r* and the *p*-level based on a specific *Fisher z* value and sample size (*N*).

After performing the selected calculations, you can send the results of the computation to a report. These values will be sent to a report when you click the *Compute* button, and not when you change the parameters with the microscrolls. Therefore, you can edit the parameters using the microscrolls and review the results in the boxes in this dialog, then when you are ready, simply click the *Compute* button to send the resulting values to a report.

Six Sigma Calculator. Select the *Six Sigma Calculator* from the *Statistics - Probability Calculator* menu to display the *Six Sigma Calculator* dialog. The *Six Sigma Calculator* is an extension or special case of the *Probability Calculator* for the normal distribution; instead of the normal distribution z and p, the *Six Sigma Calculator* expects as input either *Sigma* or *DPMO* (defects per million opportunities), and will compute each quantity from the other.

Standard Toolbar

The *Standard* toolbar, which accompanies every window in *STATISTICA* (e.g., spreadsheet, report, macro, workbook, or graph), offers a variety of tools for accessing and manipulating the various *STATISTICA* documents (i.e., creating, saving, opening, and printing files; find and replace features, and adding files to workbooks or reports).

It also contains a variety of Clipboard options and access to the *STATISTICA Electronic Manual*. For information on customizing this and other toolbars, see the *Toolbar Overview* (page 56).

☐ New File Button

Click the ☐ button (or press CTRL+N) to display the *Create New Document* dialog. This dialog is used to create a new spreadsheet, report, macro, or workbook.

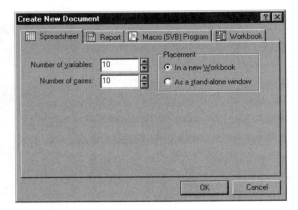

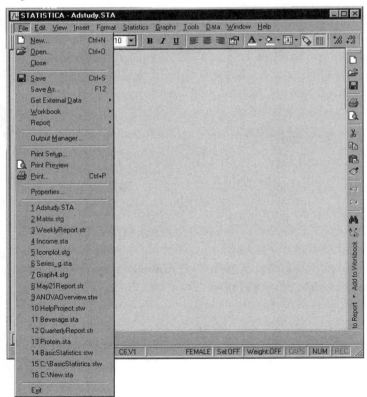 Open File Button

Click the 📂 button (or press CTRL+O) to open the file selection window, which is used to select any *STATISTICA* file or other commonly used files (Excel files, Text files, HTML files, etc.).

STATISTICA files can also be opened directly from Windows *Explorer* by double-clicking on the respective file name.

Note that a list of recently used files is automatically appended to the *File* menu, as shown above. *STATISTICA* keeps track of recently opened spreadsheets, reports, graphs, and workbooks. The length of this list can be configured on the *General* tab of the *Options* dialog, accessed by selecting *Options* from the *Tools* menu.

Save File Button

Click the button (or press CTRL+S) to save the current file under the same or different file name using the appropriate *STATISTICA* file extension. Before overwriting an existing file, *STATISTICA* asks if you want to overwrite the existing file.

Print Button

Click the button to print the current file. This is a shortcut method to send the current document to the default printer. Unlike the *Print* (or CTRL+P) command available in the *File* menu, clicking this shortcut button will not display the *Print* dialog. Instead, the entire document (or a selected section of the document) will be sent to the printer, following the current *Print* settings.

Printing documents in workbooks. Note that if you click the print button when a workbook is active, the selected document within the workbook will be sent to the printer (e.g., a spreadsheet or graph). It is not possible to print a section of a document from within a workbook. To print the entire workbook, select *Print* from the *File* menu to display the *Print* dialog, and select the *All* option button in the *Print range* group.

Print Preview Button

Click the button to display the *Print Preview* dialog. This dialog is used to see how the selected document (e.g., report, spreadsheet, graph) will look when it is printed.

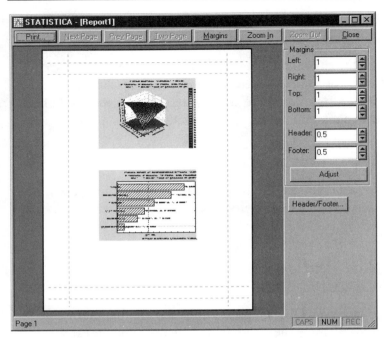

Use the controls in this dialog to modify the **Header** and **Footer** placement as well as the **Left**, **Right**, **Top**, and **Bottom** margins.

✂ Cut (Block) Button

Click the ✂ button (or press CTRL+X) to cut (remove) the contents of the currently highlighted selection and copy it to the Clipboard. When this option is used in a spreadsheet, the cleared cells are replaced with the missing data values of the respective variables.

Clipboard operations vs. global operations on cases or variables. Note that unlike the global operations performed on ranges of cases or variables and treating them as logical units (see Cases ▾ *Cases Button* and Vars ▾ *Variables Button* in Chapter 4 - *STATISTICA Spreadsheets*), the Clipboard operations follow standard conventions and apply only to the current selection. They depend entirely on the current (highlighted) block and cursor position.

For example, note that the global operation of deleting or moving variables affects (removes) not only the contents of the respective columns but also the columns themselves; thus these operations change the structure of the datafile (and those operations are always performed on entire variables regardless of whether all cases or only a subset of cases is currently highlighted in the selected variables). In the case of Clipboard operations, however, only the segment of

data that is highlighted will be cut, and (following the common spreadsheet conventions) pasting will always begin from the current cursor position, and proceed down.

Thus, for example, if in order to move a variable you (a) highlighted and cut an entire column, (b) highlighted another (entire) column, and then (c) pasted the Clipboard contents to that new location (intending to "replace" the previous values), the operation will be performed as intended only if you have placed the cursor at the top of the new column. If you have placed it somewhere in the middle, then the pasting will start from that point down.

🖹 Copy (Block) Button

Click the 🖹 button (or press CTRL+C) to copy the highlighted selection of the current document to the Clipboard.

Copy (Block) in spreadsheets. Click this button to copy the contents in the highlighted block of cells, but not the case names (or numbers) and variable names associated with the highlighted cells. Use **Copy with Headers** from the **Edit** menu to copy the highlighted block of cells with the corresponding case and variable names; for further details, see *What Is the Difference between Copy and Copy With Headers?* in Chapter 4 - *STATISTICA Spreadsheets*. Note that you can select an entire spreadsheet by clicking on the upper-left corner of the spreadsheet (in the **Info Box**).

Technical note on spreadsheets. In order to increase compatibility with other applications, *STATISTICA* automatically produces a variety of special Clipboard formats. For example, spreadsheet formats are produced and will be used when you paste the contents of the Clipboard (copied from *STATISTICA*) into a spreadsheet document in Microsoft Excel or other applications.

🖹 Paste (Block) Button

Click the 🖹 button (or press CTRL+V) to paste the current contents of the Clipboard into the *STATISTICA* document starting at the current cursor position. See *Cut (Block) Button* (page 104) for a description of differences between global operations on cases and variables (which treat cases and variables as logical units, see Cases ▾ *Cases Button* and Vars ▾ *Variables Button* in Chapter 4 - *STATISTICA Spreadsheets*), and Clipboard operations, which follow the standard spreadsheet Clipboard conventions. For example, when you intend to copy (via the Clipboard) an entire variable to a new location, make sure that the cursor is placed at the top of the destination column, because the data will always be pasted from the cursor down (even if the entire destination column is highlighted).

Clipboard operations in the spreadsheet. When you copy or move a block in the spreadsheet (e.g., via drag-and-drop), the values that are copied depend on the display mode of the spreadsheet. If the spreadsheet displays numeric values when the block is copied to the Clipboard, then only those numeric values will be copied. If the spreadsheet displays text labels when the block is copied to the Clipboard, then not only are the text labels copied to the Clipboard, but also the corresponding numeric values. This can result in the assignment of text labels to numeric values that did not previously have text label equivalents (see below).

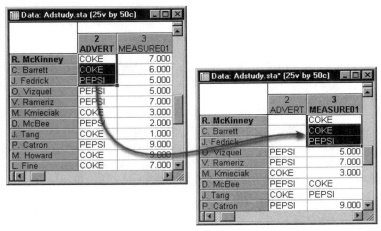

Note that in the illustration above even though only the (highlighted) block was moved, other numeric values in the target variable (in this case *1*'s and *2*'s), acquired new text identities.

Display formats in spreadsheets. When you copy or move a block in a spreadsheet, the display format for that block is also copied.

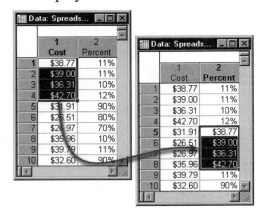

This means that the display format of the copied or moved block will overwrite the display format for the block into which it is copied. For example, if you copy a block of values that are formatted as currency and paste them into a variable that is formatted as percentage, the block will remain formatted as currency even though the rest of the values in that variable are formatted as percentage (see illustration above).

Format Painter Button

Click the button to copy character and paragraph formats. To use this button, select the text that has the formatting you want to copy, then click the **Format Painter** button. The cursor will change to a plus sign with a paintbrush. Now, select the text to which you want to apply the formatting.

Note that to apply the format to more than one cell in the spreadsheet, simply select the block of cells in the spreadsheet. You can apply formatting to more than one paragraph in a report by selecting all the paragraphs in the same manner.

Undo Button

Click the button (or press CTRL+Z) in order to reverse (undo) the last specific command or action in the document (such as editing, moving, or copying blocks; random fill; recoding or ranking variables; etc.). *STATISTICA* supports multi-level undo (with 32 buffers); therefore, you can undo multiple actions by selecting this option consecutively (up to 32 times).

Redo Button

Click the button (or press CTRL+Y) in order to reverse (redo) the last specific undo command or action in the document (such as undo editing, undo moving, or undo copying blocks; undo random fill; undo recoding or undo ranking variables; etc.). *STATISTICA* supports multi-level redo (with 32 buffers); therefore, you can redo multiple actions by selecting this option consecutively (up to 32 times).

Find Button

Click the button (or press CTRL+F) to display the **Find** dialog. The **Find** dialog is used to search the active document for words or phrases. Type the word or phrase that you are looking for in the **Find what** box. To start the search, click the **Find Next** button. *STATISTICA* will select the first word(s) that match your search criteria. Continue clicking **Find Next** until *STATISTICA*

completes the search throughout the entire document. You can stop the search at any time by clicking the *Cancel* button.

Find dialog for reports. When a report is the active document, the dialog shown below is displayed.

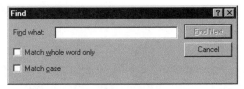

Select the *Match whole word only* check box to restrict the search to the exact word in the *Find what* box. Select the *Match case* check box to conduct a case-sensitive search for the word or phrase in the *Find what* box.

Find dialog for spreadsheets. When a spreadsheet is the active document, the dialog shown below is displayed.

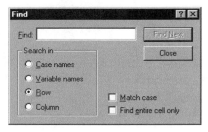

You can specify the areas of the spreadsheet to search in and whether or not to limit the search to the entire contents of the cell. You can also conduct a case-sensitive search for a word or phrase using the *Match case* check box. For more details, see the spreadsheet *Find* dialog in the *Electronic Manual*.

Find dialog for macros. When a macro is the active document, the dialog shown below is displayed.

You can specify the direction to search in your document (either *Up* or *Down*). Select the *Match whole word only* check box to restrict the search to the exact word in the *Find what* box or

select the **Match case** check box to conduct a case-sensitive search for the word or phrase in the **Find what** box.

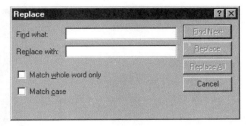 Replace Button

Click the button (or press CTRL+H) to display the **Replace** dialog. The **Replace** dialog is used to search the active document for words or phrases and replace them with different words or phrases. Specify a word or phrase for which to search (in the **Find what** box) as well as a word or phrase with which to replace it (in the **Replace with** box). To start the search, click the **Find Next** (or **Find**) button. *STATISTICA* selects the first word(s) that match your search criteria. You can choose to replace the text on an individual basis (click the **Replace** button) or globally (click the **Replace All** button). Continue clicking **Find Next** (or **Find**) until *STATISTICA* completes the search throughout the entire document. You can stop the search at any time by clicking the **Cancel** button.

Replace dialog for reports. When a report is the active document, the dialog shown below is displayed.

Select the **Match whole word only** check box to restrict the search to the exact word in the **Find what** box. Select the **Match case** check box to conduct a case-sensitive search for the word or phrase in the **Find what** box.

Replace dialog for spreadsheets. When a spreadsheet is the active document, the dialog shown below is displayed.

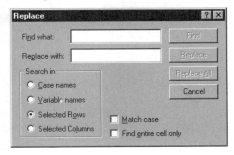

You can specify the areas of the spreadsheet to search in and whether or not to limit your search to the contents of an entire cell. You can also conduct a case-sensitive search using the **Match case** check box. For more details, see the spreadsheet **Replace** dialog in the *Electronic Manual*.

Replace dialog for macros. When a report is the active document, the dialog shown below is displayed.

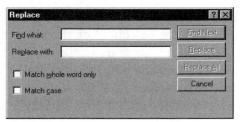

Select the **Match whole word only** check box to restrict the search to the exact word in the **Find what** box or select the **Match case** check box to conduct a case-sensitive search for the word or phrase in the **Find what** box. For more details, see the information on the *STATISTICA* Visual Basic editor in the *Electronic Manual*.

Add to Workbook Button

Click the `Add to Workbook ▼` button to add the current document to the open workbook, a previously saved workbook, or a new workbook. A menu will be displayed in which you can specify to which workbook you would like to add the document.

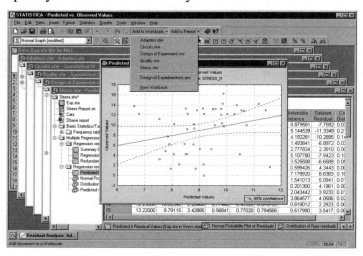

Select **New Workbook** from the menu to create a new workbook and add the current document to it. Note that if the document cannot be sent to a workbook (e.g., the document is a workbook), this button will be dimmed.

Add to Report Button

Click the `Add to Report ▼` button to add the current document to the open report, a previously saved report, or a new report. A menu will be displayed in which you can specify to which report you would like to add the document.

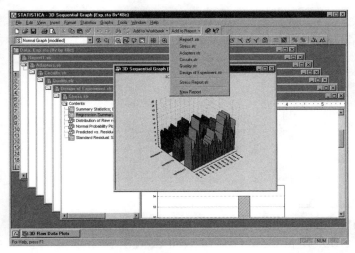

Select **New Report** from the menu to create a new report and add the current document to it. Note that if the document cannot be sent to a report (e.g., the document is a report or a workbook), this button will be dimmed.

Help Topics Button

Click the ⊘ button to access the *STATISTICA* on-line *Electronic Manual*. *STATISTICA* provides comprehensive, context-sensitive, on-line documentation for all program procedures and all options that is available by pressing the F1 key or by clicking the help button ? on the toolbar or the caption bar of all dialog boxes (there is a total of over 100 megabytes of compressed documentation included). Due to its dynamic hypertext organization (and various facilities allowing you to customize the help system), it is usually faster to use the on-line documentation (*Electronic Manual*) than to look for information in the printed manuals. Note that the status bar also displays short explanations of the menu commands or toolbar buttons,

available when a command is highlighted or a button clicked. In addition, you can press the help key (F1) when the command is highlighted for more information on any menu command.

Statistical Advisor. A *Statistical Advisor* facility is built into the *Electronic Manual*.

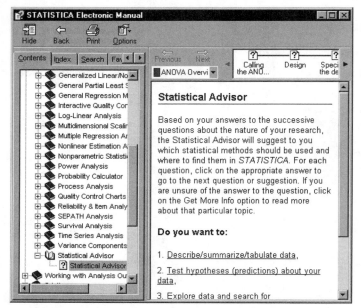

When you select *Statistical Advisor* from the *Help* menu, *STATISTICA* asks you a set of simple questions about the nature of the research problem and the type of your data. It then suggests the statistical procedures that appear most relevant and tells you where to look for them in the *STATISTICA* system.

☝ Context Sensitive Help Button

Click the ☝ button (or press SHIFT + F1) to access the on-line *Electronic Manual* in a context sensitive manner. When you click this button, the mouse pointer changes to an arrow with a question mark beside it ☝. Click the toolbar button or area of *STATISTICA* that you are interested in learning more about. The appropriate *Electronic Manual* topic is then displayed. If an appropriate topic is not available, the *Contents* tab of the *Electronic Manual* is displayed.

Statistics Toolbar

This non-default toolbar is only available if you choose to view it from the **Statistics** option of the **View - Toolbars** menu or from the **Toolbars** tab of the **Customize** dialog. It provides access to all the analysis types in *STATISTICA* as well as to the *STATISTICA* Visual Basic editor. For information on customizing toolbars, see the *Toolbar Overview* (page 56).

Resume Button

Click the ⏎ button to continue the most recently displayed analysis or graph. This option is particularly useful if you want to display the previous dialog after you have created a results spreadsheet or graph.

Macro Button

Click the ▣ button to display the *STATISTICA* Visual Basic code corresponding to all design specifications as well as output options that you have selected for the current analysis or graph. The **New Macro** dialog will first be displayed in which you can type the macro's name and description.

Commonly Used Modules Buttons

Click any of the buttons for commonly used modules to display the startup panel associated with that module. For example, click the ▣ button to display the **Basic Statistics and Tables** (Startup Panel). The table below displays the buttons and their respective modules. Note that each of these modules is discussed in the *Statistics Menu* topic (page 67)

▣ **Basic Statistics and Tables**   ▣ **Nonparametric Statistics**

▣ **Multiple Regression Analysis**   ▣ **Distribution Fitting**

▣ **ANOVA**

📑 ✳ ⊕ ✱ 🗐 ⩕ ▦ ✳ 📳 🔲 Multivariate Exploratory Techniques Buttons

Click any of the multivariate exploratory techniques buttons to display the startup panel associated with that module. For example, click the 📑 button to display the **Cluster Analysis** (Startup Panel). The table below displays the buttons and their respective modules. Note that each of these modules is discussed in the *Multivariate Exploratory Techniques* topic (page 79)

📑 **Cluster Analysis**	⩕ **Classification Trees**
✳ **Factor Analysis**	▦ **Correspondence Analysis**
⊕ **Principal Components & Classification Analysis**	✳ **Multidimensional Scaling**
✱ **Canonical Analysis**	📳 **Discriminant Analysis**
🗐 **Reliability/Item Analysis**	🔲 **General Discriminant Analysis**

🔲GLM 🔲GLZ 🔲RM 🔲PLS 🔲 📉 📈 📈 📈 📉? 🔲 Advanced Linear/Nonlinear Models Buttons

Click any of the advanced linear/nonlinear models buttons to display the startup panel associated with that module. For example, click the 🔲GLM button to display the **General Linear Models (GLM)** (Startup Panel). The table below displays the buttons and their respective modules. Note that each of these modules is discussed in the *Advanced Linear/Nonlinear Models* topic (page 71)

🔲GLM **General Linear Models**	📉 **Nonlinear Estimation**
🔲GLZ **Generalized Linear/Nonlinear Models**	📈 **Fixed Nonlinear Regression**
🔲RM **General Regression Models**	🔲 **Log-Linear Analysis**
🔲PLS **Partial Least Squares Models**	📉? **Time Series Analysis**
🔲 **Variance Components Analysis**	🔲 **Structural Equation Modeling (SEPATH)**
📉 **Survival Analysis**	

🔲 🔲 🔲 🔲 Industrial Statistics Buttons

Click any of the industrial statistics buttons to display the startup panel associated with that module. For example, click the 🔲 button to display the **Process Analysis** (Startup Panel). The table below displays the buttons and their respective modules. Note that each of these modules is discussed in the *Industrial Statistics* topic, beginning on page 86; the *Six Sigma (DMAIC) Shortcuts* are briefly discussed starting on page 89.

▤ *Quality Control Charts*
▥ *Process Analysis*

▥ *Design of Experiments (DOE)*
▥ *Six Sigma*

▥ Power Analysis Button

Click the ▥ button to display the **Power Analysis and Interval Estimation** (Startup Panel). Note that this module is discussed in the *Power Analysis* topic (page 94).

▥ ▥ ▥ ▥ ▥ Data-Mining Buttons

Click any of the data mining buttons to display the startup panel associated with that module. For example, click the ▥ button to display the **General Slicer/Dicer Explorer with OLAP** (Startup Panel). The table below displays the buttons and their respective modules. Note that each of these modules is discussed in the *Data-Mining* topic (page 94)

▥ *General Slicer/Dicer Explorer with OLAP*
▥ *General Classifier (Trees and Clusters)*
▥ *General Modeler and Multivariate Explorer*

▥ *General Forecaster*
▥ *General Neural Network Explorer*

Six Sigma Toolbar

This non-default toolbar is only available if you choose to view it from the **Six Sigma** option of the **View - Toolbars** menu or from the **Toolbars** tab of the **Customize** dialog. It provides access to all the Six Sigma shortcuts in *STATISTICA*. For information on customizing toolbars, see the *Toolbar Overview* (page 56).

▥ ▥ ▥ ▥ ▥ ▥ Six Sigma (DMAIC) Shortcuts Buttons

 The Six Sigma shortcuts provide fast access to the options and statistical analysis procedures that are commonly applied in Six Sigma quality control programs. Click any of the **D**, **M**, **A**, **I**, or **C** buttons to display a menu with shortcuts to various modules. In the listing below, these modules are identified in parentheses where applicable. Some of the shortcuts will display dialogs to specify graphs; where ambiguities exist, the type of graph that that can be specified via the respective option is also noted in parentheses. For more information, see also the *Six Sigma (DMAIC) Shortcuts* topic discussed starting on page 89. Note that all of these shortcuts

may not be available to you, depending on your configuration (installed modules) of
STATISTICA.

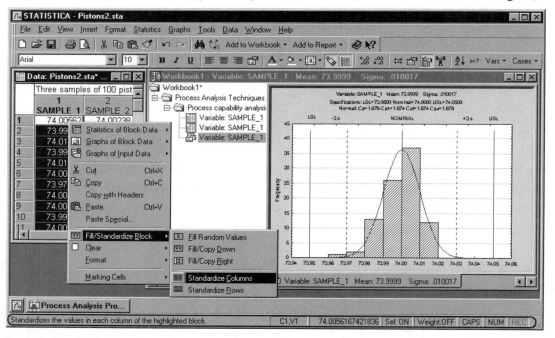

Define   **Improve**

Measure   **Control**

Analyze   **Six Sigma Calculator**

Status Bar

The *status bar* is located at the bottom of the *STATISTICA* application window. It is used to
display short help messages and explanations, display progress bars, and provide quick access
to some of the most commonly used system facilities (case selection, case weights).

It also displays the status of the *STATISTICA* Macro recording or execution. All program
control facilities accessible via the status bar shortcuts are also available from menus and
via keyboard.

Progress Bar

The status bar changes into a *progress bar* whenever data are processed. The **Cancel** button at the end of the bar is used to interrupt the current processing.

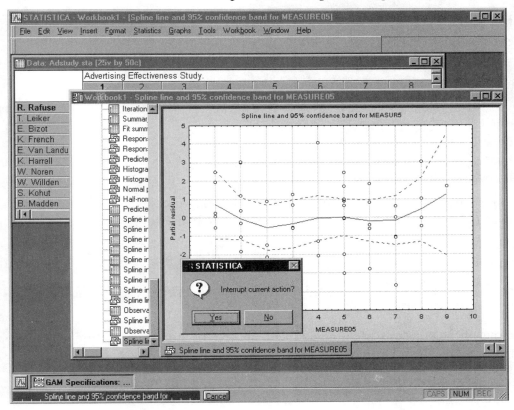

After you click the **Cancel** button, a warning message is displayed asking if you want to interrupt the current action. Click the **No** button to resume the current operation.

Multitasking. *STATISTICA* supports multitasking (between its instances or other applications).

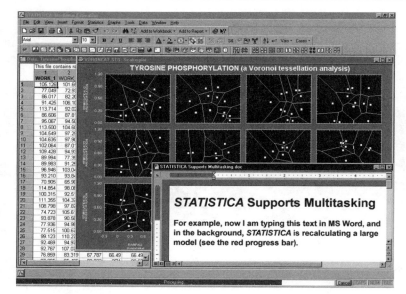

In cases when an unusually large datafile or analysis design is being processed, you can switch to another Windows application and the processing will continue in the background.

Message Area

The *Message Area* of the status bar serves as a miniature help topic. This area is used to display:

- Brief explanations of the currently highlighted menu choices or the toolbar buttons that are pointed to with the mouse pointer;

- Status information about the currently performed operation (e.g., *Extracting factors...* or *Computing residuals...*);

- Brief instructions relevant to the current stage of analysis or the operation that is being performed.

Name Box

When a spreadsheet is active, the *Name Box* indicates the case and variable number of the currently selected cell. If a block of cells is selected, the *Name Box* indicates the case and variable number of the cell that was first selected.

Show Field

The **Show Field** displays the value of the currently highlighted cell at a higher precision than will fit in the respective columns of the spreadsheet. The width of the **Show Field** can be adjusted (toggled) by clicking on it.

Note that this field is only available when a spreadsheet is being edited.

Case Selection Conditions

The clickable **Sel: Off** (or **Sel: On**) field displays the current status of the **Spreadsheet Case Selection Conditions**, that is, optional user-defined conditions (or "filters") that can be used to select a particular subset of cases for an analysis.

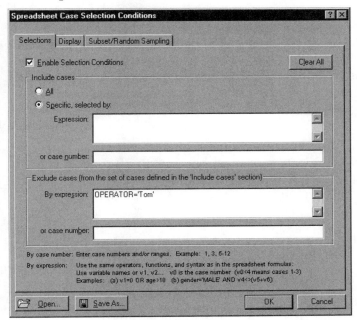

Normally, all cases encountered in the datafile are processed (as long as they do not have missing data). However, you can define temporary subsets of data and temporarily limit an

analysis to those subsets only (e.g., only females older than 60 who either have a high cholesterol level or high blood pressure). The ***Sel: Off*** (or ***Sel: On***) field also acts as a button: by clicking it, you can display the ***Spreadsheet Case Selection Conditions*** dialog, which shows the currently specified conditions (if any were defined). This is only available if the current input spreadsheet is selected. Note that case selection conditions can also be displayed using the ***Tools*** menu.

Note that unless you are at the beginning of an analysis or graph, a warning message is displayed asking if you want to start a new analysis or graph because changing selection conditions in the middle of an analysis or graph can lead to inconsistent results (i.e., different stages of computations would be based on different subsets of data). However, if you click this status bar field at any point when the conditions can be adjusted (e.g., before an analysis is started), the ***Spreadsheet Case Selection Conditions*** dialog is displayed allowing you to enter or edit the text of the conditions.

The syntax of those conditions is very simple (note the examples in the help area of the dialog). You can refer to variables by their numbers (e.g., ***v1***, ***v2***, ***v3***, ...; note that ***v0*** is the case number) or names (e.g., ***Income***, ***Profit***). Thus, for example, the expression:

Include cases: ***v0<101 and v1=1***

will include in the analysis only cases from the first 100 (i.e., case number has to be less than 101) and where the value of variable number 1 (i.e., ***v1***) is 1. Note that if the name of variable number 1 was ***Gender***, and for this particular variable, 1 was equivalent to the text value ***MALE*** (see *Can I Label Numeric Values?* in Chapter 4 - *STATISTICA Spreadsheets*), the same case selection condition could be alternatively entered as:

Include cases: ***v0<101 and GENDER = 'MALE'***

The syntax of the expressions supports a wide selection of functions and operators, and allows you to specify case selection conditions of practically unlimited complexity. For further details, see *What Syntax Can Be Used to Create Case Selection/Verification/Recode Conditions?* in Chapter 4 – *STATISTICA Spreadsheets*.

The specified conditions can be preserved even if you turn off the computer (when you exit the program, *STATISTICA* will prompt you to save the changes to the current datafile. If you click the ***Yes*** button, the selection conditions will be saved with the datafile).

Note that case selection conditions can also be specified for only the current analysis or graph (and hence not be tied to the current datafile) by clicking the ▦ button to display the ***Analysis/Graph Case Selection Conditions*** dialog. This button is included on all Startup Panels and all analysis or graph definition dialogs that are displayed before the data processing begins.

The other tabs on this dialog are used to select a cell display format that will be used to mark selected cases and to access a facility to create random subsets of cases (records).

Case Weights

The clickable **Weight: Off** (or **Weight: On**) field displays the current status of the **Spreadsheet Case Weight** option. That is, an option to treat values of a selected variable as (integer) case multipliers when processing data or fractional weights, if applicable (for example, fractional weights are supported in **Basic Statistics and Tables**, **Multiple Regression**, **Variance Components**, **Correspondence Analysis**, specialized scatterplots, and other procedures). The weights can be used either for analytic purposes (e.g., some observations can be measurably more "important" and this importance can be represented by weight scores) or to economize data storage (e.g., in some large datafiles, such as a census or survey data, many cases can be identical and can be represented by one case with an appropriate case weight attached to it).

The **Weight: Off** (or **Weight: On**) field also acts as a button: Click the **Weight: Off** field to display the **Spreadsheet Case Weight** dialog, which shows the weight variable (if one was defined). This is only available if the current input spreadsheet is selected. Note that case weights can also be displayed by clicking the 🔠 **Case Weight** button or by using the **Tools** menu.

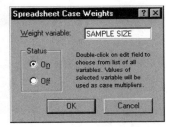

Note that unless you are at the beginning of an analysis or graph, a warning message is displayed asking if you want to start a new analysis or graph because changing case weights in the middle of an analysis or graph can lead to inconsistent results (i.e., different stages of computations would be based on different configurations of data). However, if you click this status bar field at any point when the conditions can be adjusted (e.g., before an analysis is started), the **Spreadsheet Case Weights** dialog is displayed allowing you to specify a weighting variable.

Note that case weights can also be specified for only the current analysis or graph (and hence not be tied to the current datafile) by clicking the 🔒 ⓦ button to display the **Analysis/Graph Case Weights** dialog. This button is included on all Startup Panels and all analysis or graph definition dialogs that are displayed before the data processing begins.

Caps Lock

The *CAPS* field indicates whether or not the CAPS LOCK key has been pressed on the keyboard.

Num Lock

The *NUM* field indicates whether or not the NUM LOCK key has been pressed on the keyboard.

Macro Recording Status

The *Macro* status field (the last field of the status bar) is normally not active unless the currently performed operation involves a recorded keyboard macro. When you are recording a keyboard macro (via *Tools – Macro – Record Keyboard Macro*), then the *REC* becomes active in the *Macro* status field, so you are aware that all keyboard actions are being recorded into a macro. Click the *Macro* status field at any time to display the *Macro* dialog, which is used to run, edit, and delete existing macros or create new macros.

Mouse Conventions

In addition to the standard, Windows related applications, the mouse provides shortcuts to specific features in *STATISTICA*.

When Is the Left Mouse Button Used?

The left mouse button is used for selecting (e.g., an option in a dialog), highlighting (e.g., a block of data in a spreadsheet), and dragging (e.g., dragging a block in a spreadsheet or an object in a graph; see *How Is the Mouse Used in Other Operations?*, page 131). Usually, double-clicking on an item provides a direct shortcut to some of the most commonly used dialogs in *STATISTICA* related to that item.

For example, you can double-click on a spreadsheet variable name to display a *Variable* specifications dialog that is used to edit the respective feature. Also, you can double-click on a graph object, title, point marker, etc., to display the respective dialog that is used to edit that part of the graph. In this way, you save a step whenever you want to access a commonly used option. Double-clicking on an object in a report will activate the object and initiate the in-place editing of that object.

When Is the Right Mouse Button Used?

When you right-click on an item (e.g., a cell in a spreadsheet), a shortcut menu associated with that item is displayed.

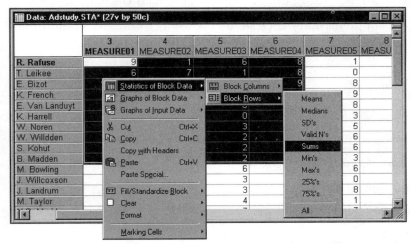

These shortcut menus allow easy access to some commonly used options in every window in *STATISTICA* (e.g., graph window, spreadsheet window, report, etc.).

For example, if you right-click on a spreadsheet cell, a shortcut menu of graph options, statistics options, Clipboard operations, etc. is displayed (see above), that are used to quickly perform the desired operation.

File Name Extensions

The default file name extensions for the types of files that *STATISTICA* uses most often are:

.sta   *STATISTICA* Spreadsheets (including datafiles and results);

.smx   *STATISTICA* Matrix Spreadsheets;

.stg   *STATISTICA* Graphics files (including all data represented in the graphs, compound documents, etc.);

.stw   *STATISTICA* Workbooks (these ActiveX containers can include spreadsheets, reports, graphs, macros, and various other objects);

.str	*STATISTICA* Report files (these ActiveX containers can include spreadsheets, reports, graphs, macros, and various other objects; note that they can be saved as standard *.rtf* files but then the *STATISTICA* tree view information is not preserved);
.svb	Macro files (recorded or written using the *STATISTICA* Visual Basic facilities);
.txt	All other non-RTF text supplementary files saved in ASCII format (e.g., multiple case selection conditions, data recoding specifications, data verification conditions, etc.);
.sel	Case selection conditions files;
.cmd	SEPATH (PATH1 language) model files;
.sqy	*STATISTICA* Query files;
.chm	*STATISTICA Electronic Manual* files;
.xml	*STATISTICA* configuration files;
.wmf	Windows graphics metafile files;
.bmp	Windows device-independent bitmap graphics files;
.jpg	Joint Photographers Experts Group highly compressed image files;
.png	Portable Network Graph highly compressed image files;
.xls	Excel worksheet files;
.dbf	dBASE, FoxPro files;
.por	SPSS portable files;
.wk?, .wq?	Lotus, Symphony, and QuattroPro worksheet files (e.g., *.wk1, .wk3*);
.txt or *.csv*	Text files;
.htm	HTML files.
.rtf	Files containing formatted (Rich Text Format) text;

Note that other file formats recognized by *STATISTICA* include all OLE/ActiveX Document compatible files (e.g., files created by MS Office). In addition, other data formats can be accessed via the *STATISTICA* Query facilities (accessible from the **Data** and **File** menus, see Chapter 10 – *STATISTICA Query*).

FAQ: SYSTEM-WIDE FEATURES

Global Control of the Program and General User Interface Conventions

How Do I Access Help for a Specific Dialog?

Quick access to the *Electronic Manual* is provided via the question mark at the right side of the caption bar of every dialog box.

Click the [?] button (or press F1) to display a **Help** window containing the description of that dialog. Note that when you highlight a menu command, you can press F1 to display its **Help** window.

What Are Microscrolls?

Numerical values in all dialogs can be changed by using the microscrolls controls: [10] [↕].

Click the microscrolls to either increment or decrement the last digit. Right-click them to either increment or decrement the next-to-last digit (e.g., clicking the up microscroll increments .15 to .16, then .17, .18, etc., right-clicking the up microscroll increments 0.15 to 0.25, then .35, .45, etc.).

The following example illustrates how the microscrolls can be controlled with the left and right mouse buttons.

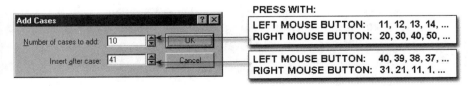

What Are ToolTips?

ToolTips are small "balloon help tips" that are displayed when the mouse pointer is on a toolbar button. ToolTips help you to quickly learn the functions of all toolbar buttons.

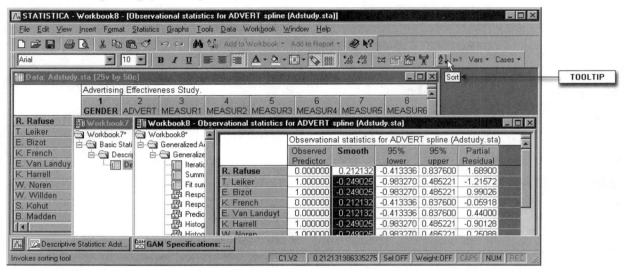

You can control the display of the ToolTips for graph objects via the **View - ToolTips on Graph Objects** menu. You can toggle the display of all other types of ToolTips on the **Options** tab of the **Customize** dialog (available by selecting **Customize** from the **Tools** menu). If a more detailed description of a button is required, click the ▶? toolbar button (the mouse pointer will change to an arrow with a question mark beside it ▶?) and then click the toolbar button. The appropriate **Help** topic will be displayed. To obtain detailed descriptions of all the buttons on the respective toolbar, review the toolbar sections in each chapter of this manual.

Can I Automate Commonly Used Procedures or Repeat Similar Tasks in *STATISTICA*?

Automation facilities in *STATISTICA* are available via *STATISTICA* Visual Basic. When you run an analytic procedure (from the **Statistics** menu) or create a graph (from the **Graphs** menu), the Visual Basic code corresponding to all design specifications as well as output options that you select are recorded. To display that code, select **Create Macro** from the 🄰 Options ▾ button drop-down list (available on any analysis or graph dialog), use the 📷 toolbar button, or select

the particular analysis or graph from the **Tools – Macro – Create Analysis/Graph Macro** menu. This code can later be executed repeatedly or edited by changing options, variables, or datafiles and optionally adding user interface, etc. Programs can also be written from scratch using the *STATISTICA* Visual Basic professional development environment featuring a convenient program editor with a powerful debugger (with breakpoints, etc.), an intuitive dialog painter, and many facilities that aid in efficient code building. To display the SVB editor, select **New** from the **File** menu to display the **Create New Document - Macro (SVB) Program** tab.

Note that also Master Macros (logs of multiple analyses) and Keyboard Macros (context sensitive recordings of sequences of keystrokes) are also available, see Chapter 11 – *STATISTICA Visual Basic*.

Also, many procedures and graphs allow you to automatically repeat the same analysis for each of a series of variables (e.g., a **DOE** analysis run on multiple dependent variables) or each level of a grouping variable (e.g., categorized graphs).

How Can I Break, Stop, or Interrupt the Current Action?

The following facilities are available to stop, break, or interrupt the current action (depending on the operation performed by *STATISTICA*).

Analysis. Click the *Cancel* button on the Progress bar to interrupt the task in progress (see *Progress Bar,* page 117).

Printing. You can interrupt the printing of spreadsheets, graphs, reports, etc., by clicking the *Cancel* button on the *Print* dialog.

STATISTICA Visual Basic. You can interrupt SVB programs in many ways. First, click the ▐▌ toolbar button to pause the currently running macro or click the ▉ toolbar button to stop running (e.g., reset) the current macro. You can also press the ESC key or CTRL+BREAK in order to interrupt the execution of the *STATISTICA* Visual Basic program.

Brushing (in a graph). You can deactivate the brushing tool by clicking on the *Selection Tool* ▶ toolbar button (see Chapter 6 – *Graphs: General Features*).

How Can I Learn What Information Is Necessary to Start an Analysis (Variables, Grouping, Codes, Options, etc.)?

Click the ? button or press the F1 key to display the relevant section of the *Electronic Manual* containing a comprehensive explanation of all options in the current dialog. However, all analysis definition screens in *STATISTICA* follow the "self-prompting" dialog conventions.

The *OK* button is never dimmed; whenever you are not sure what to select next, simply click *OK* and *STATISTICA* proceeds to the next logical step and asks you for specific input if it is necessary.

How Can I Find a Particular Statistical Procedure?

If you are not certain where to find a particular procedure within *STATISTICA*, consult the *Statistical Advisor* by selecting *Statistical Advisor* from the *Help* menu. The *Statistical Advisor* asks you a set of simple questions about the nature of the research problem and the type of your data, and then it suggests the statistical procedures that appear most relevant and tells you where to look for them in the *STATISTICA* system.

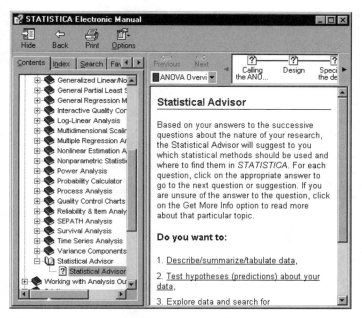

You can also search for topics in the *Electronic Manual* using the **Index** or **Search** facilities. The *Electronic Manual* is always available by pressing the F1 key.

How Can I Copy/Print Result Summaries?

Located at the top of some results dialogs is a summary box. This box contains relevant summary information for the type of analysis (e.g., descriptive statistics, design attributes, regression results, etc.). Additionally, two buttons are provided with the summary box: an expand/collapse ▲ ▼ button that is used to expand or collapse the summary box and a copy 🗈 button that is used to copy the summary results to the Clipboard.

You can copy a portion of the summary results to the Clipboard by selecting the desired text and clicking the 🗈 button. To copy all of the text, simply click the 🗈 button without selecting any text. You can then paste the text into a *STATISTICA* Report, Graph, or any word processing document (e.g., Notepad or Word) for printing.

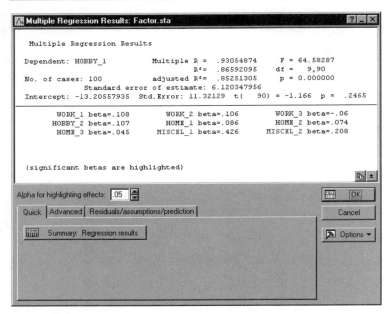

Note that the copied text retains formatting information (such as font, color, etc.).

How Do I Select Items From Multiple-Selection Lists?

You can select items in a multiple-selection list in the following manners:

- Click on an item to select (highlight) it. Click the **OK** button in the dialog to accept the selection.

- Double-click on an item to select it and accept the selection (i.e., close the dialog).

- In order to select a continuous list of items, you can (1) hold down the mouse button and drag the cursor over the items that you want to select or (2) select the first item and then while holding down the SHIFT key, click on the last item that you want to select.

- In order to select a discontinuous list of items, hold down the CTRL key and click on the desired items.

How Do I Select Items in the Workbook Tree?

You can select one or more items in the workbook tree using the standard Windows SHIFT+click and CTRL+click conventions to select ranges and discontinuous lists of variables, respectively (see above). Additionally, you can select or deselect tree items using the keyboard navigation keys (e.g., HOME, END, PAGE UP, PAGE DOWN, and arrow keys). You can delete an entire selection by pressing the DELETE key. Pressing the INSERT key will display the **Insert Item** dialog for the currently selected item.

Note that only visible items can be selected; therefore, to select the contents of a particular node, you will need to expand the node (by clicking on the plus sign adjacent to the node).

How Is the Mouse Used in Other Operations?

Besides the standard Windows SHIFT+click and CTRL+click conventions described above, you can perform the following actions using the mouse:

Drag-and-drop. Provides mouse shortcuts for moving, copying, deleting, inserting, extrapolating, etc. a block of values in the spreadsheet. See Chapter 4 - *STATISTICA Spreadsheets* for further details.

	1 Var1	2 Var2	3 Var3	4 NewVar1	5 NewVar2	6 NewVar3
1	4.000	20.000	4.000			
2	18.000	9.000	14.000			
3	15.000	3.000	10.000			
4	11.000	1.000	8.000			
5	7.000	1.000	2.000	4.000	20.000	4.000
6	1.000	8.000	13.000	18.000	9.000	14.000
7	2.000	11.000	16.000	15.000	3.000	10.000
8	8.000	12.000	17.000			

Data: Spreadsheet1* (6v by 8c)

Increase/decrease column width. Adjusts the spreadsheet column width by dragging the right column border to the desired width.

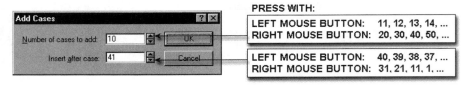

The variable header automatically expands, and the new width of the column is indicated by a dashed line (as shown in the illustration above).

Split scrolling (in spreadsheets). Splits the spreadsheet (i.e., split scrolling) by dragging the split box (the small rectangle at the top of the vertical scrollbar or to the left of the horizontal scrollbar).

Variable speed scrolling. Controls the speed at which you scroll (1 line at a time by moving the cursor a short distance away or one page at a time by moving the cursor further away) when you extend a block outside the spreadsheet.

Microscrolls ⬜ . Enable you to increase or decrease the value in a numeric edit field incrementally, by either the last digit (click the *up* microscroll, e.g., 1.11, 1.12, 1.13, ...),

StatSoft
Copyright © StatSoft, 2001

or the last digit by a factor of 10 (right-click the *up* microscroll, e.g., 1.11, 1.21, 1.31, ...; see *What Are Microscrolls?*, page 125).

Toolbar configuration. Double-click on the space between the buttons of the toolbar to undock the toolbar, allowing you to place it wherever you want. To return the toolbar to a docked setting, double-click in the title bar of the floating toolbar.

Reordering items in a list. You can reorder items in a list by selecting one or more items (in a continuous or discontinuous list) and then moving the cursor, which changes to a ↨, to the desired position.

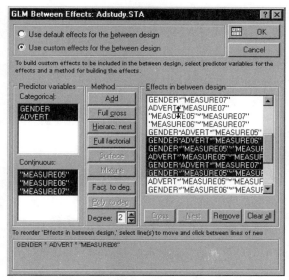

Clicking the mouse will then move the highlighted item(s) to the insertion point.

CHAPTER

STATISTICA OUTPUT MANAGEMENT

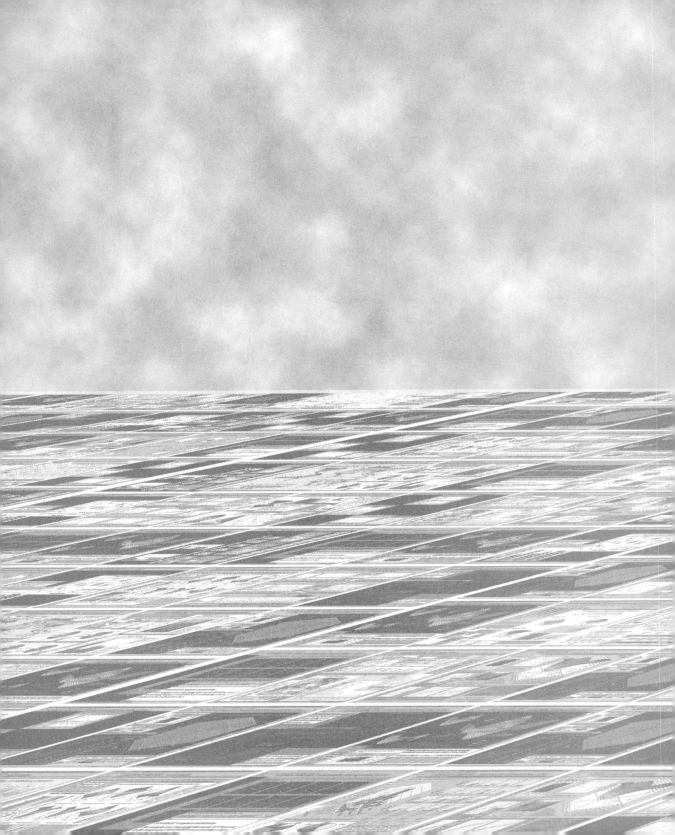

CHAPTER

STATISTICA OUTPUT MANAGEMENT

GENERAL OVERVIEWS

Three Channels for Output from Analyses

When you perform an analysis, *STATISTICA* generates output in the form of multimedia tables (spreadsheets) and graphs. There are three basic channels controlled by the **Output Manager** (accessed from the **File** menu) to which you can direct all output:

1. Workbooks,

2. Report Editors, and

3. Stand-alone Windows.

These three output channels can be used in many combinations (e.g., a workbook and report simultaneously), and each of the output channels can be customized in a variety of ways. Also, all output objects (spreadsheets and graphs) placed in each of them can contain other embedded and linked objects and documents; thus, *STATISTICA* output can be hierarchically organized in a variety of ways.

Each of the three *STATISTICA* output channels has its unique advantages:

1. Workbooks

Workbooks are the default way of managing output (for more information, see Chapter 3 – *STATISTICA Workbooks*). Each output document (e.g., a *STATISTICA* Spreadsheet or Graph, as well as a Microsoft Word or Excel document) is stored as a tab in the workbook.

Documents can be organized into hierarchies of folders or document nodes (by default, one is created for each new analysis) using a tree view, in which individual documents, folders, or entire branches of the tree can be flexibly managed.

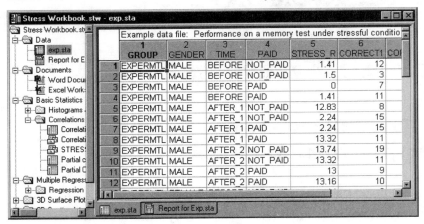

For example, selections of documents can be extracted (e.g., drag-copied or drag-moved) to the report window or to the application workspace (i.e., the *STATISTICA* application "background" where they will be displayed in stand-alone windows). Entire branches can be placed into other workbooks in a variety of ways in order to build specific folder organization, etc.

Technically speaking, workbooks are ActiveX documents (see details on ActiveX technology in Chapter 3 – *STATISTICA Workbooks*). Workbooks are compatible with a variety of foreign file formats (e.g., all the Microsoft Office documents) that can be easily inserted into workbooks and in-place edited.

User notes and comments in workbooks. Workbooks offer powerful options to efficiently manage even extremely large amounts of output, and they may be the best output handling solution for both novices and advanced users. It might appear that one of their possible drawbacks is that user comments (e.g., notes) and supplementary information cannot be as transparently inserted into the "stream" of the workbook output as they can in traditional, word processor style reports, such as *STATISTICA* Reports. However, note that:

- All *STATISTICA* documents can easily be annotated, both (a) directly, by typing text into graphs, tables, and reports, and (b) indirectly, by entering notes into the **Comments** box of the **Document Properties** dialogs (accessed from the **File – Properties** menu), and

- Formatted documents with notes and comments (in the form of text files, *STATISTICA* Report documents, WordPad or word processor documents, etc.) can easily be inserted anywhere in the hierarchical organization of output in workbooks. Moreover, such summary notes or comment documents can be made nodes for groups of subordinate objects to which the note is related to further enhance their organization.

2. Reports

STATISTICA Reports offer a more traditional way of handling output where each object (e.g., a *STATISTICA* Spreadsheet or Graph, or a Microsoft Excel spreadsheet) is displayed sequentially in a word processor style document.

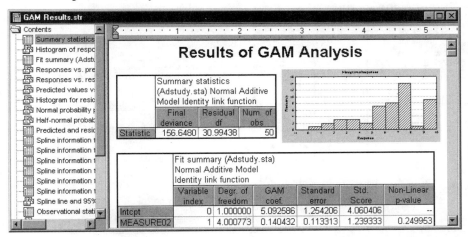

However, the technology behind this simple editor offers you very rich functionality. For example, like the workbook (see the previous section), the *STATISTICA* Report is also an ActiveX container (see the previous section; see also information on ActiveX technology in Chapter 3 – *STATISTICA Workbooks*) where each of its objects (not only *STATISTICA* Spreadsheets and Graphs, but also any other ActiveX compatible documents, e.g., Microsoft Excel spreadsheets) remains active, customizable, and in-place editable.

The obvious advantages of this way of handling output (more traditional than the workbook) are the ability to insert notes and comments "in between" the objects as well as its support for the more traditional way of quick scrolling through and reviewing the output to which some

users may be accustomed (the editor supports variable speed scrolling and other features of the IntelliMouse). Also, only the report output includes and preserves the record of the supplementary information, which contains a detailed log of the options specified for the analyses (e.g., selected variables and their labels, long names, etc.) depending on the level of supplementary information requested on the ***Output Manager*** tab of the ***Options*** dialog (assessable via the ***File – Output Manager*** menu), see page 141.

The obvious drawback, however, of these traditional reports is the inherent flat structure imposed by their word processor style format, although that is what some users or certain applications may favor.

3. Stand-Alone Windows

Finally, *STATISTICA* output documents can also be directed to a queue of stand-alone windows; the ***Queue Length*** can be controlled on the ***Output Manager*** tab of the ***Options*** dialog (assessable via the ***File – Output Manager*** menu).

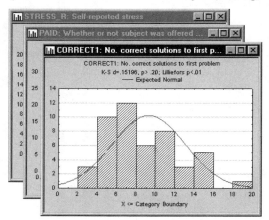
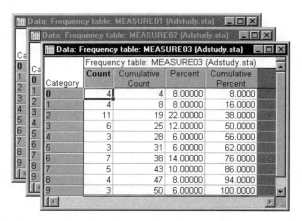

The clear disadvantage of this output mode is its total lack of organization and its natural tendency to clutter the application workspace (note that some procedures can generate hundreds of tables or graphs with a click of the button).

One of the advantages of this way of handling output is that you can easily custom arrange these objects within the *STATISTICA* application workspace (e.g., to create multiple, easy to identify "reference documents" to be compared to the new output). However, note that in order to achieve that effect, you do not need to configure the output ahead of time and generate a large number of (mostly unwanted) separate windows that can clutter the workspace. Instead, individual, specific output objects directed to and stored in the other two channels (workbooks

and reports) can easily be dragged out from their respective tree views onto the application workspace as needed.

Global Output Options – Output Manager

You can customize the output management of *STATISTICA* by selecting (or clearing) several options offered on the **Output Manager** tab of the **Options** dialog (accessible via the **Tools – Options** or the **File – Output Manager** menu). The selections that you make here will then be used as the default operating conditions whenever you open *STATISTICA*.

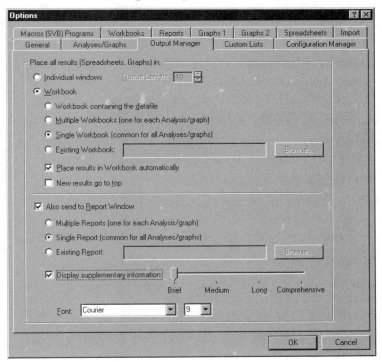

Place All Results (Spreadsheets, Graphs) in

The options in the *Place all results (Spreadsheets, Graphs) in* group allow you to determine where analysis spreadsheets and graphs will be placed upon clicking the *OK* (or *Run*) button in an analysis dialog.

Individual windows. Select the *Individual windows* option button to have each result shown in an individual window. Note that you can select either the *Individual windows* or the *Workbook* option button. You can, however, select both the *Individual windows* option button and the *Also send to Report Window* check box.

Queue length. Enter the number (or use the corresponding microscrolls) of result windows to remain on screen in the *Queue length* box.

Workbook

Select the *Workbook* option button to have each result shown in a workbook. As mentioned above, you can select either the *Individual windows* or the *Workbook* option button, but not both of them. You can, however, select both the *Workbook* option button and the *Also send to Report Window* check box. Note that this option must be selected for the next six options to be available.

Workbook containing the datafile. Select the *Workbook containing the datafile* option button to have all results included in the same workbook as the datafile. Note that if you select this option and the datafile is not included in any workbook, the results will be shown in individual windows.

If the *Place results in Workbook automatically* check box (see below) is also selected (the default setting), then results will automatically go to the workbook containing the datafile when you produce them. If the *Place results in Workbook automatically* check box is not selected, then the results will go into individual windows. Each window, however, is tagged so that if you select the  button, the tagged window will automatically go to the correct workbook containing the datafile and will be added under the appropriate workbook folder. See the *Electronic Manual* for further details.

Multiple workbooks (one for each analysis/graph). Select the *Multiple Workbooks* option button to include the results from an analysis or graph in a separate workbook for each analysis and graph.

Single workbook (common for all analyses/graphs). Select the *Single Workbook* option button to include all results in one workbook together regardless of the analysis or graph being created.

Existing workbook. Select the *Existing Workbook* option button to send all results to an existing workbook specified in the corresponding field. If a workbook has not previously been specified via the *Browse* button, then the standard *Open* dialog will be displayed when you select this option.

Browse. Select the *Browse* button to display the standard *Open* dialog, in which you can locate and specify an existing workbook.

Place results in workbook automatically. Select the *Place results in Workbook automatically* check box to include all results in a workbook automatically. Note that to use this option you must also select one of the four workbook options above.

New results go to top. Select the *New results go to top* check box to place each new result as the first child of each folder in the workbook. Note that, by default, this option is not selected, and new result objects are added at the bottom.

Also Send to Report Window

Select the *Also send to Report window* check box to show analysis and/or graph results in a report. The remaining options pertain to the use and display of the output report window. They determine what type of report is used and what amount of information is included in it. Note that this option must be selected for the remaining options to be available.

Multiple reports (one for each analysis/graph). Select the *Multiple Reports* option button to include the results from an analysis or graph in a separate report for each analysis and graph.

Single report (common for all analyses/graphs). Select the *Single Report* option button to include all results in one report together regardless of the analysis and/or graph being produced.

Existing report. Select the *Existing Report* option button to send all results to an existing report specified in the corresponding field. If a report has not previously been specified via the *Browse* button, then the standard *Open* dialog will be displayed when you select this option.

Browse. Select the *Browse* button to display the standard *Open* dialog, in which you can locate and specify an existing report.

Display Supplementary Information

Select the ***Display supplementary information*** check box, and then drag the slider at the right to select the amount of supplementary information to be included in the report.

Brief. Select ***Brief***, the most "economical" output style, to include only the selected spreadsheets and graphs (i.e., no information about the variables, case weights, or case selection conditions specified for the analyses will be output).

Medium. Select the ***Medium*** output style to include the selected spreadsheets and graphs as well as the current datafile name, information on case selection conditions and case weights (if any were specified), a list of all variables selected for each analysis, and the missing data values for each variable.

Long. Select the ***Long*** output style (which is more space consuming) to include all information from the ***Medium*** format and, additionally, the long variable labels (e.g., formulas), reserving one line of output (or more) for each variable.

Comprehensive. Select the ***Comprehensive*** output style to provide the most comprehensive information on each variable selected for analysis. In addition to all information included in the ***Long*** report format, it also includes a complete list of all of the text labels for each selected variable. For further details, see the *Electronic Manual*.

Font

Specify a ***Font*** and font size in these fields to use in the specified document.

Local Output Options –
Analysis/Graph Output Manager

Select ***Output*** from the  button (available on most dialogs in *STATISTICA*) to display the ***Analysis/Graph Output Manger*** dialog.

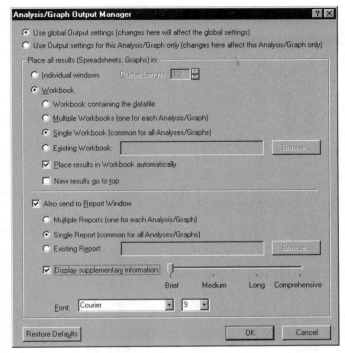

You can customize the current analysis and graph output management of *STATISTICA* by selecting (or clearing) several options offered on this dialog. Note that this dialog is identical to the **Output Manager** tab of the **Options** dialog (see above) except for the three options described below. Note that unless you select the **Use global Output settings** option button, the changes made to this dialog only affect the current analysis or graph.

Use Global Output Settings (Changes Here Will Affect the Global Settings)

Select the **Use global Output settings** option button to specify that the output management for the current analysis or graph will use the global output manager settings (specified via **Tools - Options** (or **File – Output Manager**) on the **Options - Output Manager** tab, see above).

Use Output Settings for This Analysis Only (Changes Here Affect This Analysis Only)

Select the *Use Output settings for this Analysis only* option button to use the *Analysis Output Manager* dialog to specify the output management of the current analysis or graph.

Restore Defaults

Click the *Restore Defaults* button to replace the current selection with the original default values.

Output Headers and Footers

When printing *STATISTICA* documents, you can include a header and/or footer. In addition to providing a variety of default headers and footers, *STATISTICA* contains facilities for creating custom headers and footers. To select or create headers and footers for spreadsheets, reports, or graphs, select *Header/Footer* from the *View* menu. This displays the *Modify Header/Footer* dialog. Note that the same controls described here are also available on the *Workbook Page Setup* dialog (page 152). Use this dialog to specify the header and/or footer for a workbook. Note that it is possible to apply a header to a document within a workbook (using the *View - Header/Footer* command) and to apply a different header and/or footer to that same document via the *Workbook Page Setup* dialog.

Modify Header/Footer Dialog

Select *Header/Footer* from the *View* menu to display the *Header/Footer* dialog. By default, the *Header* and *Footer* boxes provide five suggested headers and footers: *[None]*; *Page 1*; *Document Name*; *Document Name, Page 1*; and *Page 1, Document Name*. The example below shows these suggested headers for a report called *Report1*. To see how one of these headers or footers would look, select it and a preview will be displayed in the large preview box. In the example below, the *Report1, Page 1* footer has been selected and it is previewed in the *Footer* preview box.

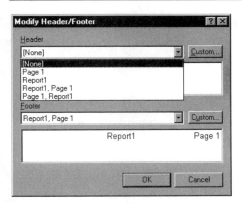

Note that once you have created a custom header or footer, that header or footer will also be displayed in the *Header* or *Footer* box, allowing you to select and preview it.

Custom Header/Custom Footer Dialog

To create a custom header or footer, click the *Header* or *Footer Custom* button to display the *Custom Header* or *Custom Footer* dialog, respectively. Both dialogs contain the same controls. The *Custom Footer* dialog is shown below.

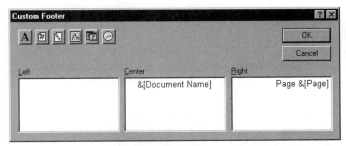

This dialog contains three boxes (*Left*, *Center*, and *Right*) in which you can enter header text or codes. The six buttons in the upper-left corner provide convenient access to certain frequently used codes and features. Once you have created the header or footer, click the *OK* button to return to the *Modify Header/Footer* dialog and preview the header or footer. Click the *Cancel* button to return to the *Modify Header/Footer* dialog without making any changes to the currently specified header or footer.

Font button. Click the button to display the standard *Font* dialog. This dialog is used to specify the font, font size, font style, and script for the selected text. It is also used to choose certain font effects (i.e., underline, strikeout).

Page number button. Click the button to insert the *&[Page]* code, which ensures that the page number is printed in the header or footer. Note that this code only prints the page number, so you may want to enter the word *Page* before this code (as shown in the example above).

Total page count button. Click the button to insert the *&[Pages]* code, which will insert the total page count at the current cursor position in the header or footer (e.g., to create the "page x of xx" format of pagination).

Document name button. Click the button to insert the *&[Document Name]* codes, which will insert the document name at the current cursor position in the header or footer.

Date button. Click the button to insert the *&[Date]* code, which will insert the current date at the current cursor position in the header or footer. Note that the date format is controlled in the Windows *Regional Settings* dialog, accessed from the Windows *Control Panel*.

Time button. Click the button to insert the *&[Time]* code, which will insert the current time at the current cursor position in the header or footer. Note that the time format is controlled in the Window*s Regional Settings* dialog, accessed from the Windows *Control Panel*.

Example: Creating a Custom Header

Use the instructions here to create a custom header (or footer) for a spreadsheet, report, or graph. You can also create custom headers (and footers) for workbooks, via the *Workbook Page Setup* dialog (page 152).

- Select *Header/Footer* from the *View* menu to display the *Modify Header/Footer* dialog.

- In the *Modify Header/Footer* dialog, click the *Custom* button that corresponds to either *Header* or *Footer*. In this example, we will create a customized header, so click the Header *Custom* button to display the *Custom Header* dialog.

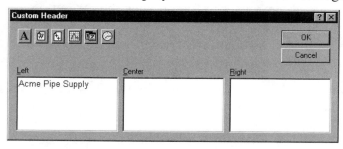

StatSoft
Copyright © StatSoft, 2001

- Enter **Acme Pipe Supply** in the **Left** box. This will insert the company logo in the left corner of the document's custom header. Note that you could apply additional formatting to the company name by selecting it and clicking the **A** (Font) button.

- In addition to manually entering text into the boxes, you can use the buttons at the top of the dialog to insert codes into the header (see *Custom Header/Custom Footer dialog*, above). To insert the current date into the **Right** box, click in the **Right** box then click the 🗓 (**Date**) button. The **&[Date]** code will insert the current date in the header of the document whenever it is viewed or printed.

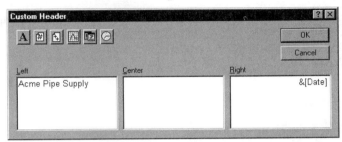

- When you are finished entering the desired text and codes into your customized header (or footer) click the **OK** button to return to the **Modify Header/Footer** dialog. You can view a sample of your customized header (or footer) in this dialog. In the example below, we see a sample of how the header with the company name and current date looks.

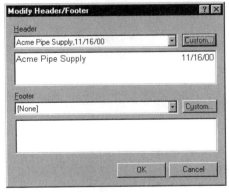

Note that you can also view the header and footer information for a specific document (e.g., a spreadsheet, report, or graph) using the **Print Preview** window (see *Using the Print Preview Window*, page 152).

Printing

Before You Print

Before you print a *STATISTICA* document, here are some options to consider:

- Do you want to create a custom header or footer for your spreadsheet or report?

- Do you want to view or change the current print settings for your spreadsheet, report, graph or workbook?

- Do you want to preview your spreadsheet or report before you print it?

Setting Basic Print Options – Spreadsheets, Reports, and Graphs

In order to view and/or change the basic print options for the active spreadsheet, graph, or report, follow these instructions:

Select **Print Setup** from the **File** menu to display the **Print Setup** dialog shown below.

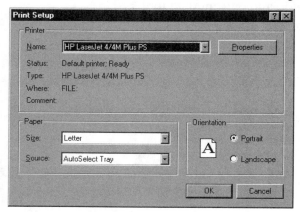

The **Print Setup** dialog provides basic information about how the active spreadsheet will print, including the designated printer, the paper size and source, and the orientation of the paper. To change the printer being used to print the spreadsheet, use the **Name** box to select from a list of available printers. To view or change the properties for the selected printer, click the **Properties** button. To change the paper size and source, select from the available options in the **Size** and

---

Source boxes. Finally, to change the orientation of the printed output, select either the **Portrait** or the **Landscape** option button in the **Orientation** group. When you have selected the desired print options, click the **OK** button.

Print Filter for spreadsheets. *STATISTICA* provides a variety of system spreadsheet layouts (on the **System** tab of the **Spreadsheet Layout Manager**). A spreadsheet layout is a convenient way to combine sets of formatting options into one collection. The **Print Filter** spreadsheet layout is used to edit the format(s) for your spreadsheet when it is printed. Formats or spreadsheet layouts are always reflected when you print your spreadsheet, with the exception of the formats or spreadsheet layouts applied to the Case Header, Variable Header, and Gridlines (unless these check boxes are cleared in the **Print Filter** spreadsheet layout). For more information on spreadsheet layouts, see Chapter 4 - *STATISTICA Spreadsheets*.

Additional options for reports. In addition to these basic print options, you can specify whether spreadsheets in reports are printed as objects or full-sized spreadsheets. These options can be set locally (for a specific report) by selecting **Print Options** from the report **Format** menu or globally (for all reports) on the **Reports** tab of the **Options** dialog. See *Global Report Options* and *Format Menu - Print Options* in Chapter 5 – *STATISTICA Reports* for more details.

Setting Basic Print Options – Workbooks

In addition to setting the options in the **Print Setup** dialog described above, you need to specify certain workbook-specific print options. To do this, select **Workbook Page Setup** from the **File** menu (when a workbook document is selected) to display the **Workbook Page Setup** dialog.

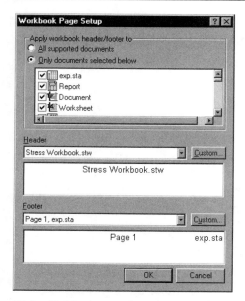

This dialog is used to create a header and footer to use when printing the workbook as well as to specify which files will have a header and/or footer applied to them.

Apply workbook header/footer to. You can either choose to apply the header and footer to *All supported documents* or *Only documents selected below*. If you choose to apply the header and footer only to selected documents, you can scroll through the box provided and select which documents will have the header and footer.

Header. You can select from a list of predefined headers (using the dropdown list), or you can click the *Custom* button and create a customized header. Once you have created a custom header, it will be displayed in the view box.

Footer. You can select from a list of predefined footers (using the dropdown list), or you can click the *Custom* button and create a customized footer. Once you have created a custom footer, it will be displayed in the view box.

Using the Print Preview Window

Before you print a document, it is useful to see how it will look when you print it. To access the *Print Preview* window, select *Print Preview* from the *File* menu, or click the print preview toolbar button.

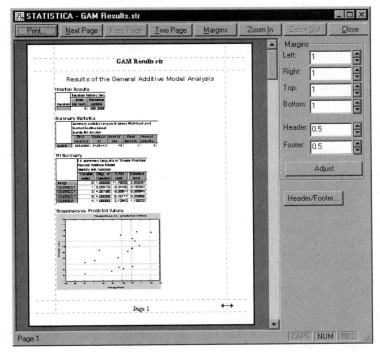

You can also access the **Print Preview** window from the **Print** dialog by clicking the **Preview** button. Note that the **Print Preview** window is only available when you are printing a single document (i.e., a spreadsheet, graph, report, or a single item within a workbook). It is not possible to preview an entire workbook when you have selected it for printing.

The way a document appears in the window depends on the resolution of your printer, and the available fonts and colors. The status bar at the bottom of the window shows the current page number.

The **Margins** group lists the margin information for **Left**, **Right**, **Top**, and **Bottom** margins, as well as how far the header and footer will print from the edge of the page. You can modify the margin information in any of these fields by entering new measurements and clicking the **Adjust** button. You can control whether the margins are measured in inches or centimeters. To measure in centimeters, select the **Use metric measurements** check box on the **General** tab of the **Options** dialog (accessed from the **Tools – Options** menu). To measure in inches, clear the **Use metric measurements** check box (the default setting). Alternatively, to adjust the margins, drag the margin lines with the mouse (the mouse pointer will change into a double-headed arrow, ↕ or ↔, see the above illustration).

The **_Header/Footer_** button accesses the header and footer information for the document (see _Example: Creating a Custom Header_, page 148).

Print. Click the **_Print_** button to access the **_Print_** dialog (see below).

Next Page. Click the **_Next Page_** button to view the next page of the active document in the **_Print Preview_** window.

Prev Page. Click the **_Prev Page_** button to view the previous page of the active document in the **_Print Preview_** window.

Two Page. Click the **_Two Page_** button to see a two-page view of the active document in the **_Print Preview_** window.

Margins. Click the **_Margins_** button to show or hide the margin information at the right side of the **_Print Preview_** window.

Zoom In. Click the **_Zoom In_** button to magnify the view of the active document.

Zoom Out. Click the **_Zoom Out_** button to decrease the magnification of the view; you can zoom out to a full-page view.

Close. Click the **_Close_** button to close the **_Print Preview_** window and return to an editable view of the active report.

Printing a Spreadsheet, Report, or Graph

To print the active spreadsheet, report, or graph, follow these instructions:

- Select **_Print_** from the **_File_** menu to display the **_Print_** dialog shown below.

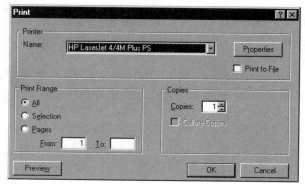

- In the **Printer** group, select the appropriate printer from the **Name** box. If desired, click the **Properties** button to view or modify printer properties. Select the **Print to File** check box to print the active document to a file, rather than queuing the print job to a printer.

- In the **Print Range** group, select from the following options: **All**, **Selection**, and **Pages**. If you select **Pages**, enter the first and last page to print in the **From** and **To** boxes.

- In the **Copies** group, you can choose to print additional copies by entering the desired number in the **Copies** field. Select the **Collate Copies** check box to ensure the printer collates the printed copies for you when you specify more than one copy.

- Click the **Preview** button to access the **Print Preview** screen.

- Click the **OK** button to print the active document with the options you selected.

Printing a Workbook

Follow these instructions to print the active workbook:

- Select **Print** from the **File** menu to display the **Print** dialog shown below.

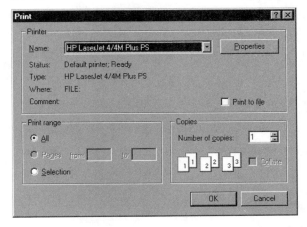

- In the **Printer** group, select the appropriate printer from the **Name** box. If desired, click the **Properties** button to view or modify printer properties. Select the **Print to file** check box to print the active report to a file, rather than queuing the print job to a printer.

- In the **Print range** group, select the **All** option button to print the entire workbook or the **Selection** option button to print only the selected document in the workbook. It is not possible to print a range of pages within the workbook.

- In the **Copies** group, you can choose to print additional copies by entering the desired number in the **Copies** field. Select the **Collate Copies** check box to ensure the printer collates the printed copies for you when you specify more than one copy.

- Click the **OK** button to print the active workbook with the options you selected.

FAQ: WORKING WITH OUTPUT

What Output Management Options Are Available in *STATISTICA*?

You can customize the way in which the output is managed in *STATISTICA*. When you perform an analysis, *STATISTICA* generates output in the form of its multimedia tables (spreadsheets) and graphs. There are three basic channels, controlled by the **Output Manager** (accessible from the **File** or **Tools** menu), to which you can direct all output: workbooks, reports, and stand-alone windows. These three output channels can be used in many combinations (e.g., a workbook and report simultaneously), and each of the output channels can be customized in a variety of ways. Also, all output objects (spreadsheets and graphs) placed in each of them can contain other embedded and linked objects and documents so *STATISTICA* output can be hierarchically organized in numerous ways. For more information on each type of document, see *Three Channels for Output from Analyses* (page 137). See also, Chapter 3 – *STATISTICA Workbooks* and Chapter 5 – *STATISTICA Reports*.

How Do I Print Spreadsheets?

The simplest way to print a spreadsheet is to click the **Print** button 🖨 on the toolbar. *STATISTICA* then sends the current spreadsheet to the printer specified in the **Print** dialog. No other intermediate option dialogs are displayed. If a block is selected in the spreadsheet, then only that block is sent to the output destination; otherwise, the entire spreadsheet is sent to the output. More options are available when you select **Print** from the **File** menu (or CTRL+P). The **Print** dialog is displayed, allowing you to customize various aspects of the printing. See also *How Can I Suppress the Printing of Gridlines in Spreadsheets*, page 159, for further details.

Automatic reports. Note that you can keep a complete log of all spreadsheets (and/or graphs) that are displayed on the screen without having to remember to individually transfer them to the report window or to print them. To do this, select the **Single Report (common for all Analyses/graphs)** option button and the **Also send to Report Window** check box on the **Output Manger** tab of the **Options** dialog (accessible via **File - Output Manager** or **Tools – Options**). This dialog also contains options for creating separate reports for each analysis. Note that this is a global option (as are all options on the **Options** dialog), and it will affect all analyses until the option is changed. To make local changes (i.e., changes for the particular analysis only),

use the ⬛ Options ▼ button on the analysis or graph definition dialog. For more information on automatic reports, see the *Analysis Report Overview* in Chapter 5 – *STATISTICA Reports*.

What Are Workbooks?

The *STATISTICA* Workbook (**.stw*) is a flexible output management facility based on the powerful ActiveX technology. Technically speaking, workbooks are "ActiveX containers" allowing you to manage all *STATISTICA* documents (e.g., tables, graphs), as well as all other ActiveX compatible documents such as Microsoft Excel worksheets or Microsoft Word documents. Each workbook contains two panels: an *Explorer*-style navigation tree on the left and a document viewer on the right. The navigation tree (workbook tree) can be hierarchically split into various nodes allowing you to organize your files in logical groupings (e.g., all analysis outputs, all macros created for a project). Tabs at the bottom of the document viewer (workbook viewer) are used to easily navigate the children of the currently selected node.

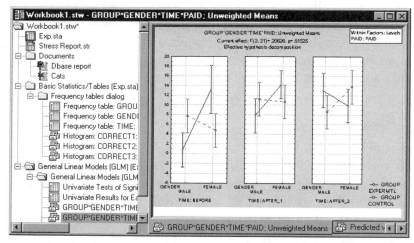

Workbooks help to organize sets of output files (e.g., spreadsheets, graphs, reports, macros, non-*STATISTICA* files, etc.) that have been created or used (e.g., reviewed) during the analysis of a datafile. Refer to the *Workbook Overview* in Chapter 3 – *STATISTICA Workbooks* for more information. See also, *Three Channels for Output from Analyses* (page 137).

How Do I Print Previously Saved Results?

There are several options available for printing previously saved results:

1. You can open each spreadsheet (and/or graph) and print it by selecting **Print** from the **File** menu (CTRL+P) as described in *How Do I Print Spreadsheets* (page 157).

2. You can open each spreadsheet (and/or graph), insert them into a report, and print the report. Note that this allows you to add supplementary text and comments to the analysis results.

3. You can insert all of the spreadsheets and graphs into a workbook and print the entire workbook by selecting **Print** from the workbook **File** menu. See the *Workbook Overview* in Chapter 3 – *STATISTICA Workbooks* for more information on workbooks.

How Can I Suppress the Printing of Gridlines in Spreadsheets?

To suppress the print of the gridlines in an active spreadsheet, you must make changes in two dialogs. First, change the **Style** of both the **Horizontal** and **Vertical Data Lines** to blank in the **Gridlines** dialog, accessed by selecting **Gridlines** from the spreadsheet **View** menu. (Note that you can access the spreadsheet **View** menu from within a report window or workbook by double-clicking on the spreadsheet. This gives you access to all spreadsheet editing tools.)

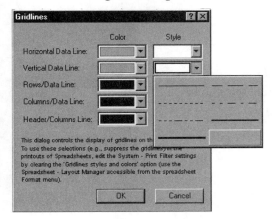

Second, clear the **Gridline styles and colors** check box on the **Edit Spreadsheet Layout: Print Filter** dialog. (To display this dialog, first select **Spreadsheet – Layout Manager** from the spreadsheet **Format** menu to display the **Spreadsheet Layouts** dialog. Then on the **System** tab, select **Print Filter** and click the **Edit** button.)

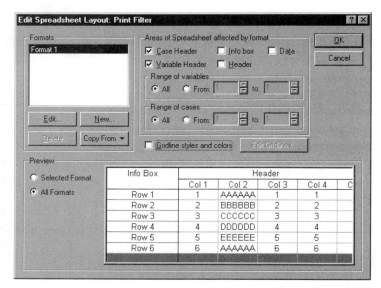

Note that clearing this check box causes *STATISTICA* to print the gridlines using the styles and colors specified in the *Gridlines* dialog rather than using a default black.

Can I Add Custom Headers
or Footers to Printed Output?

You can create a customized header or footer for a *STATISTICA* Spreadsheet, Report, Graph, or Workbook that can include information such as the date, time, page number, and name of your company.

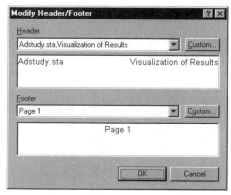

To create the header or footer for a spreadsheet, report, or graph, select **Header/Footer** from the **View** menu and use the **Modify Header/Footer** dialog to specify the custom header or footer.

Note that custom headers and footers for workbooks are created in the **Workbook Page Setup** dialog (page 152). See *Output Headers and Footers* (page 146) for more directions on creating custom headers and footers.

How Do I Change the Printer Setup?

Most options for modifying the printing specifications for a given document, including margins and customized headers and footers, can be selected from the **Print Preview** window (see *Using the Print Preview Window*, page 152). To change the printer setup for a given printer, select **Print Setup** from the **File** menu to display the **Print Setup** dialog. Then click the **Properties** button to access the printer properties dialog. For more information on printing, see the overview section on *Printing* (page 150).

CHAPTER

STATISTICA
WORKBOOKS

continued ➡

CHAPTER

STATISTICA WORKBOOKS

GENERAL OVERVIEWS

Workbook Overview

Workbooks are the default way of managing output. They store each output document (e.g., a *STATISTICA* Spreadsheet or Graph, as well as a Microsoft Word or Excel document) as a tab.

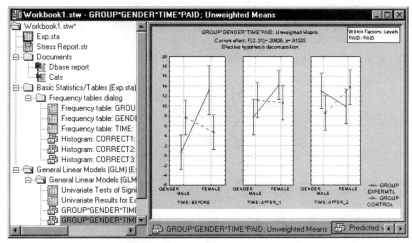

Technically speaking, *STATISTICA* Workbooks are optimized ActiveX containers that can efficiently handle large numbers of documents (see page 168 for more information). The documents can be organized into hierarchies of folders or document nodes (by default, one is created for each new analysis) using a tree view, in which individual documents, folders, or entire branches of the tree can be flexibly managed.

For example, selections of documents can be extracted (e.g., drag-copied or drag-moved) to the report window, or to the application workspace (i.e., the *STATISTICA* application "background" where they are displayed in stand-alone windows). Entire branches can be placed into other workbooks in a variety of ways in order to build a specific folder organization, etc.

Each workbook contains two panels: an *Explorer*-style navigation tree on the left and a document viewer on the right. The navigation tree (workbook tree) can be split into various nodes that are used to organize files in logical groupings (e.g., all analysis outputs or all macros created for a project). Tabs at the bottom of the document viewer (workbook viewer) are used to easily navigate the children of the currently selected node. You can easily move the tabs to the top, right, or left of the workbook viewer by right-clicking on one of the tabs and selecting a different location from the shortcut menu. One advantage of the side placement of tabs is that multiple rows (rather than one long row) are provided (as shown below). This makes it easy to select the appropriate tab.

Workbook9 - Observational statistics for ADVERT spline (Adstudy.sta)

Observational statistics for ADVERT spline (Adstudy.

	Observed Predictor	Smooth	95% lower	95% upper	Partia Residu
J. Baker	0.000000	0.017544	-0.091937	0.127025	0.411
A. Smith	1.000000	-0.020596	-0.149117	0.107925	-0.170
M. Brown	1.000000	-0.020596	-0.149117	0.107925	0.418
C. Mayer	0.000000	0.017544	-0.091937	0.127025	-0.044
M. West	0.000000	0.017544	-0.091937	0.127025	0.067
D. Young	1.000000	-0.020596	-0.149117	0.107925	0.178
S. Bird	1.000000	-0.020596	-0.149117	0.107925	0.195
D. Flynd	0.000000	0.017544	-0.091937	0.127025	0.405
J. Owen	0.000000	•0.017544	-0.091937	0.127025	0.319
H. Morrow	0.000000	0.017544	-0.091937	0.127025	0.082
F. East	0.000000	0.017544	-0.091937	0.127025	-0.296
C. Clint	1.000000	-0.020596	-0.149117	0.107925	-0.015
I. Neil	0.000000	0.017544	-0.091937	0.127025	-0.138
G. Boss	1.000000	-0.020596	-0.149117	0.107925	0.105
K. Record	0.000000	0.017544	-0.091937	0.127025	0.036
T. Bush	0.000000	0.017544	-0.091937	0.127025	-0.383
P. Squire	1.000000	-0.020596	-0.149117	0.107925	-0.658

Displaying tabs can also be suppressed to save the space. Unlike many *Explorer*-style navigation and organization applications that only allow folders to have children, the *STATISTICA* Workbook allows any item in the tree to have children. For example, you can add a spreadsheet to your workbook, and then add all the graphs produced using the data in the spreadsheet as children to the spreadsheet. A variety of drag-and-drop features and Clipboard procedures are available to aid you in organizing the workbook tree.

The workbook can hold all native *STATISTICA* documents including spreadsheets, graphs, reports, and macros. It can handle other types of ActiveX documents as well, including Excel

spreadsheets, Word documents, and others. If you want to edit these documents, you can do so using the workbook viewer pane. To edit a Microsoft Word document, double-click on the object in the workbook tree. The Word document opens in the viewer, and the workbook menu bar merges with the Microsoft Word menu bar giving you access to all of the editing features you need. Workbooks can also be used to store all output from a particular analysis.

Workbook Tree Overview

The workbook tree displays the organization of files and folders in the workbook. The files and folders are displayed in an *Explorer*-style format. Items with plus signs next to them indicate folders or files that have children associated with them. To expand the tree for a particular folder or file, click the plus sign next to it. The workbook can support an unlimited number of levels, and both individual items from the tree view and entire branches can be flexibly (interactively) managed (e.g., right-click dragging to copy or move between workbooks or reports, etc., as shown below in the second illustration).

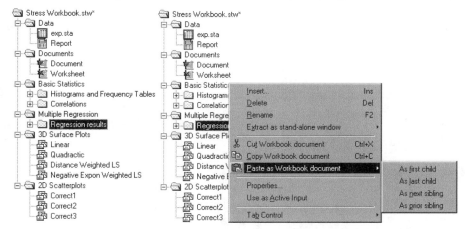

To select a workbook item for review or editing, simply locate the file in the workbook tree and double-click on its associated icon. The document will then open in the workbook viewer pane. Note that you can also navigate through the children of the currently selected node using the navigation tabs available (by default) at the bottom of the workbook viewer. As mentioned previously, you can easily move these navigation tabs to the top, right, or left of the workbook viewer by right-clicking on one of the tabs and selecting a different location from the shortcut menu or selecting the appropriate command from the **Workbook - Tab Control** menu. (see

page 197). Note that tabs at the top and bottom of the viewer scroll sideways, while multiple rows of tabs are used when tabs are placed to the left or right of the viewer.

Items in the tree are identified by the icon next to them. The 🗀 folder icon represents a folder that can contain a variety of documents and subfolders. The 🎞 spreadsheet, 📋 report, 📑 macro, and 📊 graph icons represent *STATISTICA* Spreadsheet, Report, Macro, and Graph documents, respectively.

All non-*STATISTICA* documents are represented by their respective document icons. For example, Word documents are represented by the 📘 Word icon, and Excel spreadsheet files are represented by the 📗 Excel spreadsheet icon.

The workbook tree can be organized and modified using drag-and-drop features as well as Clipboard procedures. See *Workbook Drag-and-Drop Features* (page 172) and *Workbook Clipboard Features* (page 171). Commands for inserting, extracting, renaming, and removing items from the workbook tree are available from the workbook tree shortcut menu (accessed by right-clicking anywhere in the tree) as well as from the **Workbook** menu (see page 193).

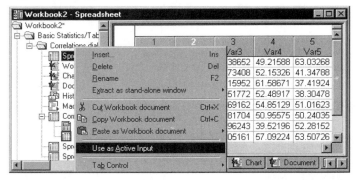

These commands are also accessible from the **Workbook** menu.

What Is ActiveX?

The term ActiveX is used in different contexts and its definitions stress different aspects of that concept. Its use within *STATISTICA*, however, can be grouped into two general categories: ActiveX objects and ActiveX documents.

ActiveX object. An ActiveX object is what was once referred to as an OLE (Object Linking and Embedding) object. At its heart is the Microsoft COM (Component Object Model) technology that allows objects to be accessed in a uniform manner. Through the use of standard

protocols, objects created in one application can be stored and edited in a different application. To support this functionality, the containing object needs to be an ActiveX object client, and the application that initially created the object needs to be an ActiveX object server. *STATISTICA* is both. As an ActiveX object client, *STATISTICA* allows you to embed and link objects from other applications in the spreadsheet, graph, and report windows. As an ActiveX object server, it allows you to embed and link spreadsheets and graphs into other applications.

ActiveX documents. ActiveX documents take the ActiveX controls one step further, in that they allow entire documents to be embedded into other applications. An ActiveX document container allows other application documents to be used within it, and an ActiveX document server allows its documents to be used within any ActiveX document container. Again, *STATISTICA* does both. *STATISTICA* Workbooks are ActiveX document containers, and allow documents from other ActiveX servers to be displayed within the workbook. Examples of this are Microsoft Word and Excel; these documents can be used directly from within a *STATISTICA* Workbook. Similarly, *STATISTICA* Spreadsheets, Graphs, and Reports are ActiveX document servers, and they also can be placed within any ActiveX document container such as Microsoft Internet Explorer and Microsoft Binder.

Analysis Workbook Overview

An analysis workbook holds the results for a given analysis. As with all *STATISTICA* Workbooks, the files in the workbook are stored as tabs, which can be easily managed. You can create an analysis workbook by selecting the **Multiple Workbooks (one for each Analysis/graph)** option button on the **Output Manager**. This option can be set globally via the **Output Manager** tab of the **Options** dialog (accessed by selecting **Output Manager** from the **File** menu) so that results for every analysis are saved in a separate workbook (for as long as the option is selected). You can also set it locally via the **Analysis/Graph Output Manager** dialog (accessed by clicking the ⬚ Options ▾ button on analysis or graph specification dialogs) so that only the results from that particular analysis are stored in a separate workbook.

Input Spreadsheet Overview

STATISTICA offers the ability to open and use many spreadsheets at the same time, allowing you to work with several different input datafiles simultaneously. In addition to storing data, *STATISTICA* uses spreadsheets to display the numeric output from its analyses and other types

of data (e.g., multimedia objects, scripts, links). Because *STATISTICA* makes no distinction in the features supported for an input spreadsheet (from which *STATISTICA* retrieves its data) and an output spreadsheet (where the results of an analysis are displayed), it is easy to use the results of one analysis as input data for further analyses.

Any spreadsheet opened from a disk file is automatically treated as an input spreadsheet, and any number of input spreadsheets can be open at a time. To avoid confusion, however, an output spreadsheet (containing the results of an analysis) is not automatically available as input data for analysis. It must first be designated as an input spreadsheet before being used for further analyses. Additionally, input spreadsheets report the number of variables and cases for that spreadsheet in the title bar.

For example, **Exp.sta (8v by 48c)** is an input spreadsheet:

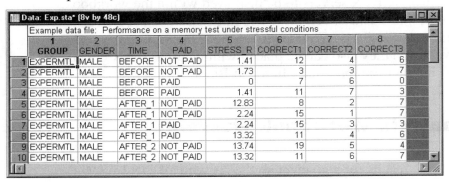

Exp.sta is not:

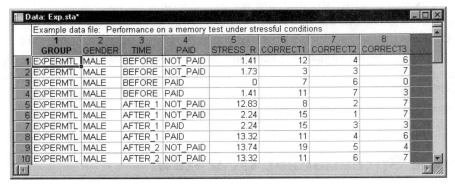

To designate an output spreadsheet as an input spreadsheet, select the spreadsheet (i.e., ensure the spreadsheet has the focus), and select **Input Spreadsheet** from the **Data** menu. Now you can begin an analysis and *STATISTICA* will use the data from the specified input spreadsheet for

the analysis. Note that if you switch back to another spreadsheet that has previously been designated as an input spreadsheet, it can still be used for analyses as well.

In a workbook, only one spreadsheet can be selected for analysis at a time, even if the workbook contains several input spreadsheets. This spreadsheet is called the **Active Input** spreadsheet, and its icon (in the workbook tree) is framed in red.

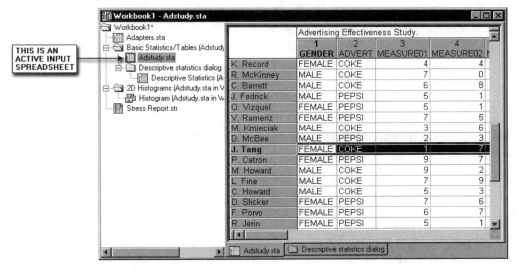

By default, when an output spreadsheet is designated as an input spreadsheet, *STATISTICA* automatically selects it as the **Active Input** spreadsheet. To select another input spreadsheet for active input, select **Use as Active Input** from the **Workbook** menu or the workbook tree shortcut menu.

It is also possible to leave a stand-alone spreadsheet open but designate it as unavailable for analysis. To do this, select the spreadsheet, and deselect (click) the **Input Spreadsheet** command on the **Data** menu. Now *STATISTICA* automatically defaults to the most recently selected input spreadsheet for analysis, ignoring all spreadsheets that are not designated as input spreadsheets.

Workbook Clipboard Features

The workbook supports copy and paste procedures available from the **Workbook** menu or by right-clicking on the file you want to relocate. These copy and paste procedures are used to

specify whether files are placed as a prior or next sibling or as a first or last child to other files in the tree.

You can copy more than one workbook item to the Clipboard at a time (e.g., either an entire node, or perhaps a discontinuous selection of items). To select a parent node and all of its children, simply click on the parent node (or press CTRL while clicking on the parent node and each of its children). To make discontinuous selections (i.e., select some of the children, but not all of them), hold down the CTRL key while clicking on each item you want to select. You can also make discontinuous selections across nodes using the CTRL key. For more information on making selections in the workbook tree, see *Can I Select Discontinuous Items in the Workbook Tree?* (page 213).

Workbook Drag-and-Drop Features

The *STATISTICA* Workbook supports an entire range of drag-and-drop features within the workbook tree. By right-clicking on an item in the workbook tree, you can drag that item to a new location within the tree. In fact, since the drag-and-drop features support parent/child relationships, you can use these features to rearrange large portions of the workbook tree.

Selecting items for drag-and-drop. The first step in using drag-and-drop features is to select the item (or items) that you want to move or copy. Afterwards, you can drag the selection (using either the left or right mouse button, see below) to its new location and drop it.

To select a single item, click on the item with either the left or right mouse button, depending on the type of drag-and-drop operation you want to perform. To select a parent node and all of its children, simply click on the parent node (or press CTRL while clicking on the parent node and each of its children). To make discontinuous selections (i.e., select some of the children, but not all of them), hold down the CTRL key while clicking on each item you want to select. You can also make discontinuous selections across nodes using the CTRL key. See also, *Can I Select Discontinuous Items in the Workbook Tree?* (page 213).

Left and right mouse button operations. *STATISTICA* supports both left and right mouse button drag-and-drop features, which means you can drag your selection using either the left or right mouse button. Left mouse button drag-and-drop features are limited to the default option of **Move as child**, while right mouse button drag-and-drop provides you with a variety of placement options.

Left mouse button drag-and-drop. To use left mouse button drag-and-drop, click on the selected item(s) and drag the selection to the desired location. When you reach the

Copyright © StatSoft, 2001

selection's new destination, release the left mouse button. The selection is automatically moved from its original location and placed as a child to the selected item in its new location.

Right mouse button drag-and-drop. To use right mouse button drag-and-drop, right-click on the selected item(s) and drag the selection to the desired location. When you reach the selection's new destination, release the right mouse button. A shortcut menu is displayed, giving you options to either copy or move the dragged item(s).

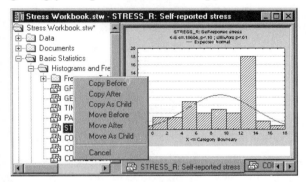

You can place the item before, after, or as a child to the newly selected item in the tree. Note that when you select more than one item, the original file structure of the dragged items will remain the same. For example, if you drag a node and all of its children and select *Copy as Child*, the original selection will be copied (not moved) to the new location with the node being placed as a child to the new selection. The children of the original node will not be placed as siblings to the original node; they will maintain their status as children. These features make it possible to customize the workbook tree to meet even the most complex of organizational structures.

Extracting Files from a Workbook

STATISTICA has options to extract files or copies of files so that they can be opened outside the workbook. This feature is useful if you need to save only a specific document from a collection in the workbook, or share it with others without sharing the entire workbook.

Inserting Files into a Workbook

To add an active file to the workbook, click the 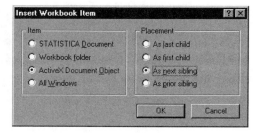Add to Workbook ▼ button, which is available on the **Standard** toolbar. You will be given the choice of adding the file to a **New Workbook** or to any existing workbooks to which you have access. Select the workbook to which you want to add the file. Once the file is added to the workbook, you can place the file anywhere in the tree using the workbook's extensive drag-and-drop facilities. Note you can also add files to an analysis workbook, a new workbook, or an open workbook using the commands on the appropriate document's **File - Workbook** menu.

It is also possible to create new *STATISTICA* files within the workbook, including spreadsheets, macros, and reports. Additionally, you can insert previously saved files and documents, as well as insert new or previously saved objects. *STATISTICA* uses a central dialog (the **Insert Workbook Item** dialog, accessible by selecting **Insert** from the **Workbook** menu or the workbook shortcut menu or by pressing the INSERT key on your keyboard) for adding items to workbooks. From this dialog you can add a *STATISTICA* document, a folder, an ActiveX document, or all open windows.

As with the copy and paste procedures, you decide where these files will be placed in the tree (either as a child or sibling to the currently selected file). See the *FAQ: Working with Workbooks* section (page 205) for a variety of examples on inserting documents via this dialog.

Output management options. Because there are many different ways in which you might want to store information, *STATISTICA* provides a variety of workbook types (e.g., a separate workbook for each analysis and/or project or one single workbook to hold all results). Additionally, these output options can be set globally or locally, allowing you to specify the best method of data storage for a given project.

By default, all results in *STATISTICA* are automatically sent to a single common workbook. However, global options available on the **Output Manager** tab of the **Options** dialog (accessible

from the **Tools - Options** menu) can be used to specify that files are sent to a separate analysis workbook for each analysis, that all files are automatically sent to the workbook containing the datafile that was used to perform the analysis, or that files are sent to a currently existing workbook.

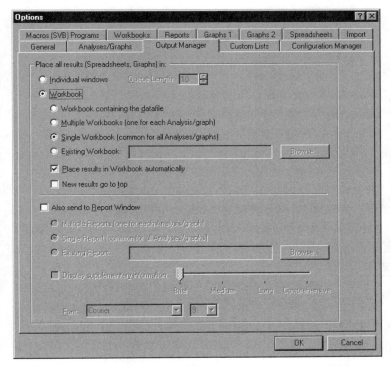

Additionally, new analyses are appended to the end of the appropriate workbook; however, you can specify that new results will always be placed at the top of the workbook tree. Note that regardless of where the results are appended, individual documents, folders, or entire branches of the tree can be flexibly managed. In fact, entire branches can be placed into other workbooks (either copy or moved) in order to build specific folder organization, etc.

Global Workbook Options

You can customize the global functionality of *STATISTICA* Workbooks using the options on the **Workbooks** tab of the **Options** dialog, accessed via the **Tools - Options** menu. The selections made on this tab determine the defaults whenever you use a workbook.

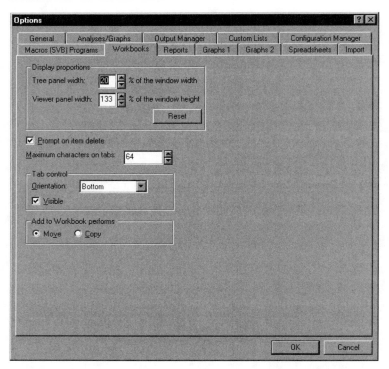

Display proportions. Use the options in the *Display proportions* group to set the global default width of the tree panel and the viewer panel. In the *Tree panel width* box, enter the percentage of the window width that the tree panel should fill. In the *Viewer panel width* box, enter the percentage of the window height that the viewer panel should occupy (e.g., *40%* and *100%*, respectively, as displayed in the illustration below).

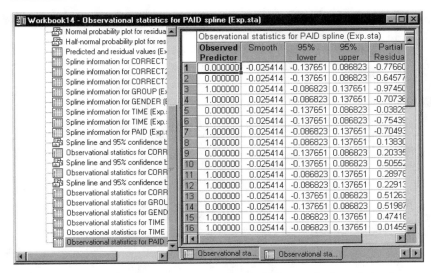

Click the **Reset** button to restore the widths to their original percentages (**20%** and **133%**, respectively).

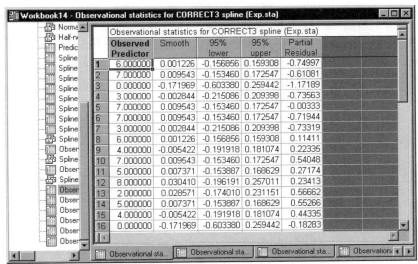

Prompt on item delete. Select the **Prompt on item delete** check box to ensure that a dialog asking you to verify the action is displayed every time you cut or delete an item from a workbook.

Maximum characters on tabs. Enter the maximum number of characters to allow on a single tab before *STATISTICA* truncates the title. The default value is **64**. Note that the

maximum character length can be set locally (for a specific workbook) by selecting *Tab Length* from the *Workbook-Tab Control* menu.

Tab control. Use the options in the *Tab control* group to set the orientation of the workbooks tabs as well as to specify whether or not they are visible. In the *Orientation* box, select *Bottom*, *Top*, *Left*, or *Right* placement. Bottom and right placement are shown below.

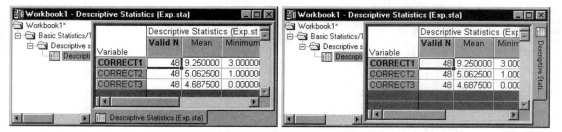

Select the *Visible* check box if you want the navigation tabs to be displayed on workbooks. Clear this check box if you do not want the tabs to be available. Note that these options can be set locally (for a specific workbook) using commands on the *Workbook - Tab Control* menu.

Add to Workbook performs. Use the options in the *Add to Workbook performs* group to specify what action is taken when you click the `Add to Workbook ▾` button. Select the *Move* option button to move the selected file to the workbook (and remove it from its stand-alone window). Select the *Copy* option button to copy the selected file to the workbook (and leave the original in its stand-alone window). By default, the *Move* option button is selected.

WORKBOOK WINDOW

The workbook menu bar contains eight menus: *File*, *View*, *Statistics*, *Graphs*, *Tools*, *Workbook*, *Window*, and *Help*. As mentioned above, items within *STATISTICA* Workbooks are in-place editable. Thus, when you select a spreadsheet within a workbook, you can expect to have the spreadsheet menu bar available. Essentially, when a document is edited within a workbook, the document's menu bar merges with the workbook menu bar. For this reason, you should expect to see more than these eight menus when editing documents within workbooks; however, those additional menus are described within the appropriate document type chapters of this manual.

File Menu

The following commands are available from the workbook *File* menu. Many of these options are also available from shortcut menus (accessed by right-clicking on an item or icon in the workbook) and from the toolbar buttons.

New

Select *New* from the *File* menu (or click the ☐ toolbar button) to display the *Create New Document* dialog, and then select the *Workbook* tab.

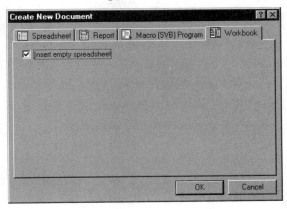

When creating a workbook, you can choose to include an empty spreadsheet.

Open

Select **Open** from the **File** menu (or click the 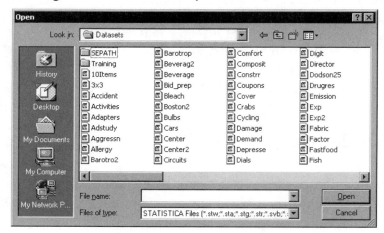 toolbar button) to display the **Open** dialog. From this dialog, you can open any document of a compatible type (e.g., a workbook). Using the **Look in** box, select the drive and directory location of the desired file. Select the file from the large box and click the **Open** button, or double-click the file name.

You can also enter the complete path of the document in the **File name** box.

Close

Select **Close** from the **File** menu to close the current document. If you have made changes to the document since you last saved it, *STATISTICA* prompts you to save changes to the document.

Save

Select **Save** from the **File** menu to save the document with its current name in the drive location you specified when you last saved it. If you have not yet saved the document, the **Save As** dialog is displayed (see below).

Save As

Select **Save As** from the **File** menu to display the **Save As** dialog. From this dialog, you can save the active document with the name of your choice by entering the name in the **File name** box. Use the **Save in** box to select the appropriate drive and folder in which to save the document, or enter the complete path in the **File name** box.

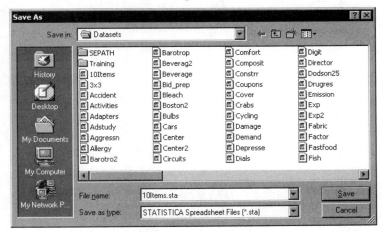

By default, *STATISTICA* recommends the file type that best fits the type of document you are saving. For example, if you are saving a workbook, the **Save as type** box contains the format **STATISTICA Workbook Files (*.stw)**.

Get External Data

The options on this menu are for use with *STATISTICA* Query. See Chapter 10 – *STATISTICA Query* for further details about creating, editing, and using queries. Note that **Edit Query**, **Properties**, **Refresh Data**, **Delete Query**, and **Cancel Query** are only available when you have run (or are currently running) a query.

Open Query from File. Select *Open Query from File* from the **File – Get External Data** menu to display the standard open file dialog, in which you can select an SQL file to run. Note that **.sqy* files are (created and) saved in *STATISTICA* Query (via the **File - Save As** command).

Create Query. Select *Create Query* from the **File - Get External Data** menu to open *STATISTICA* Query and display the **Database Connection** dialog. Once you have chosen or defined a database connection, you can use *STATISTICA* Query to write your query. For details

on defining a database connection, see the *FAQ: Working with STATISTICA Query* section in Chapter 10 – *STATISTICA Query*.

Edit Query. Select *Edit Query* from the *File – Get External Data* menu to edit the query in the active spreadsheet. Note that if you have multiple queries on the active spreadsheet, *STATISTICA* opens the query that is associated with the active cell. If you select a cell that is not connected to any query, or if you select multiple cells that involve more than one query, then the *Select Database Query* dialog is displayed and you must select the query that you want to edit before *STATISTICA* Query opens.

Properties. Select *Properties* from the *File – Get External Data* to display the *External Data Range Properties* dialog and specify options regarding the return of external data to spreadsheets. Note that if you have multiple queries on the active spreadsheet, then *STATISTICA* displays properties for the query associated with the active cell on the spreadsheet. If you select a cell that is not connected to any query, or if you select multiple cells that involve more than one query, then the *Select Database Query* dialog is displayed. On this dialog, select the query to use and click the *OK* button to display the *External Data Range Properties* dialog.

Refresh Data. Select *Refresh Data* from the *File – Get External Data* menu (or press F5) to run a query and refresh the data (retrieve the latest data from the original database). Note that if you have multiple queries on the active spreadsheet, then *STATISTICA* refreshes data for the query associated with the active cell of the spreadsheet. If you select a cell that is not connected to any query, or if you select multiple cells that involve more than one query, then the *Select Database Query* dialog is displayed so you can choose the specific query to refresh.

Delete Query. Select *Delete Query* from the *File – Get External Data* menu to delete the query from the active spreadsheet. (*STATISTICA* will always prompt you to verify that the query should be deleted.) Note that if you have multiple queries on the active spreadsheet, *STATISTICA* deletes the query associated with the active cell of the spreadsheet. If you select a cell that is not connected to any query, or if you select multiple cells that involve more than one query, then the *Select Database Query* dialog is displayed so you can choose which query to delete.

Cancel Query. Select *Cancel Query* from the *File – Get External Data* menu (or press SHIFT+F5) to stop a currently running query at any time.

Output Manager

Select *Output Manager* from the *File* menu to display the *Output Manager* tab of the *Options* dialog. Use the options on this tab to specify where to direct output files. Options for workbooks include sending all output to the workbook containing the original datafile, multiple workbooks (one for each graph and analysis), a single workbook (i.e., an analysis workbook), or an existing workbook. You can also specify whether all results are sent to the workbook automatically, and if results are placed at the top or bottom of the workbook tree. For more details, see *Three Channels for Output from Analyses* in Chapter 2 – *STATISTICA Output Management*.

Workbook Page Setup

Select *Workbook Page Setup* from the *File* menu to display the *Workbook Page Setup* dialog. Use this dialog to create a header and footer to use when printing the workbook as well as to specify which files will have a header and/or footer applied to them.

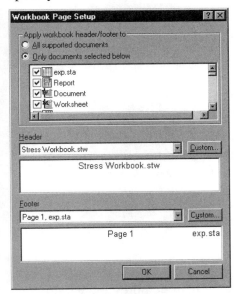

Apply workbook header/footer to. You can either apply the header/footer to *All supported documents* or *Only documents selected below*. If you choose to apply the header/footer only to selected documents, you can scroll through the list box provided and select which documents will have the header/footer.

Header, Footer. You can select from a list of predefined headers or footers (using the drop-down box), or you can click the *Custom* button and create a customized header or footer. Once you have created a custom header or footer, it is displayed in the view box. For more information on headers and footers, see *Output Headers and Footers* in Chapter 2 – *STATISTICA Output Management*.

Print Setup

Select *Print Setup* from the *File* menu to display the *Print Setup* dialog. Use this dialog to specify printing options.

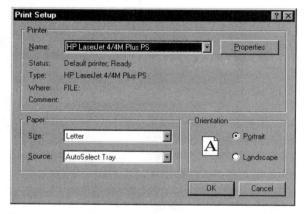

In the *Printer* group, choose a printer (using the *Name* box to view available printers). In the *Paper* group, choose the paper size (using the *Size* box to view a list of paper sizes), and the paper source (using the *Source* box to choose the printer's paper source). Additionally, choose a vertical or horizontal layout for the printed document by selecting the *Portrait* option button (for vertical) or the *Landscape* option button (for horizontal) in the *Orientation* group.

Click the *Properties* button to view options specific to the selected printer. For more information on specific printer options, consult your printer's manual.

Print Preview

Select *Print Preview* from the *File* menu to display the active document in a preview window. Before you print a document, it is useful to see how it will look when printed. The document's appearance in the print preview window depends on the resolution of your printer and the available fonts and colors. The status bar at the bottom of the window shows the current page number.

Use the options in the *Margins* group to control margin and header and/or footer placement. To modify *Left*, *Right*, *Top*, and *Bottom* margins, enter new measurements in the appropriate field, and click the *Adjust* button. You can control whether the margins are measured in inches or centimeters. To measure in centimeters, select the *Use metric measurements* check box on the *General* tab of the *Options* dialog (accessed from the *Tools – Options* menu). To measure in inches, clear the *Use metric measurements* check box (the default setting). Values in the *Header* and *Footer* boxes determine how far the header and footer are placed from the top and bottom margins, respectively. Click the *Header/Footer* button to display the *Modify Header/Footer* dialog to access the header and footer information for the document.

Print

Select *Print* from the *File* menu (or press CTRL+P) to display the *Print* dialog.

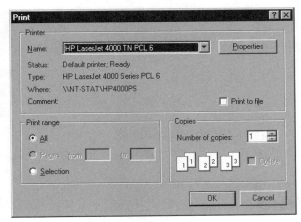

Use this dialog to execute a print command, as well as to specify printing options (see *Print Setup*, above). Verify that the printer in the *Name* box is the correct printer; if necessary, you can use the *Name* drop-down box to choose another printer. Select the *Print to File* check box if you want to print the document to a *.prn* file.

Use the options in the *Print Range* group to specify what documents in the workbook to print. To print the entire workbook, select the *All* option button. To print the selected document, select the *Selection* option button. Note that when printing workbooks, it is not possible to print a range of pages; therefore, the *Pages* option button is disabled for workbooks.

Use the options in the *Copies* group to specify the number of copies of the document to print. Enter the number of copies you require in the *Number of copies* box. Select the *Collate* check

box to ensure that your printer collates the copies of the printed document. After you have selected the desired options, click the **OK** button to print the document.

Print Active Item

This workbook printing option applies only to the currently selected (displayed) document, therefore, the option **All** in the **Print Range** group of the subsequent **Print** dialog will cause the entire document (not all documents in the workbook) to be printed, and the **Selection** option refers to the currently selected part of the document (e.g., a block of the spreadsheet).

Properties

Select **Properties** from the **File** menu to display the **Document Properties** dialog. This dialog contains two tabs: **Summary** and **Statistics**. Select the **Summary** tab to create or modify a document summary (including information about the author's name, comments about the document, etc.). Select the **Statistics** tab to view statistics about the document (including date created, who last saved the document, etc.).

Exit

Select **Exit** from the **File** menu to close *STATISTICA*. If you have made changes to your document since you last saved it, you will be prompted to save your changes.

View Menu

The following commands are available from the workbook **View** menu. Many of these options are also available from shortcut menus (accessed by right-clicking on an item or icon in the workbook) and from the toolbar buttons.

Toolbars

Use the commands on the **View - Toolbars** menu to toggle the display of various toolbars. Note that toolbars can be docked to the top, left, or right side of the workspace. They can also float (e.g., be located anywhere in the workspace). For more information on a specific toolbar, see the following:

Toolbar	Chapter
Standard	1
Statistics	1
Six Sigma	1
Spreadsheet	4
Tools	4
Report	5
Graphs	6
Graph Tools	6
Graph Data Editor	6
Macro	11

Customize. Select *Customize* from the *View - Toolbars* menu to display the *Customize* dialog. For details about customization, see Chapter 1 – *STATISTICA – A General Overview*.

Status Bar

Select *Status Bar* from the *View* menu to toggle the text and buttons that are displayed in the status bar on the bottom center of the screen. Note that the *STATISTICA* Start button ▦ is available regardless of whether the status bar is turned on or off.

Statistics Menu

The *Statistics* menu is available whenever any document is open. This menu provides access to all available analysis types within *STATISTICA* including *Basic Statistics/Tables*, *Multiple Regression*, *ANOVA*, *Nonparametrics*, *Distribution Fitting*, *Advanced Linear/Nonlinear Models*, *Multivariate Exploratory Techniques*, *Industrial Statistics & Six Sigma*, *Power Analysis*, and *Data-Mining*. Definitions for the various types of statistics are available in the *Statistics Toolbar* topic in Chapter 1 – *STATISTICA – A General Overview*. Also included on the *Statistics* menu are *Statistics of Block Data*, *STATISTICA Visual Basic*, and *Probability Calculator*.

Graphs Menu

The **Graphs** menu is available whenever any *STATISTICA* document is open. This menu provides access to all graph types in *STATISTICA* including **Histograms**, **Scatterplots**, **Means w/ Error Plots**, **Surface Plots**, **2D Graphs**, **3D Sequential Graphs**, **3D XYZ Graphs**, **Matrix Plots**, **Icon Plots**, **Categorized Graphs**, **User-defined Graphs**, **Graphs of Block Data**, and **Graphs of Input Data**. It also provides access to **Multiple Graph Layouts**. For more information on graphs, see Chapters 6, 7, 8, and 9 – *STATISTICA Graphs*.

Tools Menu

The following commands are available from the workbook **Tools** menu. Many of these options are also available from shortcut menus and from the toolbar buttons.

Analysis Bar

To take advantage of *STATISTICA*'s "multitasking" functionality, *STATISTICA*'s analyses are organized as functional units that are represented with buttons on the **Analysis** bar at the bottom of the application window (above the status bar, see the illustration below, where **Basic Statistics**, **Cluster Analysis**, and **Canonical Analysis** are running simultaneously).

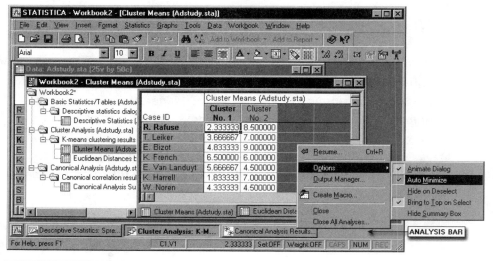

Normally, at least one analysis button is created, and consecutive buttons are added as you start new analyses. The commands on the *Tools – Analysis Bar* menu provide a variety of options for managing the *Analysis* bar.

Resume. Select *Resume* (or click the  toolbar button or press CTRL+R) from the *Tools - Analysis Bar* menu to continue the current analysis or graph. Note that you can also open the current analysis or graph by clicking on its button on the *Analysis* bar.

Select Analysis/Graph. Use the commands on the *Tools - Analysis Bar - Select Analysis/Graph* menu to select an analysis or graph from the set of active analyses and graphs. Note that you can also open an active analysis or graph by clicking on the appropriate button on the *Analysis* bar.

Options - Animate Dialog. Select *Animate Dialogs* from the *Tools - Analysis Bar - Options* menu to display animation when analysis dialogs are minimized or maximized. By default this command is checked.

Options - Auto Minimize. Select *Auto Minimize* from the *Tools - Analysis Bar - Options* menu to automatically minimize all analysis dialogs when you select another window in *STATISTICA* or another application. By default, this command is checked. When your screen is large enough to accommodate several windows, it is recommended that you clear this option. This keeps the analysis dialogs on screen while the respective output created from these dialogs is produced, thus allowing you to use the dialogs as "toolbars" from which output can be selected.

Options - Hide on Select. Select *Hide on Select* from the *Tools - Analysis Bar - Options* menu to minimize all windows associated with a particular analysis when that analysis is deselected. By default this command is cleared. Note that this command only applies when the results are sent to individual windows; see the discussion of the *Output Manager* tab of the *Options* dialog in Chapter 2 – *STATISTICA Output Management* for further details.

Options - Bring to Top on Select. Select *Bring to Top on Select* from the *Tools - Analysis Bar - Options* menu to activate (display at the top of *STATISTICA*) all windows associated with a particular analysis when that analysis is selected, replacing whatever dialogs were on top. This command also facilitates the organization of individual windows from various analyses. By default this option is checked. Note that this command only applies when the results are sent to individual windows; see the discussion of the *Output Manager* tab on the *Options* dialog in Chapter 2 – *STATISTICA Output Management* for further details.

Options - Hide Summary Box. Select *Hide Summary Box* from the *Tools - Analysis Bar - Options* menu to not display the summary box, which is located at the top of certain

results dialogs (such as *Multiple Regression - Results*) and contains basic summary information about the analysis. By default this command is not checked.

Output Manager. Select *Output Manager* from the *Tools – Analysis Bar* menu to display the *Output Manager* tab of the *Options* dialog. Use the options on this tab to specify where to direct output files. Options for workbooks include sending all output to the workbook containing the original datafile, multiple workbooks (one for each graph and analysis), a single workbook (i.e., an analysis workbook), or an existing workbook. You can also specify whether all results are sent to the workbook automatically and if results are placed at the top or bottom of the workbook tree. For more details, see *Three Channels for Output from Analyses* in Chapter 2 – *STATISTICA Output Management*.

Create Macro. Select *Create Macro* from the *Tools - Analysis Bar* menu (or click the toolbar button on the *Statistics* toolbar) to display the *New Macro* dialog, in which you can specify a name for a new macro based on the current analysis. When you run an analytic procedure (from the *Statistics* menu) or create a graph (from the *Graphs* menu), the Visual Basic code corresponding to all design specifications as well as output options that you select are recorded in the background. Thus, when you click the *OK* button on the *New Macro* dialog, the resulting macro window displays the appropriate code to recreate the current analysis.

Minimize. Select *Minimize* from the *Tools - Analysis Bar* menu to minimize the current analysis.

Close. Select *Close* from the *Tools - Analysis Bar* menu to close the current analysis.

Close All Analyses. Select *Close All Analyses* from the *Tools - Analysis Bar* menu to close all of the analyses/graphs on the *Analysis* bar.

Macro

A macro is a scripted application that extends functionality to *STATISTICA* by directly accessing *STATISTICA*'s object model and manipulating it. Macros are primarily used to automate tasks done in *STATISTICA* by harnessing its power and recording it into a *STATISTICA* Visual Basic script. Additionally, macros can be written as both stand-alone scripts and library classes to extend the statistical and mathematical capabilities of *STATISTICA*.

Macros. Select *Macros* from the *Tools - Macro* menu (or click the ▶ toolbar button on the *Macro* toolbar) to display the *Macros* dialog.

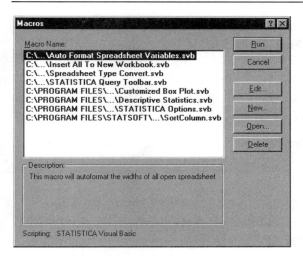

Use this dialog to run, edit, open, or delete existing macro (SVB) programs as well as to create new macros.

Start Recording Log of Analyses (Master Macro). Select *Start Recording Log of Analyses (Master Macro)* from the *Tools - Macro* menu to create a SVB program file that will include a sequence of analyses (a log of analyses) that have been performed interactively. Note that you must select this command before you start the analyses. Consult the *Electronic Manual* or the *STATISTICA Visual Basic Primer* to learn more about this powerful command and how to create recordable sequences of analyses.

Start Recording Keyboard Macro. Select *Start Recording Keyboard Macro* from the *Tools - Macro* menu to record a keyboard macro. See the *Electronic Manual* or the *STATISTICA Visual Basic Primer* for more details.

Stop Recording. Select *Stop Recording* from the *Tools - Macro* menu (or click the ■ toolbar button on the *Macro* toolbar) to stop recording a log of analyses (master macro) or a keyboard macro (see above).

Create Analysis/Graph Macro. Select *Create Analysis/Graph Macro* from the *Tools - Macro* menu to select from a list of all current analyses/graphs that are open. When you run an analytic procedure (from the *Statistics* menu) or create a graph (from the *Graphs* menu) the Visual Basic code corresponding to all design specifications as well as output options that you select are recorded. Select the appropriate analysis/graph to display that code. Note that the *New Macro* dialog will first be displayed in which you can type the macro's name and description.

Application Events - View Code. Select *View Code* from the *Tools - Macro - Application Events* menu to display the *Document Events* window, which is used to enter code to change the default behavior of application-level events. Application-level events allow you to customize the behavior of all documents of a certain type (such as all spreadsheets, all workbooks, or all reports). See *What Are Application Events and How Can They Be Controlled from STATISTICA Visual Basic?* in Chapter 11 – *STATISTICA Visual Basic* for more details.

Application Events - Autorun. Select *Autorun* from the *Tools - Macro - Application Events* menu to save your application-level event code with a specific type of document so that it will be run any time that type of document is open.

Add-Ins. Select *Add-Ins* from the *Tools - Macro* menu to display the *STATISTICA Add-Ins* dialog. Add-Ins are COM server components normally written in ATL (Active Template Library) that are used to create custom user interfaces of *STATISTICA* and/or fully functional external programs.

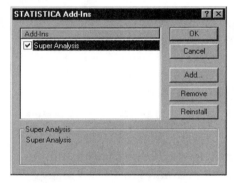

All available Add-Ins are displayed in the *Add-Ins* list. To create a new Add-In, click the *Add* button to display the *Specify Add-In to be Added* dialog, which is used to enter the program ID of the Add-In. Click the *Remove* button to delete the selected Add-In from the *Add-Ins* list. Finally, click the *Reinstall* button to register the selected Add-In to your operating system.

***STATISTICA* Visual Basic Editor.** Select *STATISTICA Visual Basic Editor* from the *Tools - Macro - Application Events* menu (or press ALT+F11) to display a *STATISTICA* Visual Basic editor (Macro Window). For more information on using *STATISTICA* Visual Basic including examples and an extensive FAQ section, see Chapter 11 - *STATISTICA Visual Basic*.

Customize

Select *Customize* from the *Tools* menu to display the *Customize* dialog, which is used to customize toolbars, menus, and keyboard hot keys with a variety of commands. For more details on the *Customize* dialog, see Chapter 1 – *STATISTICA – A General Overview*.

Options

Select *Options* from the *Tools* menu to access the *Options* dialog. In addition to general and display options, options are available for workbooks, reports, graphs, macros, file locations, custom lists, statistical analysis display, import and edit facilities, and output management. For a complete discussion of the *Options* dialog, see Chapter 2 – *STATISTICA Output Management*.

Workbook Menu

The *Workbook* menu provides access to the options commonly used to manage workbooks. Most of these options are also available from shortcut menus and from keyboard buttons.

Insert

Select *Insert* from the *Workbook* menu to display the *Insert Workbook Item* dialog. This dialog allows you to select the type of item (e.g., a *STATISTICA* document, an ActiveX document, or a folder) to insert into the workbook and to determine the placement of the item. You can also display this dialog by right-clicking on an item in the workbook tree and selecting *Insert* from the shortcut menu, or by pressing the INSERT key on your keyboard.

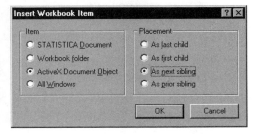

Item. Use the option buttons in the *Item* group to select what type of file to add to the workbook. Select the *STATISTICA document* option button to add a *STATISTICA* document.

This document can be new, saved, or open. Select the **Workbook folder** option button to add a new folder to the workbook. Select the **ActiveX Document Object** option button to create an ActiveX object by adding a saved ActiveX document or creating a new one. Select the **All Windows** option button to add all of the open *STATISTICA* documents to the workbook. See *ActiveX Documents* (page 169) for more information on supported document types. See also the *FAQ: Working with Workbooks* section (page 205) for extensive examples of adding files and folders to workbooks.

Placement. Use the option buttons in the **Placement** group to specify whether the item is inserted as a sibling to (i.e., on the same level as) the currently selected workbook item or as a child to (i.e., one level beneath) the currently selected workbook item. Note options for adding a file as a sibling are not available when the root folder is selected.

Delete

Select **Delete** from the **Workbook** menu to remove the selected items from the workbook tree without copying it to the Clipboard. If you want to copy the item to the Clipboard while removing it from the workbook tree, select **Cut Workbook document**, also available from the **Workbook** menu. Note that if the item has children, the children will also be deleted. To make discontinuous selections (i.e., select some of the children, but not all of them), hold down the CTRL key while clicking on each item you want to select. You can also make discontinuous selections across nodes using the CTRL key.

When you select **Delete**, a **STATISTICA Workbook** dialog is displayed, asking you to verify that you actually want to delete the selected items. Click **Yes** if you want to continue or **No** if you do not want to continue. You can turn the prompting dialog off by clearing the **Prompt on item delete** check box on the **Workbooks** tab of the **Options** dialog, which is accessed by selecting **Options** from the **Tools** menu.

You can also access the **Delete** command by right-clicking on any item in the workbook tree and selecting **Delete** from the shortcut menu, or pressing the DELETE key on your keyboard.

Rename

Select **Rename** from the **Workbook** menu to display a box that is used to rename the currently selected workbook item. You can also edit an item name by double-clicking on its name or by selecting the item and pressing F2. Note this command is also available from the workbook tree shortcut menu.

Extract As Stand-Alone Window

STATISTICA allows you to extract *STATISTICA* documents from workbooks and view them in stand-alone windows. You can then save the documents as individual files, place them in reports, or store them in other workbooks. You can also use drag-and-drop to move files from a workbook to another workbook, a report, or the workspace. When you drag-and-drop a non-*STATISTICA* file from a workbook (either a copy or the original), the appropriate program is launched and the file is opened in it.

Original. Select *Original* from the *Workbook - Extract as stand-alone window* menu to remove the selected object from the workbook and make it available in a stand-alone window. You can also access this command by right-clicking on any item in the workbook tree and selecting *Extract as stand-alone window - Original* from the shortcut menu.

Copy. Select *Copy* from the *Workbook - Extract as stand-alone window* menu to extract a copy of the selected object from the workbook and make it available as a stand-alone window. Unlike the *Original* command, this does not remove the selected file from the workbook. You can also access this option by right-clicking on any item in the workbook tree and selecting *Extract as stand-alone window - Copy* from the shortcut menu.

Cut Workbook Document

Select *Cut Workbook document* from the *Workbook* menu to remove the selected item (and it's children) from the workbook tree while copying it to the Clipboard. This feature is useful when you want to move a portion of the tree to a different location within the workbook. To paste the contents in a new location, select either *As first child*, *As last child*, *As prior sibling*, or *As next sibling* from the *Workbook - Paste as Workbook document* menu.

When you select this command, the *STATISTICA Workbook* dialog is displayed, asking you to verify that you actually want to cut the selected items. Click *Yes* if you want to continue or *No* if you do not want to continue. You can turn the prompting dialog off by clearing the *Prompt on item delete* check box on the *Workbooks* tab of the *Options* dialog, which is accessed by selecting *Options* from the *Tools* menu.

You can also access this command by right-clicking on any item in the workbook tree and selecting *Cut Workbook document* from the shortcut menu. For information on making selections in the workbook tree, see *Can I Select Discontinuous Items in the Workbook Tree?* (page 213).

Copy Workbook Document

Select **Copy Workbook document** from the **Workbook** menu to copy the selected workbook item (and it's children) to the Clipboard. For more information on making selections in the workbook tree, see *Can I Select Discontinuous Items in the Workbook Tree?* (page 213).

You can also access this option by right-clicking on any item in the workbook tree and selecting **Copy Workbook document** from the shortcut menu.

Paste As Workbook Document

To use the commands on the **Paste as Workbook document** menu, you must have an acceptable workbook item on the Clipboard (e.g., a folder, *STATISTICA* or other ActiveX document, or another workbook). If the Clipboard is empty or if you have copied something else (e.g., text from an e-mail message) to the Clipboard, these commands are dimmed.

Note that you can also access these commands by right-clicking on any item in the workbook tree and selecting the appropriate command from the **Paste as Workbook document** shortcut menu or by pressing CTRL+V.

As first child. Select **As first child** from the **Workbook - Paste as Workbook document** menu to paste the contents of the Clipboard as the first child to the currently selected item in the workbook tree.

As last child. Select **As last child** from the **Workbook - Paste as Workbook document** menu to paste the contents of the Clipboard as the last child to the currently selected item in the workbook tree.

As next sibling. Select **As next sibling** from the **Workbook - Paste as Workbook document** menu to paste the contents of the Clipboard immediately after, but on the same level as the currently selected item in the workbook tree. Note that if the top level of the workbook is selected, this option is not available as it is not possible to paste anything as a sibling to the first folder in workbook tree (i.e., on the same level as the root node).

As prior sibling. Select **As prior sibling** from the **Workbook - Paste as Workbook document** menu to paste the contents of the Clipboard immediately before, but on the same level as the currently selected item in the workbook tree. Note that if the top level of the workbook is selected, this option will not be available as it is not possible to paste anything as a sibling to the first folder in the workbook tree (i.e., on the same level as the root node).

Use As Active Input

Select *Use as Active Input* from the *Workbook* menu to toggle the use of the spreadsheet currently selected in the workbook tree as the active input spreadsheet. If you want to use the selected spreadsheet for further analysis, select this command, thus designating the spreadsheet as an active input spreadsheet. If you do not want this spreadsheet to be the active input spreadsheet, clear this command. Although any number of stand-alone spreadsheets can be specified as input spreadsheets, only one spreadsheet per workbook can be designated as the active input spreadsheet. The active input spreadsheet will be designated in the workbook tree with a red box as shown in the illustration below (*Adstudy.sta* is the designated active input spreadsheet for this workbook).

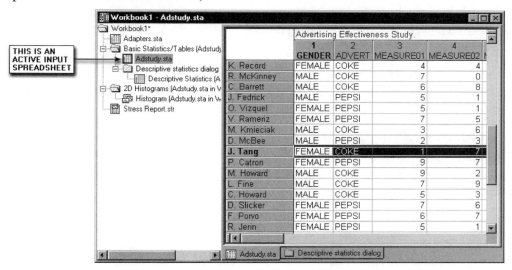

This command is not available if a spreadsheet is not selected in the workbook tree. For more details on input spreadsheets, see the *Input Spreadsheet Overview*, page 169.

Tab Control

Workbooks store each output document (e.g., a *STATISTICA* Spreadsheet or Graph, or a Word or Excel document) as a tab. You can easily navigate the children of the node currently selected in the workbook tree using the tabs provided in the workbook viewer. The commands on the *Workbook - Tab Control* menu specify where the tabs are placed and whether they are contained in a single, scrollable row or multiple rows. Note these commands are for the current workbook only. To make global changes to the workbook tab orientation and maximum

character length, use the ***Workbook*** tab of the ***Options*** dialog (accessed via the ***Tools – Options*** menu). For more details on the options available there, see *Global Workbook Options* (page 175).

Bottom (one row). Select ***Bottom (one row)*** from the ***Workbook - Tab Control*** menu to place the workbook tabs at the bottom of the workbook viewer. This produces a single, scrollable row of tabs.

Top (one row). Select ***Top (one row)*** from the ***Workbook - Tab Control*** menu to place the workbook tabs at the top of the workbook viewer. This command also produces a single, scrollable row of tabs.

Left (multiple rows). Select ***Left (multiple rows)*** from the ***Workbook - Tab Control*** menu to place the workbook tabs to the left of the workbook viewer in multiple rows.

Right (multiple rows). Select ***Right (multiple rows)*** from the ***Workbook - Tab Control*** menu to place the workbook tabs to the right of the workbook viewer in multiple rows.

Visible. Select ***Visible*** from the ***Workbook - Tab Control*** menu to display the tabs (in the position specified above). By default this command is checked. If you do not want to display the tabs, clear this command.

Tab Length. Select ***Tab Length*** from the ***Workbook - Tab Control*** menu to display the ***Workbook Tab Character Length*** dialog. Use this dialog to specify the maximum length of the tabs displayed in the workbook viewer. *STATISTICA* truncates the names on the tabs when the names are longer than the value given here. By default, the tab length is ***64***.

Events

The commands on the ***Workbook - Events*** menu apply to document-level events. Document-level events are used to customize the behavior of open documents (i.e., a single workbook or spreadsheet). Note these commands are also available from the ***Tools*** menu; however, the commands on that menu apply to application-level events. For a list of document-level events for workbooks, see *Technical Note: Changing the Workbook Document-Level Events* (page 215). See *What Are Application Events and How Can They Be Controlled from STATISTICA Visual Basic?* in Chapter 11 – *STATISTICA Visual Basic* for more details on events.

View Code. Select ***View Code*** from the ***Workbook - Events*** menu to display the ***Document Events*** window, which is used to enter code to change the default behavior of document-level events.

Autorun. Select *Autorun* from the *Workbook - Events* menu to save your document-level event code with a specific document so that it will be run any time that document is open.

Password. Select *Password* from the *Workbook - Events* menu to display the *Setup Event Password* dialog, which is used to create a password to restrict access to editing document-level events.

Window Menu

The *Window* menu is available when any document is open. It provides access to commonly used commands for organizing the workspace and switching between files.

Close All

Select *Close All* from the *Window* menu (or press CTRL+L) to close all open spreadsheets, graphs, and related windows (e.g., graph data) in *STATISTICA*. This option is useful when you need to clear the screen to start a new analysis. Note that you will be prompted to save any unsaved files before they are closed.

Cascade

Select *Cascade* from the *Window* menu to arrange the open *STATISTICA* windows in an overlapping pattern so that the title bar of each window is visible.

Tile Horizontally

Select *Tile Horizontally* from the *Window* menu to arrange the open *STATISTICA* windows in a horizontal (side by side) pattern. When you select this command, *STATISTICA* automatically optimizes the display of the open windows (with the preference given to tiling horizontally).

Tile Vertically

Select *Tile Vertically* from the *Window* menu to arrange the open *STATISTICA* windows in a vertical pattern. When you select this command, *STATISTICA* automatically optimizes the display of the open windows (with the preference given to tiling vertically).

Arrange Icons

Select *Arrange Icons* from the *Window* menu to arrange all minimized windows into rows.

Windows

Select *Windows* from the *Window* menu to display the *Windows* dialog. This dialog is used to access and manage all of the currently open windows in *STATISTICA*.

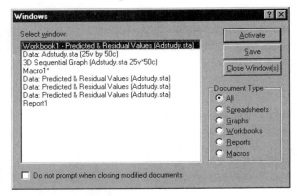

Select window. Select either a single window or multiple windows in the *Select window* box for manipulating its associated window in the *STATISTICA* environment. Note that you can use CTRL or SHIFT while selecting the items in the box to make multiple selections.

Activate. Click the *Activate* button to set the focus to the currently selected window in the *Select window* box. In other words, the associated window will be brought to the front of all of the other windows. Note that because only one window can have the focus at any given time, multiple selections for this operation will not apply.

Save. Click the *Save* button to save all of the windows that are currently selected in the *Select window* box.

Close Window(s). Click the *Close Window(s)* button to close all of the windows that are currently selected in the *Select window* box. Note that *STATISTICA* will prompt you to save any modified windows before closing them.

Document Type. Use the options in the *Document Type* group to filter all of the open windows within *STATISTICA* by their document type; the contents of the *Select window* box will be updated to reflect your filter selection. You can display all open documents in the

Select window box (select the **All** option button), or limit the display to **Spreadsheets**, **Graphs**, **Workbooks**, **Reports**, or **Macros** only by selecting the appropriate option button.

Help Menu

The **Help** menu is available when any document is open. It provides access to various types of help.

Contents and Index

Select **Contents and Index** from the **Help** menu to display the *STATISTICA Electronic Manual*. *STATISTICA* provides a comprehensive *Electronic Manual* for all program procedures and all options available in a context-sensitive manner by pressing the F1 key or clicking the help button ⬛ on the caption bar of all dialogs (there is a total of over 100 megabytes of compressed documentation included).

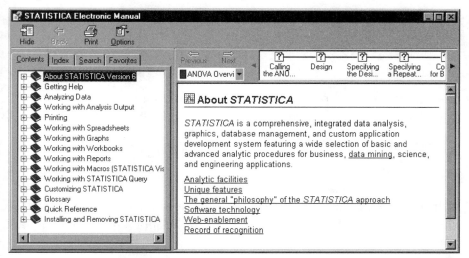

Due to its dynamic hypertext organization, organizational tabs (e.g., **Contents**, **Index**, **Search**, and **Favorites**), and various facilities allowing you to customize the help system, it is faster to use the *Electronic Manual* than to look for information in the traditional manuals. The status bar on the bottom of the *STATISTICA* window displays short explanations of the menu options or toolbar buttons when an item is highlighted or a button is clicked.

Statistical Advisor

Select **Statistical Advisor** from the **Help** menu to launch the **Statistical Advisor**.

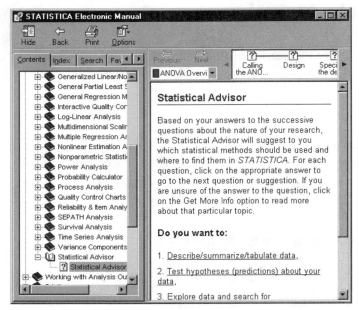

Based on your answers to the successive questions about the nature of your research, the **Statistical Advisor** suggests which statistical methods could be used and where to find them in *STATISTICA*.

Animated Overviews

Select **Animated Overviews** from the **Help** menu to display a submenu of the available animated overviews (visual overviews and specific feature help). Select from the submenu to launch one of the animated overviews.

If you select one of these commands and you have not installed that animated overview, a *STATISTICA* message dialog will prompt you for the location of that animated overview file. For example, you can run it from your *STATISTICA* CD, or if you are running a network version, the overviews can be located on the network drive. Note that if you have the *STATISTICA* CD, you can run the overviews directly from the CD without copying them to your hard drive.

StatSoft's Home Page

Select **StatSoft's Home Page** from the **Help** menu to launch the StatSoft Home Page in your default browser.

We invite you to visit the StatSoft Web site often:

- For the most recent information about *STATISTICA*, downloadable updates, new releases, new products, news about StatSoft, etc., access the *What's New...* section of the Web site.

- For a comprehensive list of *Frequently Asked Questions* (including useful tips, solutions to hardware or software compatibility problems, etc.), access the *Technical Support* section of the Web site.

- For a library of *STATISTICA* Visual Basic programs (written by users), access the *Technical Support* section of the Web site (to submit your own programs to this "user exchange forum," send an e-mail to info@statsoft.com).

Technical Support

Select **Technical Support** from the **Help** menu to launch the *Technical Support - Getting More Help* page of the StatSoft Web site in your default browser. This page contains links to download *STATISTICA* updates and links to FAQ topics on spreadsheets, graphs, printing, reports, etc. The StatSoft Technical Support Department e-mail address, phone number, and hours are also listed.

About *STATISTICA*

Select **About STATISTICA** from the **Help** menu to display the *STATISTICA* version, copyright notice, license information, and foreign office contact information.

FAQ: WORKING WITH WORKBOOKS

Can I Add a Folder to a Workbook?

By default, when results are automatically added to a workbook, they are included in a folder that identifies from what analysis they came; however, you can add as many folders (and levels) to a workbook as you want. This allows you to easily organize all of the items in your workbook, including spreadsheets, reports, and non-*STATISTICA* documents that are placed as sibling in the first level of the workbook by default. As with all items you can add to a workbook, folders can be inserted as either a sibling or a child to current workbook items.

Folders are added to the workbook tree via the **Insert Workbook Item** dialog. Right-click on the first folder in the workbook tree to display a shortcut menu, and select **Insert** to display this dialog.

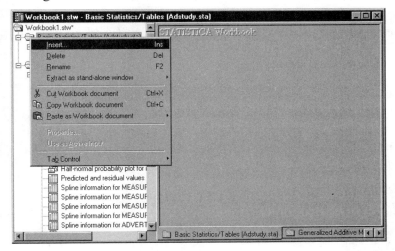

You can also select **Insert** from the **Workbook** menu or press the INSERT key on your keyboard.

When the **Insert Workbook Item** dialog is displayed, select the **Workbook folder** option button in the **Item** group and **As first child** in the **Placement** group. Note that the sibling options in this box are disabled because we are adding an item to the root level of the workbook tree and it cannot have siblings. Click the **OK** button to add the folder to the workbook tree. Although

folders can be added to any part of the workbook tree, this example adds a new folder to the first folder in the tree. Note that when the folder is initially added to the tree, its name is selected. You can rename the folder at this time by entering a new name.

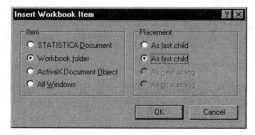

To rename the folder at a later time, double-click on the name **New Folder**, and enter a new name.

How Do I Add a New ActiveX Document to a Workbook?

New ActiveX documents (e.g., *STATISTICA* Spreadsheets and Graphs, Excel spreadsheets, Word documents) are added to the tree via the **Insert Workbook Item** dialog. To display this dialog, select **Insert** from the **Workbook** menu. You can also right-click in the tree and select **Insert** from the workbook tree shortcut menu or press INSERT on your keyboard. The example below adds a new Word document as a child to the new folder created in *Can I Add a Folder to a Workbook?* (above).

Right-click on the **New Folder** and select **Insert** from the shortcut menu. Note that you can always move or copy the item to a new location later (using drag-and-drop features), so the exact location is not critical (even for this example). On the **Insert Workbook Item** dialog, select the **ActiveX Document Object** option button in the **Item** box and the **As first child** option button in the **Placement** box. Then click the **OK** button to display the **Insert ActiveX Document Object** dialog.

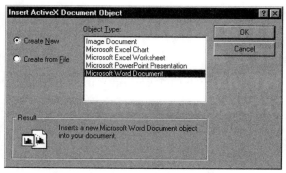

You can create your ActiveX document object using a new document or a previously saved document. For this example, select the **Create New** option button and select **Microsoft Word Document** in the **Object Type** box. Note that if you do not have Microsoft Word installed on your computer, you can select any appropriate file type from this box. Finally, click the **OK** button to add the new document to your workbook.

How Do I Add a Previously Saved Object to a Workbook?

Previously saved ActiveX documents (e.g., *STATISTICA* Spreadsheets and Graphs, Excel spreadsheets, Word documents) are added to the tree via the **Insert Workbook Item** dialog. To display this dialog, select **Insert** from the **Workbook** menu. You can also right-click in the tree and select **Insert** from the workbook tree shortcut menu or press INSERT on your keyboard. You can add a variety of file types to a workbook, a feature that allows you to easily organize all of the files associated with a topic. As with all items you can add to a workbook, files can be inserted as a sibling or as a child to current workbook items.

Follow these steps to add a previously saved file as the next sibling to the first item in the current workbook:

Select the first item in the current workbook, and then display the **Insert Workbook Item** dialog by selecting **Insert** from the **Workbook** menu.

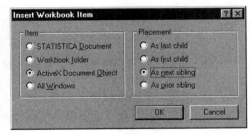

In the **Item** group, select the **ActiveX Document Object** option button, and in the **Placement** group, select the **As next sibling** option button. Click the **OK** button to display the **Insert ActiveX Document Object** dialog.

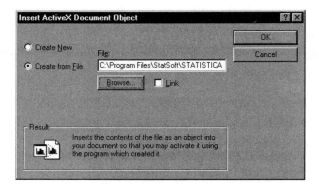

In the ***ActiveX Document Object*** dialog, select the ***Create from File*** option button and then click the ***Browse*** button to display the ***Browse*** dialog.

Use the ***Look in*** list box to locate the file you want to add. Double-click on the file to enter it in the ***File*** box on the ***Insert ActiveX Document Object*** dialog. Click the ***OK*** button to add the file to your workbook. Note that if you select a file type that is not supported by this feature, *STATISTICA* prompts you to select a different file. For information on supported file types, see the *ActiveX* document overview (page 169).

How Do I Add the Current
Item to a Workbook?

You can add any *STATISTICA* object to a workbook including graphs, reports, macros, and spreadsheets. The most direct way to add the current *STATISTICA* object (i.e., the document

---

that currently has focus) to a workbook is to click the Add to Workbook ▾ toolbar button. A list of currently available workbooks is displayed, allowing you to select in which workbook to include the item.

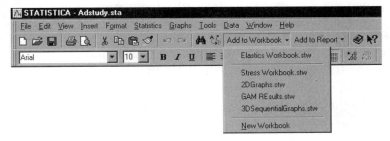

You can also add the item to a **New Workbook**.

By default, the item is appended at the end of the workbook tree one level below the root node (i.e., it is not placed as a sibling to any existing items in the workbook). You can relocate the object using the workbook's extensive drag-and-drop facilities.

You can also add currently open *STATISTICA* documents to the workbook using the **Insert** command from either the **Workbook** menu or the workbook tree shortcut menu. Select **Insert** to display the **Insert Workbook Item** dialog, and then select **STATISTICA Document** in the **Item** group of the **Insert Workbook Item** dialog.

Click the **OK** button in the **Insert Workbook Item** dialog to display the **Document type** dialog.

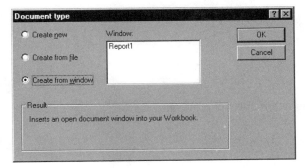

In this dialog, select the ***Create from window*** option button and specify which document to add using the ***Window*** box. Click the ***OK*** button to add the selected document to your workbook. Note that if you want to add all open *STATISTICA* documents to your workbook, select the ***All Windows*** option button on the ***Insert Workbook Item*** dialog and then click the ***OK*** button.

You can also add currently open ActiveX documents (e.g., Word documents or Excel files) to your workbook using the Clipboard. To add an open Word document, select the entire document (or a portion of it), copy it to the Clipboard (press CTRL+C), and then paste the contents of the Clipboard into the workbook (press CTRL+V or select ***Paste*** from the ***Workbook*** menu). *STATISTICA* automatically creates the appropriate type of document in the workbook and includes the contents of the Clipboard in that document.

How Do I Add a Saved *STATISTICA* Document to a Workbook?

You can add a saved *STATISTICA* document to your workbook (e.g., a spreadsheet, graph, report, or macro) by selecting ***Insert*** from the ***Workbook*** menu or workbook tree shortcut menu.

For example, to add the ***Adstudy.sta*** sample datafile to the current workbook, open a workbook and click on any item in the workbook tree. Now, press the INSERT key on your keyboard to display the ***Insert Workbook Item*** dialog. (You could also select ***Insert*** from the ***Workbook*** menu or the workbook tree shortcut menu to display this dialog.)

In the ***Insert Workbook Item*** dialog, select the ***STATISTICA Document*** option button in the ***Item*** box and the ***As next sibling*** option button in the ***Placement*** box. Click the ***OK*** button to display the ***Document type*** dialog.

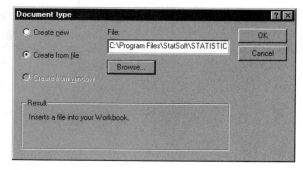

Select the ***Create from file*** option button in the ***Document type*** dialog, and then click the ***Browse*** button to display the ***Browse*** dialog.

Adstudy.sta is located in the **Examples/Datasets** subfolder of your *STATISTICA* installation folder. Use the **Look in** box to locate this file. Double-click on the file to enter it into the **File** box on the **Document type** dialog (as shown above). Click the **OK** button to add the file to your workbook.

How Do I Add a New *STATISTICA* Document to a Workbook?

In addition to adding current and previously saved *STATISTICA* documents to a workbook, you can also add new (blank) documents to a workbook, including spreadsheets, reports, and macros. When the new document is added, you must specify where to place the item in the workbook tree (either as a sibling to or as a child to another item in the workbook) and which type of item to create.

One way to add a new *STATISTICA* document is to determine an appropriate location in the workbook tree and right-click another item in that location. You can always move or copy the item to a new location later using drag-and-drop features, so the exact location is not critical.

From the workbook tree shortcut menu, select **Insert** to display the **Insert Workbook Item** dialog.

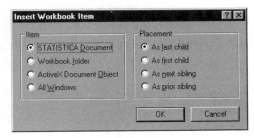

Select the **STATISTICA Document** option button in the **Item** group and the tree location (either **As last child**, **As first child**, **As next sibling**, or **As prior sibling**) option button in the **Placement** group. Click the **OK** button to display the **Document type** dialog.

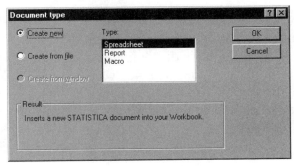

Use this dialog to create a new document, locate a previously saved document, or select an open document. For this example, select the **Select new** option button and select a **Spreadsheet**, **Report**, or **Macro** in the **Type** box. Then, click the **OK** button to add the new document to your workbook.

How Do I Delete Items from a Workbook?

Commands for deleting workbook items are available from the **Workbook** menu, as well as from the workbook tree shortcut menu (accessed by right-clicking in the workbook tree). You can also delete an item by selecting it and pressing the DELETE key on your keyboard.

Note that when you delete an item that has children, all of its children will also be deleted. For more information on making selections in the workbook tree, see *Can I Select Discontinuous Items in the Workbook Tree?* (page 213).

How Do I Rename Workbook Items?

Commands for renaming workbook items are available from the **Workbook** menu as well as from the workbook tree shortcut menu (accessibly by right-clicking in the workbook tree). You can also select a workbook item and press F2.

Can I Rearrange the Workbook Tree with Drag-and-Drop?

Yes. The *STATISTICA* Workbook supports an entire range of drag-and-drop features within the workbook tree. By right-clicking on an item in the workbook tree, you can drag that item to a new location within the tree. In fact, since the drag-and-drop features support parent/child relationships, you can use these features to rearrange large portions of the workbook tree.

To select the parent node and all of its children, simply click on the parent node (or press CTRL while clicking on the parent node and each of its children). To make discontinuous selections (i.e., select some of the children, but not all of them), hold down the CTRL key while clicking on each item you want to select. You can also make discontinuous selections across nodes using the CTRL key. For more information on making selections in the workbook tree, see *Can I Select Discontinuous Items in the Workbook Tree?* (page 213).

Follow these steps to move a folder (and its contents) to a new location in the workbook tree:

- Click on the folder to select it and all of its children.

- Right-click on the folder, and drag the folder to a new location in the tree.

- Release the right mouse button, and a shortcut menu is displayed, giving you options to either copy or move the dragged item. You can place the item before, after, or as a child to the newly selected item in the tree. For this example, select **Move After** from the shortcut menu.

For more details on drag-and-drop workbook features, see *Workbook Drag-and-Drop Features* (page 172).

How Do I Use Copy and Paste in a Workbook?

In addition to using drag-and-drop features to rearrange workbook items, you can use the **Copy** and **Paste** commands. Commands for cutting, copying, and pasting are available from the **Workbook** menu, the workbook tree shortcut menu, and the **Standard** toolbar. For more details on Clipboard functionality, see *Workbook Clipboard Features* (page 171).

Can I Select Discontinuous Items in the Workbook Tree?

You can select one or more items in the workbook tree using the standard Windows SHIFT+click and CTRL+click conventions to select ranges and discontinuous lists of variables, respectively. Additionally, you can select and deselect tree items using the keyboard navigation keys (e.g., HOME, END, PAGE UP, PAGE DOWN, and arrow keys). You can delete an entire selection by pressing the DELETE key. Press the INSERT key to display the **Insert Workbook Item** dialog for the currently selected item.

Note that to select discontinuous items of a particular node, you will need to expand the node (by clicking on the plus sign adjacent to the node).

Can I Print More Than One Item From a Workbook Without Printing the Entire Workbook?

Yes. When printing from within a workbook, only the currently displayed (active) item is printed when the **Print** 🖨 button is clicked, but if the **Print** dialog is used, multiple selections

from the workbook can be printed by using the **Selection** option on the **Print** dialog. Note that *STATISTICA* supports the standard Windows SHIFT+click and CTRL+click conventions to select ranges and discontinuous lists of items, respectively. Thus to print a range of graphs from a workbook, select the range of graphs, select **Print** from the **File** menu to display the **Print** dialog, select the **Selection** option button in that dialog, and click the **OK** button.

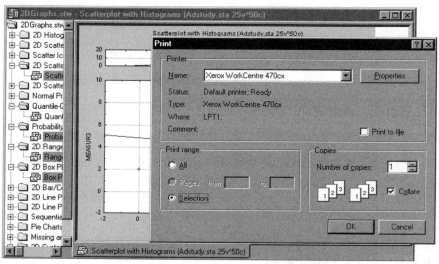

In the illustration shown above, the four selected graphs will be printed when the **OK** button is clicked on the **Print** dialog.

StatSoft
Copyright © StatSoft, 2001

TECHNICAL NOTE: CHANGING THE WORKBOOK DOCUMENT-LEVEL EVENTS (A FLEXIBLE WAY TO CUSTOMIZE *STATISTICA*)

An event is an action that is typically performed by users, such as clicking a mouse button, pressing a key, changing data, or opening a spreadsheet or workbook. In *STATISTICA*, certain events are directly accessible to customization, and you can change the default behavior of the program when the specific events are triggered (or when the events are triggered in specific parts of the documents). Using programmable events offers a particularly flexible and "far reaching" method to tailor *STATISTICA*'s behavior to your specific needs.

For example, you can prevent another user from saving changes made to a file or make specific sections of a spreadsheet dependent solely on other (e.g., automatically recalculated) values and protect them from being changed directly by editing the specific cells. You can also create a backup file of a particular workbook to a specific location whenever you save that workbook (so you won't have to save it twice). These are just a few examples of countless customizations that can easily be accomplished by customizing events.

Event Types

Events are part of a set of tools built into *STATISTICA* to make it a powerful solution building system. There are two different types of events: document-level events and application-level events.

Document-level events. Document-level events occur only for open documents and in some cases, for objects within them. They allow you to customize the behavior of open documents.

Application-level events. Application-level events occur at the level of the application itself. They are used to customize the behavior of all documents of a certain type (such as all spreadsheets, or all workbooks, or all reports, etc.). For example, you could customize *STATISTICA* so that all reports are automatically saved to a backup folder or all spreadsheets are automatically saved (without prompting) before they are closed.

Workbook Events

The following table includes the document-level events that are available for workbooks. For more information on events, see *What Are Application Events and How Can They Be Controlled from STATISTICA Visual Basic?* and *Controlling STATISTICA Events with SVB Programs* in Chapter 11 – *STATISTICA Visual Basic*. Note that application-level events are listed in Chapter 11, and document-level events for other document types are listed in the appropriate chapter (e.g., spreadsheet document-level events are listed in Chapter 4 – *STATISTICA Spreadsheets*).

Command	Action
Activate	Executes when the workbook receives the focus within the *STATISTICA* workspace (typically when you click on the workbook or a document is sent to it).
BeforeClose	Executes when you close the workbook. Before the document closes, the events within this function will first execute. An example of using this feature would be to add additional functionality, such as saving the file to multiple locations or exporting it to another application.
BeforePrint	Executes when you print the workbook. Before the workbook prints, the events within this function will first execute. Examples of using this feature would be to prevent users from printing the document or adding new functionality such as calling an alternate, customized printing dialog.
BeforeRightClick	Executes when you right-click anywhere on the workbook. Before the shortcut menu is displayed, the events within this function will first execute. An example of using this feature would be to add additional functionality to displaying shortcut menus, such as selecting the entire contents of the workbook.
BeforeSave	Executes when you save changes made to the workbook. Before the changes are actually saved, the events in this function will first execute. An example of using this feature would be to prevent users from permanently changing a workbook.
Deactivate	Executes when the workbook has lost the focus to another window within *STATISTICA*.

OnClose	Executes when the workbook is being closed.
Open	Executes when the workbook is being opened. Note, this will not run unless autorun was specified when the macro was last edited.
SelectionChanged	Executes when the focus within the workbook has moved.

4

CHAPTER

STATISTICA SPREADSHEETS

continued ➭

STATISTICA SPREADSHEETS

GENERAL OVERVIEWS

Dataset Terminology

Cases and variables. *STATISTICA* data are organized into cases and variables. If you are unfamiliar with this notation, you can think of cases as the equivalent of records in a database management program (or rows of a spreadsheet) and variables as the equivalent of fields (columns of a spreadsheet). Each case consists of a set of values of variables. For example, suppose 4 persons (cases) completed 3 tests; there can be a total of 5 variables in the datafile: *Gender* (*Male* = male subject, *Female* = female subject), *Education* (*C* = college, *H* = high school), and 3 test scores (*Test 1* through *Test 3*). Shown below is such a file.

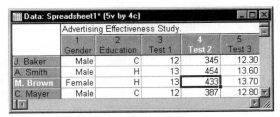

Case names. The first column in the file can (optionally) contain names of cases.

Text values. *STATISTICA* offers comprehensive support for true text values (see the next section on data types), which can be of practically unlimited length and include extensive within-cell formatting. However, for many statistical data analysis applications, it is useful to use text labels that can aid in the interpretation of their respective numeric values, as illustrated in the next paragraph.

Text labels. The two variables *Gender* and *Education* contain text labels, i.e., they are numeric variables with text labels associated with the numeric values. For example, suppose *STATISTICA* (or you) made the following assignments:

1 = *Male*, *2* = *Female* (for *Gender*); and

1 = *C*, *2* = *H* (for *Education*).

You can switch between the two views of data (numeric or text) in the spreadsheet by clicking the *Text Labels* 🖑 button on the *Spreadsheet* toolbar. After switching to numeric representation of these values, the file will look as follows:

	1 Gender	2 Education	3 Test 1	4 Test 2	5 Test 3
	Advertising Effectiveness Study.				
J. Baker	1	1	12	345	12.30
A. Smith	1	2	13	454	13.60
M. Brown	2	2	13	433	13.70
C. Mayer	1	1	12	387	12.80

Data: Spreadsheet1* (5v by 4c)

Using text labels with numeric values (summary). As shown in the example above, in *STATISTICA* each numeric value of a specific variable (e.g., *1*) can have a text label (e.g., *Male*) assigned to it. For more information, see *How Do I Enter/Edit the Assignments between Numeric Values and Text Labels?* (page 349).

Variable Types

You can specify each variable's data type via the *Variable* specifications dialog (available from the *Data* menu or by double-clicking on the column header). *STATISTICA* Spreadsheet datafiles support the four basic data types listed below (note that the spreadsheets can also contain links to other data sources, embedded multimedia objects of various types, macros, user interface, etc., however, those items will not be used as direct input for analyses):

Double. The *Double* (abbreviated Double Precision) data type is the default format for storing numeric values in *STATISTICA*. Technically, the values are stored as 64-bit floating point real numbers, with 15-digit precision (1 bit for the sign, 11 for the exponent, and 52 for the mantissa). The range of values supported by this data type is approximately $\pm 1.7*10E308$. Each numeric value can have a unique text label attached of practically unlimited length when the *Display format* is *General*. This is the only data type that allows numbers containing decimals. When your data type is *Double*, each cell takes up 8 bytes of storage (plus the optional text label). Note that for the *Double* data type, the missing data code is -9999.

Integer. If *Integer* is the data type, you can enter integers between and including -2,147,483,648 through 2,147,483,647. You cannot enter numeric values containing decimals into a variable of this type. Each numeric value can have a unique text label attached of practically unlimited length when the *Display format* is *General*. When your data type is *Integer*, each cell takes up 4 bytes of storage; hence this data type offers a more economical way of storing numbers than when *Double* is the data type and is recommended for storing integer data especially in large datafiles. Note that for the *Integer* data type, the missing data is the same as *Double*: -9999.

Byte. If *Byte* is the data type, you can enter integers between and including 0 through 255. You cannot enter numeric values containing decimals into a variable of this type. Each byte value can have a unique text label attached of practically unlimited length when the *Display format* is *General*. The advantage of specifying *Byte* as your data type is that it offers the most economical storage for values that are small integers, as each cell takes up only 1 byte of storage. Note that for the *Byte* data type, the missing data code is 255.

Text. The *Text* data type is optimized for storing sequences of any characters of practically unlimited length. Note that in *STATISTICA*, you can perform numerical analyses on text values, and in those circumstances, *STATISTICA* will assign unique numeric equivalents to all text values being processed (unlike the relation between the numeric data types listed above and their permanent text labels, the relations between text values and numbers are created *ad hoc* and are not stored by *STATISTICA*; hence, most likely different numbers will be created the next time if a text variable is included in numerical analyses). The length of a field reserved for text variable type is not constant and can be adjusted. Note that for the *Text* data type, the missing data code is always an empty string.

Why do you need different variable types? The difference between the text and the numeric types is straightforward; however, the main reason for having three types of numeric values is the storage efficiency. For most datafiles, that is not important and, thus, using the default (*Double*) data type is recommended. However, for very large datafiles, being able to switch to a 2 (or even 8) times more efficient storage (by using different data types) could make a difference between being able to perform the necessary analysis on a specific computer system or not.

Spreadsheets – Multimedia Tables

STATISTICA Spreadsheets are based on StatSoft's proprietary multimedia table technology and are used to manage both input data and the numeric or text (and optionally any other type of)

output. The basic form of the spreadsheet is a simple two-dimensional table that can handle a practically unlimited number of cases (rows) and variables (columns), and each cell can contain a virtually unlimited number of characters. Sound, video, graphs, animations, reports with embedded objects, or any ActiveX-compatible documents can also be attached.

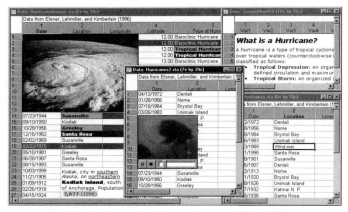

Because *STATISTICA* Spreadsheets can also contain macros and any user-defined user interface, these multimedia tables can be used as a framework for custom applications (e.g., with a list box of options or a series of buttons placed in the upper-left corner), self-running presentations, animations, simulations, etc.

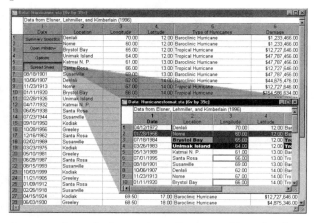

As mentioned above (see *Dataset Terminology*, page 221), *STATISTICA* data are organized into cases and variables. If you are unfamiliar with this notation, you can think of cases as the equivalent of records in a database management program (or rows of a spreadsheet), and variables as the equivalent of fields (or columns of a spreadsheet). Each case consists of a set of values of variables, and the first column in the file can (optionally) contain names of cases.

The spreadsheet window comprises several basic components.

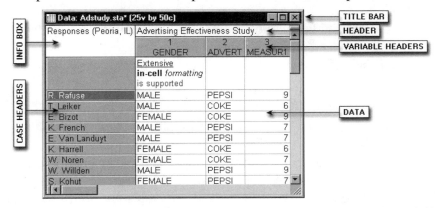

Title Bar. The *Title Bar* displays the name of the spreadsheet followed by the spreadsheet extension (*.sta*). If the spreadsheet is an input spreadsheet, the *Title Bar* also displays the number of variables by number of cases (e.g., *25v by 50c*). In the illustration shown above, the *Title Bar* contains the text *Data: Adstudy.sta (25v by 50c)*.

Info Box. Select the entire spreadsheet by clicking once in the lower-right corner (the mouse pointer will be the default arrow) of the *Info Box*, which is located in the upper-left corner of the spreadsheet window. To select the *Info Box* only (for formatting), click once in the upper-left corner of the *Info Box* (the mouse pointer will be an outlined plus sign ⊕). Double-click in the *Info Box* to enter or edit the text in the *Info Box* (e.g., additional details about the spreadsheet). In the illustration shown above, the *Info Box* contains the text *Responses (Peoria, IL)*.

Header. Located immediately above the variable headers at the top of the window, double-click in the *Header* area to enter or edit text information. To select the *Header* only (for formatting), click once in the upper-left corner (the mouse pointer will be an outlined plus sign ⊕). Press CTRL+ENTER or ALT+ENTER to enter a new line (note that you need to extend the height of the field to see new lines that you are adding). In the illustration shown above, the *Header* contains the text *Advertising Effectiveness Study*.

Case Headers. These cells, located at the far left of the window, contain header information for each case. Double-click on any *Case Header* cell to enter or edit text information. To select the *Case Header* only (for formatting) click once on the left side of the *Case Header* (the mouse pointer will be an outlined plus sign ⊕). To select the case row (for editing), click once on the middle or right side of the *Case Header* (the mouse pointer will be an outlined plus sign with an arrow ⊕). To select a block of *Case Headers*, (without selecting their respective rows), select *Select Variable Names Only* from the spreadsheet shortcut menu.

To autofit the **Case Headers**, double-click on the far-right side of any **Case Header** (the mouse pointer will be a cross with a double-headed arrow ✛). In the illustration shown above, the **Case Header** cells contain the first initials and last names of the respondents in the study. Note that **Case Headers** are optional and you can choose not to display them (toggle off **Display Case Names** from the **View** menu); if they are not displayed, the case numbers are shown.

Variable Headers. These cells, located at the top of each column, contain header information for each variable. To display details about an individual variable, double-click on the **Variable Header** cell. To select the **Variable Header** only (for formatting) click once on the upper portion of the **Variable Header** (the mouse pointer will be an outlined plus sign ⏢, instead of the ⏢). To select the variable column (for editing) click once on the lower portion of the **Variable Header** (the mouse pointer will be an outlined plus sign with an arrow ⏢). To autofit the variable column, double-click on the far-right side of the **Variable Header** (the mouse pointer will be a cross with a double-headed arrow ✛). In the illustration shown above, the first two **Variable Header** cells contain the text **Gender** and **Advert**. You have the option to change how the **Variable Header** cells display information so that they show the column number associated with the variable, the variable long name, and/or an abbreviation of the display types for the variables in the spreadsheet. Each of these options is available from the **View - Variable Headers** menu.

Data (and in-cell formatting options). The remainder of the spreadsheet contains data that pertain to the cases and variables and any optional attached or linked objects (multimedia objects, macros, custom user interface). Text in cells can be of practically unlimited length (in most *STATISTICA* configurations it is limited to 1,000 characters to protect against inadvertent pasting of unwanted large amounts of data into one cell). Text in cells can be extensively formatted including different fonts and font attributes.

Unlimited Size of Spreadsheets

An important feature of *STATISTICA* Spreadsheets (and the multimedia table technology) is their unlimited size. There are virtually no limitations in terms of the number of cases (rows), variables (columns), length of text labels and values, case names, etc. The size of your spreadsheet, its variables, and its text labels are only limited by your hardware.

The same is also true about the OLE/ActiveX and other objects that can be embedded into the spreadsheets; their size is not limited, and there can be virtually an unlimited number of them.

Importing Data

The quickest, and in many cases easiest, way to access datafiles from other Windows applications (e.g., spreadsheets) is to use the Clipboard. *STATISTICA* supports special Clipboard data formats generated by applications such as Microsoft Excel. For example, *STATISTICA* properly interprets formatted cells (such as 4/17/1999 or $10) and text values, including extensive in-cell formatting (e.g., RVS tower *120.3MHz*).

Datafiles from a wide variety of Windows and non-Windows applications can also be accessed and translated into the *STATISTICA* format (*.sta*) using the file import facilities. Select **Open** from the **File** menu to display the **Open** dialog.

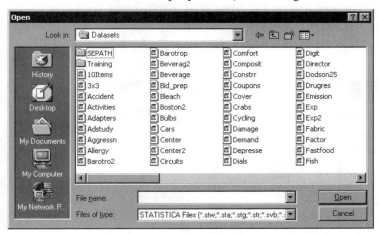

A wide variety of files is available in the **Files of type** box. Along with the numerous types of *STATISTICA* documents, Excel, dBASE, SPSS Portable, Lotus/Quattro Worksheets, Text [formatted and free format text (ASCII)], HTML, and Rich Text Files are available.

After you select the file to be opened in *STATISTICA*, you can specify the exact way in which the translation is to be performed. For instance, when you select an Excel file to be opened, you are prompted to select if *STATISTICA* should **Import all sheets to a Workbook** or **Import selected sheet to a Spreadsheet**. For more details on workbooks, see Chapter 3 – *STATISTICA Workbooks*.

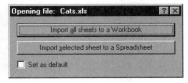

Next, you can specify the ranges in the foreign datafile to be imported and decide whether or not to import case and variable names.

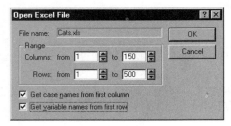

This ability to specify the exact way in which a file is to be imported is a distinct advantage of using the file import facilities instead of the Clipboard. In addition, you can access types of data that are not (or not easily) accessible to Clipboard operations.

STATISTICA Query. In addition to the file import facilities described above, *STATISTICA* provides access to virtually all databases (including many large system databases such as Oracle, Sybase, etc.) via *STATISTICA* Query, accessible from both the **File – Get External Data** and the **Data – Get External Data** menus.

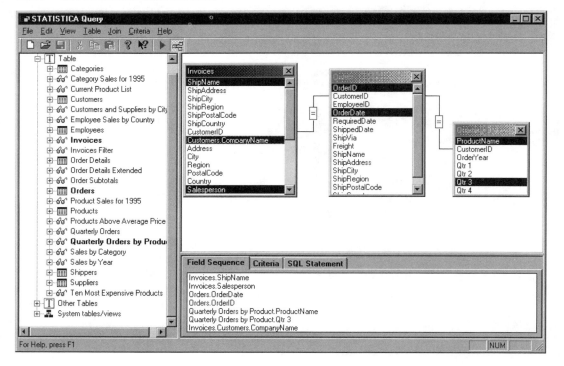

STATISTICA Query allows you to easily access data using OLE DB (Object Linking and Embedding Database) conventions. OLE DB is a database architecture [based on the Component Object Model (COM)] that provides universal data integration over an enterprise's network, from mainframe to desktop, regardless of the data type.

STATISTICA Query supports multiple database tables; specific records (rows of tables) can be selected by entering SQL statements. *STATISTICA* Query automatically builds the SQL statement for you as you select the components of the query via a simple graphical interface and/or intuitive menu options and dialogs. Hence, an extensive knowledge of SQL is not necessary in order for you to create advanced and powerful queries of data in a quick and straightforward manner. Multiple queries based on one or many different databases can also be created to return data to an individual spreadsheet; hence, you can maintain connections to multiple external databases simultaneously. For more information, see Chapter 10 - *STATISTICA Query*.

Import options. Additional import options can be selected on the *Import* tab of the *Options* dialog (assessed via the *Tools – Options* menu).

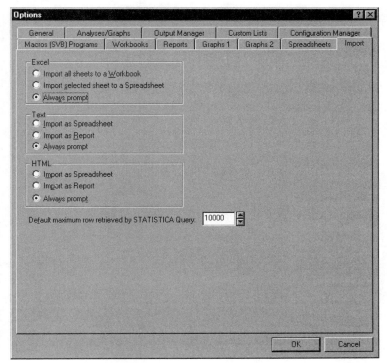

You can specify the manner in which Excel, Text, and HTML files will be imported as well as the maximum row of data that is retrieved by *STATISTICA* Query. Note that *STATISTICA* Query is capable of retrieving data larger than the value you specify here. Once you have reached the maximum row, you will be prompted to continue or stop retrieving data.

Accessing datafiles larger than the local storage. Note that enterprise versions of *STATISTICA* offer options to query and access large remote datafiles in-place (i.e., without having to import the data and create a local copy).

OLE Links

STATISTICA supports Object Linking and Embedding (OLE) conventions, allowing you to link values or text in *STATISTICA* documents (e.g., spreadsheets, reports) to documents in other (Windows) applications. Technically speaking, you can establish OLE links between a "source" (or server) file (e.g., a Microsoft Word document) and a *STATISTICA* document (the "client" file), so that when changes are made to the data in the source file, the data are automatically updated in the respective part of the *STATISTICA* document (client file). Additionally, *STATISTICA* can serve as a "source" (or server) file for other OLE compatible documents. In this way, you could link the values in a *STATISTICA* Spreadsheet to a table in a Microsoft Word document, so that the Word document updates when data in the spreadsheet are changed.

The procedure to do this is in fact much simpler than it might appear and can be easily employed without technical knowledge about the mechanics of OLE, especially when you use the **Paste Link** (instead of script-entry) method, via the **Paste Special** dialog. To create an OLE Link within a *STATISTICA* document, first copy to the Clipboard the desired text (or values) from another Windows application. Then, select **Paste Special** from the **Edit** menu to display the **Paste Special** dialog.

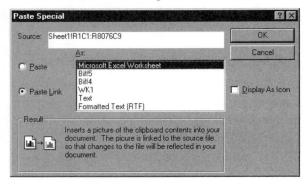

Select the **Paste Link** option button, specify the correct format in the **As** group, and click the **OK** button to establish the link using the source given in the **Source** field. See *Edit – Paste Special* for further details on the **Paste Special** dialog (page 253).

When a link is established, it can be managed using the **Links** dialog (select **Links** from the **Edit** menu). Use this dialog to update or break a link, open the source file or change to a different source file, and to specify whether updates will be automatic or manual (see *Edit – Links*, page 260).

DDE Links

STATISTICA also supports Dynamic Data Exchange (DDE) conventions (which is somewhat less flexible than OLE, but still frequently used, communication protocol). You can establish DDE links between a "source" (or server) file (e.g., a Microsoft Excel spreadsheet) and a *STATISTICA* datafile (the "client" file), so that when changes are made to the data in the source file, the data are automatically updated in the respective part of the *STATISTICA* Spreadsheet (client file). Thus, you can dynamically link a range of data in its spreadsheet to a subset of data in other (Windows) applications.

A common application for dynamically linking two files would be in industrial settings, where the *STATISTICA* datafile would be dynamically linked with a measurement device connected to the serial port (e.g., in order to automatically update specific measurements hourly). Like OLE, the procedure is in fact much simpler than it might appear and can be easily employed without technical knowledge about the mechanics of DDE.

To set up a DDE link, select **DDE Links** from the **Edit** menu to display the **Manage DDE Links** dialog. In this dialog, click the **New Link** button to display the **New DDE Link** dialog, which is used to enter the DDE link statement.

A DDE link statement can be constructed in many ways. The easiest way is to first copy to the Clipboard the desired text (or values) from another Windows application. Then, click the **From Clipboard** button on the **New DDE Link** dialog. The link statement is automatically entered in the **DDE Link** field. Alternatively, if you know the exact location of the application that will service the source file, the name of the source file, and the specific row and column ranges of the data to be linked into the client file, you can enter the link statement into the **DDE Link** field yourself. If you are unsure of the exact syntax for the link statement and you have the source file open, you can use the three columns at the bottom of the **New DDE Link** dialog to help you build the appropriate link statement.

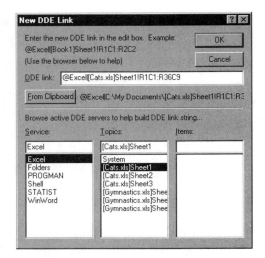

The **Service** column provides a list of all active server applications. The **Topics** column provides a list of all available objects that are associated with the **Service** you selected. The **Items** column provides a list of all available items that correspond to the **Service** and **Topic** that you selected. Note that in some cases, the **Items** column can be blank. If that is the case, to create a link to a range of cells, click in the **DDE Link** field (after the first two parts of the link statement) and enter an exclamation point followed by the desired range of cells (e.g., !R1C1:R36C9). In the example above, Excel is selected as the **Service**, Sheet 1 of Cats.xls as the **Topic**, and Row 1, Column 1 (R1C1) through Row 36, Column 9 (R36C9) as the **Items**.

Once the DDE link statement is constructed, click the **OK** button.

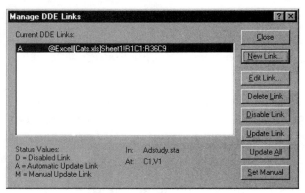

The **Manage DDE Links** dialog is displayed with the DDE link that was just created displayed in the **Current DDE Links** list. This dialog also is used to edit, update, delete, or disable any or all links. For a detailed explanation of the components of a link statement, see the *Electronic Manual*.

Exporting Data

STATISTICA can export datafiles to numerous Windows and non-Windows applications using the file export facilities. Select **Save As** from the **File** menu to display the **Save As** dialog.

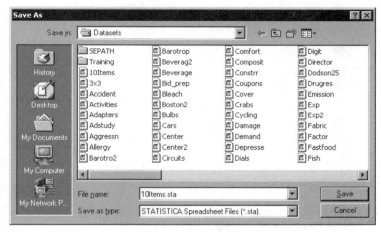

A wide variety of files is available in the **Save as type** box. *STATISTICA* Spreadsheet, *STATISTICA* Matrix, Excel, dBASE, SPSS Portable, Lotus Worksheets, Quattro Pro/DOS, Text [(formatted and free format text (ASCII)], HTML, and *STATISTICA* 5 files are available.

After you select the type of file to save the spreadsheet as, you can specify the exact way in which the translation is to be performed. For instance, if you save as an Excel Workbook, you are prompted to decide the range of the datafile and whether or not to export variable names, case names, and text labels.

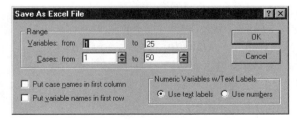

Spreadsheet Layouts

A spreadsheet layout is a convenient way to combine sets of formatting options into one collection. This feature is used to specify and apply different formats for different components of a spreadsheet. Once you have created a spreadsheet layout, it can be applied to any spreadsheet for a consistently formatted appearance.

	1 Var1	2 Var2	3 Var3	4 Var4	5 Var5
1	0.854909259	0.45663073	0.903256422	0.923182978	0.709221106
2	0.772316506	0.851738475	0.725642718	0.92539293	0.741021708
3	0.458629483	0.83194432	0.0258262829	0.94038538	0.935082895
4	0.139228945	0.513848215	0.327898993	0.721902312	0.565812094
5	0.284655199	0.241860546	0.122821887	0.958989577	0.265275867
6	0.256609706	0.842121053	0.893690437	0.425251931	0.919165991
7	0.128331067	0.990481634	0.846958357	0.956835711	0.63757508
8	0.838890554	0.390768131	0.593596071	0.17165365	0.224304197
9	0.715103764	0.193685812	0.238817803	0.164183386	0.149386736
10	0.86667626	0.874089484	0.592611911	0.35136734	0.169686784

Data: Spreadsheet4* (10v by 10c)

Note that the spreadsheet layout applied to the spreadsheet above changed the font of the data (italics, Comic Sans MS, and 12 point), the font of the case and variable headers (centered, bold, italics, 12 point, and background color), and the gridlines (thicker).

Select **Layout Manager** from the **Format - Spreadsheet** menu to display the **Spreadsheet Layouts** dialog, where there are several predefined spreadsheet layouts divided into two groups: **General** and **System** spreadsheet layouts. The **General** spreadsheet layout affects the data, header, case and variable headers, and info box of your spreadsheet. The **System** spreadsheet layouts include **Active Cell Header**, **Marked Cells**, **Print Filter**, **Selected Cells Header**, and **Selection Conditions**. These spreadsheet layouts only affect very specific characteristics associated with spreadsheet interactions.

System Default. Use this spreadsheet layout to edit the default format(s) of your spreadsheet. Whatever format(s) you specify for the **System Default** spreadsheet layout will apply to all new spreadsheets that are created.

Active Cell Header. Use this spreadsheet layout to edit the format(s) for the case name and variable name of the first selected cell.

Brushing Highlighted. Use this spreadsheet layout to edit the format(s) for points that have been highlighted via the graphic brushing mode of *STATISTICA*. For further details, see the *Electronic Manual*.

Brushing Labeled. Use this spreadsheet layout to edit the format(s) for points that have been labeled via the graphic brushing mode of *STATISTICA*. For further details, see the *Electronic Manual*.

Brushing Marked. Use this spreadsheet layout to edit the format(s) for points that have been marked via the graphic brushing mode of *STATISTICA*. For further details, see the *Electronic Manual*.

Brushing Off. Use this spreadsheet layout to edit the format(s) for points that have been turned off via the graphic brushing mode of *STATISTICA*. For further details, see the *Electronic Manual*.

Marked Cells. Use this spreadsheet layout to edit the format(s) for the marked cells in your spreadsheet. Note that you can control to which cells the *Marked Cells* spreadsheet layout will apply via the *Marking Cells* commands on the *Tools* menu. Marked cells are used in result spreadsheets. For example, you can specify the p-level to use when determining significant results. The *Marked Cells* spreadsheet layout will apply to all results less than this p-level in the result spreadsheets. Options for controlling which results are marked are typically included on the analysis dialog.

Print Filter. Use this spreadsheet layout to edit the format(s) for your spreadsheet when it is printed. If you choose to apply any format to your spreadsheet, it will always be reflected when you print your spreadsheet, with the exception of the formats applied to the *Case Header*, *Variable Header*, and *Gridlines* (unless these check boxes are cleared in the *Print Filter* spreadsheet layout). If you apply any spreadsheet layout to your spreadsheet, it will be always be reflected when you print your spreadsheet, with the exception of the *Active Cell Header* and *Selected Cells Header* spreadsheet layouts.

Selected Cells Header. Use this spreadsheet layout to edit the format(s) for the case names and variable names of the selected cells. Note that this layout applies the format(s) to all of the case names and variable names of the block of selected cells. The *Active Cell Header* (see above), on the other hand, only applies to the one case name and one variable name of your first selected cell.

Selection Conditions. Use this spreadsheet layout to edit the format(s) for the cells selected in your spreadsheet via case selection conditions. Note that you control to which cells

the **Selection Conditions** spreadsheet layout will apply when you enter selection conditions via the **Case Selection Conditions** dialog.

User-Defined Formats

A format is a group of formatting options that you can save with an assigned name for future use. Formats can be assigned to numbers as well as text. Select **Format Manager** from the **Format - Block** menu and click the **Edit** button to display the **Edit Format** dialog, which contains a wide range of font, border, alignment, and number formats that can be included in a format.

Formats incorporate maximum design power with a minimum amount of effort on your part. They can be used to apply a suite of formats without applying each formatting option separately. The use of formats also ensures that your formatting choices are consistent. Formats can be saved for future use with other spreadsheets and can be renamed at any time.

For example, if you would like to format all numbers in a specified (highlighted) group of data to be bold, red, 12 point Courier font, this would normally involve selecting the desired cells and then applying the individual formats (i.e., changing the font style to bold, changing the font color to red, changing the font size to 12 point, and changing the font to Courier). By using a format that encompasses all of these formatting preferences, you can simply apply the format to the selected cells in the spreadsheet, and each of the four distinct formatting changes are made in one step.

Input and Output Spreadsheets

STATISTICA offers the ability to open and use many spreadsheets at the same time, allowing you to work with several different input datafiles simultaneously. In addition to storing data, *STATISTICA* uses spreadsheets to display the numeric output from its analyses. Because *STATISTICA* makes no distinction in the features supported for an input spreadsheet (from which *STATISTICA* retrieves its data) and an output spreadsheet (where the results of an analysis are displayed), it is easy to use the results of one analysis as input data for further analyses.

Any spreadsheet opened from a disk file is automatically treated as an input spreadsheet, and any number of input spreadsheets can be open at a time. To avoid confusion, however, an output spreadsheet (containing the results of an analysis) is not automatically available as input

data for analysis. It must first be designated as an input spreadsheet before being used for further analyses. Additionally, input spreadsheets report the number of variables and cases for that spreadsheet in the title bar. For example, **Exp.sta (88v by 48c)** is an input spreadsheet; **Exp.sta** is not.

To designate an output spreadsheet as an input spreadsheet, select the spreadsheet (i.e., ensure the spreadsheet has the focus), and select **Input Spreadsheet** from the **Data** menu. Now you can begin an analysis, and *STATISTICA* will use the data from the specified input spreadsheet for the analysis. Note that if you switch back to another spreadsheet that has previously been designated as an input spreadsheet, it can still be used for analyses as well.

In a workbook, only one spreadsheet can be selected for analyses at a time, even if the workbook contains several input spreadsheets. This spreadsheet is called the Active Input spreadsheet, and its icon (in the workbook tree) is framed in red. By default, when an output spreadsheet is designated as an input spreadsheet, *STATISTICA* automatically selects it as the Active Input spreadsheet. To select another input spreadsheet for active input, select **Use as Active Input** on the **Workbook** menu or the workbook tree shortcut menu.

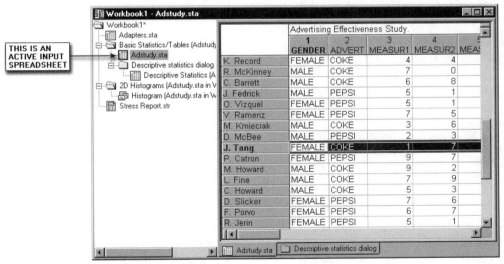

It is also possible to leave a stand-alone spreadsheet open but designate it as unavailable for analysis. To do this, select the spreadsheet, and clear the **Input Spreadsheet** command on the **Data** menu. Now *STATISTICA* automatically defaults to the most recently selected input spreadsheet for analysis, ignoring all spreadsheets that are not designated as input spreadsheets.

Global Spreadsheet Options

You can customize the global functionality of *STATISTICA* Spreadsheets using the options on the **Spreadsheet** tab of the **Options** dialog, accessed via the **Tools** menu. The selections made on this tab determine the defaults whenever you use a spreadsheet.

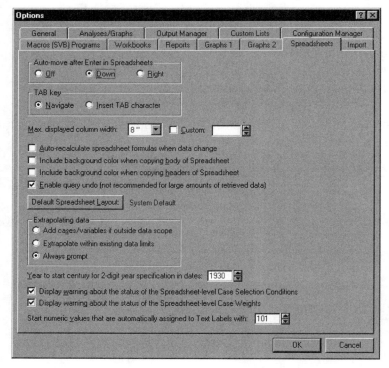

Auto-move after Enter in Spreadsheets. Use the options in the **Auto-move after Enter in Spreadsheets** group to control the movement of the cursor when you press the ENTER key after you enter or edit data in the spreadsheet. Select the **Off** option button to not move the cursor after you press ENTER. Select the **Down** option button to move the cursor down to the next cell in the column (or to the top of the next column if you are in the last cell in the column). Select the **Right** option button to move right to the next cell in the row (or to the beginning of the next row if you are in the last cell of the row).

TAB key. Use the options in the **TAB key** group to control the action that occurs when you press the TAB key after you enter or edit data in the spreadsheet. Select the **Navigate** option

button to move from cell to cell. Select the *Insert TAB character* option button to move multiple spaces within a cell.

Max. displayed column width. Use the *Max. displayed column width* drop-down box to select the maximum width of columns (variables).

Custom. Select the *Custom* check box and enter a value in the corresponding box (or use the microscrolls) to specify the maximum width of columns (variables).

Auto-recalculate spreadsheet formulas when data change. Select the *Auto-recalculate spreadsheet formulas when data change* check box to automatically recalculate all of the spreadsheet formulas when the data are changed in the spreadsheet. For further details, see *Formulas, Transformations, Recoding, Verifying, Cleaning Data* (page 361).

Include background color when copying body of Spreadsheet. Select the *Include background color when copying body of Spreadsheet* check box to include the background color of data cells that you are copying to the Clipboard. By default, the check box is cleared.

Include background color when copying headers of Spreadsheet. Select the *Include background color when copying headers of Spreadsheet* check box to include the background color of case/variable headers that you are copying to the Clipboard. By default, the check box is cleared.

Enable query undo (not recommended for large amounts of retrieved data). Select the *Enable query undo* check box to make *Undo refreshing query* from the *Edit* menu available after you retrieve external data via *STATISTICA* Query. For further details, see Chapter 10 - *STATISTICA Query*.

Default Spreadsheet Layout. Click the *Default Spreadsheet Layout* button to display the *Spreadsheet Layouts* dialog where you can select the spreadsheet layout to use whenever a new datafile is created.

Extrapolating data. Use the options in the *Extrapolating data* group to specify the action to occur when you extrapolate outside of the currently defined spreadsheet. Select *Add cases/variables if outside data scope*, *Extrapolate within existing data limits*, or *Always prompt*. If the *Always prompt* option button is selected, the *Extrapolate Values* dialog is displayed when you extrapolate outside of the currently defined spreadsheet.

Year to start century for 2-digit year specification in dates. Enter the appropriate year in the *Year to start century for 2-digit year specification in dates* box. The

default value is *1930*; hence, all values between (and including) 30 and 99 will be in the 1900s. All values between (and including) 00 and 29 will be in the 2000s.

Display warning about the status of the Spreadsheet-level Case Selection Conditions. Select the *Display warning about the status of the Spreadsheet-level Case Selection Conditions* check box to display a warning message when you change the status of the **Spreadsheet Case Selection Conditions** dialog (see 280). Briefly, any macro that is run on the spreadsheet will use the spreadsheet case selection conditions; however, if you record a macro (via the ▨ Options ▾ button available on analysis/graph specification dialogs) while analyzing this spreadsheet, the spreadsheet case selection conditions will not be recorded as part of the macro. To set selection conditions which will be recorded into a macro, you must specify them using the *Analysis/Graph Case Selection Conditions* dialog (accessible via the ▨ button available on analysis/graph specification dialogs).

Display warning about the status of the Spreadsheet-level Case Weights. Select the *Display warning about the status of the Spreadsheet-level Case Weights* check box to display a warning message when you change the status of the **Spreadsheet Case Weights** dialog (see 283). Briefly, any macro that is run on the spreadsheet will use the spreadsheet case weights; however, if you record a macro (via the ▨ Options ▾ button available on analysis/graph specification dialogs) while analyzing this spreadsheet, the spreadsheet case weights will not be recorded as part of the macro. To set weights which will be recorded into a macro, you must specify them using the *Analysis/Graph Case Weights* dialog (accessible via the ▨ w button available on analysis/graph specification dialogs).

Start numeric values that are automatically assigned to Text Labels with. Enter the value you would like assigned to the first text label in the *Start numeric values that are automatically assigned to Text Labels with* box. By default, this value is *100*. The range of valid numeric values assigned to text labels is -32,768 to 32,767, with the exception of -9999 (the default missing data code); if the maximum numeric value is reached and negative values have not yet been used, negative numeric values, starting with -32,768, are assigned.

SPREADSHEET WINDOW

File Menu

The following commands are available from the spreadsheet *File* menu. Many of these commands are also available from shortcut menus (accessed by right-clicking on an item in the spreadsheet) and from the toolbar buttons.

New

Select *New* from the *File* menu (or click the ⬚ toolbar button or press CTRL+N) to display the *Create New Document* dialog, then select the *Spreadsheet* tab. Enter the number of variables and cases to include in the spreadsheet in the *Number of variables* box and *Number of cases* box, respectively.

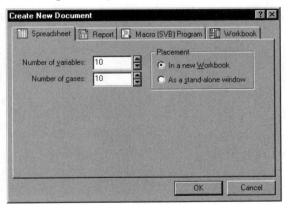

When creating a spreadsheet, you have two placement options. Select the *In a new Workbook* option button to place the newly created spreadsheet in a new *STATISTICA* Workbook, or select the *As a stand-alone window* option button to place the spreadsheet in a stand-alone window.

Open

Select **Open** from the **File** menu (or click the 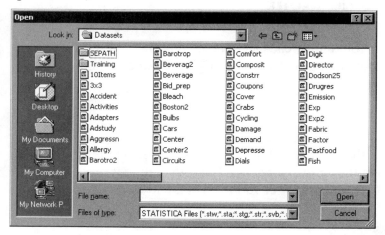 toolbar button or press CTRL+O) to display the **Open** dialog. From this dialog, you can open any document of a compatible type (e.g., a spreadsheet).

Using the **Look in** drop-down box, select the drive and directory location of the desired file. Select the file in the large box and click **Open**, or double-click on the file name. You can also enter the complete path of the document in the **File name** box.

Close

Select **Close** from the **File** menu to close the current document. If you have made changes to the document since you last saved it, *STATISTICA* prompts you to save changes to the document.

Save

Select **Save** from the **File** menu (or click the ▦ toolbar button or press CTRL+S) to save the document with the name and in the drive location you specified when you last saved it. If you have not yet saved the document, the **Save As** dialog is displayed.

Save As

Select **Save As** from the **File** menu (or press F12) to display the **Save As** dialog. From this dialog, you can save the active document with the name of your choice by typing the name in the **File name** box. Use the **Save in** box to select the appropriate drive and folder in which to save the document, or enter the complete path in the **File name** box.

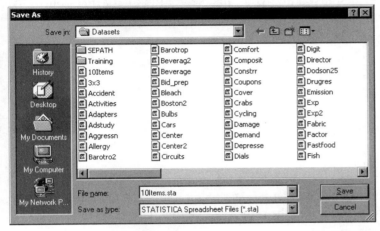

By default, *STATISTICA* recommends the file type that best fits the type of document you are saving. For example, if you are saving a spreadsheet, the **Save as type** drop-down box contains the format **STATISTICA Spreadsheet Files (*.sta)**.

Get External Data

The options on this menu are for use with *STATISTICA* Query. See Chapter 10 - *STATISTICA Query* for further details about creating, editing, and using queries. Note that **Edit Query**, **Properties**, **Refresh Data**, **Delete Query**, and **Cancel Query** are only available when you have run (or are currently running) a query.

Open Query from File. Select **Open Query from File** from the **File – Get External Data** menu to display the standard open file dialog in which you can select an SQL file to run. Note that **.sqy* files are (created and) saved in *STATISTICA* Query (via **File - Save As**).

Create Query. Select **Create Query** from the **File – Get External Data** menu to open *STATISTICA* Query and display the **Database Connection** dialog. Once you have chosen or defined a database connection, you can use *STATISTICA* Query to write your query. For details

---

on defining a database connection, see the *FAQ: Working with STATISTICA Query* in Chapter 10 – *STATISTICA Query*.

Edit Query. Select *Edit Query* from the *File – Get External Data* menu to edit the query in the active spreadsheet. Note that if you have multiple queries on the active spreadsheet, *STATISTICA* opens the query that is associated with the active cell. If you select a cell that is not connected to any query or if you select multiple cells that involve more than one query, then the *Select Database Query* dialog is displayed and you must select the query that you want to edit before *STATISTICA* Query opens.

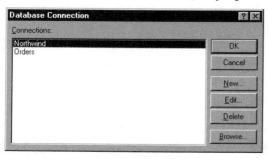

Properties. Select *Properties* from the *File – Get External Data* menu to display the *External Data Range Properties* dialog and specify options regarding the return of external data to spreadsheets.

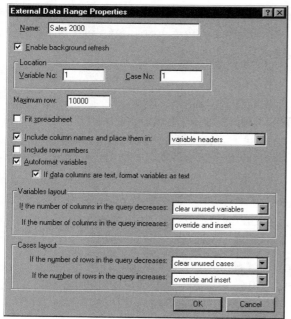

Note that if you have multiple queries on the active spreadsheet, then *STATISTICA* displays properties for the query associated with the active cell on the spreadsheet. If you select a cell that is not connected to any query or if you select multiple cells that involve more than one query, then the **Select Database Query** dialog is displayed.

On this dialog, select the query to use and click the **OK** button to display the **External Data Range Properties** dialog.

Refresh Data. Select **Refresh Data** from the **File – Get External Data** menu (or press F5) to run a query and refresh the data (retrieve the latest data from the original database). Note that if you have multiple queries on the active spreadsheet, then *STATISTICA* refreshes data for the query associated with the active cell of the spreadsheet. If you select a cell that is not connected to any query or if you select multiple cells that involve more than one query, then the **Select Database Query** dialog is displayed so you can choose the specific query to refresh.

Delete Query. Select **Delete Query** from the **File – Get External Data** menu to delete the query from the active spreadsheet. *STATISTICA* always prompts you to verify that the query should be deleted. Note that if you have multiple queries on the active spreadsheet, *STATISTICA* deletes the query associated with the active cell of the spreadsheet. If you select a cell that is not connected to any query or if you select multiple cells that involve more than one query, then the **Select Database Query** dialog is displayed so you can choose which query to delete.

Cancel Query. Select **Cancel Query** from the **File – Get External Data** menu (or press SHIFT+F5) to stop a currently running query at any time.

Workbook

Add to Analysis Workbook. Select **Add to Analysis Workbook** from the **File - Workbook** menu to add the current spreadsheet or graph to the analysis workbook. If the current spreadsheet or graph is not associated with an analysis workbook, this command is disabled. For more information on analysis workbooks, see the *Analysis Workbook Overview* in Chapter 3 – *STATISTICA Workbooks*.

Active workbooks. To add the current document to an active workbook (i.e., a workbook that is open or has recently been opened) select the workbook from the list of current workbooks shown on the *File - Workbook* menu. Note that if there are not any current workbooks at this time, the *File - Workbook* menu contains two options: *Add to Analysis Workbook* (which may be disabled) and *New Workbook*.

New Workbook. Select *New Workbook* from the *File - Workbook* menu to create a new workbook and add the current document to it. Note that in addition to adding the current document to a new workbook, you can select any currently open workbook from the list provided in the *File - Workbook* menu.

Report

Add to Analysis Report. Select *Add to Analysis Report* from the *File - Report* to add the current spreadsheet or graph to the analysis report. If the current spreadsheet or graph is not associated with an analysis report, this command is disabled. For more information, see the *Analysis Report Overview* in Chapter 5 – *STATISTICA Reports*.

Active reports. To add the current document to an active report (i.e., a report that is open or has recently been opened) select the report from the list of current reports shown on the *File - Report* menu. Note that if there are not any current reports at this time, the *File - Report* menu contains two options: *Add to Analysis Report* (which may be disabled) and *New Report*.

New Report. Select *New Report* from the *File - Report* menu to create a new report and add the current document to it.

Output Manager

Select *Output Manager* from the *File* menu to display the *Output Manager* tab of the *Options* dialog. Use the options on this tab to specify where to direct output files. The available options for reports include sending all output to multiple reports (one for each graph/analysis), a single report (i.e., an analysis report), or an existing report. You can also specify the amount of supplementary information to include when sending spreadsheets to reports.

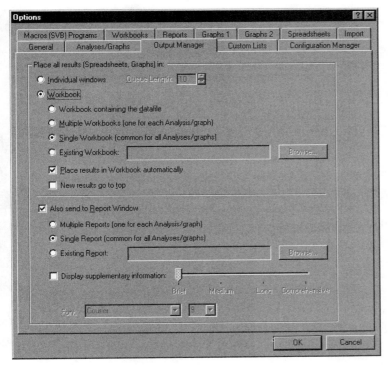

For more details, see *Three Channels for Output from Analyses* in Chapter 2 – *STATISTICA Output Management*. See also, *What Supplementary Information Is Available with Analyses and How Can I Add It to My Reports?* in Chapter 5 – *STATISTICA Reports*.

Print Setup

Select *Print Setup* from the *File* menu to display the *Print Setup* dialog. Use this dialog to specify printing options.

In the *Printer* group, you can choose a printer (using the *Name* drop-down box to view available printers). In the *Paper* group, you can choose the paper size (using the *Size* drop-down box to view a list of paper sizes), and the paper source (using the *Source* drop-down box to choose the printer's paper source). Additionally, you can choose a vertical or horizontal layout for the printed document by selecting the *Portrait* option button (for vertical) or the *Landscape* option button (for horizontal) in the *Orientation* group. Click the *Properties* button to view options specific to the selected printer.

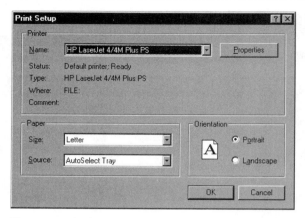

For more information on specific printer options, consult your printer's manual.

Print Preview

Select **Print Preview** from the **File** menu (or click the 🔍 toolbar button) to display the active document in a preview window. Before you print a document, it is useful to see how it will look when printed. The document's appearance in the print preview window depends on the resolution of your printer and the available fonts and colors. The status bar at the bottom of the window shows the current page number.

Use the options in the **Margins** group to control margin and header/footer placement. To modify **Left**, **Right**, **Top**, and **Bottom** margins, enter new measurements in the appropriate field, and click the **Adjust** button. You can control whether the margins are measured in inches or centimeters. To measure in centimeters, select the **Use metric measurements** check box on the **General** tab of the **Options** dialog (accessed from the **Tools – Options** menu). To measure in inches, clear the **Use metric measurements** check box (the default setting). Alternatively, to adjust the margins, drag the margin lines with the mouse (the mouse pointer will change into a double-headed arrow, ↕ or ↔, see the illustration below).

Values in the **Header** and **Footer** boxes determine how far the header and footer are placed from the top and bottom margins, respectively. Click the **Header/Footer** button to display the **Modify Header/Footer** dialog to access the header and footer information for the spreadsheet. (See *Can I Create a Custom Header or Footer for My Report* in Chapter 5 – *STATISTICA Reports* for an example of using this dialog.)

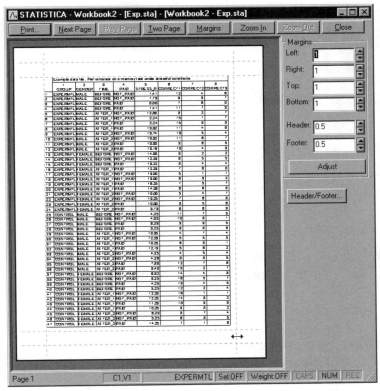

For more information on using the print preview window, see *Using the Print Preview Window* in Chapter 2 – *STATISTICA Output Management*.

Print

Select **Print** from the **File** menu (or click the 🖨 toolbar button or press CTRL+P) to display the **Print** dialog. Use this dialog to execute a print command, as well as to specify printing options (see *Print Setup*, above).

Verify that the printer in the **Name** box is the correct printer; if necessary, you can use the **Name** drop-down list box to choose another printer. Select the **Print to File** check box to print the document to a *.prn* file.

Use the options in the **Print Range** group to specify what pages to print. To print the entire document, select the **All** option button. To print a range of pages, select the **Pages** option button and specify the range using the **From** and **To** boxes. To print the selected portion of the document, select the **Selection** option button.

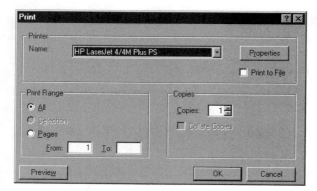

Use the options in the **Copies** group to specify the number of copies to print. Type the number of copies you require in the **Copies** box. Select the **Collate Copies** check box to ensure that your printer collates the copies of the printed document.

To review the report in a print preview window (see above), click the **Preview** button. After you have selected the desired options, click the **OK** button to print the document.

Properties

Select **Properties** from the **File** menu to display the **Document Properties** dialog. This dialog contains three tabs: **Summary, Statistics**, and **Password**. Select the **Summary** tab to create or modify a document summary (including information about the author's name, comments about the document, etc.). Select the **Statistics** tab to view statistics about the document (including date created, revision number, etc.). Finally, select the **Password** tab to password protect the current document.

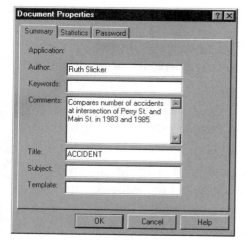

The *STATISTICA* password protection system is based on a secure advanced data encryption technology that is applied to both the text and numeric contents of *STATISTICA* datafiles. Therefore, password protected files not only cannot be opened without the knowledge of the respective password, but also the contents of their data are "scrambled" and cannot be reviewed using tools designed to inspect contents of disk files directly (such as ASCII or hexadecimal display editors, etc.). However, unlike the data, files and objects that can be attached to the *STATISTICA* Spreadsheet via either linking or embedding methods, will not be internally encrypted and thus, although they cannot be opened or accessed directly, their security potentially could be compromised by programmers familiar with OLE technologies. For that reason, the sensitive data and information should be placed in the spreadsheet directly (and not inside the objects or files that are embedded or linked to it).

Exit

Select **Exit** from the **File** menu to close *STATISTICA*. If you have made changes to your document since you last saved it, you will be prompted to save your changes.

Edit Menu

The following commands are available from the spreadsheet **Edit** menu. Many of these commands are also available from shortcut menus (accessed by right-clicking on an item in the spreadsheet) and from the toolbar buttons.

Undo

Select **Undo** from the **Edit** menu (or click the ⤺ toolbar button or press CTRL+Z) to undo your last action in the document (such as editing, moving, or copying blocks, font formatting, etc.). *STATISTICA* supports multi-level undo (with 32 buffers); therefore, you can undo multiple actions by selecting this option consecutively (up to 32 times). Note that the name of the command changes depending on which action you have performed. For example, if the last action you completed was to format a block of cells, this command will say **Undo Format**. If the last action was to fill copy down, the command will say **Undo Fill Block Down**.

Redo

Select **Redo** from the **Edit** menu (or click the  toolbar button or press CTRL+Y) to counteract an **Undo** action. For example, if your last undo action command was to undo bold formatting to the selected text, then select **Redo** to reapply bold formatting to the text. Note that the name of the option changes depending on which command you have performed.

Repeat Last Command

Select **Repeat Last Command** from the **Edit** menu to repeat the last command you performed in the spreadsheet. Note that the name of the option changes depending on which command you have performed.

Cut

Select **Cut** from the **Edit** menu (or click the toolbar button or press CTRL+X) to cut the selected text or object(s) to the Clipboard. *STATISTICA* will create in the Clipboard multiple formats of the selected object; the specific format to be used for pasting can later be selected with the **Paste Special** options in the client application.

Copy

Select **Copy** from the **Edit** menu (or click the toolbar button or press CTRL+C) to copy the selected text or object(s) to the Clipboard. **Copy** replaces the previous contents of the Clipboard with the contents of the selected cells. *STATISTICA* will create in the Clipboard multiple formats of the selected object; the specific format to be used for pasting can later be selected with the Paste Special options in the client application.

Copy with Headers

Select **Copy with Headers** from the **Edit** menu to copy the contents of the currently selected (highlighted) cells in the spreadsheet to the Clipboard. Selecting this option copies not only the contents in the selected block of cells in a spreadsheet, but also the case and variable names associated with the selected cells. (Use **Copy** to copy only the selected spreadsheet block without the row and column names.) **Copy with Headers** replaces the previous contents of the Clipboard with the contents of the selected cells. See *What Is the Difference Between Copy and Copy with Headers*, page 355, for an illustration.

Paste

Select **Paste** from the **Edit** menu (or click the 📋 toolbar button or press CTRL+V) to paste the contents of the Clipboard into the selected location in the document.

Paste Special

Select **Paste Special** from the **Edit** menu to display the **Paste Special** dialog, which contains a list of available ways to paste the contents of the Clipboard into the current document.

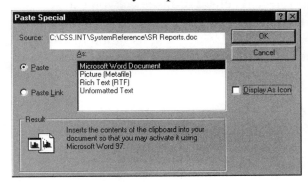

Use the options on this dialog to choose the desired format (i.e., **Rich Text**, **Unformatted Text**, or **Picture**) in which the Clipboard object will be pasted in the current document. The options available depend on the contents of the Clipboard and the options selected on this dialog. Once you have specified how to paste the Clipboard object, click the **OK** button to paste it. Click the **Cancel** button to exit this dialog without performing any action.

The **Source** field reports the source from which the current Clipboard object was taken. This information helps you determine what other options to select. For example, if the source is an Excel spreadsheet, you may want to use the **Paste Link** option button so that the selection is linked to the source file. Then, the selection is updated when the source file changes.

Select the **Paste** option button to paste the contents of the Clipboard into the document without maintaining any links to the source file. Select the **Paste Link** option button to paste the contents of the Clipboard into the document and maintain a link to the source file. Then, the selection is updated when the source file changes.

The **As** box contains a variety of formats in which the Clipboard contents can be pasted. Select the one that best suits your needs. To see what will result from selecting a specific format,

select the format and read the description in the **Result** field. Finally, you can select the **Display as icon** check box to display the Clipboard contents as an icon in the *STATISTICA* Spreadsheet.

Fill/Standardize Block

Fill Random Values. Select **Fill Random Values** from the **Edit - Fill/Standardize Block** menu to populate the selected cell(s) with random values (in the range of 0 to 1) following the Uniform distribution.

Fill/Copy Down. Select **Fill/Copy Down** from the **Edit – Fill/Standardize Block** menu to copy an identical value into a vertical block (column) of selected cells. For instance, you can select a block of cells in a spreadsheet (some cells can be blank), and use this option to copy the number from the top-most selected cell into all selected cells. Using the illustration below, the original block of selected cells includes the numbers 4, 5, 9 and 2, as well as a blank cell.

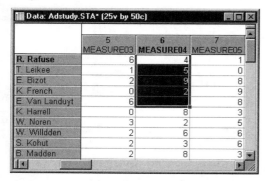

The illustration below displays the results after selecting **Fill/Copy Down** from the **Edit - Fill/Standardize Block** menu.

All selected cells have been populated with the number 4, which was the value in the top-most cell prior to completing this action.

Fill/Copy Right. Select *Fill/Copy Right* from the *Edit – Fill/Standardize Block* menu to copy an identical value into a horizontal block (row) of selected cells. For instance, you can select a block of cells in a spreadsheet (some cells can be blank), and use this command to copy the number from the left-most selected cell into all selected cells. Using the illustration below, the original block of selected cells includes the numbers 4, 9, 6 and 2, as well as a blank cell.

	3 MEASURE01	4 MEASURE02	5 MEASURE03	6 MEASURE04	7 MEASURE05
M. Bowling	4	6	6	5	6
J. Willcoxson	7	3	3	7	0
J. Landrum	6	2	3	1	8
M. Taylor	7	2	4	8	1
N.S. Madden	4	9	6	2	
K. Ridgway	3	2	5	4	4
L. Cunha	2	9	9	3	1
F. Wind	1	0	7	5	2
K. Judkasikam	0	6	2	3	2
B. Brinker	6	8	1	9	5

Data: Adstudy.STA* (25v by 50c)

The illustration below displays the results after selecting *Fill/Copy Right* from the *Edit - Fill/Standardize Block* menu.

	3 MEASURE01	4 MEASURE02	5 MEASURE03	6 MEASURE04	7 MEASURE05
M. Bowling	4	6	6	5	6
J. Willcoxson	7	3	3	7	0
J. Landrum	6	2	3	1	8
M. Taylor	7	2	4	8	1
N.S. Madden	4	4	4	4	4
K. Ridgway	3	2	5	4	4
L. Cunha	2	9	9	3	1
F. Wind	1	0	7	5	2
K. Judkasikam	0	6	2	3	2
B. Brinker	6	8	1	9	5

Data: Adstudy.STA* (25v by 50c)

All selected cells have been populated with the number *4*, which was the value in the left-most cell prior to completing this action.

Standardize Columns. Select *Standardize Columns* from the *Edit – Fill/Standardize Block* menu to standardize the values in each column of the selected block of cells. The standardized values are computed as follows:

`Std. Value = (raw value - mean of highlighted column)/std. deviation`

You can standardize the values in selected rows via *Standardize Rows* (see below).

Standardize Rows. Select *Standardize Rows* from the *Edit – Fill/Standardize Block* menu to standardize the values in each row of the selected block of cells. The standardized values are computed as follows:

```
Std. Value = (raw value - mean of highlighted row)/std. deviation
```

Clear

All. Select *All* from the *Edit - Clear* menu to delete both the contents of the selected cell(s) and the format of the selected cell(s). In the spreadsheet, the deleted values are replaced with missing data until new values are entered or pasted in. Unlike the *Cut* option (see page 252), *Clear* does not copy the text or contents of the cells to the Clipboard; therefore, it does not affect the contents of the Clipboard.

Contents. Select *Contents* from the *Edit - Clear* menu to delete only the contents of the selected cell(s) without deleting the format of the cell itself. In the spreadsheet, the deleted values are replaced with missing data until new values are entered or pasted in. Unlike the *Cut* command (see page 252), *Clear* does not copy the text or contents of the cells to the Clipboard; therefore, it does not affect the contents of the Clipboard. Keyboard Shortcut: DELETE

Formats. Select *Formats* from the *Edit - Clear* menu to remove all applied formatting options from the selected cell(s). For instance, you could use this command to remove font and color formatting from selected cells, and return the cells to the default format for their display type.

Delete

Variables. Select *Variables* from the *Edit - Delete* menu to display the *Delete Variables* dialog, in which you choose which variable(s) to delete.

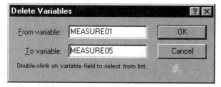

Cases. Select *Cases* from the *Edit - Delete* menu to display the *Delete Cases* dialog, in which you choose which case(s) to delete.

All Case Names. Select *All Case Names* from the *Edit - Delete* menu to replace all of the existing case names of the spreadsheet with consecutive numeric values beginning with one (i.e. 1, 2, 3, ...). This option is not available if the case names are currently labeled in this manner.

Move

Variables. Select *Variables* from the *Edit - Move* menu to display the *Move Variables* dialog, which is used to remove both the contents of the column(s) and the column(s) itself (regardless of whether or not all cases or only a subset of cases are currently selected for the respective variables) and insert the column(s) in the designated position in the spreadsheet. Therefore, the size of the file remains the same.

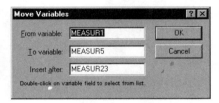

You can move one or more variables in the spreadsheet by designating the range (inclusive) of variables to be moved and the location (variable to insert after) in the *Move Variables* dialog.

Cases. Select *Cases* from the *Edit - Move* menu to display the *Move Cases* dialog, which is used to remove both the contents of the row(s) and the row(s) itself (regardless of whether all variables or only a subset of variables are currently selected for the respective cases) and insert the cases in the designated position in the spreadsheet. The size of the file remains the same.

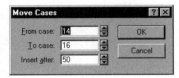

You can move one or more cases (rows) in the spreadsheet by designating the range (inclusive) of cases to be moved and the location (case to insert after) in the *Move Cases* dialog.

Select All

Select *Select All* from the *Edit* menu (or press CTRL+A) to select the entire spreadsheet. You can also select the entire spreadsheet by clicking in the lower half of the Info Box (if you click in the upper-left corner of the Info Box, the entire Info Box is selected). See *Spreadsheets - Multimedia Tables* (page 223) for further details.

Select Headers Only

Select Case Names Only. Select *Select Case Names Only* from the *Edit - Select Headers Only* menu to select only the case names associated with the currently selected cells. Note that once this option has been used, the appropriate case names will be selected, but the cells themselves will no longer be selected. To select all case names, first select *Select All* from the *Edit* menu (see above) followed by this command.

Select Variable Names Only. Select *Select Variable Names Only* from the *Edit - Select Headers Only* menu to select only the variable names associated with the currently selected cells. Note that once this option has been used, the appropriate variable names will be selected, but the cells themselves will no longer be selected. To select all variable names, first select *Select All* from the *Edit* menu (see above) followed by this command.

Find

Select *Find* from the *Edit* menu (or click the toolbar button or press CTRL+F) to display the *Find* dialog. Use this dialog to search the active document for words or phrases.

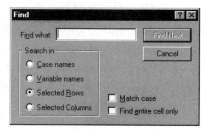

Type the word or phrase that you are looking for in the *Find what* box. To start the search, click the *Find Next* button. *STATISTICA* selects the first word(s) that match your search criteria. Continue clicking the *Find Next* button until *STATISTICA* completes the search throughout the entire document. You can click the *Cancel* button at any time to stop the search.

When searching a spreadsheet, use the *Search in* group to specify specific areas of the spreadsheet to search in and select the *Find entire cell only* check box to specify whether or not to limit the search to the entire contents of the cell. You can also conduct a case-sensitive search for a word or phrase using the *Match case* check box. For more details, see the spreadsheet *Find* dialog in the *Electronic Manual*.

Replace

Select **Replace** from the **Edit** menu (or click the  toolbar button or press CTRL+H) to display the **Replace** dialog.

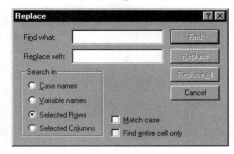

Use the **Replace** dialog to search the active document for words or phrases and replace them with different words or phrases. Specify a word or phrase for which to search (in the **Find what** box) as well as a word or phrase with which to replace it (in the **Replace with** box). To start the search, click the **Find** button. *STATISTICA* highlights the first word(s) that match your search criteria. You can replace the text on an individual basis (click the **Replace** button) or globally (click the **Replace All** button). Continue clicking **Find** until *STATISTICA* completes the search throughout the entire document. You can click the **Cancel** button at any time to stop the search.

When searching a spreadsheet, use the **Search in** group to specify specific areas of the spreadsheet to search in and select the **Find entire cell only** check box to specify whether or not to limit the search to the entire contents of the cell. You can also conduct a case-sensitive search using the **Match case** check box. For more details, see the spreadsheet **Replace** dialog in the *Electronic Manual*.

Repeat Find/Replace

Select **Repeat Find/Replace** from the **Edit** menu (or press F3) to repeat the last **Find/Replace** operation you conducted.

Go To

Select **Go To** from the **Edit** menu (or press CTRL+G) to display the **Go To** dialog, which is used to quickly move to a specific location within your spreadsheet.

Type the variable number or name in the **Variable** box. Note that you can double-click (or press the F2 key) in this box to select a variable from a list of all the available variables in the spreadsheet. Enter the desired case number or name the **Case** box. Finally, click the **Go** button to move to the selected location.

DDE Links

Select **DDE Links** from the **Edit** menu to display the **Manage DDE Links** dialog.

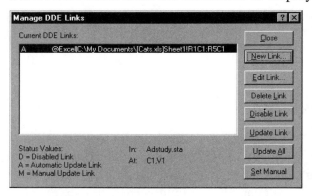

Click the **New Link** button to create a new DDE (Dynamic Data Interchange) link. You can also edit, delete, disable, or update the selected link in the **Current DDE Links** box by clicking the respective button. To update all of the DDE Links, click the **Update All** button. Click the **Set Manual/Set Auto** button to change the selected DDE link from an automatically updated link to a manually updated link or vice versa. Finally, click the **Close** button to close this dialog.

See *Import, Export, DDE, OLE* (page 389) and *How Do I Set Up DDE Links between STATISTICA and Other Windows Applications*? (page 393) for an explanation on DDE Links and how they work with spreadsheets.

Links

Select **Links** from the **Edit** menu to display the **Links** dialog, which is used to manage links between the spreadsheet and other applications.

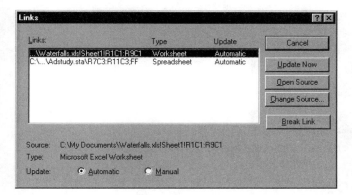

Use this dialog to update the link, open or change the associated application to which the spreadsheet is linked, or break the link. You can also specify the link as an **Automatic** or **Manual** updating link by selecting the appropriate option button. Click the **Cancel** button to close the dialog.

Object

Select **Object** from the **Edit** menu to enable editing options for the selected object in the spreadsheet. See *Insert - Object* (page 271) for further details.

Screen Catcher

Use the **Screen Catcher** commands to select any rectangular area of the screen, the entire screen, or a specific window and copy it as a bitmap to the Clipboard. The copied area can contain graphs, spreadsheets, or even windows from other applications (if they are currently displayed on the screen). The copied bitmap can then be pasted into *STATISTICA* documents or any other application that supports bitmaps.

Capture Rectangle. Select **Capture Rectangle** from the **Edit - Screen Catcher** menu (or press ALT+F3) to launch the **Screen Catcher** and capture a rectangular area of the screen.

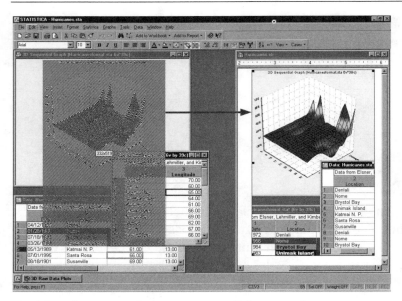

Position the cursor in one corner of the area that you want to copy. Now, drag the cursor over the area that you want to copy (you can drag across windows). When you release the mouse button, the selected area is copied to the Clipboard.

Capture Window. Select *Capture Window* from the *Edit - Screen Catcher* menu to launch the *Screen Catcher* and capture the desired window on your desktop.

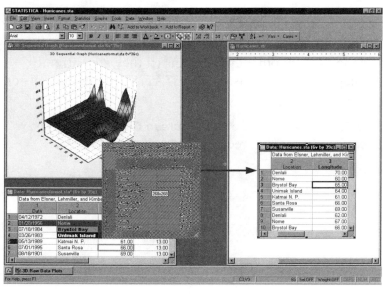

After selecting this option, move the cursor over the screen until the window you want to capture is selected. Note that *STATISTICA* gives the dimensions of each window as you pass over it. Click the window that you want to capture, and it is copied to the Clipboard.

View Menu

The following commands are available from the spreadsheet **View** menu. Many of these commands are also available from shortcut menus (accessed by right-clicking on an item in the spreadsheet) and from the toolbar buttons.

Display Text Labels

Select **Display Text Labels** from the **View** menu (or click the 🔲 toolbar button) to toggle between displaying text labels and numeric values for data in the spreadsheet. See *Using the Text Labels Editor* in the *Electronic Manual*.

Variable Headers

Display Numbers. Select **Display Numbers** from the **View - Variable Headers** menu to show the numbers in the variable headers that correspond to the variables. By default this command is selected.

Display Long Names. Select **Display Long Names** from the **View - Variable Headers** menu to show the Long Name component of variables in the spreadsheet. By default, this command is not selected. If this command is selected, the variable headers will include the Long Names that correspond to the variables. See *Tools – Variable Specs* (page 293) for further details.

Display Types. Select **Display Types** from the **View - Variable Headers** menu to show the data type in the variable headers. By default, this command is not selected and hence the display type is not shown in the variable headers. See also *Variable Types* (page 222).

Display Case Names

Select **Display Case Names** from the **View** menu (or click the 🔳 toolbar button) to toggle between displaying names or numbers for cases in the spreadsheet.

Ignore Empty Case Names

Select *Ignore Empty Case Names* from the *View* menu to toggle the display of case numbers for empty case names. If this option is checked, then *STATISTICA* displays the appropriate case number for any case that does not have a case name. If this option is cleared, then a blank appears in the case name cell for all rows that do not have a specified case name.

Display Header

Select *Display Header* from the *View* menu to toggle between displaying and not displaying the Header (located immediately above the variable headers at the top of the window) in the spreadsheet.

Gridlines

Select *Gridlines* from the *View* menu to display the *Gridlines* dialog, which is used to choose the color and format for the spreadsheet gridlines on the screen only (see the note below).

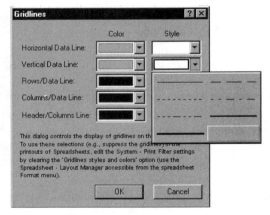

Use the *Horizontal Data Line* and *Vertical Data Line* boxes to select the color and style of the gridlines between the variables and cases. Use the *Rows/Data Line* box to select the color of the gridline between the Case Headers and the first variable in the spreadsheet. Use the *Columns/Data Line* box to select the color of the gridline between the Variable Headers and the first variable. Use the *Header/Columns Line* box to select the color of the gridline between the Variable Headers and the Header. See *Spreadsheets - Multimedia Tables*, page 223, for further details.

Note. This dialog controls the display of gridlines on the screen and does not automatically reflect the display of gridlines when the spreadsheet is printed. To use these selections in the printout of spreadsheets, edit the *Print Filter* spreadsheet layout by selecting *Layout Manager* from the *Format – Spreadsheet* menu to display the *Spreadsheet Layouts* dialog. Then select the *Print Filter* spreadsheet layout on the *System* tab and click the *Edit* button. Finally, clear the *Gridlines styles and colors* check box on the *Edit Spreadsheet Layout* dialog. See *Spreadsheet Layouts*, page 234, for further details.

Max. Displayed Column Width

Use the *Max. Displayed Column Width* options to select the maximum width of columns (variables) in the spreadsheet. You can select *8*, *6*, *4*, *3*, *2*, *1.5*, *1*, *.75*, *.5*, or *.25* inches. If you prefer, select *Customize* to specify a different width.

Customize. Select *Customize* from the *View - Max. Displayed Column Width* menu to display the *Custom Max Column Width* dialog.

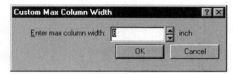

Enter a value in the *Enter max column width* box to specify the maximum width of columns (variables) in the spreadsheet. Click the *OK* button after you have specified the width or click the *Cancel* button to ignore your changes and close the dialog.

Header/Footer

Select *Header/Footer* from the *View* menu to display the *Modify Header/Footer* dialog.

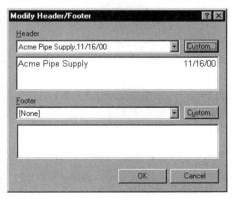

Use this dialog to view and modify the existing header and footer information. You can select a default header and footer (using the **Header** or **Footer** drop-down boxes) or you can create a custom header or footer (click the appropriate **Custom** button). For more information on using this dialog and creating custom headers and footers, see *Output Headers and Footers* in Chapter 2 – *STATISTICA Output Management*.

Display Marked Cells

Select **Display Marked Cells** from the **View** menu (or click the toolbar button) to toggle between displaying/not displaying marked cells in the spreadsheet. If you select to display marked cells, they will have a different spreadsheet layout applied to them to make them stand out from the rest of the data. By default, marked cells appear red, but this default Marked Cells Spreadsheet Layout can be changed by selecting **Layout Manager** from the **Format – Spreadsheet** menu (page 274). See *Spreadsheet Layouts* (page 234) and *Tools - Marking Cells* (page 283) for further details.

Display Selected Cases

Select **Display Selected Cases** from the **View** menu (or click the toolbar button) to toggle between displaying/not displaying the cases in the spreadsheet that are currently selected using the defined case selection conditions. If you select to display selected cases, they will have a different spreadsheet layout applied to them to make them stand out from the rest of the data. By default, selected cases appear italicized, but this default Selection Conditions Spreadsheet Layout can be changed by selecting **Layout Manager** from the **Format – Spreadsheet** menu (page 274). See *Spreadsheet Layouts* (page 234) for further details. Note that this command is only available when Selection Conditions have been turned on by toggling **Enable** from the **Tools – Selection Conditions** menu (see page 280).

Events

An event is an action that is typically performed by users, such as clicking a mouse button, pressing a key, changing data, or opening a spreadsheet or workbook. In *STATISTICA,* certain events are displayed and can be used to customize its behavior. Using programmable events, you can tailor *STATISTICA*'s behavior to your needs. For further details, see *Technical Note: Changing the Spreadsheet Document-Level Events* (page 395).

View Code. Select *View Code* from the *View - Events* menu to display the *Document Events* window, which is used to enter code to change the default behavior of document-level events. Document-level events allow you to customize the behavior of open documents.

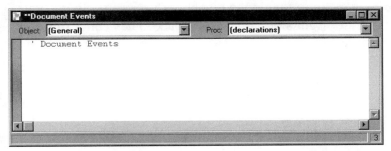

The *Object* drop-down box contains two options, *(General)* and *Document*. These options toggle between document and global declarations. If the *Object* box is set to *(General),* you are able to enter global variables, external library inclusions, and API declarations. If the *Object* box is set to *Document*, then the *Proc* box will contain a list of possible events for the current document. Select the event that you want to edit from the *Proc* box (e.g. *BeforeDoubleClick*). The corresponding function will then be written into the body of the *Document Events* window for you to edit. See Chapter 11 - *STATISTICA Visual Basic* for further details.

Autorun. Select *Autorun* from the *View - Events* menu to save your document-level event code with a specific document so that it runs any time that document is opened. Note that your code must first be run or saved before this command is applicable.

Events Password. Select *Password* from the *View - Events* menu to display the *Setup Event Password* dialog, which is used to enter a password to restrict access to editing document-level events.

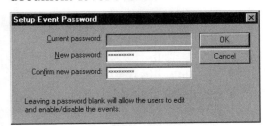

Enter a password in the *New password* box, reenter it in the *Confirm new password* box, and then click the *OK* button. The next time someone selects *View Code* from the *View - Events* menu, the *Password* dialog is displayed. The correct password must be entered before the *Document Events* window is displayed.

Toolbars

Use the commands on the *View - Toolbars* menu to toggle the display of various toolbars. Note that toolbars can be docked to the top, left, or right side of the workspace. They can also float (e.g., be located anywhere in the workspace). For more information on a specific toolbar, see the following chapters:

Toolbar	Chapter
Standard	1
Statistics	1
Six Sigma	1
Spreadsheet	4
Tools	4
Report	5
Graphs	6
Graph Tools	6
Graph Data Editor	6
Macro	11

Customize. Select *Customize* from the *View - Toolbars* menu to display the *Customize* dialog, which contains a variety of options to customize toolbars, menus, and keyboard hot keys. See *Customizing the Operation and Appearance of STATISTICA* in Chapter 1 – *STATISTICA – A General Overview* for further details.

Status Bar

Select *Status Bar* from the *View* menu to toggle between displaying and not displaying the text and buttons that are on the status bar (in the lower-center of the screen). Note that the *STATISTICA* Start button ▨ is available regardless of whether the status bar is turned on or off.

Insert Menu

The following commands are available from the spreadsheet *Insert* menu. Many of these commands are also available from shortcut menus (accessed by right-clicking on an item in the spreadsheet) and from the toolbar buttons.

Add Variables

Select *Add Variables* from the *Insert* menu to display the *Add Variables* dialog, which is used to insert variables into the spreadsheet.

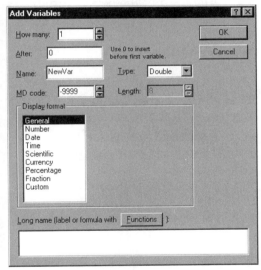

Specify how many new variables to add to the spreadsheet and where to add them. You can also specify the name, variable type (see *Variable Types,* page 222), missing data code, length, and display format of the new variable(s). You can type a more detailed name or a formula as part of the variable's definition in the *Long name* box. Click the *Functions* button to display the *Function Browser,* which is used to enter predefined functions into the formula. For further details on entering formulas into the *Long name* box, see *How Do I Calculate (Transform) Values of a Variable?* (page 361).

Copy Variables

Select **Copy Variables** from the **Insert** menu to display the **Copy Variables** dialog, which is used to make a copy of the variable(s) and insert it in the location of your choice.

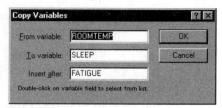

Note that unlike the Clipboard editing operations that affect blocks of data (or contents of variables), this command affects not only the contents of the variables, but the actual variables themselves.

Add Cases

Select **Add Cases** from the **Insert** menu to display the **Add Cases** dialog, which is used to insert new cases into your spreadsheet.

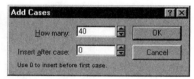

Specify how many new cases to add to the spreadsheet and where to add them.

Copy Cases

Select **Copy Cases** from the **Insert** menu to display the **Copy Cases** dialog, which is used to make a copy of the case(s) and insert it in the location of your choice.

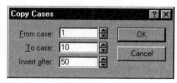

Note that unlike the Clipboard editing operations that affect blocks of data (or contents of cases), this command affects not only the contents of the cases, but the actual cases themselves.

Object

Select **Object** from the **Insert** menu (or click the 📷 toolbar button) to display the **Insert Object** dialog, which is used to insert an object into the spreadsheet.

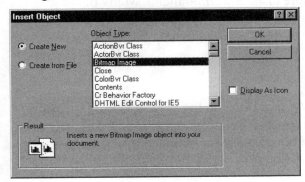

To create a new object to insert into the spreadsheet, select the **Create New** option button and choose from the list of available object types in the **Object Type** box.

To insert an existing object into the spreadsheet, select the **Create from File** option button. You can either enter a path and file name in the **File** box or click the **Browse** button to search for the desired file. If you select the **Link** check box, the inserted file is updated in both the spreadsheet and in the original location.

Select the **Display as Icon** check box to show an icon in the spreadsheet instead of the actual contents of the object. When the **Display As Icon** check box is selected, you can click the **Change Icon** button to display the **Change Icon** dialog, which is used to modify the icon that is displayed in the spreadsheet.

The **Result** field gives a brief explanation of what *STATISTICA* will do based on the current options selected in the **Insert Object** dialog. After you have made your selections, click the **OK** button to insert the object into the spreadsheet or the **Cancel** button to exit the dialog without inserting any objects. See *How Do I Insert an Object into the Spreadsheet?* (page 330) for further details.

Format Menu

The following commands are available from the spreadsheet *Format* menu. Many of these commands are also available from shortcut menus (accessed by right-clicking on an item in the spreadsheet) and from the toolbar buttons.

Cells

Select *Cells* from the *Format* menu to display the *Format Cells* dialog, which is used to specify the format for the selected cell(s). A format is a group of formatting options that you can save with an assigned name for future use. Formats can be assigned to numbers as well as text. See *User-Defined Formats* (page 236) for further details.

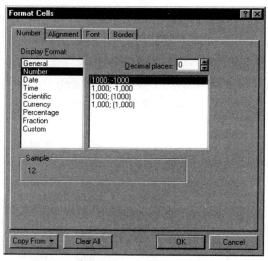

Use the *Number* tab to specify the display format of the selected cell(s). Use the *Alignment* tab to choose the vertical, horizontal, and orientation alignment of text/numerical values within each cell. Use the *Font* tab to specify the font, size, color, and certain stylist effects (i.e., bold, italic) of the selected cells. Finally, use the *Border* tab to select the style and color of the cell borders.

Variables

Width. Select *Width* from the *Format - Variables* menu to display the *Set Variable Width* dialog.

Type the desired width you want to apply to the selected variable(s) in the *Width* box and click the *OK* button. Alternatively, click the *Auto Fit* button to automatically change each variable width to accommodate the longest entry in that variable.

AutoFit. Select *AutoFit* from the *Format - Variables* menu to change the width of the selected variable(s) to accommodate the longest entry.

Cases

Height. Select *Height* from the *Format - Cases* menu to display the *Set Case Height* dialog.

Type the height you want to apply to the selected case(s) in the *Height* box and click the *OK* button. Alternatively, click the *Auto Fit* button to automatically change each case height to accommodate the tallest entry.

AutoFit. Select *AutoFit* from the *Format - Cases* menu to change the height of the selected case(s) to accommodate the tallest entry.

Block

Active formats. To apply a format to the selected cell(s) select the format from the list of current formats shown on the *Format - Block* menu. Note that if there are not any current formats at this time, the *Format - Block* menu contains four commands: *Header*, *Normal*, *New from Selection*, and *Format Manger*. See *How Do I Apply a Format?* (page 345) for further details.

Header. Select *Header* from the *Format - Block* menu to apply the Header format to the selected cell(s). The *Header* format is a default format that is always available on this menu.

Normal. Select *Normal* from the *Format - Block* menu to apply the Normal format to the selected cell(s). The *Normal* format is a default format that is always available on this menu.

New from selection. Select *New from selection* from the *Format - Block* menu to create a new format based on the format contained within the upper-left cell of the selected cells. Performing this action displays the *Format Name* dialog; enter a name for the new format and click the *OK* button. This new format is then included in the *Active formats* list (see above) and in the *Format Manager* dialog (see below).

Format Manager. Select *Format Manager* from the *Format - Block* menu to display the *Format Manager* dialog, in which you can view a list of available formats and perform various operations on formats.

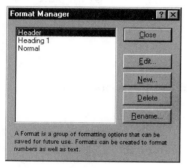

By default, the *Header* and *Normal* formats are available in this dialog. Any additional formats that you have created are also listed. Click the *Edit* button to display the *Edit Format* dialog, where you can edit the selected format. Click the *New* button to display the *Format Name* dialog, where you can specify the name of a new format. After you have specified a name, the *Edit Format* dialog is displayed where you can set the options of the new format. See *How Do I Create a New Cell/Block Format??* (page 343) for further details. Finally, you can delete or rename the selected format by clicking the respective *Delete* or *Rename* buttons.

Spreadsheet

System Default. Select *System Default* from the *Format - Spreadsheet* menu to apply the *System Default* spreadsheet layout to the entire active spreadsheet regardless of what cell(s) are selected. The *System Default* spreadsheet layout affects the data, header, case and variable

headers, and info box of your spreadsheet. See *Spreadsheet Layouts* (page 234) and *How Do I Apply a Spreadsheet Layout?* (page 349) for further details.

Layout Manager. Select *Layout Manager* from the *Format - Spreadsheet* menu to display the *Spreadsheet Layouts* dialog, in which you can view a list of available spreadsheet layouts and perform various operations on spreadsheet layouts. A spreadsheet layout is a convenient way to combine sets of formatting options into one collection. Use this feature to specify and apply different formats for different components of a spreadsheet.

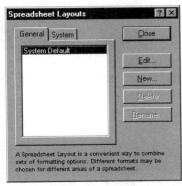

This dialog contains two tabs: *General* and *System*. Use the *General* tab to edit the *System Default* spreadsheet layout. Use the *System* tab to edit the *Active Cell Header*, *Brushing Highlighted*, *Brushing Labeled*, *Brushing Marked*, *Brushing Off*, *Marked Cells*, *Print Filter*, *Selected Cells Header*, and *Selection Condition* spreadsheet layouts. These spreadsheet layouts only affect very specific characteristics associated with spreadsheet interactions. See *Spreadsheet Layouts* (page 234) for specific details on these spreadsheet layouts.

Click the *Edit* button to display the *Edit Spreadsheet Layout* dialog, where you can edit the selected spreadsheet layout. Click the *New* button to display the *Spreadsheet Layout Name* dialog, where you can specify the name of a new spreadsheet layout. After you have specified a name, click the *OK* button to display the *Edit Spreadsheet Layout* dialog, where you can set the options of the new spreadsheet layout. See *How Do I Create a New Spreadsheet Layout*? (page 346) for further details.

Finally, you can delete or rename the selected spreadsheet layout by clicking the respective *Delete* or *Rename* buttons. Note that you can only delete or rename spreadsheet layouts that you have created. These buttons are not available when the *System Default*, *Active Cell Header*, *Brushing Highlighted*, *Brushing Labeled*, *Brushing Marked*, *Brushing Off*, *Marked Cells*, *Print Filter*, *Selected Cells Header*, or *Selection Condition* spreadsheet layout is selected.

Statistics Menu

The **Statistics** menu is available whenever any document is open. Many of these commands are also available from shortcut menus (accessed by right-clicking on an item in the spreadsheet) and from the toolbar buttons.

Resume

Select **Resume** from the **Statistics** menu (or click the  toolbar button or press CTRL+R) to continue the most recently displayed analysis or graph. This command is particularly useful if you want to display the previous dialog after you have created a results spreadsheet or graph.

Statistics

The **Statistics** menu is available whenever any document is open. This menu provides access to all available analysis types within *STATISTICA* including **Basic Statistics/Tables**, **Multiple Regression**, **ANOVA**, **Nonparametrics**, **Distribution Fitting**, **Advanced Linear/Nonlinear Models**, **Multivariate Exploratory Techniques**, **Industrial Statistics & Six Sigma**, **Power Analysis**, and **Data-Mining**. Definitions for the various types of statistics are available in the *Statistics Toolbar* topic in Chapter 1 – *STATISTICA – A General Overview*. Also included on the **Statistics** menu are **Statistics of Block Data**, **STATISTICA Visual Basic**, and **Probability Calculator**.

Graphs Menu

The **Graphs** menu is available whenever any *STATISTICA* document is open. Many of these commands are also available from shortcut menus (accessed by right-clicking on an item in the spreadsheet) and from the toolbar buttons.

Resume

Select **Resume** from the **Graphs** menu (or click the toolbar button or press CTRL+R) to continue the most recently displayed analysis or graph. This option is particularly useful if you want to display the previous dialog after you have created a results spreadsheet or graph.

Graphs

The **Graphs** menu provides access to all graph types in *STATISTICA* including **Histograms**, **Scatterplots**, **Means w/Error Plots**, **Surface Plots**, **2D Graphs**, **3D Sequential Graphs**, **3D XYZ Graphs**, **Matrix Plots**, **Icon Plots**, **Categorized Graphs**, **User-defined Graphs**, **Graphs of Block Data**, and **Graphs of Input Data**. It also provides access to **Multiple Graph Layouts**.

For more information on graphs, see Chapter 6 – *Graphs: General Features*, Chapter 7 - *Graphs: Creation and Customization*, Chapter 8 – *Graphs: Conceptual Overviews*, and Chapter 9 – *Graphs: Examples and FAQ's* .

Tools Menu

The following commands are available from the spreadsheet **Tools** menu. Many of these commands are also available from shortcut menus (accessed by right-clicking on an item in the spreadsheet) and from the toolbar buttons.

Analysis Bar

To take advantage of *STATISTICA*'s "multitasking" functionality, *STATISTICA*'s analyses are organized as functional units that are represented with buttons on the **Analysis** bar at the bottom of the application window (above the status bar, see the illustration below, where **Basic Statistics**, **Cluster Analysis**, and **Canonical Analysis** are running simultaneously). Normally, at least one analysis button is created, and consecutive buttons are added as you start new analyses.

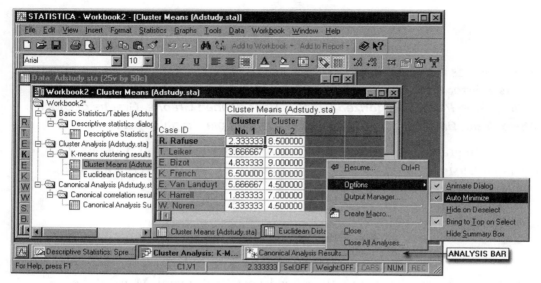

The commands on the **Tools – Analysis Bar** menu provide a variety of options for managing the **Analysis** bar.

Resume. Select *Resume* (or click the ⏎ toolbar button or press CTRL+R) from the **Tools - Analysis Bar** menu to continue the current analysis or graph. Note that you can also open the current analysis or graph by clicking on its button on the **Analysis** bar.

Select Analysis/Graph. Use the commands on the **Tools - Analysis Bar - Select Analysis/Graph** menu to select an analysis or graph from the set of active analyses and graphs. Note that you can also open an active analysis or graph by clicking on the appropriate button on the **Analysis** bar.

Options - Animate Dialog. Select *Animate Dialogs* from the **Tools - Analysis Bar - Options** menu to display animation when analysis dialogs are minimized or maximized. By default, this command is checked.

Options - Auto Minimize. Select *Auto Minimize* from the **Tools - Analysis Bar - Options** menu to automatically minimize all analysis dialogs when you select another window in *STATISTICA* or another application. By default, this command is checked. When your screen is large enough to accommodate several windows, it is recommended that you clear this option. This keeps the analysis dialogs on screen while the respective output created from these dialogs is produced, thus allowing you to use the dialogs as "toolbars" from which output can be selected.

Options - Hide on Select. Select *Hide on Select* from the *Tools - Analysis Bar - Options* menu to minimize all windows associated with a particular analysis when that analysis is deselected. By default, this command is cleared. Note that this command only applies when the results are sent to individual windows; see the discussion of the *Output Manager* tab of the *Options* dialog in Chapter 2 – *STATISTICA Output Management* for further details.

Options - Bring to Top on Select. Select *Bring to Top on Select* from the *Tools - Analysis Bar - Options* menu to activate (display at the top of *STATISTICA*) all windows associated with a particular analysis when that analysis is selected, replacing whatever dialogs were on top. This command also facilitates the organization of individual windows from various analyses. By default, this option is checked. Note that this command only applies when the results are sent to individual windows; see the discussion of the *Output Manager* tab on the *Options* dialog in Chapter 2 – *STATISTICA Output Management* for further details.

Options - Hide Summary Box. Select *Hide Summary Box* from the *Tools - Analysis Bar - Options* menu to not display the summary box, which is located at the top of certain results dialogs (such as *Multiple Regression - Results*) and contains basic summary information about the analysis. By default, this command is not checked.

Output Manager. Select *Output Manager* from the *Tools – Analysis Bar* menu to display the *Analysis/Graph Output Manager* dialog.

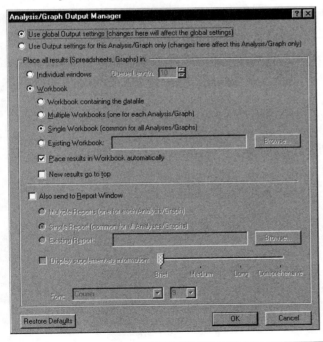

Use this dialog to specify where to direct output files for the current analysis/graph. For more details, see Chapter 2 – *STATISTICA Output Management,* particularly the topic on *Three Channels for Output from Analyses.* See also, *What Supplementary Information Is Available with Analyses and How Can I Add It to My Reports?* in Chapter 5 – *STATISTICA Reports.*

Create Macro. Select *Create Macro* from the *Tools - Analysis Bar* menu (or click the 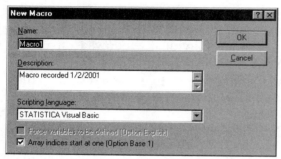 toolbar button on the *Statistics* toolbar) to display the *New Macro* dialog.

Use this dialog to specify a name, description, and scripting language for a new macro based on the current analysis. When you run an analytic procedure (from the *Statistics* menu) or create a graph (from the *Graphs* menu) the Visual Basic code corresponding to all design specifications as well as output options that you select are recorded. Thus, when you click the *OK* button in the *New Macro* dialog, the resulting macro window displays the appropriate code to recreate the current analysis.

Minimize. Select *Minimize* from the *Tools - Analysis Bar* menu to minimize the current analysis.

Close. Select *Close* from the *Tools - Analysis Bar* menu to close the current analysis.

Close All Analyses. Select *Close All Analyses* from the *Tools - Analysis Bar* menu to close the all of the analyses/graphs on the *Analysis* bar.

Selection Conditions

Selection conditions are used to include only a subset of cases in an analysis/graph (i.e., "case filters"). The conditions entered in the *Spreadsheet Case Selection Conditions* dialog are used to select cases from the current spreadsheet, and if saved to a file, they are stored along with their header in a compressed file format. The saved conditions, however, can also be used in (i.e., opened into) all those dialogs where multiple sets of case selection conditions are entered to identify multiple subsets of data (e.g., *Recode Values of Variable*, page 301, and *Verify Data*, page 292). Case selection conditions are ignored by the spreadsheet formulas.

Note that case selection conditions can also be specified locally for only the current analysis/graph (and hence not be tied to the current spreadsheet) by clicking the SELECT CASES button. This button is included on all Startup Panels and all analysis/graph definition dialogs that are displayed before the data processing begins.

Enable. Select *Enable* from the *Tools - Selection Conditions* menu (or click the toolbar button) to toggle between activating and inactivating the case selection conditions. Note that when this command is toggled on, the status bar (see page 268) displays *Sel: ON*. If it is toggled off, the status bar displays *Sel: OFF*. To view the case selection conditions, select *Edit* from the *Tools – Selection Conditions* menu (see below).

Display. Select *Display* from the *Tools - Selection Conditions* menu to toggle between applying/not applying the *Selection Conditions* spreadsheet layout to cases meeting your selection conditions. In this way, you can visually distinguish the cases that meet current selection conditions. See *Spreadsheet Layouts* (page 234) for further details.

Edit. Select *Edit* from the *Tools - Selection Conditions* menu (or click the toolbar button) to display the *Spreadsheet Case Selection Conditions* dialog.

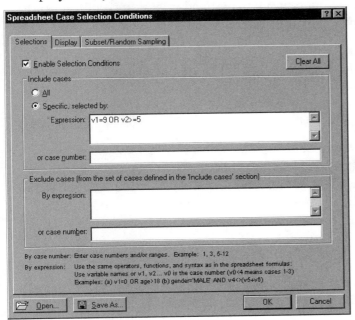

This dialog contains three tabs: *Selections*, *Display*, and *Subset/Random Sampling*. Use the *Selections* tab to create or edit selection conditions that select a subset of cases of the current

spreadsheet. Selection conditions can be entered by expression or by case number. You can enter selection conditions to either include or exclude cases.

In the **Expression** boxes, operators =, <>, <, >, <=, >=, NOT, AND, and OR can be used as well as all arithmetic operations, *STATISTICA* math and statistical functions, and *STATISTICA* distributions (see the *Function Brower* in the *Electronic Manual* for further details on *STATISTICA* functions and distributions). Variables are referred to by their names or by v1, v2, etc. Case numbers are represented by v0 (e.g., v0<=2 refers to cases 1 and 2). Finally, text must be included in single quotation marks (e.g., v1='male'). In the **or case number** boxes, case numbers (e.g. 1, 7, 9) and/or case ranges (e.g. 1-50, 75:150) separated by commas can be used. See *What Syntax Can Be Used to Create Case Selection/Verification/Recode Conditions?* (page 368).

Use the **Display** tab to apply the **Selection Conditions** spreadsheet layout to cases meeting your selection conditions. See *Spreadsheet Layouts* (page 234) for further details. Use the **Subset/Random Sampling** tab to create a new subset of the current spreadsheet either by case selection conditions or random sampling techniques.

Finally, use the **Open** and **Save As** buttons to open previously saved selection conditions or to save the current selection conditions, respectively. For further details on these tabs, see the *Electronic Manual*.

Add Selected Cases. Select **Add Selected Cases** from the **Tools - Selection Conditions** menu (or click the toolbar button) to add the selected cases in the spreadsheet to the existing case selection conditions. The selected cases are added to the **Include cases** group on the **Spreadsheet Case Selection Conditions** dialog. All other selection conditions remain the same.

Remove Selected Cases. Select **Remove Selected Cases** from the **Tools - Selection Conditions** menu (or click the toolbar button) to remove selected cases in the spreadsheet from the existing case selection conditions. The selected cases are added to the **Exclude cases** group on the **Spreadsheet Case Selection Conditions** dialog. All other selection conditions remain the same.

Replace with Selected Cases. Select **Replace with Selected Cases** from the **Tools - Selection Conditions** menu (or click the toolbar button) to add the selected cases in the spreadsheet to the **Include cases** group on the **Spreadsheet Case Selection Conditions** dialog, replacing all other selection conditions. Hence, new selection conditions are created that include only the selected cases.

Exclude Selected Cases. Select **Exclude Selected Cases** from the **Tools - Selection Conditions** menu to add the selected cases in the spreadsheet to the **Exclude cases** group on the

Spreadsheet Case Selection Conditions dialog, replacing all other selection conditions. Hence, new selection conditions are created that exclude the selected cases.

Weight

Select **Weight** from the **Tools** menu (or click the 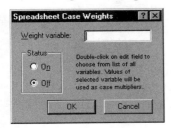 toolbar button) to display the **Spreadsheet Case Weights** dialog, which is used to specify case weights for the current spreadsheet.

Use this dialog to "adjust" the contribution of individual cases to the outcome of an analysis by "weighting" those cases in proportion to the values of a selected variable. Use the **Weight variable** box to designate the variable in the spreadsheet to be used as the weight variable. You can type the variable name (or number) or double-click on the box (or press F2 on your keyboard) to open the **Select Variable** dialog. After you have selected the weight variable, select the **On** option button in the **Status** group and click the **OK** button. Note that when **On** is selected in the **Status** group, the status bar (see page 268) displays **Weight: ON**. If **Off** is selected, the status bar displays **Weight: OFF**.

In most types of analyses, the values of the weights are rounded to the nearest integer and used as case multipliers when the datafile is processed. The statistical procedures in some analyses can use fractional weights; in those types of *STATISTICA* analyses, specific information about how weights enter into the computations is described in the help windows of the analyses.

Note that case weights can also be specified locally for only the current analysis/graph (and hence not be tied to the current spreadsheet) by clicking the button. This button is included on all Startup Panels and all analysis/graph definition dialogs that are displayed before the data processing begins.

Marking Cells

Marked cells are used to distinguish certain cases from other cases. Typically, marked cells are used in results spreadsheets. For example, you can specify the *p*-level to use when determining significant results. All results less than this *p*-level will be marked in a different spreadsheet

layout (e.g. a red color) in the results spreadsheets. Options for controlling which results are marked are typically included on the analysis dialog.

Display. Select *Display* from the *Tools – Marking Cells* menu (or click the ⊞ toolbar button) to toggle between applying/not applying the *Marked Cells* spreadsheet layout to marked cells in the spreadsheet. In this way, you can visually distinguish between the cases that are marked and cases that are not. See *Spreadsheet Layouts* (page 234) for further details.

Edit Marked Cells Layout. Select *Edit Marked Cells Layout* from the *Tools – Marking Cells* menu to display the *Edit Spreadsheet Layout* dialog, which is used to edit the *Marked Cells* spreadsheet layout.

Mark Selected Cells. Select *Mark Selected Cells* from the *Tools – Marking Cells* menu (or click the ⊞ toolbar button) to add the selected cells to the set of marked cells (i.e., the *Marked Cells* spreadsheet layout will be applied to the selected cells) in the spreadsheet. Any previously marked cells will remain marked.

Unmark Selected Cells. Select *Unmark Selected Cells* from the *Tools – Marking Cells* menu (or click the ⊞ toolbar button) to remove the selected cells from the set of marked cells in the spreadsheet (i.e. the *Marked Cells* spreadsheet layout will be removed from the selected cells). Any previously marked cells that are not selected will remain marked.

Redefine Selected Cells. Select *Redefine Selected Cells* from the *Tools – Marking Cells* menu (or click the ⊞ toolbar button) to replace the current set of marked cells with the selected cells in the spreadsheet (i.e., the *Marked Cells* spreadsheet layout will be applied to the selected cells AND removed from all unselected cells in the spreadsheet).

Unmark All. Select *Unmark All* from the *Tools – Marking Cells* menu (or click the ⊞ toolbar button) to unmark all of the cells in the spreadsheet (i.e. the *Marked Cells* spreadsheet layout will be removed from all cells).

Macro

A macro is a scripted application that extends functionality to *STATISTICA* by directly accessing *STATISTICA*'s object model and manipulating it. Macros are primarily used to automate tasks done in *STATISTICA* by harnessing its power and recording it into a *STATISTICA* Visual Basic (SVB) script. Additionally, macros can be written as both stand-alone scripts and library classes to extend the statistical and mathematical capabilities of *STATISTICA*.

Macros. Select *Macros* from the *Tools - Macro* menu (or click the 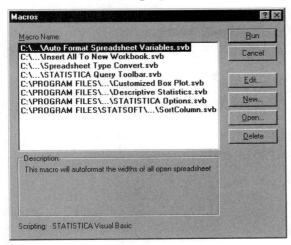 toolbar button on the *Macro* toolbar) to display the *Macros* dialog.

Use this dialog to run, edit, open, or delete existing macro (SVB) programs as well as to create new macros.

Start Recording Log of Analyses (Master Macro). Select *Start Recording Log of Analyses (Master Macro)* from the *Tools - Macro* menu to create a SVB program file that will include a sequence of analyses (a log of analyses) that have been performed interactively. Note that you must select this command before you start the analyses. Consult the *Electronic Manual* or the *STATISTICA Visual Basic Primer* to learn more about this powerful command and how to create recordable sequences of analyses.

Start Recording Keyboard Macro. Select *Start Recording Keyboard Macro* from the *Tools - Macro* menu to record a keyboard macro. See the *Electronic Manual* or the *STATISTICA Visual Basic Primer* for more details.

Stop Recording. Select *Stop Recording* from the *Tools - Macro* menu (or click the ■ toolbar button on the *Macro* toolbar) to stop recording a log of analyses (master macro) or a keyboard macro (see above).

Create Analysis/Graph Macro. Select *Create Analysis/Graph Macro* from the *Tools - Macro* menu to select from a list of all current analyses/graphs that are open. When you run an analytic procedure (from the *Statistics* menu) or create a graph (from the *Graphs* menu) the Visual Basic code corresponding to all design specifications as well as output options that you select are recorded. Select the appropriate analysis/graph to display that code. Note that the

New Macro dialog will first be displayed in which you can type the macro's name and description.

Application Events - View Code. Select *View Code* from the *Tools - Macro - Application Events* menu to display the *Document Events* window, which is used to enter code to change the default behavior of application-level events. Application-level events allow you to customize the behavior of all documents of a certain type (such as all spreadsheets, all workbooks, or all reports). See *What Are Application Events and How Can They Be Controlled from STATISTICA Visual Basic?* in Chapter 11 – *STATISTICA Visual Basic* for more details.

Application Events - Autorun. Select *Autorun* from the *Tools - Macro - Application Events* menu to save your application-level event code with a specific type of document so that it will be run any time that type of document is open.

Add-Ins. Select *Add-Ins* from the *Tools - Macro* menu to display the *STATISTICA Add-Ins* dialog. Add-Ins are COM server components normally written in ATL (Active Template Library) that are used to create custom user interfaces of *STATISTICA* and/or fully functional external programs.

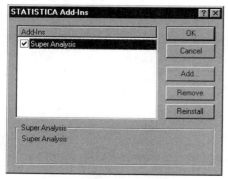

All available Add-Ins are displayed in the *Add-Ins* list. To create a new Add-In, click the *Add* button to display the *Specify Add-In to be Added* dialog, which is used to enter the program ID of the Add-In. Click the *Remove* button to delete the selected Add-In from the *Add-Ins* list. Finally, click the *Reinstall* button to register the selected Add-In to your operating system.

STATISTICA Visual Basic Editor. Select *STATISTICA Visual Basic Editor* from the *Tools - Macro - Application Events* menu (or press ALT+F11) to display a *STATISTICA* Visual Basic editor (Macro Window). For more information on using *STATISTICA* Visual Basic including examples and an extensive FAQ section, see Chapter 11 – *STATISTICA Visual Basic.*

Customize

Select **Customize** from the **Tools** menu to display the **Customize** dialog, which is used to customize toolbars, menus, and keyboard hot keys with a variety of commands. See *Customizing the Operation and Appearance of STATISTICA* in Chapter 1 – *STATISTICA – A General Overview* for further details.

Options

Select **Options** from the **Tools** menu to access the **Options** dialog. In addition to general and display options, options are available for reports, file locations, custom lists, workbooks, macros, statistical analysis display, import and edit facilities, and output management. For a discussion of the **Spreadsheets** tab of the **Options** dialog, see *Global Spreadsheet Options* (page 238).

Data Menu

The following commands are available from the **Data** menu. Many of these commands are also available from shortcut menus (accessed by right-clicking on an item or icon in the spreadsheet) and from the toolbar buttons.

Input Spreadsheet

Select **Input Spreadsheet** from the **Data** menu to toggle the designation of the active spreadsheet as an input spreadsheet. Analyses/graphs can only be performed on input spreadsheets. See *Input and Output Spreadsheets* (page 236) for further details.

Transpose

Block. Select **Block** from the **Data - Transpose** menu to transpose a square block of selected data in the spreadsheet. This command only transposes the data values defined by the selected (highlighted) data and not the entire datafile.

File. Select **File** from the **Data - Transpose** menu to turn cases into variables and variables into cases by transposing the spreadsheet datafile. (You can transpose a square block of data when you select the **Data - Transpose - Block** command.) Text label/numeric assignments in the

datafile will be lost when the file is transposed, because the text/numeric assignments are specifications of variables (i.e., columns) and not cases (i.e., rows) in *STATISTICA* datafiles. Case names (if any) from the original file will become variable names. Case names of the transposed file will be created from the former variable names.

Merge

Variables. Select *Variables* from the *Data - Merge* menu to display the *Open File to Merge* dialog. In this dialog, select the file that you would like to merge with the active spreadsheet. Next, the *Merge Files – Variables* dialog is displayed.

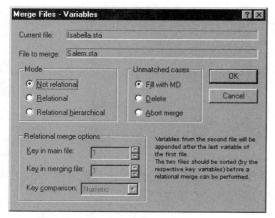

The *Current file* field displays the name of the active file, and the *File to merge* field displays the file you chose in the *Open File to Merge* dialog. Variables from the file to merge are appended after the last variable of the current file.

If the number and order of cases in the two datafiles are not precisely the same (i.e., if at least one case is missing in the second datafile or there is one case that has no equivalent in the first file), then all subsequent data would be merged with the wrong cases. To prevent this, you need to specify a mode of merging in the *Mode* group.

Select the *Not relational* option button to simply add the variables of the second file alongside those of the first (this is the default mode). Select the *Relational* option button to match the cases from the second file with those of the first file, based on the values of a specified key. When merging cases relationally from two files, specify a key (case identifier) variable in each datafile in the *Relational merge options* group; for each case, *STATISTICA* checks the values of this key in both datafiles and merges the cases only if their respective keys match.

The **Relational hierarchical** mode differs from the simple relational mode (see above) in the handling of multiple records with the same key value in either the primary or the secondary file. In the standard relational mode, successive records with identical key values will be merged. If there are uneven numbers of records with identical key values in the two files, missing data are added to "pad" the file with the lesser number of records. In contrast, in the relational hierarchical mode the file is padded with the values found in the last identical key record that was matched. Note that the sequence of cases is not changed during merging; therefore, each file should be sorted (see *Sort*, page 290) in the order of its key variable before merging.

Finally, select one of three ways of dealing with unmatched cases in the **Unmatched cases** group. Unmatched cases can result from unequal numbers of cases in the merged files, or because some of the cases do not meet the relational merge criteria. You can choose to fill the unmatched cases with missing data, delete cases from either file that cannot be matched, or display an error message and cause the merge procedure to be abandoned. See the *Electronic Manual* for further details and examples.

Cases. Select **Cases** from the **Data - Merge** menu to display the **Open File to Merge** dialog. In this dialog, select the file that you would like to merge with the active spreadsheet. The cases from the second file are appended after the cases of the active file. The number of variables in the two files must be equal.

Text Labels. Select **Text Labels** from the **Data - Merge** menu to display the **Open File to Merge** dialog. In this dialog, select the file whose text labels you would like to merge with those of the active spreadsheet. Note that only the text labels of variables with the same name in both files will be merged. If two variables (from the two files) share the same name but you do not want to merge their text labels, then temporarily change the name of the variable in one of the files until the merge procedure is completed. Next, the **Merge Files – Text Labels** dialog is displayed.

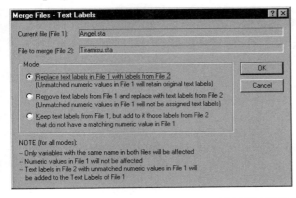

The **Current file (File 1)** field displays the name of the active file, and the **File to merge (File 2)** field displays the file you chose in the **Open File to Merge** dialog. Select one of the options in the **Mode** group to control how text labels are treated when you merge the two files and then click the **OK** button. Click the **Cancel** button to close the dialog without merging the text labels of the two files. See the *Electronic Manual* for further details and examples.

Sort

Select **Sort** from the **Data** menu (or click the ⬛ toolbar button) to display the **Sort Options** dialog, which is used to sort variables in the spreadsheet.

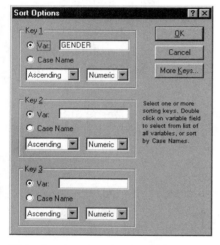

Use the **Key** groups to create custom sorting key(s). Select the **Var** option button to create the sort by variable and enter the name of the variable in the accompanying box. Note that by double-clicking in the accompanying box (or pressing F2) the **Select Variable** dialog is displayed. Optionally, select the **Case Name** option button to create the sort by case names.

Next, use the two drop-down boxes to specify the sort in **Ascending** or **Descending** order and by **Numeric** or **Text** values. If you need more than 3 keys to sort your datafile, click the **More Keys** button to access an expanded **Sort Options** dialog with the options to create seven sorting keys.

Subset/Random Sampling

Select **Subset/Random Sampling** from the **Data** menu to display the **Create a Subset/Random Sampling** dialog, which is used to extract a section (subset) of the current spreadsheet.

Click the **Variables** button to display the **Select Variables** dialog, which is used to choose the variables from the current spreadsheet to be included in the subset. Likewise, click the **Cases** button to display the **Case Selection Conditions** dialog, which is used to create conditions to define the selection of cases to be included in the subset. For more information about case selection conditions, see *Selection Conditions – Edit* (page 281). Note that the **Cases** button is only available when the **Use selection condition expression** option button is selected in the **Subset Selection Rules** group.

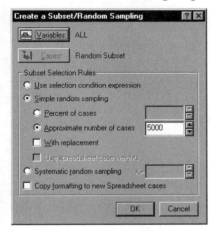

Select the **Simple random sampling** option button to obtain your subset via random sampling. You have two choices in regards to how the subset is created: either via the **Percentage of cases** within the original spreadsheet or an **Approximate number of cases**. If you select the **With replacement** check box, then once a case is selected to be included into the subset, that case will be placed back into the pool of available choices for the remaining cases in the subset (hence an individual case can appear more than once in the resulting subset). Additionally, if a weighting variable has been selected for the original spreadsheet, select the **Use spreadsheet case weights** check box to apply that case weight when the subset is being obtained.

Select the **Systematic random sampling** option button to obtain your subset via systematic random sampling. For instance, if you enter a **5** in the **K=** box, *STATISTICA* will randomly select a case within the first five cases and then finish obtaining the subset by selecting each fifth case in the spreadsheet after the originally selected case.

Finally, select the **Copy formatting to new Spreadsheet cases** check box to retain the cell format(s) of the original cases in the new subset. Note that if you do not want to overwrite the active spreadsheet with the newly created subset, you must select **Save As** from the **File** menu to save the subset of values with a different file name.

Verify Data

Verify Data. Select *Verify Data* from the *Data - Verify Data* menu to display the *Verify Data* dialog, which is used to verify the accuracy and completeness of the data in the spreadsheet.

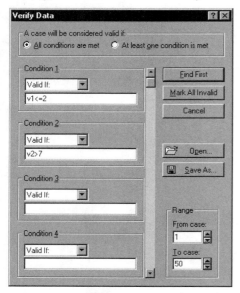

First, define that a case will be considered valid if *All conditions are met* or if *At least one condition is met* by selecting the appropriate option button. You can then create up to 16 verification conditions. To do this, select *Valid If* or *Invalid if* from the *Condition* drop-down box to determine whether the value will be considered valid or invalid if the verification condition is met. Enter the verification condition in the box below the drop-down box. For details on the syntax you can use to create verification conditions, see *What Syntax Can Be Used to Create Case Selection/Verification/Recode Conditions?* (page 368). Optionally, select *Valid cases* or *Invalid cases* from the drop-down box and enter the case number (or range of case numbers) to be considered valid/invalid.

Once you have entered your verification condition(s), click either the *Find First* button to select the first invalid case in your datafile (after this first case has been selected, you can find the next case by selecting *Find Next Invalid Case* from the *Data - Verify Data* menu) or click the *Mark All Invalid* button to mark all of the invalid cases in the datafile according to the *Marked Cells* spreadsheet layout (see page 235).

Verification conditions can be saved for future use clicking the **Save As** button. Previously saved multiple case selection conditions can be opened and used as verification conditions by clicking the **Open** button. Finally, the range of values to be verified can be defined in the **Range** group.

Find Next Invalid Case. Select **Find Next Invalid Case** from the **Data - Verify Data** menu (or press CTRL+F3) to select the next invalid case in your datafile. Note this command is only available if you have entered verification condition(s) and clicked the **Find First** button on the **Verify Data** dialog (see above).

Variable Specs

Select **Variable Specs** from the **Data** menu to display the **Variable** specifications dialog, which displays detailed information about the first selected variable. Use this dialog to enter specifications for the variable.

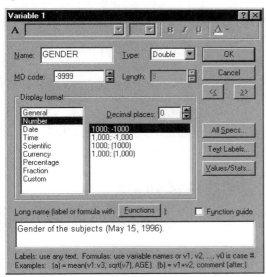

Several formatting tools, which apply to both the variable name and the long name, are available at the top of the dialog. The **Name** box contains the variable name, which is displayed in the column header in the spreadsheet. Type a new name here if you want to change the variable name. The **Type** box contains the variable data type (see *Variable Types*, page 222). The **MD code** box is used to specify a missing data code for blank cells or specific values that you intend to ignore in calculations. Note that the missing data code for a text data type is always an empty string. The **Length** box, which is only available if you have selected **Text** as

the data *Type* for the variable, is used to specify the number of characters allowed for the variable.

The *Display format* group shows the display format that accompanies the variable. When certain *Display formats* are chosen, a box to the right lists additional formats that are compatible with the selected display format. For example, if *Time* is selected as the *Display format*, the box to the right shows possible time formats (such as date and time combinations, military time options, etc.). The *Decimal places* box (which is only active when *Number*, *Scientific*, *Currency*, or *Percentage* is chosen as the *Display Format*) is used to specify the number of decimal places to be displayed.

Use the *Long name* box to type a more detailed name or a formula as part of the variable's definition. Begin the *Long name* box with an equal sign to signify if you are entering a formula instead of a name. Click the *Functions* button to display the *Function Browser,* which is used to enter predefined functions into the formula. Select the *Function guide* check box for guidance when you are entering a formula into the *Long name* field. Every time you type a letter to begin a new word (i.e. type a letter after the equal sign or after a space), the list of available *STATISTICA* functions that start with that letter will be displayed. You can then select from this list (by double-clicking on the function name) to enter the selected function into the *Long name* field.

Click the arrow buttons ⟨⟨ ⟩⟩ to browse the list of variables in the active spreadsheet; hence, you can switch between variables in the spreadsheet without closing this dialog. Click the *All Specs* button to display the *Variable Specifications Editor* dialog (see below). Click the *Text Labels* button to display the *Text Labels Editor* dialog (see page 295). Finally, click the *Values/Stats* button to display the *Values/Stats* dialog, which shows more detailed information about the selected variable, including all values in the variable and descriptive statistics.

All Variable Specs

Select *All Variable Specs* from the *Data* menu to display the *Variable Specifications Editor* dialog. This dialog, which is organized in a spreadsheet format, is used to view and edit the properties of all the variables in the active spreadsheet. It is convenient when you need to compare or edit specifications of several variables, especially when you need to copy and paste between variables (e.g., comments, formulas, or links), or extend a format definition or missing data code from one variable to subsequent variables.

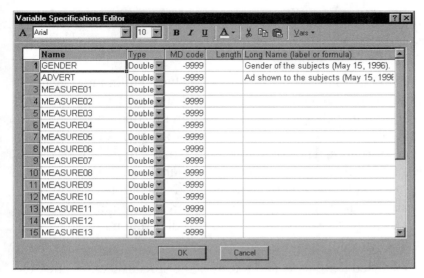

The shortcut menu contains options to **Add Vars**, **Delete Vars**, **Cut**, **Copy**, **Paste**, and **Fill/Copy Down**. See **Variable Specs** (above) for a description of the columns on this dialog.

Text Labels Editor

Select **Text Labels Editor** from the **Data** menu to display the **Text Labels Editor** dialog, which is used to create/modify text labels that accompany the selected variable(s).

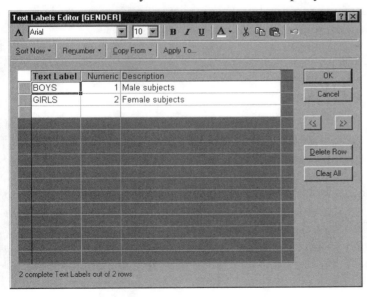

Several formatting tools are available at the top of the dialog. The text labels can be sorted in ascending/descending order in accordance with the text or numeric values. They can also be renumbered, copied from, or applied to other variables.

Text label entries (strings of characters) that label specific numeric values within the selected variable are displayed in the **Text Label** and **Numeric** columns. Note that a text label always corresponds to a numeric value. All printable characters can be used in the **Text Label** column. There is virtually no length limit (actually 10,000). A more detailed explanation of the text label can be entered in the **Description** column. The column located directly to the left of the **Text Label** column can contain descriptive information about the text label/numeric association for each row in the table. If one of the text label/numeric associations is incomplete, then this field will display a dash (**-**). If a numeric value is assigned to two different text labels, then this field will display the letter **N**. If the same text label is assigned to two different numeric values, then this field will display the letter **T**.

Click the **OK** button to accept the specified text labels and return to the spreadsheet, or click the **Cancel** button to close the dialog and ignore any changes made. Click the arrow buttons ⟨⟨ ⟩⟩ to browse the list of variables in the spreadsheet. Click the **Delete Row** button to delete the selected row. Finally, click the **Clear All** button to delete all of the values in the **Text Label**, **Numeric**, and **Description** columns.

Case Names Manager

Select **Case Names Manager** from the **Data** menu to display the **Case Names Manager** dialog, which is used to perform a variety of formatting changes on case names.

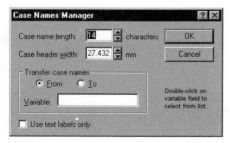

Type the desired maximum number of characters allowed for cases in the **Case name length** box. Type the number of inches desired for the width of the case headers in the **Case header width** box (see *Spreadsheets - Multimedia Tables*, page 223, for details about **Case Headers**). Use the options in the **Transfer case names** group to transfer the values **From** a specified **Variable** to create new case names or to transfer the existing case names **To** a specified **Variable**. By default, the **Use text labels only** check box is cleared, and *STATISTICA* will convert the

contents of the source variable to text (e.g., dates or numbers following the current format) and create case names from them. If the **Use text labels only** check box is selected, however, numeric values will not be converted (they will be ignored) and only text labels will be used, if the source variable contains any.

Vars

Add. Select **Add** from the **Data - Vars** menu to display the **Add Variables** dialog, which is used to insert new variables into the spreadsheet. You can specify how many new variables to add to the spreadsheet and where to add them. You can also specify the name, variable type (see *Variable Types,* page 222), missing data code, length, and display format of the new variable(s). You can type a more detailed name or a formula as part of the variable's definition in the **Long name** box. Click the **Functions** button to display the **Function Browser,** which is used to enter predefined functions into the formula.

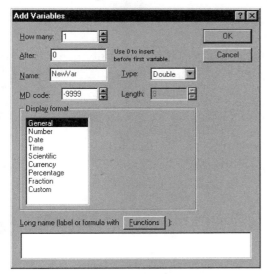

For further details on entering formulas into the **Long name** box, see *How Do I Calculate (Transform) Values of a Variable?* (page 361).

Move. Select **Move** from the **Data - Vars** menu to display the **Move Variables** dialog, which is used to remove both the contents of the variable(s) and the variable(s) itself (regardless of whether all cases or only a subset of cases are currently selected for the respective variables) and to insert the variable(s) in the designated position in the spreadsheet. Therefore, the size of the file remains the same.

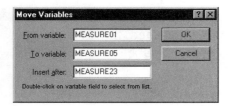

You can move one or more variables in the spreadsheet by designating the range (inclusive) of variables to be moved and the location (variable to insert after) in the **Move Variables** dialog.

Copy. Select **Copy** from the **Data - Vars** menu to display the **Copy Variables** dialog, which is used to make a copy of the variable(s) and insert it in the location of your choice.

Note that unlike the Clipboard editing operations that affect blocks of data (or contents of variables), this command affects not only the contents of the variables but the actual variables themselves.

Delete. Select **Delete** from the **Data - Vars** menu to display the **Delete Variables** dialog, in which you choose which variable(s) to delete.

Cases

Add. Select **Add** from the **Data - Cases** menu to display the **Add Cases** dialog, which is used to insert new cases into your spreadsheet.

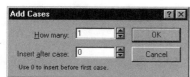

Specify how many new cases to add to the spreadsheet and where to add them.

Move. Select *Move* from the *Data - Cases* menu to display the *Move Cases* dialog, which is used to remove both the contents of the row(s) and the row(s) itself (regardless of whether all variables or only a subset of variables are selected for the respective cases) and insert the cases in the designated position in the spreadsheet. Therefore, the size of the file remains the same.

You can move one or more cases (rows) in the spreadsheet by designating the range (inclusive) of cases to be moved and the location (case to insert after) in the *Move Cases* dialog.

Copy. Select *Copy* from the *Data - Cases* menu to display the *Copy Cases* dialog, which is used to make a copy of the case(s) and insert it in the location of your choice.

Note that unlike the Clipboard editing operations that affect blocks of data (or contents of cases), this command affects not only the contents of the cases but the actual cases themselves.

Delete. Select *Delete* from the *Data - Cases* menu to display the *Delete Cases* dialog, in which you choose which case(s) to delete.

Rank

Select *Rank* from the *Data* menu to display the *Rank Order Values* dialog, which is used to control ranking information for the variables in a spreadsheet.

Click the **Variables** button to display the **Select Variables** dialog, which is used to select one or more variables to rank. Likewise, click the **Cases** button to display the **Spreadsheet Case Selection Conditions** dialog, which is used to specify a subset of cases to be ranked. The specified conditions will apply to every variable selected above. For more information about case selection conditions, see *Selection Conditions – Edit* (page 281). Before ranking, you can weight the selected variables by another variable in the spreadsheet by clicking the **Weight** button.

Various options can be selected to customize the ranking. In the **Assign rank 1 to** group, select if rank 1 should be assigned to the **smallest value** or the **largest value**. In the **Ranks for ties** group, ranks for tied values can be assigned the mean of those ranks, ranked sequentially (ignoring ties), or assigned the lowest/highest rank of the values in the tie. Finally, in the **Type of ranks** group, ranks can range from 1 to n (select **regular**), 0 to 1 (select **fractional**), or be percentages based on the fractional ranking of the values of variables (select **fractional as %**).

Recalculate

Select **Recalculate** from the **Data** menu (or click the x=? toolbar button or press SHIFT+F9) to display the **Recalculate** dialog, which is used to recalculate a variable (or group of variables).

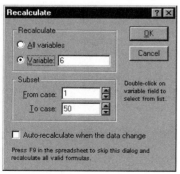

Recalculation is based on the formulas entered into the **Long name** field of the **Variable** dialog that corresponds to each variable (see *Variable Specs*, page 293).

Use the options in the **Recalculate** group to determine which variable(s) are recalculated when you click the **OK** button. By default, all cases are recalculated (the first and last case numbers in the datafile are automatically entered in the **Subset** group), or, if a block is selected in the spreadsheet, only the cases included in the block are recalculated (the first and last case numbers in the block are automatically entered in the **Subset** group). However, should you want to recalculate a block of cases that is different from the default entries, enter the case numbers in the **From case** and **To case** boxes. To automatically recalculate all of the spreadsheet formulas when the data are changed in the spreadsheet, select the **Auto-recalculate when the data change** check box.

Recode

Select **Recode** from the **Data** menu to display the **Recode Values of Variable** dialog, which is used to recode the cases in the variable.

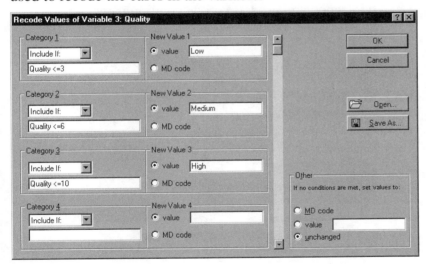

Up to 16 categories can be used to recode a variable. To specify a recode category, first select **Include cases**, **Include If**, **Exclude Cases**, or **Exclude If** from the drop-down box and then enter the case number(s)/recode condition in the field below the drop-down box. For details on the syntax you can use in recode conditions, see *What Syntax Can Be Used to Create Case Selection/Verification/Recode Conditions?* (page 368).

When you select *Include cases*, the case number (or range of case numbers) that you enter is replaced with the *New Value* (see below). *Exclude cases* will replace the original values for cases (observations) that are not included in the case number (or range of case numbers) with the *New Value*. When you select *Include if* and the recode condition you enter evaluates to true, the original values of the recoded variable for cases (observations) meeting the condition are replaced with the *New Value*. *Exclude if* assigns the *New Value* to the original values for cases (observations) that do not meet the specified recode condition.

In the *New Value* group, specify the value to replace the original value(s). You can choose either a specific text or numeric value (select the *value* option button) or the missing data value (select the *MD code* option button) as the new value.

If no conditions are met by a case (observation) when the variable is recoded, use the *Other* group to either set the original value of that case to the missing data value (select the *MD code* option button), a specific text or numeric value (select the *value* option button), or leave it as the original value (select the *unchanged* option button).

Recode conditions can be saved for future use by clicking the *Save As* button. Previously saved multiple case selection conditions can be opened and used as recode conditions by clicking the *Open* button.

Replace Missing Data

Select *Replace Missing Data* from the *Data* menu to display the *Missing Data Replacement* dialog, which is used to permanently replace missing data values for selected variables with the means of those variables.

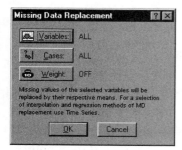

Click the *Variables* button to display the *Select Variables* dialog, which is used to select the variables in which missing data is replaced by means. By default, all cases are used in computing means for each variable. You can use case selection conditions to compute the mean for only the selected cases. Only those cases will have their missing data replaced. To do this, click the *Cases* button to display the *Spreadsheet Case Selection Conditions* dialog. Here,

you specify new case selection conditions or open existing conditions. For more information about case selection conditions, see *Selection Conditions – Edit* (page 281).

By default, each (selected) case contributes equally to the computation of the variable means. To change this, click the ***Weight*** button to display the ***Spreadsheet Case Weights*** dialog, where you select a weighting variable. By specifying one variable as containing weights, the influence of each case can be varied by an amount proportional to the value of the weight variable for the case.

Shift (Lag)

Select ***Shift (Lag)*** from the ***Data*** menu to display the ***Shift Variables*** dialog, which is used to shift selected variable(s).

Click the ***Variables*** button to select the variable(s) to be shifted. Enter a number in the ***Lag*** box to specify the magnitude of the shift (i.e., the number of cases by which to move the variable). Next, choose to either move the variable ***Forward*** (down) or ***Backward*** (up) by selecting the appropriate option button in the ***Direction*** group. After you click the ***OK*** button, *STATISTICA* shifts the selected variables and adds blank cases to the remaining variables in the file either before the first case (if the selected variable is shifted backward) or after the last case of the datafile (if the selected variable is shifted forward).

Standardize

Select ***Standardize*** from the ***Data*** menu to display the ***Standardization of Values*** dialog, which is used to standardize variables in the active spreadsheet.

Click the **Variables** button to display the **Select Variables** dialog, which is used to select the variables to be standardized. By default, all cases are used in computing means and standard deviations for each variable. However, you can use case selection conditions to compute the mean and standard deviation for a subset of the cases in the active spreadsheet. As an example of this type of subset standardization, you can standardize values once for **Males** and then for **Females** in order to obtain standardized scores according to their respective (different) group means and standard deviations. To do so, click the **Cases** button to display the **Spreadsheet Case Selection Conditions** dialog. Here, you specify new case selection conditions or open existing conditions. For more information about case selection conditions, see *Selection Conditions – Edit* (page 281).

By default, each selected case contributes equally to the computation of the variable means and standard deviations. To change this, click the **Weight** button to display the **Spreadsheet Case Weights** dialog, where you select a weighting variable. By specifying one variable as containing weights, the influence of each case can be varied by an amount proportional to the value of the weight variable for the case.

Date Operations

Select **Date Operations** from the **Data** menu to display the **Date Operations** dialog, which is used to manipulate date variables.

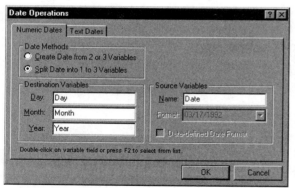

Use the **Numeric Dates** tab to create a single date variable from two or three variables (containing the month and year, day and month, day and year, or month, day, and year) or split a single date variable into one, two, or three variables (one each for the day, month, year, month and year, day and month, day and year, or month, day, and year). The **Destination Variables** and **Source Variables** groups are used to specify the new variable(s) that is (are) being created from the original date variable(s), respectively. To display a list of the current

variables in the datafile, double-click (or press F2) in **Day**, **Month**, **Year**, or **Name** boxes. When the **Create Date from 2 or 3 Variables** option button is selected, you will have the option of selecting a different format for the new variable from the **Format** box or selecting the **Data-defined Date Format** check box.

Use the **Text Dates** tab (illustrated below) to transform a date variable into a variable containing date text labels with numeric equivalents in a range that will allow *STATISTICA* to use them as codes (i.e., numeric values less than 32,000) or to transform a variable containing text labels to a date variable. The **Variables** group is used to specify both the **Source** and **Destination** variables. Note that to display a list of the current variables in the datafile double-click (or press F2) in these fields.

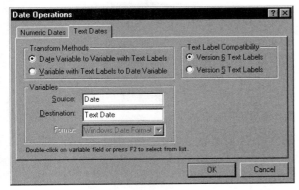

When the **Date Variable to Variable with Text Labels** option button is selected, use the **Text Label Compatibility** group to specify the creation of either Version 6 text labels (the format of the text label of the new variable will be the same as the original date variable) or Version 5 text labels (the format will always be MONDD_YY, e.g., SEP05_74). When the **Variable with Text Labels to Date Variable** option button is selected, use the **Format** drop-down box to specify the format of the new date variable.

Refer to the on-line *Electronic Manual* (press the F1 key or click 🅿️) for examples of creating date variables from numeric variables and splitting date variables into numeric variables.

Get External Data

The options on this menu are for use with *STATISTICA* Query. See Chapter 10 – *STATISTICA Query* for further details about creating, editing, and using queries. Note that **Edit Query**, **Properties**, **Refresh Data**, **Delete Query**, and **Cancel Query** are only available when you have run (or are currently running) a query.

Open Query from File. Select *Open Query from File* from the *Data – Get External Data* menu to display the standard *Open* dialog in which you can select an SQL file to run. Note that **.sqy* files are (created and) saved in *STATISTICA* Query (via the *File - Save As* command).

Create Query. Select *Create Query* from the *Data – Get External Data* menu to open *STATISTICA* Query and display the *Database Connection* dialog. Once you have chosen or defined a database connection, you can use *STATISTICA* Query to write your own query.

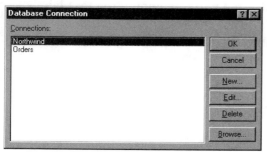

For details on defining a database connection, see the *FAQ: Working with STATISTICA Query* section in Chapter 10 – *STATISTICA Query*.

Edit Query. Select *Edit Query* from the *Data – Get External Data* menu to edit the query in the active spreadsheet. Note that if you have multiple queries on the active spreadsheet, *STATISTICA* opens the query that is associated with the active cell. If you select a cell that is not connected to any query or if you select multiple cells that involve more than one query, then the *Select Database Query* dialog is displayed and you must select the query that you want to edit before *STATISTICA* Query opens.

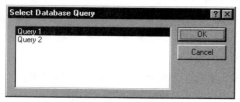

Properties. Select *Properties* from the *Data – Get External Data* menu to display the *External Data Range Properties* dialog and specify options regarding the return of external data to spreadsheets. Note that if you have multiple queries on the active spreadsheet, then *STATISTICA* displays properties for the query associated with the active cell on the spreadsheet.

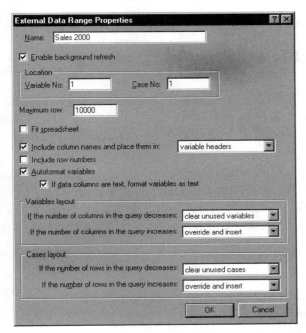

If you select a cell that is not connected to any query or if you select multiple cells that involve more than one query, then the ***Select Database Query*** dialog is displayed. On this dialog, select the query to use and click the ***OK*** button to display the ***External Data Range Properties*** dialog.

Refresh Data. Select ***Refresh Data*** from the ***Data – Get External Data*** menu (or press F5) to run a query and refresh the data (retrieve the latest data from the original database). Note that if you have multiple queries on the active spreadsheet, then *STATISTICA* refreshes data for the query associated with the active cell of the spreadsheet. If you select a cell that is not connected to any query or if you select multiple cells that involve more than one query, then the ***Select Database Query*** dialog is displayed so you can choose the specific query to refresh.

Delete Query. Select ***Delete Query*** from the ***Data – Get External Data*** menu to delete the query from the active spreadsheet. (*STATISTICA* always prompts you to verify that the query should be deleted.) Note that if you have multiple queries on the active spreadsheet, *STATISTICA* deletes the query associated with the active cell of the spreadsheet. If you select a cell that is not connected to any query or if you select multiple cells that involve more than one query, then the ***Select Database Query*** dialog is displayed so you can choose which query to delete.

Cancel Query. Select ***Cancel Query*** from the ***Data – Get External Data*** menu (or press SHIFT+F5) to stop a currently running query at any time.

Window Menu

The *Window* menu is available when any document is open. It provides access to commonly used commands for organizing the workspace and switching between files.

Close All

Select *Close All* from the *Window* menu (or press CTRL+L) to close all open spreadsheets, graphs, and related windows (e.g., graph data) in *STATISTICA*. This option is useful when you need to clear the screen to start a new analysis. Note that you will be prompted to save any unsaved files before they are closed.

Cascade

Select *Cascade* from the *Window* menu (or press SHIFT+F6) to arrange the open *STATISTICA* windows in an overlapping pattern so that the title bar of each window is visible.

Tile Horizontally

Select *Tile Horizontally* from the *Window* menu (or press ALT+F6) to arrange the open *STATISTICA* windows in a horizontal pattern. When you select this option, *STATISTICA* automatically optimizes the display of the open windows (with the preference given to tiling horizontally).

Tile Vertically

Select *Tile Vertically* from the *Window* menu (or press SHIFT+ALT +F6) to arrange the open *STATISTICA* windows in a vertical pattern. When you select this option, *STATISTICA* automatically optimizes the display of the open windows (with the preference given to tiling vertically).

Arrange Icons

Select *Arrange Icons* from the *Window* menu to arrange all minimized windows into rows.

Active Windows

Select the name of a window from the list of current windows shown on the **Window** menu to activate (bring to the top of *STATISTICA*) that window.

Windows

Select **Windows** from the **Window** menu to display the **Windows** dialog, which is used to access and manage all of the currently open windows in *STATISTICA*.

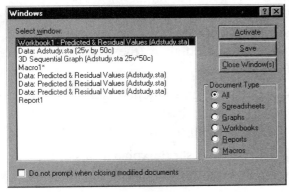

Select either a single window or multiple windows in the **Select window** box for manipulating its associated window in the *STATISTICA* environment. Note that you can use CTRL or SHIFT while selecting items to make multiple selections.

Click the **Activate** button to set the focus to the currently selected window in the **Select window** box. In other words, the associated window is brought to the front of all of the other windows. Note that because only one window can have focus at any given time, multiple selections for this operation do not apply. To save or close all of the windows that are currently selected, click the **Save** or **Close Window(s)** button, respectively. Note that *STATISTICA* prompts you to save any modified windows before closing them.

Use the options in the **Document Type** group to filter all of the open windows within *STATISTICA* by their document type; the contents of the **Select window** box are updated to reflect your filter selection. You can display all open documents in the **Select window** box (select the **All** option button), or limit the display to spreadsheets, graphs, workbooks, reports, or macros only by selecting the appropriate option button.

Help Menu

The **Help** menu provides access to various types of help. Note that the Electronic Manual is accessible at all times by pressing F1.

Contents and Index

Select **Contents and Index** from the **Help** menu (or click the 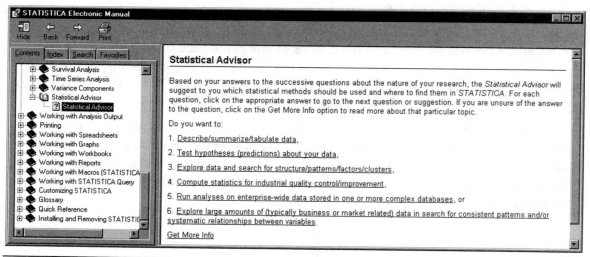 toolbar button or press F1) to display the *STATISTICA Electronic Manual. STATISTICA* provides a comprehensive electronic help for all program procedures and all options available in a context-sensitive manner by pressing F1 or clicking the help button on the caption bar of all dialogs (there is a total of more than 100 megabytes of compressed documentation included).

Due to its dynamic hypertext organization, organizational tabs (e.g., **Contents**, **Index**, **Search**, and **Favorites**), and various facilities allowing you to customize the help system, it is faster to use the *Electronic Manual* than to look for information in the traditional printed manuals.

The status bar on the bottom of the *STATISTICA* window displays short explanations of the menu options or toolbar buttons when an item is selected or a button is clicked.

Statistical Advisor

Select **Statistical Advisor** from the **Help** menu to launch the **Statistical Advisor**.

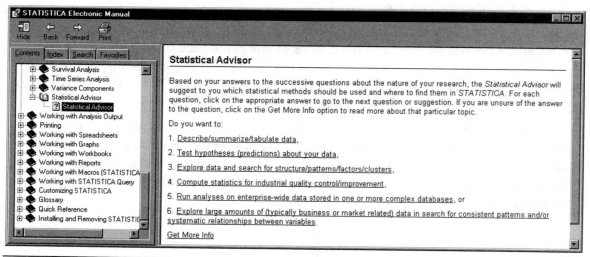

Based on your answers to the successive questions about the nature of your research, the **Statistical Advisor** suggests which statistical methods could be used and where to find them in *STATISTICA*.

Animated Overviews

Select **Animated Overviews** from the **Help** menu to display a submenu of the available animated overviews (visual overviews and specific feature help). Select from the submenu to launch one of the animated overviews.

If you select one of these commands and you have not installed that animated overview, a *STATISTICA* message dialog will prompt you for the location of that animated overview file. For example, you can run it from your *STATISTICA* CD, or if you are running a network version, the overviews can be located on the network drive. Note that if you have the *STATISTICA* CD, you can run the overviews directly from the CD without copying them to your hard drive.

StatSoft's Home Page

Select **StatSoft's Home Page** from the **Help** menu to launch the StatSoft Home Page in your default browser.

We invite you to visit the StatSoft Web page often.

- For the most recent information about *STATISTICA*, downloadable upgrades, new releases, new products, news about StatSoft, etc., access the *What's New...* section of the Web site.

- For a comprehensive list of *Frequently Asked Questions* (including useful tips, solutions to hardware/software compatibility problems, etc.), access the *Technical Support* section of the Web site.

- For a library of *STATISTICA* Visual Basic programs (written by users), access the *Technical Support* section of the Web site (to submit your own programs to this "user-exchange forum," send e-mail to *info@statsoft.com*).

Technical Support

Select **Technical Support** from the **Help** menu to launch the *Technical Support - Getting More Help* page of the StatSoft Web site in your default browser. This page contains links to download *STATISTICA* updates and links to FAQ topics on spreadsheets, graphs, printing, reports, etc. The StatSoft Technical Support Department e-mail address, phone number, and hours are also listed.

About *STATISTICA*

Select **About STATISTICA** from the **Help** menu to display the *STATISTICA* version, copyright notice, license information, foreign office contact information, and citation.

SPREADSHEET TOOLBAR

Both input data and results can be edited in a spreadsheet window. A variety of database management and data transformation/recoding operations are available as well. The **Spreadsheet** toolbar is available whenever a spreadsheet is open, and it provides quick access to the most commonly used data management and formatting facilities. All facilities accessible via the toolbar buttons are also available by using other program controls.

 Font Name and Font Size Boxes

Use the **Font Name** and **Size** boxes to change the font and font size of new or selected text in spreadsheets. Fonts, font sizes, and character attributes (i.e., bold, italic, underline, etc.) can be mixed from cell to cell of a spreadsheet.

B *I* <u>U</u> Character Formatting Buttons

These buttons control character formatting for selected text. Keyboard shortcuts are also available for formatting options as indicated below:

B   **Bold** text (keyboard: CTRL+B);

I   *Italic* text (keyboard: CTRL+I);

<u>U</u>   <u>Underline</u> text (keyboard: CTRL +U).

Paragraph Formatting Buttons

This group of buttons controls the format of new or selected paragraphs (in reports) and selected cells in spreadsheets.

   Left Justifies (left-margin aligns) text and graphs;

   Centers text or graphs between margins or within cells;

   Right Justifies (right-margin aligns) text and graphs.

Format Cells Button

Click the 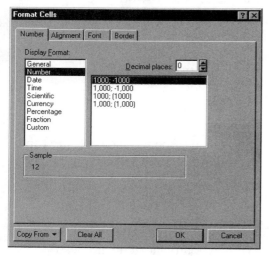 button to display the **Format Cells** dialog, which is used to specify the format for the selected cell(s). A format is a group of formatting options that you can save with an assigned name for future use. Formats can be assigned to numbers as well as text. See *User-Defined Formats* (page 236) for further details.

Use the **Number** tab to specify the display format of the selected cell(s). Use the **Alignment** tab to choose the vertical, horizontal, and orientation alignment of text/numerical values within each cell. Use the **Font** tab to specify the font, size, color, and certain stylist effects (i.e., bold, italic) of the selected cells. Use the **Border** tab to select the style and color of the cell borders.

Font Color Button

Click the arrow on the **Font Color** button to display a standard color palette, which is used to select a color for new or selected text.

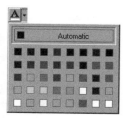

Click the letter to change the color to the default (indicated) color.

 Fill Color Button

Click the arrow on the *Fill Color* button to display a standard color palette, which is used to select a color for the background of the selected spreadsheet cells.

Click the fill bucket to change the color to the default (indicated) color.

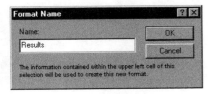 Format Menu Button

Click the arrow on the *Format Menu* button to display a format menu from which you can select a new format for the selected cells. Click the button to change the format to the most recently used (from this menu) format.

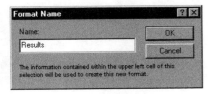

To create a new format based on the currently selected cells, select *New from Selection*. The *Format Name* dialog is displayed, which is used to specify a name for the format.

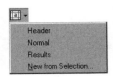

Afterwards, the new format is added to the format menu.

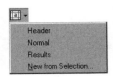

For more information on formats, see the *User-Defined Formats* (page 236).

Show/Hide Text Labels Button

Click the ⬚ button to toggle between displaying text labels in the data spreadsheet:

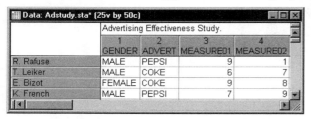

and displaying their numeric equivalents:

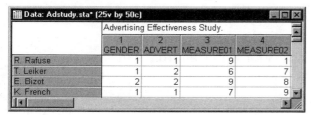

See also the *Text Labels Editor* (page 295).

Show/Hide Case Names Button

Click the ⬚ button to toggle between displaying case names in the Case Headers:

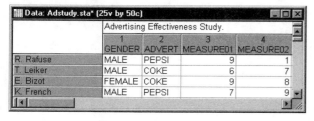

and displaying case numbers:

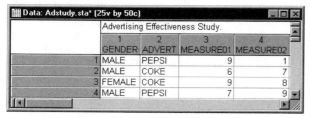

⁺⁰⁰ Increase Decimal Button

Click the ⁺⁰⁰ button to add one decimal place to the value display format of the current variable (for example, after clicking this button, the value *12.23* will be displayed as *12.234*).

⁻⁰⁰ Decrease Decimal Button

Click the ⁻⁰⁰ button to remove one decimal place from the value display format of the current variable (for example, after clicking this button, the value *12.234* will be displayed as *12.23*).

🖽 Show/Hide Marked Cells Button

Click the 🖽 button to toggle the application of the *Marked Cells* spreadsheet layout to marked cells in the spreadsheet. In this way, you can visually distinguish between the cases that are marked and cases that are not. See *Spreadsheet Layouts* (page 234) for further details.

Options for controlling which results are marked are typically included on the analysis dialog. For example, you can specify the *p*-level to use when determining significant results. All results meeting those criteria are marked in the results spreadsheet as shown below in bold, italic.

Data: Correlations (Poverty.sta)*				

Correlations (Poverty.sta)
Marked correlations are significant at p < .05000
N=30 (Casewise deletion of missing data)

Variable	POP_CHNG	N_EMPLD	PT_POOR	TAX_RATE
POP_CHNG	1.00	0.04	*-0.65*	0.13
N_EMPLD	0.04	1.00	-0.17	0.10
PT_POOR	*-0.65*	-0.17	1.00	0.01
TAX_RATE	0.13	0.10	0.01	1.00
PT_PHONE	*0.38*	0.36	*-0.73*	-0.04
PT_RURAL	-0.02	*-0.66*	*0.51*	0.02
AGE	-0.15	*-0.36*	0.02	-0.05

📷 Show Selection Conditions Button

Click the 📷 button to toggle the application of the *Selection Conditions* spreadsheet layout to cases meeting your current case selection conditions. In this way, you can visually distinguish the cases that meet current selection conditions. See *Spreadsheet Layouts* (page 234) for further details. Note that this button is only active when selection conditions are on (see below).

🖼 Selection Conditions On/Off Button

Click the 🖼 button to toggle the application of the case selection conditions. Note that when this command is toggled on, the status bar (see page 268) displays *Sel: ON*. If it is toggled off, the status bar displays *Sel: OFF*. To view or edit the case selection conditions, double-click this field on the status bar.

🏋 Weight Button

Click the 🏋 button to display the *Spreadsheet Case Weights* dialog.

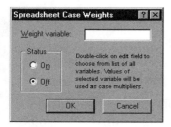

Defining weights allows you to adjust the contribution of individual cases to the outcome of an analysis by weighting those cases in proportion to the values of a selected variable. See *Weight* (page 283) for further details on this dialog.

⬇ Sort Button

Click the ⬇ button to display the *Sort Options* dialog.

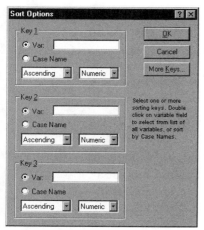

Use this dialog to specify how to sort variables or cases using up to three different sorting keys. To sort using more keys, click the ***More Keys*** button. See *Sort* (page 290) for further details on this dialog.

⊞=? **Recalculate Button**

Click the ⊞=? button to recalculate the current variable or, optionally, all variables defined by formulas in the current datafile (you can also press F9 when the spreadsheet is active to open this dialog and recalculate all valid formulas).

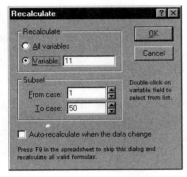

Those formulas can be entered as part of the variable specifications. To enter or edit the specifications, double-click on the variable name in the spreadsheet. See *Variable Specs* (page 293) for further details.

Note that if the intended result of the transformation is recoding of values (rather than performing arithmetic operations), there is a designated recoding facility available from the spreadsheet ⬚ Vars ▾ button (see below). See also *Formulas, Transformations, Recoding, Verifying, Cleaning Data* (page 361) for more information on data transformations in *STATISTICA*.

⬚ Vars ▾ **Variables Button**

Click the ⬚ Vars ▾ button to display a menu of global ***Variable*** editing and restructuring options: ***Add***, ***Move***, ***Copy***, ***Delete***, edit ***Specs*** (see page 293), edit ***All Specs*** (see page 294), edit ***Text Labels*** (see page 295), ***Create Subset/Random Sampling*** (see page 290), ***Verify Data*** (see page 292), ***Rank*** (see page 299), ***Recalculate*** (see above), ***Recode*** (see page 301), ***Replace Missing Data*** (see page 302), ***Shift(Lag)*** (see page 303), ***Standardize*** (see page 303), and ***Date Operations*** (see page 304).

Global vs. Clipboard operations on variables. Unlike the Clipboard editing operations that affect blocks of data (or contents of entire columns, that is, variables), these global operations affect not only the contents of the variables, but the variables themselves. For example, the **Delete** operation removes the contents of the selected range of variables and the variables themselves.

Also, note the difference between these global operations performed on variables (treated as logical units of the *STATISTICA* datafiles) and all Clipboard spreadsheet operations, which work the same way as in all standard spreadsheets (e.g., Microsoft Excel). For example, the global operations of copying, deleting, or moving variables available from this menu do not depend on the current location of the cursor or selected block. The operations are always performed on all cases of selected variables, regardless of whether or not all cases or only a subset of cases are currently selected for the respective variables.

On the other hand, in the case of Clipboard operations, only the segment of data that is selected will be copied, and (following the common spreadsheet conventions) pasting always begins from the current cursor position and proceeds down.

Thus, even when you (a) select and copy an entire variable, (b) select another (entire) variable and then (c) intend to paste the Clipboard content to that new location ("replacing" the previous values), the operation will be performed as intended only if you have placed the cursor at the top of the new variable. If you placed it somewhere in the middle, then the pasting will start from that point down.

Drag-and-drop operations. Note that all drag-and-drop with **Insert** (press the SHIFT key) operations change the size of the datafile; the same is true of simply dragging blocks outside the boundaries of the current datafile. However, dragging (without **Insert**) within the current datafile boundaries produces results identical to using the Clipboard (except that the

Clipboard is not used). For more information, see *What Are the Drag-and-Drop Facilities?* (page 352).

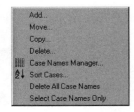 Cases Button

Click the Cases ▾ button to display a menu of global case editing and restructuring options: **Add**, **Move**, **Copy**, **Delete**, **Case Names Manager** (see page 296), **Sort Cases** (see page 290), **Delete All Case Names**, and **Select Case Names Only**.

Global vs. Clipboard operations on cases. Unlike the Clipboard editing operations that affect blocks of data (or contents of entire rows, that is, cases), these global operations affect not only the contents of the cases, but the cases themselves. For example, the **Delete** operation removes the contents of the selected range of cases and the cases themselves.

See *Variables Button* (page 319) for more explanation and examples of the differences between the global and Clipboard operations and the drag-and-drop operations.

Tools Toolbar

The **Tools** toolbar provides access to commonly used options for marking/unmarking cells in the spreadsheet and for updating and replacing selection conditions. This toolbar is available by selecting **Tools** from the **View - Toolbars** menu or on the **Toolbars** tab of the **Tools - Customize** dialog. For information on customizing toolbars, see Chapter 1 – *STATISTICA – A General Overview*.

Mark (Selected Cells) Button

Click the button to add the selected cells to the set of all marked cells (i.e., the **Marked Cells** spreadsheet layout is applied to the selected cells) in the spreadsheet. Any previously marked cells remain marked. See *Spreadsheet Layouts* (page 234) for further details.

Cells marked by *STATISTICA*. Many procedures in *STATISTICA* automatically mark specific cells or blocks in spreadsheets in order to "highlight" some results (e.g., unusually high frequencies in a frequency table, statistically significant correlation coefficients in a correlation matrix, or statistically significant effects in an ANOVA table of all effects). For example, in the following spreadsheet, note that (among others) the correlation between **POP_CHNG** and **PT_POOR** and between **POP_CHNG** and **PT_PHONE** are shown in bold, italic indicating significance at p<.05.

Data: Correlations (Poverty.sta)*				
Correlations (Poverty.sta) Marked correlations are significant at p < .05000 N=30 (Casewise deletion of missing data)				
Variable	POP_CHNG	N_EMPLD	PT_POOR	TAX_RATE
POP_CHNG	1.00	0.04	**-0.65**	0.13
N_EMPLD	0.04	1.00	-0.17	0.10
PT_POOR	**-0.65**	-0.17	1.00	0.01
TAX_RATE	0.13	0.10	0.01	1.00
PT_PHONE	*0.38*	0.36	*-0.73*	-0.04
PT_RURAL	-0.02	*-0.66*	*0.51*	0.02
AGE	-0.15	*-0.36*	0.02	-0.05

Such spreadsheets usually offer an option to change the criterion used to select the cells to be highlighted (e.g., the *p*-level for correlation coefficients) via the analysis definition dialog. For example, the **Options** tab of the **Product-Moment and Partial Correlations** dialog contains an option for specifying the **p-level for highlighting** in results spreadsheets.

Highlighted values in the Graph Data Editor. The **Graph Data Editor** is a special type of spreadsheet associated with each graph, which contains the specific values displayed in the graph. The values highlighted in that **Editor** (i.e., displayed in a different color) are those that are selected using the interactive **Brushing** tool ⊕ in the graph.

Note that the term highlighted has a specific meaning when used in the context of brushing data points in the graph. There, data points that are highlighted (in the graph) are those that have only been selected with the mouse pointer but still not **Marked**, **Labeled**, or **Turned OFF** as long as a brushing **Action** (e.g., **Labeling** or **Marking**) is not executed. See the *Brushing Toolbar Button* ⊕ in Chapter 6 – *Graphs: General Features* for more details.

⊞ Unmark (Selected Cells) Button

Click the ⊞ button to remove the selected cells from the set of marked cells in the spreadsheet (i.e. the **Marked Cells** spreadsheet layout is removed from the selected cells). Any previously marked cells that are not selected remain marked. See the **Mark (Selected Cell)** button, above, for information on cells marked by *STATISTICA* and cells highlighted in the **Graph Data Editor**.

New Mark Button

Click the 斑 button to replace the current set of marked cells with the selected cells in the spreadsheet (i.e., the **Marked Cells** spreadsheet layout is applied to the selected cells AND is removed from all unselected cells in the spreadsheet). See *Mark (Selected Cells) Button* (page 321) for information on cells marked by *STATISTICA* and cells highlighted in the **Graph Data Editor**.

Unmark All Button

Click the 斑 button to unmark all of the cells in the spreadsheet (i.e. the **Marked Cells** spreadsheet layout is removed from all cells). See *Mark (Selected Cells) Button* (page 321) for information on cells marked by *STATISTICA* and cells highlighted in the **Graph Data Editor**.

Select - Include Button

Click the 斑 button to add the selected cases in the spreadsheet to the existing case selection conditions. The selected cases are added to the **Include cases** group on the **Spreadsheet Case Selection Conditions** dialog. All other case selection conditions remain the same. See *Selection Conditions – Edit* (page 281) for details on defining case selection conditions.

Select - Exclude Button

Click the 斑 button to remove selected cases in the spreadsheet from the existing case selection conditions. The selected cases are added to the **Exclude cases** group on the **Spreadsheet Case Selection Conditions** dialog. All other case selection conditions remain the same. See *Selection Conditions – Edit* (page 281) for details on defining case selection conditions.

New Selection Conditions Include Button

Click the 斑 button to add the selected cases in the spreadsheet to the **Include cases** group on the **Spreadsheet Case Selection Conditions** dialog, replacing all other case selection conditions. Hence, new case selection conditions are created that include only the selected cases. See *Selection Conditions – Edit* (page 281) for details on defining case selection conditions.

FAQ: WORKING WITH SPREADSHEETS

General

Can I Open More Than One Input Datafile Simultaneously?

Yes. Moreover, you can simultaneously run different analyses on the same or different datafiles. To open (and use) more than one input datafile simultaneously, open multiple datafiles and make sure each is specified as an input datafile (by selecting **Input Spreadsheet** on the **Data** menu) as you start the respective analyses.

For example, open the file **Adstudy.sta** and select **Basic Statistics/Tables** from the **Statistics** menu. When the **Basic Statistics and Tables** (Startup Panel) is displayed, double-click **Descriptive Statistics** on the **Quick** tab and calculate summary statistics for the variable **Measure01**.

Next, open the file **Aggressn.sta** and once again select **Basic Statistics/Tables** from the **Statistics** menu. This time, double-click **t-test, independent, by groups** and perform an independent *t*-test with **Gender** as the grouping variable and **Aggressn** as the dependent variable.

You will notice that the analyses stay associated with the different datafiles when you switch between them (using the buttons next to the *STATISTICA* Start button, above the status bar, see page 268). Note that you can also perform different analyses on the same or different datafiles. For example, you could also perform analysis of variance on **Adstudy.sta** or on some other datafile.

While you can specify any number of spreadsheets as input datafiles, only one file (per workbook) can be designated as the active input datafile.

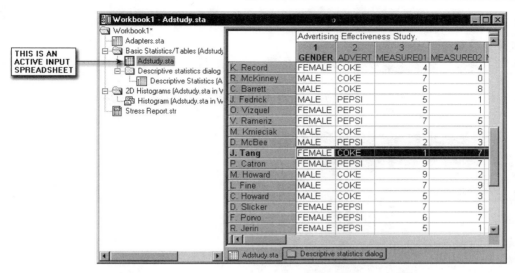

This means that if all your spreadsheets are opened within one workbook, only one of them can be used for input data at a time. The active input data spreadsheet is highlighted in the workbook tree with a red square, as shown above.

Can I Use the Results of One Analysis to Perform Another Analysis?

Yes. To designate a spreadsheet (i.e., the results from one analysis) as an input spreadsheet, select the spreadsheet (e.g., ensure the spreadsheet has the focus) and select **Input spreadsheet** from the **Data** menu. Now you can specify an analysis, and *STATISTICA* uses the selected spreadsheet as the input spreadsheet. Note that if you switch back to another spreadsheet that has previously been designated as an input spreadsheet, it can still be used for analyses as well.

While you can specify any number of spreadsheets as input datafiles, only one file (per workbook) can be designated as the active input datafile. This means that if all your spreadsheets are opened within one workbook, only one of them can be used for input data at a time. The active input data spreadsheet is highlighted in the workbook tree with a red square as shown in *Can I Open More Than One Datafile Simultaneously?* (page 324).

Can I Create a Custom List to Use for Extrapolation?

Yes. Select the **Custom Lists** tab of the **Options** dialog (accessible via **Tools - Options**) to edit or create custom lists for *STATISTICA* to use.

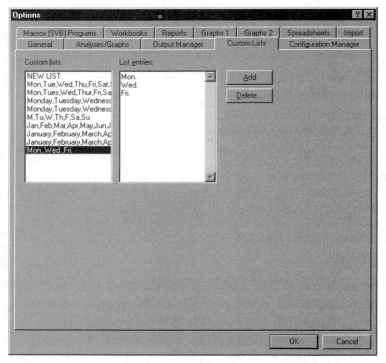

The **Custom lists** box reports all the custom lists currently available in *STATISTICA*. To see the entries included in a custom list, select it here and the entries are shown in the **List entries** box. To create a new list, select **NEW LIST** and enter the list items for it in the **List entries** box. Once you have added all the items for the new list, click the **Add** button and the list is added to those in the **Custom lists** box (see **Mon., Wed., Fri.** in the above illustration). When entering data in spreadsheets, you can enter the first few items in the list, and then use extrapolation to fill in the remaining cases in the spreadsheet (as shown below).

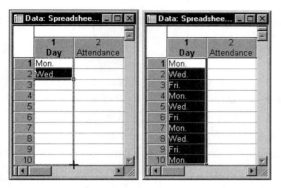

See *How Does the Extrapolation of Blocks (AutoFill) Work?* (page 380) for further details.

What Is a Matrix Spreadsheet?

A *STATISTICA* Matrix Spreadsheet (**.smx*) is a spreadsheet file with a strictly defined format that can be used as input in several modules of *STATISTICA* (e.g., **Cluster Analysis**, **Multidimensional Scaling**, **SEPATH**). Matrix spreadsheets have to meet the following conditions:

- The number of cases (rows) = the number of variables (columns) + 4.

- The matrix must be a square matrix and the case names should be the same as the variable names.

- The last four cases contain the following case names and information:

Means: The mean of each variable is given in this row; this case can be left empty (i.e., do not enter anything in this row) for Similarities and Dissimilarities matrices.

Std.Dev.: The standard deviation of each variable is given in this row; this case can be left empty (i.e., do not enter anything in this row) for Similarities and Dissimilarities matrices.

No.Cases: This required entry is the number of cases from which the matrix was produced, not the number of cases (rows of data) in this matrix file.

Matrix: This required entry represents the type of matrix file; 1 = Correlation, 2 = Similarities, 3 = Dissimilarities, and 4 = Covariance.

When entering these last four cases into the matrix file manually, be sure to spell the case names exactly as they appear above (i.e., *Means*, *Std.Dev.*, *No.Cases*, and *Matrix*).

While a matrix spreadsheet looks identical to a regular spreadsheet, analyses that require matrix input (such as Multidimensional Scaling) only consider input spreadsheets that are also matrix spreadsheets.

You can create a matrix spreadsheet from any place that allows you to create a matrix of results [e.g., **Basic Statistics - Correlations**, **Multiple Regression**, **Cluster Analysis**, **Multidimensional Scaling**, and **Structural Equation Modeling (SEPATH)**]. To create the matrix, click the ![Matrix] button. Note that clicking this button does not cause *STATISTICA* to save the matrix file to disk, rather it causes *STATISTICA* to create the special matrix spreadsheet. This matrix spreadsheet is always displayed in an individual window (regardless of the settings on the **Options - Output Manager** tab or the **Analysis/Graph Output Manager** dialog). You can, however, add the spreadsheet to a workbook or report using the ![Add to Workbook ▾] or ![Add to Report ▾] button, respectively.

When you save a matrix spreadsheet (using **Save** or **Save As** from the **File** menu), *STATISTICA* detects that it is in this special matrix form and sets the default file extension to **.smx*.

Does *STATISTICA* Support Drag-and-Drop Opening of Spreadsheets and Graphs?

Yes. In Windows *Explorer,* you can simply drag any *STATISTICA* Spreadsheet (file name extension **.sta*) or graphics file (file name extension **.stg*) onto the *STATISTICA* icon to automatically open the file in *STATISTICA*.

How Much Disk Space Is Necessary to Perform Database Management Operations?

In order to allow you to revert to the original datafile after file editing, *STATISTICA* creates temporary and backup files. Thus, in order to edit a datafile, *STATISTICA* needs at least twice as much free disk space as the size of the file to be edited. More disk space is necessary if you perform large-scale editing operations (e.g., changing large blocks) on very large datafiles. To allow for the **Undo** function, *STATISTICA* needs to store multiple copies (up to 32) of the modified sections of data. Some operations (e.g., import/export) use intermediate scratch files, so they may need additional space.

Can I Undo Editing Operations?

You can undo most operations such as editing, moving, or copying blocks; random fill; recoding or ranking variables; etc. by selecting **Undo** from the spreadsheet **Edit** menu or pressing CTRL+Z. Multi-level undo is supported (with 32 buffers), so you can undo up to 32 of the most recent spreadsheet operations.

What Is the Difference between the File Header and a Customized Header?

File Header. The file header is located immediately above the variable headers at the top of the spreadsheet window. You can use this header to include a summary title or other identification of the datafile (or results). Double-click in the header area (or press F2) to edit/enter information.

To select the header only (for formatting), click once in the upper-left corner (the mouse pointer will be an outlined plus sign ✛). Press CTRL+ENTER or ALT+ENTER to enter a new line (note that you need to extend the height of the field to see new lines that you are adding). For more information on the spreadsheet window, see *Spreadsheets - Multimedia Tables* (page 223).

In addition to adding new lines to the file header, *STATISTICA* Spreadsheets (and all other *STATISTICA* documents) can easily be annotated by entering notes into the **Comments** area of the **Document Properties** (on the **Summary** tab) dialog, which is accessed by selecting **Properties** from the **File** menu. The **Comments** area can hold any type of comments or supplementary information.

Customized header. A customized header is used when printing spreadsheets (or other *STATISTICA* documents, e.g., reports, graphs, etc.). This header can include a title, page numbers, time of printing, date, the name of the file, and any custom text you might want to include. To create a header, select **Header/Footer** from the **View** menu. For more information, see the **Modify Header/Footer** dialog in the *Electronic Manual*. Note that while the file header is associated only with spreadsheets, any printed *STATISTICA* document can have a customized header.

What Graphs Are Available from Spreadsheets?

A variety of graphs is available from the spreadsheet shortcut menu (accessed by right-clicking in any cell or selected block of the spreadsheet). These graphs belong to two broad categories, *Graphs of Block Data* and *Graphs of Input Data*, and they produce either summary graphs from rows or columns of the currently highlighted block, or graphs as the original input data identified by the currently selected row and/or column of the spreadsheet (respectively).

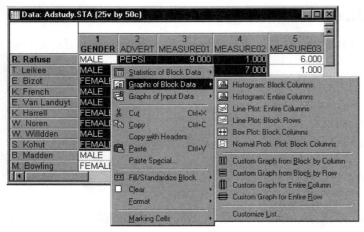

For more information of types of graphs available in *STATISTICA*, see Chapter 6 – *STATISTICA Graphs: General Features*.

How Do I Insert an Object into the Spreadsheet?

You can insert either a new or an existing object into the spreadsheet. To do this, select *Object* from the *Insert* menu to display the *Insert Object* dialog. To insert a new object, select the *Create New* option button, select the type of object to be created from the *Object Type* box, and click the *OK* button. In the following example, a *Bitmap Image* will be inserted into the spreadsheet.

Note that *STATISTICA* will now emulate Microsoft Paint, and you can edit the new bitmap image using Paint utilities (as shown below).

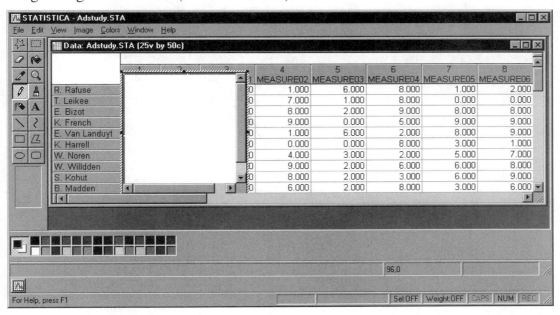

To return to the standard *STATISTICA* user interface, click a cell in the spreadsheet. (Note that if the **Display As Icon** check box is selected on the **Insert Object** dialog, then the object is displayed as an icon in the spreadsheet, and when you double-click the icon, the application in which it was created will be opened.)

To insert an existing object into the spreadsheet, select the **Create from File** option button on the **Insert Object** dialog. Next, click the **Browse** button to display the **Browse** dialog and select the file to be displayed in the spreadsheet. You can also select the **Display As Icon** check box to display the selected file as the icon that is pictured below the **Display As Icon** check box. Click

the **Change Icon** button to specify a new icon. To link the object to the original file (so that changes in that file will automatically be reflected in the object that is displayed in *STATISTICA*), select the **Link** check box.

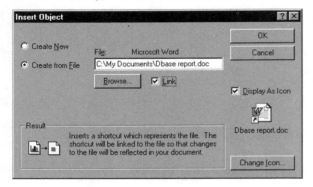

Finally, click the **OK** button.

	1 GENDER	2 ADVERT	3 MEASURE01	4 MEASURE02	5 MEASURE03	6 MEASURE04	7 MEASURE05	8 MEASURE06
R. Rafuse			9.000	1.000	6.000	8.000	1.000	2.000
T. Leikee			6.000	7.000	1.000	8.000	0.000	0.000
E. Bizot	Dbase report.doc		9.000	8.000	2.000	9.000	8.000	8.000
K. French			7.000	9.000	0.000	5.000	9.000	9.000
E. Van Landuyt	MALE	PEPSI	7.000	1.000	6.000	2.000	8.000	9.000
K. Harrell	FEMALE	COKE	6.000	0.000	0.000	8.000	3.000	1.000
W. Noren	FEMALE	COKE	7.000	4.000	3.000	2.000	5.000	7.000
W. Willdden	MALE	PEPSI	9.000	9.000	2.000	6.000	6.000	8.000
S. Kohut	FEMALE	PEPSI	7.000	8.000	2.000	3.000	6.000	9.000
B. Madden	MALE	PEPSI	6.000	6.000	2.000	8.000	3.000	6.000

Data: Adstudy.STA (25v by 50c)

Double-click on the object to edit it. It will be opened in the application in which it was created (unless both the **Link** and the **Display As Icon** check boxes are cleared, in that case, *STATISTICA* will emulate the application in which the object was created, as shown in the insertion of the **Bitmap Image** above).

Cases, Variables, Formats

What Are Cases and Variables?

STATISTICA datafiles are organized into cases and variables. If you are not familiar with this notation, you can think of cases as the equivalent of records in a database management

program (or rows of a spreadsheet) and variables as the equivalent of fields (columns of a spreadsheet). Each case consists of a set of values of variables.

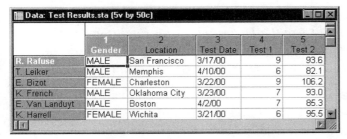

See the *STATISTICA Spreadsheets – General Overviews* (page 221) for information on the organization of datafiles in *STATISTICA*.

Why Do You Need Different Variable Types?

The difference between the text and the numeric types is straightforward; however, the main reason for having three types of numeric values is the storage efficiency. For most datafiles, that is not important and, thus, using the default (***Double***) data type is recommended. However, for very large datafiles, being able to switch to a 2 (or even 8) times more efficient storage (by using different data types) could make a difference between being able to perform the necessary analysis on a specific computer system or not. For further details, see *Variable Types* (page 222).

How Do I Add/Delete Variables (Columns of Data)?

Perhaps the easiest way to add variables to an existing datafile is to expand the spreadsheet beyond the existing columns and double-click on the gray area of the spreadsheet. This will display the ***Add Cases and/or Variables*** dialog.

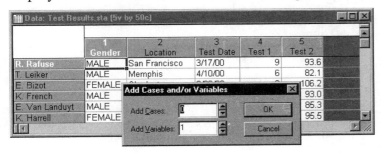

Specify the number of variables to add and click the **OK** button. The new variables are added to the far right of the datafile.

Variables can also be added to and/or deleted from a datafile either by (1) using the drag-and-drop operation (insert to add columns), see *What Are the Drag-and-Drop Facilities?* (page 352), or (2) selecting the appropriate option from the spreadsheet menu of global operations on variables, accessed by clicking the ⊻ars ▾ button on the **Spreadsheet** toolbar (see page 319).

How to add variables before variable 1. The quickest way to add variables before variable 1 is to select **Add Variables** from the spreadsheet **Insert** menu. In the **After** box of the **Add Variables** dialog, enter **0**, which references the variable number. If the variable number is **0**, it effectively means that you are inserting before the first variable.

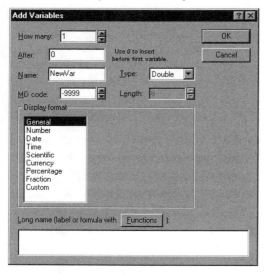

Once you click the **OK** button, a new variable is inserted before the first variable of your spreadsheet.

	1 NewVar	2 GENDER	3 ADVERT	4 MEASURE01	ME
	Advertising Effectiveness Study.				
R. Rafuse		MALE	PEPSI	9	
T. Leiker		MALE	COKE	6	
E. Bizot		FEMALE	COKE	9	
K. French		MALE	PEPSI	7	
E. Van Landuyt		MALE	PEPSI	7	
K. Harrell		FEMALE	COKE	6	
W. Noren		FEMALE	COKE	7	
W. Willden		MALE	PEPSI	9	
S. Kohut		FEMALE	PEPSI	7	

Note that the name of the new variable is the name specified in the **Name** box of the **Add Variables** dialog (in this example, **NewVar**).

How Do I Add/Delete Cases (Rows of Data)?

Perhaps the easiest way to add cases to an existing datafile is to double-click on the gray area of the spreadsheet. This displays the **Add Cases and/or Variables** dialog.

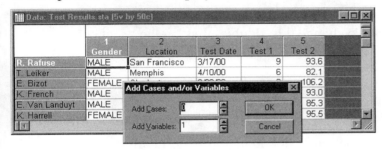

Specify the number of cases you want to add and click the **OK** button. The new cases are added to the bottom of the datafile.

Cases can also be added to and/or deleted from a datafile either by (1) using the drag-and-drop operation (insert to add cases), see *What Are the Drag-and-Drop Facilities?* (page 352), or (2) selecting the appropriate option from the spreadsheet menu of global operations on cases, accessed by clicking the Cases ▼ button on the **Spreadsheet** toolbar (see page 321).

How to add cases before case 1. The quickest way to add cases before case 1 is to select **Add Cases** from the spreadsheet **Insert** menu. In the **Insert after case** box of the **Add Cases** dialog, enter **0**, which references the case number. If the case number is **0**, it effectively means that you are inserting before the first case.

After you click the **OK** button, a new case is inserted before the first case of your spreadsheet.

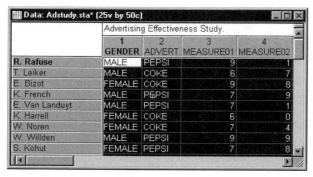

Note that the name of the new case is *1*. To change this, double-click on the *1* and enter a new name.

Can I Select (Highlight) the Entire Spreadsheet?

Yes. Following standard spreadsheet conventions, click in the upper-left corner of the spreadsheet (in the *Info Box*, see *Spreadsheets - Multimedia Tables*, page 223). This shortcut is useful, for example, when you intend to copy the entire file to the Clipboard or reformat the entire spreadsheet. You can also press CTRL+A or select *Select All* from the spreadsheet *Edit* menu or shortcut menu (accessed by right-clicking in the *Info Box*).

Note that by clicking twice in the *Info Box*, you deselect (i.e., select and then deselect) the entire spreadsheet.

Can I Edit the Contents of a Cell (and Not Overwrite It)?

Yes. Follow the standard spreadsheet (e.g., Microsoft Excel) conventions, specifically:

Double-click on the cell. To avoid clearing the contents of the cell at the point when you start entering a correction, double-click on the cell before you start typing. This action enters the edit mode and positions the cursor within the cell.

Press F2. Alternatively, you can follow the "old" spreadsheet convention and press the F2 key, which also enters the edit mode for the currently highlighted cell.

How Are Dates Represented in *STATISTICA* Datafiles?

Date values of variables are stored internally in Julian format, that is, as a single integer value that represents the number of days that have passed since January 1, 1900. For example, a date entered and displayed as *1/21/1968* is stored as the Julian date *24858*; the (optional) decimals are interpreted as time (see *How Is Time Represented in STATISTICA Datafiles?*, page 338). Date values stored in this manner can be used in subsequent analyses (e.g., in **Survival Analysis** in order to calculate survival times, see below) and transformed using arithmetic operations; at the same time, they can be displayed as dates in reports or graphs (e.g., used to label scale values).

Julian date values can be displayed in the spreadsheet in numeric (Julian) format or in one of several predefined date display formats (e.g., *1/6/64*, *6-Jan-64*, *Jan-1964*, *01/06/64*, *01/06/1964*, *6-Jan*).

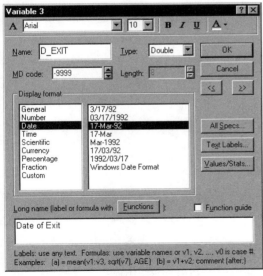

To change the date display format, select **Date** in the **Display format** group of the **Variable** specifications dialog (accessible by double-clicking on the variable name in the spreadsheet

or selecting **Variable Specs** from the **Tools** menu) and choose one of the predefined display formats.

When you enter data into a new variable using a format that is recognized as valid in the **Display format** group of the **Variable** specifications dialog (e.g., **Time**, **Date**, **Currency**, etc.), *STATISTICA* displays the **Auto Format Cells** dialog. In this dialog, you can either apply the format to the individual cell or to the whole column.

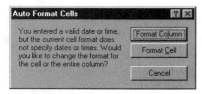

You can create a single date variable from two (month, year or day, month) or three (day, month, year) variables as well as split a single date variable into two or three variables via the **Date Operations** dialog (accessed from the **Data - Date Operations** menu - see page 304 for further details on the **Date Operations** dialog).

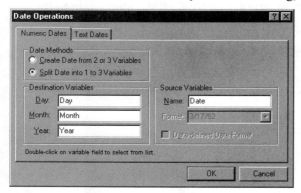

Refer to the on-line *Electronic Manual* (press the F1 key or click ) for examples of creating date variables from numeric variables and splitting date variables into numeric variables.

How Is Time Represented in *STATISTICA* Datafiles?

Time values of variables are stored as (optional) decimal values representing the fraction of the day since midnight; for example, 6:00 A.M. is stored as 0.25. Time values stored in this manner can be used in subsequent analyses and transformed using arithmetic operations; at the same time, they can be displayed as times in reports or graphs (e.g., used to label scale values).

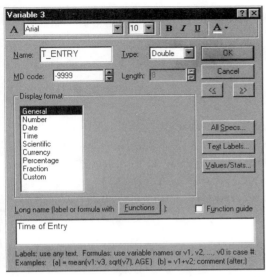

STATISTICA formats the display of time values according to the current settings of the Windows *Control Panel*.

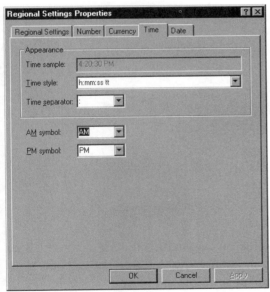

See the previous topic (*How Are Dates Represented in STATISTICA Datafiles*, page 337) for a discussion of the *Auto Format Cells* and the *Date Operations* dialogs. Refer to the on-line *Electronic Manual* (press the F1 key or click **?** in the *Date Operations* dialog) for examples of

creating time variables from numeric variables and splitting time variables into numeric variables.

How Do I Convert Date-Values into Text Labels and Vice-Versa?

In some circumstances, it can be useful to create text labels with date information [e.g., when using a date variable as a coding variable with codes greater than 32,000. In this case, you can transform the date variable into a variable containing date text labels with numeric equivalents in a range that allows them to be used as codes by all procedures of *STATISTICA* (i.e., numeric values less than 32,000). Use the **Dates Operations –Text Dates** tab to convert dates into codes.

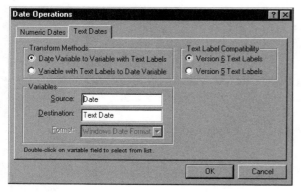

This dialog is accessed by selecting **Date Operations** from the **Data** menu. See *How Are Dates Represented in STATISTICA Datafiles?*, page 337, or *Date Operations,* page 304, for further details.

How Do I Review and Edit Variable Specifications (Names, Formats, Notes, Formulas, etc.)?

Specifications of a single variable. Double-click on a variable name in a spreadsheet to display the **Variable** specifications dialog where you can change the variable name, font format, display format, missing data value, long label, or formula for the current variable. (As in most other facilities commonly used in *STATISTICA*, this dialog can also be accessed from the toolbar, shortcut menus, and menus.)

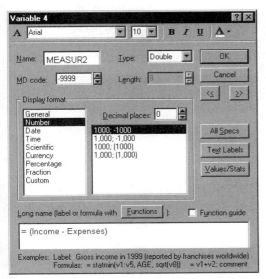

This dialog can also be used to access descriptive statistics and a listing of sorted numeric values and text labels for the current variable (click the **Values/Stats** button). From this dialog, you can also access the **Text Labels Editor** (click the **Text Labels** button) to review and change the assignments between the text and numeric values for the current variable. Note that you can click the ⸨ ⸩ buttons to scroll through each variable in your datafile. See *Variables Specs* (page 293) for further details on this dialog.

Specifications of all variables. You can also click the **All Specs** button in the **Variable** specifications dialog to display an editable, combined table of specifications of all variables in the current datafile. Note that selecting **All Variable Specs** from the **Data** menu or **All Specs** from the ▾Vars ▾ button on the **Spreadsheet** toolbar can also access this dialog.

The table format is convenient when you need to compare or edit specifications of several variables, especially when you need to paste and copy between variables (e.g., comments, formulas, or links), or extend a format definition or missing data code from one variable to subsequent variables.

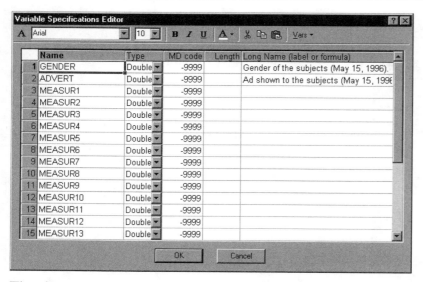

The shortcut menu contains commands to **Add Vars**, **Delete Vars**, **Cut**, **Copy**, **Paste**, and **Fill/Copy Down**.

How Are Missing Data Handled in *STATISTICA*?

Specifying missing data codes. A missing data code, that is, a value that signifies that there is no data for a particular case and variable (displayed as a blank cell in the spreadsheet), can be specified separately for each variable. To change the code, double-click on a variable name in the spreadsheet to access the **Variable** specifications dialog for that variable. Alternatively, click the Vars ▾ toolbar button and select **All Specs** from the resulting menu to display the **Variable Specifications Editor** dialog for all variables. The default **Missing Data code** in *STATISTICA* (used when new files are created, new variables are added, or data are imported) depends on the variable type being used. For more details, see *Variable Types* (page 222).

Processing cases with missing data. The way in which missing data are handled when processing data can be adjusted individually for each analysis (see the **MD deletion** group in most analysis definition dialogs). Whenever applicable, you have the choice to eliminate them from calculations in a casewise or pairwise manner, substitute them with means, or reconstruct or interpolate them (e.g., in the **Time Series** module).

Press the F1 key or click the help button ❓ in the respective analysis-definition dialog to learn about the specific missing data handling options available for the procedures of interest.

How Do I Change the Missing Data Code for Individual Variables?

The value used to designate missing data values for individual variables can be changed in all **Variable** specifications dialogs [see *How Do I Review and Edit Variable Specifications (Names, Formats, Notes, Formulas, etc.)?*, page 340].

How Do I Edit Case Names?

In order to enter or edit the current case names, double-click on any case name in the spreadsheet. To manage (e.g., change the width, copy from a variable, etc.) case names, right-click on any case name and select **Case Names Manager** from the resulting shortcut menu to display the **Case Names Manager** dialog. The **Case Names Manager** command is also available from the Cases ▾ toolbar button and the **Tools** menu.

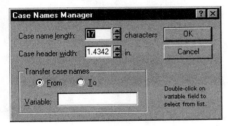

See *Case Names Manager* (page 296) for further details on this dialog. If the current datafile contains no case names, you can create them by double-clicking on the case number column or resizing the case number column with the mouse pointer.

How Do I Create a New Cell/Block Format?

A format is a group of formatting options that you can save with an assigned name for future use. Formats can be assigned to numbers as well as text. To create a new format in a spreadsheet, follow these instructions:

Select **Format Manager** from the **Format - Block** menu to display the **Format Manager** dialog.

In this dialog, click the **New** button to display the **Format Name** dialog.

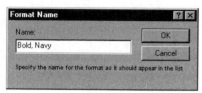

Enter a name in the **Name** box, and then click the **OK** button to display the **Edit Format** dialog.

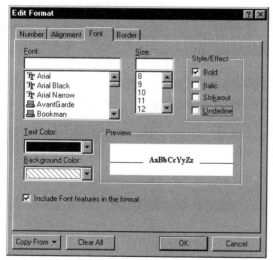

This dialog contains four tabs (**Number**, **Alignment, Font**, and **Border**) that are used to specify the formatting components for the new format. Select the desired formatting options for your new format. In the illustration above, the **Bold** check box has been selected in the **Style/Effect** group and the **Text Color** box has been changed to navy.

Click the **OK** button to close the **Edit Format** dialog. The format that you just created is added to the list of formats available on the **Format Manager** dialog and the **Format – Block** menu.

You can now apply this format to new selections (see the next section).

How Do I Apply a Format?

Formats can be applied quickly and consistently to segments of data. Once a format has been created (see *How Do I Create a New Cell/Block Format?,* page 343), it can be applied to any part (or all) of the data in a spreadsheet.

To apply an existing format, first select the cell(s) that you want to format in the spreadsheet. Then select **Block** from the **Format** menu to display the list of available formats. (Optionally, right-click on the selected cell(s) and choose **Format** from the shortcut menu, as displayed below.) Finally, select the name of the format to apply to the selected cells.

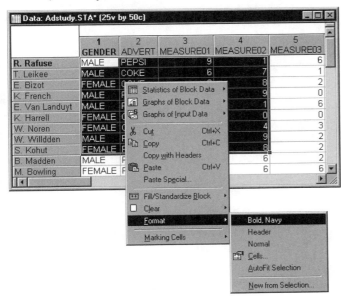

How Do I Create a New Spreadsheet Layout?

A spreadsheet layout is a convenient way to combine sets of formatting options into one collection. This feature can be used to specify and apply different formats for different components of a spreadsheet. Once you have created a spreadsheet layout, it can be applied to any spreadsheet for a consistently formatted appearance (see *How Do I Apply a Spreadsheet Layout*, page 349). Follow these instructions to create a new spreadsheet layout:

Select **Layout Manager** from the **Format – Spreadsheet** menu to display the **Spreadsheet Layouts** dialog.

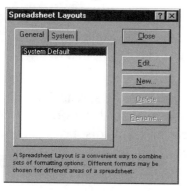

In this dialog, click the **New** button to display the **Spreadsheet Layout Name** dialog.

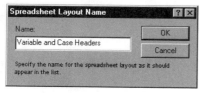

Enter a name in the **Name** box, and then click the **OK** button to display the **Edit Spreadsheet Layout** dialog.

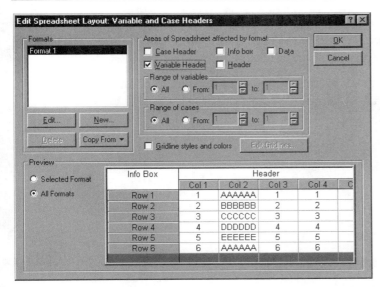

In the **Formats** group, a default **Format 1** will be displayed. Select the area of the spreadsheet to be affected by **Format 1** in the **Areas of Spreadsheet affected by format** group. In the above illustration, only the **Variable Header** check box is selected. Note that you can control the range of variable headers to be affected by **Format 1** by using the **Range of variables** group.

Next, click the **Edit** button in the **Format** group to display the **Edit Format: Format 1** dialog.

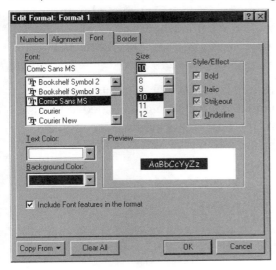

Select any formatting options to be associated with **Format 1**. Above, **Comic Sans MS** has been chosen as the **Font**, **10** for the **Size**, **white** for the **Text Color**, and **blue** for the **Background color**. Note that the **Preview** group is used to view your specified format.

Click the **OK** button to accept the settings and return to the **Edit Spreadsheet Layout** dialog.

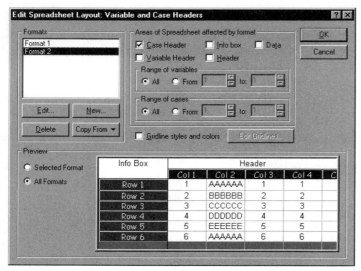

To define additional formats for other areas of the spreadsheet, click the **New** button to display the **New Format: Format 2** dialog and select any formatting options to be associated with **Format 2**. In this manner, you can add additional formats to the spreadsheet layout as necessary. In the illustration above, **Format 2 (Font - Comic Sans MS**, **Size - 10**, **Text Color – white**, and **Background color - blue)** has been applied to the Case Headers.

When you are satisfied with the appearance of your spreadsheet layout, which is displayed in the **Preview** group, click the **OK** button. The spreadsheet layout that you just created is added to the list of formats available on the **Spreadsheet Layouts** dialog and the **Format – Spreadsheet** command.

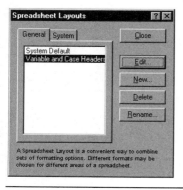

You can now apply this spreadsheet layout to new selections (see below).

How Do I Apply a Spreadsheet Layout?

You can apply any existing spreadsheet layout to the active spreadsheet. To do this, select **Spreadsheet** from the **Format** menu to display the list of available spreadsheet layouts. Then, select the name of the spreadsheet layout to apply to the spreadsheet.

Text Labels

Can I Label Numeric Values?

Yes, in *STATISTICA*, each numeric value can have two identities: the actual numeric value (e.g., *1*) and a text label (e.g., **Male**) assigned to it. For more information, see *How Do I Enter/Edit the Assignments between Numeric Values and Text Labels?* (below).

Note that in addition to text labels for numeric values, *STATISTICA* also supports text variables (see *Variable Types*, page 222).

How Do I Enter/Edit the Assignments between Numeric Values and Text Labels?

Normally, the assignments between text labels and numeric values are handled automatically as you enter or edit data in the spreadsheet (see *Can I Label Numeric Values?*, above). However, in some circumstances, you may want to review, edit, or restructure all assignments of values for a particular variable or copy text/numeric assignments from one variable to another. These operations can be performed in the **Text Labels Editor** dialog, accessible by selecting **Text Labels** from the `Vars ▾` toolbar button menu or by clicking the **Text Labels** button in the **Variable** specifications dialog (see page 293) for a particular variable.

Use the **Text Labels Editor** dialog to sort the assignments by text labels or numeric values, perform the **Fill Down** operation (to automatically reassign numeric values to text labels), **Copy** numeric/text assignments from other variables or from case names, and perform other operations.

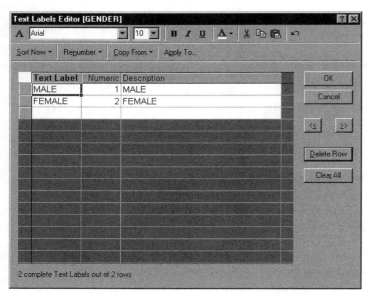

See *Text Labels Editor* (page 295) for further details on this dialog.

Can I Switch between Displaying Text Labels and Numeric Values in the Data Spreadsheet?

Yes. Each value of a particular variable can simultaneously have a numeric and text identity. You can toggle the display of text labels or numeric values by selecting *Display Text Labels* from the spreadsheet *View* menu or by clicking the ▧ toolbar button.

Can I Copy a Set of Text Labels, Numeric Values, and Descriptions to Other Variables and Files?

Yes, you can. Click the *Copy From* button in the *Text Labels Editor* (see above) to copy to the current variable the text labels, numerical values, and descriptions from other variables or from the case names.

When you need to copy the text labels, numerical values, and descriptions from one file to another, select *Text Labels* from the *Data - Merge* menu. This allows you to concatenate or

replace text labels with labels from another file (see *Can I Merge the Text Labels/Numeric Values from Two Files?*, page 361).

Can Clipboard and Drag-and-Drop Operations in the Spreadsheet Affect Values Outside the Range that is Being Pasted or Dropped?

Text labels. When you copy or move a block in the spreadsheet, the values copied to the Clipboard depend on the display mode of the spreadsheet. If the spreadsheet displays numeric values when the block is copied, then only those numeric values are copied to the Clipboard. If the spreadsheet displays text labels when the block is copied, then not only are the text labels copied to the Clipboard, but also the corresponding numeric values.

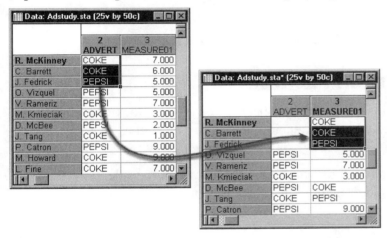

This can result in the assignment of text labels to numeric values that did not previously have text label equivalents. Note that in the illustration above, even though only the (highlighted) block was moved, other numeric values in the target variable (in this case *1*'s and *2*'s), acquired new text identities.

Display formats. When you copy or move a block in a spreadsheet, the display format for that block is also copied. This means that the display format of the copied or moved block overwrites the display format for the block into which it is copied. For example, if you copy a block of values that are formatted as currency and paste them into a variable that is formatted as percentage, the block remains formatted as currency even though the rest of the values in that variable are formatted as percentage.

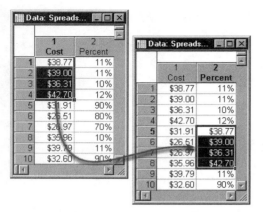

In the example above, the four currency values copied from **Cost** are still formatted as currency even though they were pasted into the variable **Percent**.

Operations on Blocks of Values

What Are the Drag-and-Drop Facilities?

STATISTICA Spreadsheets support all standard spreadsheet (Excel-style) block drag-and-drop operations (copy, move, insert, etc.) and block extension operations, as well as mouse pointer-controlled column width, variable-speed scrolling, split scrolling, etc.

Moving a block. You can move a block by pointing to the border of the selection (the mouse pointer changes to an arrow) and dragging it to a new location.

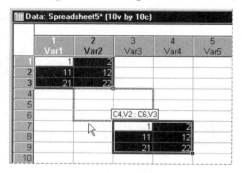

Copying a block. To copy a block, point to the border of the selection (the mouse pointer changes to an arrow), then drag the selection to a new location while pressing the CTRL key.

Note that when you are dragging the selection, a plus sign is displayed next to the cursor to indicate you are copying the text rather than moving it (see the cursor in the illustration below).

Inserting a block. To insert a block between columns or rows, point to the border of the selection (the mouse pointer changes to an arrow), then drag the selection while pressing the SHIFT key. If you point between rows, an insertion bar appears between the rows, and when you release the mouse button, the block is inserted between those two rows [creating new case(s)]. If you point the cursor between columns, an insertion bar appears between the columns, and when you release the mouse button, the block is inserted between those two columns [creating new variable(s)]. Note that if you also press the CTRL key while you are dragging the selection, the block will be copied and inserted instead of moved and inserted; a plus will appear next to the arrow-cursor, as shown in the illustration below.

Note that all SHIFT + drag-and-drop operations change the size of the datafile; the same is true of simply dragging blocks outside the boundaries of the current datafile. However, dragging (without SHIFT) within the current datafile boundaries produces results identical to using the Clipboard (except that the Clipboard is not used).

Clearing a block. To clear the contents of a block of cells, drag the block Fill Handle (a small solid square located on the lower-right corner of the block) within the selected block. After selecting the block, point at the lower-right corner of the block displaying the Fill

Handle. When the mouse pointer changes to a plus sign **+**, drag within the selected block. As you are dragging, the values within the block are dimmed. When you release the left mouse button, the dimmed values are deleted.

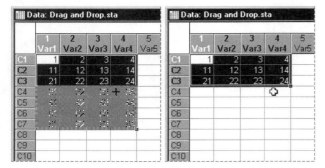

Extrapolating a block (AutoFill). Values within a block can be extrapolated by using intelligent, Excel-style, series extension facilities. After selecting the block, point at the lower-right corner of the block displaying the Fill Handle. When the mouse pointer changes to a plus sign **+**, drag to the last cell(s) to which you want to extrapolate the data.

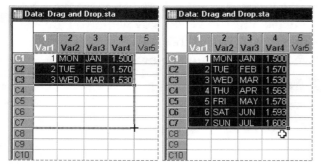

Numeric values are extrapolated using linear regression; text values are extrapolated following meaningful sequences (e.g., Q1, Q2, Q3, ..., or JAN, FEB, MAR, ...). See *How Does the Extrapolation of Blocks (AutoFill) Work?* (page 380) for more details. See also, *Can I Create a Custom List to Use for Extrapolation?* (page 326).

Can I Rearrange Blocks of Data or Ranges of Cases and Variables in a Datafile?

There are three types of operations that you can use to rearrange blocks of data or ranges of cases and variables in a datafile: Clipboard-based, global, and drag-and-drop-based; they operate differently and can produce different effects.

Clipboard-based operations. The standard Clipboard-based operations of cut, copy, and paste (implemented via the standard Clipboard keyboard, toolbar, or menu commands) affect only the contents of blocks of data, rows, or columns. They do not influence the overall size of the datafile (e.g., they can empty a column but will not remove the column from the spreadsheet).

Global operations. Global operations (accessible from the ▼ Vars and ▼ Cases *Spreadsheet* toolbar buttons and the spreadsheet *Insert* menu) are performed on entire rows or columns as "units"; for example, they move or delete entire columns, not only their contents. Hence, the size of the datafile increases or decreases when you use these options. Refer to *Variables Button* (page 319) for more information on those differences.

Drag-and-drop operations. *STATISTICA* supports standard (Microsoft Excel-style) drag-and-drop facilities, allowing you to easily perform both Clipboard-style (but without using the Clipboard) and global (if the insert mode is implemented by pressing the SHIFT key) operations; for details, see *What Are the Drag-and-Drop Facilities?* (page 352).

What Is the Difference Between Copy and Copy with Headers?

The example below shows how the selected portion of the spreadsheet will be copied to the Clipboard when using *Copy with Headers* or *Copy* (both available from the *Edit* menu).

	4 MEASURE02	5 MEASURE03	6 MEASURE04	7 MEASURE05	8 MEASURE06	9 MEASURE07
W. Noren	4	3	2	5	7	1
W. Willdden	9	2	6	6	8	7
S. Kohut	8	2	3	6	9	1
B. Madden	6	2	8	3	6	4
M. Bowling	6	6	5	6	8	7
J. Willcoxson	3	3	7	0	6	5
J. Landrum	2	3	1	8	1	4
M. Taylor	2	4	8	1	2	6
N.S. Madden	2	7	5	7	2	5
K. Ridgway	2	5	4	4	4	3
L. Cunha	9	9	3	1	4	5
F. Wind	0	7	5	2	4	2

Data: Adstudy.STA* (25v by 50c)

Copy with Headers:

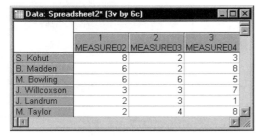

Copy:

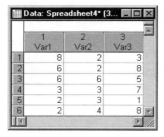

Do *STATISTICA* Spreadsheets
Support Split Display?

Yes. Spreadsheets can be split into up to four sections (panes) by dragging the split box (the black rectangle at the top of the vertical scrollbar or to the left of the horizontal scrollbar). This is useful if you have a large amount of information and you want to review results from different parts of the spreadsheet. When you guide the mouse pointer to the split box, the pointer changes to ⬌ (or ⬍). Now, to position the split, drag to the desired position.

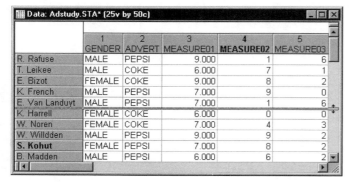

You can change the position of the split by dragging the split box (now located between panes) to a new position. Note that vertically split panes scroll together when you scroll vertically; horizontally split panes scroll together when you scroll horizontally.

Data: Adstudy.STA* (25v by 50c)							
	1 GENDER	**2** ADVERT	**3** MEASURE01	**22** MEASURE20	**23** MEASURE21	**24** MEASURE22	**25** MEASURE23
R. Rafuse	MALE	PEPSI	9	2.000	3.000	3.000	7.000
T. Leikee	MALE	COKE	6	4.000	8.000	0.000	7.000
E. Bizot	FEMALE	COKE	9	1.000	0.000	2.000	6.000
K. French	MALE	PEPSI	7	0.000	3.000	7.000	6.000
E. Van Landuyt	MALE	PEPSI	7	4.000	7.000	9.000	9.000
K. Harrell	FEMALE	COKE	6	7.000	4.000	4.000	8.000
W. Noren	FEMALE	COKE	7	2.000	9.000	4.000	3.000
W. Willdden	MALE	PEPSI	9	8.000	0.000	7.000	8.000
S. Kohut	FEMALE	PEPSI	7	9.000	3.000	8.000	5.000
B. Madden	MALE	PEPSI	6	0.000	9.000	4.000	4.000

How Can I Expand a Block in the Spreadsheet Outside the Current Screen?

Blocks can be selected by (1) selecting a block with the mouse pointer, (2) clicking in one corner of the block to be selected, scrolling to the desired opposite corner (the original cell remains selected), and SHIFT+click in that cell, or (3) pressing the SHIFT key while using the cursor keys on the keyboard. To expand a previously selected block, you can use the SHIFT+cursor key, or scroll the display with the mouse pointer and SHIFT+click in the desired corner of the block. To select a large block in "split-pane" mode (as shown below), click in a cell in one pane, and then scroll (in the same pane) to display the diagonally opposite corner of another pane and SHIFT+click to select the block.

Data: Factor.sta* (10v by 100c)						
	This file contains random varial			This file contains random variable		
	1 WORK_1	**2** WORK_2	**3** WORK_3	**4** HOBBY_1	**5** HOBBY_2	**6** HOME_1
1	105.126	101.659	115.060	100.998	95.184	100.281
2	77.049	72.933	77.485	72.744	61.563	93.854
3	86.017	82.206	78.889	77.951	91.705	86.773
4	91.425	106.107	95.640	90.901	111.466	100.248
96	94.618	117.866	103.155	113.343	122.243	118.594
97	84.764	97.653	91.380	99.431	103.378	116.611
98	138.876	117.427	112.185	132.825	128.104	99.538
99	95.370	89.835	107.429	105.645	106.761	92.319
100	106.050	120.708	119.815	101.847	94.961	75.864

Variable speed of block highlighting. Note that you can control the speed at which you scroll when you extend a block outside the current display window. By moving the cursor a short distance away from the spreadsheet, you can scroll one line at a time when a block is selected; you can scroll one page at a time by moving the cursor farther away from the spreadsheet.

Can I Expand the Datafile Using Drag-and-Drop?

Yes. You can use drag-and-drop operations to increase the size of the spreadsheet by moving or copying a block [either all or some of the variable(s) or case(s)] beyond the current spreadsheet boundaries.

	1 Var1	2 Var2	3 Var3	4 NewVar1	5 NewVar2	6 NewVar3
1	4.000	20.000	4.000			
2	18.000	9.000	14.000			
3	15.000	3.000	10.000			
4	11.000	1.000	8.000			
5	7.000	1.000	2.000	4.000	20.000	4.000
6	1.000	8.000	13.000	18.000	9.000	14.000
7	2.000	11.000	16.000	15.000	3.000	10.000
8	8.000	12.000	17.000			

Data: Spreadsheet1* (6v by 8c)

You can also expand the spreadsheet using the *Insert* operation (see *What Are the Drag-and-Drop Facilities?*, page 352). This operation inserts the block between cases or variables, thereby increasing the size of the spreadsheet. Note that whether you move, copy, or insert new cases or variables, when the block contains only part of a variable or case, then *STATISTICA* fills the remaining values in the variable/case with missing data.

Splitting and Merging Files

How Do I Split a *STATISTICA* Datafile into Smaller Files?

Select *Subset/Random Sampling* from the *Data* menu to display the *Create a Subset/Random Sampling* dialog, where you can create a subset of your datafile based on the specified variables and either case selection conditions or random sampling.

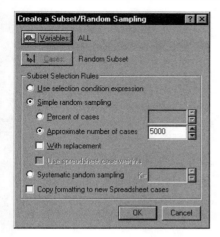

Once your selections have been made, a new spreadsheet with the requested subset is created, and you can continue the analyses using the subset of data. See *Subset/Random Sampling* for further details (page 290).

Saving subsets. When you intend to save the subset, select either *Save As* or *Save* from the *File* menu to save the new spreadsheet. The original file is not in danger of being overwritten unless you use the same name.

Permanent vs. temporary subsets. Subset creation is designed mostly to create permanent subset files, because in order to temporarily select a subset of data to be analyzed, the on-line *Case Selection Conditions* are more convenient to use (see *Selection Conditions*, page 280, for further details).

How Do I Merge Two *STATISTICA* Datafiles?

Select *Merge* from the *Data* menu to display a submenu of merge options.

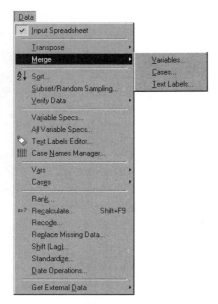

Either cases (rows of data) or variables (columns of data) can be merged. In other words, either the second file is appended to the "bottom" of the first one, or it is appended to the "right side" of the first one.

If you select **Merge – Variables...** then additional options are available.

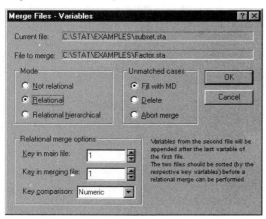

For example, you can select **Relational** or **Relational hierarchical** option buttons, for which a key variable (in each file) is used to "match" cases based on the corresponding values of the key. See *Merge - Variables* (page 288) for further details on this dialog.

Can I Merge the Text Labels/Numeric Values from Two Files?

Yes. Select *Text Labels* from the *Data - Merge* menu to display the *Merge Files - Text Labels* dialog.

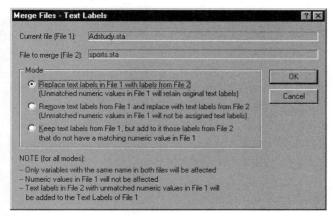

In this dialog, you can select the way in which the text values from the two files are to be merged. See *Merge – Text Labels* (page 289) for further details on this dialog.

Formulas, Transformations, Recoding, Verifying, Cleaning Data

How Do I Calculate (Transform) Values of a Variable?

Spreadsheet formulas. To perform data transformation and recoding operations on single variables using one (as opposed to a set of transformation formulas), you can use the data spreadsheet formulas. Double-click on the variable name in a spreadsheet variable that you want to transform to display the *Variable* specifications dialog (see below) where a data transformation or recoding formula can be typed directly into the *Long name (label or formula with Functions)* box.

Following the Windows spreadsheet formula conventions (e.g., Microsoft Excel), start the formula with an "=" sign (otherwise *STATISTICA* will not recognize that the text is to be

interpreted as a formula). For example, enter *=(v1+v2)/2*. Variables can be referenced by their names (e.g., *Income*, *profit*, *TEST1*) or numbers (e.g., *v1*, *v2*, *v3*, ...); *v0* is the case number. Logical operators can be used to define conditional transformation expressions (e.g., *= ScoreA + ScoreB*). Note that <, <=, >, >=, and <> can be used to specify logical conditions.

Variable names can include special characters (e.g., spaces, plus or minus signs). Variable names that include special characters must be enclosed in quotation marks (e.g., *= 'Score A' + 'Score B'*). If a quotation mark itself occurs in the variable name, use double quotation marks instead (e.g., *= "A's Score" + "B's Score"*). Note that if double quotation marks are used in the name, then the variable name must be enclosed in single quotation marks.

Following the logic of the concept of "missing data," the formula will evaluate to missing data if any variable used in the formula is missing. However, several functions are supported that calculate specific statistics for ranges and/or lists of data "adjusting" for missing values. For example, the formula *= mean (v1:v5, v7, sqrt(v9), TIME)* will calculate the mean of variables *v1*, *v2*, *v3*, *v4*, *v5*, *v7*, square root of *v9*, and variable *TIME*, even if some of them may have missing values.

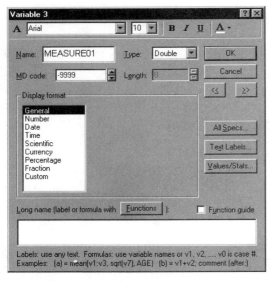

Click the *Functions* button on the *Variable* specifications dialog to access the *Functions Browser* facility that can be used to select functions and other elements of the syntax (for the formulas) and insert them into the formula editor.

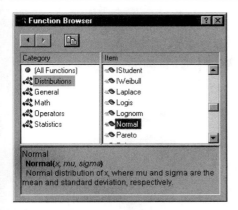

The **Function Browser** also is used to quickly access information on any of the functions and the complete on-line syntax documentation. After entering a formula and clicking the **OK** button on the **Variable** specifications dialog, you are given the option to recalculate the variable now. If you choose not to recalculate at this time, you can do so later by clicking the **Spreadsheet** toolbar **Recalculate** button ⬛, selecting **Recalculate** from the **Data** menu, or pressing F9.

The *STATISTICA* Visual Basic programming language. If you need to write more complex data transformation programs than those that can be entered via spreadsheet formulas, the integrated programming language *STATISTICA* Visual Basic can be used. The industry standard *STATISTICA* Visual Basic language offers incomparably more than just a "supplementary application programming language" that can be used to write custom extensions. *STATISTICA* Visual Basic takes full advantage of the object model architecture of *STATISTICA* and allows you to access programmatically every aspect and virtually every detail of the functionality of the program. *STATISTICA* Visual Basic adds an arsenal of more than 10,000 new functions to the standard comprehensive syntax of Microsoft Visual Basic, thus comprising one of the largest and richest development environments available.

STATISTICA Visual Basic programs can be executed from within *STATISTICA*, but because of the industry standard compatibility of *STATISTICA* Visual Basic, you can also execute its programs from any other compatible environment (e.g., MS Excel, MS Word, or a stand-alone Visual Basic language).

The *STATISTICA* Visual Basic environment includes a flexible program editor and powerful debugging tools. To display the *STATISTICA* Visual Basic editor, select **New** from the **File** menu to display the **Create New Document - Macro (SVB) Program** tab. When editing macro programs by typing in general Visual Basic commands or program commands specific to

STATISTICA Visual Basic, the editor displays type-ahead help to illustrate the appropriate syntax.

```
OpenFiles.svb
Object: (General)                    Proc: Main
  Dim ResultsWorkbook As Workbook
  Dim i As Integer

  'Create a new Workbook
  Set ResultsWorkbook = Workbooks.New
  ResultsWorkbook.Visible = True
  'This is is previously saved and
  'must be in this location
  Sname = "j:\STATISTICA 6\Examples\DataSets\Descriptives.sta"
  'Open the Spreadsheet
  Set SS = Spreadsheets.Open(Sname)
  'Place it in the workbook
  ResultsWorkbook.InsertObject(
                    InsertObject( ByVal Object As Unknown , [ItemReference As STATISTICA.WorkbookItem =
  For i = 1 To 4   0], [ByVal Placement As STATISTICA.WorkbookPlacement =
      'This is is  scWorkbookDefaultPlacement] ) As STATISTICA.WorkbookItem
      'must be in
      GName = "j:\STATISTICA 6\Examples\DataSets\Histo" + Str(i) + ".stg"
      'Open the previously saved graph
      Set G = Graphs.Open(GName)
      'Place it into the Workbook
      ResultsWorkbook.InsertObject(G, ResultsWorkbook.Root)
  Next i
```

Help on the members and functions for each class (object) is also provided in-line (see also *The STATISTICA VB Object Model* in the *Electronic Manual*).

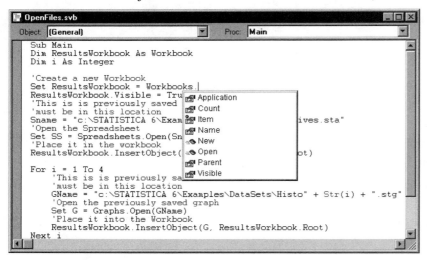

You can access examples and concise syntax summaries by pressing the F1 key or clicking the help button ▣ to display the *Electronic Manual*. Note that to learn more about a particular Visual Basic function, select the function in the *STATISTICA* Visual Basic editor and then press F1 on your keyboard. This will call the Interactive Syntax Help Engine and display the help for that keyword.

When executing a program, you can set breakpoints in the program, step through line by line, and observe and change the values of variables in the macro program as it is running.

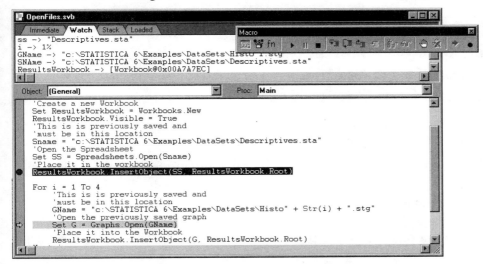

To summarize, *STATISTICA* Visual Basic is not only a powerful programming language, but it represents a very powerful professional programming environment for developing simple macros as well as complex custom applications. For more information, refer to Chapter 11 - *STATISTICA Visual Basic*.

Can *STATISTICA* Automatically Recalculate All Spreadsheet Formulas When the Data Change?

Yes. Select the *Auto-recalculate when the data change* check box in the *Recalculate* dialog (accessed by clicking the ▣=? button on the toolbar or selecting *Recalculate* from the *Data* menu).

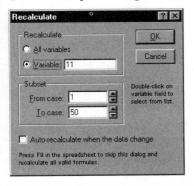

STATISTICA will automatically recalculate all spreadsheet formulas when data are changed in the spreadsheet. Alternatively, you can press F9 to manually recalculate the formulas in the spreadsheet.

What Is the Simplest Way to Recode Values of a Variable (e.g., Split a Continuous Variable into Categories)?

Recoding functions of practically unlimited complexity can be custom-defined in *STATISTICA* Visual Basic and used repeatedly in your data transformation programs.

However, a quick on-line recoding facility can also be accessed directly from the spreadsheet at any point. Select **Recode** from the **Spreadsheet** Vars ▾ toolbar button to display the **Recode Values** dialog. You can scroll the **Recode Values** dialog to define new values (text or numeric) of the current variable (see the groups **New Value 1**, **New Value 2**, **New Value 3**, etc., below) depending on the specific conditions that you define (see the **Category 1**, **Category 2**, **Category 3**, etc., groups below).

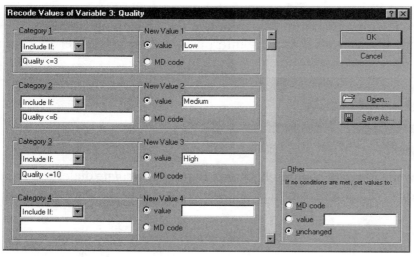

For example, the recoding conditions specified above would "translate" the values less than or equal to 3 of the current variable (**Quality**) into **Low**, the values greater than 3 and less than or equal to 6 into **Medium**, and the values greater than 6 and less than or equal to 10 into **High**. Note that any other values that occur in the **Quality** variable would be "translated" into the missing data code because the **MD Code** option button is selected in the **Other** group. Note also that *STATISTICA* first finds all cases that meet **Category 1** and recodes them; then *STATISTICA*

only searches through the remaining cases that have not be recoded to see if they meet **Category 2**, etc. Hence, you do not have to specify **Quality > 3 AND Quality <= 6** as **Category 2**.

When specifying conditions, follow the standard syntax conventions common in *STATISTICA* to all those procedures that involve any operation of "selecting cases" based on their values (see *What Syntax Can Be Used to Create Case Selection/Verification/Recode Conditions?*, page 368). You can also save the current recoding conditions to a text file or open a file with previously saved conditions. See *Data - Recode* (page 301) for further details on this dialog.

Note that recoding conditions can be much more complex (see the *Electronic Manual* by pressing the F1 key or clicking ▐?▌ in this dialog), and they can be defined such that the new values of the current variable do not depend on the old values of that variable, but only on values of some other variables in the datafile. Thus, this facility can be used not only to recode existing data, but also to create values of a new variable based on conditions met by other variables (as illustrated in *How Do I Create Values of a New Variable Based on Conditions Met by Other Variables?*, below).

How Do I Create Values of a New Variable Based on Conditions Met by Other Variables?

You can use either of the data transformation facilities: spreadsheet formulas or *STATISTICA* Visual Basic. However, often the quickest way to create values of a new variable based on conditions met by other variables is to use the on-line data recoding facility (described in *What Is the Simplest Way to Recode Values of a Variable?*, above), which is accessible at any point from the spreadsheet by clicking the ▐Vars ▼▌ button and selecting **Recode** or selecting **Recode** from the **Data** menu. Note that the currently selected variable does not even have to be included in the text of the recoding conditions. Thus, you can use this facility to create values of a variable based on conditions met by other variables.

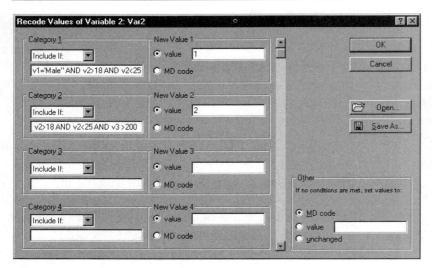

For example, you can add a new (empty) variable to the datafile, and then use this facility to create the new values. For instance, the recoding conditions could be used to assign 1's to the new variable for all "male subjects 18 to 25 years old with cholesterol levels below 200," 2's to "male subjects 18 to 25 years old with cholesterol levels above 200," and assign the missing data value to all other subjects.

What Syntax Can Be Used to Create Case Selection/Verification/Recode Conditions (or to Calculate Variable Values Using Spreadsheet Formulas)?

Recoding variable values, performing case selection for analyses, or verifying quality of data involve a comparison of conditions or values encountered in the spreadsheet with conditions specified by an expression to determine a further action for each case (i.e., respectively assigning a new variable value, including the case in an analysis, or verifying that the data associated with a case are correct). Assigning variable values via a spreadsheet formula involves case-by-case calculation of values for a variable based directly on the values of other spreadsheet variables and/or parameters not included in the spreadsheet. For data transformations more extensive or more complex than can be accomplished using the variable recode and spreadsheet formula facilities provided, use the *STATISTICA* Visual Basic language (see Chapter 11).

- Refer to variables by either their numbers (e.g., ***v1 = 1***) or their names (e.g., ***Gender = 1***). Note that you can type variable names in either upper case or lower case letters (i.e., ***"GENDER"*** is equivalent to ***"gender"***). Note also that ***v0*** refers to the case number when used in expressions.

- In expressions, enclose text labels of a variable in single quotation marks (e.g., ***v1 = 'MALE'***). Note that you can type text labels in either upper case or lower case letters (i.e., ***'YES'*** is equivalent to ***'yes'***).

- In expressions or spreadsheet formulas, enclose variable names containing special characters (e.g., spaces, plus or minus signs) in single quotation marks. If the single quotation mark itself occurs in the variable name, use double quotation marks instead (e.g., ***"A's Score"***). Note that if double quotation marks are used in the name, then the variable name must be placed in single quotation marks.

Syntax conversions for spreadsheet formulas. Spreadsheet formulas (specified on the ***Variable Specifications*** dialog accessed via the ***Data – Variable Specs*** menu) must start with an equal sign. When you enter a label that starts with an equal sign, *STATISTICA* will assume that it is a formula and will verify it for formal correctness. Note that a semicolon after a formula starts a comment: e.g., ***= v1 + v2; this is a comment***.

Missing values. The ***IsMD(x)*** function will return a value of ***true (1)*** if the passed expression is a missing data value.

Operators. A number of arithmetic, relational, and logical operators are available for creating expressions for recoding, case selection, and data verification or for creating spreadsheet formulas.

Arithmetic: ***+, -, *, /, ***** or ***^*** (exponentiation), ***( )***

Relational:

      = (equal to)

      <>, >< (not equal to)

      < (less than)

      > (greater than)

      <= (less than or equal to)

      >= (greater than or equal to)

Logical:

> **AND** (equivalent to **&**)
>
> **OR** (equivalent to **!**)
>
> **NOT** (equivalent to **~**)

Note that a common error is caused by omitting parentheses needed to adjust for the precedence of operators; for example, the expression **x > 0** and **x < 1** is incorrect and needs parentheses: **(x > 0) and (x < 1)** because relational operators (**>**, **<**) have a lower precedence than the conjunction (**and**).

Math functions. Math functions can be used in expressions for recoding, case selection or data verification as well as in spreadsheet formulas for calculating variable values. If the value of any variable used in the expression or formula is missing (in the current case), then the expression evaluates to missing data (for the current case).

> **Abs(x)** - absolute value of x
>
> **Arccos(x)** - arc cosine of x
>
> **Arcsin(x)** - arc sine of x
>
> **Arctan(x)** - arc tangent of x
>
> **Cos(x)** - cosine of x
>
> **CosH(x)** - hyperbolic cosine of x
>
> **Exp(x)** - e to the power of x
>
> **Hypot(x,y)** - returns hypotenuse of x and y (square root($x^2 + y^2$))
>
> **Log(x)** - natural logarithm of x
>
> **Log2(x)** - binary logarithm of x
>
> **Log10(x)** - common logarithm of x
>
> **Max(x,y)** - returns the greater of x and y
>
> **Min(x,y)** - returns the lesser of x and y
>
> **Rnd(x)** - random number in the range of 0 to x
>
> **RndNormal(x)** - normalized random value in range of 0 to x
>
> **RndPoisson(x)** - Poisson random value in range 0 to x
>
> **Sign(x)** - sign of x: if x>0 then +1, if x<0 then -1, if x = 0 remains 0
>
> **Sin(x)** - sine of x

SinH(x) - hyperbolic sine of x

Sqrt(x) - square root of x

Tan(x) - tangent of x

TanH(x) – hyperboloic tangent of x

Trunc(x) - truncate x to an integer "towards zero"

Uniform(x) – random value in range 0 to x (same as Rnd(x))

Function names are not case sensitive, i.e., *Log(x)* is the same as *log(x)* or *LOG(x)*. As indicated in the function list above, math functions accept one to two arguments depending on the function. The position of the function in the expression or spreadsheet formula will be replaced by the return value of the function. Numeric values (e.g, *Sqrt(155)*), variable names (e.g., *Max(SCORE1,SCORE9)*), or variable numbers (e.g., *Log(v8)*) are acceptable arguments. Additional arguments acceptable to math functions are expressions that evaluate to a number (e.g., *Max(v7,(v5+v8-BASELINEVALUE)/3)*) or functions that return a numeric result (e.g., *Sin(Sqrt(v5))*). Some commonly used constants can also be specified in expressions and formulas by reference: e.g., *Pi* = 3.14... *Euler (e)* = 2.71...

Statistics functions. Nine statistics functions that accept lists of values and/or ranges and arguments, and adjust to missing data, are also available for use in recoding, case selection, and verification expressions as well as in spreadsheet formulas:

Mean(x1, x2,..xn) – mean of n arguments

Median(x1, x2,..xn) – 50th percentile of n arguments

Perc25(x1, x2,..xn) – 25th percentile of n arguments

Perc75(x1, x2,..xn) – 75th percentile of n arguments

Statmax(x1, x2,..xn) – maximum of n arguments

Statmin(x1, x2,..xn) – minimum of n arguments

Stdev(x1, x2,..xn) – standard deviation of n arguments

Sum(x1, x2,..xn) – sum of n arguments

Validn(x1, x2,..xn) – number non-missing of n arguments

Note that the *Statmin* and *Statmax* functions have the stat prefix to distinguish them from the arithmetical *Max* and *Min* (math) functions discussed above. Like math functions, the names of statistics functions are not case sensitive when used in expressions and formulas.

All statistics functions ignore arguments that contain missing values, basing their results on non-missing data only. Unlike math functions, statistics functions will not evaluate to a missing result unless all arguments are missing values or an argument is encountered that evaluates to an undefined value (e.g., square root of a negative number or division by zero). Also unlike math functions, each statistics function accepts any number of arguments placed in parentheses and separated by commas.

The following are acceptable statistics function arguments:

> Numbers: e.g., ***Mean(18,80,120,68,40)***

> Variable names: e.g., ***Mean(SALARY,1050,BONUS,500)***

> Variable numbers: e.g., ***Stdev(120,v3,v2,v8,255)***

> Ranges of variables designated by name or number (use colons to define ranges): e.g., ***Sum(v54,COST1:COST5,2550,v23:v35,1575,OVERHEAD)***

> Expressions that evaluate to a number: e.g., ***Mean(500,(v6+456)/v3),TRAVELEXPENSE,v8)***

> Functions that return a numeric result: e.g., ***Mean(220,Sqrt(v8+v7),v12)***

Distribution functions and their integrals. *STATISTICA* provides a predefined broad selection of distribution functions, their integrals and inverse distribution functions that can be used in spreadsheet formulas and in recoding, case selection, and verification expressions like all other functions.

Below is a listing of all available distributions (parameters are given in parenthesis):

Distribution	Density/ Probability Function	Distribution Function	Inverse Distribution Function
Beta	*beta(x,n,w)*	*ibeta(x,n,w)*	*vbeta(x,n,w)*
Binomial	*binom(x,p,n)*	*ibinom(x,p,n)*	
Cauchy	*cauchy(x,h,q)*	*icauchy(x,h,q)*	*vcauchy(x,h,q)*
Chi-square	*chi2(x,n)*	*ichi2(x,n)*	*vchi2(x,n)*
Exponential	*expon(x,l)*	*iexpon(x,l)*	*vexpon(x,l)*
Extreme	*extreme(x,a,b)*	*iextreme(x,a,b)*	*vextreme(x,a,b)*
F	*F(x,n,w)*	*iF(x,n,w)*	*vF(x,n,w)*
Gamma	*gamma(x,c)*	*igamma(x,c)*	*vgamma(x,c)*

Geometric	geom(x,p)	igeom(x,p)	
Laplace	laplace(x,a,b)	ilaplace(x,a,b)	vlaplace(x,a,b)
Logistic	logis(x,a,b)	ilogis(x,a,b)	vlogis(x,a,b)
Lognormal	lognorm(x,m,s)	ilognorm(x,m,s)	vlognorm(x,m,s)
Normal	normal(x,m,s)	inormal(x,m,s)	vnormal(x,m,s)
Pareto	pareto(x,c)	ipareto(x,c)	vpareto(x,c)
Poisson	poisson(x,l)	ipoisson(x,l)	
Rayleigh	rayleigh(x,b)	irayleigh(x,b)	vrayleigh(x,b)
Student's t	student(x,df)	istudent(x,df)	vstudent(x,df)
Weibull	weibull(x,b,c,q)	iweibull(x,b,c,q)	vweibull(x,b,c,q)

See the *Electronic Manual* for detailed discussion of the distribution types and the parameters that are required by each function.

How Can I Verify and "Clean" Data?

Select **Verify** from the **Data** menu to access an interactive data-verification and cleaning facility. Use the **Verify Data** dialog to enter the conditions to be met by the data.

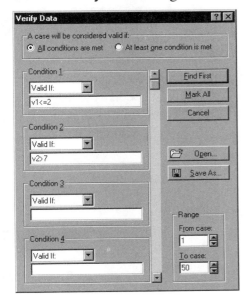

Follow the standard syntax conventions common in *STATISTICA* to all those procedures that involve any operation of "selecting cases" based on their values (see *What Syntax Can Be Used to Create Case Selection/Verification/Recode Conditions?*, page 368). You can also save the current verification condition to a text file or open a file with previously saved conditions. See *Data - Verify Data* (page 292) for further details on this dialog.

The verification can be as simple as checking whether values in a variable are "legal" (e.g., only *1* and *2* might be allowed for *Gender*) or whether they fall within allowed ranges of values (e.g., *Age* must be more than *0* and less than *200*). It can also be as complex as checking multiple logical conditions that some values must meet in relation to other values.

Consider the following example of conditional verification:

If a person is a male or less than 10 years old, then the number of pregnancies for that person cannot be more than zero.

In order to apply these conditions, you would specify (for example):

Invalid if: *(v1='MALE' or AGE<10) and PREGN>0*

Once you have entered your verification condition(s), click either the *Find First* button to select the first invalid case in your datafile (after this first case has been selected, you can find the next case by selecting *Find Next Invalid Case* from the *Data - Verify Data* menu) or click the *Mark All* button to mark all of the invalid cases in the datafile according to the *Marked Cells* spreadsheet layout (see page 234).

How Do I Perform a Multiple Sort?

Select *Sort* from the *Data* menu, *Sort Cases* from the Cases button, or click the toolbar button. All of these actions display the *Sort Options* dialog where you can specify the key variables and the type of sort. See *Data - Sort* (page 290) for further details on this dialog.

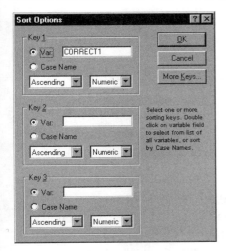

If you need to sort based on more than three keys, click the **More Keys** button to switch to a larger dialog.

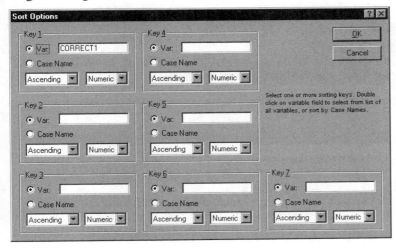

How Do I Rank-Order Values of a Variable (Replace Values with Their Ranks)?

Select **Rank** from the Vars ▾ button on the **Spreadsheet** toolbar or from the **Data** menu to display the **Rank Order Values** dialog.

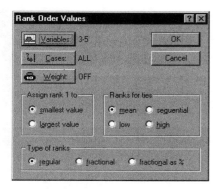

Then click the **Variables** button to select the variables to be ranked. Note that this operation replaces the original values with their respective ranks based on the sorted series. Optionally, you can also specify a subset of cases to be affected by the operation (click the **Cases** button), use case weights (click the **Weight** button), and set a number of options to perform particular (non-default) types of ranking. See *Data - Rank* (page 299) for further details on this dialog.

How Do I Transpose Data (Convert Cases Into Variables)?

Select **Transpose** from the **Data** menu, then select either **Block** or **File**.

Transposing a block of data. The block transposing command affects only the contents of cells in the block currently selected in the spreadsheet (the block must be square); the variable names and case names will not be affected. For example, the **Transpose - Block** command executed on the following square block of data:

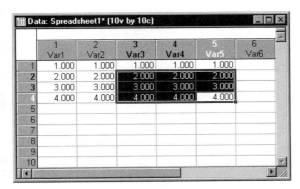

produces the following result:

Transposing a datafile. Datafile transposing restructures the entire file. For example, transposing the datafile shown above (before the block was transposed), changes the number of cases and variables in the datafile:

Note that when you transpose a file, case names become variable names and variable names become case names.

How Do I Automatically Fill
Ranges of Data in the Spreadsheet?

In addition to the custom-defined operations of filling specific ranges of data with values available via the data transformation options [see *How Do I Calculate (Transform) Values of a Variable?*, page 361], *STATISTICA* can fill selected blocks with random values. *STATISTICA* also supports standard (Microsoft Excel-style) block extrapolation operations, which can be implemented by dragging a border of the block onto the adjacent area of the spreadsheet that is to be filled following the extrapolation rules.

Random value fill. Select a block, then right-click and select *Fill/Standardize Block - Fill Random Values* from the spreadsheet shortcut menu. The currently selected block will be filled with random values (following a uniform distribution) in the range of 0 to 1. An error message is displayed if a block is not selected.

Data: Spreadsheet1* (4v by 10c)			
1 Var1	2 Var2	3 Var3	4 Var4
0.070053	0.834943	0.225343	
0.685111	0.840637	0.060952	
0.749607	0.58757	0.866258	
0.28634	0.432971	0.718811	
0.342844	0.397755	0.584113	
0.533606	0.535986	0.780004	
0.898758	0.730164	0.606118	

Extrapolation of values (AutoFill). A selected series of values (consisting of at least two values) will be extended if a meaningful pattern is detected. For example, the following numbers: *1.00*, *2.00*, and *3.00* would be extended to *1.00*, *2.00*, *3.00*, *4.00*, *5.00*, *6.00*, *7.00*, and *8.00* as illustrated in the following spreadsheets:

Data: Spreadsheet2* (10v by 10c)			
1 Var1	2 Var2	3 Var3	4 Var4
1.00	1.00	1.51	
2.00	3.00	1.53	
3.00	5.00	1.52	

Data: Spreadsheet2* (10v by 10c)			
1 Var1	2 Var2	3 Var3	4 Var4
1.00	1.00	1.51	
2.00	3.00	1.53	
3.00	5.00	1.52	
4.00	7.00	1.53	
5.00	9.00	1.54	
6.00	11.00	1.54	
7.00	13.00	1.54	
8.00	15.00	1.55	

If no meaningful pattern can be detected, predicted values from a linear trend regression will be used for the extrapolation. If a single value (or a repeated value) is selected, the value will be copied into the extrapolated block; also, sequences consisting of names of months, days, or quarters can be automatically extended. See *How Does the Extrapolation of Blocks (AutoFill) Work?* (page 380) fur further details.

Fill down or right. Finally, simple spreadsheet-style "fill block" (***Down*** or ***Right***) options are available that fill the currently selected block by duplicating the first row or column (respectively) of the block. To use this facility, first select a block to be filled in the spreadsheet. Then select ***Fill/Standardize Block*** from the spreadsheet ***Edit*** menu (or the spreadsheet shortcut menu, shown below). Use the hierarchical menu to select either ***Fill/Copy Down*** or ***Fill/Copy Right***.

The former copies (duplicates) the first row of the block to all remaining rows of the block; the latter copies the first column of the block to the remaining columns of the block. Both options work in a manner similar to Microsoft Excel's Fill Right and Fill Down facility (also available in Excel from the Edit menu).

How Does the Extrapolation of Blocks (*AutoFill*) Work?

A horizontal or vertical series in a block can be extrapolated by dragging the block Fill Handle (a small solid square located on the lower-right corner of the block). After selecting the block, point at the Fill Handle. When the mouse pointer changes to a plus sign **+**, drag to the last cell(s) to which you want to extrapolate the data. *STATISTICA* can create series of values such as sequential numbers, linear extrapolations, and dates (e.g., you can extend a series such as 1, 2, 3 to include 4, 5, 6).

You can extrapolate (*AutoFill*) a block in the following ways:

- If the initial selection contains repeated values, these values are duplicated in the extended block.

- Non-identical values are extended by linear trend regression to compute extrapolated values for the series.

- If the block contains the same text labels (i.e., all the cells have the text label '*Gender*'), then this text label is copied into the extended block. If the block contains several different text labels, then the text labels (beginning with the numeric value of the last text label in the block) are extended by linear regression to compute extrapolated values for the series.

- If a variable in the block contains dates (i.e., you selected **Date** in the **Display format** group on the **Variable** specifications dialog, see page 293), then based on the pattern of dates, *STATISTICA* extrapolates the dates appropriately.

- If a variable in the block contains text names (i.e., not dates) of months (e.g., Jan, Feb, Mar,), days (e.g., Mon, Tue, Wed,), or quarters (e.g., Q1, Q2,), then *STATISTICA* extrapolates either up, down, or to the left or right of the block the rest of the names in the series (e.g., Apr, May, ..., Dec; Thr, Fri, Sat, Sun; or Q3, Q4, respectively). The syntax for specifying these names is given below (note that it does not matter whether you use upper case or lower case letters, however you need to be consistent throughout the series; e.g., use Jan, Feb, Mar, ..., not Jan, feb, Mar, ...):

Months. The names of months can be spelled out or abbreviated in the following manner:

- January, February, March, April, May, June, July, August, September, October, November, December

- Jan, Feb, Mar, Apr, May, Jun, Jul, Aug, Sep, Oct, Nov, Dec

Days. The names of the days can be spelled out or abbreviated in the following manner:

- Monday, Tuesday, Wednesday, Thursday, Friday, Saturday, Sunday

- Mon, Tue, Wed, Thu, Fri, Sat, Sun

- M, Tu, W, Th, F, Sa, Su

Quarters. The names of quarters of the year can be specified as follows in the variable:

- Q1, Q2, Q3, Q4

Custom lists. You can also create your own list of values to use in extrapolating values. See *Can I Create a Custom List to Use for Extrapolation?* (page 326).

Extrapolating (auto filling) a block upwards or to the left. In the same manner as extrapolating a block down or right, a block can be extrapolated by dragging the Fill Handle up or to the left past the original start of the block.

If you drag the Fill Handle up or to the left and stop within the original selection without going past the top or left side of the selection, you will delete data within the selection (data to be deleted are indicated in gray as you drag within the selection). You can use the *Fill/Standardize Block* commands on the spreadsheet *Edit* menu to copy to adjacent cells within the currently selected block.

How Do I Standardize Values in a Block?

Select a block of values in the spreadsheet, right-click, and select either *Fill/Standardize Block - Standardize Columns* or *Fill/Standardize Block - Standardize Rows* from the shortcut menu. The standardized values are computed as follows:

```
Standard Value = (raw value - mean of highlighted row/column)/standard
deviation
```

In the following illustration, the columns of the selected block have been standardized by columns:

	1 Var1	2 Var2	3 Var3	4 Var4	5 Var5
1	4.000	4.000	6.000	3.000	6.000
2	8.000	4.000	6.000	1.000	4.000
3	5.000	1.000	2.000	3.000	7.000
4	7.000	2.000	4.000	6.000	2.000
5	4.000	3.000	5.000	7.000	7.000
6	9.000	4.000	7.000	10.000	7.000
7	9.000	9.000	6.000	0.000	6.000
8	5.000	2.000	2.000	2.000	1.000

Data: Spreadsheet2* (5v by 8c)

	1 Var1	2 Var2	3 Var3	4 Var4	5 Var5
1	4.000	4.000	6.000	3.000	6.000
2	8.000	4.000	6.000	1.000	4.000
3	5.000	-1.162	-1.201	-1.212	7.000
4	7.000	-0.387	-0.240	-0.173	2.000
5	4.000	0.387	0.240	0.173	7.000
6	9.000	1.162	1.201	1.212	7.000
7	9.000	9.000	6.000	0.000	6.000
8	5.000	2.000	2.000	2.000	1.000

Data: Spreadsheet2* (5v by 8c)

You can also standardize selected variables by selecting *Standardize* from the *Data* menu (this procedure works independent of the currently selected block, but takes into account the current case selection conditions and weights). See *Data – Standardize* (page 303) for further details.

Statistical Analyses

How Do I Select Variables for an Analysis?

Every analysis definition dialog in *STATISTICA* contains at least one *Variables* button that is used to specify variables to be analyzed. You can click the button (or press v). If you forget to specify variables and click the *OK* button to start the analysis, *STATISTICA* prompts you to specify the variables to be analyzed. The variable selection dialog that will be displayed supports various ways of selecting variables, and it offers various shortcuts and options to review the contents of the datafile (for more information and illustrations, see the *Example 1: Correlations (Introductory Overview)* in Chapter 1 – *STATISTICA – A General Overview*).

Can Variables Be Selected for Analyses by Highlighting Them in the Spreadsheet?

Yes. If you select a block in the spreadsheet, the variables included in the block are automatically preselected for the next analysis. This shortcut is designed to limit the chance of producing unintended results in the following ways:

- The preselection of variables by marking a block in the spreadsheet works only as long as you have not selected a specific list of variables for the analysis (i.e., it never overwrites your previous choices; *STATISTICA* remembers your previous choices).

- If the variables from the block are not what you intend to analyze, you do not need to "undo" the selection. When you display the variable selection dialog, the list of preselected variables is selected; thus, the first click of the mouse clears the previous range (unless you keep the CTRL key pressed). If you prefer to use the keyboard to specify the list, then the first (non-cursor moving) key you press deletes the previous entry in the variable selection edit field.

How Do I Select a Subset of Cases (Observations) to Be Included in an Analysis?

Before an analysis begins (i.e., before the data are processed), you can instruct *STATISTICA* to select only cases (i.e., rows in the spreadsheet) that meet some specific selection criteria. A facility to define and manage case selection conditions can be accessed from the **Tools - Selection Conditions - Edit** menu (see page 280) or by clicking on the status bar field **Sel**, which shows the current status of the **Case Selection Conditions**. Note that when case selection conditions are disabled, this field reads **Sel: OFF**, when they are enabled, it reads **Sel: ON**).

Note that there are two types of case selection conditions in *STATISTICA* – permanent (saved with the datafile) and temporary (associated with the current analysis only). For more information, see *Case Selection Conditions* in Chapter 1 – *STATISTICA – A General Overview*. See also the next topic.

How Are Case Selection Conditions Stored and Saved?

Global case selection conditions (definitions of subsets of data) are attached to a spreadsheet and are called Spreadsheet Case Selection Conditions. They are accessed from the **Tools - Selection Conditions** menu (see page 280) or by clicking the status bar field **Sel**. Analysis/Graph Case Selection Conditions can also be locally connected to an analysis or graph. They are accessed by clicking the **Select Cases** button that is available on all analysis/graph definition dialogs. You can save and open either type of selection conditions (and maintain libraries of case selection conditions) using the **Open** or **Save As** buttons in the appropriate **Case Selection Conditions** dialog. Note that case weights are specified in a similar manner.

Can I Select Random Subsets of Data?

Yes, you can. Subsets of data can be created via both simple random sampling and systematic random sampling. Select **Subset/Random Sampling** from the **Data** menu to display the **Create a Subset/Random Sampling** dialog.

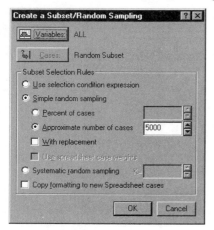

Select the **Simple random sampling** option button in the **Subset Selection Rules** group to obtain your subset via random sampling. You have two choices in regards to how the subset is created: either via the **Percentage of cases** within the original spreadsheet or an **Approximate number of cases**. If you select the **With replacement** check box, then once a case is selected to be included into the subset, that case will be placed back into the pool of available choices for the remaining cases in the subset (hence an individual case can appear more than once in the resulting subset). Additionally, if a weighting variable has been selected for the original spreadsheet, select the **Use spreadsheet case weights** check box to apply that case weight when the subset is being obtained.

Select the **Systematic random sampling** option button to obtain your subset via systematic random sampling. For instance, if you enter a **5** in the **K=** box, *STATISTICA* will randomly select a case within the first five cases and then finish obtaining the subset by selecting each fifth case in the spreadsheet after the originally selected case.

See *Subset/Random Sampling* for further details on this dialog (page 290).

What Is the Quickest Way to Review Basic Descriptive Statistics for a Variable?

You can either select **Descriptive Statistics** from the **Basic Statistics and Tables** (Startup Panel) (accessible from the **Statistics** menu) or click the **Values/Stats** button in the **Variable** specifications dialog, as illustrated below. To display the **Variable** specifications dialog, double-click on the variable name in the spreadsheet. Alternatively, right-click on the variable name and select **Variable Specs** from the shortcut menu.

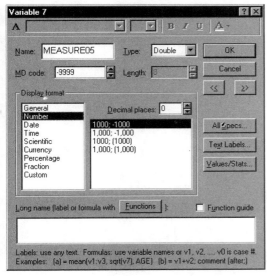

After you click the **Values/Stats** button, the resulting dialog displays information about the selected variable, a sorted list of its values, and descriptive statistics. The descriptive statistics can be copied to the Clipboard by clicking the **Copy** button 🗐 (see the lower part of the window in the illustration below).

For information about producing descriptive statistics for columns or rows of the currently selected block of values, see *What Are Statistics of Block Data?*, page 387.

What Is the Quickest Way to Review a Sorted List of All Unique Values of a Variable?

Click the **Values/Stats** button in the **Variable** specifications dialog to display a sorted list of all unique numeric values and corresponding text labels (if there are any) for that variable. The **Variable** specifications dialog is accessed by double-clicking the variable name in the spreadsheet. See *What Is the Quickest Way to Review Basic Descriptive Statistics for a Variable*, page 385, for further details on this dialog.

Note that the assignments between the numeric values and text labels for the variable can be edited in the **Text Labels Editor** (see *Tools – Text Labels Editor*, page 295, for further details).

What Are Statistics of Block Data?

Statistics for each row or column in a selected block can be computed and added to the spreadsheet (i.e., appended at the end) by selecting the desired **Statistics of Block Data** from the shortcut menu.

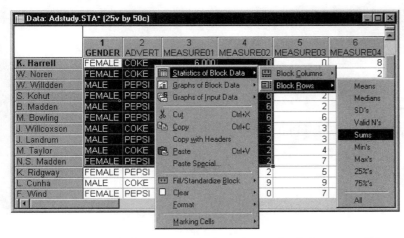

The selected statistics are added to the end of the spreadsheet as additional rows or columns. For example, the selection shown above computes the sums for each row in the selected block

and appends them to the datafile as a new variable **Sum**. Note that in the illustration below the new variable **Sum** has been moved next to **Measure02**.

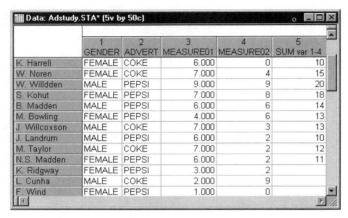

If you select **Block Columns**, the results of the selected statistic are appended as a new case, and the case name contains the name of the selected statistic and the range description, as shown below.

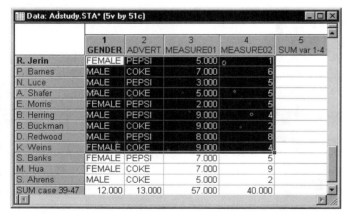

In addition, a (customizable) list of predefined graphs is also available from this shortcut menu.

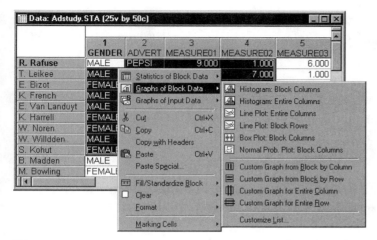

These **Graphs of Block Data** allow you to visualize the data in the selected block either by row or by column.

Can Matrix Data Be Used for Input Instead of Raw Data?

Yes. *STATISTICA* supports a variety of matrix file types as input data (e.g., for Multiple Regression, Factor Analysis, Reliability/Item Analysis, and other modules). Matrix files can be edited in the spreadsheet (as if they were raw datafiles); however, in order to be properly interpreted as matrices, they need to meet specific content and format conditions (depending on the matrix type). *STATISTICA* also supports the multiple matrix datafile format for input in procedures (such as Structural Equation Modeling) that supports matrix data for groups. Note that matrix data that meet the specific content and format conditions are automatically saved in *STATISTICA* Matrix Spreadsheet (**.smx*) format. For more details, see *What Is a Matrix Spreadsheet? (*page 327).

Import, Export, DDE, OLE

How Can I Access Data from Excel and Other Foreign Datafiles?

Clipboard. The quickest, and in many cases easiest, way to access datafiles from other Windows applications (e.g., spreadsheets) is to use the Clipboard, which in *STATISTICA*

supports special Clipboard data formats generated by applications such as Microsoft Excel. For example, *STATISTICA* properly interprets formatted cells (such as 4/17/1999 or $10) and text values (including extensive in-cell formatting, e.g., RVStower*120.3MHz*).

File import facilities. Datafiles from a wide variety of Windows and non-Windows applications can be accessed and translated into the *STATISTICA* format using the file import facilities (available by opening the file using the **File - Open** command). You can even access formatted and free format text (ASCII) files.

The main advantages of using the file import facilities (over the Clipboard) are that:

- You can specify the exact way in which the translation is to be performed (e.g., access named ranges in the foreign datafiles; decide whether or not to import variable names, text labels, and case names; and how to interpret them), and

- You can access types of data that are not (or not easily) accessible to Clipboard operations (such as missing data codes).

See *Importing Data* (page 227) for further details.

STATISTICA Query. In addition to the file import facilities described above, *STATISTICA* provides access to virtually all databases via *STATISTICA*'s flexible *STATISTICA* Query, accessible from both the **File – Get External Data** and the **Data – Get External Data** menus. For more information, see Chapter 10 - *STATISTICA Query*.

OLE links. *STATISTICA* supports the Object Linking and Embedding (OLE) conventions allowing you to link values or text in *STATISTICA* documents (e.g., spreadsheets, reports) to documents in other (Windows) applications. The procedure is in fact much simpler than it might appear and can be easily employed without technical knowledge about the mechanics of OLE, especially when you use the **Paste Link** (instead of script-entry) method. See *How Do I Set Up OLE Links between STATISTICA and Other Windows Applications?* (page 393) and *OLE Links* (page 230) for further details.

DDE links. Finally, *STATISTICA* supports the Dynamic Data Exchange (DDE) conventions; thus, you can dynamically link a range of data in a spreadsheet to a subset of data in other (Windows) applications. Like OLE, the procedure is in fact much simpler than it might appear, and can be easily employed without technical knowledge about the mechanics of DDE. To set up a DDE Link, select **DDE Links** from the **Edit** menu. See *How Do I Set Up DDE Links between STATISTICA and Other Windows Applications?* (page 393) and *DDE Links* (page 231) for further details.

Accessing datafiles larger than the local storage. Note that enterprise versions of *STATISTICA* offer options to query and access large remote datafiles in-place (i.e., without having to import the data and create a local copy).

How Do I Export Data from *STATISTICA* to Excel and Other Foreign Datafiles?

STATISTICA can export datafiles to a wide variety of Windows and non-Windows applications via the **Save As** command accessible from the **File** menu. You can even save spreadsheets as formatted and free format text (ASCII) files. These facilities allow you to specify the exact way in which the translation is to be performed (e.g., decide whether or not to export variable names, text labels, and case names). Also, *STATISTICA* Spreadsheets are OLE servers and clients, thus other applications can link to them. See *Exporting Data* (page 233) for further details.

Can *STATISTICA* Access Data from Enterprise Databases?

Yes. *STATISTICA* Query (including options to combine fields from multiple tables) is provided via both the **File – Get External Data** and the **Data – Get External Data** menus (see Chapter 10 – *STATISTICA Query*). As long as the database format supports a 32-bit ODBC driver or an OLE DB provider, accessing such data should present no problem.

Note that you can either take advantage of *STATISTICA* Query's built-in graphical user interface for querying:

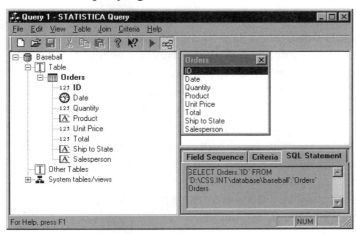

or write SQL statements in a text editor.

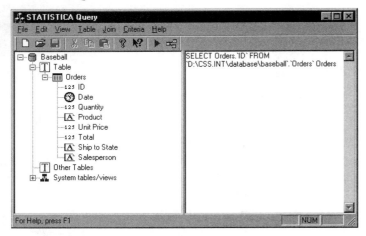

For further details about setting up connections to external databases, consult the database vendor and/or your database administrator.

Can *STATISTICA* Process Files That Are Larger Than the Local Hard Drive?

Yes, the enterprise versions of *STATISTICA* offer options to query and access large remote datafiles in-place (i.e., without having to import the data and create a local copy).

Does *STATISTICA* Support ODBC?

Yes, via *STATISTICA* Query, an external data querying facility available from the **Data – Get External Data** menu and the **File – Get External Data** menu. The *STATISTICA* Query interface includes options to combine fields from multiple tables and provides flexible access to a wide variety of database management files, including the major enterprise database formats such as Microsoft SQL Servers and Oracle. Specific records (rows of tables) can be selected by entering SQL statements. *STATISTICA* Query automatically builds the SQL statement for you as you select the components of the query via a simple graphical interface and/or intuitive menu options and dialogs. Hence, an extensive knowledge of SQL is not necessary in order for you to create advanced and powerful queries of data in a quick and straightforward manner. Multiple queries based on one or many different databases can also be created to return data to an individual spreadsheet. Therefore, you can maintain connections to multiple external databases simultaneously (see Chapter 10 – *STATISTICA Query*).

How Do I Set Up OLE Links between *STATISTICA* and Other Windows Applications?

You can establish OLE (Object Linking and Embedding) links between a "source" (or server) file (e.g., a Microsoft Word document) and a *STATISTICA* document (the "client" file) so that when changes are made to the data in the source file, the data are automatically updated in the respective part of the *STATISTICA* document (client file). Additionally, *STATISTICA* can serve as a "source" (or server) file for other OLE compatible documents. In this way, you could link the values in a *STATISTICA* Spreadsheet to a table in a Microsoft Word document so that the Word document updates when data in the spreadsheet are changed.

OLE links are established using the quick, "paste-like" **Paste Link** option in the **Paste Special** dialog. To create an OLE Link within a *STATISTICA* document, first copy to the Clipboard the desired text (or values) from another Windows application. Then (in *STATISTICA*), select **Paste Special** from the **Edit** menu to display the **Paste Special** dialog. Select the **Paste Link** option button, specify the correct format in the **As** group, and click the **OK** button to establish the link using the source given in the **Source** field. See *Edit – Paste Special* for further details on the *Paste Special* dialog (page 253).

When a link is established, it can be managed using the **Links** dialog (select **Links** from the **Edit** menu). Use this dialog to update or break a link, open the source file or change to a different source file, and determine whether updates will be automatic or manual (see *Edit – Links*, page 260).

How Do I Set Up DDE Links between *STATISTICA* and Other Windows Applications?

You can establish DDE (Dynamic Data Exchange) links between a "source" (or server) file (e.g., a Microsoft Excel spreadsheet) and a *STATISTICA* datafile (the "client" file) so that when changes are made to the data in the source file, the data are automatically updated in the respective part of the *STATISTICA* Spreadsheet (client file).

A common application for dynamically linking two files would be in industrial settings, where the *STATISTICA* datafile would be dynamically linked with a measurement device connected to the serial port (e.g., in order to automatically update specific measurements hourly).

DDE links are established in the **Manage DDE Links** dialog (select **DDE Links** from the **Edit** menu). To create a new link, click the **New Link** button in this dialog. The dialog also is used to edit, update, delete, or disable any or all links.

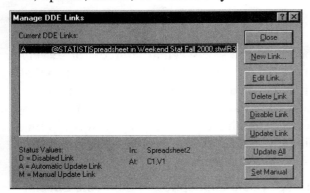

See *DDE Links* (page 231) and *Edit – DDE Links* (page 260) for further details.

TECHNICAL NOTE: CHANGING THE SPREADSHEET DOCUMENT-LEVEL EVENTS (A FLEXIBLE WAY TO CUSTOMIZE *STATISTICA*)

An event is an action that is typically performed by users, such as clicking a mouse button, pressing a key, changing data, or opening a spreadsheet or workbook. In *STATISTICA*, certain events are directly accessible to customization and you can change the default behavior of the program when the specific events are triggered (or when the events are triggered in specific parts of the documents). Using programmable events offers a particularly flexible and "far reaching" method to tailor *STATISTICA*'s behavior to your specific needs.

For example, you can prevent another user from saving changes made to a file or make specific sections of a spreadsheet dependent solely on other (e.g., automatically recalculated) values and protect them from being changed directly via editing the specific cells. You could also create a backup file of a particular workbook to a specific location whenever you save that workbook (that way you wouldn't have to save it twice). Those are just a few examples of countless customizations that can easily be accomplished by customizing events.

Event Types

Events are part of a set of tools built into *STATISTICA* to make it a powerful solution building system. There are two different types of events: document-level events and application-level events.

Document-level events. Document-level events occur only for open documents and in some cases, for objects within them. They allow you to customize the behavior of open documents. Commands for editing, running, and password protecting document-level events are available from the *View - Events* menu (page 266).

Application-level events. Application-level events occur at the level of the application itself. They allow you to customize the behavior of all documents of a certain type (such as all spreadsheets, or all workbooks, or all reports, etc.). For example, you could customize *STATISTICA* so that all reports are automatically saved to a backup folder or all spreadsheets are automatically saved (without prompting) before they are closed. Commands for editing and

running application level events are available from the **Tools - Macro – Application Events** menu (page 286).

Spreadsheet Events

The following table includes the document-level events that are available for spreadsheets. For more information on events, see *What Are Application Events and How Can They Be Controlled from STATISTICA Visual Basic?* and *Controlling STATISTICA Events with SVB Programs* in Chapter 11 – *STATISTICA Visual Basic*. Note that application-level events are listed in Chapter 11, and document-level events for other document types are listed in the appropriate chapter (e.g., report document-level events are listed in Chapter 5 – *STATISTICA Reports*).

Command	**Action**
Activate	Executes when the spreadsheet receives the focus within the *STATISTICA* workspace (typically - when you click on the spreadsheet).
BeforeClose	Executes when you close the spreadsheet. Before the document closes, the events within this function will first execute. An example of using this feature would be to add additional functionality, such as saving the file to multiple locations or exporting it to another application.
BeforeDoubleClick	Executes when you double-click anywhere on the spreadsheet. Before the clicked area goes into edit mode, the events within this function will first execute. Examples of using this feature would be to prevent users from altering a spreadsheet's content or to add additional functionality such as applying a custom format to the clicked area.
BeforePrint	Executes when you select a print option. Before the spreadsheet prints, the events within this function will first execute. Examples of using this feature would be to prevent users from printing the document or adding new functionality such as calling an alternate, customized printing dialog.
BeforeRightClick	Executes when you right-click anywhere on the spreadsheet. Before the shortcut menu is displayed, the events within this function will first execute. An example of using this feature would be to add additional functionality to displaying shortcut menus, such as highlighting the entire contents of the spreadsheet.

BeforeSave	Executes when you save changes made to the spreadsheet. Before the changes are actually saved, the events in this function will first execute. An example of using this feature would be to prevent users from permanently changing a spreadsheet.
DataChanged	Executes when data within a cell or block of cells is altered. This will also execute when a variable's Data type or Display format is changed.
Deactivate	Executes when the spreadsheet has lost the focus to another window within *STATISTICA*.
OnClose	Executes when the spreadsheet is being closed.
Open	Executes when the spreadsheet is being opened. Note that this will not run unless autorun was specified when the macro was last edited.
SelectionChange	Executes when the focus within the spreadsheet has moved.
StructureChanged	Executes when the size of the spreadsheet is changed. Examples of events that trigger this include the addition or deletion of variables/cases and Block statistics being added to the spreadsheet.

5

STATISTICA REPORTS

continued ➥

STATISTICA REPORTS

GENERAL OVERVIEWS

Report Overview

STATISTICA Reports offer a more traditional way of handling output (compared to workbooks) as each object (e.g., a *STATISTICA* Spreadsheet or Graph, or a Microsoft Excel spreadsheet) is displayed sequentially in a word processor style document.

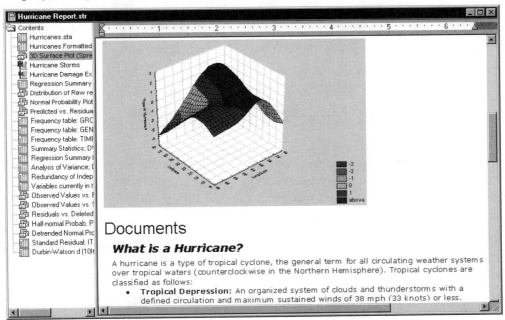

However, the technology behind this simple report offers you rich functionality. For example, like the workbook, each *STATISTICA* Report is also an ActiveX container where each of its objects (not only *STATISTICA* Spreadsheets and Graphs, but also any other ActiveX-compatible documents, e.g., Microsoft Word documents and bitmap images) is active, customizable, and in-place editable. Reports are stored in the STR file format, which is a StatSoft extension of the Microsoft RTF (Rich Text Format, *.rtf*) format. STR files share the RTF formatting information and additionally they include the tree view information (which cannot be stored in the standard RTF files). Hence, report files are by default saved with the file name extension *.str*, but they can also be saved as standard RTF files (in which case the tree information will not be preserved).

The obvious advantages of this way of handling output (more traditional than the workbook) are the ability to insert notes and comments "in between" the objects as well as its support for the more traditional way of quickly scrolling through and reviewing the output to which some users may be accustomed. (Note that the editor supports variable speed scrolling and other features of the IntelliMouse.) Also, only the report output includes and preserves a record of the supplementary information, which contains a detailed log of the options specified for the analyses (e.g., selected variables and their labels, long names, etc.) depending on the level of supplementary information requested on the **Output Manager** tab of the **Options** dialog (assessable via the **File – Output Manager** menu), see *Chapter 2 – STATISTICA Output Management*.

The obvious drawback, however, of these traditional reports is the inherent flat structure imposed by their word processor style format, though that is what some users of certain applications may favor.

Report Tree Overview

The report tree displays the organization of files in the report. The files are displayed in an *Explorer*-style format; objects are displayed sequentially and represented by their respective icons. This tree view offers a convenient way to locate a specific object or document, rearrange objects and documents (by dragging or via the Clipboard), etc.

You can embed any type of *STATISTICA* document in a report, including spreadsheets, graphs, and analyses. In addition to *STATISTICA* document types, you can embed other types of ActiveX/OLE objects in a report, including Excel spreadsheets, Word documents, bitmap images, and others. To edit one of these types of embedded documents, double-click on the document. The file opens in the viewer, and the **Report** toolbar merges with the toolbar from

the embedded file's native application, giving you access to all of the editing features you need.

Items in the tree are identified by the icon next to them. The ▦ spreadsheet, ▤ macro, and 🔲 graph icons represent *STATISTICA* Spreadsheet, Macro, and Graph documents, respectively. All non-*STATISTICA* documents are represented by their document icons. For example, Word documents are represented by the 🔲 Word icon, and Excel spreadsheet files are represented by the 🔲 Excel spreadsheet icon.

The report tree can be organized and modified using drag-and-drop features as well as Clipboard procedures.

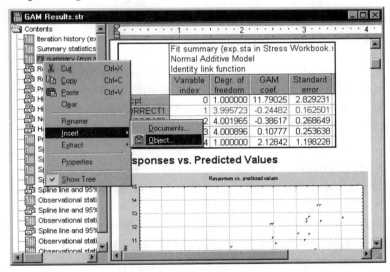

Commands for inserting, extracting, renaming, and removing items from the report tree are available from the report tree shortcut menu (accessed by right-clicking anywhere in the tree, as shown above).

Analysis Report Overview

An analysis report holds the results for a given analysis. As with all *STATISTICA* Reports, the files in the report are easily managed via the report tree. You can create an analysis report by selecting the *Multiple Reports (one for each Analysis/graph)* option button on the *Output Manager*. This option can be set globally via the *Output Manager* tab of the *Options* dialog (accessed via the *Tools – Options* menu) so that results for every analysis are saved in a

separate report (for as long as the option is selected). You can also set it locally via the **Analysis/Graph Output Manager** dialog (accessed by selecting the [🖪 Options ▼] button on an analysis or graph specification dialog) so that only the results from that particular analysis are stored in a separate report. For more information on report options, see *Output Manager* (page 412). See also Chapter 2 – *STATISTICA Output Management*.

Adding Items to a Report

Reports are ideal document containers for other *STATISTICA* documents, as well as objects from other applications. When adding items to a report, it is not necessary that the item itself be open. Nor is it necessary that you manually open the report; however, you must have either a report or an appropriate *STATISTICA* document (e.g., a spreadsheet, graph, or macro) open in order to access the options described here. To add any active *STATISTICA* document (except a workbook) to a report, use the **File - Report** commands (page 412) or the **Add to Report** toolbar button (see Chapter 1 – *STATISTICA – A General Overview*). When a report is active, you can add documents and objects to it using the commands on the report **Insert** menu (or the report shortcut menu) as well.

Insert an active *STATISTICA* document. To add a *STATISTICA* document that is open in your current *STATISTICA* session, select **Documents** from the **Insert** menu. Note that if you do not have a *STATISTICA* document open in your current session, this command is dimmed. This command displays the **Insert Document** dialog, which lists all active (open) *STATISTICA* documents. Select the document to insert, and click the **OK** button. The selected document is inserted at the cursor point, and the document icon is added to the navigation tree at the left side of the screen.

Insert documents from Workbooks. To insert documents from a workbook, select them in the tree pane of the workbook, copy (e.g., press CTRL+C), and then paste them all into the desired place in the report (note that multiple items can be selected in the workbook tree panel using the standard Windows CTRL and SHIFT conventions).

Insert a new or previously saved document. To add a new or saved document (e.g., a *STATISTICA* Spreadsheet or Graph, or an Excel spreadsheet) to a report, select **Objects** from the **Insert** menu. This displays the **Insert Object** dialog, which you can then use to create a new document or locate a previously saved document. The specified document is inserted at the cursor point and added to the navigation tree at the left side of the screen.

Add to Report toolbar button. Click the `Add to Report ▾` button (on the **Standard** toolbar, see Chapter 1 – *STATISTICA – A General Overview*) to add the current document to the open report, a previously saved report, or a new report. After you click this button, a menu is displayed allowing you to specify to which report you would like to add the document. Click the arrow to immediately display the menu and select a report to which the document will be added. Select **New Report** from the menu to create a new report and add the current document to it. Note that if the document cannot be sent to a report (e.g., the document is a report or a workbook), the **Add to Report** button is dimmed.

Global Report Options

You can customize the global functionality of *STATISTICA* Reports using the options offered on the **Reports** tab of the **Options** dialog, accessed via the **Tools - Options** menu. The selections made on this tab determine the defaults whenever you use a report.

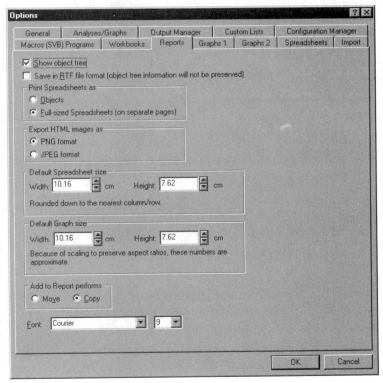

Show object tree. Select the *Show object tree* check box if you want to show the navigation tree by default whenever a report is opened. Note that if you clear this check box and want to view the tree for a particular report, you can display the navigation tree by selecting *Show Tree* from the *View* menu.

Save in RTF file format. Select the *Save in RTF file format (object tree information will not be preserved)* check box to save the report in RTF (Rich Text Format, *.rtf*) format. By default, *STATISTICA* Reports are saved (and viewed) using the *.str* file format. Using this format allows you to preserve the tree information with object names (when the report is opened in *STATISTICA*), so you can use the navigation tree to locate specific items in the report. By default, the *Save in RTF file format* check box is cleared.

Print Spreadsheets as. Use the options in the *Print Spreadsheets as* group to determine whether spreadsheets are printed as objects or full-sized spreadsheets when the report is printed. If you select the *Objects* option button, the spreadsheet is printed exactly as it is shown in the report, using the default dimensions. This means that only a portion of the spreadsheet is printed. If you select the *Full-sized Spreadsheets (on separate pages)* option button, then the entire spreadsheet is printed using separate pages.

Export HTML images as. Use the options in the *Export HTML images as* group to specify a format for saving graphs when reports are saved in HTML format. Select the *PNG format* option button to save images in PNG format. Select the *JPEG format* option button to save images in JPG format. For more information on these image formats, see *What Is the PNG (Portable Network Graphics) Files Format?* and *What Is the JPEG Graphics Format?* in Chapter 9 – *STATISTICA Graphs: Examples and FAQs.*

Default Spreadsheet size. Use the options in the *Default Spreadsheet size* group to set the default *Width* and default *Height* of all spreadsheets that are sent to reports. Note that the values entered in the *Width* and *Height* boxes are rounded down to the nearest column/row. You can control whether the width and height are measured in inches or centimeters. To measure in centimeters, select the *Use metric measurements* check box on the *General* tab of the *Options* dialog. To measure in inches, clear the *Use metric measurements* check box (the default setting).

Default Graph size. Use the options in the *Default Graph size* group to set the default *Width* and default *Height* of all graphs that are sent to reports. Note that because of scaling to preserve aspect ratios, these numbers are approximate. You can control whether the width and height are measured in inches or centimeters. To measure in centimeters, select the *Use metric measurements* check box on the *General* tab of the *Options* dialog. To measure in inches, clear the *Use metric measurements* check box (the default setting).

Add to Report performs. Use the options in the *Add to Report performs* group to specify what action is taken when the `Add to Report ▼` button is used. Select the *Move* option button to move the selected file to the report (and remove it from its stand-alone window) when the `Add to Report ▼` button is clicked. Select the *Copy* option button to copy the selected file to the report (the original is left in its stand-alone window). By default, the *Move* option button is selected.

Font. Specify a *Font* and font size in these fields to use in the specified report.

REPORT WINDOW

File Menu

The following commands are available from the report *File* menu. Many of these commands are also available from shortcut menus (accessible by right-clicking on an item or icon in the report) and from the toolbar buttons.

New

Select *New* from the *File* menu (or click the ☐ toolbar button) to display the *Create New Document* dialog, and then select the *Report* tab.

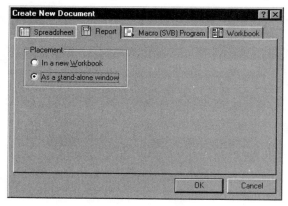

When creating a report, you have two placement options. Select the *In a new Workbook* option button to place the newly created report in a new *STATISTICA* Workbook, or select the *As a stand-alone window* option button to place the report in an individual report window.

Open

Select *Open* from the *File* menu (or click the ☐ toolbar button) to display the *Open* dialog. From this dialog, you can open any document of a compatible type (e.g., a report). Using the *Look in* box, select the drive and directory location of the desired file. Select the file in the

large box and click **Open**, or double-click on the file name. You can also enter the complete path of the document in the **File name** box.

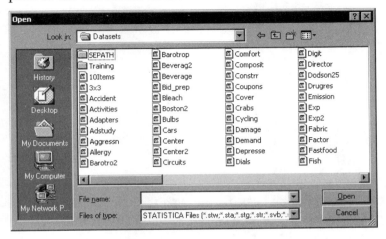

Close

Select **Close** from the **File** menu to close the current document. If you have made changes to the document since you last saved it, *STATISTICA* prompts you to save changes to the document.

Save

Select **Save** from the **File** menu to save the document with the name and in the drive location you specified when you last saved it. If you have not yet saved the document, the **Save As** dialog is displayed (see below).

Save As

Select **Save As** from the **File** menu to display the **Save As** dialog. From this dialog, you can save the active document with the name of your choice by entering the name in the **File name** box. Use the **Save in** box to select the appropriate drive and folder in which to save the document, or enter the complete path in the **File name** box.

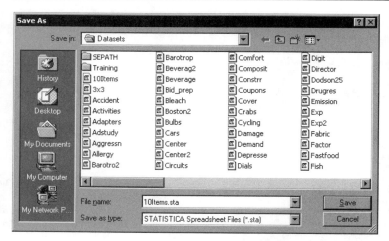

By default, *STATISTICA* recommends the file type that best fits the type of document you are saving. For example, if you are saving a report, the **Save as type** box contains the format **STATISTICA Report Files (*.str)**. Note that you can save reports in the standard RTF (Rich Text Format, *.rtf*) files, but then the tree information will not be preserved.

Get External Data

Options on this menu are for use with *STATISTICA* Query. See Chapter 10 - *STATISTICA Query* for further details about creating, editing, and using queries. Note that **Edit Query**, **Properties**, **Refresh Data**, **Delete Query**, and **Cancel Query** are only available when you have run (or are currently running) a query.

Open Query from File. Select **Open Query from File** from the **File – Get External Data** menu to display the standard open file dialog in which you can select an SQL file to run. Note that *.sqy* files are (created and) saved in *STATISTICA* Query (via the **File - Save As** command).

Create Query. Select **Create Query** from the **File – Get External Data** menu to open *STATISTICA* Query and display the **Database Connection** dialog. Once you have chosen or defined a database connection, you can use *STATISTICA* Query to write your query. For details on defining a database connection, see the *FAQ: Working with STATISTICA Query* section in Chapter 10 - *STATISTICA Query*.

Edit Query. Select **Edit Query** from the **File – Get External Data** menu to edit the query in the active spreadsheet. Note that if you have multiple queries on the active spreadsheet, *STATISTICA* opens the query that is associated with the active cell. If you select a cell that is not connected to any query or if you select multiple cells that involve more than one query,

then the **Select Database Query** dialog is displayed, and you must select the query that you want to edit before *STATISTICA* Query opens.

Properties. Select **Properties** from the **File – Get External Data** menu to display the **External Data Range Properties** dialog and specify options regarding the return of external data to spreadsheets. Note that if you have multiple queries on the active spreadsheet, then *STATISTICA* displays properties for the query associated with the active cell on the spreadsheet. If you select a cell that is not connected to any query or if you select multiple cells that involve more than one query, then the **Select Database Query** dialog is displayed. On this dialog, select the query to use and click the **OK** button to display the **External Data Range Properties** dialog.

Refresh Data. Select **Refresh Data** from the **File – Get External Data** menu (or press F5) to run a query and refresh the data (retrieve the latest data from the original database). Note that if you have multiple queries on the active spreadsheet, then *STATISTICA* refreshes data for the query associated with the active cell of the spreadsheet. If you select a cell that is not connected to any query or if you select multiple cells that involve more than one query, then the **Select Database Query** dialog is displayed so you can choose the specific query to refresh.

Delete Query. Select **Delete Query** from the **File – Get External Data** menu to delete the query from the active spreadsheet. (*STATISTICA* will always prompt you to verify that the query should be deleted.) Note that if you have multiple queries on the active spreadsheet, *STATISTICA* deletes the query associated with the active cell of the spreadsheet. If you select a cell that is not connected to any query or if you select multiple cells that involve more than one query, then the **Select Database Query** dialog is displayed so you can choose which query to delete.

Cancel Query. Select **Cancel Query** from the **File – Get External Data** menu (or press SHIFT+F5) to stop a currently running query at any time.

Workbook

Add to Analysis Workbook. Select **Add to Analysis Workbook** from the **File - Workbook** menu to add the current spreadsheet or graph to the analysis workbook. If the current spreadsheet or graph is not associated with an analysis workbook, this command is dimmed. For more information on analysis workbooks, see the *Analysis Workbook Overview* in Chapter 3 – *STATISTICA Workbooks*.

New Workbook. Select **New Workbook** from the **File - Workbook** menu to create a new workbook and add the current document to it. Note that in addition to adding the current

document to a new workbook, you can select any currently open workbook from the list provided in the *File - Workbook* menu.

Active workbooks. To add the current document to an active workbook (i.e., a workbook that is open or has recently been opened) select the workbook from the list of available workbooks shown on the *File - Workbook* menu. Note that if there are not any active workbooks at this time, the *File - Workbook* menu contains two options: *Add to Analysis Workbook* (which may be disabled) and *New Workbook*.

Report

Add to Analysis Report. Select *Add to Analysis Report* from the *File - Report* menu to add the current spreadsheet or graph to the analysis report. If the current spreadsheet or graph is not associated with an analysis report, this command is dimmed. For more information on analysis reports, see the *Analysis Report Overview*, page 403.

New Report. Select *New Report* from the *File - Report* menu to create a new report and add the current document to it.

Active Reports. To add the current document to an active report (i.e., a report that is open or has recently been open) select the report from the list of available reports shown on the *File - Report* menu. Note that if there are not any active reports at this time, the *File - Report* menu contains two options: *Add to Analysis Report* (which may be disabled) and *New Report*.

Output Manager

Select *Output Manager* from the *File* menu to display the *Output Manager* tab of the *Options* dialog. Use the options on this tab to specify where to direct output files. Options for reports include: sending all output to multiple reports (one for each graph/analysis), a single report (i.e., an analysis report), or an existing report. You can also specify the amount of supplementary information to include when sending spreadsheets to reports. For more details, see *Three Channels for Output from Analyses* in Chapter 2 – *STATISTICA Output Management*. See also, *What Supplementary Information Is Available with Analyses and How Can I Add It to My Reports?* (page 447).

Print Setup

Select *Print Setup* from the *File* menu to display the *Print Setup* dialog. Use this dialog to specify printing options.

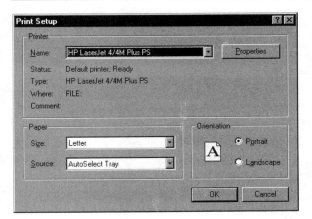

In the *Printer* group, choose a printer (using the *Name* box to view available printers). In the *Paper* group, choose the paper size (using the *Size* box to view a list of paper sizes), and the paper source (using the *Source* box to choose the printer's paper source). Additionally, choose a vertical or horizontal layout for the printed document by selecting the *Portrait* option button (for vertical) or the *Landscape* option button (for horizontal) in the *Orientation* group of the dialog. Click the *Properties* button to view options specific to the selected printer. For more information on specific printer options, consult your printer's manual.

Print Preview

Select *Print Preview* from the *File* menu to display the active document in a preview window. Before you print a document, it is useful to see how it will look when you print it. The document's appearance in the print preview window depends on the resolution of your printer, and the available fonts and colors. The status bar at the bottom of the window shows the current page number.

Use the options in the *Margins* group (see illustration below) to control margin and header/footer placement. To modify *Left*, *Right*, *Top*, and *Bottom* margins, enter new measurements in the appropriate box, and click the *Adjust* button. You can control whether the margins are measured in inches or centimeters. To measure in centimeters, select the *Use metric measurements* check box on the *General* tab of the *Options* dialog (accessed from the *Tools – Options* menu). To measure in inches, clear the *Use metric measurements* check box (the default setting). Margins can also be adjusted by dragging the margin lines with the mouse pointer (as shown below).

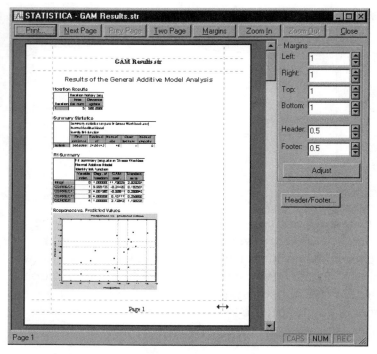

Values in the **Header** and **Footer** boxes determine how far the header and footer are placed from the top and bottom margins, respectively. Click the **Header/Footer** button to display the **Modify Header/Footer** dialog and access the header and footer information for the report. [See *Can I Create a Custom Header or Footer for My Report?* (page 446) for an example of using that dialog.] For more information on using the print preview window, see *Using the Print Preview Window* in Chapter 2 – *STATISTICA Output Management.*

Print

Select **Print** from the **File** menu (or press CTRL+P) to display the **Print** dialog. Use this dialog to execute a print command, as well as to specify printing options (see *Print Setup*, on page 412).

Verify that the printer in the **Name** box is the correct printer; if necessary, you can use the **Name** box to choose another printer. Select the **Print to File** check box if you want to print the document to a *.prn* file.

Use the options in the **Print Range** group to specify what pages to print. To print the entire report, select the **All** option button. To print a range of pages, select the **Pages** option button

and specify the range using the **From** and **To** boxes. To print the selected portion of the report, select the **Selection** option button.

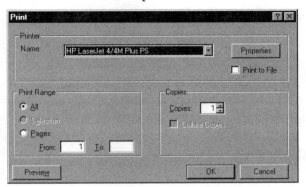

Use the options in the **Copies** group to print more than one copy of the document. Enter the number of copies you require in the **Copies** box. Select the **Collate Copies** check box to ensure that your printer collates the copies of the printed document.

To review the report in a print preview window (see above), click the **Preview** button. After you have selected the desired options, click the **OK** button to print the document.

Properties

Select **Properties** from the **File** menu to display the **Document Properties** dialog. This dialog contains two tabs: **Summary** and **Statistics**.

Select the **Summary** tab to create or modify a document summary (including information about the author's name, comments about the document, etc.). Select the **Statistics** tab to view statistics about the document (including date created, revision number, etc.).

Exit

Select **Exit** from the **File** menu to close *STATISTICA*. If you have made changes to your document since you last saved it, you will be prompted to save your changes.

Edit Menu

The following commands are available from the report **Edit** menu. Many of these commands are also available from shortcut menus (accessible by right-clicking on an item or icon in the report) and from the toolbar buttons.

Undo

Select **Undo** from the **Edit** menu (or press CTRL+Z) to undo your last action in the document (such as editing, moving, or copying blocks; font formatting; paragraph formatting, etc.). *STATISTICA* supports multi-level undo (with 32 buffers); therefore, you can undo multiple actions by selecting this option consecutively (up to 32 times).

Redo

Select **Redo** from the **Edit** menu (or press CTRL+Y) to counteract an **Undo** action. For example, if your last undo action command was to undo bold formatting to the selected text, then select **Redo** to reapply the bold formatting to the text.

Cut

Select **Cut** from the **Edit** menu (or press CTRL+X) to cut the selected text or object(s) to the Clipboard. *STATISTICA* will create in the Clipboard multiple formats of the selected object; the specific format to be used for pasting can later be selected with the **Paste Special** options in the client application.

Copy

Select **Copy** from the **Edit** menu (or press CTRL+C) to copy the selected text or object(s) to the Clipboard. *STATISTICA* will create in the Clipboard multiple formats of the selected object; the specific format to be used for pasting can later be selected with the **Paste Special** options in the client application.

Paste

Select **Paste** from the **Edit** menu (or press CTRL+V) to paste the contents of the Clipboard into the selected location in the document.

Paste Special

Select **Paste Special** from the **Edit** menu to display the **Paste Special** dialog, which contains a list of available ways of pasting the contents of the Clipboard into the current document.

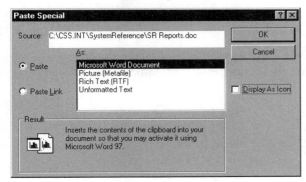

Use the options on this dialog to choose the desired format (i.e., Rich Text Format, Unformatted Text, or Picture) in which the Clipboard object will be pasted in the current document. The options available depend on the contents of the Clipboard and the options selected on this dialog. Once you have specified how to paste the Clipboard object, click the **OK** button to paste it. Click the **Cancel** button to exit this dialog without performing any actions.

Source. The **Source** box reports the source from which the current Clipboard object was taken. This information can help you in determining what options to select below. For example, if the source is an Excel spreadsheet, you may want to select the **Paste Link** option

button so that the selection is linked to the source file. In this way, the selection can be updated when the source file changes.

Paste. Select the *Paste* option button to paste the contents of the Clipboard into the document without maintaining any links to the source file.

Paste Link. Select the *Paste Link* option button to paste the contents of the Clipboard into the document and maintain a link to the source file. In this way, the selection can be updated when the source file changes.

As. The *As* box contains a variety of formats in which the Clipboard contents can be pasted. Select the one that best suits your needs. To see what will result from selecting a specific format, select the format and look at the *Results* box (below).

Display as icon. Select the *Display as icon* check box to display the Clipboard contents as an icon.

Result. The *Result* box gives a detailed description of what will happen when you paste the Clipboard contents into your document using the format selected in the *As* box.

Clear

Select *Clear* from the *Edit* menu (or press SHIFT+DELETE) to erase the current selection. Note this option is only available when something in the report has been selected.

Select All

Select *Select All* from the *Edit* menu (or press CTRL+A) to select the entire report.

Find

Select *Find* from the *Edit* menu (or press CTRL+F) to display the *Find* dialog. The *Find* dialog is used to search the active document for words or phrases.

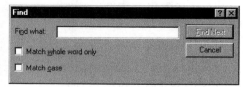

Type the word or phrase that you are looking for in the *Find what* box. To start the search, click the *Find Next* button. *STATISTICA* selects the first word(s) that match your search criteria.

Continue clicking **Find Next** until *STATISTICA* completes the search throughout the entire document. You can stop the search at any time by clicking the **Cancel** button.

When searching a report, select the **Match whole word only** check box to restrict the search to the exact word in the **Find what** box. Select the **Match case** check box to conduct a case-sensitive search for the word or phrase in the **Find what** box. Note that a  **Find** button is also available on the **Standard** toolbar (see Chapter 1 – *STATISTICA – A General Overview*).

Replace

Select **Replace** from the **Edit** menu (or press CTRL+H) to display the **Replace** dialog.

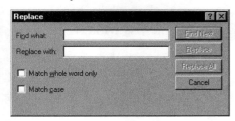

The **Replace** dialog is used to search the active document for words or phrases and replace them with different words or phrases. Specify a word or phrase for which to search in the **Find what** box, as well as a word or phrase with which to replace it in the **Replace with** box. To start the search, click the **Find Next** button. *STATISTICA* selects the first word(s) that match your search criteria. You can replace the text on an individual basis (click the **Replace** button) or globally (click the **Replace All** button). Continue clicking **Find Next** until *STATISTICA* completes the search throughout the entire document. You can stop the search at any time by clicking the **Cancel** button.

When searching a report, select the **Match whole word only** check box to restrict the search to the exact word in the **Find what** box. Select the **Match case** check box to conduct a case-sensitive search for the word or phrase in the **Find what** box. Note that a **Replace** button is also available on the **Standard** toolbar (see Chapter 1 – *STATISTICA – A General Overview*).

Repeat Find/Replace

Select **Repeat Find/Replace** from the **Edit** menu (or press F3) to repeat the last **Find/Replace** operation you conducted.

Screen Catcher

The *Screen Catcher* utility is used to select any rectangular area of the screen, the entire screen, or a specific window and copy it as bitmap to the Clipboard. The copied area can contain graphs, spreadsheets, or even screens from other applications (if they are currently displayed on the screen). The copied bitmap can then be pasted into *STATISTICA* documents or any other application that supports bitmaps.

Capture Rectangle. Select *Capture Rectangle* from the *Edit - Screen Catcher* menu (or press ALT+F3) to launch the *Screen Catcher* and capture a rectangular area of the screen. After selecting this command, position the cursor in one corner of the area that you want to copy. Now, drag the cursor over the area that you want to copy (you can drag across windows).

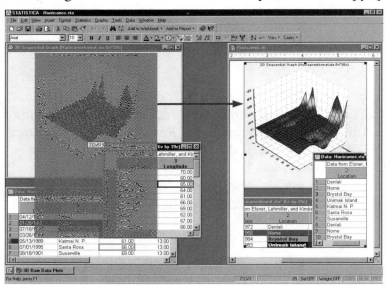

When you release the mouse button, the selected area is copied to the Clipboard.

Capture Window. Select *Capture Window* from the *Edit - Screen Catcher* menu to launch the *Screen Catcher* and capture the desired window on your desktop. After selecting this command, move the cursor over the screen until the window you want to capture is selected. Note that *STATISTICA* gives the dimensions of each window as you pass over it.

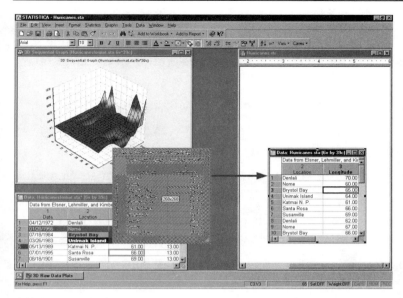

Click the window you want to capture, and it is copied to the Clipboard.

Links

Select **Links** from the **Edit** menu to display the **Links** dialog, which is used to manage links between the report and other applications. Note that this command is not available unless you have already created a link to the report (see the *Electronic Manual* for further details on creating links). All the links for the active document are listed in the large box. Select the link you want to manage, and then use the options below. Note that the **Source** to which the link is made, the **Type** of associated application and **Update** status are reported in the large box as well as beneath it.

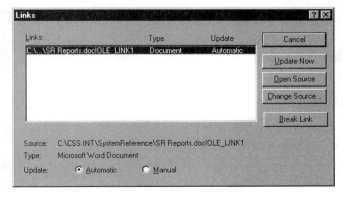

Automatic. Select the *Automatic* option button to automatically update the report when changes to the source file are made. This is the default link status. Note that you can toggle between *Automatic* and *Manual* (see below) by clicking on the *Update* portion of the large box as well.

Manual. Select the *Manual* option button to update the report manually by clicking the *Update Now* button.

Update Now. Click the *Update Now* button to update the report manually.

Open Source. Click the *Open Source* button to open the associated application to which the report is linked.

Change Source. Click the *Change Source* button to display the *Change Source* dialog, a standard file selection dialog that is used to change the file to which the report is linked.

Break Link. Click the *Break Link* button to permanently break the link between the report and file.

Object Properties

Select *Object Properties* from the *Edit* menu to display the *Object Properties* dialog. This dialog contains tabs that allow you to view and edit properties for the selected object in the report, such as a spreadsheet, a linked object, or a graph. The options available in this dialog depend on the type of object you are viewing. For example, when a spreadsheet is selected, you can view *General* properties and *View* properties (i.e., how the spreadsheet is displayed in the report). For linked objects, you can also view link information (i.e., whether the link is updated manually or automatically). Note that this command is also available from the shortcut menu, accessible by right-clicking on an object in the report. For more details on specific tabs, click the ? button in the dialog or press F1 to access the *Electronic Manual*.

View Menu

The following commands are available from the report *View* menu. Many of these commands are also available from shortcut menus (accessible by right-clicking on an item or icon in the report) and from the toolbar buttons.

Show/Hide Tree

Select **Show Tree** from the **View** menu to toggle the display of the pane on the left side of the screen. This pane contains a list of each of the spreadsheets and other embedded objects present in the report. When the **Show Tree** command is checked, the tree pane is displayed. When the command is cleared, the tree pane is hidden. See *Report Tree Overview* (page 402) for more details on using the report tree (or navigation tree).

Toolbars

Use the commands on the **View - Toolbars** menu to toggle the display of various toolbars. Note that toolbars can be docked to the top, left, or right side of the workspace. They can also float (e.g., be located anywhere in the workspace). For more information on a specific toolbar, see the following:

Toolbar	Chapter
Standard	1
Statistics	1
Six Sigma	1
Spreadsheet	4
Tools	4
Report	5
Graphs	6
Graph Tools	6
Graph Data Editor	6
Macro	11

Customize. Select **Customize** from the **View - Toolbars** menu to display the **Customize** dialog. For details about customization, see Chapter 1 – *STATISTICA – A General Overview*.

Status Bar

Select **Status Bar** from the **View** menu to toggle the text and buttons that are displayed in the status bar in the bottom center of the screen. Note that the *STATISTICA* Start button ▦ is available regardless of whether the status bar is turned on or off.

Header/Footer

Select **Header/Footer** from the **View** menu to display the **Modify Header/Footer** dialog. This dialog is used to view and modify the existing header and footer information.

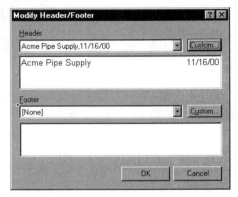

You can select a default header and footer (using the **Header** or **Footer** box), or you can create a custom header or footer (click the appropriate **Custom** button). For more information on using this dialog and creating custom headers and footers, see *Output Headers and Footers* in Chapter 2 – *STATISTICA Output Management*.

Events

An event is an action that is typically performed by users, such as clicking a mouse button, pressing a key, changing data, or opening a spreadsheet or workbook. In *STATISTICA* certain events are displayed and can be used to customize its behavior. Using programmable events, you can tailor *STATISTICA*'s behavior to your needs. Events are part of set of tools built into *STATISTICA* to make it a powerful solution building system. There are two different types of events: document-level events and application-level events.

Document-level events occur for only open documents and in some cases, for objects within them. They allow you to customize the behavior of open documents. Application-level events occur at the level of the application itself. They allow you to customize the behavior of all documents of a certain type (such as all spreadsheets, or all workbooks, or all reports, etc.). See *What Are Application Events and How Can They Be Controlled from STATISTICA Visual Basic* and the *Controlling STATISTICA Events with SVB Programs* in Chapter 11 – *STATISTICA Visual Basic* for more details.

View Code. Select *View Code* from the *View - Events* menu to display the *Document Events* window, which is used to enter code to change the default behavior of document-level events. Note that it may take a few seconds for the *Document Events* window to appear. Document-level events allow you to customize the behavior of open documents.

Autorun. Select *Autorun* from the *View - Events* menu to save your document-level event code with a specific document so that it runs any time that document is open.

Password. Select *Password* from the *View - Events* menu to display the *Setup Event Password* dialog, which is used to create a password to restrict access to editing document-level events.

Insert Menu

The following commands are available from the report *Insert* menu. Many of these commands are also available from shortcut menus (accessible by right-clicking on an item or icon in the report) and from the toolbar buttons.

Documents

Select *Documents* from the *Insert* menu to display the *Insert Document* dialog. Use this dialog to insert an active document (e.g., graph, spreadsheet, macro) into the report window. Double-click on the document to activate it. To deactivate it, click on the report window outside the document frame. While the document is active, you have access to all document-related menus and toolbars. Note you can also add an active document to the report using the *File - Report* menu (page 412) or the Add to Report ▼ toolbar button.

Objects

Select *Object* from the *Insert* menu to display the *Insert Object* dialog, which is used to insert an object into the report. You can create a new object or insert an object from a file. For example, you can insert a previously saved *STATISTICA* Spreadsheet or a new Microsoft Word document. For more information on inserting files into a report, see *Adding Items to a Report* (page 404).

Page Break

Select *Page Break* from the *Insert* menu to insert a manual page break into your report.

Format Menu

The following commands are available from the report *Format* menu. Many of these commands are also available from shortcut menus (accessible by right-clicking on an item or icon in the report) and from the toolbar buttons.

Font

Select *Font* from the *Format* menu to display the *Font* dialog. The *Font* dialog offers a choice of font and font size, style/effect, text and background colors, as well as a preview of the selected font options. Note you can also access this dialog using the **A** toolbar button (page 442).

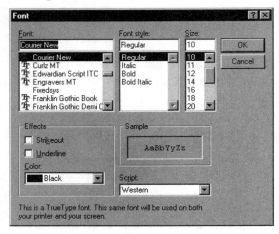

Font. Use the *Font* box and the accompanying list to select the desired font.

Font style. Select a font style using the options in the *Font style* box. Typical font styles are *Regular*, *Italic*, *Bold*, and *Bold Italic*.

Size. Use the *Size* box and the accompanying list to select the desired font size, in points. Note that 72 points equals 1 inch.

Effects. Select the ***Strikeout*** and/or ***Underline*** check boxes to apply a specific effect to the text.

Color. Use the ***Color*** box to select the text color from a palette of available colors.

Sample. The ***Sample*** box displays sample text that is formatted according to the choices you have made in the ***Font*** dialog.

Script. Use the ***Script*** box to select from the available script types for the font specified in the ***Font*** box.

Bullet Style

Select ***Bullet Style*** from the ***Format*** menu to toggle the use of bullets with the currently selected paragraphs. Note you can also use the ▤ toolbar button.

Paragraph

Select ***Paragraph*** from the ***Format*** menu to display the ***Paragraph Format*** dialog. This dialog controls three aspects of paragraph formatting within the document: ***Indentation***, ***Alignment***, and ***Spacing***. Once you have formatted the paragraphs, click the ***OK*** button. Click the ***Cancel*** button to exit without making any changes.

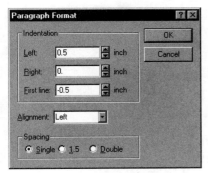

Indentation. The options in the ***Indentation*** group are used to control the indention of the selected paragraph(s). You can control whether the indentation is measured in inches or centimeters. To measure in centimeters, select the ***Use metric measurements*** check box on the ***General*** tab of the ***Options*** dialog (accessed from the ***Tools – Options*** menu). To measure in inches, clear the ***Use metric measurements*** check box (the default setting). In the ***Left*** box and ***Right*** box, type the measure by which to indent the first paragraph from the left and right margins, respectively. Use the ***First line*** box to specify the measure by which to indent the first

line from the left margin. For all three boxes, you can enter the text directly or use the accompanying microscrolls.

Alignment. Use the *Alignment* box to choose *Left*, *Center*, or *Right* alignment for the selected paragraph(s). Note that you can also use the paragraph formatting toolbar buttons ▤ ▤ ▤ to specify alignment.

Spacing. Use the options in the *Spacing* group to control the line spacing of the selected paragraph(s). Select the *Single* option button to apply single spacing to the selected paragraph(s). Select the *1.5* option button to apply one-and-a-half line spacing to the selected paragraph(s). Select the *Double* option button to apply double spacing to the selected paragraph(s).

Tabs

Select *Tabs* from the *Format* menu to display the *Tabs* dialog. By default, a report contains tab stops every half-inch (see below for information on how to change the settings from inches to centimeters). However, you can set custom tab stops by using this dialog. Once you have set the tab stops, click the *OK* button. Click the *Cancel* button to exit without setting or removing any tab

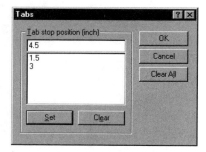

Tab stop position. Use the options in the *Tab stop position* group to enter a tab stop position as measured from the left margin. You can control whether the distance between the tab and the left margin is measured in inches or centimeters. To measure in centimeters, select the *Use metric measurements* check box on the *General* tab of the *Options* dialog (accessed from the *Tools – Options* menu). To measure in inches, clear the *Use metric measurements* check box (the default setting). To set a new tab, enter the tab position in the upper box, and then click the *Set* button to add the new tab stop to the list of tab stops in the lower box. Click the *Clear* button to delete the selected tab stop from the list of tab stops, or delete all tab stops from the list of tab stops by clicking the *Clear All* button.

Print Options

Select **Print Options** from the **Format** menu to display the **Print Options** dialog.

Use the options in the **Print Spreadsheets as** group to specify whether to print the spreadsheets as objects (select the **Objects** option button) or full-sized spreadsheets [select the **Full sized Spreadsheets (on separate pages)** option button].

Statistics Menu

The **Statistics** menu is available whenever any document is open. This menu provides access to all available analysis types within *STATISTICA* including **Basic Statistics/Tables**, **Multiple Regression**, **ANOVA**, **Nonparametrics**, **Distribution Fitting**, **Advanced Linear/Nonlinear Models**, **Multivariate Exploratory Techniques**, **Industrial Statistics & Six Sigma**, **Power Analysis**, and **Data-Mining**. Definitions for the various types of statistics are available in the *Statistics Toolbar* topic in Chapter 1 – *STATISTICA – A General Overview*. Also included on the **Statistics** menu are **Statistics of Block Data**, **STATISTICA Visual Basic**, and **Probability Calculator**.

Graphs Menu

The **Graphs** menu is available whenever any *STATISTICA* document is open. This menu provides access to all graph types in *STATISTICA* including **Histograms**, **Scatterplots**, **Means w/ Error Plots**, **Surface Plots**, **2D Graphs**, **3D Sequential Graphs**, **3D XYZ Graphs**, **Matrix Plots**, **Icon Plots**, **Categorized Graphs**, **User-defined Graphs**, **Graphs of Block Data**, and **Graphs of Input Data**. It also provides access to **Multiple Graph Layouts**. For more information on graphs, see Chapters 6, 7, 8, and 9 – *STATISTICA Graphs*.

Tools Menu

The following commands are available from the report **Tools** menu. Many of these commands are also available from shortcut menus (accessible by right-clicking on an item or icon in the report) and from the toolbar buttons.

Analysis Bar

To take advantage of *STATISTICA*'s "multitasking" functionality, *STATISTICA*'s analyses are organized as functional units that are represented with buttons on the **Analysis** bar at the bottom of the application window (above the status bar, see the illustration below, where **Basic Statistics**, **Cluster Analysis**, and **Canonical Analysis** are running simultaneously). Normally, at least one analysis button is created, and consecutive buttons are added as you start new analyses.

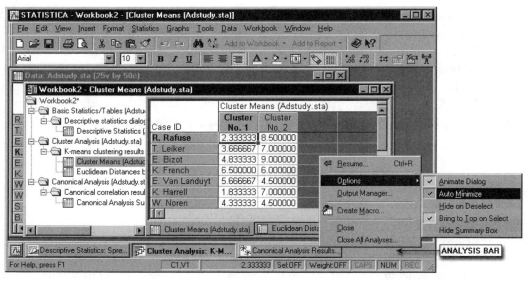

The commands on the **Tools – Analysis Bar** menu provide a variety of options for managing the **Analysis** bar.

Resume. Select **Resume** (or click the  toolbar button or press CTRL+R) from the **Tools - Analysis Bar** menu to continue the current analysis or graph. Note that you can also open the current analysis or graph by clicking on its button on the **Analysis** bar.

Select Analysis/Graph. Use the commands on the *Tools - Analysis Bar - Select Analysis/Graph* menu to select an analysis or graph from the set of active analyses and graphs. Note that you can also open an active analysis or graph by clicking on the appropriate button on the *Analysis* bar.

Options - Animate Dialog. Select *Animate Dialogs* from the *Tools - Analysis Bar - Options* menu to display animation when analysis dialogs are minimized or maximized. By default, this command is checked.

Options - Auto Minimize. Select *Auto Minimize* from the *Tools - Analysis Bar - Options* menu to automatically minimize all analysis dialogs when you select another window in *STATISTICA* or another application. By default, this command is checked. When your screen is large enough to accommodate several windows, it is recommended that you clear this option. This keeps the analysis dialogs on screen while the output created from these dialogs is produced, thus allowing you to use the dialogs as "toolbars" from which output can be selected.

Options - Hide on Select. Select *Hide on Select* from the *Tools - Analysis Bar - Options* menu to minimize all windows associated with a particular analysis when that analysis is deselected. By default, this command is cleared. Note that this command only applies when the results are sent to individual windows; see the discussion of the *Output Manager* tab of the *Options* dialog in Chapter 2 – *STATISTICA Output Management* for further details.

Options - Bring to Top on Select. Select *Bring to Top on Select* from the *Tools - Analysis Bar - Options* menu to activate (display at the top of *STATISTICA*) all windows associated with a particular analysis when that analysis is selected, replacing whatever dialogs were on top. This command also facilitates the organization of individual windows from various analyses. By default, this option is checked. Note that this command only applies when the results are sent to individual windows; see the discussion of the *Output Manager* tab on the *Options* dialog in Chapter 2 – *STATISTICA Output Management* for further details.

Options - Hide Summary Box. Select *Hide Summary Box* from the *Tools - Analysis Bar - Options* menu to not display the summary box, which is located at the top of certain results dialogs (such as *Multiple Regression - Results*) and contains basic summary information about the analysis. By default, this command is not checked.

Output Manager. Select *Output Manager* from the *Tools – Analysis Bar* menu to display the *Output Manager* tab of the *Options* dialog. Use the options on this tab to specify where to direct output files. Options for reports include sending all output to multiple reports (one for each graph and analysis), a single report (i.e., an analysis report), or an existing report. You can also specify how much supplementary information is displayed in the report. For more details, see *Three Channels for Output from Analyses* in Chapter 2 – *STATISTICA Output Management*.

Create Macro. Select *Create Macro* from the *Tools - Analysis Bar* menu to display the *New Macro* dialog and specify a name for a new macro based on the current analysis. When you run an analytic procedure (from the *Statistics* menu) or create a graph (from the *Graphs* menu) the Visual Basic code corresponding to all design specifications as well as output options that you select are recorded. Thus, when you click the *OK* button in the *New Macro* dialog, the resulting macro window displays the appropriate code to recreate the current analysis.

Minimize. Select *Minimize* from the *Tools - Analysis Bar* menu to minimize the current analysis.

Close. Select *Close* from the *Tools - Analysis Bar* menu to close the current analysis.

Close All Analyses. Select *Close All Analyses* from the *Tools - Analysis Bar* menu to close the all of the analyses/graphs on the *Analysis* bar.

Macro

A macro is a scripted application that extends functionality to *STATISTICA* by directly accessing *STATISTICA*'s object model and manipulating it. Macros are primarily used to automate tasks done in *STATISTICA* by harnessing its power and recording it into a *STATISTICA* Visual Basic script. Additionally, macros can be written as both stand-alone scripts and library classes to extend the statistical and mathematical capabilities of *STATISTICA*.

Macros. Select *Macros* from the *Tools - Macro* menu (or click the ▶ toolbar button on the *Macro* toolbar) to display the *Macros* dialog.

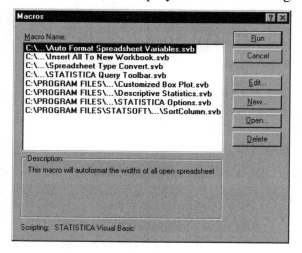

Use this dialog to run, edit, open, or delete existing macro (SVB) programs as well as to create new macros.

Start Recording Log of Analyses (Master Macro). Select *Start Recording Log of Analyses (Master Macro)* from the *Tools - Macro* menu to create a SVB program file that will include a sequence of analyses (a log of analyses) that have been performed interactively. Note that you must select this command before you start the analyses. Consult the *Electronic Manual* or the *STATISTICA Visual Basic Primer* to learn more about this powerful command and how to create recordable sequences of analyses.

Start Recording Keyboard Macro. Select *Start Recording Keyboard Macro* from the *Tools - Macro* menu to record a keyboard macro. See the *Electronic Manual* or the *STATISTICA Visual Basic Primer* for more details.

Stop Recording. Select *Stop Recording* from the *Tools - Macro* menu (or click the ■ toolbar button on the *Macro* toolbar) to stop recording a log of analyses (master macro) or a keyboard macro (see above).

Create Analysis/Graph Macro. Select *Create Analysis/Graph Macro* from the *Tools - Macro* menu to select from a list of all current analyses/graphs that are open. When you run an analytic procedure (from the *Statistics* menu) or create a graph (from the *Graphs* menu) the Visual Basic code corresponding to all design specifications as well as output options that you select are recorded. Select the appropriate analysis/graph to display that code. Note that the *New Macro* dialog will first be displayed in which you can type the macro's name and description.

Application Events - View Code. Select *View Code* from the *Tools - Macro - Application Events* menu to display the *Document Events* window, which is used to enter code to change the default behavior of application-level events. Application-level events allow you to customize the behavior of all documents of a certain type (such as all spreadsheets, all workbooks, or all reports). See *What Are Application Events and How Can They Be Controlled from STATISTICA Visual Basic?* in Chapter 11 – *STATISTICA Visual Basic* for more details.

Application Events - Autorun. Select *Autorun* from the *Tools - Macro - Application Events* menu to save your application-level event code with a specific type of document so that it will be run any time that type of document is open.

Add-Ins. Select *Add-Ins* from the *Tools - Macro* menu to display the *STATISTICA Add-Ins* dialog. Add-Ins are COM server components normally written in ATL (Active Template Library) that are used to create custom user interfaces of *STATISTICA* and/or fully functional external programs.

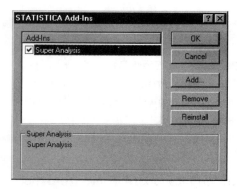

All available Add-Ins are displayed in the **Add-Ins** list. To create a new Add-In, click the **Add** button to display the **Specify Add-In to be Added** dialog, which is used to enter the program ID of the Add-In. Click the **Remove** button to delete the selected Add-In from the **Add-Ins** list. Finally, click the **Reinstall** button to register the selected Add-In to your operating system.

STATISTICA* Visual Basic Editor.** Select ***STATISTICA Visual Basic Editor from the **Tools - Macro - Application Events** menu (or press ALT+F11) to display a *STATISTICA* Visual Basic editor (Macro Window). For more information on using *STATISTICA* Visual Basic, including examples and an extensive FAQ section, see Chapter 11 – *STATISTICA Visual Basic*.

Customize

Select **Customize** from the **Tools** menu to display the **Customize** dialog, which is used to customize toolbars, menus, and keyboard hot keys with a variety of commands. For more information, see Chapter 1 – *STATISTICA – A General Overview*.

Options

Select **Options** from the **Tools** menu to display the **Options** dialog. In addition to general and display options, options are available for reports, file locations, custom lists, workbooks, graphs, macros, statistical analysis display, import and edit facilities, and output management. For a discussion of the **Reports** tab on the **Options** dialog, see *Global Report Options* (page 405).

Window Menu

The *Window* menu is available when any document is open. It provides access to commonly used commands for organizing the workspace and switching between files.

Close All

Select *Close All* (or press CTRL+L) from the *Window* menu to close all open spreadsheets, graphs, and related windows (e.g., graph data) in *STATISTICA*. This command is useful when you need to clear the screen to start a new analysis. Note that you will be prompted to save any unsaved files before they are closed.

Cascade

Select *Cascade* from the *Window* menu to arrange the open *STATISTICA* windows in an overlapping pattern so that the title bar of each window is visible.

Tile Horizontally

Select *Tile Horizontally* from the *Window* menu to arrange the open *STATISTICA* windows in a horizontal (side by side) pattern. When you select this option, *STATISTICA* automatically optimizes the display of the open windows (with the preference given to tiling horizontally).

Tile Vertically

Select *Tile Vertically* from the *Window* menu to arrange the open *STATISTICA* windows in a vertical pattern. When you select this command, *STATISTICA* automatically optimizes the display of the open windows (with the preference given to tiling vertically).

Arrange Icons

Select *Arrange Icons* from the *Window* menu to arrange all minimized windows into rows.

Windows

Select **Windows** from the **Window** menu to display the **Windows** dialog.

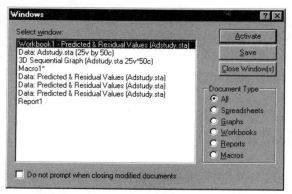

This dialog is used to access all of the currently open windows in *STATISTICA* and manage them.

Select Window. Select either a single window or multiple windows in the **Select Window** box to manipulate its associated window in the *STATISTICA* environment. Note that you can use CTRL or SHIFT while highlighting the items in the list box to make multiple selections.

Activate. Click the **Activate** button to set the focus to the currently selected window in the **Select Window** box. In other words, the associated window will be brought to the front of all of the other windows. Note that because only one window can have the focus at any given time, multiple selections for this operation do not apply.

Save. Click the **Save** button to save all of the windows that are currently selected in the **Select Window** box.

Close Windows. Click the **Close Windows** button to close all of the windows that are currently selected in the **Select Window** box. Note that *STATISTICA* will prompt you to save any modified windows before closing them.

Document Type. Use the options in the **Document Type** group to filter all of the open windows within *STATISTICA* by their document type; the contents of the **Select Window** box will be updated to reflect your filter selection. You can display all open documents in the **Select Window** box (select the **All** option button), or limit the display to spreadsheets, graphs, workbooks, reports, or macros only by selecting the appropriate option button.

Help Menu

The *Help* menu is available when any document is open. It provides access to various types of help.

Contents and Index

Select *Contents and Index* from the *Help* menu to display the *STATISTICA Electronic Manual*. *STATISTICA* provides a comprehensive *Electronic Manual* for all program procedures and all options available in a context sensitive manner by pressing the F1 key or clicking the help button **?** on the caption bar of all dialogs (there is a total of over 100 megabytes of compressed documentation included).

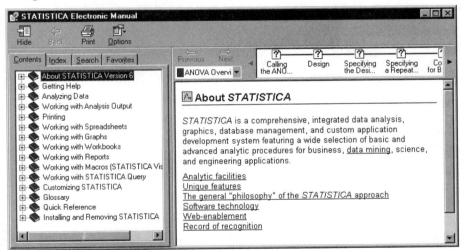

Due to its dynamic hypertext organization, organizational tabs (e.g., *Contents*, *Index*, *Search*, and *Favorites*), and various facilities allowing you to customize the help system, it is faster to use the *Electronic Manual* than to look for information in the traditional manuals. The status bar on the bottom of the *STATISTICA* window displays short explanations of the menu options or toolbar buttons when an item is highlighted or a button is clicked.

Statistical Advisor

Select **Statistical Advisor** from the **Help** menu to launch the **Statistical Advisor**. Based on your answers to the successive questions about the nature of your research, the **Statistical Advisor** suggests which statistical methods can be used and where to find them in *STATISTICA*.

Animated Overviews

Select **Animated Overviews** from the **Help** menu to display a submenu of the available animated overviews (visual overviews and specific feature help). Select from the submenu to launch one of the animated overviews.

If you select one of these commands and you have not installed that animated overview, a *STATISTICA* message dialog will prompt you for the location of that animated overview file. For example, you can run it from your *STATISTICA* CD, or if you are running a network version, the overviews can be located on the network drive. Note that if you have the *STATISTICA* CD, you can run the overviews directly from the CD without copying them to your hard drive.

StatSoft's Home Page

Select **StatSoft's Home Page** from the **Help** menu to launch the StatSoft Home Page in your default browser. We invite you to visit the StatSoft Web page often:

- For the most recent information about *STATISTICA*, downloadable updates, new releases, new products, news about StatSoft, etc., access the *What's New...* section of the Web site.

- For a comprehensive list of *Frequently Asked Questions* (including useful tips, solutions to hardware/software compatibility problems, etc.), access the *Technical Support* section of the Web site.

- For a library of *STATISTICA* Visual Basic programs (written by users), access the *Technical Support* section of the Web site (to submit your own programs to this "user exchange forum," send an e-mail to info@statsoft.com).

Technical Support

Select **Technical Support** from the **Help** menu to launch the *Technical Support - Getting More Help* page of the StatSoft Web site in your default browser. This page contains links to

download *STATISTICA* updates and links to FAQ topics on spreadsheets, graphs, printing, reports, etc. The StatSoft Technical Support Department e-mail address, phone number, and hours are also listed.

About *STATISTICA*

Select **About STATISTICA** from the **Help** menu to display the *STATISTICA* version, copyright notice, license information, and foreign office contact information.

REPORT TOOLBAR

The **Report** toolbar provides quick access to the most commonly used text output management facilities such as character formatting and paragraph formatting.

All of the options accessible via the toolbar buttons are also available from menus and via the keyboard. For information on customizing toolbars, see the *Toolbar Overview* in Chapter 1 – *STATISTICA – A General Overview*.

Arial ▾ 10 ▾ **Font Name and Font Size Boxes**

Use the Arial ▾ and 10 ▾ boxes to change the font and font size of new or selected text in reports.

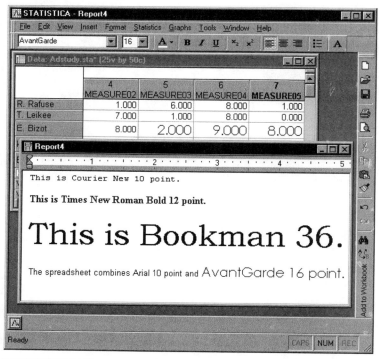

Fonts, font sizes, and character attributes (i.e., bold, italics, underline, etc.) can be mixed within each line of text.

B *I* **U** x₂ x² **Character Formatting Buttons**

Click the character formatting buttons to set formats for selected text. Keyboard shortcuts are also available for formatting options where indicated below.

B   Bold text (keyboard: CTRL+B);

I   *Italic* text (keyboard: CTRL+I);

U   Underline text (keyboard: CTRL+U);

x₂   *Subscript* text ($_{\text{sub}}$script);

x₂   *Superscript* text (^{super}script);

☰ ☰ ☰ ☷ **Paragraph Formatting Buttons**

Click the paragraph formatting buttons to control the format of new or selected paragraphs. Note that these commands are also available from the **Format** menu; see *Paragraph* (page 427) and *Bullet Style* (page 427).

☰   **Left justifies** (left margin aligns) text and graphs;

☰   **Centers** text or graphs between margins or within cells;

☰   **Right justifies** (right margin aligns) text and graphs;

☷   Toggles the use of **bullets** for the selected paragraphs.

A▾ Font Color Button

Use the **A▾** button to change the color of the selected text in the report. Click the arrow to display a standard color palette permitting the selection of color for new or selected text.

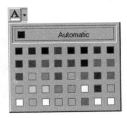

Click the letter to change the color to the default color (indicated by the color of the bar on the button).

 Fill Color Button

Use the button to change the background color of the selected text in the report. Click the arrow to display a standard color palette to select a color for the background.

Click the fill bucket to change the color to the default color (indicated by the color of the bar on the button).

A Font Button

Click the **A** button to display the *Font* dialog, which is used to select from a variety of font names, styles, sizes, colors, scripts, and effects.

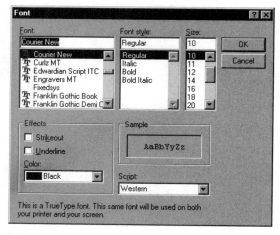

You can also access this dialog from the **Format** menu; see *Font* for more details (page 426).

FAQ: WORKING WITH REPORTS

How Can I Print Text/Graphics Reports from Analyses?

Combined text and graphics reports can be created in the report window by creating a new report (using the **Create New Document – Report** tab, accessed from the **File – New** menu) and clicking the **Add to Report** toolbar button to add individual spreadsheets and graphs. You can also select the **Also send to Report Window** check box and the **Single Report (common for all Analyses/graphs)** option button on the **Output Manager** tab of the **Options** dialog (accessible via the **Tools – Options** menu) to create an automatic report of all output created during your analyses.

To print the report, select **Print** from the **File** menu (or press CTRL+P) to display the **Print** dialog. Use the options in the **Print Range** group to specify what pages to print. To print the entire report, select the **All** option button. To print a range of pages, select the **Pages** option button and specify the range using the **From** and **To** boxes. To print the selected portion of the report, select the **Selection** option button.

What Is the Difference between *RTF* and *STR* (*STATISTICA* Report) Format?

RTF (Rich Text Format) is a Microsoft standard method of encoding formatted text and graphics for easy transfer between applications. When reports are saved in Rich Text Format (**.rtf*), all file formatting is preserved so that it can be read and interpreted by other RTF-compatible applications (e.g., Microsoft Word).

The *STATISTICA* Report format (*.str*) adheres to RTF conventions; however, saving reports in the default *STATISTICA* Report format (**.str*) ensures that reports will be opened in *STATISTICA,* giving you complete access to the report tree. In order to open the report in an RTF-compatible application, select **Save As** from the **File** menu to save the report as an RTF file. You can then open it in any RTF-compatible application.

Can I Combine Text, Tables, and Graphs in One Report?

Yes. You can use *STATISTICA* Reports to combine text and graphics by creating a new report (using the **Create New Document – Report** tab, accessed from the **File – New** menu) and clicking the **Add to Report** toolbar button to add individual spreadsheets and graphs to that report. Note that this is only one way you can create reports of your analyses in *STATISTICA*. For more information, see *Adding Items to a Report* (page 404).

How Do I Adjust Margins in the Reports?

Use the **Print Preview** dialog, accessible by selecting **Print Preview** from the **File** menu when the report is active or clicking the button.

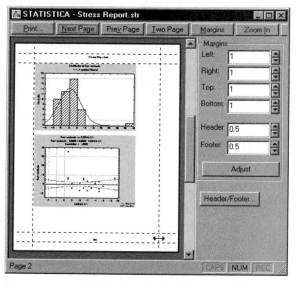

You can position the desired text on the printed page by adjusting the margins (specify the **Left**, **Right**, **Top**, and **Bottom** margin widths, or drag the margin lines with the mouse pointer, as shown above). You can also create a header and footer for the document. For more information, see *Using the Print Preview Window* in Chapter 2 – *STATISTICA Output Management*.

How Can I Automatically Print (or Save to Reports) All Spreadsheets and/or All Graphs from an Analysis?

To create an analysis report, select the *Also send to Report Window* check box and the *Multiple Reports (one for each Analysis/graph)* option button on the *Output Manager* dialog (accessible from the *File - Options* menu). When used together, these two options automatically generate a report of all spreadsheets and graphs associated with a single analysis.

Can I Append Output from Multiple Sessions to the Same Report?

Yes. To automatically send all results to the same report, select the *Also send to Report Window* check box on the *Output Manager* tab of the *Options* dialog (accessible by selecting *Output Manager* from the *File* menu). Then select the *Single Report (common for all Analyses/graphs)* option button. When used together, these two options automatically generate a report that contains all of the results and graphs created in *STATISTICA* (from the time the options are specified).

Can I Save My Reports in HTM Format?

Yes. Select *Save As* from the *File* menu to display the *Save As* dialog. To save the file using an *.htm* extension, select *HTML Files* from the *Save as type* group.

Note that any graphs in the report are saved as *.png* files in the same folder as the HTM file using the following naming convention: *reportname_pict0001.png*, *reportname_pict0002.png*, etc. You can save graphs as JPG files, instead. To do this, select *Options* from the *Tools* menu to display the *Options* dialog. Select the *Reports* tab, and then select the *JPEG format* option button in the *Export HTML images as* group. For more information on the *Options* dialog, see Chapter 2 - *STATISTICA Output Management*.

How Do I Create a New Report?

The most direct way to create a new report is to select *New* from the *File* menu or click the 🗋 toolbar button. In the *Create New Document* dialog, select the *Report* tab and specify whether you want to create the report as part of a new workbook or a stand-alone report. Alternatively, *STATISTICA* will create a new report for you (and add the active document to it) when you

select **New Report** from the **File - Report** menu or the  toolbar button. For more details, see *Adding Items to a Report* (page 404) and *New* (page 408).

Can I Create a Custom Header or Footer for My Report?

Yes. You can create a customized header or footer for a *STATISTICA* Report (or any other document) that can include information such as the date, time, or the name and logo of your company. To create a header or footer, select **Header/Footer** from the **View** menu. This displays the **Modify Header/Footer** dialog in which you can select a default header or footer or create a custom header or footer. For more information, see *Header/Footer* (page 424). See also, *Example: Creating a Custom Header* in Chapter 2 – *STATISTICA Output Management*.

Can I Rename an Item in a Report?

Yes. When you add items to a report, they are inserted with generic document names (i.e., spreadsheet). However, you can rename report items (using a variety of Windows standard renaming conventions) so that they are more descriptive. For example, you can click to select an item in the report tree and press F2. Then enter the new name for the item and press ENTER.

Can I Use Drag-and-Drop to Rearrange Items in the Report Tree?

Yes. The *STATISTICA* Report supports an entire range of drag-and-drop features within the report tree. By right-clicking on an item in the report tree, you can drag that item to a new location within the tree. For more information on standard drag-and-drop features, see *What Are the Drag-and-Drop Facilities?* in Chapter 4 – *STATISTICA Spreadsheets*.

Can I Use Clipboard Functions to Rearrange Items in the Report Tree?

Yes. In addition to using drag-and-drop features to rearrange report items, you can use cut (CTRL+X), copy (CTRL+C), and paste (CTRL+V) functions. For example, if you have a report that contains two spreadsheets and a graph, you can rearrange the order of the spreadsheets by using cut and paste functions. To move (cut) a spreadsheet (or other document) in a report, right-click on the spreadsheet and select **Cut** from the shortcut menu (or press CTRL+X or click

the ✄ button). This action places a copy of the item on the Clipboard while removing it from its current location in the report.

To paste the spreadsheet into a new position in the report, right-click in the report where you want to place the spreadsheet, and select **Paste** from the shortcut menu. Alternatively, you can click the 📋 toolbar button or press CTRL+V. This action places the spreadsheet (that you had cut to the Clipboard) in the desired location in the report. Note that the pasted object is placed above the object that you right-click.

What Supplementary Information Is Available with Analyses and How Can I Add It to My Reports?

STATISTICA provides a variety of supplementary information that can be included with graphs or spreadsheets from specific analyses when they are sent to a report. You can specify on the **Output Manager** tab of the **Options** dialog (select **Options** from the **Tools** menu) how much information to include.

Displaying supplementary information. Select the **Display supplementary information** check box on the **Output Manager** tab of the **Options** dialog and then drag the slider to the right to select the amount of supplementary information to be included in the report. Note that the **Display supplementary information** check box is only available if the **Also send to Report Window** check box is selected.

- **Brief.** Select **Brief**, the most "economical" output style, to include only the contents of the selected spreadsheets (i.e., no information about the variables or conditions specified for the analyses will be output).

- **Medium.** Select the **Medium** output style to include the contents of the selected spreadsheets as well as the current datafile name, information on Case Selection Conditions and Case Weights (if any were specified), the Output Header, a list of all variables selected for each analysis, and the missing data values for each variable.

- **Long.** Select the **Long** output style (which is more space consuming) to include all information from the **Medium** format and, additionally, the long variable names (e.g., labels, formulas), reserving one line of output (or more) for each variable.

- **Comprehensive.** Select the **Comprehensive** output style to provide the most comprehensive information on each variable selected for analysis. In addition to all information included in the **Long** format, it also includes a complete list for each selected variable of all its values that have text label descriptions.

TECHNICAL NOTE: CHANGING THE REPORT DOCUMENT-LEVEL EVENTS (A FLEXIBLE WAY TO CUSTOMIZE *STATISTICA*)

An event is an action that is typically performed by users, such as clicking a mouse button, pressing a key, changing data, or opening a spreadsheet or workbook. In *STATISTICA*, certain events are directly accessible to customization and you can change the default behavior of the program when the specific events are triggered (or when the events are triggered in specific parts of the documents). Using programmable events offers a particularly flexible and "far reaching" method to tailor *STATISTICA*'s behavior to your specific needs.

For example, you can prevent another user from saving changes made to a file or make specific sections of a spreadsheet dependent solely on other (e.g., automatically recalculated) values and protect them from being changed directly via editing the specific cells. You could also create a backup file of a particular workbook to a specific location whenever you save that workbook (that way you wouldn't have to save it twice). Those are just a few examples of countless customizations that can easily be accomplished by customizing events.

Event Types

Events are part of a set of tools built into *STATISTICA* to make it a powerful solution building system. There are two different types of events: document-level events and application-level events.

Document-level events. Document-level events occur only for open documents and in some cases, for objects within them. They are used to customize the behavior of open documents. Commands for editing and running document-level events are available from the *View - Events* menu (page 424).

Application-level events. Application-level events occur at the level of the application itself. They allow you to customize the behavior of all documents of a certain type (such as all spreadsheets, or all workbooks, or all reports, etc.). For example, you could customize *STATISTICA* so that all reports are automatically saved to a backup folder or all spreadsheets are automatically saved (without prompting) before they are closed. Commands for editing and

running application level events are available from the **Tools - Macro – Application Events** menu (page 433).

Report Events

The following table includes the document-level events that are available for reports. For more information on events, see *What Are Application Events and How Can They Be Controlled from STATISTICA Visual Basic?* and *Controlling STATISTICA Events with SVB Programs* in Chapter 11 – *STATISTICA Visual Basic*. Note that application-level events are listed in Chapter 11, and document-level events for other document types are listed in the appropriate chapter (e.g., spreadsheet document-level events are listed in Chapter 4 – *STATISTICA Spreadsheets*).

Command	Action
Activate	Executes when the report receives the focus within the *STATISTICA* workspace (typically when you click on the report or a document is sent to it).
BeforeClose	Executes when you close the report. Before the document closes, the events within this function will first execute. An example of using this feature would be to add additional functionality, such as saving the file to multiple locations or exporting it to another application.
BeforePrint	Executes when you print a report. Before the report prints, the events within this function will first execute. Examples of using this feature would be to prevent users from printing the document or adding new functionality such as calling an alternate, customized printing dialog.
BeforeRightClick	Executes when you right-click anywhere on the report. Before the shortcut menu is displayed, the events within this function will first execute. An example of using this feature would be to add additional functionality to displaying shortcut menus, such as selecting the entire contents of the report.
BeforeSave	Executes when you save changes made to the report. Before the changes are actually saved, the events within this function will first execute. An example of using this feature would be to prevent users from permanently changing a report.
Deactivate	Executes when the report has lost the focus to another window within *STATISTICA*.

OnClose   Executes when the report is being closed.

Open   Executes when the report is being opened. Note that this will not run unless autorun was specified when the macro was last edited.

SelectionChanged   Executes when the focus within the report has moved.

6

CHAPTER

STATISTICA GRAPHS: GENERAL FEATURES

continued

CHAPTER

STATISTICA GRAPHS:
GENERAL FEATURES

GRAPHS IN *STATISTICA*

The most common application of graphs is to efficiently present and communicate information (typically, numerical data). However, graphical techniques also provide powerful analytical tools for the exploration of data and verification of hypotheses.

A broad selection of graphics options. *STATISTICA* includes a comprehensive selection of graphical methods for both data analysis and the presentation of results. All graphs in *STATISTICA* include a broad selection of built-in, interactive analytic techniques and extensive customization tools allowing you to interactively control virtually all aspects of the display. Also, flexible multi-graphics management facilities are available that are used to integrate various graphical displays and to build dynamic links between applications (e.g., using OLE-Object Linking and Embedding).

Comprehensive support for Visual Basic and other languages. *STATISTICA* graphical options can also be accessed programmatically (using the built in *STATISTICA* Visual Basic or other compatible languages), which creates practically unlimited possibilities to produce highly customized graphical displays. These custom graphs can later be permanently added to *STATISTICA*'s user interface (e.g., assigned to buttons on toolbars or added to the menus).

General categories of graphs. The *STATISTICA* system offers a variety of methods in which graphs can be requested or defined. These methods (constituting broad categories of graphs such as input data, block data and specialized) are reviewed in *General Categories of Graphs* on page 461; they complement each other, providing a high level of integration

between numbers (such as raw data, intermediate results, or final results) and graphical displays. For example, specialized graphs can be requested as part of the automatic output from statistical procedures, but they can also be requested via integrated tools to visualize virtually any combination of numbers (and/or labels) that are displayed or generated by *STATISTICA*.

Customization of Graphs

Interactive graph customization. The customization options in *STATISTICA* graphics include hundreds of features and tools that can be used to adjust every detail of the display and associated data processing. However, these options are arranged in a hierarchical manner, so those used most often are accessible directly via shortcuts by double-clicking or right-clicking on the respective element of the graph.

Permanent settings and automation options. The initial (default) settings of all of these features can be easily adjusted so that even the default appearance and behavior of *STATISTICA* Graphs will match your specific needs and/or will require very little intervention on your part. There are at least four different ways to make these adjustments:

1. **Options dialog.** Perhaps the most straightforward way to adjust the default appearance of graphs is by using the *Graphs 1* and *Graphs 2* tabs of the *Options* dialog (accessible via the *Tools – Options* menu). Most commonly used settings can be easily adjusted here, and the results will be reflected in the default styles (see point 2 below) that will be used by the system and as such, they will be automatically saved in the *STATISTICA* configuration file (e.g., different settings can be used for different projects). For further details, see the *Configuration Manager* tab of the *Options* dialog in the *Electronic Manual*.

2. **Graph style system.** All of the numerous features that affect the appearance of the graph (from as elementary as the color of the font in the footnote to as general as the global features of the graph document) can be saved as individual "styles." These styles can be given custom names and later be reapplied using simple shortcuts (such as pressing a specific key combination or clicking a button on a custom toolbar). An intelligent system internally manages these thousands of styles and their combinations in *STATISTICA* and helps you achieve your customization objectives with a minimum amount of effort. All user-defined or modified styles will be saved automatically in the *STATISTICA* configuration file (e.g., different sets or systems of styles can be used for different projects). For further details on *STATISTICA* graph styles, see Chapter 7 – *STATISTICA Graphs: Creation and Customization* or the *Electronic Manual*; for information on saving

configurations of *STATISTICA*, see the **Configuration Manager** tab of the **Options** dialog in the *Electronic Manual*.

3. User-defined graphs. New types of graphs can be defined in a variety of ways and they can be added to the menus, dialogs, or toolbars. If a custom graph that you intend to use repeatedly is not built "from scratch" (e.g., drawn using the graphics primitives) but is based on one of the **Graphs** menu graphs and is produced by some combination of the existing graph customization options, then adding it to the **Graphs** menu as a new type of graph is as simple as clicking the **Add As User-defined Graph to Menu** button on the **Options 2** tab of the graph definition dialog. All user-defined graph specifications will be saved automatically in the *STATISTICA* configuration file (e.g., different sets of custom graphs can be used for different projects). For further details, see the **Configuration Manager** tab of the **Options** dialog in the *Electronic Manual*.

4. STATISTICA Visual Basic. Finally, note that there are no limits to how "deeply customized" your *STATISTICA* custom graphs can be, because *STATISTICA* Visual Basic (with all its powerful custom drawing tools as well as the *STATISTICA*-based library of graphics procedures) can be used to produce virtually any graphics or multimedia output supported by the contemporary computer hardware. Those custom developed displays or multimedia output can be assigned to *STATISTICA* toolbars, menus, or dialogs and become a permanent part of "your" *STATISTICA* application.

As mentioned previously (and discussed in detail on page 461), there are various methods to request *STATISTICA* Graphs. You could say that these methods represent different types of "interfaces" between numbers and graphs.

For example, the numbers represented in a pie chart can simply depict values of a spreadsheet column (e.g., variable **Sales**) in the consecutive cases of the spreadsheet (e.g., cases labeled: **Year 2002**, **Year 2003**, **Year 2004**, ..., etc.). The numbers in a similar pie chart, however, can also represent results of some calculations. For example, the slices of the pie can represent relative frequencies of observations that belong to certain categories calculated by one of the histogram or frequency categorization procedures (e.g., numbers of years when the **Sales** were below $10 million, between $10 and $20 million, and above $20 million).

Regardless of the method used to create a graph (i.e., regardless of where the numbers represented in the graph were obtained or how they were calculated), all *STATISTICA* Graph customization and multigraphics management facilities can be used to change the appearance of the graph or integrate it with other graphs or documents.

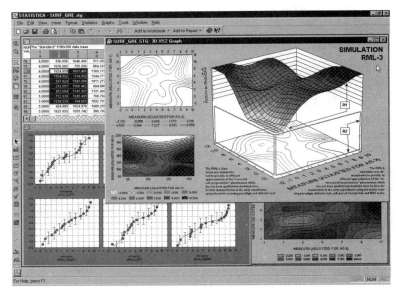

Also, all integrated analytic facilities that are accessible from within graphs in *STATISTICA* (such as function fitting, smoothing, rotation, brushing, analytical zooming, etc.) are available and can be applied to the graph regardless of the source of the numbers in the graph or the method that was used to create it.

The graph editing facilities offered in *STATISTICA* allow you to create not only highly customized scientific and technical publication-ready displays:

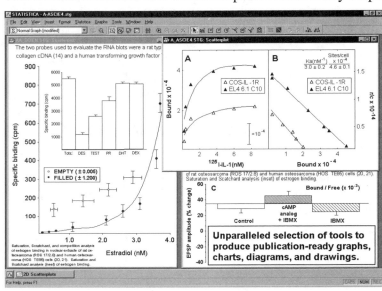

and precise drawings:

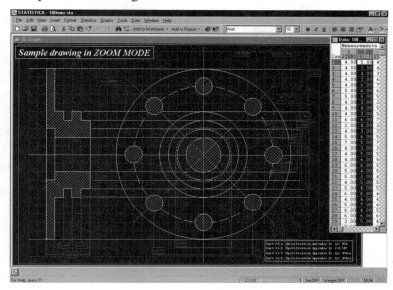

but also presentation-quality diagrams, posters, business charts, and other displays:

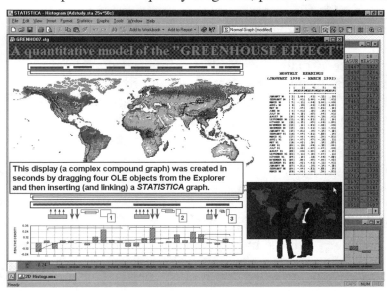

that are designed to communicate information in an effective and attractive manner.

Graphs that are saved into files or that in any other way have been temporarily detached from the *STATISTICA* application (e.g., copied to the Clipboard or linked to a document in another

application) are complete "objects" (technically speaking, ActiveX objects, see Chapter 3) that contain not only all customization features and other embedded objects, but also all data that are necessary to continue editing all aspects of the display or the analysis of its contents (fitting, smoothing, etc.).

Because *STATISTICA* Graphs are ActiveX objects, they can easily be linked to or embedded into other compatible documents (e.g., Excel or Word documents), where they can be in-place edited by double-clicking on them. *STATISTICA* Graphs are also ActiveX containers and, therefore, can contain a wide variety of embedded or linked documents such as Visio drawings, Adobe illustrations, Excel spreadsheets, or Word documents. Moreover, *STATISTICA* supports hierarchies of embedded objects up to four levels, which means that it can manage "documents, containing documents, containing documents, which contain documents."

Creating Graphs via
STATISTICA Visual Basic

STATISTICA graphical options can also be accessed programmatically using the built-in *STATISTICA* Visual Basic or other compatible languages. Therefore, there are no limits to how "deeply customized" your *STATISTICA* custom graphs can be, because *STATISTICA* Visual Basic (with all its powerful custom drawing tools as well as the *STATISTICA*-based library of graphics procedures) can be used to produce virtually any graphics or multimedia output supported by the contemporary computer hardware. These custom developed displays or multimedia output can be assigned to *STATISTICA* toolbars, menus, or dialogs and become a permanent part of "your" *STATISTICA* application.

An application written in *STATISTICA* Visual Basic can operate on graphs in three ways:

- make a new graph and then modify, print, or save it, etc.;
- access an existing graph window and then modify it;
- open an existing graph file and then modify, print, or save it, etc.

Every graph available in *STATISTICA* can be produced by *STATISTICA* Visual Basic and then customized using *STATISTICA* procedures or general options offered in this comprehensive language.

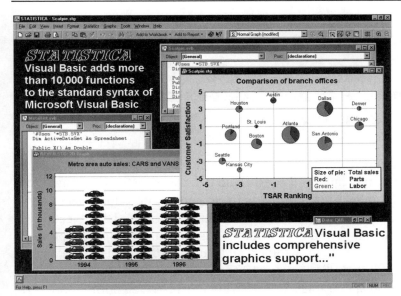

As with all other functions of *STATISTICA* Visual Basic, functions to access the graphics library of *STATISTICA* can be easily incorporated into *STATISTICA* Visual Basic programs via a hierarchically organized **Function Browser**, which contains short descriptions of all functions and options to insert them directly into the source code of your program (i.e., into the *STATISTICA* Visual Basic Editor).

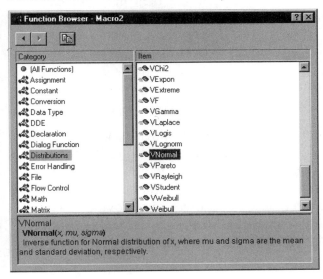

For more information on accessing the graphics libraries of *STATISTICA* via the *STATISTICA* Visual Basic programming language, refer to the *Electronic Manual*.

References, Acknowledgments, Standards

The *Electronic Manual* contains topics devoted to specific categories of graphs, includes conceptual overviews and examples of typical applications, and discusses distinctive functional properties of the respective types of graphs. More comprehensive general introductions to the science (and art) of visualization of data can be found in numerous sources on the subject. Some of the recommended sources that address a broad range of issues in statistical graphing are Buja and Tukey (1991), Chambers, Cleveland, Kleiner, and Tukey (1983), Cleveland (1984, 1985), Kolata (1984), Tufte (1983, 1990), Tukey (1977), and Tukey and Tukey (1981). Representative collections of articles on specific issues, covering a diverse selection of areas of visualization of statistical data, can be found in the proceedings of the *Annual Meetings of the Section on Statistical Graphics of the American Statistical Association* (published by the American Statistical Association).

The default settings of most statistical graphs offered in *STATISTICA* follow the established conventions that are either explicitly described in the literature on statistical and technical graphing, or they represent standards that are commonly accepted by major scientific journals (e.g., *SCIENCE*). However, practically all default settings of *STATISTICA* can be customized to meet specific requirements of unusual applications. *STATISTICA*'s graphics facilities were designed to play the role of flexible "tools," capable of producing effects that go far beyond established patterns and templates. Moreover, these tools can be customized, new tools can be designed, and both can be added to toolbars or menus for repeated use (see page 454).

In addition to a comprehensive selection of standard statistical and technical graphs, *STATISTICA* includes numerous unique types of graphs and graph customization facilities. While StatSoft statisticians designed most of them, it is important to say that our users have played an important role in their creation. In fact, the final selection of graphics options included in *STATISTICA* is the result of input from thousands of users who provided their comments and "wish lists" in response to StatSoft's inquiries. Many unique facilities of *STATISTICA* Graphs (e.g., the multiple subset selection facility, on-line categorization options, and others) were introduced in response to users' ideas and requests. We at StatSoft are very grateful for the input from our users.

General Categories of Graphs

In addition to the specialized statistical graphs that are available from the output dialogs in all statistical procedures (see page 468), there are two general categories or classes of graphs both accessible from the **Graphs** menu, **Graphs** toolbar, shortcut menus, and the *STATISTICA* Start button ⊞ menu:

- Input data graphs (**Graphs of Input Data**, see page 462, and **Graphs** menu graphs, see page 466) and

- **Graphs of Block Data** (see page 464).

The most important difference between these two general categories lies in the data that the graph types utilize for generating plots:

Input data graphs. *Graphs of Input Data* and their expanded version in the **Graphs** menu produce statistical summaries or other representations of the raw data in the current input data spreadsheet (typically for the entire variable(s) or its subsets if case selection conditions are used). Note that if graphs of this general category are produced using a shortcut menu from within a spreadsheet of results that does not contain the actual data (e.g., a correlation matrix), *STATISTICA* will still reach to the respective input (raw) data in order to produce the graph (e.g., a scatterplot of the variables identified by the selected cell in the correlation matrix from which the shortcut menu was opened).

Graphs of Block Data. *Graphs of Block Data*, on the other hand, are entirely independent of the concept of "input data" or "datafile." They provide a general tool to visualize numeric values in the currently highlighted block of any spreadsheet (which can contain values from custom defined subsets of numerical output or arbitrarily selected subsets of raw data).

Common features of the two categories of graphs. Note that these two general categories of graphs offer the same customization options and the same selection of types of graphs. For example, you can create the same, highly specialized categorized ternary graph from the input (raw) dataset, and from a custom defined block of values representing results of a particular test.

These two general categories of graphs will be briefly discussed in the next two sections, followed by a section on the **Graphs** menu, which contains an exhaustive selection of all graphs from the first (input data graphs) category (often referred to as **Graphs** menu graphs), although it also gives access to **Graphs of Block Data** and other options.

Graphs of Input Data

The **Graphs of Input Data** command is available from the shortcut menu of all spreadsheets and it offers quick and simplified access to the most commonly used types of graphs that are based on the current input dataset.

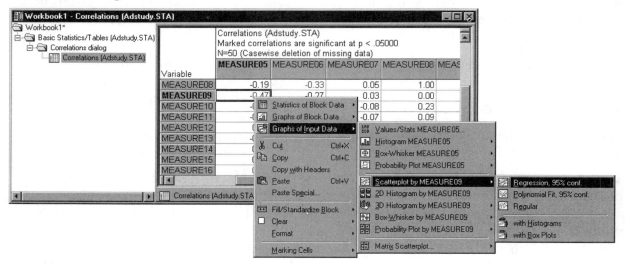

Note that all these graphs are also available from the **Graphs** menu, the *STATISTICA* Start button ▦ menu on the status bar, or by clicking the **Graphs Gallery** ▦ button on any graph specification dialog. **Graphs of Input Data** do not offer as many options as the corresponding **Graphs** menu graphs; however, they are quicker to select because unlike **Graphs** menu graphs:

- **Graphs of Input Data** can be called directly from the spreadsheet shortcut menus,

- **Graphs of Input Data** do not require you to select variables (the variable selection is determined by the current cursor position within a spreadsheet), and

- **Graphs of Input Data** do not require you to select options from any intermediate dialogs (default formats of the respective graphs are produced).

Graphs of Input Data process data directly from the current input datafile and they only take their cues as to which variables to use from the current cursor position (in any type of spreadsheet).

For example, if you right-click a single correlation in a results spreadsheet and create a **Scatterplot by...** graph, *STATISTICA* generates a 2D scatterplot using the original raw values of

---

the two variables represented by that correlation (see the *Introductory Example* in Chapter 1 – *STATISTICA – A General Overview* for a more detailed example).

Although the most convenient (and you could say most logical) way to select **Graphs of Input Data** is via the spreadsheet shortcut menu, you can also select them from the **Graphs** menu or the *STATISTICA* Start button 🖼 menu. Either method will display a submenu from which you can choose one of the statistical graphs applicable to the current variable (i.e., to the variable indicated by the current cursor position in the spreadsheet).

If the spreadsheet has a matrix format or a format where a cursor position indicates not one but two variables (as in the illustration showing a correlation matrix, below), then predefined bivariate graphs for the specified pair of variables will be directly available from the **Graphs of Input Data** menu.

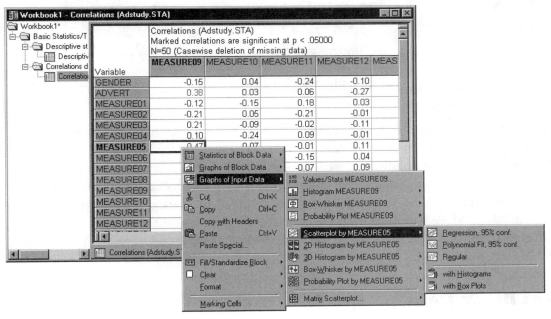

Otherwise (that is, when the current cursor position indicates only one variable as in a table of descriptive statistics, as shown in the illustration below), if you select any of the bivariate graphs in the menu, *STATISTICA* will prompt you to select the second variable. For example, in the illustration below, if you select **Scatterplot by**, *STATISTICA* would ask you by which variable **Measure05** is going to be plotted.

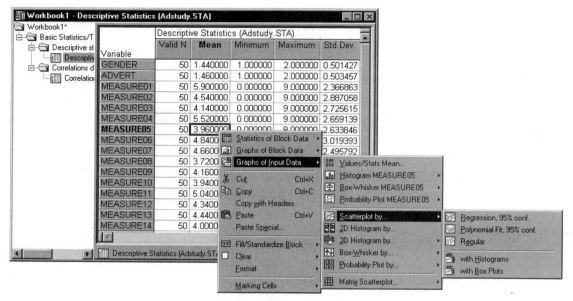

If more than one variable is indicated by a highlighted section (i.e., when a block is selected), then the **Graphs of Input Data** menu will apply to the first selected variable.

When generating **Graphs of Input Data**, *STATISTICA* takes into account the current case selection and weighting conditions for the variables that are being plotted. Note, however, that the case selection or weighting conditions need to be specified for the current spreadsheet (i.e., using the **Tools – Selection Conditions – Edit** and the **Tools – Weight** menu commands) and not only "locally" for an analysis (i.e., selected from the respective analysis/graph specification dialogs using the ▣ and ▣ buttons). The latter conditions will be ignored by the **Graphs of Input Data**. For more information on specific types of **Graphs of Input Data**, see *Graphs of Input Data* (page 527) or the *Electronic Manual*.

Graphs of Block Data

Unlike **Graphs of Input Data**, **Graphs of Block Data** use the currently selected (continuous) block of data in the active spreadsheet to specify input data for the graph.

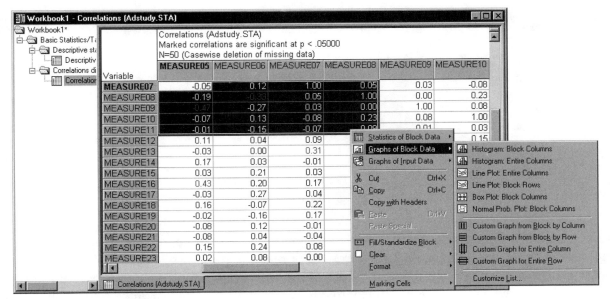

Note that these graphs are entirely independent from the concept of "input data." They process values (numbers) from whatever is currently selected in the block and ignore the "meaning" of those numbers (e.g., the numbers can be raw data or values of correlation coefficients). These graphs offer an effective means of visualizing, exploring, and efficiently summarizing numeric output from analyses displayed in results spreadsheets (e.g., histograms of Monte Carlo output scores in the **SEPATH** module, or a box plot of aggregated means from a multivariate multiple classification table in the **ANOVA** module).

Although the most convenient (and you could say most logical) way to select **Graphs of Block Data** is via the shortcut menu associated with the respective block selected in a spreadsheet, **Graphs of Block Data** are also available from the **Graphs** menu or the *STATISTICA* Start button ▦ menu. When creating **Graphs of Block Data**, you can select from default graphs (e.g., **Histogram: Block Columns** or **Line Plot: Block Rows**), or you can create your own custom graphs for either the selected cells in the rows or columns or of all cells in the selected rows or columns (i.e., going beyond the values that are selected in the block).

Default graphs. The default graphs (the first six commands in the illustration above) allow you to create specified graphs with a single click. For specific information on default graphs, refer to the *Graphs Menu* topic on *Graphs of Block Data* (page 525).

Custom graphs. Select one of the four **Custom Graph** commands to display the **Select Graph** dialog, which provides a variety of options for creating customized graphs.

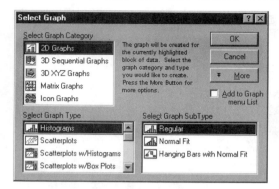

For specific information on a custom graph, refer to the *Graphs Menu* topic on *Graphs of Block Data* (page 525).

Customizing graphs. As with most features of *STATISTICA*, **Graphs of Block Data** are fully customizable. Select **Customize List** from the **Graphs of Block Data** menu to display the **Customize Graph Menu** dialog, which provides options to remove, rename, or edit the currently listed graphs as well as to add new (user-defined) graphs to the **Graphs of Block Data** menu.

For example, if you want to include a normal fit on the histograms created using **Histogram: Block Columns**, select **Histogram: Block Columns** in the **Customize Graph Menu** dialog, click the **Edit** button, and switch the **Graph SubType** to **Normal Fit**. All subsequently created **Histogram: Block Columns** plots will include a normal fit to the data.

Graphs Menu Graphs

Commands available on the **Graphs** menu provide a complete selection of all statistical graphs and all optional customizations available in *STATISTICA*. They are available from not only the **Graphs** menu, but also the *STATISTICA* Start button 🔲 menu and offer hundreds of types of graphical representations and analytic summaries of data.

Note that unlike the **Graphs of Block Data** (which are also included in this menu to offer a full complement of all graphical options accessible from a single control), all other graph types from the **Graphs** menu are not limited to the values in the current output spreadsheet. Instead, they process data directly from the current input spreadsheet in the same way the (previously discussed) **Graphs of Input Data** do. They represent either standard methods to graphically summarize raw data (e.g., various scatterplots, histograms, or plots of central tendencies such as medians) or standard graphical analytic techniques (e.g., categorized normal probability plots, detrended probability pots, or plots of confidence intervals of regression lines). When

generating these graphs, *STATISTICA* takes into account the current case selection and weighting conditions for the variables selected to be plotted.

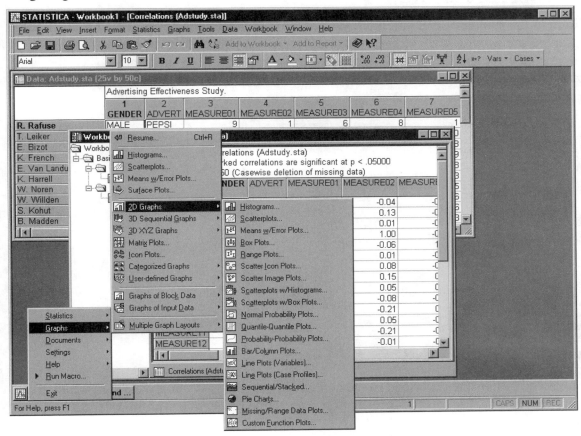

Graphs menu graphs include **2D Graphs**, **3D Sequential Graphs**, **3D XYZ Graphs**, **Matrix Plots**, **Icon Plots**, **Categorized Graphs**, and **User-Defined Graphs**. Note that the top portion of that menu includes four or five of the most commonly used types of graphs (**Histograms**, **Scatterplots**, **Mean/Error Plots**, etc.), and the lower portion contains the comprehensive list of all graph types. Like all menus in *STATISTICA*, it can be easily customized (use the **Menu** tab of the **Customize** dialog, accessed from the **Tools - Customize** menu) to position the most commonly used options in the most convenient locations. See *Types of Graphs Menu Graphs* in Chapter 7 – *STATISTICA Graphs: Creation and Customization*; see also the *Electronic Manual.*

Other Specialized Graphs

In addition to the standard selection of *Graphs of Input Data*, *Graphs of Block Data*, and *Graphs* menu graphs (see above), other specialized statistical graphs that are related to a type of analysis (e.g., cluster analysis results) are accessible directly from results dialogs (i.e., the dialogs that contain output options from the current analysis).

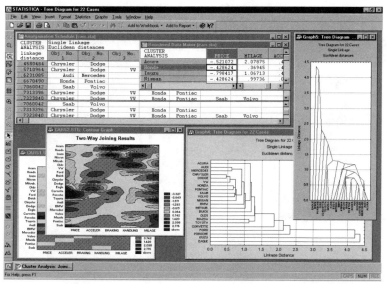

The specialized graphs are described in the context of the respective analyses from which they can be produced; thus, for specific information refer to the respective sections of the *Electronic Manual*).

GRAPH WINDOW

File Menu

The following commands are available from the graphs *File* menu. Many of these options are also available from shortcut menus (accessed by right-clicking on an item in the graph) and from the toolbar buttons.

New

Select *New* from the *File* menu (or click the ☐ toolbar button) to display the *Create New Document* dialog. You can create a new spreadsheet, report, macro (SVB) program, or workbook from this dialog by selecting the appropriate tab.

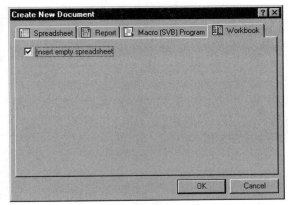

You can use the commands on the *Graphs* menu (from the menu bar or the *STATISTICA* Start button 📊 menu to create new graphs).

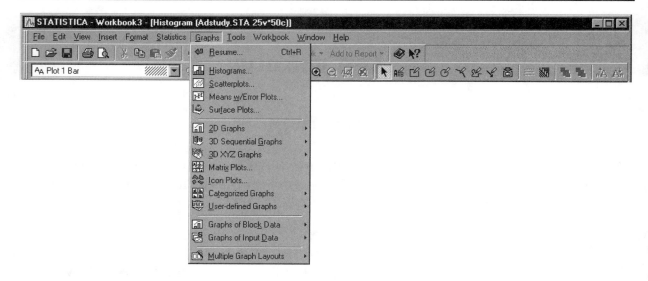

Open

Select **Open** from the **File** menu (or click the 🗁 toolbar button) to display the **Open** dialog. From this dialog, you can open any document of a compatible type (e.g., a graph). Using the **Look in** box, select the drive and directory location of the desired file. Select the file in the large box and click **Open**, or double-click the file name. You can also enter the complete path of the document in the **File name** box.

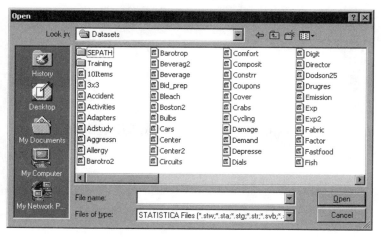

Close

Select **Close** from the **File** menu to close the current document. If you have made changes to the document since you last saved the document, *STATISTICA* prompts you to save changes to the document.

Save

Select **Save** (or press CTRL+S) from the **File** menu to save the current graph in the *STATISTICA* Graph format (file name extension *.stg*). Unlike other graphics formats such as JPEGs, metafiles, bitmaps, and PNG files, this format not only stores the "image" of the current graph but also all information necessary to continue graph customization or graphical data analysis (including all data represented in the graph, fitted equations, linked artwork, etc.). Graphs stored in the *STATISTICA* format can be opened later, allowing you to continue graphical data analysis or graph customization. This format is recommended whenever you intend to use the graph again in the *STATISTICA* system or link it to other application documents using Microsoft Windows OLE conventions. For more information on the *STATISTICA* graphics format, see *What Is the Native STATISTICA Graphics Format?* in Chapter 9 - *STATISTICA Graphs: Examples and FAQ's*.

Note that if you have not yet saved the graph, the **Save As** dialog is displayed (see below).

Save As

Select **Save As** from the **File** menu to display the **Save As** dialog.

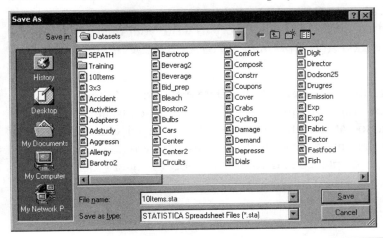

From this dialog, you can save the active document with the name of your choice by typing the name in the **File** name box. Use the **Save in** box to select the appropriate drive and folder in which to save the document. By default, *STATISTICA* recommends the file type that best fits the type of document you are saving. For example, if you are saving a graph, the **Save as type** box contains the format **STATISTICA Graph Files** (**.stg*).

Other graphics formats. To save graphs in other formats (e.g., bitmap, metafile, JPEG, or PNG formats), select the desired format in the **Save as type** box on the **Save As** dialog. *STATISTICA* provides support for each of the formats mentioned above, for more information, see *In What Formats Can I Save STATISTICA Graphs?* in Chapter 9 - *STATISTICA Graphs: Examples and FAQ's*. For information on the different formats, see *What Is the Windows Metafile Graphics Format?*, *What Is the Bitmap Graphics Format?*, *What Is the JPEG Graphics Format?*, and *What Is the PNG (Portable Network Graphics) File Format?* also in Chapter 9 - *STATISTICA Graphs: Examples and FAQ's*.

Get External Data

The options on this menu are for use with *STATISTICA* Query. See Chapter 10 - *STATISTICA Query* for further details about creating, editing, and using queries. Note that **Edit Query**, **Properties**, **Refresh Data**, **Delete Query**, and **Cancel Query** are only available when you have run (or are currently running) a query.

Open Query from File. Select **Open Query from File** from the **File - Get External Data** menu to display the standard open file dialog in which you can select an **SQL** file to run. Note that **.sqy* files are (created and) saved in *STATISTICA* Query (via the **File - Save As** option).

Create Query. Select **Create Query** from the **File - Get External Data** menu to open *STATISTICA* Query and display the **Database Connection** dialog. Once you have chosen or defined a database connection, you can use *STATISTICA* Query to write your query. For details on defining a database connection, see the *FAQ: Working with STATISTICA Query* section in Chapter 10.

Edit Query. Select **Edit Query** from the **File - Get External Data** menu to edit the query in the active spreadsheet. Note that if you have multiple queries on the active spreadsheet, *STATISTICA* opens the query that is associated with the active cell. If you select a cell that is not connected to any query or if you select multiple cells that involve more than one query, then the **Select Database Query** dialog is displayed and you must select the query that you want to edit before *STATISTICA* Query opens.

Properties. Select *Properties* from the *File - Get External Data* to display the *External Data Range Properties* dialog and specify options regarding the return of external data to spreadsheets. Note that if you have multiple queries on the active spreadsheet, then *STATISTICA* displays properties for the query associated with the active cell on the spreadsheet. If you select a cell that is not connected to any query or if you select multiple cells that involve more than one query, then the *Select Database Query* dialog is displayed. On this dialog, select the query to use and click the *OK* button to display the *External Data Range Properties* dialog.

Refresh Data. Select *Refresh Data* from the *File - Get External Data* menu (or press F5) to run a query and refresh the data (retrieve the latest data from the original database). Note that if you have multiple queries on the active spreadsheet, then *STATISTICA* refreshes data for the query associated with the active cell of the spreadsheet. If you select a cell that is not connected to any query or if you select multiple cells that involve more than one query, then the *Select Database Query* dialog is displayed so you can choose the specific query to refresh.

Delete Query. Select *Delete Query* from the *File - Get External Data* menu to delete the query from the active spreadsheet. (*STATISTICA* will always prompt you to verify that the query should be deleted.) Note that if you have multiple queries on the active spreadsheet, *STATISTICA* deletes the query associated with the active cell of the spreadsheet. If you select a cell that is not connected to any query or if you select multiple cells that involve more than one query, then the *Select Database Query* dialog is displayed so you can choose which query to delete.

Cancel Query. Select *Cancel Query* from the *File - Get External Data* menu (or press SHIFT+F5) to stop a currently running query at any time.

Workbook

Add to Analysis Workbook. Select *Add to Analysis Workbook* from the *File - Workbook* menu to add the current spreadsheet or graph to the analysis workbook. If the current spreadsheet or graph is not associated with an analysis workbook, this command is disabled. For more information on analysis workbooks, see the *Analysis Workbook Overview* in Chapter 3.

New Workbook. Select *New Workbook* from the *File - Workbook* menu to create a new workbook and add the current document to it. Note that in addition to adding the current document to a new workbook, you can select any currently open workbook from the list provided in the *File - Workbook* menu.

Active Workbooks. To add the current document to an active workbook (i.e., a workbook that is open or has recently been open), select the workbook from the list of available workbooks shown on the *File - Workbook* menu. Note that if there are not any active workbooks at this time, the *File - Workbook* menu contains two options: *Add to Analysis Workbook* (which may be disabled) and *New Workbook*.

Report

Add to Analysis Report. Select *Add to Analysis Report* from the *File - Report* menu to add the current spreadsheet or graph to the analysis report. If the current spreadsheet or graph is not associated with an analysis report, this command is disabled. For more information on analysis reports, see the *Analysis Report Overview* in Chapter 5.

Add New Report. Select *New Report* from the *File - Report* menu to create a new report and add the current document to it.

Active Reports. To add the current document to an active report (i.e., a report that is open or has recently been open), select the report from the list of available reports shown on the *File - Report* menu. Note that if there are not any active reports at this time, the *File - Report* menu contains two options: *Add to Analysis Report* (which may be disabled) and *New Report*.

Output Manager

Select *Output Manager* from the *File* menu to display the *Output Manager* tab of the *Options* dialog. Use the options on this tab to specify where to direct output files. *Graphs* menu graphs, *Graphs of Input Data*, and *Graphs of Block Data* can be directed to individual windows, workbooks, and/or reports dependent on selections made on the *Output Manager* dialog (for more details, see *Three Channels for Handling Output* in Chapter 2 – STATISTICA Output Management.).

Print Setup

Select *Print Setup* from the *File* menu to display the *Print Setup* dialog. Use this dialog to specify printing options. In the *Printer* group, you can choose a printer (using the *Name* box to view available printers). In the *Paper* group, you can choose the paper size (using the *Size* box to view a list of paper sizes), and the paper source (using the *Source* box to choose the printer's paper source). Additionally, you can choose a vertical or horizontal layout for the printed document by selecting the *Portrait* option button (for vertical) or the *Landscape* option button (for horizontal) in the *Orientation* group of the dialog.

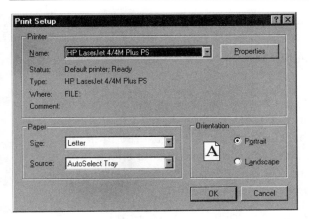

Click the **Properties** button to view options specific to the selected printer. For more information on specific printer options, consult your printer's manual.

Print Preview

Select **Print Preview** from the **File** menu to display the active document in a preview window.

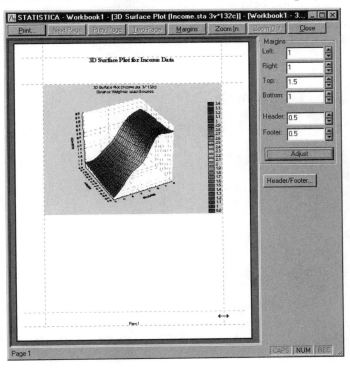

Before you print a document, it is useful to see how it will look when you print it. The document's appearance in the print preview window depends on the resolution of your printer, and the available fonts and colors. The status bar at the bottom of the window shows the current page number.

Use the options in the **Margins** group to control margin and header and/or footer placement. To modify **Left**, **Right**, **Top**, and **Bottom** margins, enter new measurements in the appropriate field, and click the **Adjust** button. You can control whether the margins are measured in inches or centimeters. To measure in centimeters, select the **Use metric measurements** check box on the **General** tab of the **Options** dialog (accessed from the **Tools – Options** menu). To measure in inches, clear the **Use metric measurements** check box (the default setting).

You can also adjust the margins with the mouse pointer by dragging the margin lines (as shown above).

Values in the **Header** and **Footer** boxes determine how far the header and footer are placed from the top and bottom margins, respectively. Click the **Header/Footer** button to display the **Modify Header/Footer** dialog and access the header and footer information for the graph (or other document). See *Can I Create a Custom Header or Footer?* in Chapter 5 - *STATISTICA Reports*, for an example of using that dialog.) For more information on using the print preview window, see *Using the Print Preview Window* in Chapter 2 - *Output Management*.

Print

Select **Print** from the **File** menu (or press CTRL+P) to display the **Print** dialog.

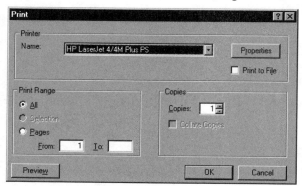

Use this dialog to execute a print command, as well as to specify printing options (see *Print Setup*, above). Verify that the printer in the **Name** box is the correct printer; if necessary, you can use the **Name** box to choose another printer. You can select the **Print to File** check box if you want to print the document to a **.prn* file.

Use the options in the **Print range** group to specify what to print. Note that for graphs in stand-alone windows, only the **All** option button is active. This allows you to print the entire graph. When printing graphs from within a workbook, you can select one or more graphs to print using the **Selection** option button.

Use the options in the **Copies** group to print more than one copy of the document. Enter the number of copies you require in the **Copies** box. Select the **Collate Copies** box to ensure that your printer collates the copies of the printed document.

To review the report in a print preview window (see above), click the **Preview** button. After you have selected the desired options, click **OK** to print the document.

Printing from workbooks. When printing from within a workbook, only the currently displayed (active) graph is printed when the **Print** 🖨 button is clicked, but if the **Print** dialog is used, multiple selections from the workbook can be printed by using the **Selection** option on the **Print** dialog.

Note that *STATISTICA* supports the standard Windows SHIFT+click and CTRL+click conventions to select ranges and discontinuous lists of items in the workbook, respectively. Thus to print a range of graphs from a workbook, select the range of graphs, select **Print** from the **File** menu to display the **Print** dialog, select the **Selection** option button in that dialog, and click the **OK** button. For more information on a variety of printing issues, see *Printing Graphs* in the FAQ section in Chapter 9 - *STATISTICA Graphs: Examples and FAQ's*.

Properties

Select **Properties** from the **File** menu to display the **Document Properties** dialog. This dialog contains two tabs: **Summary** and **Statistics**.

Select the **Summary** tab to create or modify a document summary (including information about the author's name, comments about the document, etc.). Select the **Statistics** tab to view statistics about the document (including date created, who last saved the document, etc.).

Exit

Select **Exit** from the **File** menu to close *STATISTICA*. If you have made changes to your document since you last saved it, you will be prompted to save your changes.

Edit Menu

The following commands are available from the graph *Edit* menu. Many of these options are also available from shortcut menus (accessible by right-clicking on an item in the graph) and from the toolbar buttons.

Undo

Select *Undo* from the *Edit* menu (or press CTRL+Z) to undo your last action in the document (such as modifying point markers, area patterns, line properties, etc.). *STATISTICA* supports multi-level undo (with 32 buffers); therefore, you can undo multiple actions by selecting this option consecutively (up to 32 times).

Redo

Select *Redo* from the *Edit* menu (or press CTRL+Y) to counteract an *Undo* action. For example, you if your last undo action command changed a histogram to a pie chart, then select *Redo*, and *STATISTICA* changes the pie chart back into a histogram.

Cut

Select *Cut* from the *Edit* menu (or press CTRL+X) to cut the selected text or object(s) to the Clipboard. *STATISTICA* will create in the Clipboard multiple formats of the selected object; the specific format to be used for pasting can later be selected with the Paste Special options in the client application.

Copy

Select *Copy* from the *Edit* menu (or press CTRL+C) to copy the selected text or object(s) to the Clipboard. *STATISTICA* will create in the Clipboard multiple formats of the selected object; the specific format to be used for pasting can later be selected with the Paste Special options in the client application.

Paste

Select **Paste** from the **Edit** menu (or press CTRL+V) to paste the contents of the Clipboard into the selected location in the document.

Paste Special

Select **Paste Special** from the **Edit** menu to display the **Paste Special** dialog, which provides you with a list of available ways of pasting the contents of the Clipboard into the current document.

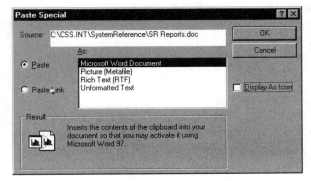

Use the options on this dialog to choose the desired format (i.e., Rich Text Format, text, or picture) in which the Clipboard object will be pasted in the current document. The options available depend on the contents of the Clipboard and the options selected on this dialog. Once you have specified how to paste the Clipboard object, click the **OK** button to paste it. Click the **Cancel** button to exit this dialog without performing any actions.

Source. The **Source** box reports the source from which the current Clipboard object was taken. This information can help you in determining what other options to select. For example, if the source is an Excel spreadsheet, you may want to use **Paste Link** so that the selection is linked to the source file. In this way, the selection can be updated when the source file changes.

Paste. Select the **Paste** option button to paste the contents of the Clipboard into the document without maintaining any links to the source file.

Paste Link. Select the **Paste Link** option button to paste the contents of the Clipboard into the document and maintain a link to the source file so that the selection can be updated when the source file changes.

As. The *As* box contains a variety of formats in which the Clipboard contents can be pasted. Select the one that best suits your needs. To see what will result from selecting a specific format, select the format and look at the *Results* box (below).

Display as icon. Select the *Display as icon* check box if you want the Clipboard contents to be displayed as an icon.

Result. The Result box gives a detailed description of what will happen when you paste the Clipboard contents into your document using the format selected in the *As* box.

Delete

Select *Delete* from the graph *Edit* menu to remove (delete) the selected object (e.g., arrow, rectangle, oval, embedded or linked object, etc.) from the graph. To delete an object, click on the object to select it and then select *Edit - Delete*.

Links

Select *Links* from the *Edit* menu to display the *Links* dialog, which is used to manage links between the graph and other applications. All the links for the active document are listed as well as the *Source* to which the link is made, the *Type* of associated application, and *Update* status.

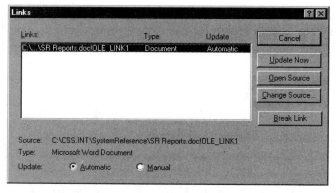

Select the link you want to manage, and then choose any other options desired.

Automatic. Select the *Automatic* option button to automatically update the graph when changes to the source file are made. This is the default link status. Note that you can toggle between *Automatic* and *Manual* (see below), by clicking in the *Update* portion of the large box as well.

Manual. Select the *Manual* option button to update the graph manually by clicking the *Update Now* button.

Update Now. Click the *Update Now* button to update the graph manually.

Open Source. Click the *Open Source* button to open the associated application to which the graph is linked.

Change Source. Click the *Change Source* button to display the *Change Source* dialog, a standard file selection dialog that allows you to change the file to which the graph is linked.

Break Link. Click the *Break Link* button to permanently break the link between the graph and file.

Screen Catcher

The *Screen Catcher* utility allows you to select any rectangular area of the screen, the entire screen, or a specific window and copy it as bitmap to the Clipboard. The copied area can contain graphs, spreadsheets, or even screens from other applications (if they are currently displayed on the screen). The copied bitmap can then be pasted into *STATISTICA* documents, or any other application that supports bitmaps.

Capture Rectangle. Select *Capture Rectangle* from the *Edit - Screen Catcher* menu (or press ALT+F3) to launch the *Screen Catcher* in order to capture a rectangular area of the screen.

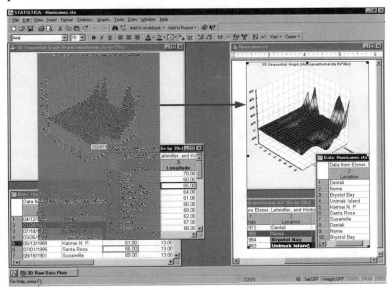

Then, position the cursor in one corner of the area that you want to copy. Now, drag the cursor over the area that you want to copy (you can drag across windows). When you release the mouse button, the selected area is copied to the Clipboard.

Capture Window. Select *Capture Window* from the *Edit - Screen Catcher* menu to launch the *Screen Catcher* to capture the desired window on your desktop.

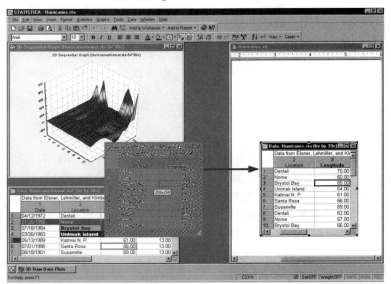

Move the cursor over the screen until the window you want to capture is selected. Note that *STATISTICA* gives the dimensions of each window as you pass over it. Click the window that you want to capture, and it is copied to the Clipboard (as shown above).

View Menu

The following commands are available from the graph *View* menu. Many of these options are also available from shortcut menus (accessed by right-clicking on an item in the graph) and from the toolbar buttons.

Graph Data Editor

Select *Graph Data Editor* from the *View* menu to display the Graph Data Editor. This spreadsheet contains the variables that have been specified for the current graph.

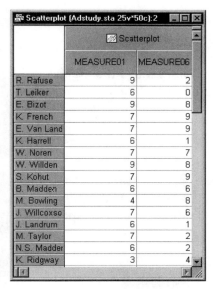

Using the Graph Data Editor, you can perform many editing operations on the data as well as brushing of the data. See *Graph Data Editor* (page 570) for more details.

Rotate

Select **Rotate** from the **View** menu or click the ![icon] toolbar button to display the **Point of View** dialog. Use this dialog to change the point of view for 3D plots by rotating them on their axes. For more information, see *Special Topics in 3D Graph Customization* (Chapter 7 - *STATISTICA Graphs: Creation and Customization*) and the 3D *Rotation Control Button* (page 545).

Fit to Window

Select **Fit to Window** from the **View** menu or click the ![icon] toolbar button to fit the graph to the current default graph window size so that as you resize the window, the graph proportions are not distorted. For more details on this selection, see *Display Graph Fit to Window Button* (page 547).

Original Size

Select *Original Size* from the *View* menu or click the  toolbar button to move the graph position within the graph area. For additional details on this selection, see *Display Graph at Actual Size Button* (page 547).

Document Size and Scaling

Select *Document Size and Scaling* from the *View* menu or click the toolbar button to display the *Document Size and Scaling* dialog. Use this dialog to "stretch" or "squeeze" the graph in either direction, changing the proportions between the X and Y coordinates of 2D displays and other relations between graph components. For more information, see *Adjust the Graph Actual Size/Scaling Button* (page 548).

Zoom Subgraph

Select *Zoom subgraph* from the *View* menu to zoom-in on the selected portion of the graph. This command is only available when a subgraph has been selected in a compound graph (see *What Are Compound Graphs?* in Chapter 9 - *STATISTICA Graphs: Examples and FAQ's*) for more details.

Brushing

Select *Brushing* from the *View* menu or click the toolbar button to activate the brushing mode and display the interactive *Brushing* dialog. When this mode is activated, the cursor changes to a "gunsight-style" cross-hair, and you can highlight data points from the graph by clicking on them with the cross-hair, enclosing them in a box, lasso, or cube (3D plots, only). You can also slice graphs along the x-, y-, or z-axis (3D plots, only). *STATISTICA* features a comprehensive selection of brushing facilities, including a variety of tools for manipulation of data points on screen, as well as the management of brushed (selected) data points (see also *Graph Data Editor*, page 570, for more details.). For more details on the interactive *Brushing* dialog, see the *Brushing Tool Button* (page 545) and the *Electronic Manual*.

Graph Placement

Select *Graph Placement* from the *View* menu or click the ▣ toolbar button to set the graph area either by selecting the appropriate graph area (using the graph placement selection tool) or by using the resizing squares. For more details on this selection, see *Set Graph Area Button* (page 549).

Alignment Grid

Select *Alignment Grid* from the *View* menu (or press CTRL+G) to toggle the display of the graph's alignment grid (see below). This grid is a visible representation of the grid specified via the *Snap to Grid* option, and the spacing of the gridlines is dependent on the *Horizontal* and *Vertical Spacing* settings in the *Snap to Grid* dialog. The alignment grid aids in the placement and alignment of graphic objects (e.g., added text, embedded or linked objects, arrows, freehand drawings such as the rectangles shown below, a previously saved graph, etc.). Note that the grid is not printed when you print the graph. For more information, see *Snap to Grid* (below) and *Snap to Grid Button* (page 551).

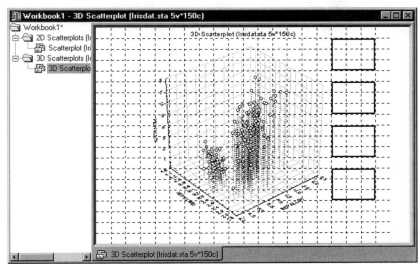

Snap to Grid

Select **Snap to Grid** from the **View** menu or click the toolbar button to display the **Snap to Grid** dialog, in which you can enable (and customize) an invisible drawing grid to help you align drawn objects (e.g., lines, rectangles, inserted graphics or text objects).

By default, the object is pulled into alignment with the nearest intersection of gridlines (as if the resolution of the screen were "decreased"). This is the snap-to-grid effect.

The display of the alignment grid itself in the graph can be toggled on and off by pressing CTRL+G, or selecting **Alignment Grid** from the **View** menu. For more details, see *Snap to Grid Button* (page 551).

Header/Footer

Select **Header/Footer** from the **View** menu to display the **Modify Header/Footer** dialog. This dialog allows you to view and modify the existing header and footer information. You can select a default header and footer (using the **Header** or **Footer** box), or you can create a custom header or footer (click the appropriate **Custom** button).

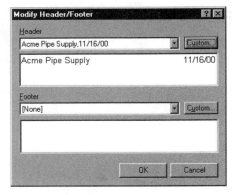

For more information on using this dialog and creating custom headers and footers, see *Output Headers and Footers*, in Chapter 2.

Events

An event is an action that is typically performed by users, such as clicking a mouse button, pressing a key, changing data, or opening a spreadsheet or workbook. In *STATISTICA*, certain events are visible to the outside world and can be used to customize its behavior. Using programmable events, you can tailor program behavior to your needs. Events are part of set of

tools built into *STATISTICA* to make it a powerful solution building system. There are two different types of events: document-level events and application-level events.

Document-level events occur for only open documents, and in some cases, for objects within them. They allow you to customize the behavior of open documents. Application-level events occur at the level of the application itself. They allow you to customize the behavior of all documents of a certain type (such as all spreadsheets, or all workbooks, or all reports, etc.). See *What Are Application Events and How Can They Be Controlled from STATISTICA Visual Basic?* and the *STATISTICA Events Overview* in Chapter 11 for more details.

View Code. Select *View Code* from the *View - Events* menu to display the *Document Events* window, which is used to enter code to change the default behavior of document-level events. Document-level events allow you to customize the behavior of open documents.

Autorun. Select *Autorun* from the *View - Events* menu to save your document-level event code with a specific document so that it runs any time that document is opened.

Events Password. Select *Password* from the *View - Events* menu to display the *Setup Event Password* dialog, which is used to create a password to restrict access to editing document-level events.

Toolbars

Use the commands on the *View - Toolbars* menu to toggle the display of various toolbars. Note that toolbars can be docked to the top, left, or right side of the workspace. They can also float (e.g., be located anywhere in the workspace). For more information on a specific toolbar, see the following:

Toolbar	Chapter
Standard	1
Statistics	1
Six Sigma	1
Spreadsheet	4
Tools	4
Report	5
Graphs	6
Graph Tools	6
Graph Data Editor	6
Macro	11

Customize. Select *Customize* from the *View – Toolbars* menu to display the *Customize* dialog. For details about customization, see Chapter 1 – *STATISTICA – A General Overview.*

Status Bar

Select *Status Bar* from the *View* menu to toggle the text and buttons that are displayed in the status bar in the bottom center of the screen. Note that the *STATISTICA* Startup button is available regardless of whether the status bar is turned on or off.

ToolTips on Graph Objects

Select *ToolTips on Graph Objects* from the *View* menu to toggle the display of ToolTips with graph objects. These ToolTips report the name of the currently selected object (e.g., subtitle, custom text) and, in 2D displays, the true, that is, *dynamic* coordinates (see the FAQ section in Chapter 9 for more details on fixed vs. dynamic coordinates) of the cursor position if the cursor is positioned within the plot area. The latter feature is useful when you need to identify the exact coordinates of a specific data point, especially in such multi-graphics displays as matrix plots where there is no space for the display of the scale values.

Note that in 3D displays (where the *dynamic coordinates* cannot be identified using the regular mouse pointer (see 3D brushing tools in Chapter 9 for information on advanced interactive tools to accomplish that goal), fixed coordinates of the current cursor position are displayed, where the middle of the graph is set to have coordinates of 0,0. (See the FAQ section in Chapter 9 for more details on fixed and dynamic placement of graph objects.)

Clone Graph on Zoom

Select *Clone Graph on Zoom* from the *View* menu to automatically send the graph to a new window when using the *Zoom-in* toolbar button.

Automatically Update Style

Select *Automatically update style* from the *View* menu to allow a newly modified default style to take effect (i.e., redraw with new appearance) on all currently open graphs that are using that default style.

Insert Menu

The following commands are available from the graph *Insert* menu. Many of these options are also available from shortcut menus and from the toolbar buttons.

Plots Legend

Select *Plots Legend* from the *Insert* menu to display the *Add Legend* dialog. Use this dialog to add a plot legend to the current graph.

Type. In the *Type* box, select the type of legend to add to the plot. *STATISTICA* provides two types of legends: *Fixed (title)* or *Floating*. If you select *Fixed (title)*, the legend is added as a title at the top, left, right or bottom of the graph depending on the selection in the *Position* box. If you select, *Floating* in the *Type* box, a floating legend is created and positioned in one of the four corners of the graph depending on the selection in the *Position* box.

Position. In the *Position* box, select the position for the initial placement of the newly created legend. Fixed titles (see *Type*, above) can be placed at the top of the graph (select either *Title* or *Subtitle*), to the left of the graph (select *Left*), to the right of the graph (select *Right*), or at the bottom of the graph (select *Footnote*). Floating legends can be placed in one of the four corners of the graph window by selecting *Left-up*, *Left-down*, *Right-up*, or *Right-down*.

Note that this option is for creation of the legend only. To add text or objects, customize fonts, etc., use the *Graph Titles/Text* tab of the *All Options* dialog (see Chapter 7 - *STATISTICA Graphs: Creation and Customization*).

Fits Legend

Select *Fits Legend* from the *Insert* menu to display the *Add Legend* dialog. You can use this dialog to add a fit legend to the current graph. For more information on the *Add Legend* dialog, see *Plots Legend*, above. Note that this option is for creation of the legend only. To add text or objects, customize fonts, etc., use the *Graph Titles/Text* tab of the *All Options* dialog (see Chapter 7 – *STATISTICA Graphs: Creation and Customization*).

Surface Legend

Select **Surface Legend** from the **Insert** menu to display the **Add Legend** dialog. You can use this dialog to add a surface legend to the current graph. For more information on the **Add Legend** dialog, see **Plots Legend**, above. Note that this option is for creation of the legend only. To add text or objects, customize fonts, etc., use the **Graph Titles/Text** tab of the **All Options** dialog (see Chapter 7 - *STATISTICA Graphs: Creation and Customization*).

Statistics

Select **Statistics** from the **Insert** menu to display the **Add Statistics** dialog. You can use this dialog to add appropriate statistics as fixed or floating titles to the current graph.

Type. Use the **Type** box to specify whether the statistics are added as movable (**Floating**) or fixed (**Fixed (title)**) text.

Position. Use the **Position** box to specify the placement of the statistics. If you have selected a **Floating** type, you can choose from **Left-up**, **Left-down**, **Right-up**, or **Right-down**. If you have selected to add the statistics to a **Fixed (title)** position, you can select from one of the five default title positions: **Title**, **Subtitle**, **Footnote**, **Left**, or **Right**. For more on titles in graphs, refer to the **Graph Titles/Text** tab of the **All Options** dialog (Chapter 7 - *STATISTICA Graphs: Creation and Customization*).

Edit statistics. Click the **Edit statistics** button to display the **Statistics** dialog.

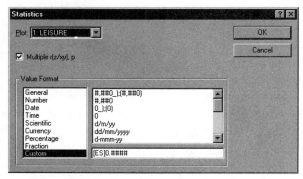

Using this dialog, you can specify the plot to use for the statistics, the appropriate statistics for the plot type (e.g., for a 3D surface plot, you can select **Multiple r(z/xy), p** as shown above), and the **Value Format** to use.

Graph Drawing Tools, OLE Insert Object

In addition to the options for inserting legends (above), the graphs **Insert** menu provides access to a variety of drawing tools for graphs as well as the OLE **Insert Object** dialog. All of these drawing tools are also available from the **Graph Tools** toolbar. For more information on a given menu option, see the relative discussion in the *Graph Tools Toolbar* section: *Selection Tool Button*, page 553; *Text Button*, page 554; *Rectangle Button*, page 555; *Rounded Rectangle Button*, page 556; *Ellipse Button*, page 557; *Arc Button*, page 557; *Polygon Button*, page 558; *Arrow Button*, page 560; and *Insert OLE Object Button*, page 562.

Format Menu

The following commands are available from the graph **Format** menu. Many of these options are also available from shortcut menus (accessed by right-clicking on an item in the graph) and from the toolbar buttons.

Selection

Select **Selection** from the **Format** menu to display the most commonly used formatting options for the currently selected graph item (e.g., when a title is selected, this displays the **Graph Titles/Text** dialog). Selecting this command will display the same customization dialog that is displayed when the item is double-clicked.

All Options

Select **All Options** from the **Format** menu to display the **All Options** dialog. This dialog provides access to all the customization options available for the currently active graph. For more discussion of this dialog, see graph customization dialogs in Chapter 7 – *STATISTICA Graphs: Creation and Customization*.

Graph Data Editor

Select *Graph Data Editor* from the *Format* menu to display the *Graph Data Editor*. This spreadsheet contains the selected variables for the graph (i.e., it contains a subset of the spreadsheet from which the variables were selected). Using the *Graph Data Editor*, you can perform many editing operations on the data as well as brushing of the data. See *Graph Data Editor* (page 570) for more details.

Line Color, Fill Color

Select *Line Color* or *Fill Color* from the *Format* menu to specify the color for line attributes and fill attributes, respectively. Note that these options are also available as buttons on the *Graph Tools* toolbar. See the relative discussions in the *Graph Tools Toolbar* section for more details: *Line/Pattern Color Button* (page 564) and *Area Pattern/Color Button* (page 565).

Move to Front, Move to Back

Select *Move to Front* or *Move to Back* from the *Format* menu to move the currently selected object to the front of the graph or to the back of the graph, respectively. Note that these options are also available as buttons on the *Graph Tools* toolbar. For more details see the appropriate discussions in the *Graph Tools Toolbar* section: *Move to Front Button* (page 565) and *Move to Back Button* (page 566).

Move Forward, Move Back

Select *Move Forward* or *Move Back* from the *Format* menu to move the selected object one spot forward or one spot backwards, respectively, in a group of overlaid objects (e.g., arrows, polygons, arcs). To move an object (i.e., an arrow) to the first position in a group of overlaid objects, use *Move to Front* (page 565). To move an object to the last position in a group of overlaid objects, use *Move to Back* (page 566).

Statistics Menu

The *Statistics* menu is available whenever any document is open. This menu provides access to all available analysis types within *STATISTICA* including *Basic Statistics/Tables*, *Multiple*

Regression, *ANOVA*, *Nonparametrics*, *Distribution Fitting*, *Advanced Linear/Nonlinear Models*, *Multivariate Exploratory Techniques*, *Industrial Statistics & Six Sigma*, *Power Analysis*, and *Data-Mining*. Definitions for the various types of statistics are available in the *Statistics Toolbar* topic in Chapter 1 – *STATISTICA – A General Overview*. Also included on the *Statistics* menu are *Statistics of Block Data*, *STATISTICA Visual Basic*, and *Probability Calculator*.

Graphs Menu

The following commands are available from the *Graphs* menu. The following sections include brief overviews of the respective graph types; for more information, see Chapter 8 - *STATISTICA Graphs: Conceptual Overviews* or the *Electronic Manual*.

Many of these options are also available from shortcut menus (accessed by right-clicking on selected cells in a spreadsheet) and from the toolbar buttons. The *Graphs* menu is available whenever any *STATISTICA* document is open.

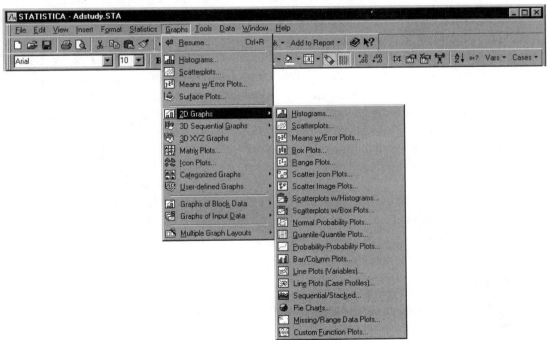

The top portion of the *Graphs* menu includes several most commonly used graph types (which are also included in the submenus of the respective general types of graphs, see page 496). As

mentioned before (Chapter 1), it is very easy in *STATISTICA* to customize the menus and, for example, add to this list additional graphs that you use most often.

The middle portion of the **Graphs** menu includes a set of hierarchical options representing all types of graphs of input data available in *STATISTICA*. (Note that some of these are also included in the top portion for easy access, see below.)

Resume

Select **Resume** from the **Graphs** menu (or press CTRL+R) to restore the current graph specification dialog.

Histograms

Select **Histograms** from the **Graphs** menu (or the **Graphs - 2D Graphs** menu) to display the **2D Histogram** dialog. Two-dimensional histograms (the term was first used by Pearson, 1895) present a graphical representation of the frequency distribution of the selected variable(s) in which the columns are drawn over the class intervals and the heights of the columns are proportional to the class frequencies.

STATISTICA offers several types of 2D histograms: **Regular** for single variables, **Double-Y** for plotting two variables\histograms on the same graph each with a separate frequency scale, **Multiple** for plotting multiple variables\histograms on the same graph with a common frequency scale, and **Hanging Bars** for hanging the frequency columns of a variable from the normal curve. The latter type provides a "visual test of normality" to identify areas of a variable's distribution that show discrepancies between observed and predicted normal frequencies. Any of the distributions specified in **Distributions and Fits** (Chapter 7 - *STATISTICA Graphs: Creation and Customization*) can be displayed on **Regular**, **Double-Y**, or **Multiple** histogram graphs for visual comparison. Histogram intervals or categorization can be defined by any of the options discussed in *Methods of Categorization* in Chapter 7 - *STATISTICA Graphs: Creation and Customization*.

Scatterplots

Select **Scatterplots** from the **Graphs - 2D Graphs** menu (or the **Graphs - 2D Graphs** menu) to display the **2D Scatterplots** dialog. Two-dimensional scatterplots visualize a relation (correlation) between two variables X and Y (e.g., weight and height). Individual data points are represented in two-dimensional space, where axes represent the variables (X on the

horizontal axis and Y on the vertical axis). The two coordinates (X and Y) that determine the location of each point correspond to its specific values on the two variables.

Numerous types of X-Y scatterplots are available: **Regular**, **Double-Y**, **Multiple**, **Frequency**, **Bubble**, **Voronoi**, and **Quantile**. **Regular** scatterplots are simple scatterplots of one set of X-Y variables. **Double-Y** plots involve one *X* variable and two *Y* variables. One *Y* variable is plotted against the left y-axis scale and one is plotted against a separate right y-axis scale. Different point markers are used for the two plots. **Multiple** scatterplots have one *X* variable and multiple *Y* variables. All *Y* variables are fit to a single y-axis scale, but each variable is represented on the graph by a different point marker. **Frequency** plots are regular X-Y scatterplots in which the size of point markers is proportional to the number of superimposed points at that set of X-Y coordinates. **Bubble** plots are regular X-Y scatterplots in which the size of point markers reflects the value of a third (weighting) variable. In **Voronoi** plots, the space between each plotted point marker on a regular scatterplot is divided by straight lines that define the areas surrounding each point such that any set of X-Y coordinates within the defined space is closer to its center point than to any other point. In **Quantile** scatterplots, the quantiles of the *X* variable are plotted against the quantiles of the *Y* variable. If the resulting data points fall on a straight line, the two variables follow the same distribution.

STATISTICA provides a variety of fit types for the graph: **Linear**, **Polynomial**, **Logarithmic**, **Exponential**, **Distance Weighted LS**, **Neg Expon Weighted LS**, **Spline**, and **Lowess**. See *2D Fit Lines* (Chapter 7 - *STATISTICA Graphs: Creation and Customization*) for more details. Note that the selected function is applied to each set of X-Y variables (e.g., a *Double-Y* plot has two fit lines applied, one for each y-variable. Options to mark selected subsets of the data are available for **Regular**, **Quantile**, and **Voronoi** type plots. See *Mark Selected Subsets* in the *Electronic Manual* for a discussion of subset definition.

Means w/Error Plots

Select **Means w/Error Plots** from the **Graphs** menu (or the **Graphs - 2D Graphs** menu) to display the **Means with Error Plots** dialog. This plot visualizes means of the dependent variable(s) for subsets of cases specified by the user. Each mean marker is accompanied by a set of whiskers representing error bars that is the confidence intervals around the mean. These plots are typically used to compare marginal means across groups and visualize the reliability of these means.

Surface Plots

Select **Surface Plots** from the **Graphs** menu (or the **Graphs - 3D XYZ Graphs** menu) to display the **3D Surface Plots** dialog. Use this dialog to create a surface plot in which a surface is fitted

(defined by a smoothing technique) to the data (variables corresponding to sets of XYZ coordinates). The resulting plot is a three-dimensional representation of three variables with the x-, y-, and z-axes placed along the edges of a cube. The specified surface is fit to the values of the variable represented by the z-axis and is shaded in colors corresponding to the z-axis values. In order to change the range of values and number of levels used to define the surface, see *How Do I Define a Custom Selection of Levels for a Contour Plot or Surface Plot?* in Chapter 9 – *STATISTICA Graphs: Examples and FAQ's*. Equations that can be fit to surface graphs include **Linear**, **Quadratic**, **Distance Weighted LS**, **Neg Expon Weighted LS**, and **Spline**. These fit types are discussed in *3D Graph Surfaces* in Chapter 7 - *STATISTICA Graphs: Creation and Customization*.

2D Graphs

Histograms. Select *Histograms* from the *Graphs - 2D Graphs* menu (or the *Graphs* menu) to display the *2D Histogram* dialog. Two-dimensional histograms (the term was first used by Pearson, 1895) present a graphical representation of the frequency distribution of the selected variable(s) in which the columns are drawn over the class intervals and the heights of the columns are proportional to the class frequencies.

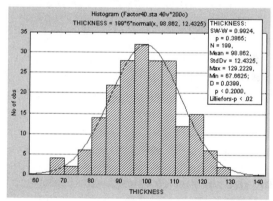

See the *Histograms* discussion (page 494) for more details on this type of graph.

Scatterplots. Select *Scatterplots* from the *Graphs - 2D Graphs* menu (or the *Graphs* menu) to display the *2D Scatterplots* dialog. Two-dimensional scatterplots visualize a relation (correlation) between two variables X and Y (e.g., weight and height). Individual data points are represented in two-dimensional space, where axes represent the variables (X on the horizontal axis and Y on the vertical axis). The two coordinates (X and Y) that determine the location of each point correspond to its specific values on the two variables.

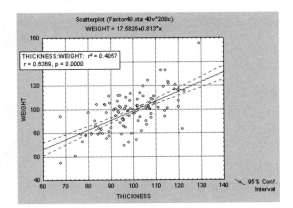

See the *Scatterplots* discussion (page 494) for more details on this type of graph.

Means with Error Plots. Select *Means w/Error Plots* from the *Graphs - 2D Graphs* menu (or the *Graphs* menu) to display the *Means with Error Plots* dialog. This plot visualizes means of the dependent variable(s) for subsets of cases specified by the user. Each mean marker is accompanied by a set of whiskers representing error bars that is the confidence intervals around the mean.

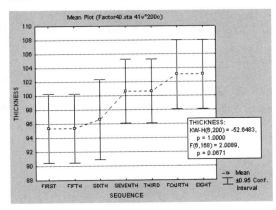

These plots are typically used to compare marginal means across groups and visualize the reliability of these means.

Box Plots. Select *Box Plots* from the *Graphs - 2D Graphs* menu to display the *2D Box Plots* dialog. In box plots (this term was first used by Tukey, 1970), ranges or distribution characteristics of values of a selected variable (or variables) are plotted separately for groups of cases defined by values of a categorical (grouping) variable. The central tendency (e.g., median or mean), and range or variation statistics (e.g., quartiles, standard errors, or standard deviations) are computed for each group of cases and the selected values are presented in the

selected box plot style. Outlier data points can also be plotted (see *Outliers and Extremes* in the *Electronic Manual*).

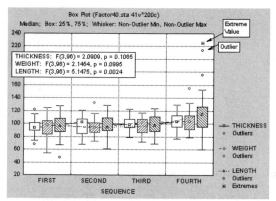

Box plots can be specified to display the desired variability data using **Whiskers**, **Boxes**, **Box-Whiskers**, **Columns**, or **High-Low Close** (a type of whisker display) formats. Multiple box plots of a single box plot variable can be produced by the categories of a grouping variable (see *Methods of Categorization* in Chapter 7 - *STATISTICA Graphs: Creation and Customization*), and function fits can be applied to the resulting series of box plots. Function fits available include: **Linear**, **Polynomial**, **Logarithmic**, **Exponential**, **Distance Weighted LS**, **Neg Expon Weighted LS**, **Spline**, and **Lowess** (see *2D Fit Lines* in Chapter 7 – *STATISTICA Graphs: Creation and Customization*). Single or multiple box plot variables can be selected.

Range Plots. Select *Range Plots* from the *Graphs - 2D Graphs* menu to display the *2D Range Plots* dialog. Range plots display ranges of values or error bars related to specific data points in the form of boxes or whiskers.

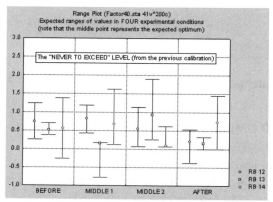

Unlike the standard box plots or means with error plots, the ranges or error bars are not calculated from data but defined by the raw values in the selected variables. One range or error bar is plotted for each case.

In the simplest instance, three variables need to be selected, one representing the mid-points, one representing the upper limits and one representing the lower limits, as illustrated in the example below:

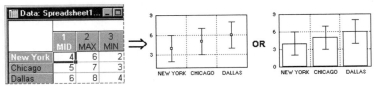

The range variables can be interpreted either as absolute values or values representing deviations from the midpoint. Simple or multiple variables can be represented in the graph. To produce simple graphs of sequences of values (without range or error bars) for either style or multiple variables use bar/column plots.

Scatter Icon Plots. Select *Scatter Icon Plots* from the **Graphs - 2D Graphs** menu to display the **Scatter Icon Plots** dialog. A scatter icon plot is a scatterplot whose point markers are icons that represent for each specific data point (defined by the scatterplot X and Y coordinates) the variability within a set of other variables chosen for the icon. Additionally, a weight variable can be selected and used to scale the entire icon representing the point.

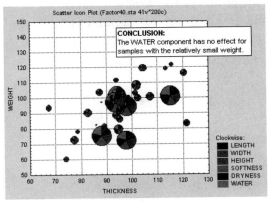

For a detailed overview of scatter icon plots, see *Scatter Icon Plots Overview* in the *Electronic Manual*. To create simple icon plots, select **Icon Plots** from the **Graphs** menu.

Scatter Image Plots. Select *Scatter Image Plots* from the **Graphs - 2D Graphs** menu to display the **Image Plots** dialog. Like scatter icon plots, a scatter image plot is a scatterplot

whose point markers are images. However, these images are not icons defined by specific variables, but user-selected graphics files (bitmap, metafile, PNG, and JPG formats are supported). Different image files can be attached to different data points (defined by the scatterplot *X* and *Y* coordinates), and a weight variable can be used to scale specific image markers accordingly.

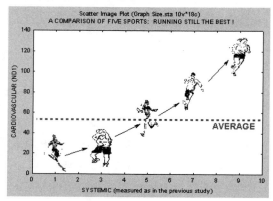

For a detailed overview of scatter image plots, see *Scatter Image Plots Overview* in the *Electronic Manual*. To create icon plots, select **Icon Plots** from the **Graphs** menu.

Scatterplots w/Histograms. Select **Scatterplots w/Histograms** from the **Graphs - 2D Graphs** menu to display the **2D Scatterplots with Histograms** dialog. This type of graph combines features of the regular X-Y scatterplot with histograms of the X and Y variables. Histograms are plotted parallel to the x- and y-axes of the graph, the X-histogram at the top and the Y-histogram at the right. Each component graph in this compound graph can be edited by clicking on it.

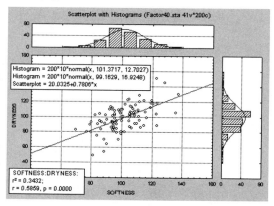

The function fits available for the scatterplot (i.e., **Linear**, **Polynomial**, **Logarithmic**, **Exponential**, **Distance Weighted LS**, **Neg Expon Weighted LS**, **Spline**, and **Lowess**) are discussed in *2D Fit Lines* and distributions available for plotting on the histograms are discussed in *Fitted Functions for Histograms* in Chapter 7 - *STATISTICA Graphs: Creation and Customization*. Histogram interval definition can be specified separately for both the x- and y-axes by any of the options discussed in *Methods of Categorization* in Chapter 7 - *STATISTICA Graphs: Creation and Customization*.

▦ Scatterplots w/Box Plots. Select **Scatterplots w/Box Plots** from the **Graphs - 2D Graphs** menu to display the **2D Scatterplots with Box Plots** dialog. This type graph combines features of the regular X-Y scatterplot with box plots of the *X* and *Y* variables. Box plots are plotted parallel to the x- and y-axes of the graph, the *X*-box plot at the top and the *Y*-box plot at the right. Each component graph in this compound graph can be edited by clicking on it.

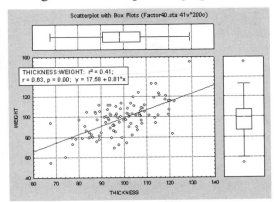

The function fits available for application to the scatterplot (i.e., **Linear**, **Polynomial**, **Logarithmic**, **Exponential**, **Distance Weighted LS**, **Neg Expon Weighted LS**, **Spline**, and **Lowess**) are discussed in *2D Fit Lines* (Chapter 7 - *STATISTICA Graphs: Creation and Customization*). Complete options for specifying box plot parameters are also available for this graph type.

▦ Normal Probability Plots. Select **Normal Probability Plots** from the **Graphs - 2D Graphs** menu to display the **Normal Probability Plots** dialog. This type of graph is used to evaluate the normality of the distribution of a variable, that is, whether and to what extent the distribution of the variable follows the normal distribution. The selected variable will be plotted in a scatterplot against the values "expected from the normal distribution."

The standard normal probability plot is constructed as follows. First, the deviations from the mean (residuals) are rank ordered. From these ranks *STATISTICA* computes z values (i.e., standardized values of the normal distribution) based on the assumption that the data come from a normal distribution (see *Computation Note* in the *Electronic Manual*). These z values

are plotted on the y-axis in the plot. If the observed residuals (plotted on the x-axis) are normally distributed, then all values should fall onto a straight line. If the residuals are not normally distributed, then they will deviate from the line. Outliers can also become evident in this plot. If there is a general lack of fit, and the data seem to form a clear pattern (e.g., an S shape) around the line, then the variable may have to be transformed in some way (e.g., a log transformation to "pull-in" the tail of the distribution, etc.).

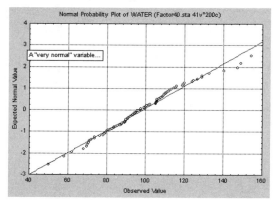

In addition to the **Normal** probability plot described above, **Half-Normal** and **Detrended** normal probability plots are also available. **Half-Normal** probability plots employ only the positive half of the normal curve in the analysis, and **Detrended** normal probability plots remove the linear trend of a normal probability plot to make lack-of-fit more apparent.

Quantile-Quantile Plots. Select **Quantile-Quantile Plots** from the **Graphs - 2D Graphs** menu to display the **Quantile-Quantile Plots** dialog. You can visually check for the fit of a theoretical distribution to the observed data by examining the quantile-quantile (or Q-Q) plot. In this plot, the observed values of a variable are plotted against the theoretical quantiles. A good fit of the theoretical distribution to the observed values would be indicated by this plot if the plotted values fall onto a straight line. To produce a Q-Q plot, *STATISTICA* first sorts the n observed data points into ascending order, so that:

$$x_1 \leq x_2 \leq \ldots \leq x_n$$

These observed values are plotted against one axis of the graph; on the other axis the plot will show:

$$F^{-1}((i-r_{adj}) \ / \ (n+n_{adj}))$$

where i is the rank of the respective observation, r_{adj} and n_{adj} are adjustment factors (≤ 0.5) and F^{-1} denotes the inverse of the probability integral for the respective standardized

distribution. The resulting plot is a scatterplot of the observed values against the (standardized) expected values, given the respective distribution.

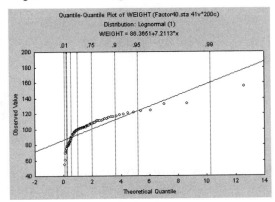

Note that, in addition to the inverse probability integral value, *STATISTICA* also shows the respective cumulative probability values on the opposite axis, that is, the plot shows not only the standardized values for the theoretical distribution, but also the respective *p*-values. Note also that the adjustment factors r_{adj} and n_{adj} ensure that the *p*-value for the inverse probability integral will fall between 0 and 1, but not including 0 and 1 (see Chambers, Cleveland, Kleiner, and Tukey, 1983; in *STATISTICA*, the default value for both adjustment factors is 1/3=.333).

You can choose a variety of distributions to use in creating the Q-Q plot. For a list of the theoretical distributions available, see *Quantile-Quantile Plots - Advanced Tab* in the *Electronic Manual*.

▢ **Probability-Probability Plots.** Select *Probability-Probability Plots* from the *Graphs - 2D Graphs* menu to display the *Probability-Probability Plots* dialog. You can visually check for the fit of a theoretical distribution to the observed data by examining the probability-probability plot. In probability-probability plots (or P-P plots) the observed cumulative distribution function is plotted against the theoretical cumulative distribution function. As in the quantile-quantile plot, the values of the respective variable are first sorted into ascending order. The *i*th observation is plotted against one axis as *i/n* (i.e., the observed cumulative distribution function), and against the other axis as $F(x_{(i)})$, where $F(x_{(i)})$ stands for the value of the theoretical cumulative distribution function for the respective observation $x_{(i)}$. If the theoretical cumulative distribution approximates the observed distribution well, then all points in this plot should fall onto the diagonal line. You can choose a variety of distributions to use in creating the P-P plot.

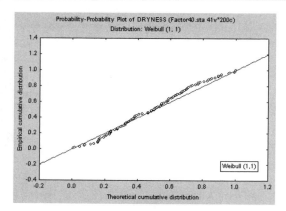

For a list of the theoretical distributions available, see *Probability-Probability Plots - Advanced Tab* in the *Electronic Manual*.

Bar/Column Plots. Select ***Bar/Column Plots*** from the ***Graphs - 2D Graphs*** menu to display the ***2D Bar/Column Plots*** dialog. Bar/column plots represent simple sequences of values (one case is represented by one bar/column).

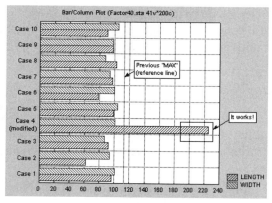

If more than one variable is selected, then either multivariate clusters of bars/columns are plotted (one cluster per case) or each variable is represented in a separate graph, depending on whether a multiple or regular graph is chosen. To produce column, bar, or box plots with error (or range) bars, use either ***Range Plots*** or ***Box Plots***.

Line Plots (Variables). Select ***Line Plots (Variables)*** from the ***Graphs - 2D Graphs*** menu to display the ***2D Line Plots*** dialog. In line plots, individual data points are plotted as one line in the sequence they are encountered in the input datafile, providing a simple way to visually present a sequence of values. Line plots can also be used to plot continuous functions, theoretical distributions, etc.

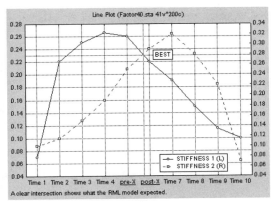

STATISTICA offers five types of line plots: ***Regular***, ***Multiple***, ***Double-Y***, ***XY Trace***, and ***Aggregated***. ***Regular*** line plots produce plots of single variables. ***Double-Y*** line plots produce two line plots with the first variable plotted against a scale displayed on the left y-axis and the second variable plotted against a scale displayed on the right y-axis. ***Multiple*** line plots produce plots of multiple variables against a common y-axis scale for all. When ***XY Trace*** plots are made, a scatterplot of two variables is first created; then the individual data points are connected with a line (in the order in which they are read from the datafile). In this sense, trace plots visualize a "trace" of a sequential process (movement, change of a phenomenon over time, etc.). ***Aggregated*** line plots display a sequence of means (i.e., a moving average) for consecutive subsets of a selected variable. You can select the number of consecutive observations from which the mean will be calculated and if desired, the range of values in each subset will be displayed as well by whisker-type range bars.

Fitted functions can be applied to line plots. The function fits available for application to line plots (i.e., ***Linear***, ***Polynomial***, ***Logarithmic***, ***Exponential***, ***Distance Weighted LS***, ***Neg Expon Weighted LS***, ***Spline***, and ***Lowess***) are discussed in *2D Fit Lines* (Chapter 7 - *STATISTICA Graphs: Creation and Customization*).

Line Plots (Cases). Select *Line Plots (Cases)* from the ***Graphs - 2D Graphs*** menu to display the ***2D Line Plots – Case Profiles*** dialog. As described above, in regular line plots the values of one variable are plotted as one line (individual data points are connected by a line). In case profile line plots, the values of the selected variables in a case (row) are plotted as one line (i.e., one line plot will be generated for each of the selected cases). Case profile line plots provide a simple way to visually present the values in a case (e.g., test scores for several tests).

Note that you can use ***Statistics of Block Data/Block Columns*** from the spreadsheet shortcut menu to perform some statistics on the cases [e.g., sum the values or take the mean of the values for each test (variable)]. You can then plot those statistics along with the cases in the block for comparison purposes.

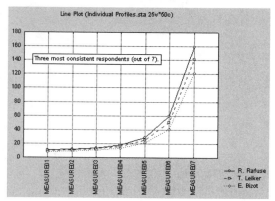

STATISTICA offers two types of case profile line plots: **Regular** and **Multiple**. **Regular** case profile line plots produce a sequence of graphs (one for each case). **Multiple** case profile lineplots produce a line plot for each case in one graph with a common y-axis scale.

Fitted functions can be applied to line plots. The function fits available for application to line plots (i.e., **Linear**, **Polynomial**, **Logarithmic**, **Exponential**, **Distance Weighted LS**, **Neg Expon Weighted LS**, **Spline**, and **Lowess**) are discussed in *2D Fit Lines* (Chapter 7 - *STATISTICA Graphs: Creation and Customization*).

Sequential/Stacked Plots. Select *Sequential/Stacked* from the *Graphs - 2D Graphs* menu to display the *Sequential/Stacked Plots* dialog. Sequential/stacked plots are used to present sequences of values of selected variables. The stacked feature of this type of graph is specifically designed to represent the broad category of datasets in which consecutive variables represent parts ("portions") of one whole. Display options include a variety of combinations of lines, stepped lines, areas, and columnar displays.

Pie Charts. Select *Pie Charts* from the *Graphs - 2D Graphs* menu to display the *Pie Charts* dialog. Pie charts (the term first used by Haskell, 1922) are useful for representing proportions. Individual data values of the X variable are represented as the "wedges" of the pie. In *STATISTICA*, circular or elliptical, 2D or 3D pie charts can be made to display either frequency data or data values of a single variable. Frequency data can be calculated in intervals defined in several different ways as discussed in *Methods of Categorization* (Chapter 7 - *STATISTICA Graphs: Creation and Customization*).

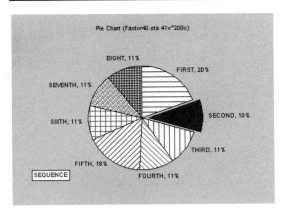

Labels on pie charts can be specified as values, percents, text labels, text labels and values, or text labels and percents.

Missing and 'Out of Range' Data Plots. Select *Missing/Range Data Plots* from the *Graphs - 2D Graphs* menu to display the *Missing and 'Out of Range' Data Plots* dialog. Use this graph to examine the pattern or distribution of missing and/or user-specified "out of range" data points in the current dataset (or in a subset of variables and cases).

This graph is useful in exploratory data analysis to determine the extent of missing (and/or "out of range") data and whether the patterns of those data occur more or less randomly.

Custom Function Plots. Select *Custom Function Plots* from the *Graphs - 2D Graphs* menu to display the *2D Custom Function Plots* dialog. This dialog provides access to two types of custom plots: functions and parametric curves. Enter a custom function to produce a graph based not on values of variables in a datafile but on a user-defined formula (custom function), for example: *y = x2 + x/2*. To plot a parametric curve, enter a function for y and a function for x. For example, enter *y(t)=sin(2*pi*t)* and *x(t)=cos(2*pi*t)* to plot a circle.

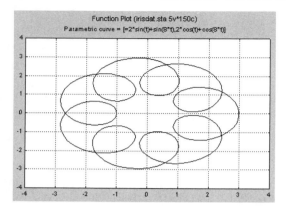

For more on parametric curves, see Chapter 9 – *STATISTICA Graphs: Examples and FAQ's.*

3D Sequential Graphs

Raw Data Plots. Select *Raw Data Plots* from the *Graphs - 3D Sequential Graphs* menu to display the *3D Raw Data Plots* dialog. Raw data plots allow you to plot sequences of raw data from selected variables in a three-dimensional display. The selected variables are represented on the y-axis, the consecutive cases on the x-axis, and the values are plotted against the z-axis.

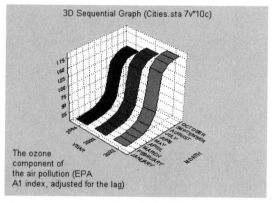

The contour and surface types integrate all individual plots into one structure. Display options include lines, columns, ribbons, blocks, and spikes as well as contour and surface plot options.

Bivariate Histograms. Select *Bivariate Histograms* from the *Graphs - 3D Sequential Graphs* menu to display the *3D Bivariate Histograms* dialog. Three-dimensional histograms are used to visualize crosstabulations of values in two variables. They are considered a conjunction of two simple (i.e., univariate) histograms, combined such that the frequencies of co-

occurrences of values on the two analyzed variables can be examined. In the most common format of this graph, a 3D bar is drawn for each "cell" of the crosstabulation table and the height of the bar represents the frequency of values for the respective cell of the table.

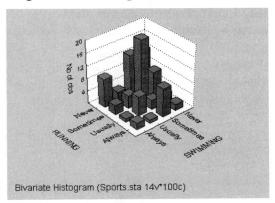

Bivariate Histogram (Sports.sta 14v*100c)

Histogram intervals or categorization can be defined for each variable by any of the options discussed in *Methods of Categorization* (Chapter 7 - *STATISTICA Graphs: Creation and Customization*).

Range Plots. Select *Range Plots* from the *Graphs - 3D Sequential Graphs* menu to display the *3D Range Plots* dialog. 3D range plots display ranges of values or error bars related to specific data points.

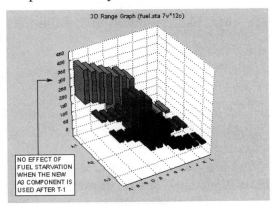

The ranges or error bars in this plot are not calculated from data (as they are in box plots, see below) but defined by the raw values in the selected variables. One range or error bar is plotted for each case. The range variables can be interpreted as either absolute values or values representing deviations from the midpoint. Simple or multiple variables can be represented in the graph. Display options include simple points or error bar styles, ribbons, boxes, or blocks.

Box Plots. Select *Box Plots* from the *Graphs - 3D Sequential Graphs* menu to display the *3D Box Plots* dialog. In 3D box plots, the ranges of values of a selected variable are plotted separately for groups of cases defined by values of a categorical (grouping) variable. The central tendency (e.g., median or mean), and range or variation statistics (e.g., quartiles, standard errors, or standard deviations) are computed for each group of cases, and the selected values are presented in the selected style.

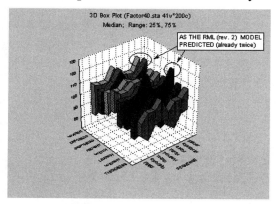

3D Range Plots differ from *3D Box Plots* in the way in which ranges are calculated. For range plots, the dataset provides the ranges (e.g., one variable contains the minimum range values and another variable contains the maximum range values), while the ranges for box plots are calculated using the observed (raw) data (e.g., standard deviations, standard errors, or min-max value are calculated from the variable). As with *3D Range Plots*, display options include simple points or error bar styles, ribbons, boxes, or blocks.

3D XYZ Graphs

Scatterplots. Select *Scatterplots* from the *Graphs - 3D XYZ Graphs* menu to display the *3D Scatterplots* dialog. 3D scatterplots visualize a relationship between three or more variables, representing the X, Y, and one or more Z (vertical) coordinates of each point in three-dimensional space. You can select one of several types of graphs: *Scatterplot*, *Space Plot*, *Spectral Plot*, *Deviation Plot*, or *Trace Plot*. Use *Scatterplots* to visualize the relationship between three (or more) variables using *X*, *Y*, and *Z* coordinates in a three-dimensional representation with the x-, y-, and z-axes at the edges of a cube.

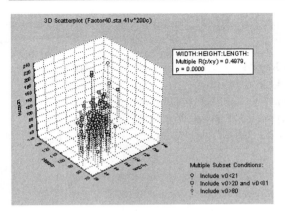

Space Plots move the z-axis to the center of the plot and allow the x-y plane to be placed at a specified point on the z-axis. Use ***Spectral Plots*** to explore relationships among three variables by compressing the data into a specified number of planes. Values of variables X and Z are interpreted as the x- and z-axis coordinates of each point respectively; values of variable Y are clustered into equally spaced values corresponding to the locations of the consecutive spectral planes. In ***Deviation Plots***, data points are represented in 3D space as "deviations" from a specified base level of the z-axis. In ***Trace Plots***, individual data points are connected with a line (in the order in which they were read from the datafile), visualizing a "trace" of sequential values.

 **Scatter Image Plots.** Select ***Scatter Image Plots*** from the ***Graphs - 3D XYZ Graphs*** menu to display the ***3D Scatter Image Plots*** dialog. An XYZ scatter image plot is a scatterplot whose point markers are images represented by user-selected graphics files (bitmap, metafile, JPG, and PNG formats are supported). Different image files can be attached to different data points (defined by the scatterplot X, Y, and Z coordinates), and a weight variable can be used to scale specific image markers accordingly.

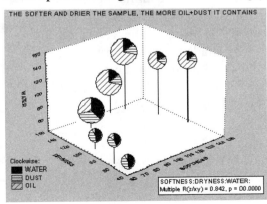

For a detailed overview of scatter image plots, see *Scatter Image Plots Overview* in the *Electronic Manual*.

Surface Plots. Select ***Surface Plots*** from the ***Graphs - 3D XYZ Graphs*** menu (or the ***Graphs*** menu) to display the ***3D Surface Plots*** dialog. Use this dialog to create a surface plot in which a surface is fitted (defined by a smoothing technique) to the data (variables corresponding to sets of XYZ coordinates). The specified surface is fit to the values of the variable represented by the z-axis and is shaded in colors corresponding to the z-axis values.

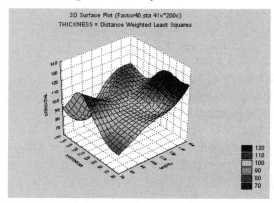

In order to change the range of values and number of levels used to define the surface, see *How Do I Define a Custom Selection of Levels for a Contour Plot or Surface Plot?* in Chapter 9 - *STATISTICA Graphs: Examples and FAQ's*.

Equations and smoothing techniques that can be fit to surface graphs include ***Linear***, ***Quadratic***, ***Distance Weighted LS***, ***Neg Expon Weighted LS***, and ***Spline***. These fit types are discussed in *3D Graph Surfaces* (Chapter 9 - *STATISTICA Graphs: Examples and FAQ's*).

Contour Plots. Select ***Contour Plots*** from the ***Graphs - 3D XYZ Graphs*** menu to display the ***3D Contour Plots*** dialog. A contour plot is the projection of a three-dimensional surface onto a two-dimensional plane. The values of the fitted surface in terms of the Z variable are depicted either by variously colored and patterned lines or by shades of color (areas) on an X-Y scatterplot. For a discussion on changing the range of values and number of levels used to define the contour or surface, see *How Do I Define a Custom Selection of Levels for a Contour Plot or Surface Plot?* in Chapter 9 - *STATISTICA Graphs: Examples and FAQ's*.

Equations and smoothing techniques that can be fit to contour plots include ***Linear***, ***Quadratic***, ***Distance Weighted LS***, ***Neg Expon Weighted LS***, and ***Spline***.

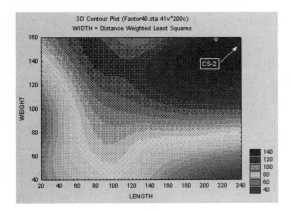

These fit types are discussed in *3D Graph Surfaces* (Chapter 7 - *STATISTICA Graphs: Creation and Customization*).

Ternary Plots. Select ***Ternary Plots*** from the ***Graphs - 3D XYZ Graphs*** menu to display the ***Ternary Graphs*** dialog. A ternary plot can be used to examine relations between four dimensions where three of those dimensions represent components of a mixture (i.e., the relations between them is constrained such that the values of the three variables add up to the same constant). One typical application of this graph is when the measured response(s) from an experiment depends on the relative proportions of three components (e.g., three different chemicals) that are varied in order to determine an optimal combination of those components (e.g., in mixture designs).

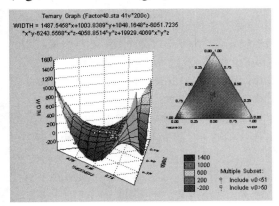

For plotting three (*X*, *Y*, and *Z*) or four (*X*, *Y*, *Z*, and *V*) variables, ternary plot display options include triangular shaped ***2D Scatterplots*** or ***Contour*** plots with contours defined by either lines or areas. Triangular (prism) shaped plots are available for plotting four variables (***Surface Plots***) or for more than four variables (i.e., *X*, *Y*, *Z*, and *V1...Vn*) using ***3D Scatterplots*** or ***Trace***

Plots. *Linear* and *Quadratic* as well as *Full Cubic* and *Special Cubic* fits can be applied to all ternary plots except the *2D Scatterplot* and *Trace* ternary plot types.

▩ Categorized XYZ Plots. Select *Categorized XYZ Plots* from the *Graphs - 3D XYZ Graphs* menu to display the *3D Categorized Plots* dialog. *STATISTICA* supports a variety of 3D graph types (i.e., *Scatterplots*, *Surface Plots*, *Contour Plots*, *Trace Plots*, *Deviation Plots*, *Space Plots*, *Spectral Plots*). Using categorized plots you can categorize 3D plots by the specified categories of a selected variable. One component graph is produced for each level of the grouping variable (or user-defined subset of data) and all the component graphs are arranged in one display to allow for comparisons between the subsets of data (categories).

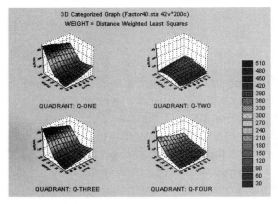

For a detailed discussion of categorized graphs, see *What Are Categorized Graphs?* (Chapter 8 - *STATISTICA Graphs: Conceptual Overviews*).

▨ Categorized Ternary Plots. Select *Categorized Ternary Plots* from the *Graphs - 3D XYZ Graphs* menu to display the *Categorized Ternary Graphs* dialog. In categorized ternary graphs, the points representing the proportions of the component variables (*X*, *Y*, and *Z*) are plotted in a two-dimensional display for each level of the grouping variable (or user-defined subset of data). One component graph is produced for each level of the grouping variable (or user-defined subset of data), and all the component graphs are arranged in one display to allow for comparisons between the subsets of data (categories).

Use categorized ternary plots to examine relationships between three or more dimensions where three of those dimensions represent components of a mixture for each level of a grouping variable. One typical application of this graph is when the measured response(s) from an experiment depends on the relative proportions of three components (e.g., three different chemicals) that are varied in order to determine an optimal combination of those components (e.g., in mixture designs).

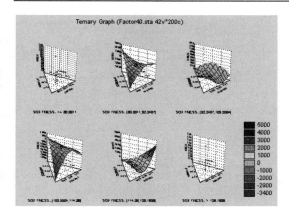

In a categorized ternary plot, one component graph is produced for each level of the grouping variable (or user-defined subset of data) and all the component graphs are arranged in one display to allow for comparisons between the subsets of data (categories).

Custom Function Plots. Select *Custom Function Plots* from the *Graphs - 3D XYZ Graphs* menu to display the *3D Custom Function Plots* dialog.

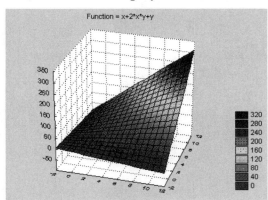

This graph produces a 3-dimensional plot based not on values of variables in a datafile but on a user-defined formula (custom function), for example: $z = x + 2 * x * y + y$.

Matrix Plots

Select *Matrix Plots* from the *Graphs* menu to display the *Matrix Plots* dialog. Matrix plots summarize the relationships between several variables in a matrix of X-Y plots. The most common type of matrix plot is the scatterplot matrix, which can be considered the graphical equivalent of the correlation matrix.

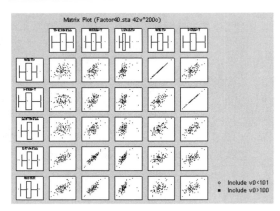

If one list of variables is selected when defining the matrix plot, then a square matrix plot is produced and the histograms of frequency distributions of the respective variables are plotted on the diagonal. If you select variables in two lists, then a rectangular matrix plot is produced. In this style of matrix plot, the histograms for the respective variables are displayed in the first row and column of the matrix.

Icon Plots

Select **Icon Plots** from the **Graphs** menu to display the **Icon Plots** dialog. Icon graphs represent cases or units of observation as multidimensional symbols, and they offer a powerful exploratory technique.

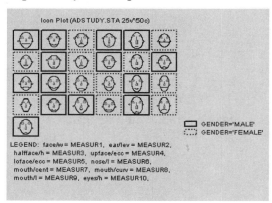

The general idea behind this method capitalizes on the human ability to "automatically" spot complex (sometimes interactive) relations between multiple variables if those relations are consistent across a set of instances (in this case "icons"). Sometimes the observation (or a "feeling") that certain instances are "somehow similar" to each other comes before the

observer (in this case an analyst) can articulate which specific variables are responsible for the observed consistency (Lewicki, Hill, & Czyzewska, 1992). However, further analysis that focuses on such intuitively spotted consistencies can reveal the specific nature of the relevant relations between variables.

The basic idea of icon plots is to represent individual units of observation as particular graphical objects where values of variables are assigned to specific features or dimensions of the objects (usually one case = one object). The assignment is such that the overall appearance of the object changes as a function of the configuration of values.

Thus, the objects are given visual "identities" that are unique for configurations of values and that can be identified by the observer. Examining such icons can help to discover specific clusters of both simple relations and interactions between variables

STATISTICA provides a variety of graph types for icon plots including **Chernoff faces**, **Pies**, **Stars**, **Sunrays**, **Polygons**, **Columns**, **Lines**, and **Profiles**. For more information on these types of graphs, see *Taxonomy of Icon Plots* in the *Electronic Manual* or in Chapter 8 - *STATISTICA Graphs: Conceptual Overviews*.

Categorized Graphs

In addition to the many graph types listed above, *STATISTICA* offers a variety of categorized graphs in which values of the selected variable(s) are displayed broken down by values of another (categorized) variable. These graphs offer a selection of categorization methods (including user-defined intervals, multiple subsets defined by logical expressions, and others). See *Methods of Categorization* (Chapter 7 - *STATISTICA Graphs: Creation and Customization*) for more details. For a detailed discussion of categorized graphs, see *What Are Categorized Graphs?* (Chapter 8 - *STATISTICA Graphs: Conceptual Overviews*).

Histograms. Select **Histograms** from the **Graphs - Categorized Graphs** menu to display the **2D Categorized Histograms** dialog. Categorized histograms offer an extension of the regular histograms, allowing you to examine the distribution of values of the selected variable(s) broken down (categorized) by some other grouping or independent variable(s). One or two grouping variables can be represented on one graph (i.e., a graph can represent a one-way or two-way categorization of cases). If two grouping variables are selected, a "crosstabulation of graphs" is produced.

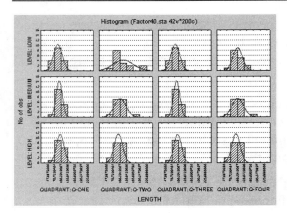

Note that you can choose to have separate plots for each category or to overlay the categorized plots in one graph using different colors and/or patterns for the individual plots.

Scatterplots. Select *Scatterplots* from the *Graphs - Categorized Graphs* menu to display the *2D Categorized Scatterplots* dialog. Categorized scatterplots plot the values of a selected variable against the values of another selected variable broken down (categorized) by some other grouping or independent variable(s).

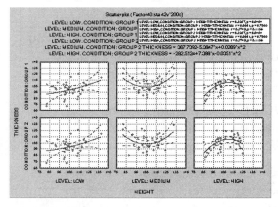

One or two grouping variables can be represented on one graph (i.e., a graph can represent a one-way or two-way categorization of cases). If two grouping variables are selected, a "crosstabulation of graphs" is produced. Note that you can choose to have separate plots for each category or to overlay the categorized plots in one graph using different colors and/or patterns for the individual plots.

STATISTICA provides a variety of fit types for the scatterplots: *Linear*, *Polynomial*, *Logarithmic*, *Exponential*, *Distance Weighted LS*, *Neg Expon Weighted LS*, *Spline*, and *Lowess*. See *2D Fit Lines* (Chapter 7 - *STATISTICA Graphs: Creation and Customization*) for more

details. Note that the selected function is applied to each set of X-Y variables (e.g., a **Double-Y** plot has two fit lines applied, one for each Y variable).

Means w/Error Plots. Select *Means w/Error Plots* from the *Graphs - Categorized Graphs* menu to display the *Categorized Means with Error Plots* dialog.

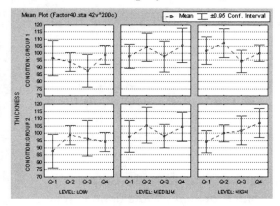

When you create categorized means with error plots, a series of standard means with error plots, one for each category of cases identified by the *X* or *X* and *Y* category variables (or identified by the multiple subset criteria) is produced.

Box Plots. Select *Box Plots* from the *Graphs - Categorized Graphs* menu to display the *Categorized Box Plots* dialog. When you create categorized box plots, a series of standard 2D box plots, one for each category of cases identified by the *X* or *X* and *Y* category variables (or identified by the multiple subset criteria) is produced.

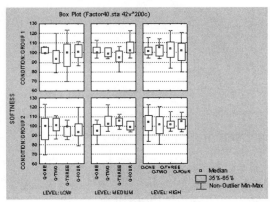

The categorized box and whisker plot typically summarizes the distribution of a variable broken down (categorized) by some other grouping or independent variable(s), by three components:

1. A central line to indicate central tendency or location;

2. A box to indicate variability around this central tendency;

3. Whiskers around the box to indicate the range of the variable (or another measure of variability).

For more information on the types of box and whisker plots, see *Box-Whisker Type* in the *Electronic Manual*.

One, two, or three grouping variables can be represented on one graph (i.e., a graph can represent a one-way, two-way, or three-way categorization of cases). If two or three grouping variables are selected, a "crosstabulation of graphs" is produced. These graphs allow the user to plot multiple dependent variables in one graph, support different graph layout formats, and offer a selection of categorization methods (including user-defined intervals, multiple subsets defined by logical expressions, and others).

Normal Probability Plots. Select *Normal Probability Plots* from the *Graphs - Categorized Graphs* menu to display the *Categorized Normal Probability Plots* dialog. When you create categorized probability plots, a series of standard probability plots, one for each category of cases identified by the X or X and Y category variables (or identified by the multiple subset criteria) is produced.

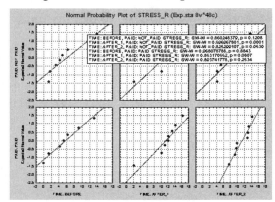

As with *2D Normal Probability Plots* (page 501), you can create normal, half-normal, and detrended normal probability plots.

Quantile-Quantile Plots. Select *Quantile-Quantile Plots* from the *Graphs - Categorized Graphs* menu to display the *Categorized Quantile-Quantile Plots* dialog. Use categorized quantile-quantile plots to visually check for the fit of a theoretical distribution to the observed data by examining each quantile-quantile (or Q-Q) plot for the respective level of the grouping variable (or user-defined subset of data).

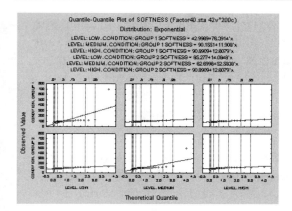

In a Q-Q plot, the observed values of a variable are plotted against the theoretical quantiles. When the plotted values fall onto a straight line, a good fit of the theoretical distribution to the observed values is indicated. One component graph is produced for each level of the grouping variable (or user-defined subset of data), and all the component graphs are arranged in one display to allow for comparisons between the subsets of data (categories). See *2D Graphs* (page 496) for more details on Q-Q plots.

Probability-Probability Plots. Select *Probability-Probability Plots* from the *Graphs - Categorized Graphs* menu to display the *Categorized Probability-Probability Plots* dialog. Use probability-probability (P-P) plots to check for the fit of a theoretical distribution to the observed data by examining each probability-probability plot for the respective level of the grouping variable (or user-defined subset of data).

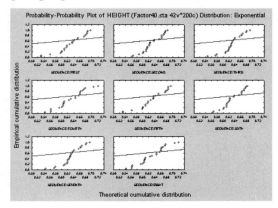

In P-P plots, the observed cumulative distribution function is plotted against the theoretical cumulative distribution function. As in the categorized quantile-quantile plot, the values of the respective variable are first sorted into ascending order. The ith observation is plotted against

one axis as *i/n* (i.e., the observed cumulative distribution function), and against the other axis as $F(x_{(i)})$, where $F(x_{(i)})$ stands for the value of the theoretical cumulative distribution function for the respective observation $x_{(i)}$. If the theoretical cumulative distribution approximates the observed distribution well, then all points in this plot should fall onto the diagonal line. One component graph is produced for each level of the grouping variable (or user-defined subset of data), and all the component graphs are arranged in one display to allow for comparisons between the subsets of data (categories).

⊞ Line Plots. Select *Line Plots* from the *Graphs - Categorized Graphs* menu to display the *Categorized Line Plots* dialog. When you create categorized line plots, a series of standard 2D line plots, one for each category of cases identified by the *X* or *X* and *Y* category variables (or identified by the multiple subset criteria) is produced.

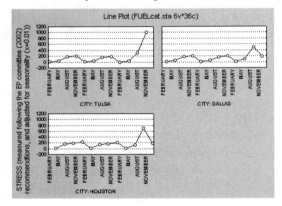

See *Line Plots (Variables)* on page 504 for more details on line plots.

⠿ Pie Charts. Select *Pie Charts* from the *Graphs - Categorized Graphs* menu to display the *Categorized Pie Charts* dialog.

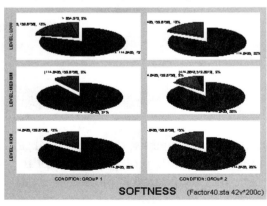

When you create categorized pie charts, a series of standard pie charts, one for each category of cases identified by the *X* or *X* and *Y* category variables (or identified by the multiple subset criteria) is produced.

Missing and 'Out of Range' Data Plots. Select *Missing/Range Data Plots* from the *Graphs - Categorized Graphs* menu to display the *Categorized Missing and 'Out of Range' Data Plots* dialog.

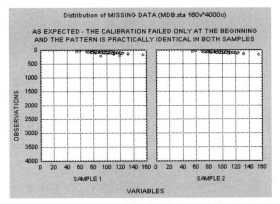

When you create categorized missing data/range plots, a series of standard 2D missing data/range plots, one for each category of cases identified by the *X* or *X* and *Y* category variables (or identified by the multiple subset criteria) is produced.

3D XYZ Plots. Select *3D XYZ Plots* from the *Graphs - Categorized Graphs* menu to display the *3D Categorized Plots* dialog (see the illustration on page 514). *STATISTICA* supports a variety of 3D graph types (i.e., *Scatterplots*, *Surface Plots*, *Contour Plots*, *Trace Plots*, *Deviation Plots*, *Space Plots*, *Spectral Plots*). Using categorized plots you can categorize 3D plots by the specified categories of a selected variable.

One component graph is produced for each level of the grouping variable (or user-defined subset of data) and all the component graphs are arranged in one display to allow for comparisons between the subsets of data (categories).

3D Ternary Plots. Select *3D Ternary Plots* from the *Graphs - Categorized Graphs* menu to display the *Categorized Ternary Graphs* dialog (see the illustration on page 515). In categorized ternary graphs, the points representing the proportions of the component variables (*X*, *Y*, and *Z*) are plotted in a two-dimensional display for each level of the grouping variable (or user-defined subset of data).

System Reference – 523

One component graph is produced for each level of the grouping variable (or user-defined subset of data), and all the component graphs are arranged in one display to allow for comparisons between the subsets of data (categories).

You can use categorized ternary plots to examine relationships between three or more dimensions where three of those dimensions represent components of a mixture for each level of a grouping variable. One typical application of this graph is when the measured response(s) from an experiment depends on the relative proportions of three components (e.g., three different chemicals) that are varied in order to determine an optimal combination of those components (e.g., in mixture designs). In a categorized ternary plot, one component graph is produced for each level of the grouping variable (or user-defined subset of data) and all the component graphs are arranged in one display to allow for comparisons between the subsets of data (categories).

User-Defined Graphs

Use *User-Defined Graphs* to save all of the options specified during the initial graph specification.

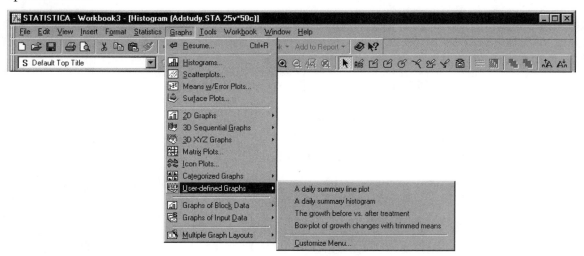

This graph can then be used with the same or a different input datafile to create similar analytic or presentation output with a minimum of required input. User-defined graphs can be named and saved by clicking the *Add as User-defined Graph to Menu* button on the *Options 2* tab of any graph specification dialog. Once saved, these "templates" can be accessed by name from the *Graphs - User-Defined Graphs* menu.

Select **Customize Menu** from the **Graphs - User-Defined Graphs** menu to display the **Customize User Graph Menu** dialog. You can use this dialog to review or modify the list of user-defined graphs. See the *Electronic Manual* for more details on using and managing user-defined graphs.

Graphs of Block Data

Graphs of Block Data process data from whatever is currently highlighted in the block and ignore the "meaning" of those numbers (e.g., the numbers can be raw data or values of correlation coefficients).

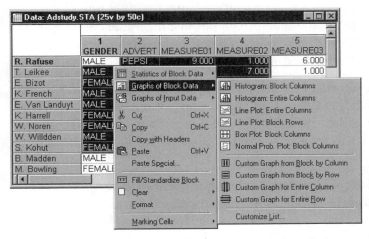

These graphs offer an effective means of visualizing, exploring, and efficiently summarizing numeric output from analyses displayed in results spreadsheets (e.g., histograms of Monte Carlo output scores in the **SEPATH** module, or a box plot of aggregated means from a multivariate multiple classification table in the **ANOVA** module). For more details on these types of graphs, see *Graphs of Block Data* in the *Overview* section (page 464).

Histogram: Block Columns. Select **Histogram: Block Columns** from the **Graphs - Graphs of Block Data** menu (or the spreadsheet shortcut menu) to create a multiple histogram with a fitted normal curve for each column of the selected block. This histogram allows for comparisons between the frequency distributions of values in block columns. To create a histogram for each row of the block, use the **Custom Graph from Block by Row** option.

Histogram: Entire Columns. Select **Histogram: Entire Columns** from the **Graphs - Graphs of Block Data** menu (or the spreadsheet shortcut menu) to create a multiple histogram for each variable (column) represented in the selected block (e.g., not just the selected values but all the values in the entire column). This histogram allows for comparisons between the

---

frequency distributions of values in block columns. To create a histogram for each row (i.e., case) represented in the block, use the *Custom Graph for Entire Row* option.

⊠ Line Plot: Block Rows. Select *Line Plot: Block Rows* from the *Graphs - Graphs of Block Data* menu (or the spreadsheet shortcut menu) to create a multiple line plot for the values in each row of the selected block. To create a line plot for the values in each column of the block, use the *Custom Graph from Block by Column* option. This multiple line plot is helpful in comparing values in block rows.

⊠ Line Plot: Entire Columns. Select *Line Plot: Entire Columns* from the *Graphs - Graphs of Block Data* menu (or the spreadsheet shortcut menu) create a multiple line plot that includes each variable (column) represented in the block (e.g., not just the values highlighted in the block but all the values in the entire column). To create a line plot for each row (i.e., case) represented in the block, use the *Custom Graph for Entire Row* option.

⊞ Box Plot: Block Columns. Select *Box Plot: Block Columns* from the *Graphs - Graphs of Block Data* menu (or the spreadsheet shortcut menu) to create a box plot of medians (and min/max values and 25th and 75th percentiles) for each of the columns in the block. To create a box plot of medians for each of the rows in the block, use the *Custom Graph from Block by Row* option.

▦ Normal Probability Plot: Block Columns. Select *Normal Probability Plot: Block Columns* from the *Graphs - Graphs of Block Data* menu (or the spreadsheet shortcut menu) to create a multiple normal probability plot for the values in the selected columns. This multiple plot allows for comparisons between the normality of the distribution of values in the block columns.

To create a normal probability plot for each of the rows in the block, use the *Custom Graph from Block by Row* option. Note that if the data in columns (or rows) that are compared in the graph have different ranges, then the comparison will be biased, as the range will affect the slope of the fit. To avoid this bias, bring the data to a common range (e.g., via standardization or other transformations) before creating this graph.

▦ Custom Graph from Block by Column. Select *Custom Graph from Block by Column* from the *Graphs - Graphs of Block Data* menu (or the spreadsheet shortcut menu) to display the *Select Graph* dialog. Use this dialog to create customized graphs of any kind (e.g., *2D Graphs, 3D Sequential Graphs, 3D XYZ Graphs, Matrix Graphs, Icon Graphs*) using the values in each column of the selected block.

After selecting the *Graph Category*, *Type*, and *SubType*, click the *OK* button to create the selected graph. To add your custom graph to the *Graphs - Graphs of Block Data* menu, select *Add to Graph menu List*.

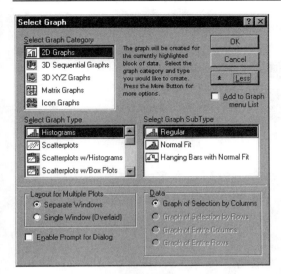

Access additional options by clicking the **More** button.

▤ Custom Graph from Block by Row. Select *Custom Graph from Block by Row* from the *Graphs - Graphs of Block Data* menu (or the spreadsheet shortcut menu) to display the *Select Graph* dialog (described above). You can use this dialog to create customized graphs using the values in each row of the selected block.

▥ Custom Graph for Entire Column. Select *Custom Graph for Entire Column* from the *Graphs - Graphs of Block Data* menu (or the spreadsheet shortcut menu) to display the *Select Graph* dialog (described above). You can use this dialog to create customized graphs using each variable (column) represented in the block (e.g., not just the values selected in the block but all the values in the entire column).

▤ Custom Graph for Entire Row. Select *Custom Graph for Entire Row* from the *Graphs - Graphs of Block Data* menu (or the spreadsheet shortcut menu) to display the *Select Graph* dialog (described above). You can use this dialog to create customized graphs using each case (row) represented in the block (e.g., not just the values selected in the block but all the values in the entire row).

Graphs of Input Data

Graphs of Input Data offer quick and simplified access to the most commonly used types of graphs. The most common way to call these graphs is directly from the spreadsheet shortcut menu (here in the *Graphs* menu, they are included mostly for users who prefer to use the

---

 StatSoft

keyboard and/or to simplify the process of recording keyboard macros). **Graphs of Input Data** do not require that you specify variables or options.

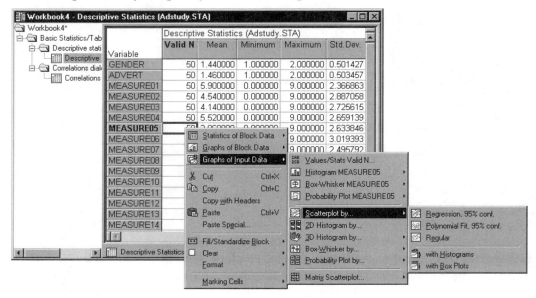

For more details on this type of graph, see *Graphs of Input Data* (page 462).

Value/Stats. Select *Value/Stats* from the **Graphs - Graphs of Input Data** menu to display the **Values/Stats** dialog and view the values of the currently selected variable.

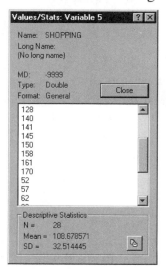

Additionally, you can view the variable name, missing data value, format and long variable name (if any) and a selection of descriptive statistics (*n*, mean, and standard deviation) for the variable. Note that you can click the 📑 button to copy the descriptive statistics to the Clipboard.

Histogram - Regular. Select *Regular* from the *Graphs - Graphs of Input Data - Histogram* menu to create a histogram (without a normal fit to the data) of the values of the currently highlighted variable without a normal fit to the data. If the selected variable is categorical (usually 10 or less categories), then *STATISTICA* will use those categories as the histogram intervals. If the variable has either more than 10 categories or the values are non-integers (continuous), then *STATISTICA* will divide the values into about 10 to 12 intervals based on the range of the values. These intervals are then used in the histogram.

Histogram - Normal Fit. Select *Normal Fit* from the *Graphs - Graphs of Input Data - Histogram* menu to create a histogram (with a normal fit to the data) of the values of the currently highlighted variable. If the selected variable is categorical (usually 10 or less categories), then *STATISTICA* will use those categories as the histogram intervals. If the variable has either more than 10 categories or the values are non-integers (continuous), then *STATISTICA* will divide the values into about 10 to 12 intervals based on the range of the values. These intervals are then used in the histogram.

Box-Whisker - Median/Quart./Range. Select *Median/Quart./Range* from the *Graphs - Graphs of Input Data - Box-Whisker* menu to create a box plot that describes the central tendency of the selected variable in terms of the median of the values (represented by the smallest box in the plot). The spread (variability) in the variable values are represented in this plot by the quartiles (the 25th and 75th percentiles, larger box in the plot) and the minimum and maximum values of the variable (the "whiskers" in the plot).

Box-Whisker - Mean/SD/SE. Select *Mean/SD/SE* from the *Graphs - Graphs of Input Data - Box-Whisker* menu to create a box plot in which the smallest box in the plot represents the mean (central tendency) of the variable, while the dispersion (variability) is represented by ± 1 times the standard error (large box) and ± 1 times the standard deviation about the mean (whiskers).

Box-Whisker - Mean/SD/1.96*SD. Select *Mean/SD/1.96*SD* from the *Graphs - Graphs of Input Data - Box-Whisker* menu to create a box plot that shows the mean (small box in the plot) of the variable surrounded by a larger box (± 1 times the standard deviation). If the distribution is normal, then the whiskers in this plot represent a 95% confidence interval defined as the variable mean ± 1.96 times the variable standard deviation.

Box-Whisker - Mean/SE/1.96*SE. Select *Mean/SE/1.96*SE* from the *Graphs - Graphs of Input Data - Box-Whisker* menu to create a box plot that shows the mean (small box

in the plot) of the variable surrounded by a larger box. However, in this plot, the larger box represents ± 1 times the standard error. If the distribution is normal, then the whiskers in this plot represent a 95% confidence interval defined as the variable mean ± 1.96 times the variable standard error.

Probability Plot - Normal Probability. Select *Normal Probability* from the *Graphs - Graphs of Input Data - Probability Plot* menu to create a normal probability plot for the indicated variable. This plot (described in *2D Graphs*, beginning on page 496) allows you to visually verify the normality of the data.

Probability Plot - Half-Normal Prob. Select *Half-Normal Prob* from the *Graphs - Graphs of Input Data - Probability Plot* menu to create a half-normal probability plot for the indicated variable. The half-normal probability plot is constructed in the same way as the standard normal probability plot, except that only the positive half of the normal curve is considered. Consequently, only positive normal values will be plotted on the y-axis.

Probability Plot - Detrended. Select *Detrended* from the *Graphs - Graphs of Input Data - Probability Plot* menu to create a detrended probability plot for the indicated variable. This plot is constructed in the same way as the standard normal probability plot, except that before the plot is generated, the linear trend is removed. This often "spreads out" the plot, thereby allowing the user to detect patterns of deviations more easily.

Scatterplot by - Regression, 95% Conf. Select *Regression, 95% Confidence* from the *Graphs - Graphs of Input Data - Scatterplot by* menu to plot the values of the currently highlighted variable against the values of another variable (specified in the *Select Variable* dialog). When this command is selected, a linear regression line is also fitted to the data and a 95% confidence interval for that line is shown.

Scatterplot by – Polynomial Fit, 95% Conf. Select *Polynomial, 95% Confidence* from the *Graphs - Graphs of Input Data - Scatterplot by* menu to create a scatterplot of the selected variable with another variable (specified in the *Select Variable* dialog). When this command is selected, the data are fitted with a polynomial fit (by default, of the 2nd degree) and a 95% confidence interval is shown for that line.

Scatterplot by - Regular. Select *Regular* from the *Graphs - Graphs of Input Data - Scatterplot by* menu to create a scatterplot of the selected variable with another variable (specified in the *Select Variable* dialog). The regular scatterplot visualizes a relation between two variables X and Y (e.g., weight and height). Individual data points are represented by point markers in two-dimensional space, where axes represent the variables. The two coordinates (X and Y) that determine the location of each point, correspond to its specific values on the two variables. If the two variables are strongly related, then the data points form a systematic shape

(e.g., a straight line or a clear curve). If the variables are not related, then the points form an irregular "cloud."

Scatterplot by - with Histograms. Select *with Histograms* from the *Graphs - Graphs of Input Data - Scatterplot by* menu to create a scatterplot with histograms of the selected variable with another variable (specified in the *Select Variable* dialog). Scatterplots with histograms contain a regular scatterplot of the variables with frequency distribution histograms for each variable displayed along the x- and y-axis of the scatterplot, respectively.

Scatterplot by - with Box Plots. Select *with Box Plots* from the *Graphs - Graphs of Input Data - Scatterplot by* menu to create a scatterplot with box plots of the selected variable with another variable (specified in the *Select Variable* dialog). Scatterplots with box plots contain a regular scatterplot of the variables with box plots for each variable displayed along the x- and y-axis of the scatterplot, respectively.

2D Histogram by. Select *2D Histogram by* from the *Graphs - Graphs of Input Data* menu to create a categorized histogram that allows you to examine the distribution of values of the currently selected variable broken down (categorized) by some other grouping or independent variable. For example, you may want to break down the distribution of test scores by the gender of the students.

In some results spreadsheets, the selection of the grouping variable is predetermined by some global grouping variable (i.e., the grouping variable designated in an analysis dialog). When this is the case, *STATISTICA* will incorporate the (predetermined) grouping variable name in the appropriate *Graphs of Input Data* option (e.g., *2D Histogram by Gender*). In most other cases, *STATISTICA* will prompt you to select the grouping variable via the *Select Variable* dialog.

When you are prompted to select a grouping variable that will be used to categorize the values of the currently highlighted variable, the *Select Variable* dialog will be displayed. Once you have selected the grouping variable from this dialog, another dialog will be displayed (*Codes* dialog) prompting you to enter the values (codes) for the previously specified grouping variable (if you leave the list of codes empty, all distinctive values of the grouping variable will automatically be used). After the codes have been entered, a plot displaying separate histograms for each category of the grouping variable will be produced.

3D Histogram by - Continuous by Continuous. Select *Continuous by Continuous* from the *Graphs - Graphs of Input Data - 3D Histogram by* menu to create a bivariate histogram for two continuous variables or if you want *STATISTICA* to perform an automatic categorization on the values of the variables for this bivariate histogram. Both selected variables will be divided into categories for this bivariate histogram. When a selected variable is categorical (usually 10 or fewer categories), *STATISTICA* will use those categories

as the histogram intervals for that variable. If a variable has more than 10 categories or its values are non-integers (continuous), then *STATISTICA* will divide the values into approximately 10 to 12 intervals based on the range of the values.

3D Histogram by - Discrete by Discrete. Select *Discrete by Discrete* from the *Graphs - Graphs of Input Data - 3D Histogram by* menu to create a bivariate histogram when you have highlighted a discrete variable in the spreadsheet and you want to plot another discrete variable on the second axis. After you have selected the second variable, the *Select Codes* dialog is displayed in which you will designate the codes for each variable. Click the *OK* button to accept all of the codes for both variables.

3D Histogram by - Continuous by Discrete. Select *Continuous by Discrete* from the *Graphs - Graphs of Input Data - 3D Histogram by* menu to plot a continuous variable (the currently highlighted variable) against a discrete variable in a bivariate histogram. The values of the first variable will be categorized according to the *STATISTICA* automatic categorization conventions (described above in *3D Histograms by - Continuous by Continuous*); however, you will be given the option to specify the codes for the discrete variable in the *Select Codes* dialog.

3D Histogram by - Discrete by Continuous. Select *Discrete by Continuous* from the *Graphs - Graphs of Input Data - 3D Histogram by* menu to plot a discrete (or to-be-categorized automatically) variable (the currently highlighted variable) against a continuous variable in a bivariate histogram. The values of the continuous variable will be categorized according to the *STATISTICA* automatic categorization conventions (described above in *3D Histograms by - Continuous by Continuous*); however, you will be given the option to specify the codes for the discrete variable in the *Select Codes* dialog.

Box-Whisker by - Median/Quart./Range. Select *Median/Quart./Range* from the *Graphs - Graphs of Input Data - Box-Whisker by* menu to create a box-whisker plot that represents the distribution of categorized values of the currently highlighted variable. When selected, the resulting box plots represent the currently highlighted variable's central tendency and variability for each category of some grouping variable.

In some results spreadsheets, the selection of the grouping variable is predetermined by some global grouping variable (i.e., the grouping variable designated in an analysis dialog). When this is the case, *STATISTICA* will incorporate the (predetermined) grouping variable name in the appropriate *Graphs of Input Data* option (e.g., *Box-Whisker by Gender*). In most other cases, *STATISTICA* will prompt you to select the grouping variable via the *Select Variable* dialog.

When you are prompted to select a grouping variable that will be used to categorize the values of the currently highlighted variable, the *Select Variable* dialog is displayed. Once you have

selected the grouping variable from this dialog, another dialog is displayed (**Codes** dialog) prompting you to enter the values (codes) for the previously specified grouping variable (if you leave the list of codes empty, all distinctive values of the grouping variable will automatically be used). After the codes have been entered, a plot displaying separate box and whisker plots for each category of the grouping variable will be produced.

For this particular type of categorized box plots, the central tendency of the categorized variable is shown in terms of the medians of the categorized values (represented by the smallest boxes in the plot). The spread (variability) in the categorized variable values is represented in this plot by the quartiles (the 25th and 75th percentiles, larger box in the plots) and the minimum and maximum values of the variable (the whiskers in the plots).

Box-Whisker by - Mean/SD/SE. Select *Mean/SD/SE* from the *Graphs - Graphs of Input Data - Box-Whisker by* menu to create a categorized box plot in which the smallest box in the plot represents the mean (central tendency) of the variable, while the dispersion (variability) is represented by ± 1 times the standard error (large box) and ± 1 times the standard deviation about the mean (whiskers). Note that the grouping variable for this plot is selected as described above in *Box-Whisker by - Mean/Quart./Range*.

Box-Whisker by - Mean/SD/1.96*SD. Select *Mean/SD/1.96*SD* from the *Graphs - Graphs of Input Data - Box-Whisker by* menu to create a categorized box plot that shows the mean (small box in the plot) of the variable surrounded by a larger box (± 1 times the standard deviation). If the distribution is normal, then the whiskers in this plot represent a 95% confidence interval defined as the variable mean ± 1.96 times the variable standard deviation. Note that the grouping variable for this plot is selected as described above in *Box-Whisker by - Mean/Quart./Range*.

Box-Whisker by - Mean/SE/1.96*SE. Select *Mean/SE/1.96*SE* from the *Graphs - Graphs of Input Data - Box-Whisker by* menu to create a categorized box plot that shows the mean (small box in the plot) of the variable surrounded by a larger box. However, in this plot, the larger box represents ± 1 times the standard error. If the distribution is normal, then the whiskers in this plot represent a 95% confidence interval defined as the variable mean ± 1.96 times the variable standard error. Note that the grouping variable for this plot is selected as described above in *Box-Whisker by - Mean/Quart./Range*.

Probability Plot by - Normal Probability. Select *Normal Probability* from the *Graphs - Graphs of Input Data - Probability Plot by* menu to produce standard normal probability plots for each category of the currently highlighted variable. In some spreadsheets, the selection of the grouping variable is predetermined by some global grouping variable (i.e., the grouping variable designated in an analysis dialog). When this is the case, *STATISTICA*

will incorporate the (predetermined) grouping variable name in the appropriate *Graphs of Input Data* option (e.g., *Probability Plot by Gender*). In most other cases, *STATISTICA* will prompt you to select the grouping variable via the *Select Variable* dialog.

When you are prompted to select a grouping variable that will be used to categorize the values of the currently highlighted variable, the *Select Variable* dialog is displayed. Once you have selected the grouping variable from this dialog, another dialog is displayed (*Codes* dialog) prompting you to enter the values (codes) for the previously specified grouping variable (if you leave the list of codes empty, all distinctive values of the grouping variable will automatically be used). After the codes have been entered, a plot displaying separate probability plots for each category of the grouping variable will be produced.

▦ Probability Plot by - Half-Normal Prob.

Select *Half-Normal Prob.* from the *Graphs - Graphs of Input Data - Probability Plot by* menu to produce half-normal probability plots for each category of the currently highlighted variable. See *Probability Plot by - Normal Probability* (above) for a discussion of how the grouping variable is specified. For more information on how the half-normal probability plot is constructed, see *Normal Probability Plots* (page 501).

▦ Probability Plot by - Detrended.

Select *Detrended* from the *Graphs - Graphs of Input Data - Probability Plot by* menu to produce detrended probability plots for each category of the currently highlighted variable. See *Probability Plot by - Normal Probability* (above) for a discussion of how the grouping variable is specified. For more information on how the detrended probability plot is constructed, see *Normal Probability Plots* (page 501).

▦ Matrix Scatterplot - Casewise MD Deletion.

Select *Casewise MD Deletion* from the *Graphs - Graphs of Input Data - Matrix Scatterplot* menu to produce a visual (scatterplot) representation of the correlations between several variables in a matrix format using casewise missing data deletion. In *Matrix Scatterplots*, one or two lists of variables can be selected in the *Select Variables for Matrix* dialog, and one scatterplot is generated for each pair of variables. If you select all of the variables in only the first list, then a square matrix of scatterplots is produced. You can generate a rectangular matrix of scatterplots by selecting only specific variables in the first list and then a different set of variables in the second list. When the scatterplots are generated, the cases that have missing data in at least one of the selected variables, will be excluded from each scatterplot. With this type of missing data deletion, all scatterplots will be based on the same set of observations. However, you could easily end up with no valid cases if the missing data are randomly distributed across cases, because each of them might have at least one missing data value for one of the selected variables.

 **StatSoft**
Copyright © StatSoft, 2001

▦ Matrix Scatterplot - Pairwise MD Deletion. Select *Pairwise MD Deletion* from the *Graphs - Graphs of Input Data - Matrix Scatterplot* menu to produce a visual (scatterplot) representation of the correlations between several variables in a matrix format (either square or rectangular as described above) using pairwise missing data deletion. When pairwise deletion of missing data is used, each individual scatterplot in the matrix is based on cases that have valid data for that pair of variables.

Multiple Graph Layouts

In addition to the compound graphs created by *STATISTICA* via results dialogs and the other graph types described above (e.g., categorized graphs, matrix plots, scatterplots with box plots), *STATISTICA* provides two methods for creating user-defined compound or multiple graphs. For more details on using these options, see *Creating User-Defined Compound Graphs* (Chapter 7 - *STATISTICA Graphs: Creation and Customization*).

▥ Wizard. Select *Wizard* from the *Graphs - Multiple Graph Layouts* menu to display the *AutoLayout Wizard – Step 1* dialog. Use this dialog to gather the graphs for the multi-graph layout you are designing. Once you have gathered graphs for the layout, click *OK* on this dialog to proceed to the *AutoLayout Wizard - Step 2* dialog. This dialog allows you to pick an appropriate (for the number of selected graphs) layout for your multiple graph. Note that both of these dialogs are discussed in detail in the *AutoLayout Wizard* topic (Chapter 7 - *STATISTICA Graphs: Creation and Customization*).

▥ Templates. Select *Templates* from the *Graphs - Multiple Graph Layouts* menu to display the *Graph Template* dialog and specify the number of blank graphs to use in your compound graph template. After specifying the number of graphs, click *OK* and proceed to the *AutoLayout Wizard - Step 2* dialog to pick an appropriate layout for your multiple graph. Note that this option is discussed in detail in the *Template* topic in Chapter 7 - *STATISTICA Graphs: Creation and Customization*.

▱ Blank Graph. Select *Blank Graph* from the *Graphs - Multiple Graph Layouts* menu to sproduce a blank graph; use drawing tools to create highly customized charts, maps, etc.

Tools Menu

The following commands are available from the graph *Tools* menu. Many of these options are also available from shortcut menus (accessed by right-clicking on an item or icon in the report) and from the toolbar buttons.

Analysis Bar

To take advantage of *STATISTICA*'s "multitasking" functionality, *STATISTICA* analyses are organized as functional units that are represented with buttons on the **Analysis** bar at the bottom of the application window (above the status bar, see the illustration below, where **Basic Statistics**, **Cluster Analysis**, and **Canonical Analysis** are running simultaneously). Normally, at least one analysis button is created, and consecutive buttons are added as you start new analyses.

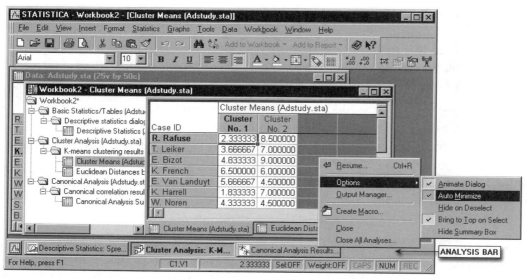

The commands on the **Tools – Analysis Bar** menu provide a variety of options for managing the **Analysis** bar.

Resume. Select **Resume** from the **Tools - Analysis Bar** menu (or click the toolbar button or press CTRL+R) to continue the current analysis or graph. Note that you can also open the current analysis or graph by clicking on its button on the **Analysis** bar.

Select Analysis/Graph. This command is not shown in the illustration above because it is accessible only when this menu is accessed from the **Tools** menu and not when you access it by right-clicking on a specific analysis button on the **Analysis** bar. Use the commands on the **Select Analysis/Graph** menu to select an analysis or graph from the set of active analyses and graphs.

Options - Animate Dialog. Select *Animate Dialogs* from the *Tools - Analysis Bar - Options* menu to display animation when analysis dialogs are minimized or maximized. By default this command is checked.

Options - Auto Minimize. Select *Auto Minimize* from the *Tools - Analysis Bar - Options* menu to automatically minimize all analysis dialogs when you select another window in *STATISTICA* or another application. By default this command is checked. When your screen is large enough to accommodate several windows, it is recommended that you clear this option. This keeps the analysis dialogs on screen while the respective output created from these dialogs is produced, thus allowing you to use the dialogs as "toolbars" from which output can be selected.

Options - Hide on Select. Select *Hide on Select* from the *Tools - Analysis Bar - Options* menu to minimize all windows associated with a particular analysis when that analysis is deselected. By default this command is cleared. Note that this command only applies when the results are sent to individual windows; see the discussion of the *Output Manager* tab of the *Options* dialog in Chapter 2 – *STATISTICA Output Management* for further details.

Options - Bring to Top on Select. Select *Bring to Top on Select* from the *Tools - Analysis Bar - Options* menu to activate (display at the top of *STATISTICA*) all windows associated with a particular analysis when that analysis is selected, replacing whatever dialogs were on top. This command also facilitates the organization of individual windows from various analyses. By default this option is checked. Note that this command only applies when the results are sent to individual windows; see the discussion of the *Output Manager* tab on the *Options* dialog (Chapter 2) for further details.

Options - Hide Summary Box. Select *Hide Summary Box* from the *Tools - Analysis Bar - Options* menu to not display the summary box, which is located at the top of certain results dialogs (such as *Multiple Regression - Results*) and contains basic summary information about the analysis. By default this command is not checked.

Output Manager. Select *Output Manager* from the *Tools – Analysis Bar* menu to display the *Output Manager* tab of the *Options* dialog. Use the options on this tab to specify where to direct output files. For more details, see *Three Channels for Handling Output* in Chapter 2.

Create Macro. Select *Create Macro* from the *Tools - Analysis Bar* menu to display the *New Macro* dialog and specify a name for a new macro based on the current analysis. When you run an analytic procedure (from the *Statistics* menu) or create a graph (from the *Graphs* menu) the Visual Basic code corresponding to all design specifications as well as output options that you select are recorded. Thus, when you click the *OK* button in the *New Macro* dialog, the resulting macro window displays the appropriate code to recreate the current analysis.

Minimize. Select *Minimize* from the *Tools - Analysis Bar* menu to minimize the current analysis.

Close. Select *Close* from the *Tools - Analysis Bar* menu to close the current analysis.

Close All Analyses. Select *Close All Analyses* from the *Tools - Analysis Bar* menu to close the all of the analyses/graphs on the *Analysis* bar.

Macro

A macro is a scripted application that extends functionality to *STATISTICA* by directly accessing *STATISTICA*'s object model and manipulating it. Macros are primarily used to automate tasks done in *STATISTICA* by harnessing its power and recording it into a *STATISTICA* Visual Basic script. Additionally, macros can be written as both stand-alone scripts and library classes to extend the statistical and mathematical capabilities of *STATISTICA*.

Macros. Select *Macros* from the *Tools - Macro* menu (or click the ▶ toolbar button on the *Macro* toolbar) to display the *Macros* dialog.

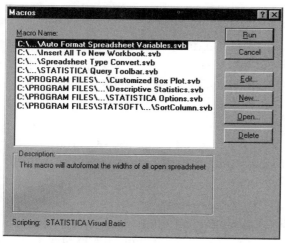

Use this dialog to run, edit, open, or delete existing macro (SVB) programs as well as to create new macros.

Start Recording Log of Analyses (Master Macro). Select *Start Recording Log of Analyses (Master Macro)* from the *Tools - Macro* menu to create a SVB program file that will include a sequence of analyses (a log of analyses) that have been performed interactively. Note that you must select this command before you start the analyses. Consult the *Electronic*

Manual or the *STATISTICA Visual Basic Primer* to learn more about this powerful command and how to create recordable sequences of analyses.

Start Recording Keyboard Macro. Select *Start Recording Keyboard Macro* from the *Tools - Macro* menu to record a keyboard macro. See the *Electronic Manual* or the *STATISTICA Visual Basic Primer* for more details.

Stop Recording. Select *Stop Recording* from the *Tools - Macro* menu (or click the ■ toolbar button on the *Macro* toolbar) to stop recording a log of analyses (master macro) or a keyboard macro (see above).

Create Analysis/Graph Macro. Select *Create Analysis/Graph Macro* from the *Tools - Macro* menu to select from a list of all current analyses/graphs that are open. When you run an analytic procedure (from the *Statistics* menu) or create a graph (from the *Graphs* menu) the Visual Basic code corresponding to all design specifications as well as output options that you select are recorded. Select the appropriate analysis/graph to display that code. Note that the *New Macro* dialog will first be displayed in which you can type the macro's name and description.

Application Events - View Code. Select *View Code* from the *Tools - Macro - Application Events* menu to display the *Document Events* window, which is used to enter code to change the default behavior of application-level events. Application-level events allow you to customize the behavior of all documents of a certain type (such as all spreadsheets, all workbooks, or all reports). See *What Are Application Events and How Can They Be Controlled from STATISTICA Visual Basic?* in Chapter 11 for more details.

Application Events - Autorun. Select *Autorun* from the *Tools - Macro - Application Events* menu to save your application-level event code with a specific type of document so that it will be run any time that type of document is open.

Add-Ins. Select *Add-Ins* from the *Tools - Macro* menu to display the *STATISTICA Add-Ins* dialog.

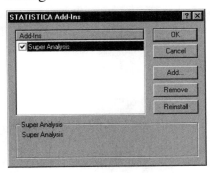

Add-Ins are COM server components normally written in ATL (Active Template Library) that are used to create custom user interfaces of *STATISTICA* and/or fully functional external programs. All available Add-Ins are displayed in the ***Add-Ins*** list. To create a new Add-In, click the ***Add*** button to display the ***Specify Add-In to be Added*** dialog, which is used to enter the program ID of the Add-In. Click the ***Remove*** button to delete the selected Add-In from the ***Add-Ins*** list. Finally, click the ***Reinstall*** button to register the selected Add-In to your operating system.

STATISTICA* Visual Basic Editor.** Select ***STATISTICA Visual Basic Editor from the ***Tools - Macro - Application Events*** menu (or press ALT+F11) to display a *STATISTICA* Visual Basic editor (Macro Window). For more information on using *STATISTICA* Visual Basic including examples and an extensive FAQ section, see Chapter 11.

Customize

Select ***Customize*** from the ***Tools*** menu to display the ***Customize*** dialog, which is used to customize toolbars, menus, and keyboard hot keys with a variety of commands.

Options

Select Select ***Options*** from the ***Tools*** menu to access the ***Options*** dialog. In addition to general and display options, options are available for graphs, reports, file locations, custom lists, workbooks, macros, statistical analysis display, import, and edit facilities, and output management. For a discussion of the ***Graphs 1*** and ***Graphs 2*** tab of the ***Options*** dialog, see global graph options in Chapter 7 - *STATISTICA Graphs: Creation and Customization*.

Window Menu

The ***Window*** menu is available when any document is open. It provides access to commonly used commands for organizing the workspace and switching between files.

Close All

Select ***Close All*** (or press CTRL+L) from the ***Window*** menu to close all open spreadsheets, graphs, and related windows (e.g., graph data) in *STATISTICA*. This option is useful when you need to clear the screen to start a new analysis. Note that you will be prompted to save any unsaved files before they are closed.

Cascade

Select *Cascade* from the *Window* menu to arrange the open *STATISTICA* windows in an overlapping pattern so that the title bar of each window is visible.

Tile Horizontally

Select *Tile Horizontally* from the *Window* menu to arrange the open *STATISTICA* windows in a horizontal (side by side) pattern. When you select this option, *STATISTICA* automatically optimizes the display of the open windows (with the preference given to tiling horizontally).

Tile Vertically

Select *Tile Vertically* from the *Window* menu to arrange the open *STATISTICA* windows in a vertical pattern. When you select this option, *STATISTICA* automatically optimizes the display of the open windows (with the preference given to tiling vertically).

Arrange Icons

Select *Arrange Icons* from the *Window* menu to arrange all minimized windows into rows.

Windows

Select *Windows* from the *Window* menu to display the *Windows* dialog. This dialog allows you to access all of the currently open windows in *STATISTICA* and manage them.

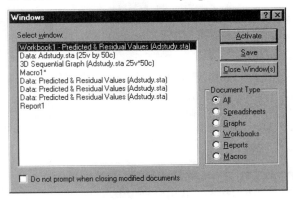

Select Window. Select either a single window or multiple windows in the **Select Window** list box for manipulating its associated window in the *STATISTICA* environment. Note that you can press CTRL or SHIFT while highlighting the items in the list box to make multiple selections.

Activate. Click the **Activate** button to set the focus to the currently selected window in the **Select Window** box. In other words, the associated window will be brought to the front of all of the other windows. Note that because only one window can have the focus at any given time, multiple selection for this operation will not apply.

Save. Click the **Save** button to save all of the windows that are currently selected in the **Select Window** list box.

Close Windows. Click the **Close Windows** button to close all of the windows that are currently selected in the **Select Window** list box. Note that *STATISTICA* will prompt you to save any modified windows before closing them.

Document Type. Use the options in the **Document Type** group to filter all of the open windows within *STATISTICA* by their document type; the contents of the **Select Window** box will be updated to reflect your filter selection. You can display all open documents in the **Select Window** box (select the **All** option button), or limit the display to spreadsheets, graphs, workbooks, reports, or macros only, by selecting the appropriate option button.

Help Menu

The **Help** menu is available when any document is open. It provides access to various types of help.

Contents and Index

Select **Contents and Index** from the **Help** menu to display the *STATISTICA Electronic Manual*. *STATISTICA* provides comprehensive on-line documentation for all program procedures and all options available in a context-sensitive manner by pressing the F1 key or clicking the help button ? on the caption bar of all dialogs (there are over 100 megabytes of compressed documentation included).

Due to its dynamic hypertext organization, organizational tabs (e.g., **Contents**, **Index**, **Search**, and **Favorites**), and various facilities allowing you to customize the help system, it is often faster to use the *Electronic Manual* than to look for information in the traditional manuals. The

status bar on the bottom of the *STATISTICA* window displays short explanations of the menu options or toolbar buttons when an item is highlighted or a button is clicked.

Statistical Advisor

Select **Statistical Advisor** from the **Help** menu to launch the **Statistical Advisor**. Based on your answers to the successive questions about the nature of your research, the **Statistical Advisor** suggests which statistical methods could be used and where to find them in *STATISTICA*.

Animated Overviews

Select **Animated Overviews** from the **Help** menu to display a submenu of the available animated overviews (visual overviews and specific feature help). Select from the submenu to launch one of the animated overviews.

If you select one of these commands and you have not installed that animated overview, a *STATISTICA* message dialog will prompt you for the location of that animated overview file. For example, you can run it from your *STATISTICA* CD, or if you are running a network version, the overviews can be located on the network drive. Note that if you have the *STATISTICA* CD, you can run the overviews directly from the CD without copying them to your hard drive.

StatSoft's Home Page

Select **StatSoft's Home Page** from the **Help** menu to launch the StatSoft home page in your default browser. We invite you to visit the StatSoft Web site often:

- For the most recent information about *STATISTICA*, downloadable upgrades, new releases, new products, news about StatSoft, etc., access the *What's New...* section of the Web site.

- For a comprehensive list of *Frequently Asked Questions* (including useful tips, solutions to hardware and software compatibility problems, etc.), access the *Technical Support* section of the Web site.

- For a library of *STATISTICA* Visual Basic programs (written by users), access the *Technical Support* section of the Web site (to submit your own programs to this "user-exchange forum," send an e-mail to *info@statsoft.com*).

Technical Support

Select **Technical Support** from the **Help** menu to launch the *Technical Support - Getting More Help* page of the StatSoft Web site in your default browser. This page contains links to download *STATISTICA* updates and links to FAQ topics on spreadsheets, graphs, printing, reports, etc. The StatSoft Technical Support Department e-mail address, phone number, and hours are also listed. Note that most chapters of this manual also contain FAQ sections.

About *STATISTICA*

Select **About STATISTICA** from the **Help** menu to display the *STATISTICA* version, copyright notice, license information, and foreign office contact information.

GRAPH TOOLBARS

Graph Tools Toolbar

The **Graph Tools** toolbar (displayed when a graph window is active) offers a selection of drawing, graph customization, and multigraphics management options (e.g., linking, embedding graphs and artwork).

Most of the graph control and customization facilities accessible via the toolbar buttons listed below are also available from menus. All toolbars can be customized to adjust the number and order of buttons in the toolbar.

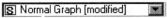

Graphic Styles Box

Use the 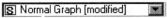 box to select a style for the currently highlighted object (e.g., grid lines, point markers, axis line) or the entire graph. For more on styles, see the *Graph Styles* in Chapter 7 - *STATISTICA Graphs: Creation and Customization*.

3D Rotation Control Button

Click the button to change the perspective on 3D graphs, specifically graphs that actually portray three dimensions such as 3D XYZ Scatterplots. Clicking this button displays the **Point of View Settings and Exploratory Spin** dialog on which you can rotate the graph about horizontal and vertical axes and change the viewpoint distance or the virtual distance from which the graph is viewed. For a detailed discussion of this dialog, see *Viewing 3D Graphs* (Chapter 7 - *STATISTICA Graphs: Creation and Customization*).

Brushing Tool Button

Click the button to activate the **Brushing** mode and display the interactive **Brushing** dialog. When this mode is activated, the cursor changes to a "gunsight-style" cross-hair (several other brushes are also available, see the **Brushing** dialog). You can then highlight data points from the graph by clicking on them with the cross-hair (or enclosing them in the rectangle or lasso, or for 3D displays with a cube or a custom 3D slice).

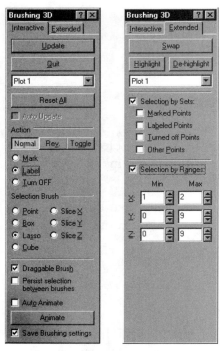

STATISTICA features a comprehensive selection of brushing facilities, including a unique set of interactive tools to make precise selections of subsets in 3D displays using the simple mouse movements and a variety of tools for manipulation of data points on screen, as well as the management of brushed (selected) data points.

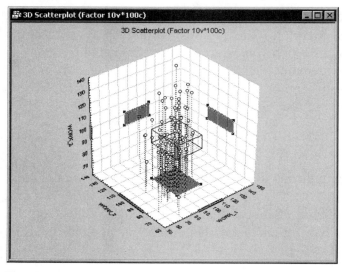

For more information on brushing, see the *Electronic Manual*. See also, *Brushing, Interactive Analysis* (Chapter 9 - *STATISTICA Graphs: Examples and FAQ's*).

Display Graph Fit to Window Button

When this button is selected (a default setting), the graph window is resized while maintaining the aspect ratio (ratio of vertical to horizontal dimensions) of the graph.

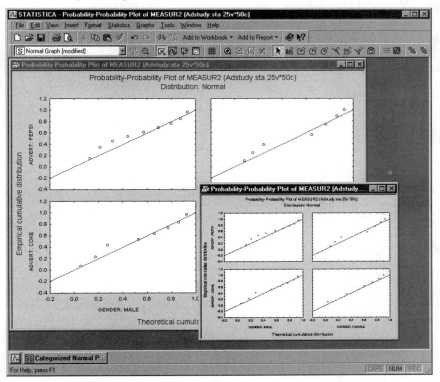

The default graph aspect ratio. The global default graph aspect ratio can be modified on the *Graphs* tab of the *Options* dialog (accessible via the *Tools* menu). For the currently active graph, the default graph aspect ratio can be modified in the *Document Size and Scaling* dialog, accessible from the *View* menu or from the button (see below).

Display Graph at Actual Size Button

When you click the button, scroll bars are displayed around the graph, allowing you to view the graph at its actual size.

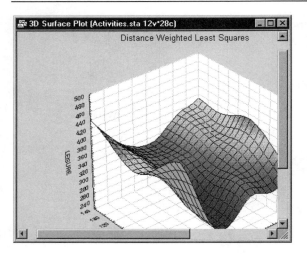

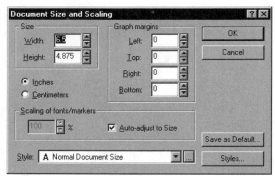 Adjust the Graph Actual Size/Scaling Button

Click the button to display the *Document Size and Scaling* dialog

and change the current graphs aspect ratio (i.e., the ratio of its vertical to horizontal dimensions). Note that the global default graph aspect ratio can be modified on the *Graphs* tab of the *Options* dialog (accessible via the *Tools* menu).

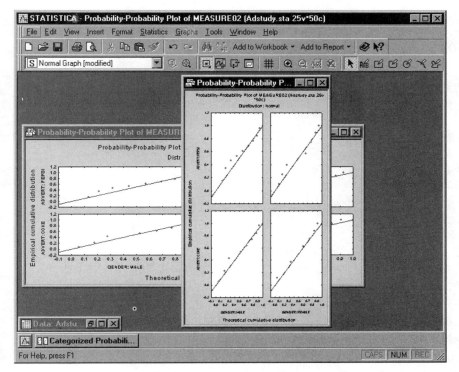

The modified proportions of the graph will also be reflected in the printout (as can be examined using the ***Print Preview*** facility, use the ***Print Preview*** 🔍 toolbar button or select ***Print Preview*** from the ***File*** menu).

🔳 Set Graph Area Button

Click the 🔳 button (or select ***Graph placement*** from the ***View*** menu) to specify the placement of the plot(s) within the graph window. There are two methods to do it interactively using this tool (in addition to making the adjustment in the ***Document Size and Scaling*** dialog accessible by clicking the 🔁 button; see above).

You can increase (or decrease) the margin around the plot by clicking this button and dragging the resizing squares that are displayed around the edges of the graph window. As you drag the buttons towards the plot, the margins around the plot increase (as illustrated below).

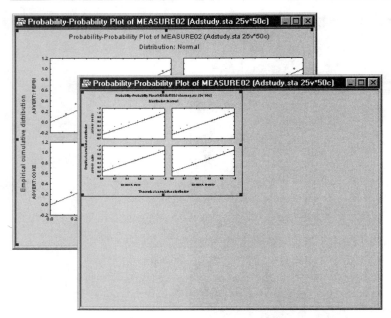

You can also increase the margins by simply drawing a rectangle in the place where you want the graph to be. After clicking the 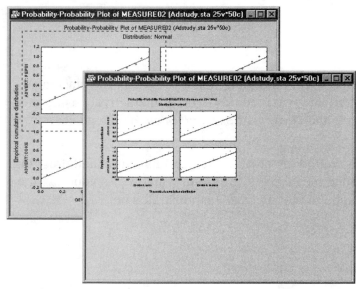 button, drag the mouse pointer across the graph to "draw" a rectangle in the graph window (see the dashed line in the illustration below). When you release the mouse button, the plot will be redrawn in the area specified by the rectangle and the margins will have been increased (or decreased) as shown below.

You can use the **Snap to Grid** option (see below), to align the area of the graph with other objects, e.g., other graphs to be pasted later in the blank area of the display.

⊞ Snap to Grid Button

Click the ⊞ button to display the **Snap to Grid** dialog in which you can enable (and customize) an invisible drawing grid to help you align drawn objects (e.g., lines, rectangles, inserted graphics or text objects).

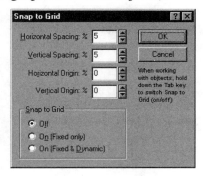

When **Snap to Grid** is enabled, an object that is being drawn or placed in the graph is pulled into alignment with the nearest intersection of gridlines (as if the resolution of the screen were "decreased"). This is referred to as the snap-to-grid effect.

The display of the alignment grid itself in the graph can be toggled on and off by pressing CTRL+G or selecting **Alignment Grid** from the **View** menu.

If the **Snap to Grid** option is turned off, the grid no longer affects drawn objects. The snap-to-grid facility can be toggled while placing individual objects by holding down the TAB key (e.g., even if the **Snap to Grid** facility is turned off, it can be enabled by holding down the TAB key).

Applications. A typical application of the **Snap to Grid** facility is when you need to align objects (e.g., a series of custom text objects, independent segments of a legend, or embedded graphs); it is also useful while drawing diagrams, flow charts, etc.

⊕ Zoom In Button

When you click the ⊕ button, the mouse pointer becomes a magnifier, enabling you to proportionally enlarge the current graph and redraw it in the new ranges of coordinates either in the same window, or in a separate window (CTRL+⊕ to create a new graph). Place the magnifier over the area of the graph that you want to view and click. This results in a

magnified view of the specified area; the point where you click (the focal point of magnification) becomes the center of the new graph window. Each time you click the *Zoom In* tool, the graph area is magnified by a factor of approximately 2; you can click it up to 5 times to achieve a magnification ratio of approximately 32:1.

Alternatively, you can click and drag the mouse pointer to define ("draw") a rectangular area of the graph to enlarge.

This tool offers more than just "mechanical stretching" of the picture. Specifically, it provides a logical magnification of the selected area while maintaining the sizes of point markers, fonts, and width of all lines. Thus, the *Zoom In* tool effectively increases the "functional resolution" of the display, allowing you to inspect areas of the graph that in normal mode were not readable due to overlapping markers or point labels.

Reversing the Zoom In operation. You can also decrease the graph size (in the same increments) using the *Zoom Out* tool ⊖ (see discussion below). In order to quickly restore the original graph magnification, you can use the *Reset Zoom* ⊗ button (see discussion below).

⊖ Zoom Out Button

This button is active only when you have previously "zoomed-in" on the current graph. When you click the ⊖ button, the mouse pointer becomes a magnifier, enabling you to proportionally "shrink" the current graph and redraw it in the new ranges of coordinates either in the same window, or in a separate window (CTRL+⊖ to create a new graph). Place the magnifier over the area of the graph that you want to reduce and single-click. This results in a reduced view of the specified area; the point where you click (the focal point of reduction) becomes the center of the new graph window. Each time you click the *Zoom Out* tool, the graph area is reduced by a factor of approximately 2. Note that you are not able to zoom out beyond the scaling of the graph.

Reversing the Zoom In operation. If you previously magnified a portion of the graph using the *Zoom In* ⊕ tool (see discussion above), the original location of the graph in the window will not be restored unless you select exactly the same focal points for the *Zoom Out* operation as those you had used for the *Zoom In*. In order to quickly restore the original graph magnification, you can use the *Reset Zoom* ⊗ button (see discussion below).

▦ Zoom Pan Button

Use the ▦ button to move the focus of an area of the graph that has been enlarged via the *Zoom In* ⊕ button. After the *Zoom Pan* button has been selected, click and drag the mouse

pointer in the direction you would like the focus to shift. An arrow will be displayed on the graph to indicate the direction and magnitude of the shift that will be applied.

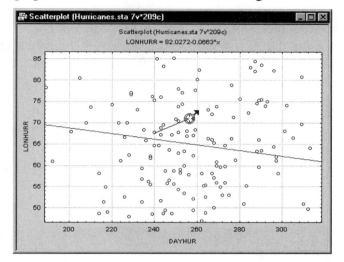

When the mouse button is released, the graph will be redrawn at the same magnification with the focus in the shifted position.

Reset Zoom Button

Click the ⊗ button to reverse the effects of any enlargement or shrinking of the graph as a result of using the *Zoom In* or *Zoom Out* options discussed above. Clicking this button will restore the graph to its original magnification.

Selection Tool Button

Click the ⬉ button to enable the selection mode, which is the default mode for all graph windows (indicated when the button is depressed). This mode will allow you to use the left-click to select graph objects (e.g., arrows, embedded or linked objects, etc.) or different parts of the graph (e.g., titles, scales, gridlines, etc.) for customization, editing, or interactive analysis. You will temporarily exit from this selection mode whenever you choose one of the interactive graph customization options (e.g., brushing, add text, draw arrows and other objects, embed graphs). Once you have completed drawing the selected object, you will automatically return to this mode.

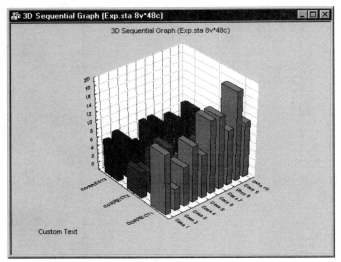 Text Button

Click the ![] button (or select **Text** from the **Insert** menu) to place custom text objects on the graph at any location. After clicking the **Text** button, the mouse pointer will turn to a cross-hair shape when it is passed over the graph. Clicking at any location on the graph places a graphic text box downward and to the right of the cursor location (as shown below).

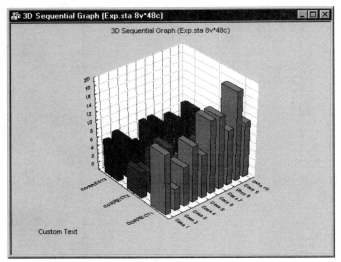

Click to activate this text object and drag it to a new position. Double-click the object to display the **Graph Titles/Text** dialog where the default text (i.e., **Custom Text**) can be replaced and customized.

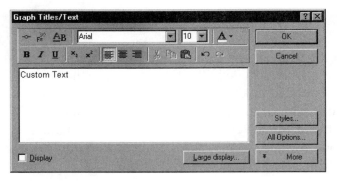

Enter the desired text in the edit box; then use the buttons on the mini-formatting toolbar above the box to customize the text. A listing and brief discussion of each of these mini-formatting toolbar options are available in the mini-formatting toolbar topic in the *Electronic Manual*.

Click the **OK** button to place the text on the graph. Note that all custom text is initially inserted as static text (i.e., its position in the graph window is independent of the graph coordinates and scaling). It can be changed to **Dynamic** in the **Text Object Properties** dialog (shown below); see *What Is the Difference between the Dynamic and Fixed Status of Custom Objects (e.g., Arrows, Custom Text, Drawings) in Graphs?* (Chapter 9 - *STATISTICA Graphs: Examples and FAQ's*) for information about the static and dynamic status of objects in graphs. To access this dialog, click the **More** button on the **Graph Titles/Text** dialog, then click the **Text Object Properties** button.

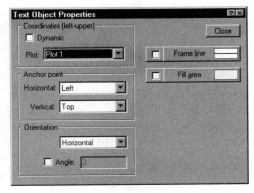

This dialog also provides access to customization options for other features such as frames, background patterns, and colors.

Note that you can also edit custom graph text on the **Graph Titles/Text** tab of the **All Options** graph customization dialog (see Chapter 7 - *STATISTICA Graphs: Creation and Customization*). If custom text needs to be aligned with other custom objects on the graph, consider using the **Snap to Grid** feature (see page 551).

Rectangle Button

Click the button (or select **Rectangle** from the **Insert** menu) to draw a rectangle. When this button is clicked, the cursor changes into a cross-hair allowing you to draw rectangles by dragging the cross-hair on the graph.

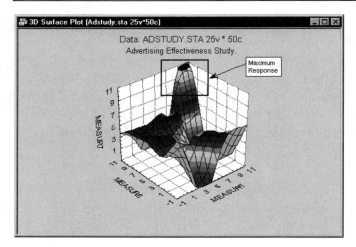

Drawing squares. If you hold down the SHIFT key while drawing, a square is produced (equal height and width proportions are maintained).

Exit. To exit from the drawing mode, click the selection tool ▶ button or press ESC.

Moving or redimensioning rectangles. Click on the perimeter of a rectangle or square to select the object for moving or redimensioning. To move the object, place the mouse pointer inside the object and drag it with the left mouse button depressed. To redimension the object, drag the black selection squares that surround the object. If a rectangle or square needs to be aligned with other custom objects on the graph, consider using the *Snap to Grid* feature (page 551).

Customizing properties of rectangles. To further specify properties of newly created rectangles or to make changes to properties of existing rectangles, double-click on the perimeter of the object to display the *Shape Object Properties* dialog. You can use this dialog to specify line and fill patterns and colors as well as to define whether the anchor points (upper-left and lower-right corners) of the object are to be positioned in *Dynamic* or *Fixed* (the default) mode. In *Dynamic* mode, the corner(s) of the object will be positioned with respect to scale values of the graph axes. In this mode, the object will move if the scaling on the graph changes. In *Fixed* mode, the positioning will be with respect to the dimensions of the graph window and will not change if scale values change. See *What Is the Difference between the Dynamic and Fixed Status of Custom Objects in Graphs?* (Chapter 9 - *STATISTICA Graphs: Examples and FAQ's*).

✍ Rounded Rectangle Button

Click the ✍ button (or select *Rounded Rectangle* from the *Insert* menu) to draw a rounded rectangle. When this button is clicked, the cursor changes into a cross-hair allowing you to

draw rectangles with rounded corners by dragging the cross-hair on the graph. (For details on operation of this feature, see the summary on drawing and customizing rectangles, above.)

Drawing rounded squares. If you hold down the SHIFT key while drawing, a rounded square is produced (equal height and width proportions are maintained).

Ellipse Button

Click the button (or select **Ellipse** from the **Insert** menu) to draw an oval. When this button is clicked, the cursor changes into a cross-hair allowing you to draw ellipses or circles by dragging the cross-hair on the graph. (For details on operation of this feature, see the summary on drawing and customizing rectangles, above.)

Drawing circles. If you hold down the SHIFT key while drawing, a circle is produced (equal height and width proportions are maintained).

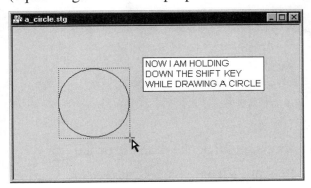

Arc Button

Click the button to draw an arc. When this button is clicked, the cursor changes into a cross-hair allowing you to draw arcs by dragging the cross-hair on the graph. The arc is drawn as one-quarter of an ellipse and can be controlled as illustrated below.

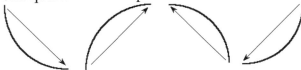

Specifically, the proportions of the ellipse depend on the angle of mouse movement (45° will produce a quarter-circle arc; alternatively, hold down the SHIFT key while drawing to draw a circular arc). The orientation of the arc (concave vs. convex) will depend on the direction of the

movement (up or down, respectively). For details on operation of this feature, see the summary on drawing and customizing rectangles, above.

Polygon Button

Click the button to draw a freehand/polyline/polygon drawing mode. When this button is clicked, the cursor changes into a "pencil" with a pointed tip (see below).

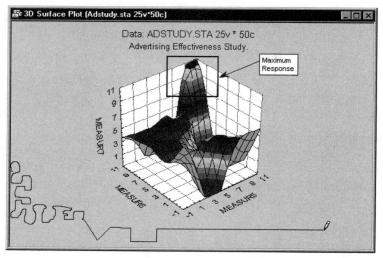

Freehand lines. In order to produce freehand drawings, click and drag the pencil.

Straight lines. To draw a straight segment of a line, place the tip of the pencil on the starting point of the segment and click the mouse, then (after releasing the button), move the tip of the brush to the endpoint of the segment, and click again. To finish drawing the object, double-click.

Filled shapes. Filled shapes are produced when the fill pattern is enabled (see the fill pattern customization button); note that the shape of the line or polyline does not have to be closed in order to be filled as long as it forms at least one cove (i.e., when it is not straight), as shown below.

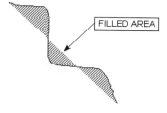

FILLED AREA

Proportional vs. nonproportional resizing. In addition to the standard graphic object customization facilities (see the summary on drawing and customizing rectangles, ✐ button), several line-drawing specific features are supported. When the object is selected, it is surrounded by eight black handles. Dragging a corner handle will proportionately resize the drawing, whereas dragging by the side handles will resize (i.e., stretch or squeeze) the object in only one direction.

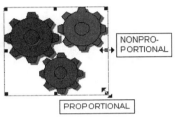

The arrow shape of the cursor (when you move it over the handles) indicates the resizing mode. For example, in the illustration above, the cursor is positioned over the lower-right handle. Its diagonal, double-headed appearance indicates that dragging this handle will proportionately resize the object.

Line shape editing. Additionally, when you select a line drawing, you can edit the details of the shape of the line. A handle will be displayed in every point of the line drawing where the line changes its direction (see the illustration below).

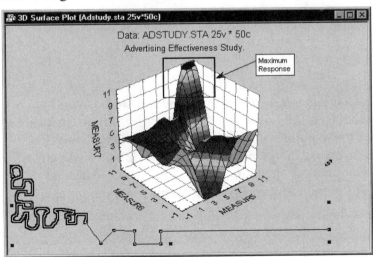

You can drag any one of them with the mouse to modify the local shape of the drawing.

Editing in zoom mode. True freehand drawings will be densely covered by the handles because they consist of many very short segments. In order to edit such densely curved shapes, zoom-in on the drawing (see 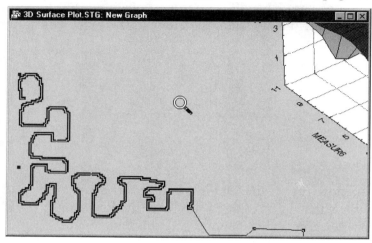 *Zoom In Button*, page 551).

Drawings can consist of large numbers of independent or overlapping segments and the order in which they are redrawn (and, consequently, overwrite each other) can be controlled using *Redraw* order control tools (*Bring to Front* and *Send to Back* buttons). Drawing and all other graph customization options (e.g., embedding) can be performed in every graph resizing mode, including zoom-in mode (which can be used to increase the drawing precision, as shown in the illustration above).

Arrow Button (Error Bar Button)

Click the button (or select *Arrow* from the *Insert* menu) to enter the arrow and error bar drawing mode. You can use the *Arrow* tool to place arrows and error bars of various styles on a graph. When this button is clicked, the mouse pointer changes to a small vertical arrow. To draw an arrow, place the small vertical arrow on the graph to position the default-style arrow head. Then drag the arrow away from that point to create the arrow. The tail's length and position is determined by the location of the mouse pointer on the graph when the mouse button is released.

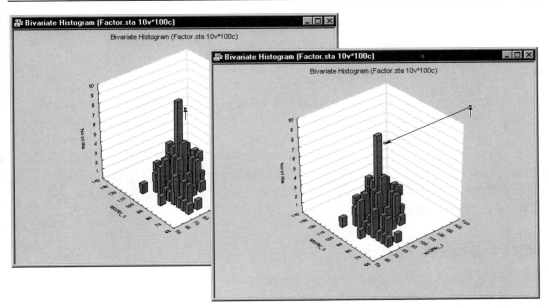

The arrow style (including selecting error bar options) as well as the size and location of the arrow can be adjusted later.

Moving or redimensioning arrows. Click on the head or tail of the arrow to select the object for moving or redimensioning (a handle will be displayed at the head end and at the tail end of the arrow). To move the arrow, place the mouse pointer on the arrow shaft, click and drag it. To redimension or change the orientation of the arrow, drag either of the handles at the ends of the arrow.

Customizing properties of arrows. To further specify properties of newly created arrows or to make changes to properties of existing arrows, double-click on either the head or the tail of the arrow to display the *Arrow Object Properties* dialog.

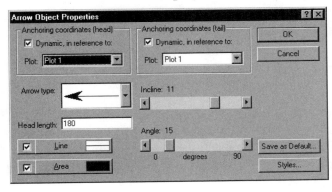

Use this dialog to specify a number of options for customizing the appearance of arrow heads and their fill characteristics as well as specify line colors and patterns for both arrow and error bars.

You can also specify whether the anchor points (head or first-defined end and tail or last-defined end) of the arrow are positioned in *Dynamic* or *Fixed* (the default) mode. In *Dynamic* mode, the specified end of the object is positioned with respect to scale values of the graph axes. In this mode, the head or tail moves if the scaling on the graph changes. In *Fixed* mode, the positioning is with respect to the dimensions of the graph window and arrow position does not change if scale values change. See *What Is the Difference between the Dynamic and Fixed Status of Custom Objects (e.g., Arrows, Custom Text, Drawings) in Graphs?* (Chapter 9 - *STATISTICA Graphs: Examples and FAQ's*).

Insert OLE Object Button

Click the button (or select *OLE Object* from the *Insert* menu) to enter the object embedding or linking mode. When this button is clicked, the *Insert Object* dialog is displayed, allowing you to combine different objects (text, spreadsheets, metafiles, bitmaps, *STATISTICA* Graphs, etc.) in the graph window.

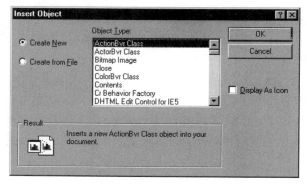

Object types. When the *Create New* option button is selected, you can choose the type of object to be created from a list of Windows applications that support OLE. After you select the object type, click the *OK* button and the selected application opens allowing you to create the new object.

When the *Create from File* option button is selected, you can specify the file to be added; a *Browse* button is available to help you in locating the file. Note that the properties of the object are customizable, including whether or not the object has dynamic or fixed coordinates. See Chapter 9 for more details on dynamic versus fixed placement of graph objects.

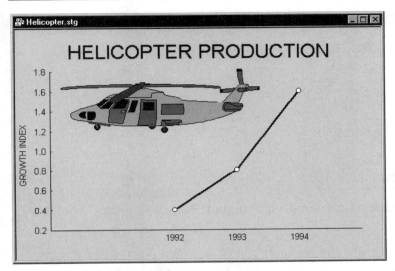

Linking vs. embedding (overview). *STATISTICA* supports OLE (Object Linking and Embedding) as both a server and a client. That is, not only can *STATISTICA* Graphs be dynamically updated in other applications (server mode), but foreign (OLE-compliant) objects (e.g., graphs, worksheets) and *STATISTICA*'s own objects can be built into *STATISTICA* Graphs and later be dynamically updated.

You can incorporate external objects in *STATISTICA* Graphs not only by pasting them (see the 📋 button), but also by accessing the objects directly from files on your computer (e.g., by dragging them with the mouse directly from Windows *Explorer*, across application windows, and placing them onto *STATISTICA* Graphs). *STATISTICA* supports nested compound documents (up to fourth order), that is, *STATISTICA* documents with embedded documents can be embedded in other *STATISTICA* documents.

Note that each of the two methods of incorporating artwork from external sources into *STATISTICA* Graphs (linking and embedding) has its advantages and disadvantages.

Linked objects. Graphs with linked objects are redrawn somewhat slower because their redrawing can involve updating links to external files; however, updating such graphs can be automatic (the status of links can be adjusted in the **Links** dialog accessible from the graph **Edit** menu).

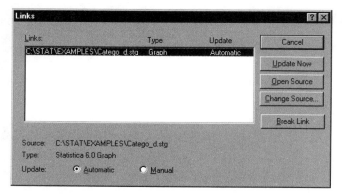

This allows you to easily create compound documents that include "always current" contents of other files.

Embedded objects. Graphs with embedded objects can be redrawn faster (than graphs with linked objects) because they include no links to external files that need to be updated. The server application is called when you double-click on the object; however, embedded objects can be updated only by manually replacing or modifying them.

All characteristics of the foreign objects (whether linked or embedded) and their relation to other components of the current graph can be adjusted later from the *Edit* menu (or by right-clicking on them, which will display a shortcut menu of available object customization choices). The only exception is the mode for incorporating an object (linking or embedding), which can be determined at the point of incorporating the file [later, only linked objects can be changed into embedded, using the *Break Link* option on the *Links* dialog, see *Links* (page 480)].

Line Pattern/Color Button

Click the button to select the line style (pattern, color combination, size, measurement units and mode) for the currently selected custom graphic object. In order to adjust the patterns for the new (subsequent) objects to be entered, use the style management facilities (accessible by clicking the *Styles* button on any graph customization dialog), or double-click on any existing object, change its properties and click the *Save as Default* button to make the style permanent (i.e., applicable to subsequent objects of the same type).

Special line customization features. In addition to the common line customization options, this dialog offers some unique features, such as support for two-color line patterns (e.g., select any of the noncontinuous line patterns, set the line to nontransparent, and choose a desired background color that will fill the breaks in the line pattern). The line thickness can be adjusted in device-independent point units (1 point = 1/72 of an inch) in arbitrary increments,

including fractions of a point (only if the output device supports such fine adjustments). Note that to achieve the desired effects when hard copies are produced, and to avoid the common "too thin" one-pixel lines printed by default on laser printers, a global minimum width for all printed lines can be adjusted in the **Print Options** dialog (see the ▤ button).

▧ Area Pattern/Color Button

Click the ▧ button to select the fill pattern style (pattern design, color combination, and mode) for the currently selected object. In order to adjust the patterns for the new (subsequent) objects to be entered, use the style management facilities (accessible by clicking the **Styles** button on any graph customization dialog), or double-click on any existing object, change its properties and click the **Save as Default** button to make the style permanent (i.e., applicable to subsequent objects of the same type).

Special area customization features. In addition to the common area customization options, this dialog offers some unique features, such as "transparent hatch" fill patterns or support for two-color fill patterns (e.g., set the fill pattern to nontransparent and not solid, then choose a desired background color that will fill the breaks in the pattern).

▣ Move to Front Button

Click the ▣ button (or select **Move to Front** from the **Format** menu) to bring the currently selected custom graphic object to the front. The **Move to Front** operation allows you to place a custom graphic object on top of all others by changing the object redrawing order such that the current object is redrawn last and thus will remain uncovered by any other objects (see also the next button, *Move to Back* ▣).

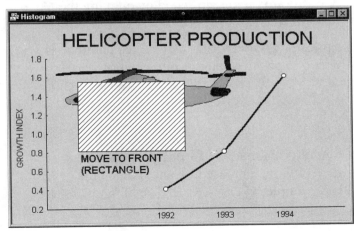

For example, if you intended to place a text on top of some artwork that was drawn (or embedded) later, you would select the text and click the ⬜ *Move to Front* button (alternatively, you could select the background object and click the 🔳 *Move to Back* button.

If a series of objects are present, you can use the *Move Forward* and *Move Back* commands available on the *Format* menu to place an object one position later or one position earlier (respectively) in the drawing queue.

🔳 Move to Back Button

Sends the currently selected custom graphic object to the back. In other words, it will be redrawn first, so that it can be covered by other custom graphic objects that are redrawn later.

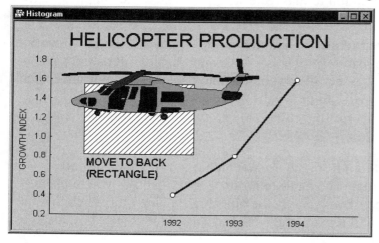

This *Move to Back* operation is similar (but opposite) to the one initiated by clicking the *Move to Front* ⬜ button.

If a series of objects are present, you can use the *Move Forward* and *Move Back* commands available on the *Format* menu to place an object one position later or one position earlier (respectively) in the drawing queue.

🔠 Increase Font Button

Click the 🔠 button to increase the selected font or point markers by one point. If nothing is selected in the graph, all fonts and point markers will be increased by one point. This general function of the *Increase Font* tool is useful, for example, in cases when a small graph is to be

produced (e.g., for publication) and when the size of all fonts relative to the body of the graph needs to be different (typically much larger) than in graphs of standard size.

Decrease Font Button

Click the button to decrease the selected font or point markers by one point. As with the *Increase Font* button (above), if nothing is selected in the graph, all fonts and point markers will be increased by one point.

Graphs Toolbar

This toolbar is only available if you choose to view it by selecting **Graphs** from the **View - Toolbars** menu or from the **Toolbars** tab of the **Customize** dialog. It allows quick access to the various types of **Graphs** menu graphs available in *STATISTICA*. Note that clicking any of these buttons automatically launches the appropriate graph specification dialog. The **Graphs** menu is available from the menu bar and from the *STATISTICA* Start button menu. For information on customizing toolbars, see the *Toolbar Overview* in Chapter 1.

Resume Button

Click the button (or press CTRL+R) to restore the current graph specification dialog and resume the analysis.

2D Graph Buttons

Click any of the 2D graph buttons to display the graph specification dialog associated with that button. For example, click the button to display the **2D Histogram** dialog. The table below displays the buttons and their respective graph types. Note that each of these graphs is discussed in the *Graphs Menu - 2D Graphs* topic (page 496), for overviews of these conceptual types of graphs, see Chapter 8 - *STATISTICA Graphs: Conceptual Overviews* and the *Electronic Manual*.

2D Histograms	**Normal Probability Plots**
2D Scatterplots	**Quantile-Quantile Plots**
Mean with Error Plots	**Probability-Probability Plots**
2D Box Plots	**2D Bar/Column Plots**

▪ **2D Range Plots**
▪ **Scatter Icon Plots**
▪ **Scatter Image Plots**
▪ **2D Scatterplots with Histograms**
▪ **2D Scatterplots with Box Plots**

▪ **2D Line Plots**
▪ **2D Line Plots – Case Profiles**
▪ **Sequential/Stacked Plots**
▪ **Pie Charts**
▪ **Missing and 'Out of Range' Data Plots**

▪ ▪ ▪ ▪ 3D Sequential Graph Buttons

Click any of the 3D sequential graphs buttons to display the graph specification dialog associated with that button. For example, click the ▪ button to display the **3D Raw Data Plots** dialog. The table below displays the buttons and their respective graph types. Note that each of these graphs is discussed in the *Graphs Menu - 3D Sequential Graphs* topic (page 508), for overviews of these conceptual types of graphs, see Chapter 8 - *STATISTICA Graphs: Conceptual Overviews* and the *Electronic Manual*.

▪ **3D Raw Data Plots**
▪ **3D Bivariate Histograms**

▪ **3D Range Plots**
▪ **3D Box Plots**

▪ ▪ ▪ ▪ ▪ 3D XYZ Graph Buttons

Click any of the 3D XYZ graphs buttons to display the graph specification dialog associated with that button. For example, click the ▪ button to display the **3D Scatterplots** dialog. The table below displays the buttons and their respective graph types. Note that each of these graphs is discussed in the *Graphs Menu - 3D XYZ Graphs* topic (page 510), for overviews of these conceptual types of graphs, see Chapter 8 - *STATISTICA Graphs: Conceptual Overviews* and the *Electronic Manual*.

▪ **3D Scatterplots**
▪ **3D Scatter Image Plots**
▪ **3D Surface Plots**

▪ **3D Contour Plots**
▪ **Ternary Graphs**

▪ Matrix Plots Button

Click the ▪ button to display the **Matrix Plots** dialog. For more details on matrix plots, refer to the *Graphs Menu - Matrix Plots* topic (page 515), for overviews of these conceptual types of graphs, see Chapter 8 - *STATISTICA Graphs: Conceptual Overviews* and the *Electronic Manual*.

⚫⚫ Icon Plots Button

Click the ⚫⚫ button to display the **Icon Plots** dialog. For more details on icon plots, refer to the *Graphs Menu - Icon Plots* topic (page 516), for overviews of these conceptual types of graphs, see Chapter 8 - *STATISTICA Graphs: Conceptual Overviews* and the *Electronic Manual*.

▦ ▦ ▦ ▦ ▦ ▦ ▦ ▦ ▦ ▦ ▦ ▦
Categorized Graphs Buttons

Click any of the categorized graphs buttons to display the graph specification dialog associated with that button. For example, click the ▦ button to display the **2D Categorized Histograms** dialog. The table below displays the buttons and their respective graph types. Note that each of these graphs is discussed in the *Graphs Menu - Categorized Graphs* topic (page 517), for overviews of these conceptual types of graphs, see Chapter 8 - *STATISTICA Graphs: Conceptual Overviews* and the *Electronic Manual*.

▦ **2D Categorized Histograms**

▦ **2D Categorized Scatterplots**

▦ **Categorized Means with Error Plots**

▦ **Categorized Box Plots**

▦ **Categorized Normal Probability Plots**

▦ **Categorized Quantile-Quantile Plots**

▦ **Probability-Probability Plots**

▦ **Categorized Line Plots**

▦ **Categorized Pie Charts**

▦ **Categorized Missing & 'Out of Range' Data Plots**

▦ **3D Categorized Plots**

▦ **Categorized Ternary Graphs**

GRAPH DATA EDITOR

Overview

The **Graph Data Editor** is a special version of the spreadsheet that allows you to review directly the data that are plotted in the graph. This tool is useful for a variety of analytic applications, such as brushing or other forms of identifying specific data points. It also offers (sometimes) the only way to access data from those graphs that do not plot raw values but values that have been derived, transformed, or result from specific analytic calculations, as well as values of the fitted functions. It can also be used to add additional plots of a compatible type to an existing graph. This editor is available from the graph **View** menu, the graph **Format** menu, and the general graph shortcut menu (accessible by right-clicking on the background of any graph).

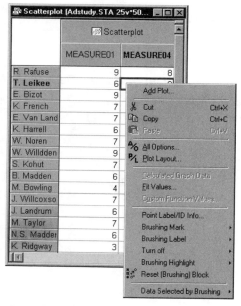

Columns in the Graph Data Editor. Each *STATISTICA* Graph carries with it a copy of all relevant data from the input dataset or spreadsheet upon which the graph is based. The type and amount of data associated with a particular graph is a function of the type of graph as well as the specific options chosen in making the graph. For example, a simple histogram carries with it only a copy of the variable upon which the histogram is based. A simple two-variable scatterplot

(shown in the illustration above) contains copies of both variables involved. Graphs with multiple plots contain copies of the relevant variables for each plot on the graph. As illustrated below, the **Graph Data Editor** for the bar/column plot contains three plots, with each plot having two variables: the numeric value for the x-axis (representing the locations of the columns on the x-axis) and the actual values for the variable (**Measure01**, **Measure02**, **Measure03**, respectively).

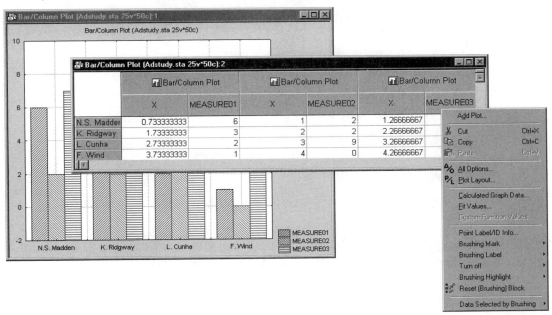

Supplementary data values. Additional variables used in categorization of plots are also stored with the graph when it is created. Also, when graphs are made using the **Multiple Subsets** options for data division (see Chapter 7), a copy of all variables in the dataset will be stored with the graph to allow for subsequent interactive modification of the multiple subset conditions - which may potentially involve any variables in the data set. (Note that this example illustrates well the function of the **Graph Data Editor** - namely, it contains all data that allow you to change the layout and analytic characteristics of the graph even if the original data set is not available, e.g., when the graph file is sent to a different computer without the original data set.)

Data in the Graph Data Editor. The data displayed in the **Graph Data Editor** are the raw data values for each plot. Data are organized in the editor under headers identifying the specific plot type to which the data apply. For plot types (such as histograms) in which the raw data are not directly plotted on the graph, select **Calculated Graph Data** from the **Graph Data Editor** shortcut menu to display the calculated graph data (for example, the heights of columns - i.e., counts - representing specific frequencies in a histogram, as shown below).

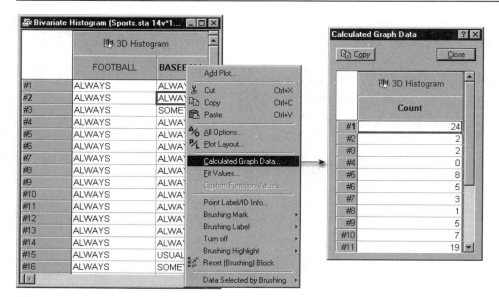

These data points cannot be edited in place; however, they can be copied and graphed separately or added to other graphs.

Fit values in the Graph Data Editor. If the graph contains a fit line or multiple fit lines, the coordinates of the segments of each line can be viewed by selecting *Fit Values* from the shortcut menu.

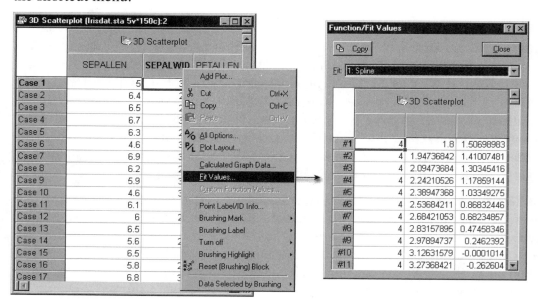

Use the *Fit* box at the top of the *Function/Fit Values* dialog to select the appropriate fit (if there is more than one fit in the graph), and again the data can be copied for separate analytic operations or addition to other graphs. The number of segments in a given fit line can be altered using the *Resolution* option found on the *Plot: Fitting* tab of the *All Options* dialog (see Chapter 7 - *STATISTICA Graphs: Creation and Customization*). Custom functions that have been applied to a graph can be viewed/copied in a similar fashion by selecting *Custom Function Values* from the shortcut menu.

Changing data values. Raw data can be changed, deleted, copied from, or pasted into the columns of the *Graph Data Editor*. Changes in data values are immediately reflected in the active graph. The *Graph Data Editor* can be used both independently and in conjunction with the graph brushing tools to explore relationships among graph data or to identify specific data points or subsets of data points. Convenient data exploration options are available from the shortcut menu (shown above) and from toolbar buttons when the *Graph Data Editor* is open. The options include marking, labeling, and turning off specific points on the plot.

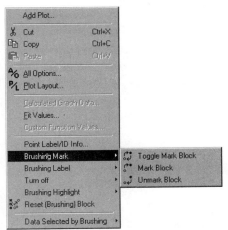

An additional option for highlighting specific points is available when the *Brushing* dialog is open. Each of these options can be toggled, turned on, or turned off for points selected in the *Graph Data Editor*.

Appearance of data values. Data for selected cases or blocks of cases when designated as marked (either by direct action from within the *Graph Data Editor* or by using the brushing tools), will be displayed as bold in the *Graph Data Editor* and the corresponding point markers on the graph will be changed. If cases are selected for labeling, the data in the *Graph Data Editor* will be displayed in italics, and the selected cases will be labeled on the graph. On the shortcut menu a *Point Label/ID Info* option allows editing the labels to be posted. Cases selected

to be turned off will be displayed in the *Graph Data Editor* with a darkened background and the points will be removed from the graph, causing any fit lines applied to the plot to be recalculated. If the *Brushing* dialog is open, cases selected and designated for highlighting will be highlighted by use of a different point marker on the graph and their data displayed in the *Graph Data Editor* in red. Cases in the *Graph Data Editor* that fall under more than one of the categories described above will have multiple overlaid formats applied, (e.g., a case that is both labeled and marked will be displayed in bold italics).

Additional toolbar buttons and shortcut options. A *Reset Brushing Block* feature is available on the toolbar and shortcut menu to simultaneously remove all marked, labeled, and turned-off designations from the selected points. Toolbar buttons and shortcut menu options are also provided to quickly access the *All Options* dialog and the *Plot Layout* dialog from the *Graph Data Editor*.

Adding new plots. Click the Add Plot... button on the toolbar (see page 576) to display the *New plot(s)* dialog and define a new plot for the graph.

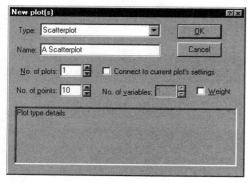

Once you have specified the what type of plot to add, the appropriate number of columns will be created in the *Graph Data Editor*, and you can enter the values to use in the plot.

Graph Data Editor Toolbar

The *Graph Data Editor* toolbar (displayed when the *Graph Data Editor* is active) offers a selection of brushing tools as well as access to various dialogs.

Most of the features available via the toolbar buttons listed below are also available from the **Graph Data Editor** shortcut menu. The **Graph Data Editor** itself can be access from the graph **View** menu, the graph **Format** menu, and the main graph shortcut menu. All toolbars can be customized to adjust the number and order of buttons in the toolbar.

Many of the toolbar buttons described below are used to assign brushing attributes. For more information on brushing in the **Graph Data Editor**, see *Brushing, Interactive Analysis* in Chapter 9 or the *Electronic Manual*.

Open the All Options Dialog

Click the button to display the **All Options** dialog. This dialog offers convenient access to most of the commonly used graph customization features organized into tabs. For more information on this dialog and the options available on each tab, see *The All Options Dialog* in Chapter 7. See also, *What Is the All Options Dialog?* in Chapter 9.

Open the Plot Layout Dialog

Click the button to open the **Plot Layout** dialog. This dialog contains most of the commonly used graph customization features for plots. For more information on this dialog and the options available on each of its tabs, see *Plot Options* in Chapter 7. See also, *What Is the Plot Layout Dialog?* in Chapter 9.

Calculated Graph Data... Display Calculated Graph Data Associated with this Plot

Click the Calculated Graph Data... button to display the **Calculated Graph Data** dialog. The data displayed in the **Graph Data Editor** are the raw data values for each plot. Data are organized in the editor under headers identifying the specific plot type to which the data apply. For plot types (such as histograms) in which the raw data are not directly plotted on the graph, clicking this button (or selecting **Calculated Graph Data** from the graph data editor shortcut menu) will display the calculated graph data (for example, the heights of columns - i.e., counts - representing specific frequencies in a histogram, as shown above).

These data points cannot be edited in place; however, they can be copied and graphed separately or added to other graphs.

Add Plot... **Add a New Plot**

Click the Add Plot... button to display the **New plot(s)** dialog and define a new plot for the graph.

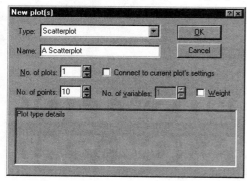

Once you have specified the what type of plot to add, the appropriate number of columns will be created in the **Graph Data Editor**, and you can enter the values to use in the plot.

Type. Use the **Type** box to select a graph type to use for the new plot. Note that the plot types available in this list are dependent on the types of plots, which already exist in the graph. For example, if you have created a 3D surface plot, you can only add additional 3D surface plots, but if you have added a 2D scatterplot, you can choose from a variety of two-dimensional plot types, including scatterplots, histograms, and bar plots.

Name. Enter a name for the new plot in the *Name* box. *STATISTICA* assigns the name to the plot number. This option is useful if you have several plots on one graph. Note that if you enter a name here and create more than one plot (by entering 2 or more in the *No. of plots* box), *STATISTICA* will use this name for each of the new plots and number them consecutively. For example, if you create three plots named *A Scatterplot*, the actual names will be *A Scatterplot (1)*, *A Scatterplot (2)*, and *A Scatterplot (3)*. You can also assign names to plots using the *Name* box on the *Plot: General* tab of the *All Options* (or *Plot Layout*) dialog.

No. of plots. Enter the number of plots to add in the *No. of plots* box.

No. of points. Enter the number of points (cases) to add for each plot in the *No. of points* box.

Connect to current plot's settings. Select the *Connect to current plot's settings* check box if you want to use the same scale values and labeling options as the current plot.

No. of variables. Enter the number of variables to use in the *No. of variables* box.

Weight. Select the *Weight* check box if you want to include a weight variable for the newly created plot.

▣ Toggle Mark Points (Brushing)

Click the ▣ button to toggle the marking of points in the graph data editor. When this button is clicked the selected points will be marked (bolded) if they were not marked and will be represented by a different point marker in the graph. They will be unmarked if they have been previously marked using the brushing tools, or the options in the *Graph Data Editor*.

▣ Mark Data Points (Brushing)

Click the ▣ button to mark the currently selected points. Marked data points are bolded in the *Graph Data Editor* and a different point marker is used in the graph.

▣ Unmark Data Points (Brushing)

Click the ▣ button to unmark the currently selected points in the *Graph Data Editor*. This means the points will no longer be bolded in the graph data editor, and when the graph is redrawn, these unmarked points will be displayed using the original (or default) point markers.

Toggle Label Data Points (Brushing)

Click the button to toggle the labeling of data points in the *Graph Data Editor*. When this button is clicked, the selected points will be labeled in the graph if they were not labeled. They will be unlabeled if they have been previously labeled, using the brushing tools or the options in the graph data editor. If cases are selected for labeling, the data in the *Graph Data Editor* will be displayed in italics. You can select *Point Label/ID Info* from the shortcut menu to edit the labels to be posted.

Label Data Points (Brushing)

Click the button to italicize the selected points in the *Graph Data Editor*, and identify the points with the respective data point label (e.g., the case name) in the graph. Note that the selected points will be labeled with the values specified on the *Plot: Point Labels* tab of the *All Options* dialog. When points are labeled via the brushing tool (or this button), they are the only points labeled in the plot. Data point labels displayed by brushing can only be turned off via the brushing tool.

Unlabel Data Points (Brushing)

Click the button to unlabel the currently selected points in the graph data editor. This means the points will no longer be italicized in the *Graph Data Editor*, and when the graph is redrawn, these points will not be labeled.

Toggle Data Points (Temporarily) On/Off (Brushing)

Click the button to toggle the display of the selected data points. When this button is clicked the selected points will be turned off (removed) in the graph if they were turned on. They will be turned on if they have been previously turned off, using the brushing tools or the options in the *Graph Data Editor*. Cases that have been turned off are shown on a darkened background in the *Graph Data Editor*.

Turn Off (Temporarily) Data Points (Brushing)

Click the button to display the selected cases with a darkened background in the graph data editor and temporarily remove them from the display and fitted function calculations in the graph.

Turn Back On Data Points (Brushing)

Click the button to turn back on the selected data points, which have been previously turned off using the brushing tools or the *Graph Data Editor* toolbar buttons. When points are turned back on, they are once again shown in the display and used for all fitted function calculations in the graph.

Toggle Highlight Data Points (Brushing)

Click the button to toggle the highlighted data points. When this button is clicked the selected points will be highlighted in the graph (when the *Brushing* dialog is opened) if they were not previously highlighted. They will not be highlighted if they are currently highlighted in the graph. Highlighted cases are shown in red in the *Graph Data Editor*. Note this option is only valid if the *Brushing* dialog is opened.

Highlight Data Points (Brushing)

Click the button to display the selected points in red in the *Graph Data Editor* and to highlight those points in the graph if the *Brushing* dialog is open.

Dehighlight Data Points (Brushing)

Click the button to change the color of the selected points from red to black in the *Graph Data Editor* and to unhighlight those points in the graph if the *Brushing* dialog is open.

Reset Data Points (Brushing)

Click the button remove all previously assigned attributes to the selected points.

STATISTICA GRAPHS: CREATION AND CUSTOMIZATION

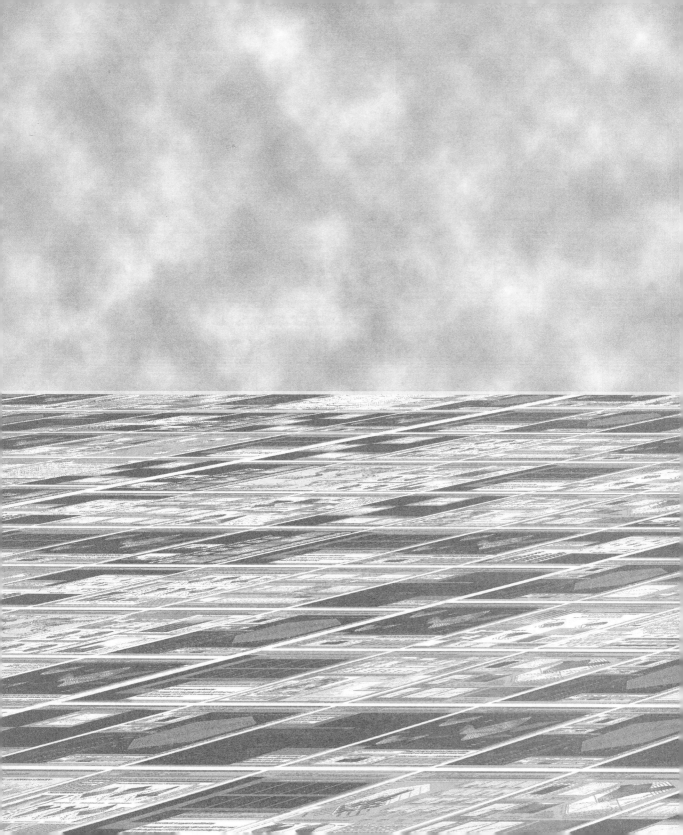

CHAPTER

STATISTICA GRAPHS:
CREATION AND
CUSTOMIZATION

GRAPH CREATION
(GRAPHS MENU GRAPHS)

STATISTICA offers facilities to convert any set of values into a wide variety of graphical representations via a large selection of graph specification and customization options. *Graphs* menu graphs offer the largest selection of options among *STATISTICA* Graphs (see *Graphs in STATISTICA* in Chapter 6 – *STATISTICA Graphs: General Features*) for defining the most appropriate data presentations for both data analysis and display. Thousands of different combinations of options can be applied to achieve the desired analytic or exploratory goals, as well as to create the attractive and concise graphics necessary for clear communication of results. Graph specification dialogs are available from the *Graphs* menu or the *STATISTICA* Start button ▦.

Types of Graphs Menu Graphs

Major *Graphs* menu graph types include *2D Graphs*, *3D Sequential Graphs*, *3D XYZ Graphs*, *Matrix Plots*, *Icon Plots*, *Categorized Graphs*, and *User-Defined Graphs*. In addition to the brief descriptions given here, Chapter 6 - *STATISTICA Graphs: General Features* and the *Electronic Manual* provide detailed discussions for each graph type listed and their numerous subtypes.

Additionally, Chapter 8 – *STATISTICA Graphs: Conceptual Overviews* contains conceptual overviews of many of the graph types discussed here.

2D Graphs

2D graphs include 2D displays of values from single or multiple variables. These plots include a very wide variety of both common and unique graph types. For more information on a specific type of 2D graph, see the overviews below as well as Chapter 8 – *STATISTICA Graphs: Conceptual Overviews* and *Graphs Menu* in Chapter 6 – *STATISTICA Graphs: General Features*.

2D Histograms. Two-dimensional histograms (the term was first used by Pearson, 1895) present a graphical representation of the frequency distribution of the selected variable(s) in which the columns are drawn over the class intervals and the heights of the columns are proportional to the class frequencies.

2D Scatterplots. Two-dimensional scatterplots help you visualize a relation (correlation) between two variables X and Y (e.g., weight and height). Individual data points are represented in two-dimensional space, where axes represent the variables (x on the horizontal axis and y on the vertical axis). The two coordinates (x and y) that determine the location of each point correspond to its specific values on the two variables.

2D Means with Error Plots. Plots of marginal means with error bars have long been used in science and industrial settings to visualize the differences in means with their respective confidence intervals. 2D means with error plots visualize means of the dependent variable(s) for subsets of cases specified by the user. Each mean marker is accompanied by a set of whiskers representing error bars that is the confidence intervals around the mean. These plots are typically used to compare marginal means across groups and visualize the reliability of these means.

2D Scatterplots with Box Plots. A scatterplot with box plots is a compound graph containing a regular scatterplot of the specified two (*X* and *Y*) variables and box plots for each variable displayed along the x- and y-axis of the scatterplot, respectively. Each component graph in this compound graph can be edited by clicking on it.

2D Range Plots. Range plots display ranges of values or error bars related to specific data points in the form of boxes or whiskers. Unlike the standard box plots, the ranges or error bars are not calculated from data but defined by the raw values in the selected variables. One range or error bar is plotted for each case. In the simplest instance, three variables need to be

selected, one representing the mid-points, one representing the upper limits and one representing the lower limits, as illustrated in the example below:

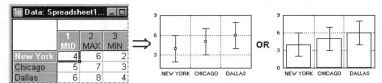

The range variables can be interpreted either as absolute values or values representing deviations from the midpoint. Simple or multiple variables can be represented in the graph. To produce simple graphs of sequences of values (without range or error bars) for either style or multiple variables use bar/column plots.

Scatter Icon Plots. A scatter icon plot is a scatterplot whose point markers are icons that represent for each specific data point (defined by the scatterplot X and Y coordinates) the variability within a set of other variables chosen for the icon. Additionally, a weight variable can be selected and used to scale the entire icon representing the point. For a detailed overview of scatter icon plots, see *Scatter Icon Plots Overview* in the *Electronic Manual*. To create simple icon plots, select **Icon Plots** from the **Graphs** menu.

Scatter Image Plots. Like scatter icon plots, a scatter image plot is a scatterplot whose point markers are images. However, these images are not icons defined by specific variables but user-selected graphics files (bitmap, metafile, PNG, and JPG formats are supported). Different image files can be attached to different data points (defined by the scatterplot X and Y coordinates), and a weight variable can be used to scale specific image markers accordingly. For a detailed overview of scatter image plots, see *Scatter Image Plots Overview* in the *Electronic Manual*. To create icon plots, select **Icon Plots** from the **Graphs** menu.

2D Scatterplots with Histograms. A scatterplot with histograms is a compound graph containing a regular scatterplot of the specified two (X and Y) variables and histograms for each variable displayed along the x- and y-axis of the scatterplot, respectively. Each component graph in this compound graph can be edited by clicking on it.

2D Box Plots. In box plots (this term was first used by Tukey, 1970), ranges or distribution characteristics of values of a selected variable (or variables) are plotted separately for groups of cases defined by values of a categorical (grouping) variable. The central tendency (e.g., median or mean), and range or variation statistics (e.g., quartiles, standard errors, or standard deviations) are computed for each group of cases and the selected values are presented

in the selected box plot style. Outlier data points can also be plotted (see *Outliers and Extremes* in the *Electronic Manual*).

Normal Probability Plots. This type of graph is used to evaluate the normality of the distribution of a variable, that is, whether and to what extent the distribution of the variable follows the normal distribution.

Quantile-Quantile Plots. You can visually check for the fit of a theoretical distribution to the observed data by examining the quantile-quantile (or Q-Q) plot. In this plot, the observed values of a variable are plotted against the theoretical quantiles. A good fit of the theoretical distribution to the observed values would be indicated by this plot if the plotted values fall onto a straight line.

Probability-Probability Plots. You can visually check for the fit of a theoretical distribution to the observed data by examining the probability-probability plot. In probability-probability plots (or P-P plots) the observed cumulative distribution function is plotted against the theoretical cumulative distribution function.

2D Bar/Column Plots. Bar/Column plots represent simple sequences of values (one case is represented by one bar/column). If more than one variable is selected, then either multivariate clusters of bars/columns are plotted (one cluster per case) or each variable is represented in a separate graph, depending on whether a multiple or regular graph is chosen. To produce column, bar, or box plots with error (or range) bars, use 2D range plots or 2D box plots.

2D Line Plots. In line plots, individual data points are connected by a line. Line plots provide a simple way to visually present a sequence of values. *XY* trace-type line plots can be used to display a trace (instead of a sequence). Line plots can also be used to plot continuous functions, theoretical distributions, etc.

2D Line Plots - Case Profiles. Unlike regular line plots that plot the values of one variable as one line (individual data points are connected by a line), case profile line plots plot the values of the selected variables in a case (row) as one line (i.e., one line plot will be generated for each of the selected cases). Case profile line plots provide a simple way to visually present the values in a case (e.g., test scores for several tests). Note that you can use *Statistics of Block Data/Block Columns* on the spreadsheet shortcut menu to perform some statistics on the cases [e.g., sum the values or take the mean of the values for each test (variable)] and then plot those statistics along with the cases in the block for comparison purposes.

Sequential/Stacked Plots. These types of graphs are used to present sequences of values of selected variables. The stacked feature of this type of graph is specifically designed to represent the broad category of datasets in which consecutive variables represent parts ("portions") of one whole.

Pie Charts. Pie charts (the term first used by Haskell, 1922) are useful for representing proportions. Individual data values of the *X* variable are represented as the "wedges" of the pie.

2D Missing and 'Out of Range' Data Plots. This graph allows you to examine the pattern or distribution of missing and/or user-specified "out of range" data points in the current dataset (or in a subset of variables and cases). This graph is useful in exploratory data analysis to determine the extent of missing (and/or "out of range") data and whether the patterns of those data occur more or less randomly.

Custom Function Plots. You can create to two types of custom plots: functions and parametric curves. Enter a custom function to produce a graph based not on values of variables in a datafile but on a user-defined formula (custom function), for example: *y = x2 + x/2*. To plot a parametric curve, enter a function for y and a function for x. For example, enter *y(t)=sin(2*pi*t)* and *x(t)=cos(2*pi*t)* to plot a circle. For more information on parametric curves, see Chapter 9 – *STATISTICA Graphs: Examples and FAQ's*.

3D Sequential Graphs

3D sequential graphs are a unique subset of 3D graphs. Like the more traditional 3D graphs (see *3D XYZ Graphs* below), they display information from multiple variables in graphs that portray three dimensions; however, they do not represent 3D (*XYZ*) coordinates of specific data points, but only show 3D representations of multiple sequences of values and/or their variability. They are especially designed for quick (and often more attractive than multiple 2D graphs) comparison of range and variability of multiple variables.

3D Raw Data Plots. Raw data plots allow you to plot sequences of raw data from selected variables in a 3-dimensional display. The selected variables are represented on the y-axis, the consecutive cases on the x-axis, and the values are plotted against the z-axis. The contour and surface types integrate all individual plots into one structure.

3D Bivariate Histograms. Three-dimensional histograms are used to visualize crosstabulations of values in two variables. They are considered to be a conjunction of two simple (i.e., univariate) histograms, combined such that the frequencies of co-occurrences of values on the two analyzed variables can be examined. In a most common format of this graph,

a 3D bar is drawn for each "cell" of the crosstabulation table and the height of the bar represents the frequency of values for the respective cell of the table.

3D Range Plots. 3D range plots display ranges of values or error bars related to specific data points. Unlike in 3D box plots (see below), the ranges or error bars in this plot are not calculated from data (as they are in box plots, see below) but defined by the raw values in the selected variables. One range or error bar is plotted for each case. The range variables can be interpreted either as absolute values or values representing deviations from the midpoint. Simple or multiple variables can be represented in the graph.

3D Box Plots. In 3D box plots, the ranges of values of a selected variable are plotted separately for groups of cases defined by values of a categorical (grouping) variable. The central tendency (e.g., median or mean), and range or variation statistics (e.g., quartiles, standard errors, or standard deviations) are computed for each group of cases and the selected values are presented in the selected style. 3D range plots differ from 3D box plots in that for range plots, the ranges are the values of the selected variables (e.g., one variable contains the minimum range values and another variable contains the maximum range values) while the ranges are calculated from variable values (e.g., standard deviations, standard errors, or min-max value) for box plots.

3D XYZ Graphs

3D XYZ Graphs contain options to produce a variety of types of 3D graphs where the location of data points is determined by their three (and in the case of ternary plots, four) coordinates.

3D Scatterplots. 3D scatterplots visualize a relationship between three or more variables, representing the X, Y, and one or more Z (vertical) coordinates of each point in 3-dimensional space.

Scatter Image Plots. A scatter image plot is a scatterplot whose point markers are images represented by user-selected graphics files (bitmap, metafile, PNG, and JPG formats are supported). Different image files can be attached to different data points, and a weight variable can be used to scale specific image markers accordingly. For a detailed overview of scatter image plots, see *Scatter Image Plots Overview* in the *Electronic Manual*.

3D Surface Plots. In this plot a surface is fitted (using a smoothing technique) to the data (representing sets of XYZ coordinates). In order to change the range of values and number of levels used to define the contour or surface, see *Specify Contour Levels* in the *Electronic Manual*.

3D Contour Plots. A contour plot is the projection of a 3-dimensional surface onto a 2-dimensional plane. See also, *Calculating Contour Levels and Distance* in the *Electronic Manual.* In order to change the range of values and number of levels used to define the contour or surface, see *Specify Contour Levels* in the *Electronic Manual.*

Ternary Plots. A ternary plot can be used to examine relations between four or more dimensions where three of those dimensions represent components of a mixture (i.e., the relations between them is constrained such that the values of the three variables add up to the same constant). One typical application of this graph is when the measured response(s) from an experiment depends on the relative proportions of three components (e.g., three different chemicals) which are varied in order to determine an optimal combination of those components (e.g., in mixture designs).

Custom Function Plots. This graph produces a 3-dimensional plot based not on values of variables in a datafile but on a user-defined formula (custom function), for example: $z = x^2 + y/2$.

Matrix Plots

Matrix plots produce a two dimensional matrix of scatterplots or *XY* line plots from either one or two variable lists. The single graph containing the matrix of scatterplots or *XY* line plots can also include a histogram or a box plot representing the variability of each variable.

Icon Plots

Icon plots are a powerful exploratory data analysis technique, representing cases or units of observation as multidimensional symbols or "icons." The general idea behind this graphic analytical method capitalizes on the human ability to "automatically" spot complex (sometimes interactive) relations between multiple variables if those relations are consistent across a set of instances represented by icons. Values of variables are assigned to specific features or dimensions of the icons. The assignment is such that the overall appearance of the icon changes as a function of the configuration of values.

Categorized Graphs

Many of the graph types discussed above under 2D Graphs, 3D Sequential Graphs, and 3D XYZ Graphs can be created in multi-graph displays defined by the categories of an additional one or two variables to subdivide the originally selected data into the desired subsets ("categories").

One component graph is produced for each level of the grouping variable (or a combination of levels of two grouping variables or user-defined subset of data) and all the component graphs are arranged in one display to allow for comparisons between the subsets of data (categories).

Categorized graphs can be specified either from the **Categorized** tab of the graph specification dialog (see *Categorized Tab*, page 592) for the respective graph types or by selecting **Categorized Graphs** from the **Graphs** menu. Categorized graphs are available for the following graph types:

- Histograms;
- Scatterplots;
- Means with Error Plots;
- Box plots;
- Normal probability, quantile-quantile, and probability-probability plots;
- Line plots;
- Pie charts;
- Missing data and range violation plots;
- 3D XYZ plots; and
- 3D Ternary plots.

For more information on these graph types, see *Graphs Menu - Categorized Graphs* in Chapter 6 – *STATISTICA Graphs: General Features*.

🔲 **User-Defined Graphs**

Use **User-Defined Graphs** to save all of the options specified during the initial graph specification in the form of a new graph "type." This graph can then be used with the same or a different input datafile to create similar analytic or presentation output with a minimum of required input. User-defined graphs can be named and saved by clicking the **Add as User-defined Graph to Menu** button on the **Options 2** tab of any graph specification dialog. Once saved, these "templates" can be accessed by name from the **Graphs - User-Defined Graphs** menu.

Select **Customize Menu** from the **Graphs - User-Defined Graphs** menu to display the **Customize User Graph Menu** dialog (at least one user-defined graph needs to be placed in that menu before this is available). You can use this dialog to review or modify the list of user-defined graphs. See the *Electronic Manual* for more details on using and managing user-defined graphs.

Graph Specification Dialogs

The definition dialog boxes for most *Graphs* menu graphs are very similar, consisting of five or six tabs containing all of the most relevant features and options for creating each graph as shown in this example.

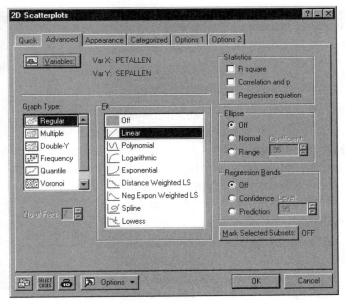

These tabs employ the same logic in variable and feature selection used to specify analytical procedures throughout *STATISTICA*.

Quick Tab

The *Quick* tab, available for nearly every graph type, contains a limited assortment of the options that are most commonly used for the respective graph type. This tab offers optimized ("quick") access to the most commonly used options. On this tab, the options available will always include variable selection appropriate for the graph type and sufficient options selected by default to produce a graph once variable selection has been completed. For a few graph types, such as 2D custom function graphs, no *Quick* tab is present.

Advanced Tab

The **Advanced** tab contains a wider set of options that are applicable for defining the detailed architecture of the graph type being created. Options to select graph subtypes are, in general, more numerous on the **Advanced** tab than on the **Quick** tab. This tab also contains variable selection options and, like the **Quick** tab, will have sufficient options selected by default to produce a graph.

STATISTICA will remember which of the two tabs (**Quick** or **Advanced**) you used last and will display the respective tab the next time you display this dialog.

Appearance Tab

Typically, the **Appearance** tab contains options for graph subtype selection and for style selection. You can select a saved **Graph Style** to use as a "template" in creating the graph, and you can select a predefined **Document Style** to govern the graph's general printed format (e.g., page orientation, paper size differences, slide proportions and formatting). The **Document Style** selected affects the appearance of the graph in workbooks and reports as well as in printed hardcopy (see *Document Styles*, page 617). See also the *Electronic Manual* for more information on document styles and graph styles.

Categorized Tab

Many **Graphs** menu graphs dialogs contain a **Categorized** tab. On this tab, defined categories of an additional one or two variables can be used to subdivide the originally selected variable data for presentation on categorized graph displays. See *Categorized Graphs* (page 589) for details on applicable graph types. Either one or two variables (or more than two if **Multiple subsets** categorization options are chosen) can be used to define the categories corresponding to the individual plots on the categorized graph to be produced. Note that in order to specify categories, you must first select the **On** check box in the **X Categories** (and **Y Categories**) group. The categorization methods and options presented on this tab are available in a wide variety of contexts throughout *STATISTICA* and are discussed in detail under *Methods of Categorization* (page 599).

Options 1 Tab

The **Options 1** tab provides access to a number of less frequently used options that govern graph appearance and behavior. This tab is the same no matter what type of **Graphs** menu graph is being created, and the selected options are only in effect until the graph specification dialog is closed. Many of the options on this tab and the **Options 2** tab (see below) can also be set globally on the **Graphs 2** tab of the **Options** dialog (see global graph options in Chapter 6 – *STATISTICA Graphs: General Features*). Note that global options will be over-ridden by options specified on any individual specification dialog as long as the definition dialog is in use.

Titles. Select the **Display default title** check box to display the default title for this graph. *STATISTICA* Graph default titles are always displayed at the top of the graph. In addition to the default title, you can include your own custom default title. To include a custom title, enter the title in the box provided, and select either **Show on top**, **Show on right**, or **Show on left**. Now, when a graph is created the custom title is shown as well as the default title. Note that if you prefer to show only the custom title, clear the **Display default title** check box. If you clear the **Display default title** check box and do not enter a custom title, created graphs will not have titles (other than axis titles).

Footnotes. To display a custom footnote, select *Display footnote* and enter the text for the footnote in the box provided.

Case labels. Use the *Case labels* box in the *Display Options* group box to specify what values are used to label individual points on the graph. By default, individual points are not labeled (i.e., *Off* is selected); however, you can label points with *Case names* or the values of another *Variable*. If *Variable* is selected, you can select the variable for labeling by clicking the *Variable* button in the *Display Options* box.

Display text labels (or dates) as axis values. Select the *Display text labels (or dates) as axis values* check box in the *Display Options* group box to use text (or dates) category labels on the graph axes. If this check box is cleared, then the equivalent numeric values (or Julian date values) are used as category labels.

Auto Update. Use the options in the *Auto Update* group box to control the links between graphs and their input datafiles. A link established between an input datafile and a graph will result in the graph being updated when a value within one of the graphed variables changes in value. The data links established on this panel will remain in effect only as long as the specification dialog is displayed and will be effective only if the graphs are produced to separate windows (see Chapter 2 - *STATISTICA Output Management* for more details on output options).

When selected, the *Update existing graph* check box can be used to change an existing graph (previously produced to an individual window) after one or more data values have been changed in the input datafile. This will update the previous graph rather than produce a new graph. When the graph addressed is contained in a workbook or a report, a new graph displaying the updates will be produced to an individual window.

If the *Update when input data change* check box is also selected (either when creating a graph to an individual window or updating it after creation), a link is established between the input datafile and the graph that will result in instant updating of the graph whenever the data in the datafile change.

Scaling. Use the options in the *Scaling* group box to specify a scale type for the x-, y-, z-, or v-axis. Select the appropriate axis in the *Axis* box, then select a *Linear*, *Logarithmic*, *Logit*, *Probability*, or *Weibull* scale type in the *Type* box. For examples of each scale type, see the *Electronic Manual*.

Automatically close this dialog after graph is created. Select the *Automatically close this dialog after Graph is created* check box to close the definition dialog

after the graph is produced. If this check box is cleared, the dialog is minimized as a button on the **Analysis** bar.

It is recommended to select this option (and close the dialog after the graph is produced) when you seldom continue to produce multiple graphs with different options specified in these dialogs, but instead use these dialogs typically to produce one graph (or an automatic sequence of graphs) and later customize them using tools available directly from the graph.

Options 2 Tab

Select the **Options 2** tab on the graph specification dialog to display an additional set of options that are always available to **Graphs** menu graphs. Note, however, that some of these options are not applicable to all graph types. For example, 3D plots cannot use polar coordinates (see below).

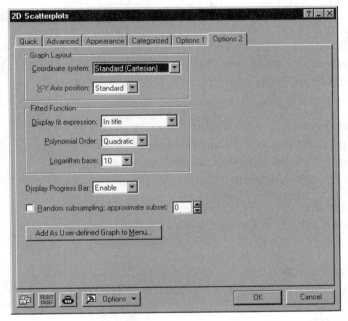

Coordinate system. Use the **Coordinate system** box in the **Graph Layout** group box to select a coordinate system to use with 2D graphs. Select **Standard (Cartesian)** to use standard X vs. Y (Cartesian) coordinates. Select **Polar** to convert the X-Y data to polar coordinates based on r and θ (theta). In the polar coordinate system, r represents the distance from the polar axis (center of the circle) along a straight line and θ represents the angle of the straight line measured in radians.

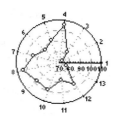

Polar plots are automatically scaled from 0 to 2π (pi) radians and are circular in appearance, with the X variable plotted around the circumference and the Y variable plotted radially. Polar coordinates are not an available option for 3D and some other graph types.

X-Y Axis position. Use the *X-Y Axis position* box in the *Graph Layout* group box to specify the placement of X and Y variables (selected on the *Quick* or *Advanced* tab) on the x- and y-axes. Select *Standard* to place the X variable on the x-axis and the Y variable on the y-axis. Select *Reversed* to reverse the axis positions of the X and Y variables.

Display fit expression. Use the *Display fit expression* box in the *Fitted Function* group box to display fitted function equations as fixed lines in the title of the graph (select *In title*) or as separate movable objects placed on the graph as custom text (select *As custom text*). Select *Off* if you do not want to display the fitted function. Note that this option can be set globally on the *Analyses/Graphs* tab of the *Options* dialog, accessible from the *Tools* menu.

Polynomial Order. Use the *Polynomial Order* box in the *Fitted Function* group box to select the order of polynomial fit to use when a polynomial fit has been specified for the graph. You can choose *Quadratic* (2nd order), *Cubic* (3rd order), *Quartic* (4th order), or *Quintic* (5th order).

Logarithm base. Use the *Logarithm base* box in the *Fitted Function* group box to apply the logarithm base when a logarithmic fit has been specified for the graph. You can choose *10* (for base 10) or *Euler* (for Euler's e, approximately 2.71828).

Display Progress Bar. Use the *Display Progress Bar* box to enable or disable the red progress bar on the status bar during graph creation. If *Enable* is selected, *STATISTICA* always displays the progress bar during creation. If *Disable* is selected, *STATISTICA* does not display the progress bar. If *Auto* is selected, *STATISTICA* decides whether or not to display the progress bar (based on its own estimate of the time it will take to draw the graph).

Random subsampling. Select the *Random subsampling* check box to plot a random subset of points from those available. If a large number of points are selected for plotting on a graph, drawing time can be effectively reduced and the relationship between the data points effectively preserved by plotting a random subset of points from those available. If the *Random*

subsampling check box is selected, enter the approximate number of points to display in the *approximate subset* box. Pleas note that this option (as well as the logic of its operation) is different from the random sampling options available from the *Subset/Random Sampling* dialog accessible from the *Data* menu (see Chapter 4 - *STATISTICA Spreadsheets*).

Add as User-Defined Graph to Menu. Click the *Add as User-Defined Graph to Menu* button to display the *New User-Defined Graph* dialog. This dialog provides options to name and save the specifications for this graph as a user-defined graph. This saved template can be used to quickly re-create the specified graph in the future using different input datafiles and/ or different input variables by selecting the name of the template from the *User-Defined Graphs* menu. See *User-Defined Graphs* (page 590) for further discussion. See also the *Electronic Manual*.

Distributions and Fits

A variety of predefined fits and functions can be fitted or used with graphs. For example, you can fit and superimpose a normal (or *Beta*, or *Gamma*, or many others) distribution over a 2D Histogram. In addition to the brief discussion give here, these distributions, fits, and functions are discussed in detail in the glossary of the *Electronic Manual*.

Fitted Functions for Histograms

When creating a 2D Histogram, you can fit an equation to the histogram by selecting one of twelve predefined functions. (The maximum likelihood method is used to estimate the distribution parameters.) The twelve functions are: *Normal*, *Beta*, *Exponential*, *Extreme*, *Gamma*, *Geometric*, *Laplace*, *Logistic*, *Lognormal*, *Poisson*, *Rayleigh*, and *Weibull*. You can refer to the glossary of the *Electronic Manual* for a definition of a particular distribution.

Note that for the *Beta* distribution, the data are required to be in the valid range of $0<x<1$. For the *Gamma*, *Lognormal*, *Weibull*, *Extreme*, and *Rayleigh* distributions, the valid range is $x>0$. If the data fall outside of the valid range, then *STATISTICA* will give an error message ("Fit not drawn because of invalid range of values") in the title of the graph.

2D Fit Lines

You can fit an equation to the points in the line plots by selecting one of the predefined functions described below:

Linear. Select *Linear* to fit a linear function of the form:

```
Y = a + bX
```

to the points in the 2D scatterplot. Confidence bands can be requested in the **2D Scatterplots - Advanced** tab.

Polynomial. Select *Polynomial* to fit to the data a polynomial function of the form:

$$y = b_0 + b_1x + b_2x^2 + b_3x^3 + \ldots + b_nx^n$$

where *n* is the order of the polynomial ($1<n<6$). The order of the polynomial function can be changed in the **Fitting** dialog.

Logarithmic. Select *Logarithmic* to fit to the data, a logarithmic function of the form:

```
y = q*[logn(x)]+b
```

where the logarithm base (*n*) is selected by the user (by default, base 10 is selected). The base of the log function can be changed globally on the **Graphs 2** tab of the **Options** dialog, accessible from the **Tools - Options** menu (see page 614). The log function can be changed for an individual graph on the **Options 2** tab during graph creation (see page 595).

Exponential function. Select *Exponential* to fit to the data an exponential function of the form:

```
y = b*exp(q*x)
```

where *b* and *q* are constants.

Least squares. Select *Least Squares* to fit a curve to the X-Y coordinate data according to the distance-weighted least squares smoothing procedure (the influence of individual points decreases with the horizontal distance from the respective points on the curve; see the *Electronic Manual*).

Negative Exponential. Select *Negative Exponential* to fit a curve to the X-Y coordinate data according to the negative exponentially weighted smoothing procedure (the influence of individual points decreases exponentially with the horizontal distance from the respective points on the curve). For more information, see *Spline Fitting* in the *Electronic Manual*.

Spline. Select *Spline* to fit a curve to the X-Y coordinate data using the bicubic spline smoothing procedure. For more information, see *Spline Fitting* in the *Electronic Manual*.

Lowess smoothing. Select *Lowess* to use the Lowess method of smoothing data (pairs of X-Y data) in which a local regression model is fit to each point and the points close to it. The method is also sometimes referred to as robust locally weighted regression. The smoothed data

usually provide a clearer picture of the overall shape of the relationship between the X and Y variables. For more information, see also Cleveland (1979, 1985). Note that the **Lowess** function in *STATISTICA* Visual Basic will also produce a table of residuals.

3D Graph Surfaces

You can fit one of several equations to the data or use one of the data smoothing procedures in the 3D surface plot by selecting one of the predefined functions described below:

Linear. Select *Linear* to fit a linear surface (e.g., $Z = a + bX + cY$) to the points in the 3D scatterplot.

Quadratic. Select *Quadratic* to fit a second-order polynomial surface to the points in the 3D scatterplot.

Least squares. Select *Least squares* to fit a surface to the XYZ coordinate data according to the distance-weighted least squares smoothing procedure (the influence of individual points decreases with the horizontal distance from the respective points on the surface); the stiffness of the fit can be controlled in the *Fitting* dialog. For more information, see *Distance-Weighted Least Squares* in the *Electronic Manual*.

Negative exponential. Select *Negative exponential* to fit a surface to the XYZ coordinate data according to the negative exponentially weighted smoothing procedure (the influence of individual points decreases exponentially with the horizontal distance from the respective points on the surface); the stiffness of the fit can be controlled on the *Plot: Fitting* tab of the *All Options* dialog, accessible from the *Format* menu. For more information, see *Negative Exponentially Weighted Fitting* in the *Electronic Manual*.

Spline. Select *Spline* to fit a surface to the XYZ coordinate data using the bicubic spline smoothing procedure. For more information, see *Spline Fitting* in the *Electronic Manual*.

Methods of Categorization

Categorization (or interval definition) is used in two classes of graphs in *STATISTICA*: categorized graphs (e.g., categorized scatterplots) and graphs that include grouping or categorized variables (e.g., 2D histograms, or 2D box plots). You can choose from several methods of categorization that are described below:

- Integer Mode
- Categories
- Boundaries
- Codes
- Multiple Subsets

Integer Mode

When you select this method, *STATISTICA* will truncate each encountered value of the selected variable to an integer value, and create one category (or graph in the case of categorized graphs) for each integer. When you select this option, the **Change Variable** button in this group box allows you to select a different variable whose values will be categorized in the graph. Note that if the number of integer categories exceeds 256, then *STATISTICA* will automatically change the method of categorization to **Categories** (see below) with 16 categories.

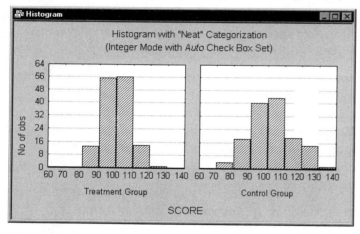

The Auto option. In the case of histograms (2D or 3D), the **Integer** categorization method offers an additional option designed to automatically identify the most informative method of categorization if the encountered distribution of values does not lend itself to a simple integer categorization (e.g., when it contains fractions or wide ranges of values). The specific manner in which this method will categorize the respective variable depends on whether or not the **Auto** check box is selected. In general, if you want to force single integer categories (as described under the simple **Integer Mode** above; e.g., if you want to produce a histogram for 100 different integer values), deselect the **Auto** check box. If you want "neat" categorization, then leave the **Auto** check box selected. When you select the **Auto** check box, *STATISTICA* will

perform integer categorization as described above, unless one of the following conditions is met: (1) if the range of values is greater than 20, or (2) if there are any non-integer values in the respective variable. In that case, *STATISTICA* will create a "neat" categorization for the histogram. In this context, "neat" categories means that the program will choose a step size with the only significant digit being 1, 2, or 5 (e.g., possible step sizes could be .1, .05, 50, 200, etc.). The minimum will be chosen (1) so that it includes the smallest value in the respective variable, and (2) so that it will produce "neat" scale values, that is, scale values with the last significant digit of either 0, 2, or 5 (e.g., 10.5, .002, .004, 1000, etc.).

Categories

Enter the desired number of categories in the box to the right of the **Categories** option button. *STATISTICA* divides the entire range of values of the selected variable (from minimum to maximum) into the requested number of equal length intervals. Cases with values of the selected variable that belong to each interval will compose one category.

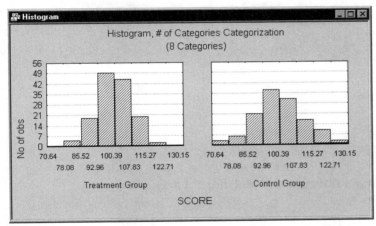

When you select the **Categories** option button, the **Change Variable** button in this group box allows you to select a different variable whose values will be categorized in the graph.

Boundaries

After you select this option button, click the **Specify Boundaries** button and enter a list of the desired boundaries for the selected variable in the **Specify Boundaries for** dialog.

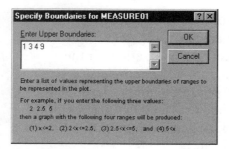

For example, if you enter **1 3 4 9**, then the following five ranges (i.e., categories) of values of the selected variable will be created: (1) x<=1, (2) 1<x<=3, (3) 3<x<=4, (4) 4<x<=9, and (5) x>9.

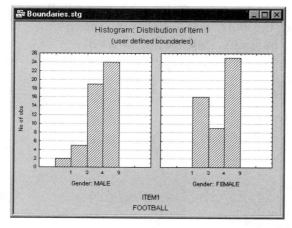

As you can see from this example, the intervals do not have to be of equal width, but the boundaries do need to be of consecutive numbers (e.g., **1 3 9**, not **1 3 1 9**).

Codes

Select the **Codes** option button if the selected variable contains codes from which you want to specify the categories.

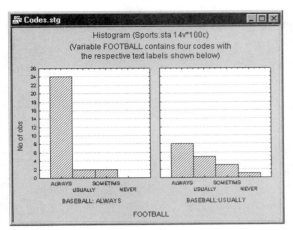

After you select this option, click the **Specify Codes** button and enter the desired codes in the **Category Codes** dialog.

Multiple Subsets

When you select this option button, click the **Specify Subsets** button to display the **Specify Multiple Subsets** dialog in which you can define the selection conditions that define specific categories.

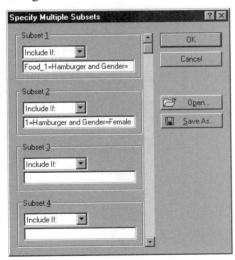

This method allows you to custom-define the categories and enables you to use more than one variable to define the category. See *Specify Multiple Subsets* in the *Electronic Manual* for more information on this method of categorization.

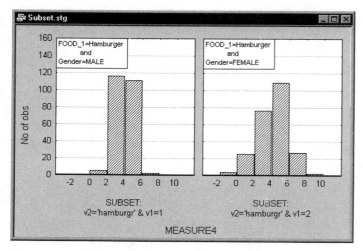

Note that when this categorization method is selected, the current selection of the respective categorical variable is ignored because several different variables can be used to define the categories.

COMPOUND/MULTIPLE GRAPHS IN *STATISTICA*

Compound graphs include more than one graph positioned in the same display. In *STATISTICA* compound graphs include:

- **Graphs** menu graphs specified using the **Categorized** tab from the graph specification dialogs or using **Categorized Graphs** from the **Graphs** menu.

- Certain types of **Graphs** menu graphs that display different graph types on a single graph, such as scatterplots with box plots, scatterplots with histograms, and matrix plots.

- Certain specialized graphs made from the dialogs of statistical procedures (e.g., X-bar and R Charts in the **Quality Control** module).

- User-defined compound graphs specified using either the **Multiple Graph Layouts Wizard** or **Template** (see the following section).

There are two classes of compound graphs in *STATISTICA*: multiple graphs (see the first two bullets, above) and graphs that comprise a collection of independent graphs (see the last two bullets). For multiple graphs, there is one **All Options** dialog for all plots of the same type (e.g., all the scatterplots in a matrix plot are controlled via the same **All Options** dialog, so if you increase the font size on the x-axis title in one plot, the font size on the x-axis will be increased in all plots). For compound graphs made via the **AutoLayout Wizard**, **Template**, or from the dialogs of statistical procedures, the individual graphs can be accessed and edited individually (see *Compound Graph Customization*, page 608).

Creating User-Defined Compound Graphs

In addition to the compound graphs created by *STATISTICA* (see the first three bullets above), *STATISTICA* offers several methods to create custom compound graphs. The simplest method is by copying and pasting graphs manually and arranging them interactively on-screen. *STATISTICA* Visual Basic can also be used to create highly customized compound graphs. In

addition, *STATISTICA* provides two wizard-based methods for creating user-defined compound or multiple graphs.

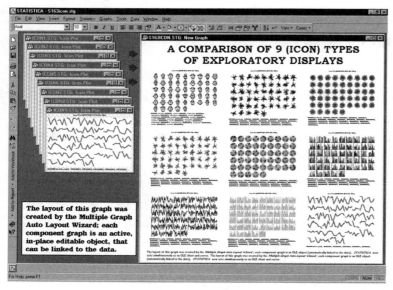

To access these methods select *Multiple Graph Layouts* from the *Graphs* menu or click the *STATISTICA* Start button ▣ on the *STATISTICA* *Analysis* bar and select *Graphs - Multiple Graph Layouts*.

AutoLayout Wizard

Selecting *Wizard* from the *Graphs - Multiple Graph Layouts* menu displays the *AutoLayout Wizard – Step 1* dialog. Use this dialog to gather the graphs for the multi-graph layout you are designing.

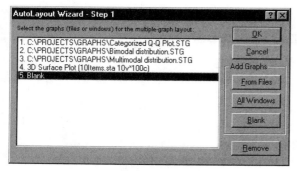

Click the **From Files** button to add saved graphs to the list. Clicking the **All Windows** button adds all currently displayed graphs in individual windows to the list (graphs stored in active workbooks or reports are not included). Graphs can be removed from the list by selecting them and clicking the **Remove** button. To add a blank graph to the list, click the **Blank** button. These blank graphs can serve as "placeholders" in setting up the layout. Additional *STATISTICA* Graphs or other graph objects (i.e., text) can later fill the openings reserved by the placeholder blank graphs. The number of graph items (i.e., graphs plus blanks) on the completed list determines the number of items used in creating the graph layout. The order in which the items are displayed can be adjusted on the following dialog.

Click **OK** in the **AutoLayout Wizard – Step 1** dialog to display the **AutoLayout Wizard – Step 2** dialog. Use this dialog to determine the actual layout and appearance of the compound graph.

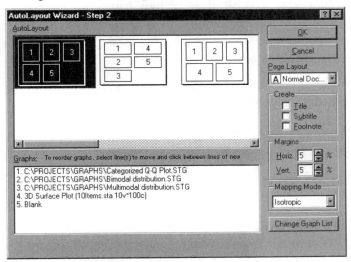

The **AutoLayout** box contains a variety of layout options. These options correspond to the number of graph items you specified in **Step 1** above (e.g., if you selected five graphs, then each of the layout arrangements contains placeholders for five graphs). The order of items in the **Graphs** box can be changed to correspond to the positions you want the graphs to occupy in the layout option selected.

Other options available on this dialog address changing the list of graph items (via returning to **Step 1** dialog), addition of titles to the compound graph, and adjusting margins and overall proportions of the graph. For specific details on these options, press the F1 key when the **AutoLayout Wizard - Step 2** dialog is displayed.

Template

Selecting **Template** from the **Graphs - Multiple Graph Layouts** menu displays the **Graph Template** dialog, in which you can specify the number of blank graphs to use in your template.

Unlike the **Wizard**, **Template** does not allow specifying the inclusion of saved graphs or graphs from individual windows. Rather, only blank graphs can be specified. The blank graphs can be used later as containers to hold other graph objects such as bitmaps, or they can serve as "placeholders" in setting up the layout. Additional *STATISTICA* Graphs or other graph objects (i.e., text) can later fill the openings reserved by the placeholder blank graphs. A maximum of 25 blank graphs can be specified in the **Number of blank graphs to arrange** box.

Click the **OK** button on the **Graph Template** dialog to display the **AutoLayout Wizard – Step 2** dialog (discussed above) and select the layout option you want.

Compound Graph Customization

Collections of independent graphs. Individual graphs within certain types of compound graphs (e.g., those made via the **AutoLayout Wizard**, **Multiple Graph Templates**, or specialized graphs from statistical analyses) can be selected and customized individually.

To select a single graph in a compound graph, click outside the plot area of the individual graph (i.e., on the outer background of the graph). The graph will be surrounded by an animated border (as shown in the illustration below), and you can select specific features of that graph (e.g., point markers, fit lines) by double-clicking on them.

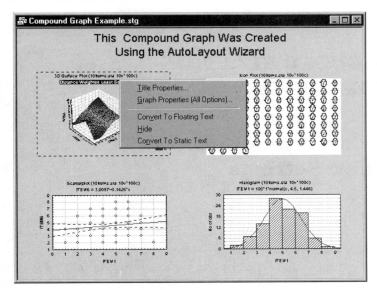

To select the entire plot for customization via the ***All Options*** dialog, double-click outside the plot area of the individual graph. See *Common Customization Features* (page 610) for more details.

Double-clicking on the common background of the compound graph (or right-clicking on the background and using the shortcut menu) provides access to customizing the compound graph's features independently from the contained graphs.

Multiple graphs. Graphs of the same type within a single compound graph created via *STATISTICA*'s ***Graphs*** menu graphs procedures (e.g., categorized graphs, matrix plots, scatterplots with histograms, or scatterplots with box plots) allow customization of features on all of the like graph types simultaneously. That is, on a graph that has four histograms and twelve scatterplots, the twelve scatterplots can be customized independently. Double-clicking on a specific feature (such as an axis) will display a dialog to quickly address altering that feature on all the scatterplots at once. A single-click selects the feature to be customized on all the like graphs (e.g., single-click the x-axis in one scatterplot, and the x-axis in all other scatterplots is selected as well). Double-clicking on the common outer background of the compound graph (or double-clicking on the inner background of any one of the graphs) displays the ***All Options*** dialog for customizing features of all the scatterplots simultaneously as well as for customizing the container graph.

COMMON CUSTOMIZATION FEATURES

Graph Customizations Default Settings

In order to custom-tailor graphics options in *STATISTICA* and limit the need for repeating the same graph customization, the default settings of practically all features of *STATISTICA* graphs can be adjusted by the users. This flexible system customizability is possible because all features of the appearance of *STATISTICA* graphs are driven by the same, comprehensive system of internal "graphics styles."

Graphics styles – a quick overview. Although most users may not need to use that advanced functionality and may not be interested in the internal working of the program, practically all features of *STATISTICA* graphs (thousands of combinations of settings) are internally represented by "styles" (see *Graphics Styles*, page 655). You can define these styles in a very simple way by assigning custom names to your own selections of features in the respective graph customization dialogs. Later you can quickly re-use these styles by selecting them by name, instead of repeating the tedious sequences of customizations. For example, a custom style of a specific arrow with a wide, filled head and a thick red tail can be called *A red fat arrow* and later selected with a single click of the mouse. Because the internal system of styles drives the management of practically all graph customizations, also all program defaults (i.e., the appearance of graphs when you do not make any customizations) are stored as styles and can be easily modified by the users.

Changing defaults in specific dialogs. Many graph customization dialogs (e.g., all dialogs used to customize lines, rectangles, circles, arrows, etc.) feature a *Save as Default* button (see an example below).

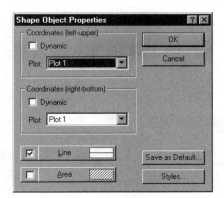

Clicking this button will (after you confirm your intentions) overwrite the previous default settings of that dialog with the settings you have just selected. Technically speaking, the program will internally modify the contents of the default style for that particular object.

Most graph customization dialogs offer access to the internal organization of styles (by pressing the **Styles** button).

The quickest method to modify the default settings of the most commonly used graph customization options. Many of the commonly used features of graphs have been collected in one, easy to find location, namely in the **Graphs 1** and **Graphs 2** tabs of the **Options** dialog (available from the **Tools** menu), where their default settings can be adjusted by the user.

Note that all changes that you will make in those dialogs will affect directly the respective internal styles (and advanced users may later fine tune these changes by accessing the structure of the corresponding styles directly, see *Graphics Styles*, page 655).

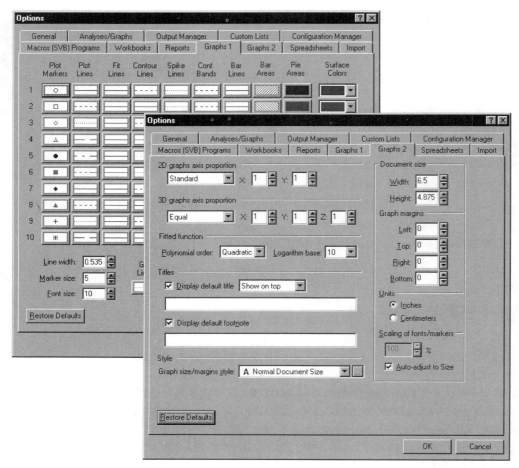

The options here will be reflected in the respective styles of graphs that govern the defaults for **Graphs of Input Data**, **Graphs of Block Data**, and **Graphs** menu graphs, but not for graphs created from statistical analyses. Upon exiting from this dialog, all your adjustments will be stored as part of the current configuration of *STATISTICA*. To save the settings on these tabs into a specific non-default set of configurations, use **Configuration Manager** tab of the **Options** dialog.

Graphs 1 Tab

Select the **Graphs 1** tab of the **Options** dialog to make changes to the default color and size of various graph objects (e.g., markers, lines, areas, surface colors).

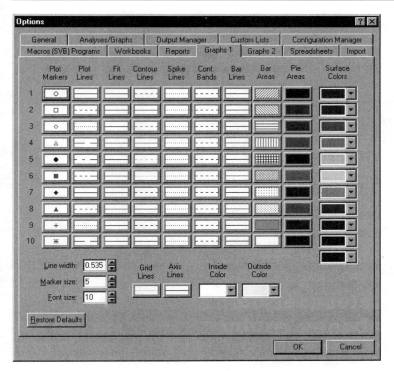

Plot Defaults. Graphs in *STATISTICA* consist of one or more plots (i.e., they may represent one or more series/sets of data). By default, a different set of colors and patterns is used for each consecutive plot. You can select for each plot (of the first 10 plots), the default point, line, area colors and patterns of specific plot components. Click on a (numbered) plot component to display the appropriate dialog (i.e., *Line Properties*, *Marker Properties*, *Area Properties*, *Color*) from which you can select the default color and pattern for the component.

Line width. In the *Line width* box, enter the default line width (in points, 1 point = 1/72 of an inch) to use when plotting lines on graphs. This setting will affect the width of all lines in the graph including the *Grid Lines* and *Axis Lines* (see below), which can then be modified to different settings without affecting the global line width.

Marker size. In the *Marker size* box, enter the default point marker size (in points, 1 point = 1/72 of an inch) to use when plotting point markers on graphs.

Font size. In the *Font size* box, enter the default font size (in points, 1 point = 1/72 of an inch) to use when placing text on graphs.

Grid Lines, Axis Lines. Click the *Grid Lines* button or the *Axis Lines* button to display the *Line Properties* dialog in which you can specify various features about the default grid lines

or axis lines, respectively. See the **Line Properties** dialog in the *Electronic Manual* for more information on using this dialog. See also *Line width* setting above.

Outside Color, Inside Color. Click either the **Outside Color** box or the **Inside Color** drop-down box to display a variety of background color options for the graph. In addition to a selection of predefined colors, you can choose a **Transparent background** or an **Automatic background** (the Windows System color). You can also choose from a list of background color styles. Select **Style Name** to view the list of styles. Select **More Colors** to display the **Color** dialog from which you can select a custom color.

Graphs 2 Tab

Select the **Graphs 2** tab of the **Options** dialog to set global defaults for the general appearance of the graph (e.g., fitted functions, axis proportions, document styles).

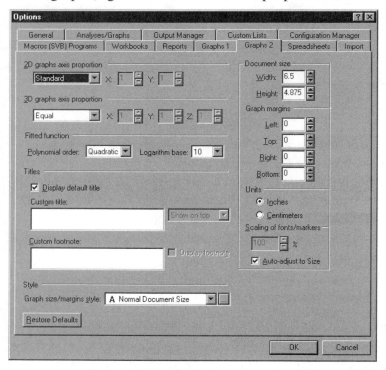

2D graphs axis proportion. Use the options in the **2D graphs axis proportion** section to set the default proportions of the axes in any 2D graph in order to adjust the frame of reference for the graph (axes) to its shape (i.e., in order to expose or explore a particular aspect of the data). In the drop-down box, select **Standard**, **Equal**, **Same Scale Units**, or **Custom**.

- **Standard.** Select **Standard** to scale the x- and y-axes independently of each other. This option generates standard rectangular proportions.

- **Equal.** Select **Equal** to scale the x- and y-axes to the same (1:1) length (i.e., the proportions of the *x* and *y* sides are equal and form a regular square).

- **Same scale units.** Select **Same scale units** to generate axis proportions based on the relative scale values on the x- and y-axes. If the x-axis values range from 0 to 3 and the y-axis values range from 0 to 1, the x-axis will be three times the length of the y-axis.

- **Custom defined.** Select **Custom defined** to set the axes to any proportion you want. The *X* and *Y* boxes to the right determine the relative length of the axes. By changing the respective values, you can "flatten" or "stretch" the 2D box in any direction.

3D graphs axis proportion. Use the options in the **3D graphs axis proportion** section to set the default proportions of the axes in any 3D graph in order to adjust the frame of reference for the graph (axes) to its shape (i.e., in order to expose or explore a particular aspect of the data). In the list box, select **Equal**, **Same Scale Units**, or **Custom**.

- **Equal.** Select **Equal** to scale the x-, y-, and z-axes to the same (1:1:1) length (i.e., the proportions of the *X*, *Y*, and *Z* sides are equal and form a regular cube).

- **Scale units.** Select **Scale units** to generate axis proportions based on the relative scale values on the x-, y-, and z-axes. If the x and y-axis values range from 0 to 3 and the z-axis values range from 0 to 1, the x- and y-axes will be three times the length of the z-axis.

- **Custom defined.** Select **Custom defined** to set the axes to any proportion you want. The *X*, *Y*, and *Z* boxes to the right determine the relative length of the axes. By changing the respective values, you can "flatten" or "stretch" the 3D box in any direction.

Fitted function. When you fit a function to the data in a *STATISTICA* Graph, several options are available to customize the fit. You can set global defaults for both the polynomial fit and the logarithm base in the **Fitted function** section. Use the **Polynomial order** drop-down box to set a default polynomial fit of either a second (**Quadratic**), third (**Cubic**), fourth (**Quartic**) or fifth (**Quintic**) order polynomial. Use the **Logarithm base** drop-down box to set the default logarithm base to either the natural log (**Euler**) or the base 10 (**10**) logarithmic function.

Titles. Use the options in the **Titles** section to toggle the display of default titles and footnotes as well as to specify custom default titles and footnotes.

- **Titles.** All *STATISTICA* Graphs have a default title that is shown at the top of the graph; select the **Display default title** check box to show the standard default title for each graph. In addition to the standard default title, you can include your own default title. To include

your own title, enter the title in the box provided, and select either **Show on top**, **Show on left**, or **Show on right** in the drop-down box. Now, when a graph is created the custom title is shown as well as the standard default title. Note that if you prefer to show only the custom title, clear the **Display default title** check box. If you clear the **Display default title** check box and do not enter a custom title, created graphs will not have titles (other than axis titles).

- **Footnotes.** Graphs do not have standard default footnotes; however, you can create and include a default footnote in a similar manner as the custom title described above. To display a default footnote, select the **Display default footnote** check box and enter the text for the footnote in the box provided. By default, this text will be displayed as a footnote to all graphs.

Graph size/margins style. Use the **Graph size/margins style** list box to select from a variety of styles for document sizes and margins. Note that when you select a style from this list, the appropriate boxes (e.g., **Width**, **Height**, **Left**, **Top**, **Right**, **Bottom**) are modified to fit the chosen style. To save the current set of defaults as a new style, click the ▦ button. See the *Electronic Manual* for more information on style management options, including a description of the various default document styles.

Document size. Use the **Width** and **Height** boxes in the **Document size** group box to specify the width and height (in either inches or centimeters, see **Units**) of the graph window (or graph in a workbook).

Graph margins. Use the options in the **Graph Margins** group to specify the margins (in either inches or centimeters, see **Units**). Note that the sum of horizontal or vertical margins cannot be larger than the width and height specified above.

Units. Use the options in the **Units** group to specify whether graph margins and document sizes are measured in inches (select the **Inches** option button) or centimeters (select the **Centimeters** option button).

Scaling of fonts/markers. Use the options in the **Scaling of fonts/markers** group to specify how fonts and point markers will be scaled as the document size and margins are modified. By default, the **Auto-adjust to Size** check box is selected permitting *STATISTICA* to determine appropriate scaling percentages. To specify your own percentages, clear the **Auto-adjust to Size** check box and enter the percentage in the **%** box.

Restore Defaults. Click the **Restore Defaults** button to return all the options to their original (default) values.

Document Styles

Document styles are a set of predefined graph size and margin specifications that allow you to print (or display) your graph with settings optimized for a particular layout (for more information on *Graphics Styles*, see page 655). For example, if you select the **Landscape Printer Page Size** style, the graph size and margins will be adjusted so that when the graph is printed in landscape mode, it will fill the entire page.

You can set a global document style for graphs on the **Graphs 2** tab of the **Options** dialog (accessible from the **Tools** menu). You can also select a new document style, on the **Graph Window** tab of the **All Options** dialog (for all graph types), or on the **Appearance** tab of any graph specification dialog (for **Graphs** menu graphs).

For more on document styles, including an overview of the various default styles, refer to the *Electronic Manual*.

Interactive Graph
Customization

The customization options in *STATISTICA* Graphs include hundreds of features and they are used to customize every detail of the display and associated data processing so that even the default appearance and behavior of *STATISTICA* Graphs will match your specific needs and/or will require very little intervention on your part (see page 610 for information on changing default settings of graph customization).

All graph customizations can be divided into two groups: before and after creation of the graph. Those custom characteristics that can be defined prior to graph creation include numerous defaults on the **Options** dialog and the use of graph styles. See *Graph Customization Default Settings* (page 610) for more details. See also, *Graphics Styles* (page 655). The customization features that are available after graph creation include clicking on graph components (described below) and the **All Options** dialog (see page 620).

Customizing the structural components of a graph. The components of *STATISTICA* Graphs are readily customizable after creation. The options controlling nearly all aspects of the appearance of *STATISTICA* Graphs are interactively accessible in a variety of ways. They are arranged in a hierarchical manner, so those used most often are accessible directly via shortcuts by clicking on the respective element of the graph. You can also double-

click on the graph feature or component you want to change, right-click the graph component and select **Properties** (or other appropriate options) from the shortcut menu, or access the **All Options** graph customization dialog. (See below for a discussion of the **All Options** dialog.) Only a few, very fundamental graph features, specifically those that control how data are extracted and processed from the data source (e.g., general graph type, variable selections, case selection, and weighting) cannot be altered through this on-screen customization after a graph has been produced.

The same plot customization options are available for all graph types (see Chapter 6 – *STATISTICA Graphs: General Features* for a discussion of these main graph types and their differences). In general, all graph customization options available in *STATISTICA* can be used to customize a graph, regardless of how that graph was requested or defined. For example, you can use the same set of options to customize a histogram made via **Graphs of Block Data - Histogram: Block Columns** and a histogram made via **Graphs - 2D Graphs - Histogram**. The exceptions to this general rule are certain highly specialized and unusual graph types produced as options from buttons on statistical analysis dialogs. The latter graphs are still customizable in most respects, but can contain unique features that cannot be altered.

Graph customization dialogs. *STATISTICA* provides direct access to common customization facilities for each graph component: Double-click on the component to display the most commonly used options for that component. For example, double-click on the x-axis of a scatterplot to display the **Scaling** dialog, in which you can adjust scaling options.

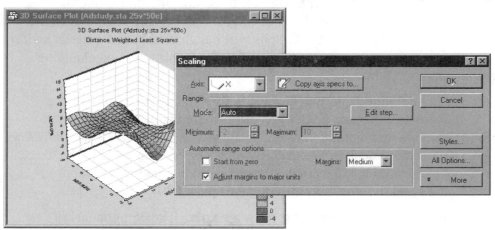

Click the **All Options** button on the **Scaling** dialog to display the **Axis Layout** dialog, which contains seven tabs offering a more diverse selection of axis-related options. Click the **All Options** button on the **Axis Layout** dialog to display the **All Options** dialog.

The *All Options* dialog contains 15 to 20 tabs that contain all of the relevant customizable features for the particular graph. The tabs are grouped in clusters containing logically related items. The options on the *All Options* dialog are an all-inclusive "superset" of options accessed by double-clicking on specific graph features. You can access the *All Options* dialog directly by selecting *All Options* from the graph *Format* menu, by double-clicking on the background (e.g., the area outside the axes) on a graph, or by selecting *Graph Properties (All Options)* from any graph shortcut menu.

Adding/editing custom graphic objects. You can access the tools to add and modify custom graphics objects to a current (active) graph (such as drawing, managing, and changing objects; pasting; embedding; linking; etc.) from the *Graph Tools* toolbar or the *Insert* menu. Objects that can be added include text, rectangles, rectangles with rounded corners, circles or ovals, arcs, polygons or freehand lines, arrows, and linked or embedded objects. To add a custom object (e.g., a drawing, arrow, or rectangle) to the graph, click a custom graphic object button on the *Graph Tools* toolbar and then click on the appropriate spot on the graph. Double-click on an existing object to open it for customization. For more details on specific tools, see the *Graph Tools Toolbar* in Chapter 6 – *STATISTICA Graphs: General Features*.

THE ALL OPTIONS DIALOG

The **All Options** dialog is accessible by double-clicking the outside background area of the graph, or by selecting **All Options** from the graph shortcut menu (accessible by right-clicking anywhere on the graph).

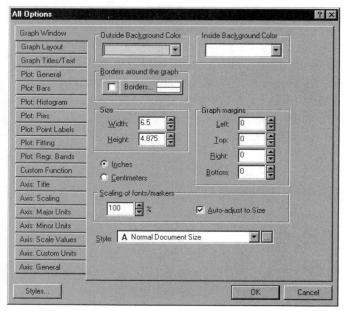

This dialog offers convenient access to most of the commonly used graph customization features organized into tabs.

General Customizing Options

This section describes the general customization features available on the tabs of the **All Options** dialog for most graphs. Each specific graph type can have one or more additional tabs that include the specific properties of that particular graph type or can contain additional options on the tabs discussed here. Also, for many graph types, special options allow graphs to

be converted from one specific type to another within a general graph type (e.g., conversion of histograms to pie charts). The axis and plot related options are discussed below.

Note that most tabs of the **All Options** dialog are also available as individual dialogs or tabs on other dialogs. For example, the options on the **Graph Titles/Text** tab are also available on the **Graph Titles/Text** dialog (accessible by double-clicking a graph title). For more information on the tabs of the **All Options** dialog, refer to the *Electronic Manual*. (Press F1 when the **All Options** dialog is displayed.)

Graph Window Tab

Select the **Graph Window** tab to specify background color, graph margins, graph borders, and other graph window properties.

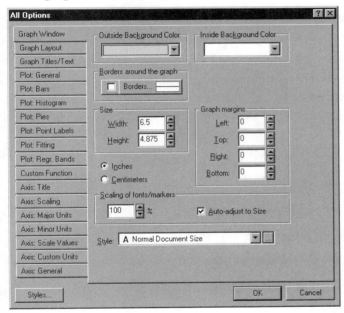

Outside Background Color, Inside Background Color. Click either the **Outside Background Color** box or the **Inside Background Color** box to display a variety of background color options for the graph. You can also access the appropriate background color dialog by double-clicking in either the outside background or inside background of the graph. In addition to a selection of predefined colors, you can choose a **Transparent background** or an **Automatic background** (the Windows System color). You can also choose from a list of background color

styles. Select **Style Name** to view the list of styles. Select **More Colors** to display the **Color** dialog from which you can select a custom color.

Borders around the graph. Select the **Borders** check box to include a border around the graph window. Click the **Borders** button to display the **Line Properties** dialog, in which you can choose the color, pattern, and thickness of that border. Note this border encircles the entire graph window and not the plot itself.

Size. Use the **Width** and **Height** boxes in the **Size** group box to specify the width and height (in either inches or centimeters, see **Units**) of the graph window (or graph in a workbook).

Graph margins. Use the options in the **Graph Margins** group to specify the margins (in either inches or centimeters, see **Units**). Note that the sum of horizontal or vertical margins cannot be larger than the width and height specified above.

Units. Use these options to specify whether graph margins and document sizes are measured in inches (select the **Inches** option button) or centimeters (select the **Centimeters** option button).

Scaling of fonts/markers. Use the options in the **Scaling of fonts/markers** group to specify how fonts and point markers will be scaled as the document size and margins are modified. By default, the **Auto-adjust to Size** check box is selected permitting *STATISTICA* to determine appropriate scaling percentages. To specify your own percentages, clear the **Auto-adjust to Size** check box and enter the percentage in the **%** box.

Style. Use the **Style** list box to select from a variety of styles for document sizes and margins. Note that when you select a style from this list, the appropriate boxes (e.g., **Width**, **Height**, **Left**, **Top**, **Right**, **Bottom**) are modified to fit the chosen style. To save the current set of defaults as a new style, click the ▦ button. See *Graphics Styles* on page 655 and the *STATISTICA Electronic Manual* for more information on style management options, including a description of the various default document styles.

Graph Layout Tab

Select the **Graph Layout** tab of the **All Options** dialog to access options that affect the graph layout. In general, the options on this tab are plot specific options used for re-creating the plot. As an example, for 2D scatterplots, options are available to reverse the x- and y-axes, use a polar coordinate system instead of the standard Cartesian coordinate system, or set the relative proportion of the x-axis and y-axis dimensions. The **Graph Layout** tab also contains a zoom feature for magnifying a specific area of the plot. Many of these options can also be set globally via the **Options** dialog (accessible from the **Tools** menu).

Note that the options available on the **Graph Layout** tab for each type of graph (e.g., 2D, 3D, Matrix Graphs, etc.) are different. For more specific details on the layout settings for a particular graph type, select **All Options** from the **Format** menu when the appropriate graph type is displayed. Then select the **Graph Layout** tab and press F1. For additional axis-related features, see also discussions on the **Axis: Title** tab, **Axis: Scaling** tab, **Axis: Major Units** tab, **Axis: Minor Units** tab, and **Axis: Scale Values** tab (below).

Graph Titles/Text Tab

Select the **Graph Titles/Text** tab from the **All Options** dialog to access options for formatting, deleting, and adding graph titles. You can add titles to the top, right, left, and bottom (footnote) of any graph. Note that the top title can also have a subtitle. You can also use the options described below to edit all titles, including those generated by *STATISTICA* during graph creation. Note that movable (floating) legends placed on graphs during graph creation are customizable using options on this tab; see *Graph Customization - Titles, Legends, Custom Text* in Chapter 9 – *STATISTICA Graphs: Examples and FAQ's*. To edit axis titles, use the **Axis Title** tab (page 635).

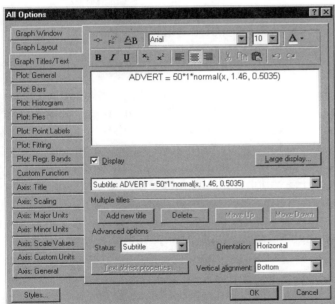

Mini-formatting toolbar. Use the buttons on the **Mini-formatting** toolbar near the top of the **Graph Titles/Text** tab to quickly specify font type and style, add superscripts and subscripts, change the font attributes to bold or italics, etc. A listing and a brief description of each of

these **Mini-formatting** toolbar buttons are available in the **Mini-formatting** toolbar topic in the *Electronic Manual.*

Title. By default, you can enter up to five titles in the active graph (although more can be added using the **Multiple titles** options below). Each of these titles can be formatted using the **Mini-formatting** toolbar. To edit/create a title, pick the title in the list box below, and enter the text into the large box. You can add legends and fits/functions to the titles using buttons in the **Mini-formatting** toolbar. To actually display the selected title in the graph, you must select the **Display** check box. Click the **Large display** button to display an expanded **Titles** dialog for this particular title.

The five available titles are **Title** (shown at the top of the graph), **Subtitle** (directly beneath the title), **Footnote**, **Left Title**, and **Right Title**. Note that unlike axis titles, which move with the axis, these titles are stationary.

Multiple titles. Use the buttons in the **Multiple titles** group box to manage existing titles or create new ones. Click the **Add New** button to add a new title of the type selected in the list box above. Click the **Delete** button to delete the currently selected title. Click **Move Up** or **Move Down** to reposition the selected title within titles of similar type (i.e., you can rearrange footnotes with these buttons, but you cannot promote a footnote to a subtitle other than by copying its contents to a title). Note that these buttons are disabled when they cannot be used. For example, you cannot move the top footnote any higher than it already is; hence, the **Move Up** button is disabled when **Footnote 1** is selected.

Advanced options. The text of floating legends (i.e., movable object legends) can also be edited using the options above. These legends (generated by *STATISTICA*) are used to identify multiple plots, multiple fits, surface or contour intervals on 3D plots, and statistics in text boxes. Convert movable legends to fixed titles by selecting the movable legend, then changing its status to either **Title**, **Subtitle**, **Left**, **Right**, or **Footnote** in the **Status** box in the **Advanced options** group box. The same conversion option is available on the **Insert** menu (see Chapter 6 – *STATISTICA Graphs: General Features*) through the **Plots Legend**, **Fits Legend**, **Surface Legend**, and **Statistics** options. Note that the color or pattern of points, lines, areas etc. appearing in the legend, must be edited and changed on the graph itself. Note that you can also change fixed titles to movable text by selecting the fixed title and changing its **Status** to **Floating**.

You can use other options in the **Advanced options** group to set the orientation and vertical alignment of the title. You can set properties for any text objects by clicking the **Text object properties** button, also in the **Advanced options** group.

Plot Options

As mentioned before, *STATISTICA* graphs consist of one or more plots. For example, a simple histogram or scatterplot consists of one plot, whereas a multiple histogram or a double-y scatterplot contains two or more plots. As with customization options for any specific graph object, the most direct way to access customization options for a specific plot object (e.g., fit lines, point markers) is to double-click on that item. This displays a dialog with the most relevant options for that graphic item (e.g., **Fitting** dialog, **General** dialog). Plot related options are also contained in the **Plot** dialog (accessible by clicking the **All Options** button on any plot specific option dialog). The tabs on the **Plot** dialog are also available on the **All Options** dialog.

When accessing the plot options as part of the **Plot** or **All Options** dialog, you must first select which plot you want to customize (assuming that more than one plot is included in the graph). Although all of the options are the same for each plot, you can make changes to each plot independently of the others. A **Plot** box is provided at the top of each tab; some tabs also provide another feature specific box. For example, when customizing fit options, you can choose to customize a particular fit in a particular plot, thus, you must specify the plot number and the fit number (see illustration below).

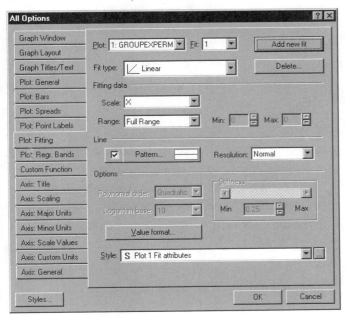

If you have several plots in one graph, you can assign names to each plot using the **Name** box on the **Plot: General** tab (see below).

Plot: General tab

Select the **Plot: General** tab of the **All Options** dialog to access the individual plots on a graph. Use the options here to toggle the display of individual plots and to customize point markers, lines, and area patterns that are integral parts of each plot. Note that if multiple graphs of a single type are displayed on the graph (as in the case of categorized scatterplots), the **Plot** box contains only one item. New plots can also be added to the graph from this tab.

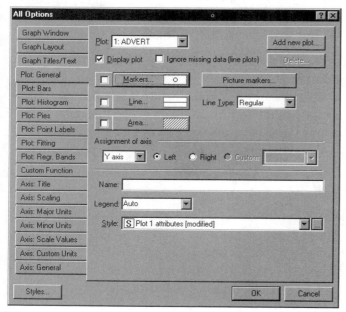

Add new plot, Delete. Click the **Add new plot** button to display the **New plot(s)** dialog and define a new plot for the graph. To delete a plot, select the plot you want to delete in the **Plot** box, and click the **Delete** button. Note if there is only one plot in the graph, you cannot delete it (the **Delete** button is disabled). However, you can choose not to display the plot by clearing the **Display plot** check box. To add or delete custom functions, use the **Custom Function** tab (described on page 633).

Display plot. Select the **Display plot** check box to display the currently selected plot in the graph. If this check box is cleared, the currently selected plot is not displayed in the graph.

Ignore missing data (line plots). Select the *Ignore missing data (line plots)* check box to ignore the missing data in the graph. This option, applicable only to line plots, allows you to create line plots when gaps in the data sequence will not break the line (the line will bridge all gaps).

Markers. Select the *Markers* check box to display point markers in the plot. Click the *Markers* button to display the *Marker Properties* dialog in which you can customize the appearance of the point markers. The current point marker specifications are shown in the view area of the *Markers* button.

Line. Select the *Line* check box to connect the data points in the plot with a line. If you choose this option, you will need to specify a *Line type*. Click the *Line* button to customize the appearance of the line in the *Line Properties* dialog. The current line specifications are shown in the view area on the *Line* button.

Area. Select the *Area* check box to display the area under the points in the plot. Click the *Area* button to customize the appearance of the area in the *Area Properties* dialog. The current area specifications are shown in the view area on the *Area* button. Note that the area patterns of specialized plots like histograms are accessed on other tabs (e.g., for histograms, see the *Plot: Bars* tab).

Name. Enter a name for the currently selected plot in the *Name* box. *STATISTICA* assigns the name to the plot number. This option is useful if you have several plots on one graph.

Legend. Select the display property of the legend in the *Legend* drop-down box. Select *Auto* to allow *STATISTICA* to control when the legend is displayed, select *Off* to turn off the display of the legend, and select *On* to turn on the display of the legend.

Styles. Instead of using the options on this tab (e.g., *Marker*, *Area*, *Line*, etc.), you can select the style you wish to use for this plot from the *Style* drop-down box.

Additional Options. Additional customization options may be displayed on the *Plot: General* tab as a function of the graph type. For example, 2D scatterplot graphs present options related to display of lines connecting point markers as either *Regular* or *Step type* lines, display of *Voronoi* lines, or options allowing plots to be assigned to different axes in the graph. 3D graphs may offer options on spike lines etc. For more details on these general plot options, press F1 on the *Plot: General* tab of the *All Options* dialog when the appropriate graph type is displayed.

Plot: Point Labels Tab

Select the **Plot: Point Labels** tab of the **All Options** dialog to plot labels (either the coordinates of the points or text labels) over individual data points in the graph.

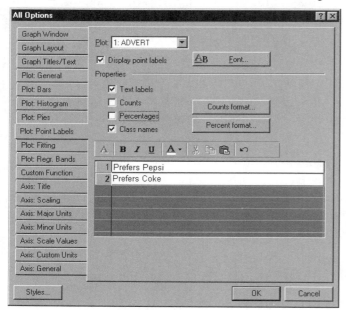

Display point labels. Select the **Display point labels** check box to use point labels (shown in the spreadsheet below) in the currently specified plot. Clear this check box if you do not wish to use point labels.

Font. Click the **Font** button to display the **Font** dialog from which you can choose a different font (e.g., size, type, etc.) for all of the point labels. For more information on the **Font** dialog, see Chapter 6 – *STATISTICA Graphs: General Features* or the *Electronic Manual*.

Properties. Select the **Text labels** check box in the **Properties** group to use strings of text for the point labels. By default, the strings of text used are either the numbered cases (e.g., **#1**, **#2**, . . .) or case names (e.g, **J. Baker**, **A. Smith**, …) in the plot. In the spreadsheet below, you can edit the text labels for the plot so that they are more meaningful. You can also show counts, percentages, or class names by selecting the **Counts**, **Percentages**, or **Class names** check boxes, respectively. Note that you can specify value (number) formats for the counts or percentages by clicking the **Count format** or **Percent format** button. For more information on these number formats see *Number Formats in STATISTICA* in the *Electronic Manual*.

Mini-formatting toolbar. Use the buttons on the *Mini-formatting* toolbar above the Point label spreadsheet to quickly specify font type and style, add superscripts and subscripts, change the font attributes to bold or italics, etc. A listing and a brief description of each of these *Mini-formatting* toolbar buttons are available in the *Mini-formatting* toolbar topic in the *Electronic Manual*.

Point label editor. The current point labels are shown in this box, you can edit individual point labels by typing directly into the spreadsheet and using the *Mini-formatting* toolbar (see above). Note that changes made here override any general formats applied using the *Font* dialog above.

Plot: Fitting Tab

Select the *Plot: Fitting* tab of the *All Options* dialog to add, delete, and/or customize any number of predefined fits to individual 2D and 3D plots selected from the *Plot* box at the top of the tab. Use the *Fit type* box to select from a wide variety of fits appropriate to the plot being customized. The range of *X* values (or *X* and *Y* values in the case of 3D plots) over which the fit is to be calculated can also be specified.

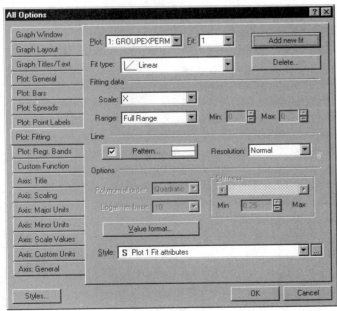

Add new fit, Delete. Click the *Add new fit* button to add a new fit of the type currently selected in the *Fit type* box to the selected plot. To delete an existing fit, select the plot that has

the fit in the **Plot** field and the fit you wish to delete in the **Fit** field, and then click the **Delete** button. Note that the options here are for deleting fits, not plots. To delete plots, use the **Plot: General** tab.

Fit type. Use the **Fit type** box to select an appropriate fit type for the plot you are editing. Depending on the type of plot you have, you can fit a variety of distributions or one of several equations to the data in the plot by selecting one of the predefined functions in this dialog. Fit options available for a 2D scatterplot include **Linear**, **Polynomial** (of specified order 2 through 5), **Logarithmic** (Euler and base 10), **Exponential**, **Distance Weighted Least Squares**, **Negative Exponential Weighted Least Squares**, **Lowess**, and **Spline**. Fits available for a 2D histogram include the above plus the following distribution fits: **Normal**, **Beta**, **Exponential**, **Extreme**, **Gamma**, **Geometric**, **Laplace**, **Logistic**, **Poisson**, **Rayleigh**, and **Weibull**. Surface fits available for 3D graphs include **Linear**, **Quadratic**, **Distance Weighted Least Squares**, **Negative Exponential Weighted Least Squares**, and **Spline**. For more information on distribution and function fitting options available, see *Distributions and Fits* (page 597).

Fitting data. In the **Scale** box of the **Fitting data** group select the scale (axis values) to use in fitting the data. Use the **Range** box to choose one of three set ranges over which to plot the function. Choose **Full range** to calculate the selected fit using the entire range of data. Select **Axis Range** to calculate the fit using the entire range of the axis (i.e., using all points shown on the axis). For example, if the plotted variable contains values from 0 to 9, but your axis only shows values from 3-8, you can select **Axis Range** to calculate a fit using only the values between 3 and 8. To use all the values (including those which are not shown on the axis), you would select **Full Range**. Select **Custom Range** to calculate the fit using the range of values specified in the **Min** and **Max** fields.

Additional options. Depending on the graph and fit types created, a variety of additional options is available to customize the fit. These options include line properties, polynomial order, logarithm base, and stiffness parameters. For more information on these options (and any other options available on the **Plot: Fitting** tab), refer to the appropriate **Plot: Fitting tab** topic in the *Electronic Manual* by pressing F1 when the **Plot: Fitting** tab is displayed in *STATISTICA*.

Plot: Ellipse Tab

Available only for 2D graphs that display bivariate data, the **Plot: Ellipse** tab provides addition or customization of any desired number of normal or range type data ellipses to each plot on the graph. These plots can serve as aids to visualize potential outliers, provide a prediction interval in which additional sample points might be expected to fall, or to display the relative

ranges of values for a variable pair. The options described here are for creating new ellipses and/or customizing the ellipses previously plotted in a graph including the type of ellipse and the data used in calculating the ellipse.

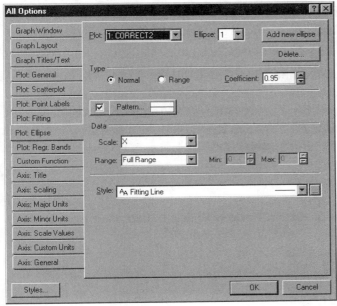

Add new ellipse, Delete. Click the *Add new ellipse* button to add a new ellipse to the selected plot. To delete an existing ellipse, select the plot that has the ellipse in the *Plot* field and the ellipse you wish to delete in the *Ellipse* field, and click the *Delete* button. Note that the options here are for deleting ellipses, not plots. To delete plots, use the *Plot: General* tab.

Normal. Select the *Normal* option button in the *Type* group to produce an ellipse based on the assumption that the two variables follow the bivariate normal distribution. The orientation of the ellipse is determined by the sign of the linear correlation between the two variables (the longer axis of the ellipse is superimposed on the regression line). The *Normal* ellipse (also referred to as a prediction interval ellipse; see the glossary in the *Electronic Manual*) shows the prediction interval for a single new observation, given the parameter estimates and the given *n*. Parameter estimates (means, standard deviations, covariance) for the bivariate distribution are computed from the data. The coefficient (or confidence level) for this ellipse is entered in the *Coefficient* box.

Note that if the number of observations in the scatterplot is small, then the prediction interval can be very large, exceeding the area shown in the graph for the default scaling of the axes.

Thus, in some cases (with small **n**) you may not see the **Normal** ellipse on the default graph. To fix this problem, change the scaling to show larger intervals for the two variables in the plot.

Range. Select the **Range** option button to produce a range ellipse. This option produces a fixed size ellipse such that the length of its horizontal and vertical projection onto the x- and y-axis (respectively) is equal to the mean ± (Range/2 * I) where the mean and range refer to the **X** or **Y** variable, and **I** is the current value in the **Coefficient** box.

Additional options. A variety of additional options are available to customize the appearance of the ellipse. These options include line properties (patterns) and the range of values to use in the ellipse. For more information on these options (and other options available on the **Plot: Ellipse** tab), refer to the appropriate **Plot: Ellipse Tab** topic in the *Electronic Manual* by pressing F1 when the **Plot: Ellipse** tab is displayed in *STATISTICA*.

Plot: Regr. Bands Tab

Select the **Plot: Regr. (Regression) Bands** tab of the **All Options** dialog to access options for plotting either confidence or prediction limits (or both) in the graph.

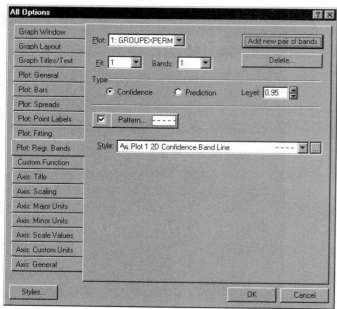

Bands. Any number of regression bands (e.g., 90%, 95%, 98%) can be added or edited on each fit on each plot. The **Plot, Fit,** and **Bands** fields work together. For example, to edit the third regression band on the second fit on plot 1, set **Plot** to **1, Fit** to **2,** and **Bands** to **3**.

Add new pair of bands, Delete. Click the *Add new pair of bands* button to add a new regression band to the selected fit and selected plot. To delete an existing band, select the appropriate fit and plot in the *Fit* and *Plot* boxes and the regression band in the *Bands* box, then click the *Delete* button. Note that the options here are for deleting regression bands, not plots or fits. To delete plots, use the *Plot: General* tab (page 626). To delete fits, use the *Plot: Fitting* tab (page 629).

Confidence. Select the *Confidence* option button in the *Type* group to create a confidence band around the regression line using the confidence level you specify in the *Level* box. This confidence level represents the probability that the "true" fitted line (in the population) falls between the bands. For a more precise definition of confidence levels, see *How Can I Interpret a 100(1-alpha)% Confidence Interval?* in Chapter 9 – *STATISTICA Graphs: Examples and FAQ's*).

Prediction. Select the *Prediction* option button in the *Type* group to create a prediction band (or prediction interval) around the individual values using the confidence level you specify in the *Level* box. This confidence level represents the probability that additional sample points will fall within the prediction interval. For a more precise definition of confidence levels, see *How Can I Interpret a 100(1-alpha)% Confidence Interval?* in Chapter 9 – *STATISTICA Graphs: Examples and FAQ's*).

Pattern. Select the *Pattern* check box if you wish to show the regression band on the graph. Click the *Patterns* button to display the *Line Properties* dialog and customize the appearance of the bands. The current specifications for the bands are shown in the view area on the *Pattern* button. See the *Electronic Manual* to learn more about the *Line Properties* dialog.

Custom Function Tab

Select the *Custom Function* tab of the *All Options* dialog to add any number of user-specified functions (e.g., $Y = X^2 - 2*X$ or $Z = 3*X^2 - 2*Y^2 + 48$) to the graph. Each function can be applied over a specified range of *X* values (or *X* and *Y* values in the case of 3D plots). The function equations are automatically added to the graph as subtitles. The options here can be used to modify or delete an existing custom function or add a new one to the current graph.

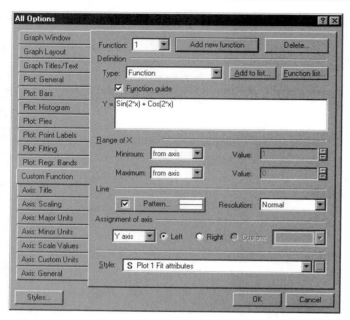

Function, Add new function, Delete. The *Function* box provides a list of custom functions in the current graph. To edit an existing custom function plot, select the function number and use the options described below. If no functions are available, or you want to create a new function, click the *Add new function* button. To delete an existing function, specify the function number in the *Function* box and click *Delete*.

Type. For 2D graphs, use the *Type* box to select either a function or a parametric curve (defined by two equations such as a circle). Any number of functions (lines) can be displayed simultaneously in 2D graphs. In 3D graphs, the parametric curve option is not available, and only one function (surface) at a time can be displayed.

After selecting the plot type, enter the appropriate functions in the boxes provided (*Y=* box for 2D custom functions, *X(t)* and *Y(t)* boxes for parametric curves, *Z(x,y)=* box for 3D custom functions, and *v(x, y, z)=* for ternary plots). Note that as you enter the function, a menu with possible characters or functions is displayed if the *Function guide* check box is selected. You can choose from any item on the menu or enter the function yourself. See *How Do I Plot a Parametric Curve?* in Chapter 9 – *STATISTICA Graphs: Examples and FAQs* for more details on parametric curves, which, for example, allow you to plot an ellipse or a circle.

Add to list. Click the *Add to list* button to display the *New Function* dialog and save a function for reuse at a later time. Afterwards, the saved function can be selected and either

renamed, deleted, or edited in the *Function Name Manager* dialog (accessible clicking the *Function list* button).

Range of. Use the boxes in the *Range of* group box to specify the range (i.e., minimum and maximum values) for the custom function or parametric curve. You can either base the range on the current axis (select *From axis* for both the minimum and maximum values), or you can specify a custom range (select *Custom* for both values). Note that it is also possible to select *From axis* for one and use a *Custom* range for the other (e.g., *From axis* for the *Minimum* and *Custom* for the *Maximum*). If you choose a *Custom* range, specify the *Minimum* and/or *Maximum* value in the appropriate *Value* box.

Line pattern and resolution. Select the *Pattern* check box to plot a line for the custom function. Click the *Pattern* button to customize the appearance of the line in the *Line Properties* dialog. The current specifications for the line are shown in the view box on the *Pattern* button. Use the *Resolution* box to control how the fit is approximated by points on the x-axis. For more details on specific resolution settings, see *Can I Control the Resolution of Fit Lines?* in Chapter 9 - STATISTICA Graphs: Examples and FAQ's. Note that the default (*Normal*) setting will produce smooth function lines in most cases; the higher settings will make a difference only when you either zoom in on part of the graph or use logarithmic scaling (to "stretch" a part of the scale). These options are only available for 2D custom functions.

Assignment of axis. Use the options in the *Assignment of axis* group box to specify the axis against which the function will be plotted. If you select *Y axis*, you can either assign the custom function to the *Left*, *Right*, or *Custom* axis. If you select *X axis*, you can either assign the custom function to the *Top*, *Bottom*, or *Custom* axis. Note that if a custom axis is not being used in the graph, the *Custom* option is dimmed. Additionally, assigning the custom function to the *Top* or *Right* axis when neither axis is being used in the graph will not automatically cause scale values to be displayed on that newly assigned axis. To edit or add scale values, use the *Scale Values* tab of the *All Options* dialog.

Axis Options

As mentioned above, the most direct way to access customization options for a specific graphic object (e.g., axis title, major units) is to double-click on that item. This displays a dialog with the most relevant options for that graphic item (e.g., *Title* dialog, *Major Units* dialog). Axis related options are also contained in the *Axis* dialog (accessible by clicking the *All Options* button on any axis specific option dialog). The tabs on the *Axis* dialog are also available on the *All Options* dialog.

When accessing the axis options as part of the **Axis** or **All Options** dialog, you must first select which axis you wish to customize. In the **Axis** box, select the axis to which you would like to make changes: **X**, **Y-left**, **Y-right**, or **Top** axis for 2D graphs; **X**, **Y**, or **Z** axes for 3D graphs; **X**, **Y**, **Z**, or **V** axes for 3D ternary graphs; or any axes that you may have added to the graph. Axes can be customized independently. Each tab also contains the **Copy axis specs to** button. Click this button to display the **Copy axis specs to** dialog, which includes all axis specification options that can be automatically copied to the remaining axes. This makes it possible to automatically format several axes with the same configurations.

Axis: Title Tab

Select the **Axis: Title** tab of the **All Options** dialog to customize axis titles for all axes defined on the graph.

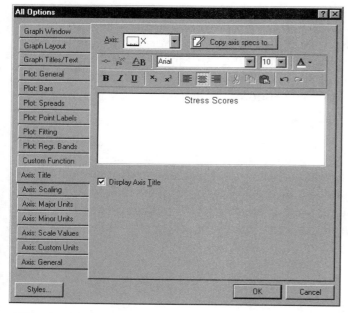

All axis titles, including those generated by *STATISTICA* during graph creation can be edited in terms of fonts, font size, color, and centering characteristics. Special symbols can be added and display formats such as bold, italics, underlines, subscripts, and superscripts can be applied using the mini-formatting toolbar.

Mini-formatting toolbar. Use the buttons on the **Mini-formatting** toolbar above the title box to quickly specify font type and style, add superscripts and subscripts, change the font

attributes to bold or italics, etc. A listing and a brief description of each of these **Mini-formatting** toolbar buttons are available in the **Mini-formatting** toolbar topic in the *Electronic Manual*.

Title. Enter the axis title in the **Title** box. Any legends, fits or functions added to the title will be shown in this box as will any formatting changes you make using the options on the **Mini-formatting** toolbar.

Display Axis Title. Select the **Display Axis Title** check box to display the axis title (shown in the **Title** box) in the graph. If you do not want to display the selected title, clear this box.

Axis: Scaling Tab

Select the **Axis: Scaling** tab on the **All Options** dialog to access options for complete and independent customization of major scale parameters for all axes present on the graph.

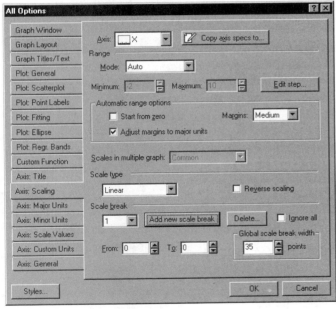

This tab provides selection of automatic versus manual scaling modes, major features such as minimum and maximum values, step size, selection of special scale types such as logarithmic, probability, etc., addition of scale breaks, and reversal of scales.

Mode. Use the **Mode** box in the **Range** group to choose one of two scaling methods. The method you choose here determines whether the minimum, step, and maximum scaling values of each axis will be set automatically by the program (**Auto**) or manually by the user (**Manual**).

Minimum, Maximum. Use the *Minimum* and *Maximum* boxes in the *Range* group to enter the minimum and maximum scaling values for the current axis. These boxes are available only if you choose the *Manual* scaling option (see above).

Edit Step. Click the *Edit Step* button to display the *Axis: Major Units* tab (see below) and edit the step size. The step parameter determines the distance in scale units between major tickmarks for the axis selected.

Automatic range options. The *Automatic range options* group contains options that apply only when you select *Auto* in the *Mode* box of the *Range* group. Select the *Start from zero* check box to determine the steps by "counting" from 0, even if 0 is not in the current range between the minimum and maximum scaling values. Select the *Adjust margins to major units* check box to ensure your graph margin begins and ends on a major unit. You can also specify the width of your margin (whether the margins of the graph are extremely close to the actual data points or very far away from them) using the *Margins* box. You can choose from *Narrow*, *Medium*, *Wide* or *Extreme*.

Scales in multiple graph. Use the *Scales in multiple graph* box when a multiple graph such as a categorized scatterplot is selected. You can specify a *Common* scale for the selected axis on all graphs (e.g., all the x-axes would have the same scale) or use *Independent* scaling for each graph.

Scale type. Use the *Scale type* box to choose between five types of scaling (*Linear*, *Logarithmic*, *Logit*, *Probability* or *Weibull*). Note that if the *Logarithmic*, *Logit*, *Probability* or *Weibull* scale is selected, then the scale minimum for that axis must be greater than 0. For examples of each scale type, see the *Electronic Manual*.

Reverse scaling. Select the *Reverse scaling* check box to reverse the scale values (technically: apply the current *Maximum* value (see above) as a minimum and the current *Minimum* value as a maximum and set the step size to a respective negative value) for the currently selected axis.

Scale break. Use the options in the *Scale break* group to place a 'scale break' mark across the range specified in the *From* and *To* (see below) boxes for the specified axis. As shown in the illustration below, several scale breaks can be specified on the axis.

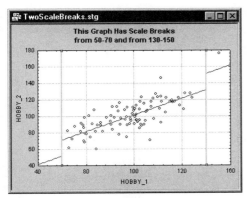

Note that if the options in this box are disabled, click the **Add new scale break** button to add your first scale break. Note that true (multiple) scale breaks are supported in *STATISTICA*, that is the plotting area can be arbitrarily "segmented" by removing segments specified by the break(s).

Add new scale break, Delete. Click the **Add new scale break** button to add scale breaks. *STATISTICA* adds another number to the list box at left each time this button is clicked. To delete the currently selected scale break, click the **Delete** button.

Ignore all. Select the **Ignore all** check box to suppress plotting of all scale breaks on the specified axis. If this box is cleared, *STATISTICA* plots each scale break.

From, To. Use the **From** and **To** boxes to specify where the scale break should begin and end. For example, if you want the scale break to begin at 40 and end at 60, enter **40** in the **From** box and **60** in the **To** box (the area from 40 to 60 will not be displayed in the graph).

Global scale break width. Use the **Global scale break width** box to specify the physical width (in points) for display of all scale breaks. If you enter **From** and **To** values, the graph above and below the break will be rescaled. You can place labels on scale breaks using text via the **Custom Units** tab, where you can also suppress an unwanted scale value (such as the first scale value if you are placing a break near the left side of the axis by entering a "blank" as text at that scale value.

Axis: Major Units Tab

The **Axis: Major Units** tab provides options for changing the units defining the major tickmark intervals on every graph axis. By default, scale value labels will be applied at intervals defined by **Major Units** parameters.

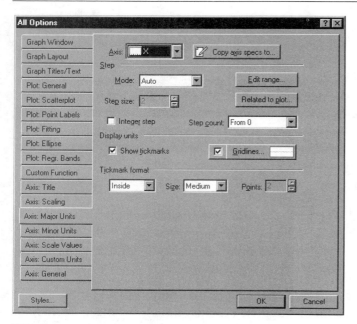

Mode. Use the **Mode** box to specify one of two possible methods for determining step size. You can have *STATISTICA* determine the step size for you based on the current range (select **Auto**) or specify the step size yourself (select **Manual**).

Edit range. Click the **Edit range** button to switch to the **Axis: Scaling** tab, where you can view or edit the range of values for the selected axis.

Related to plot. Click the **Related to plot** button to display the **Plot Connection** dialog.

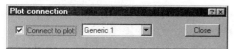

STATISTICA graphs often contain more than one plot. If the text labels for the various plots are different (i.e., based on different variables), you will need to specify which plot to use when applying scale labels to the selected graph axis. The effect of this option is most noticeable when text labels are associated with variable values, but the option also determines which variable plotted on the graph will influence other selected axis labeling options. Use the list provided to select the plot to which to "connect" the chosen axis-related scale labeling options. Note that when the **Connect to plot** check box is cleared, no text labels are used on the axis and default numeric scale labeling is applied.

Step size, Step count, Integer step. When you have selected *Manual* mode, use the *Step size* and *Step count* boxes to specify step size and step count, respectively. Enter the desired step size in the *Step size* box. If you enter a 5 in this field, major tickmarks will be shown every five units. Use the *Step count* box to specify where the step count should begin. You can choose to begin counting at zero or at a specified minimum. Select *From 0* to start the major tickmark count at 0. Select *From minimum* to start the calculation of major tickmarks at the current *Minimum* range value for the axis. Note that the minimum range value is set on the *Axis: Scaling* tab (above), available by clicking the *Edit range* button. If you have selected *Auto* mode, select the *Integer step* check box to calculate the major tickmarks at integer values only.

Display units. Check the *Show tickmarks* check box in the *Display units* group to display a tickmark at each major unit. Select the *Gridlines* check box to display a gridline at each tickmark. Click the *Gridlines* button to display the *Line Properties* dialog and specify the pattern, color, and other gridline properties. The current gridline properties are displayed in the view area on the *Gridlines* button. Note that the style and size of the tickmark are determined by the specifications in the *Tickmark format* group box.

Tickmark format. Use the options in the *Tickmark format* group to customize the style and size of major tickmarks for each of the axes. The color and thickness of the tickmarks are determined by the current specifications of the respective *Axis Line* (see the *Axis: General Tab*, page 646).

Setting of format. Three format settings are provided. Select either *Inside*, *Outside*, or *Crossed* format settings to display tickmarks on the inside, outside, or both sides of the axis (respectively). To remove tickmarks from the axis, clear the *Show tickmarks* check box in the *Display units* group (above).

Size. In the *Size* box, select *Small*, *Medium*, *Large*, or *Custom* sized tickmarks. If you select *Custom*, then specify the length (from 0 to 720) of the tickmarks in the *Points* box to the right.

Axis: Minor Units Tab

Select the *Axis: Minor units* tab from the *All Options* dialog to define the number and display of subintervals within the intervals defined on the *Axis: Major units* tab (see above).

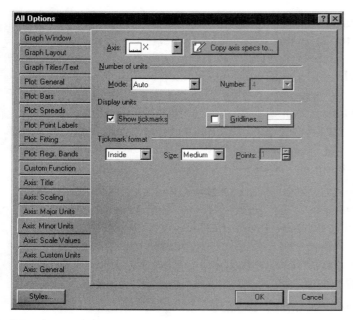

Mode. Use the *Mode* box of the *Number of units* group to specify one of two possible methods for determining how many minor units to include. You can have *STATISTICA* determine the number of minor units to include (select *Auto*), or you can specify the number yourself (select *Manual*). If you select *Manual*, you will need to specify a number between 1 and 15 in the *Number* box. Note that this number defines the number of subintervals that is one more than the number of tickmarks placed between major unit tickmarks.

Display units. Select the *Show tickmarks* check box in the *Display units* group to display a tickmark at each minor unit. Select the *Gridlines* check box to display a gridline at each tickmark. Click the *Gridlines* button to display the *Line Properties* dialog and specify the pattern, color, and other gridline properties. The current gridline properties are displayed in the view box on the *Gridlines* button. Note that the style and size of the tickmark are determined by the specifications in the *Tickmark format* group.

Tickmark format. Use the options in the *Tickmark format* group to customize the style and size of minor tickmarks for each of the axes. The color and thickness of the tickmarks are determined by the current specifications of the respective *Axis Line* (see the *Axis: General Tab*, page 646).

Setting of format. Three format settings are provided. Select *Inside*, *Outside*, or *Crossed* format settings to display tickmarks on the inside, outside, or both sides of the axis

(respectively). To remove tickmarks from the axis, clear the **Show tickmarks** check box in the **Display units** group (above).

Size. In the **Size** box, select **Small**, **Medium**, **Large**, or **Custom** sized tickmarks. If you select **Custom**, then specify the length (from 0 to 720) of the tickmarks in the **Points** box to the right.

Axis: Scale Values Tab

Select the **Axis: Scale Values** tab on the **All Options** dialog to specify the posting of scale values on any graph axis. A variety of options exist for controlling the display of scale values. For a discussion of scale values/labels associated with scale breaks, see the **Axis: Scaling** tab.

Display scale values. The **Display scale values** group contains a variety of options for controlling the display of scale values. Note that if you do not want to show any scale values, you can clear all available check boxes in this group. Note that the items in this group function in a hierarchical manner. That is, when a scale value could be labeled with more than one type of value (e.g., automatic, data value, or custom value), *STATISTICA* will display custom labels instead of data values or automatic values (if the **Custom labels** check box is selected), and it will display data values over automatic values (if the **Data values** check box is selected).

Automatic - at major tickmarks. Select the *Automatic - at major tickmarks* check box to have *STATISTICA* automatically format and display numeric scale values for you. If you clear this box, numeric scale values will not be included in your graph.

Use text labels from data set. Select the *Use text labels from data set* check box to display the dataset's text labels on the selected axis.

Data values. Select the *Data values* check box to display data values on the selected axis. You can click the *Related to plot* button to display a dialog in which you select the plot from which to get data values.

Related to plot. Click the *Related to plot* button to display the *Plot Connection* dialog.

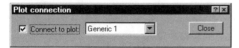

STATISTICA graphs often contain more than one plot. If the text labels for the various plots are different (i.e., based on different variables), you will need to specify which plot to use when applying scale labels to the selected graph axis. The effect of this option is most noticeable when text labels are associated with variable values, but the option also determines which variable plotted on the graph will influence other selected axis labeling options. Use the list provided to select the plot to which to "connect" the chosen axis-related scale labeling options. Note that when the *Connect to plot* check box is cleared, no text labels are used on the axis and default numeric scale labeling is applied.

Custom labels. Select the *Custom labels* check box to display custom labels on the selected axis. To create custom labels, click the *Edit custom labels* button, which will display the *Axis: Custom Units* tab.

Skip values. Select the *Skip values* check box to skip labels, displaying only the text and/or numeric labels specified in the *Show every X label* box (e.g., every fourth label). This option is useful when the scale values (numbers, dates, or labels) are so close together on the axis that they are hard to distinguish.

Value format. Use the options in the *Value format* box to select a format to use for numeric scale values. You can choose from *General*, *Number*, *Date*, *Time*, *Scientific*, *Currency*, *Percentage*, *Fraction* or *Custom*. For more information on these number formats, see the *Electronic Manual*.

Layout. Use the *Layout* box in the *Options* group to specify the layout of the scale values with respect to their axis. Select *Perpendicular* to position the scale values perpendicular (at a right angle) to their respective axis. Select *Parallel* to position the scale values parallel to their

respective axis. Select **Alternate** to position the scale values in an alternating pattern (one above, one below) along the axis. When you select **Auto**, *STATISTICA* determines whether the **Parallel**, **Perpendicular**, or **Alternate** layout is appropriate for the current graph.

Font. Click the **Font** button to display the **Font** dialog, in which you can choose a different font for the scale values of the currently selected axis; see the *Electronic Manual* for more details on the **Font** dialog.

Axis: Custom Units Tab

Select the **Axis: Custom Units** tab to plot numeric or text labels at any position on the selected axis of the graph. Special formatting can be applied to these values or labels, and each can be accompanied by an added tickmark and/or graph gridline. In the simplest case, these features can be used to add labels at custom locations on the graph axes, but the same features can be used to completely customize the graph axis units if desired.

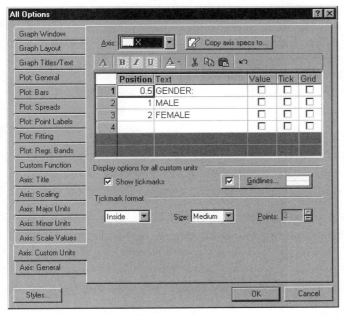

Mini-formatting toolbar. Use the buttons on the **Mini-formatting** toolbar above the custom unit spreadsheet to quickly specify font type and style, add superscripts and subscripts, change the font attributes to bold or italics, etc. A listing and a brief description of each of these **Mini-formatting** toolbar buttons are available in the **Mini-formatting** toolbar topic in the *Electronic Manual*.

 StatSoft
Copyright © StatSoft, 2001

Custom unit editor. Use the custom unit editor to create (or modify) text or numeric custom labels to use with the specified axis. In the *Position* cell, enter the axis position (a number, e.g., 5.3) for the label. In the *Text* cell, enter the text you want to use in that position. Note that if you leave the *Text* cell blank and select the *Value* check box, *STATISTICA* displays the numeric value corresponding to the axis position. You can also display a gridline and/or a tickmark for individual labels by selecting the appropriate check boxes (*Grid* and *Tick*, respectively). The text of the custom unit can by formatted using the *Mini-formatting* toolbar above the editor. Note that if the text you have entered is not displayed on the graph, you should verify that the *Custom labels* check box is selected on the *Axis: Scale Values* tab.

Display options for all custom units. Select the *Show tickmarks* check box in the *Display options for all custom units* group to display a tickmark at each custom unit when a tickmark has been selected in the custom unit editor (see above). Select the *Gridlines* check box to display a gridline at each tickmark (when gridlines and tickmarks have been selected in the custom unit editor). Click the *Gridlines* button to display the *Line Properties* dialog and specify the pattern, color, and other gridline properties. The current gridline properties are displayed in the view box on the *Gridlines* button. Note that the style and size of the tickmark are determined by the specifications in the *Tickmark format* group.

Tickmark format. Use the options in the *Tickmark format* group to customize the style and size of custom unit tickmarks for each of the axes. The color and thickness of the tickmarks are determined by the current specifications of the respective *Axis Line* (see *Axis: General Tab*, below).

Type. Three types of tickmarks are provided. Select either *Inside*, *Outside*, or *Crossed* to display tickmarks on the inside, outside, or both sides of the axis (respectively). To remove tickmarks from the axis, clear the *Show tickmarks* check box in the *Display options for all custom units* group (above).

Size. In the *Size* box, select *Small*, *Medium*, *Large*, or *Custom* sized tickmarks. If you select *Custom*, then specify the length (from 0 to 720) of the tickmarks in the *Points* box to the right.

Axis: General Tab

Select the *Axis: General* tab on the *All Options* dialog to access options that control the general appearance of each individual axis including color, line pattern and thickness, and position of the axis with respect to the plot area of the graph.

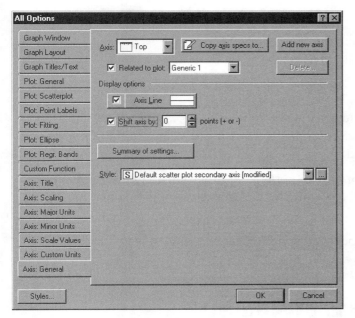

Add new axis, Delete. Click the *Add new axis* button to add a new axis to the plot. To delete an existing axis, select the appropriate axis in the *Axis* box, and then click the *Delete* button.

Related to plot. Use the *Related to plot* check box and drop-down list to specify which plot data to use (when more than one plot is included on the graph) when applying scale labels to the selected graph axis. The effect of this option is most noticeable when text labels are associated with variable values, but the option also determines which variable plotted on the graph will influence other selected axis labeling options. Use the list provided to select the plot to which to "connect" the chosen axis-related scale labeling options. Note that when the *Related to plot* check box is cleared, no text labels are used on the axis and default numeric scale labeling is applied.

Axis line. Select the *Axis Line* check box to include an axis line in the graph. Click the *Axis Line* button to display the *Line Properties* dialog. You can use that dialog to customize the appearance of the current axis line. The current specifications for the axis line are shown in the view box on the *Axis Line* button. To turn off the axis line, clear the *Axis Line* check box.

Shift axis by. Select the *Shift axis by* check box to shift the current axis position by the number of points specified in the *X points (+ or -)* box.

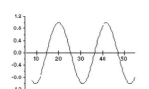 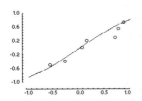

To shift the axis into the graph plot area, enter negative values in the **X points (+ or -)** box. To shift the axis away from the graph plot area, enter positive/unsigned values in the **X points (+ or -)** box. Note that there are 72 points per inch.

Summary of settings. Click the **Summary of settings** button to display the **Summary of settings** dialog. Use this dialog to view selected information about the axis line and title, gridlines, tickmarks, and scale values.

Style. Use the **Style** box to select a pre-specified style for the currently selected axis. For more information on styles, see the *Electronic Manual*.

Special Topics in 3D
Graph Customization

Viewing 3D Graphs

You can use the **Point of View** tab on the **All Options** dialog to change the perspective on 3D graphs, specifically graphs that actually portray 3 dimensions, such as 3D XYZ Scatterplots. These options are also available by clicking the 3D rotation control 🔲 toolbar button on the **Graph Tools** toolbar when a graph is active (see Chapter 6 – *STATISTICA Graphs: General Features*).

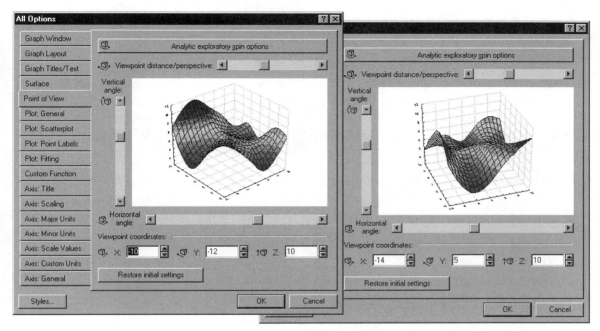

The window portrays the graph in the current perspective. This allows you to monitor the viewpoint distance/perspective and horizontal and vertical angles as the graph spins.

X, Y, and Z boxes. In the *X*, *Y*, and *Z* boxes, you can enter eye position parameters (i.e., X, Y, Z coordinates) to expose specific aspects of the 3D display for analytic purposes and also to create special presentation effects by changing the perspective of the graph. Use these eye position parameters to separately determine the location of the "view point" of the graph for each of the three axes X, Y, and Z. The absolute value of each parameter determines the distance between the eye and the respective axis (i.e., the degree of the perspective transformation for the respective plane). The sign of the parameter determines the side from which the axis is viewed (i.e., a value of 0 represents the "facing position" and no perspective transformation; negative and positive values represent left and right or up and down shifts of the viewpoint, respectively).

Analytic exploratory spin options. Click the *Analytic exploratory spin options* button to rotate the graph about a vertical axis as indicated by the movement of the *Horizontal angle* scrollbar during rotation. After the *Analytic exploratory spin options* button is clicked, it is replaced by a series of four buttons. Click the << or >> buttons to change the direction of rotation. Click the *Manual* button to temporarily stop the rotation and allow settings to be adjusted manually. Once you have clicked the *Manual* button, you can use either the << or the >> button to re-start the spin action. Click the *Stop* button to stop the rotation of the graph and

replace these four buttons with the *Analytic exploratory spin options* button. You can click the *OK* button at any time (even while the graph is spinning) to redraw the actual graph in the current perspective.

Viewpoint distance/perspective. Use the *Viewpoint distance/perspective* scrollbar to determine the virtual distance from which the 3D graph is being viewed. If the scrollbar is moved to the left, the distance is effectively reduced and the perception is given of seeing "inside" the volume of the graph cube. If the scrollbar is moved to the right, the cube is viewed from a more external perspective.

Horizontal angle. Use the *Horizontal angle* scrollbar to control the rotation about the vertical line of the graph. Moving the scrollbar from extreme left to extreme right rotates the graph one complete rotation (360°).

Vertical angle. Use the *Vertical angle* scrollbar to control rotation of the graph up to 180° about a horizontal line parallel to the plane of the display. Moving the scrollbar to the extreme top allows direct viewing of the bottom of the graph; moving the scrollbar to the extreme bottom allows direct viewing of the top of the graph.

Changing 3D XYZ Graphic Display Types

When any 3D graph type is active, the options on the *Graph Layout*, *Surface*, *Plot: General*, and *Plot: Fitting* tabs of the *All Options* dialog can be used to convert the graph from one basic type to another and access the major features of surfaces, contours, and point markers for customization.

The *Graph Layout* tab contains a *Graph type* box with *Standard*, *2D Projection*, *Spectral*, and *Space Plot* options. So, for example, a *Standard* scatterplot:

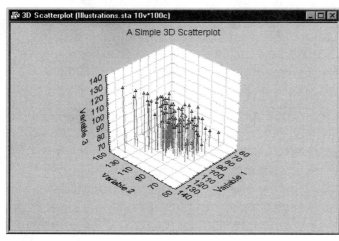

could be changed to a *Spectral* plot (that might reveal a pattern of data that was entirely hidden in the simple XYZ scatterplot display):

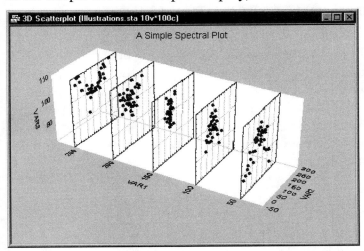

a *Space* plot:

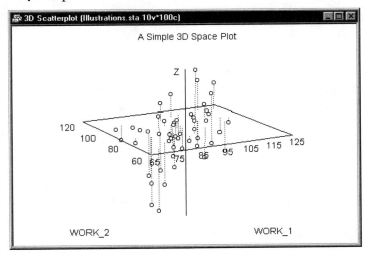

or a *2D Projection*.

When opened, the *Graph Layout* tab displays the current graph type in the window. *Standard* corresponds to the normal 3D scatterplot graph type (or 3D surface plot type if a fit is specified). *2D projection* corresponds to a 2D contour plot, a plot in which a specified 3D fit surface is projected onto the X-Y plane resulting in a 2D graph. The *Spectral* and *Space* plots are special instances of 3D scatterplots. Note that point markers, spike lines, and trace lines

(lines connecting point markers) can be specified on the **Plot General** tab. You can also access these marker and line features by double-clicking them on the graph (or selecting **Properties** from the shortcut menu after right-clicking them) to display the **General** dialog.

To remove, add, or change a fit type (i.e., a 3D surface plot or 2D contour plot), select the **Plot: Fitting** tab. Note that the same functionality is available on the **Fitting** dialog obtained by double-clicking on the fitted surface on a 3D surface plot or a contour on a 2D contour plot. You can also access the **Fitting** dialog by right-clicking on a surface or contour and selecting **Properties** from the shortcut menu. Here, the type of fit can be specified in the **Fit type** box (i.e., **Linear**, **Quadratic**, **Distance Weighted Least Squares**, or **Negative Exponential Weighted Least Squares**). The selected fit can be turned on or off via the **Display fit** checkbox. Note that only one fit at a time can be displayed on a 3D graph.

Customizing 3D Fitted Surfaces or Contours

To customize colors used on 3D fitted surfaces or 2D contours, use the **Surface Specs** button on the **Plot: Fitting** tab (page 629) of the **All Options** dialog. Note that the same **Surface Specs** button is also on the **Fitting** dialog accessed by double-clicking the fitted surface on a 3D surface plot or a contour on a 2D contour plot. Note also that you can access the **Fitting** dialog by right-clicking on a surface or contour and selecting **Properties** from the shortcut menu. Clicking the **Surface Specs** button displays the **Surface Specifications** dialog. This dialog provides access to the method of determining the number of intervals to be displayed on the surface or contour plot, the colors to be used for the intervals (area colors for surface plots, line colors and styles for contour plots), and other appearance aspects of the intervals defined. (For details, see *How Do I Define a Custom Selection of Levels for a Contour Plot or Surface Plot?* and *How Do I Define a Custom Palette for a Contour Plot or Surface Plot?* in Chapter 9 - *STATISTICA Graphs: Examples and FAQ's*). Note that changing the definition of intervals and/or colors on the graph also changes the graph legend. See *Graph Titles/Text Tab* (page 623) and *Graph Customization - Titles, Legends, Custom Text* in Chapter 9 - *STATISTICA Graphs: Examples and FAQ's* for other legend-related customization topics).

Use the options on the **Surface** tab of the **All Options** dialog to customize other properties of 3D surfaces or 2D contours.

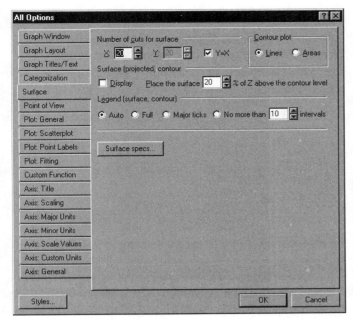

Number of cuts for surface. Use the options in the *No. of cuts for surface* group to enter the number of cuts on a 3D fitted surface (or 2D contour) for both the *X* and *Y* directions. If you want the same number of cuts for both axes, you can enter the number in the *X* box and then select the *Y=X* check box to copy that number to the *Y* box. The number of cuts affects the smoothness of the surface or contour (i.e., the larger the number, the smoother and more detailed the surface).

Contour plot. If a contour plot is already present or has been specified for the graph, select the *Lines* option button in the *Contour* plot group to display 2D contour plots with lines. Select the *Areas* option button to display the 2D contour plots with areas.

Surface (projected) contour. Use the options in the *Surface (projected) contour* group to project an already specified 3D fitted surface onto an x-y plane as a 2D contour plot at a specified location on the z-axis. Select the *Display* check box to display the 2D contour plot. Specify the location using the *Distance* box. This produces the 2D contour plot in addition to the 3D fitted surface plot.

Legend (surface, contour). Use the options in the *Legends (surface, contour)* group to control how many boxes are to be displayed in legends for surface and contour plots. Note that these options have no effect on the actual number of intervals displayed on the contour surface, just the number of legend boxes displayed. By default the *Auto* option button is selected and *STATISTICA* decides how many boxes to use. If the *Full* option button is selected, the number

of boxes displayed will match the number of surface intervals defined (via minor axis tics as default, which will occasionally cause the legend to run off the graph at top). Select the **Major axis** option button to reduce the displayed intervals to correspond to the number of major axis tickmarks. Select the **No more than** option button to use a user-specified maximum number of legend boxes.

Surface Specs. Click the **Surface Specs** button to display the **Surface Specifications** dialog.

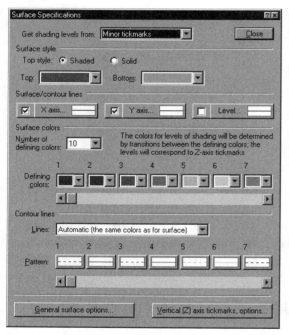

Note that in order to determine the colors (and shades) of the surface, *STATISTICA* surface plots use the concept of "defining colors" and the intermediate colors (shades) are created automatically by *STATISTICA* to fill the gaps (with appropriate color gradient) between the defining colors, following the user-requested number of steps (depending either on major or minor tickmarks). For example, if you have specified only three defining colors: yellow, red, and green, and the number of levels (resulting from your selection of scaling, e.g., major tickmarks, on the z-axis) is 13, *STATISTICA* will create five gradient colors to create a "smooth" transition between yellow and red, and five others to create a transition from red to green.

GRAPHICS STYLES

Using Styles

STATISTICA graphs consist of many components and literally thousands of different characteristics (e.g., line colors, patterns, settings, etc.). Styles are collections of settings that you can apply to a particular graph or a specific aspect or part of a graph, e.g., the font of the top title. You can also configure the styles you create or modify as defaults that will be automatically applied to all new graphs that you create. *STATISTICA* allows you to create, manage, and reuse styles pertaining to aspects of the graphs themselves, such as line patterns, hatch patterns, sizes and colors of arrows, etc.; these are called **Graphics styles**. In addition, you can create, manage, and reuse styles pertaining to the size and proportions of graphs; these are called **Document styles** (see page 617).

Graphics styles can be edited and created via the **All Options** dialogs (accessible by selecting **All Options** from the graph **Format** menu); **Document styles** can be edited via the **Graphs2** tab of the **Options** dialog (accessible via the **Tools - Options** menu). Note that when styles are applied to graphs, they modify the graphs themselves, and those modifications will be saved along with the respective graphs. Thus, for example, if you e-mail to a colleague a graph customized by applying a particularly elaborate and customized graphics style, then that graph will appear on your colleague's computer customized in the same manner as you intended, even though on your colleague's installation of *STATISTICA*, the styles used to achieve the customization may not be defined.

Overview

Although most users may not need to use the advanced functionality and may not be interested in the internal working of *STATISTICA*, practically all features of *STATISTICA* graphs (thousands of combinations of settings) are internally represented by "styles." You can define these styles in a very simple way – by assigning custom names to your own selections of features in the respective graph customization dialogs. Later you can quickly re-use these styles by selecting them by name instead of repeating the tedious sequences of customizations. For example, a custom style of a specific arrow with a wide, filled head and a thick red tail can be called **A red fat arrow** and later selected with a single click of the mouse. Because the internal

system of styles drives the management of practically all graph customizations, also all program defaults (i.e., the appearance of graphs when you do not make any customizations) are stored as styles and can be easily modified.

In order to custom-tailor graphics options in *STATISTICA* and limit the need for repeating the same graph customization, the default settings of practically all features of *STATISTICA* Graphs can be adjusted and saved. This flexible system customizability is possible because all features of the appearance of *STATISTICA* Graphs are driven by the same, comprehensive system of styles.

A graphics style is a convenient way to combine sets of formatting and specification options into one collection. This feature allows you to specify and apply specific settings and formats for different components of a graph. Once you have created a graphics style, it can be applied to not only the current graph, but also to future graphs (of the same type).

There are several predefined, modifiable graphics styles to choose from, but the power and flexibility of creating entirely new graphics styles is at your fingertips. Graphics styles follow a hierarchical organization and higher-level styles consist of other (lower-level) styles. Therefore, styles can be created and applied on the component level (e.g. background color), or they can be used to encapsulate all of the component styles into a single graphics style that can be applied to the graph as a whole.

Exchanging graphs that have styles applied to them. Once a style has been applied to a graph (either during creation of the graph or after the graph has been created), the graph features included in this particular style are determined by the style. The specifications of the resulting graph, however, are stored in the graph not only in terms of references to the respective styles, but also their absolute values: colors, sizes, patterns. This means that a graph with an applied style can be sent to another *STATISTICA* user, and its features will be correctly interpreted even though the other user has not defined an equivalent graph style.

Saving styles. Graphics styles (along with complete sets of globally defined graph options) can be saved, exported, or imported through the **Configuration Manager**, available by selecting **Options** from the **Tools** menu. Styles can also be imported or exported using a simple shortcut method, specifically the method of *Using a graph as a carrier of a graphics style from one system to another* (see page 661).

Contents of Graphics Styles

Graphs are composed of collections of properties of graph objects; that is, they are collections of labels, axes, fit lines, point markers, etc., as well as separate added objects such as circles,

Copyright © StatSoft, 2001

text boxes, or arrows (e.g., size, color, thickness, and pattern of lines; size, shape, and color of point markers; colors and patterns used for definition of areas; size, color, and fonts for labels, titles, and scales; etc.). Graphics styles are not used to add objects to graphs, but to specify the properties of objects already added to or specified for the graph. Styles can also be used to specify properties of objects that can be added to a graph later (e.g., arrows, rectangles, text, etc.).

If you are new to graphics styles, at this point you may want to make a graph in *STATISTICA* (any graph), double click on the graph background to display the **All Options** dialog, and then click the **Styles** button typically located in the lower-left corner of the **All Options** dialog. Click the **Styles** button to display the **Graphics Style** dialog, which is used to edit the current graphics style (i.e., the style that is the collection of all attributes and properties for the current graph). Then click the **More** button to display the full **Graphics Style** dialog (and thus expose attributes and properties) as shown in the illustration below.

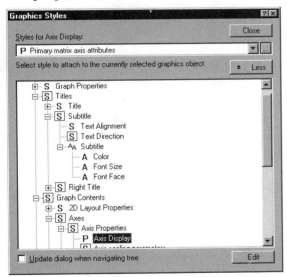

Attributes and properties. Technically speaking, there are two classes of items for which a style can be created: attributes and properties. In practically all cases, there are no differences between the two in the manner in which you manage (e.g., edit) attributes and properties; however, the distinction can be useful to allow you to keep track of the various aspects of the graph that you may want to customize.

An attribute (designated with an icon **A**) is an object that affects the simple appearance of the graph such as colors, line patterns, font sizes, font names, etc. Attributes typically have simple values, such as particular colors, line patterns, point patterns, or font sizes. A property

(designated with an icon **P**) of a graph is an aspect of the graph that usually cannot be represented or summarized by a simple value, such as the **Axis Display** property. To edit this property (see *Managing and Editing Graphics Styles* below), select the **Axis Display** property and then click the **Edit** button. A (somewhat) complex dialog is displayed with various settings that describe this aspect of the graph.

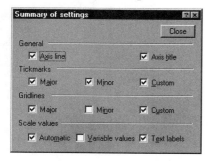

Clearly, the **Axis Display** property cannot be summarized by assigning to it a single value, and the individual check boxes on this dialog are not associated with separate styles; they can only be saved collectively (as a property).

Style collections. There are two additional classes of styles: collections of attributes (ᴬᴬ) and collections of properties and/or attributes (**S**). All the elements of a collection of attributes are simple attributes, **A** . All the elements of a collection of properties and/or attributes are either properties or collections of attributes, **P** and ᴬᴬ. Again, this distinction (and the respective icons used to reinforce this distinction) is not important with respect to how you manage (e.g., edit) these aspects of the style, but is mostly used here to help you keep track of the very large number of aspects of the graphics that can be customized and managed via styles.

Managing and Editing Graphics Styles

Look back at the **Graphics Styles** dialog (page 657), which lists all graphics styles. You can see how the various aspects of the graph are hierarchically organized into a tree structure. For example, the **Titles** collection contains various collections of other attributes and properties, etc. To reiterate, practically every aspect of the graph can be customized, and the customization saved as a style that can be reused in subsequent graphs. Note that you can edit some specific elements of the style by clicking the **Edit** button on the **Graphics Styles** dialog; the **Edit** button, however, will only be available (not dimmed) if the respective style property or attribute is applicable and available in the current graph.

Editing the current graphics style. When you click the *Edit* button, the focus will change to the respective option on the *All Options* dialog where you can make changes to the respective aspect of the graph. After a property or attribute was modified (from the default configuration), a red "box" will be drawn around it, as shown in the *Graphics Styles* illustration on page 657. Note that many properties and attributes may be "circled red" the first time you create the graph, because of the specific requirements of the particular graph.

Saving the current graph configurations as a style. You also save all customizations by right-clicking on the *Graphics Styles* box on the *Graph Tools* toolbar. Many edit fields, drop-down boxes, etc. available on the various tabs of the *All Options* dialog or sub dialogs (e.g., the *Line properties* dialog) also contain a *Style* box; e.g., a *Line style* box as shown in the illustration below.

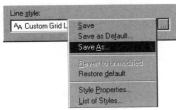

If you click the ▢ button to the right of the *Line styles* box, or right-click on the *Line styles* box, a menu is displayed containing the commands to *Save*, *Save As Default*, *Save As*, etc. Using these commands, you can make the respective customization (e.g., line pattern) the default for future graphs (select *Save as Default*), save it in the current style, or use *Save As* to create a new style that you can later apply to your new graphs.

Applying a Graphics Style

Suppose you have created a graphic style by specifying various non-default point markers for scatterplots. You have used the *Save As* command to save the style under the name *MyScatterPlotStyle*. There are various ways to retrieve the style and apply it to selected newly created scatterplots.

Selecting the style on the Appearance tab. Begin by selecting *Scatterplots* from the *Graphs* menu to display the *2D Scatterplots* dialog. Select the variables for the plot, etc., and then click on the *Appearance* tab.

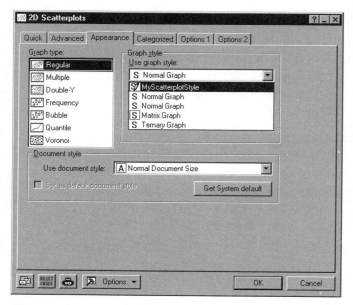

The newly created style can now be selected in the **Graph style** box of this tab, as shown above.

Selecting the style from the Graphics Style box of the Graph Tools toolbar. You can also apply the newly created style after the graph has been created. When the currently active window is a graph, then by default (unless you disabled that toolbar) the **Graph Tools** toolbar will be displayed. This toolbar contains a drop-down box with all graphics styles.

Select **MyScatterplotStyle** to apply all specifications in that style to the current (existing) graph.

Selecting the style from the Graphics Styles dialog. You can also double-click anywhere on the graph background to display the **All Options** dialog. In the lower-left corner is the **Styles** button. Click the **Styles** button to display the **Graphics Styles** dialog. Then click the **More** button to see all properties and attributes of the current style. Finally, collapse the root **Graph** directory of the tree browser, as shown below.

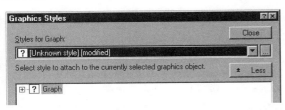

Now, click on the **Styles for Graph** drop-down box to see all currently available styles.

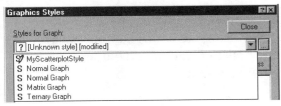

Select **MyScatterplotStyle** to apply all specifications in that style to the current (existing) graph.

Importing and Exporting Styles

Suppose you have created a complex graphics style that you want to use on another computer, for example, on your computer at home (and you have created the style with *STATISTICA* installed on the computer at your office). There are basically two ways to make graphics styles "portable": By using a graph itself as the carrier of the style or via the **Configuration Manager**.

Using a graph as a carrier of a graphics style from one system to another. This method involves using the graph itself as a "carrier" of the graphics style. First apply the desired graphics style to the graph. Next, save the graph and then open it on the other system. Click the background of the graph and then right-click on the **Graphics Styles** box on the **Graph Tools** toolbar. In the shortcut menu that is displayed, select **Save As** to display the **Save As** dialog. Specify the name for the new graphics style and click the **OK** button. Now the graph style is part of the other system's graphics library. This is the traditional and recommended means of porting a graphics style.

Configuration manager. You can also save and retrieve (export and import) the complete configuration of your *STATISTICA* installation, along with all graphics styles. To use this method of transferring graphics styles (along with other *STATISTICA* system configurations), select **Options** from the **Tools** menu to display the **Options** dialog, and click on the **Configuration manager** tab. Next, click the **Export** button to display the **Select Configurations** dialog. Then, select the active configuration and click the **OK** button. Finally, use the **Save As** dialog to save the **.xml* file to a location where the other system can retrieve it and later import it into their system.

To import the configuration, click the **Import** button on the **Configuration Manager** tab of the **Options** dialog to display the **Open** dialog. Next, select the **.xml* file that you have previously saved and click the **Open** button to display the **Select Configurations** dialog. Here, select the configuration with the graphics styles that you need to import, and click the **OK** button. After you have imported the **.xml* file, select the configuration that you have just imported on the **Configuration Manager** tab and click the **Select** button to load it. Once *STATISTICA* restarts, all of the graphics styles from the other system will now be installed on the new system.

This latter method is the most efficient means of porting numerous graphics styles; however, it may not be the one you want to use in many cases because using the configuration manager will import all aspects of a system's settings (e.g., toolbar customization and output settings), not just its graphics styles. If you only need to import graphics styles, then use the previous "graph-as-a-carrier" method.

Examples

Graphics styles are designed for users who are familiar with the specification and customization options available on *STATISTICA*'s individual graph creation dialogs and the **All Options** customization dialog (and its subsidiary dialogs). Property options associated with the graph objects available on the **Graph Tools** toolbar can also be captured as graphics styles. The following simple examples illustrate the use of graphics styles.

Example 1: Styles that Apply to the Entire Graph

If a variety of customizations has been applied to a graph using the tabs of the **All Options** dialog, the graphic style can be saved and then applied with new data to completely reproduce the appearance of the customized graph. As an example, use the **Adstudy.sta** file to make a 2D scatterplot with the default linear fit applied from the **Graphs** menu (**Measure01** and **Measure02** were used in the graph below).

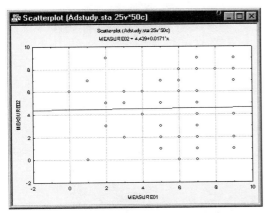

- Next, double-click on the graph background (i.e., outside the graph axes) to display the **All Options** dialog. On the **Graph Window** tab, change the **Inside Background Color** to light yellow and the **Outside Background Color** to light blue.

- On the **Plot: General** tab, change the point marker color and type by clicking the **Markers** button. Select a solid square marker type and set the marker color to black.

- Finally, click the **Axis: Major Units** tab and change (for both the **X** and **Y** axes) the **Gridlines** to a solid red line of width 1.5 points.

- Click the **OK** button to redraw the graph.

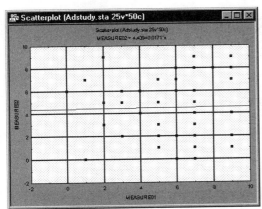

- Next, right-click on the **Graphics Styles** box on the **Graph Tools** toolbar, select **Save As**, and give the style a distinctive or descriptive name (e.g., **NewColorScatterplot**).

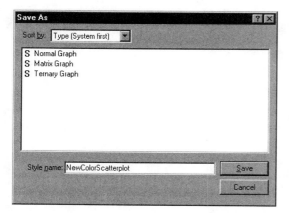

The saved graphic style incorporates both the creation and customization features of the graph.

- Now display a new **2D Scatterplot** creation dialog by selecting **Scatterplots** from the **Graphs** menu, and select the **Appearance** tab.

- Use the **Use graph style** box to select the newly defined style (**NewColorScatterplot**).

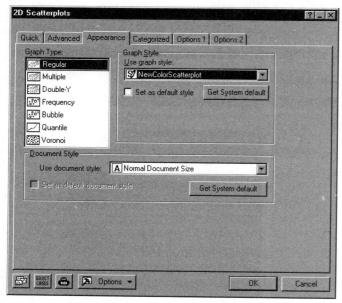

When you click the **OK** button, *STATISTICA* will ask you to select variables for the graph, and then will produce a new scatterplot with the background colors, marker shape and color, and gridline type, thickness, and color defined by the saved style.

Note that this graph style can be instantly applied to future scatterplots after creation by selecting the saved *NewColorScatterplot* style from the *Graphics Styles* box on the *Graph Tools* toolbar with the graph active.

Note also that a graph style specified in the *Use graph style* box on the *Appearance* tab of a graph creation dialog will be saved with the graph creation specifications if the graph is saved as a user-defined graph via the *Add As User-defined Graph to Menu* button on the *Options 2* tab of the graph creation dialog.

Example 2: Adding Styles for Graph Objects

Properties of graphic objects such as arrows, rectangles, ovals, etc. and other features/ properties addressed by options available from the *Graph Tools* toolbar can also be specified with graphics styles.

Separate styles applied to individual objects. When you select a specific component of the graph (e.g., an arrow or the top title), the name of the current style of that component is displayed in the *Graphics Styles* box on the *Graph Tools* toolbar (in this case, initially, most likely AA Default arrow or S Default Top Title , respectively). Now you can click on the *Graphics Styles* box and scroll down to the custom style that you need to apply (if there is one already created). When you select that style (e.g., a previously created arrow style Red Arrow or title style PPP), the respective object will instantly switch to the selected style.

Styles of specific objects contained in higher-level styles. If, before saving the graph style in the example above, a red arrow with a solid-filled head is placed on the graph and the *Increase Font* toolbar button is clicked several times to increase the size of graph fonts and point markers, these object properties will be appended to the graph style. Upon creating a new graph using the saved style (i.e., specifying it on the *Appearance* tab of the creation dialog) or applying the saved style to a previously created graph (via selecting the style from the *Graphics Styles* box on the *Graph Tools* toolbar when the graph is active), the increased font and marker sizes are applied immediately in addition to the other features specified in *Example 1* above. No arrow will be added automatically to the graph, but if the *Arrow* tool is selected from the *Graphs Tools* toolbar, any arrow you draw on the graph will have the solid head as specified by the style and will be red.

Example 3: Styles for Individual Features

Advanced users can define and apply graphics styles for altering individual graph features or hierarchically related groups of features via the tabs of the **All Options** customization dialog. If, for example, the font of the x-axis label is increased to 20 points, bolded, and changed to red (using the **Font** button) on the **Axis: Scale Values** tab and the color of the x-axis line changed to blue on the **Axis: General** tab, these axis-related features can then be saved as a single graph style on the **Axis: General** tab by clicking the **Style Ellipsis** button (![]). The defined axis style can be applied to subsequent graphs after creation by selecting the saved style on the **Axis: General** tab of the **All Options** dialog. (Upon clicking the **OK** button, the axis style will be applied to whichever axis is selected in the **Axis** box at the top of the **Axis: General** tab.)

Note that if a style for a particular feature or hierarchically related group of features is saved as the default style, all subsequent graphs made will reflect the redefined feature or features addressed by that style.

8

CHAPTER

STATISTICA GRAPHS: CONCEPTUAL OVERVIEWS

continued ➡

STATISTICA GRAPHS: CONCEPTUAL OVERVIEWS

This chapter includes general discussions of features and applications of only the most often used types of graphs found in *STATISTICA*. For information about the specialized types of graphs associated with particular analyses, please refer to the documentation on the respective modules in the *Electronic Manual*.

2D GRAPHS

2D Histograms

The 2D Histograms procedure contains a selection of univariate and multivariate (noncategorized) histograms.

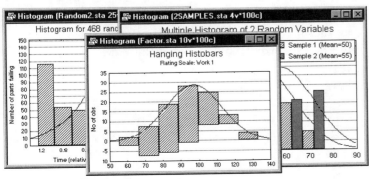

Histograms (the term was first used by Pearson, 1895) are used to examine frequency distributions of values of variables. For example, the frequency distribution plot shows which specific values or ranges of values of the examined variable are most frequent, how differentiated the values are, whether most observations are concentrated around the mean, whether the distribution is symmetrical or skewed, whether it is multimodal (i.e., has two or more peaks) or unimodal, etc. Histograms also allow you to evaluate the similarity of an observed distribution to theoretical or expected distributions.

There are two major reasons why frequency distributions are of interest:

- You may learn from the shape of the distribution about the nature of the examined variable (e.g., a bimodal distribution may suggest that the sample is not homogeneous, and consists of observations that belong to two populations that are more or less normally distributed).

- Many statistics are based on assumptions about the distributions of analyzed variables; histograms help you to test whether those assumptions are met.

Often the first step in the analysis of a new dataset is to run histograms on all variables.

Note that if more than one variable is selected for any type of univariate *Histogram* (see below), then a series of graphs, one for each variable, will be created and can be reviewed in a results workbook, report, or cascade of stand-alone windows; see Chapter 2 – *STATISTICA Output Management*.

Histograms vs. Descriptive Statistics

Histograms provide information similar to descriptive statistics (e.g., mean, median, minimum, maximum, differentiation of values, etc.). Although specific (numerical) descriptive statistics are easier to read in a table, the overall shape and global descriptive characteristics of a distribution are much easier to examine in a graph.

Moreover, the graph provides qualitative information about the distribution that cannot be fully represented by any single index. For example, the overall skewed distribution of income may indicate that the majority of people have an income that is much closer to the minimum than maximum of the range of income. Although this information will be contained in the index of skewness, when presented in the graphical form of a histogram, the information is usually much more easily recognized and remembered.

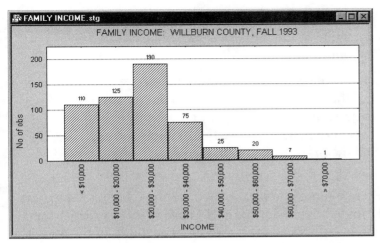

The histogram may also reveal "bumps" that may represent important facts about the specific social stratification of the investigated population or anomalies in the distribution of income caused by a recent tax reform.

Categorization of Values

All **Graphs** menu graphs histogram procedures offer the standard, comprehensive selection of categorization methods (see Chapter 7 – *STATISTICA Graphs: Creation and Customization* for more details).

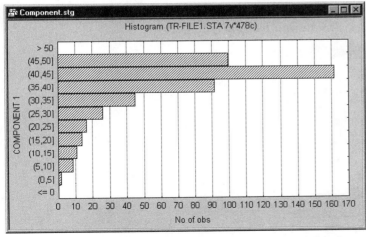

These categorization methods divide the entire range of values of the examined variable into a number of categories or sub-ranges for which frequencies are counted and presented in the plot

as individual columns or bars (horizontal bars are plotted when the **X-Y Axis Position** box on the **Options 2** tab of the graph specification dialog is set to **Reversed**, see Chapter 7 – *STATISTICA Graphs: Creation and Customization*).

For example, you can create a histogram where each column represents a range of 10 units on the scale used to represent the variable; if the minimum value is 0 and the maximum is 120, then 12 columns would be created. Alternatively, you can request that the entire range of values of the variable be divided into a specific number of equal size intervals (e.g., 10); in the latter case, if the minimum value is 0 and the maximum is 120, then each interval would be equal to 12 units of the scale.

There are also options supported to generate more complex categorizations. For example, you can create uneven ranges by custom defining boundaries for each range (e.g., in order to create more interpretable ranges or to concatenate outliers and increase the readability of the middle part of the histogram).

The ranges can also be created by defining specific inclusion or exclusion criteria using logical statements (e.g., the first column in the histogram could represent persons who traveled by plane more than 10 times in the last year and who are not traveling more than 50% of the time on business; etc.).

Fitting Theoretical Distributions to Observed Distributions

The distribution fitting facilities integrated with histograms allow you to compare the observed data to a selection of common distributions including **Normal** (see below),

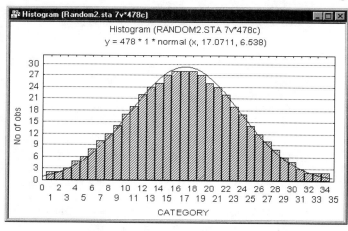

Beta, **Exponential**, **Extreme**, **Gamma**, **Geometric**, **Laplace**, **Logistic**, **Lognormal**, **Poisson**, **Rayleigh**, and **Weibull** (see Chapter 7 – *STATISTICA Graphs: Creation and Customization*).

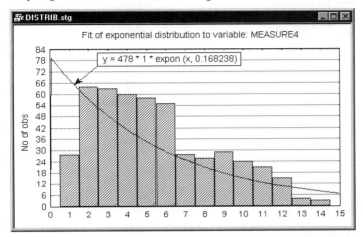

Note that *STATISTICA* also includes designated distribution fitting procedures (see **Distribution Fitting**, and **Process Analysis** in the *Electronic Manual*) featuring a comprehensive selection of theoretical distribution functions, graphs and statistical tests of differences between observed and expected distributions. See also *Quantile-Quantile* and *Probability-Probability Plots*, pages 682 and 683, respectively.

2D Scatterplots

Two-dimensional scatterplots are used to visualize relations between two variables X and Y (e.g., weight and height). Individual data points are represented by point markers in two-dimensional space, where axes represent the variables. The two coordinates (X and Y) that determine the location of each point correspond to its specific values on the two variables.

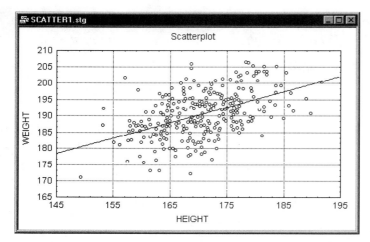

If the two variables are strongly related, then the data points form a systematic shape (e.g., a straight line or a clear curve), as shown in the example.

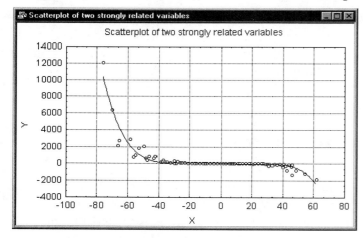

Fitting functions to scatterplot data helps identify the patterns of relations between variables.

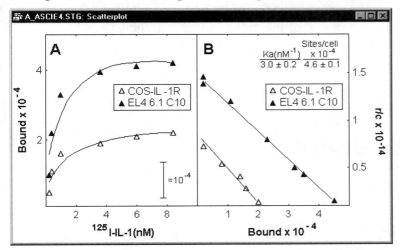

If the variables are not related, then the points form a round "cloud."

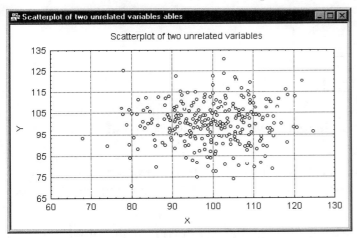

Homogeneity of Bivariate Distributions (Shapes of Relations)

Scatterplots are typically used to explore or identify the nature of relations between two variables (e.g., blood pressure and cholesterol level), because they may provide much more information than a correlation coefficient. For example, a lack of homogeneity in the sample from which a correlation was calculated may bias the value of the correlation. Imagine a case

where a correlation coefficient is calculated from data points that came from two different experimental groups, but this fact was ignored when the correlation was calculated. You can assume that the experimental manipulation in one of the groups increased the values of both correlated variables and thus the data from each group form a distinctive "cloud" in the scatterplot (as shown in the illustration below).

In this example, the high correlation is entirely due to the arrangement of the two groups, and it does not represent the "true" nature of the relation between the two variables, which is practically equal to 0 (if you looked at each group separately).

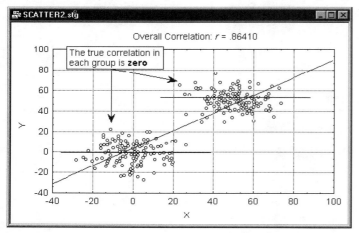

If you suspect that such a pattern may exist in your data, and you know how to identify the possible "subsets" of data, try to run the correlations separately in each subset of observations, or use the **Categorized Scatterplot** instead.

This example shows an extreme case; however, this and similar types of problems caused by the lack of homogeneity of the population (or the sample) tested are common and are often encountered in research practice.

Another aspect of relationships between variables that can be examined in scatterplots is curvilinearity. There are no "automatic" or easy-to-use tests to measure curvilinear relationships between variables: The standard Pearson r coefficient measures only linear relations; some nonparametric correlations such as the Spearman R can measure curvilinear relations, but not non-monotonous relations. Examining scatterplots allows you to identify the shape of relations, so that later an appropriate data transformation can be chosen to "straighten" the data or select an appropriate nonlinear equation to be fit.

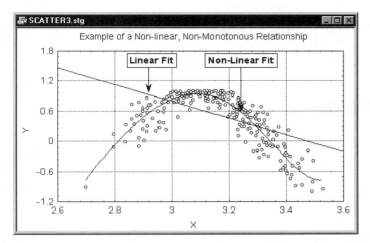

Refer to the *Electronic Manual* for more information on the **Basic Statistics and Tables**, **Distribution Fitting**, **Multiple Regression**, and **Nonlinear Estimation** modules.

Outliers

Another major advantage of scatterplots is that they allow you to identify "outliers" (atypical data points) that artificially increase or decrease ("bias") the correlation coefficient.

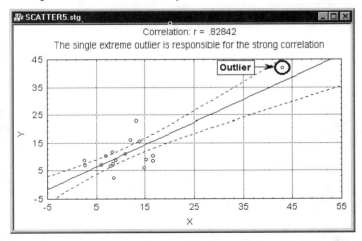

For example, a single outlier may "artificially" increase the value of a correlation between two variables to the point where it becomes highly significant. A scatterplot allows you to identify such anomalies. For example, the correlation between the two variables in the previous

illustration would have been nearly *0* without the single outlier. The presence of this outlier "artificially" increases the value of the correlation to a highly significant value.

In *STATISTICA*, the brushing tools (see Chapter 9 – *STATISTICA Graphs: Examples and FAQ's*) are particularly useful in such circumstances because they allow you

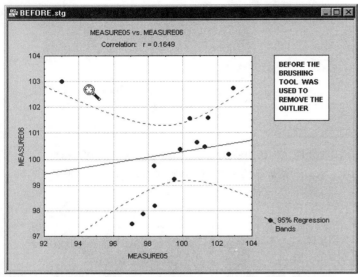

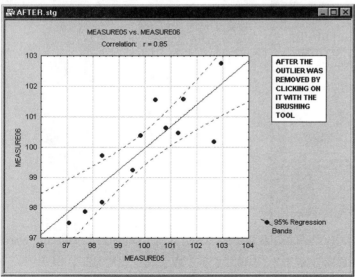

to interactively remove outliers and see how the fitted function or regression line changes, as illustrated in the example above.

Normal Probability Plots

The options in the **Normal Probability Plots** dialog allow you to produce three types of probability plots: **Normal**, **Half-Normal**, and **Detrended**. Normal probability plots provide a quick way to visually inspect to what extent the pattern of data follows a normal distribution.

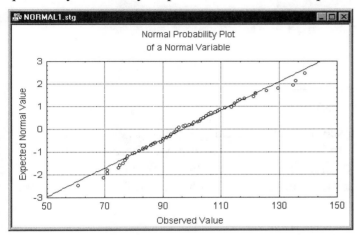

Normal Probability Plot

The way the standard normal probability plot is constructed is as follows. First the values are rank ordered. From these ranks, Z values (i.e., standardized values of the normal distribution) are computed based on the assumption that the data come from a normal distribution. Specifically, the normal probability value z_j for the j'th ordered value (rank) in a variable with n observations is computed as:

$$z_j = \Phi^{-1}[(3*j-1)/(3*N+1)]$$

where Φ^{-1} is the inverse normal cumulative distribution function (converting the normal probability p into the normal value Z).

These Z values are plotted on the y-axis in the plot. If the observed values (plotted on the x-axis) are normally distributed, then all values should fall onto a straight line in the plot. If the values are not normally distributed, they will deviate from the line.

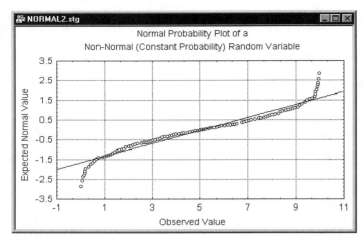

Outliers may also become evident in this plot.

If there is a general lack of fit, and the data seem to form a clear pattern (e.g., an *S* shape) around the line, then the variable may have to be transformed in some way before it can be used in a procedure that assumes normality (e.g., a log transformation is often used to "pull in" the tail of the distribution; see Neter, Wasserman, and Kutner, 1985, page 134, for a discussion of such remedies for non-normality).

Half-Normal Probability Plot

The half-normal probability plot is constructed in the same way as the standard normal probability plot, except that only the positive half of the normal curve is considered. Consequently, only positive normal values will be plotted on the y-axis.

Specifically, the half-normal probability value z_j for the j'th ordered value (rank) in a variable with n observations is computed as:

$$z_j = \Phi^{-1}[(3*N+3*j-1)/(6*N+1)]$$

where Φ^{-1} is again the inverse normal cumulative distribution function.

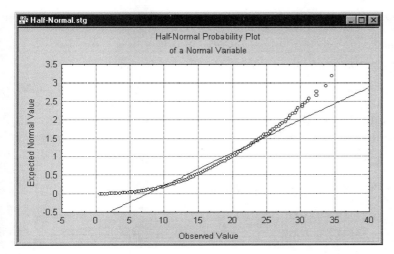

This plot is often used for examining the distribution of residuals (e.g., in **Multiple Regression**) when you want to ignore the sign of the residual, that is, when you are mostly interested in the distribution of absolute residuals, regardless of sign.

Detrended Normal Probability Plot

The detrended normal probability plot is constructed in the same way as the standard normal probability plot, except that before the plot is generated, the linear trend is removed.

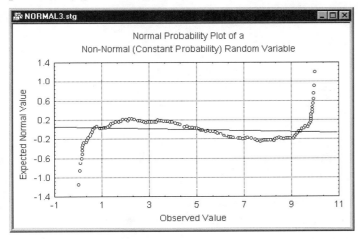

Specifically, in this plot each value (X_j) is standardized by subtracting the mean and dividing by the respective standard deviation (s). The detrended normal probability value z_j for the j'th ordered value (rank) in a variable with n observations is computed as:

```
z_j = Φ⁻¹[(3*j-1)/(3*N+1)] - (x_j-mean)/s
```

where Φ^{-1} is again the inverse normal cumulative distribution function. This often "spreads out" the plot, thereby allowing the user to detect patterns of deviations more easily.

Quantile-Quantile Plots

The Quantile-Quantile (or Q-Q) plot is useful for finding the best fitting distribution within a family of distributions.

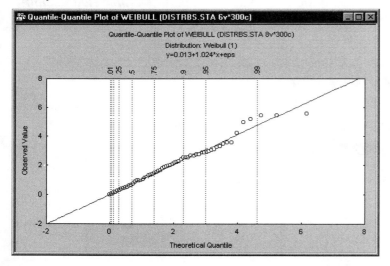

The first step, therefore, is to choose which of the theoretical distributions to fit to the data. After selecting the distribution from the list in the **Quantile-Quantile Plots** dialog, you may or may not need to specify certain parameters for that distribution.

In order to assess the fit of the theoretical distribution to the observed data, the non-missing observed values of the variable are ordered ($x_1 < ... < x_n$), and then these values (x_i) are plotted against the inverse probability distribution function denoted as F^{-1} [specifically, $F^{-1}(i - rank_{adj}/n + n_{adj})$, where F^{-1} depends on the distribution, and $rank_{adj}$ and n_{adj} are user-defined adjustments]. A regression line is then fit to the data in the resulting scatterplot. If the observed

values fall on the regression line (fitting line), then it can be concluded that the observed values follow the specified distribution. The equation of the fitting line ($Y=a + bx$, given in the third title of the resulting Q-Q plot) provides parameter estimates (a and b, where a is the **Threshold** parameter and b is the **Scale** parameter) for the best fitting distribution (see the respective distribution for more information on these parameters).

For the **Exponential**, **Extreme**, **Normal**, and **Rayleigh** distributions, the standardized distribution function is used, and no additional parameters are needed.

For the **Beta**, **Gamma**, **Lognormal**, and **Weibull** distributions, the standardized distribution with specific **Shape** parameters is used. The **Shape** parameters can be specified in one of two ways:

1. The **Shape** parameters are user defined (clear the **Compute parameters from** check box).

2. The **Shape** parameters are estimated (select the **Compute parameters from** check box) based on user-defined **Threshold** and **Scale** parameters and using either the maximum likelihood or matching moments approximation.

The **Shape** parameters are given in the second title of the graph.

Probability-Probability Plots

The Probability-Probability (or P-P) plot is useful for determining how well a specific theoretical distribution fits the observed data. In the P-P plot, the observed cumulative distribution function (the proportion of non-missing values $\leq x$) is plotted against a theoretical cumulative distribution function in order to assess the fit of the theoretical distribution to the observed data.

If all points in this plot fall onto a diagonal line (with intercept 0 and slope 1), then you can conclude that the theoretical cumulative distribution adequately approximates the observed distribution.

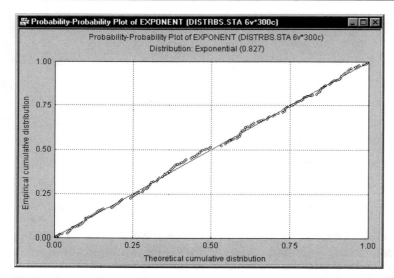

If the data points do not all fall on the diagonal line, then you can use this plot to visually assess where the data do and do not follow the distribution (e.g., if the points form an S shape along the diagonal line, then the data may need to be transformed in order to bring them to the desired distribution pattern).

In order to create this plot, the theoretical distribution function must be completely specified. Therefore, the parameters for the distribution must either be defined by the user or computed from the data (see the specified distribution in the *Electronic Manual* for more information on the respective parameters).

In general, if the observed points follow the selected distribution with the respective parameters, then they will fall onto the straight line in the P-P plot.

Note that you can also use the ***Quantile-Quantile*** plot (see page 682) to obtain the parameter estimate(s) (for the best-fitting distribution from a family of distributions) to be used here.

2D Range Plots

Graphs menu graphs range plots display ranges of values or error bars related to specific data points in the form of boxes or whiskers. Unlike the standard box plots (see page 688), the ranges or error bars are not calculated from data but defined by the raw values in the selected variables.

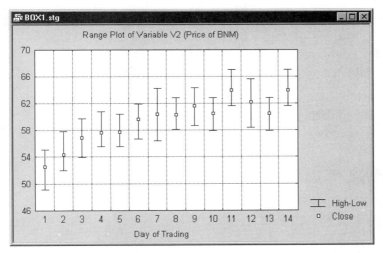

One range or error bar is plotted for each case or observation. In the simplest instance, three variables need to be selected, one representing the mid-points, one representing the upper limits and one representing the lower limits, as illustrated in the example below:

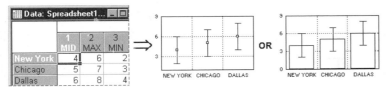

Range graphs are used to display (a) ranges of values for individual units of analysis (items, cases, samples, etc.); (b) variation of scores in individual groups or samples (this is applicable when the measures of variation are obtained from independent measurements; otherwise the box plots, see page 688, which calculate the variation for the samples represented in the graph are more appropriate);

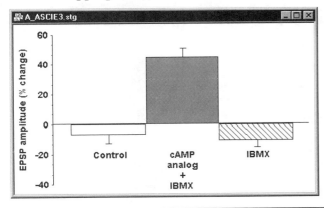

and (c) time spans or value ranges of independently measured phenomena, processes, or actions (e.g., advertising schedules). The major difference between the range and box plots is that in the range plots, all values that define the ranges ("middle points," minimum, and maximum) are not computed from the data but are provided as raw values of data variables.

Typically, horizontal range plots are used to display time span data rather than variation data; they are also recommended when the range labels are very long, because in horizontal range plots the range labels do not have to be rotated (as is the case when long labels are used on the *X*-axis).

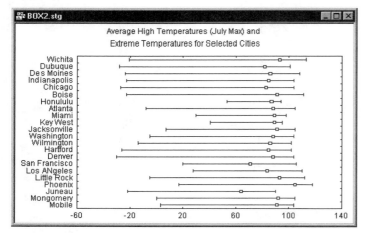

Vertical range plots are often used to represent market data, trading ranges, etc.

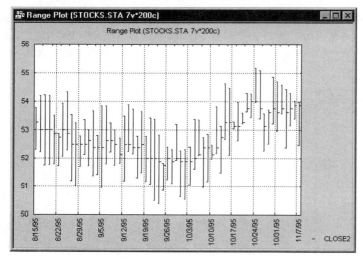

"One-Sided" Ranges or Error Bars in Range Plots

For some applications, it is necessary that the range (or error) bar extends in only one direction.

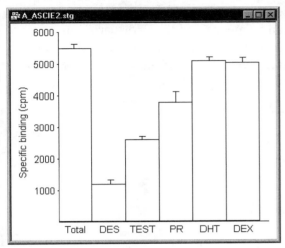

In order to display a "one-sided" range (relative to the mid-point) or an error bar that extends in only one direction, set the respective values of the variable defining the range boundary to *0* (when the **Relative to the Mid-point** mode is selected) or the mid-point (when the **Absolute** mode is selected).

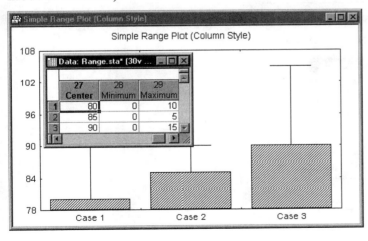

2D Box (and Means with Error) Plots

In box plots (this term was first used by Tukey, 1970), ranges of values of a selected variable (or variables) are plotted separately for groups of cases defined by values of a categorical (grouping) variable. The central tendency (e.g., median or mean) and range or variation statistics (e.g., quartiles, standard errors, or standard deviations) are computed for each group of observations, and the selected values are presented in the style specified in the **Graph Type**. Outlier data points can also be plotted (see the subsection on outliers and extremes, below).

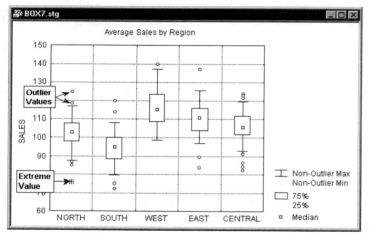

More than one dependent variable can be represented in one graph to allow for comparisons of distributions of scores of the respective measures across groups.

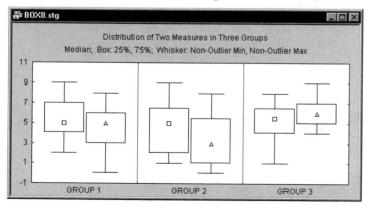

Range plots (see the previous section, page 684) differ from box plots in that for range plots, the ranges to be plotted are defined by (i.e., represent) the values of the selected variables (e.g., one variable contains the minimum range values and another variable contains the maximum range values), while for box plots, the ranges are calculated from raw values of the variable (e.g., standard deviations, standard errors, or ranges).

There are two typical applications for box plots: (a) comparing ranges of values for individual samples or categories of cases (e.g., a typical min-max plot for stocks or commodities or aggregated sequence data plots with ranges); and (b) comparing distributions or variations of scores in individual groups or samples (e.g., box and whisker plots presenting the mean for each sample as a point inside the box, standard errors as the box, and standard deviations around the mean as a narrower box or a pair of "whiskers").

The box plots, showing variation of scores, allow you to visualize and quickly evaluate the strength of the relation between the grouping and dependent variable(s). Specifically, assuming that the dependent variable is normally distributed, and knowing what proportion of observations fall within, for example, ±1 or ±2 standard deviations from the mean (see *Elementary Concepts* in the *Electronic Manual*), you can easily evaluate the results of an experiment and say that (for example) the scores in about 95% of cases in experimental group 1 belong to a different range than scores in about 95% of cases in group 3.

Analysis of outliers and extreme values. Another specific application of box plots is in the analysis of the distribution of values that deviate from central tendencies for their respective groups (so-called outlier and extreme values, see below). The implementation of box plots in *STATISTICA* offers comprehensive facilities to detect such values and to display them in a variety of ways; options are also provided to "trim the distributions" and display the respective statistical summaries based on values that fall only within requested ranges of the distributions.

Outliers and Extremes

Values that are "far" from the middle of the distribution are referred to as outliers and extreme values if they meet the conditions specified in the next three paragraphs.

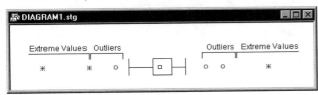

Outliers. For *Graphs* menu graphs box plots, a data point is deemed to be an outlier if the following conditions hold:

```
data point value > UBV + o.c.*(UBV - LBV)
```

or

```
data point value < LBV - o.c.*(UBV - LBV)
```

where

UBV   is the upper value of the box in the box plot (e.g., the mean + standard error or the 75th percentile).

LBV   is the lower value of the box in the box plot (e.g., the mean - standard error or the 25th percentile).

o.c.   is the outlier coefficient specified in the *Outliers* group of the box plot dialog (by default, the outlier coefficient is 1.5).

Non-outlier range. The non-outlier range is the range of values in the *2D Box Plots* (and *3D Sequential Graphs - Box Plots*, or *Categorized Box Plots*), which fall below the upper outlier limit (for example, if the outlier coefficient is 1.5, then +1.5 * the height of the box) and above the lower outlier limit (for example, -1.5 * the height of the box).

Extremes. For *Graphs* menu graphs box plots, a data point is deemed to be an extreme value if the following conditions hold:

```
data point value > UBV + 2*o.c.*(UBV - LBV)
```

or

```
data point value < LBV - 2*o.c.*(UBV - LBV)
```

where

UBV   is the upper value of the box in the box plot (e.g., the mean + standard error or the 75th percentile).

LBV   is the lower value of the box in the box plot (e.g., the mean - standard error or the 25th percentile).

o.c.   is the outlier coefficient specified in the *Outliers* group of the box plot dialog (by default, the outlier coefficient is 1.5, thus, the extreme values are those that are outside the 3 box length range from the upper and lower value of the box).

For example, the following diagram illustrates the ranges of outliers and extremes in the "classic" box and whisker plot (for more information about box plots and the topic behind these default values, see Tukey, 1977).

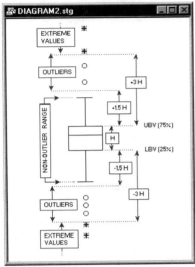

In this plot, the upper box value (UBV) is in the 75th percentile, the lower box value (LBV) is in the 25th percentile, and the outlier coefficient is equal to 1.5.

Note that in *STATISTICA* box and whisker plots, you can adjust all these parameters.

2D Line Plots (Variables)

The *2D Line Plots* dialog contains a selection of two-dimensional univariate and multivariate line plots. In line plots, individual data points are connected by a line.

Line plots provide a simple way to visually present a sequence of many values (e.g., stock market quotes over a number of days); *XY Trace* line plots (see below) can be used to display a trace (instead of a sequence).

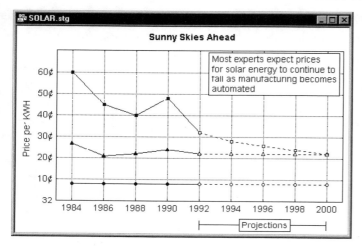

Line graphs can also be used to plot continuous functions, theoretical distributions, etc.

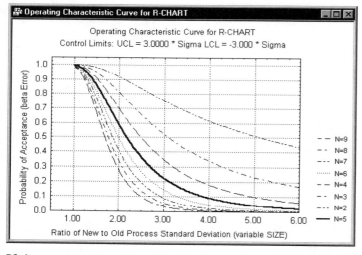

If there are only a few observations, then a vertical bar graph is usually better, although notable exceptions to this rule are plots of differences between means across a number of groups. Such plots are integrated into various modules of *STATISTICA* (e.g., integrated plots of interactions are built into the **ANOVA**, **Basic Statistics and Tables**, and other modules).

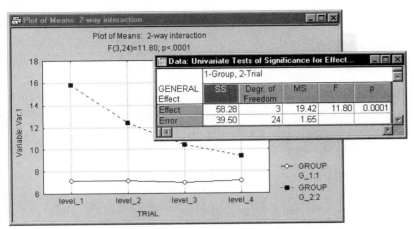

If there are very many observations in the sequence and if they are differentiated, then smoothing is necessary to uncover or expose the overall pattern of the data sequence. The simplest form of smoothing is aggregation, where instead of raw data, the means of consecutive sets of *n* observations are plotted. In aggregated line plots the ranges of values in consecutive segments are marked using whisker-type high-low marks.

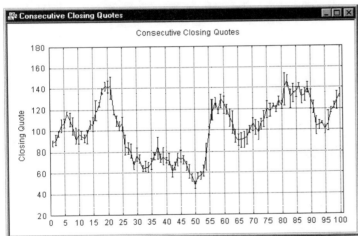

Aggregation can also be used as a data reduction technique allowing you to present in one plot more data points than could otherwise be displayed (given the resolution of the monitor or the printer). For example, on most display monitors, even in high-resolution mode (e.g., 800x600), the plotting area consists of no more than 500 vertical lines of pixels, thus series longer than 500 data points cannot be shown properly; aggregation can make long series much more readable on the screen.

XY Trace Line Plot (Drawing Based on Coordinates)

Note that unlike in the regular (sequential) line plots, in *XY Trace* line plots, individual data points are connected with lines in the sequence in which they appear in the datafile; they are not sorted in any way. An empty data cell (missing data) "breaks" the line. Therefore, line plots can be used to "draw" from coordinates stored in a file (e.g., maps, special contours, etc.).

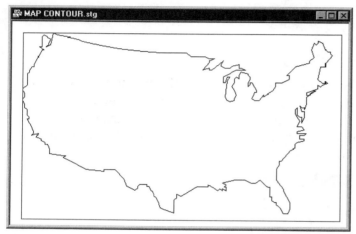

Obviously, for interactive customization of graphs, the integrated on-screen drawing tools are much more convenient.

Line plots can also be used to draw contour-like computed projections of surfaces.

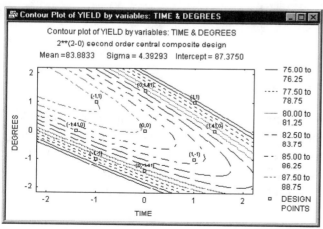

This method of generating contour plots is used in some modules of *STATISTICA*.

Sequential/Stacked Plots

All types of graphs included in this category are used to present sequences of values. In this respect, the way in which they visualize data is similar to line plots (see page 691). In fact, as long as you select only one variable to be plotted, the graphical representation of data will be identical to the one offered by line plots, although the specific graphical representations of data offered by the current graphs are more differentiated (e.g., line plots, area charts, step charts, bar charts, etc.).

The only major difference between the representation of data in those graphs as compared to line graphs appears when you select more than one variable to be plotted. In **Line Plots**, each of them would be plotted independently from the others; thus, for example, if two variables have the same values for case number 3, then at this point (case 3) the two lines will cross or overlap. However, all graphs in the current category "stack" corresponding values of consecutive variables (from the selected list).

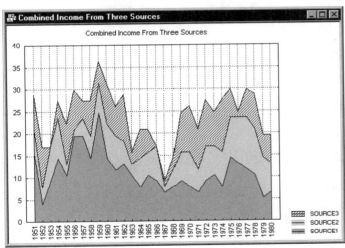

Thus, in this example, the point representing case number 3 in the second variable will be respectively higher than in the first variable. The variables are stacked in the order in which they were selected.

Due to this stacked representation of values in consecutive variables, the lines (or steps, areas, columns, etc.) of consecutive variables will never overlap as long as they are larger than 0.

This interpretation of data involves a limitation concerning missing data in the plotted dataset. Specifically, the graph location of each data point in each consecutive variable (from the selected list) is calculated as a sum of its value and the corresponding values (i.e., those for the same case) in all "preceding" variables in the list. Therefore, if any of those preceding values are missing, then the sum cannot be calculated and the graph will be "broken" at this point. Thus, there should be no missing data in the datasets selected for the stacked representation (except for the last variable).

Applications

These types of graphs are used to present sequences of values of selected variables. Sequential/stacked line plots can be used for some of the applications reviewed in the section on line plots (see page 691). However, the stacked feature of those graphs (applicable when more than one variable is selected) is specifically designed to represent the broad category of datasets in which consecutive variables represent parts ("portions") of one whole. For example, each case may represent the GNP in one fiscal year, and each of several variables may represent the dollar amount contributed by each of several industries and other sources of goods and services. If such data were represented as a stacked column graph, then the total height of each column would indicate the total GNP, and each of its stacked segments would show the relative contribution of the respective industry.

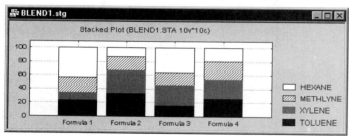

If the variables shown in the graph represent percentages and/or add up to the same value (e.g., 100%) for each "case," then the total height of the graph will be constant across cases, as shown in the example above.

Pie Charts

The pie chart (the term first used by Haskell, 1922) is one of the most common graph formats used for representing proportions or values of variables. Depending on the selected type of graph (see below), the pie chart will display either *raw values* or *counts* (i.e., frequencies) of specific categories of values (such as those that can be displayed in histograms (see page 669).

Pie Charts of Values

The sequence of values from the variable will be represented by consecutive slices of the pie; the size of each slice will be proportional to the respective value.

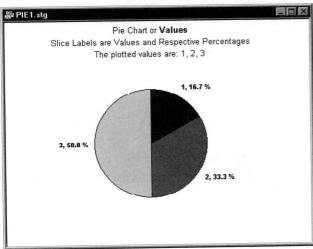

The values should be greater than *0* (*0*'s and negative values cannot be represented as slices of the pie). This simple type of pie chart (sometimes called data pie chart) interprets data in the most straightforward manner: one case = one slice.

Slices of this pie chart can be labeled with case names, case numbers, or values of specific variables, and those labels can also include numeric information (e.g., percentages, see below).

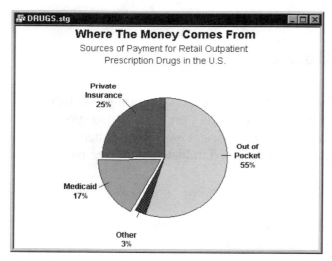

The typical application for simple (raw data) pie charts is to show how some quantity is divided (e.g., how the total amount of the national budget is spent) or how different components contribute to the total outcome (e.g., production in each of several factories that belong to a single, parent company).

Pie Charts of Counts

Unlike the simple data pie chart (see above), this type of pie chart (sometimes called frequency pie chart) interprets data like a histogram. It categorizes all values of the selected variable following the selected categorization technique and then displays the relative frequencies as pie slices of proportional sizes.

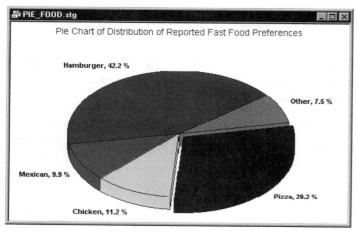

Frequency pie charts categorize the data and display the results of the categorization in terms of relative frequencies of consecutive categories. Thus, those pie charts offer an alternative method to display frequency histogram data (see the section on histograms, page 669).

Missing/Range Data Plots

The **Missing and 'Out of Range' Data Plots** allow you to examine the pattern or distribution of missing and/or user-specified "out of range" data points in the current dataset (or in a subset of variables and cases).

This graph is useful in exploratory data analysis to determine the extent of missing (and/or "out of range") data and whether the patterns of those data occur more or less at random:

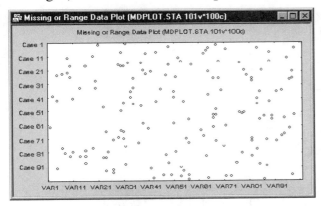

or they form an identifiable pattern:

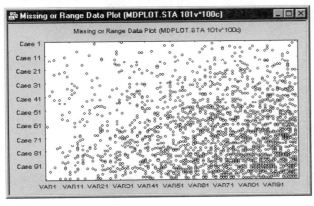

In a sense, they depict a "map" of a datafile (or its segments) and allow you to examine patterns of missing data, particularly low values, high values, etc.

There are many obvious applications for this general type of graph allowing you to quickly visualize the pattern of an entire dataset. It is also one of the graphs recommended for Six Sigma applications and it is included in the list of *STATISTICA* Six Sigma shortcuts (see Chapter 1 – *STATISTICA – A General Overview*).

The categorized format allows you to compare such patterns between particular subsets of data.

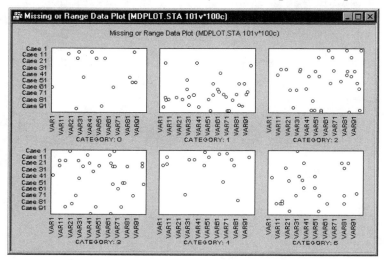

Note that a variety of methods is available in *STATISTICA* to transform and recode data (e.g., spreadsheet formulas or *STATISTICA* Visual Basic; see Chapter 4 – *STATISTICA Spreadsheets*) including menu-driven techniques to replace or convert missing values. In most analyses, missing data can be managed using either pairwise deletion or casewise deletion of missing data, or substitution by means.

3D SEQUENTIAL GRAPHS

Raw Data Plots

These graphs allow you to plot sequences of raw data (observations) for one or more selected variables in a 3-dimensional display. The selected variables are represented on the y-axis, the consecutive cases on the x-axis, and the values are plotted against the z-axis as illustrated below:

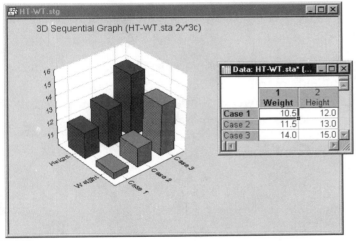

Three-dimensional raw data plots are used to visualize sequences of values. The nature of this graph is analogous to the multiple line graph, except that in the **3D Raw Data Graphs**, the ribbons, lines, boxes, or other 3D representations of values of each variable are not overlaid (as in the 2D graph), but "spaced" from each other in 3D perspective.

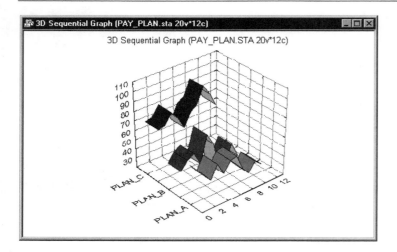

Applications

The **3D Raw Data Plots** have both presentation and analytic applications. The most common application of **3D Raw Data Plots** is to present and communicate data (e.g., price quotes, population growth, sales vs. earnings). These graphs provide a simple and often attractive way to visually present sequences of observations, such as various types of time series.

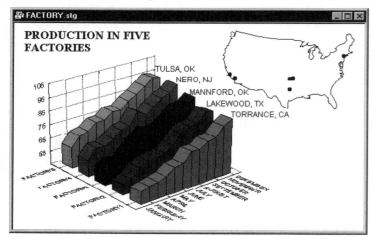

The major advantage of such 3D representations over 2D multiple line plots is that for some datasets, in 3D displays, the individual sequences of values can easily be identified; if a proper viewpoint is selected (e.g., via interactive rotation, see Chapter 7 – *STATISTICA Graphs: Creation and Customization*), then the lines can never overlap or "run into each other" as is often the case in multiple line 2D graphs.

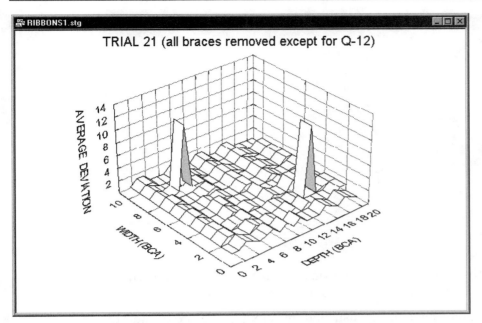

Three-dimensional raw data plots are also used for analytic purposes, when the input data are in a matrix format and the pattern of the matrix data is to be analyzed (as shown above).

Bivariate Histograms

Three-dimensional (bivariate) histograms are used to visualize crosstabulations of values in two variables. They can be considered to be a conjunction of two simple (i.e., univariate) histograms, combined such that the frequencies of co-occurrences of values on the two analyzed variables can be examined.

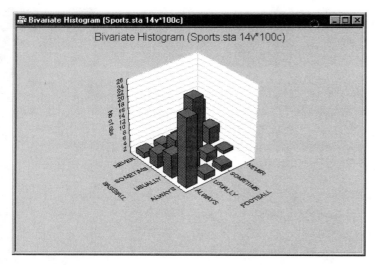

As mentioned before, there are two major reasons why frequency distributions (either univariate or bivariate, such as those visualized in 3D histograms) are of interest.

- You may learn from the shape of the distributions about the nature of the examined variables (e.g., a bimodal distribution may suggest that the sample is not homogeneous, and consists of observations that belong to two populations that are more or less normally distributed); this application is particularly relevant for smoothed *3D Bivariate Histograms* (see below).

- Many statistics are based on assumptions about the distributions of analyzed variables; *3D Bivariate Histograms* help you to test whether those assumptions are met for pairs of variables.

3D Histograms vs. Crosstabulations

3D Bivariate Histograms provide information similar to crosstabulations. Although specific (numerical) frequency data are easier to read in a table, the overall shape and global descriptive characteristics of bivariate distributions may be easier to explore in a graph. Moreover, the graph provides qualitative information about the distribution that cannot be fully represented by any single index. For example, a bivariate skewed distribution of response latencies vs. the duration of a reaction-time task (in a reaction-time experiment) may result from the changes in subjects' strategies of dealing with fatigue.

Different Categorization Methods in One Graph

Different methods of categorization can be used for each of the two variables for which the bivariate distribution is visualized in the graph, as illustrated in the following 3D Bivariate Histogram of reaction time scores by experimental conditions.

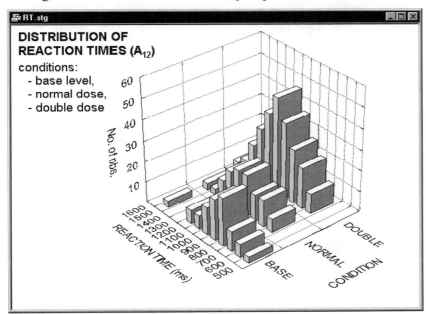

Specifically, in this graph the distribution of reaction times (a continuous variable categorized by dividing the entire range of values into 12 intervals of equal size) can be reviewed across three experimental conditions (a discrete variable with three distinctively labeled levels: *Base*, *Normal*, and *Double*).

Smoothing Bivariate Distributions

The smoothing facilities available for 3D Bivariate Histograms (see Chapter 7 – *STATISTICA Graphs: Creation and Customization*) allow you to fit surfaces to 3D representations of bivariate frequency data. Thus, every 3D histogram can be turned into a smoothed surface. This technique is of relatively little help if applied to a simple pattern of categorized data (such as the histogram that was shown above).

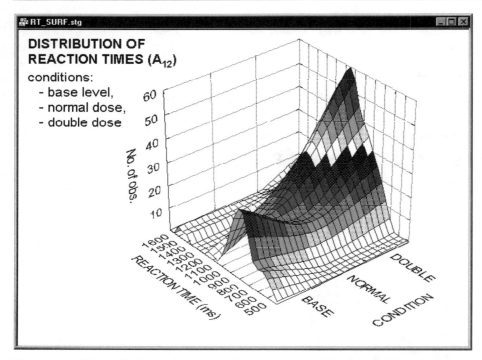

However, if applied to more complex patterns of frequencies such as this one:

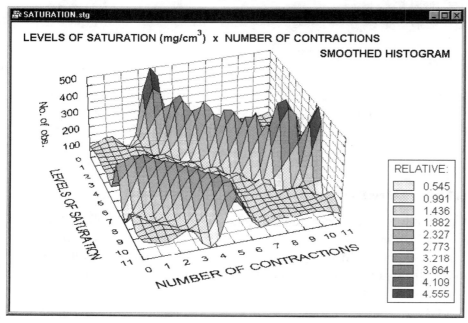

it may provide a valuable exploratory technique, allowing identification of regularities that are less salient when examining the standard 3D histogram representations (e.g., see the systematic surface "wave patterns" shown on the smoothed histogram above).

3D Range and Box Plots

Similar to **2D Range Plots** (see page 684) and to **2D Box plots** (see page 688), 3D range and 3D box plots display ranges of values or error bars related to specific measures of central tendency, but the results are displayed in three dimensions, which may produce a more attractive presentation effect and sometimes increases the readability (and informativeness) of the display, especially when the appropriate viewpoint is selected.

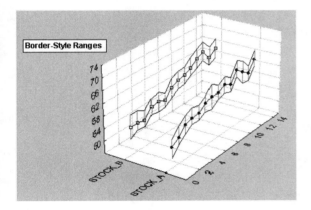

Applications

The typical applications for both range and box plots are the same (they differ only in terms of the source of data: existing (i.e., already calculated) values for central tendencies and ranges vs. values calculated by the graphing procedure directly from raw data). In general, these graphs are used to display (a) ranges of values for individual units of analysis (items, cases, samples, etc.) or (b) variation of scores in individual groups or samples.

Graph Type

STATISTICA offers a unique selection of **3D Range Plots** and **3D Box Plots** (illustrated below). When the 📁 **Point Ranges** style is used, the ranges are represented by point markers (connected by a line). One range is plotted for each case. When 📁 **Border-style Ranges** are

plotted, the ranges are represented by two continuous lines (upper and lower ranges). The mid-points are represented by point markers connected with a line. For  **Error Bar-style Ranges**, the mid-points are represented by point markers, and the ranges are represented by error bars. One error bar is plotted for each case.

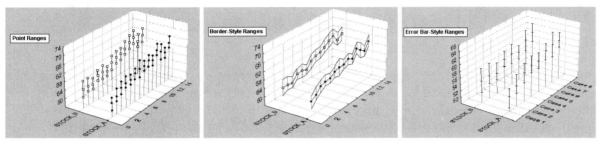

When 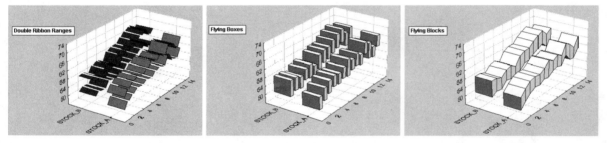 **Double Ribbon Ranges** are plotted, the ranges are represented by two ribbons (upper and lower ranges). For **Flying Boxes**, the ranges are represented by "flying" boxes, and for **Flying Blocks**, the ranges are represented by "flying" blocks.

For these three graph types (i.e., **Double Ribbon Ranges**, **Flying Boxes**, and **Flying Blocks**), the mid-points are not displayed in the graph (they can be shown when you switch to one of the first three types of range plots).

3D XYZ GRAPHS

3D Scatterplots

3D Scatterplots (also called XYZ Scatterplots) represent the most elementary form of 3D data plots. Its most typical application is to visualize relations between continuous variables.

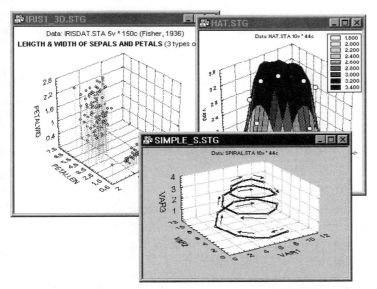

Although there may be specific reasons for using 3D Scatterplots for particular applications, one of the major general advantages of this type of plot is that it may reveal complex interactive relations between the represented variables.

Applications

Exploring interactive relations between variables. Consider the following illustrative example. Imagine that the price and supply of a commodity were registered over a period of time (at different times of the day). Plotting the values of those three variables (*Price*, *Supply*, and *Hour* of the day) may reveal a complex multivariate interactive relationship that would be almost impossible to identify in a numerical exploratory data analysis.

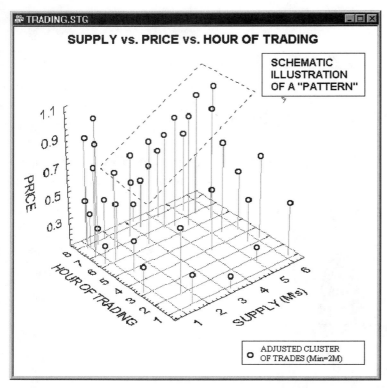

For example, it could be revealed that the strength of the relationship increases in the second half of the day (i.e., the later the hour, the closer the price is related to supply); however, the shape of the plot may also show that this relationship does not hold when the supply is very low (i.e., for low values of the variable *Supply*).

It is often easier to reveal such complex interactive relations using graphical as compared to numerical methods, especially when the relationships involved are curvilinear.

Identifying clusters and subsets in samples from non-homogenous populations. Another area of exploratory data analysis where XYZ scatterplots can be useful is when it is expected that there are distinctive clusters of observations that can be identified by examining the distribution of observations (simultaneously) on three or more variables.

For example, the following XYZ Scatterplot shows data from the "classic" iris dataset (Fisher, 1936; file *Irisdat.sta*); the observations represent different types of iris.

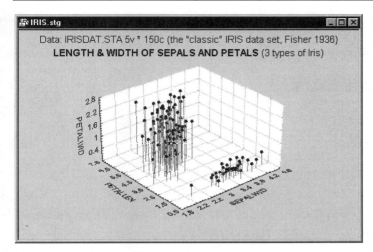

The graph shows that by plotting the width of petals against the length of petals and width of sepals, you can see that the sample is not homogenous.

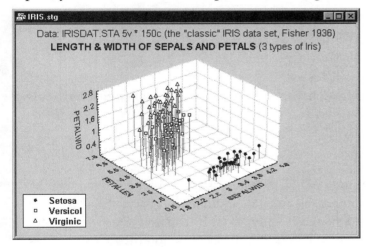

You can easily identify the distinctive types of iris in the graph above, where the subsets are marked using the multiple subsets facility (see Chapter 7 – *STATISTICA Graphs: Creation and Customization*).

Examining results from exploratory multivariate analyses. In statistics, XYZ scatterplots are often used to visualize results of multivariate exploratory techniques, such as factor analysis and multidimensional scaling.

For example, plotting labeled cases in three-dimensional space created by a three-dimensional solution of multidimensional scaling can aid in the interpretation of the dimensions, as well as classification of individual cases.

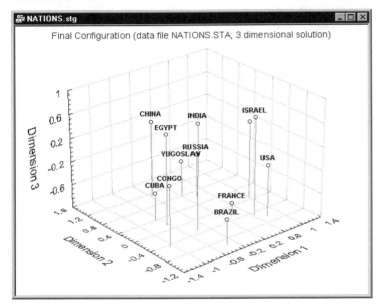

For more information, see also the discussions on *Factor Analysis* and *Multidimensional Scaling* in the *Electronic Manual*.

Rotation

One of the common problems when examining 3D Scatterplots is overlapping points, which decrease the readability of the graph. In some cases when the number of observations is very large, the graph is almost impossible to understand if you look at it from only one viewpoint. Thus, interactive on-screen rotation is particularly useful in the exploration of 3D scatterplots, as illustrated below.

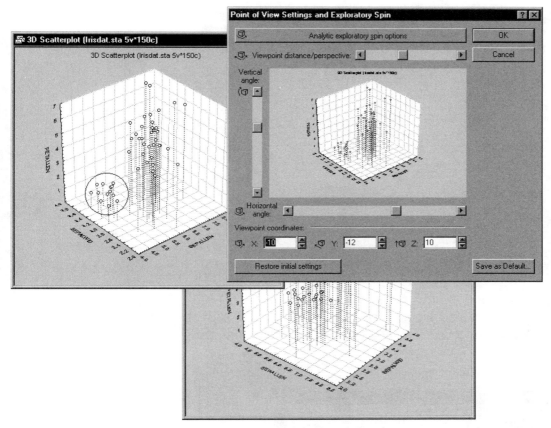

The rotation (and perspective) facilities for three-dimensional graphs are described in Chapter 7 – *STATISTICA Graphs: Creation and Customization*.

Exploration of data via 3D spinning. By clicking the *Spin* button, the display can be set into continuous rotation in clockwise or counter-clockwise directions.

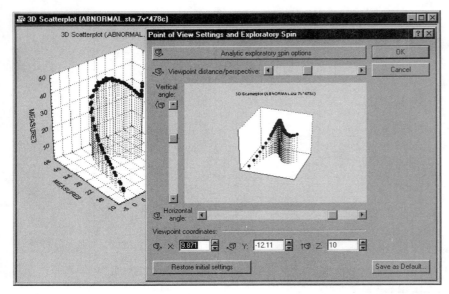

Brushing. In addition to the regular (2D) brushing, *STATISTICA* also supports "true 3D" subset selection. This can be accomplished in two ways: interactively using a unique set of tools to make precise selections of subsets in 3D displays using the simple mouse movements,

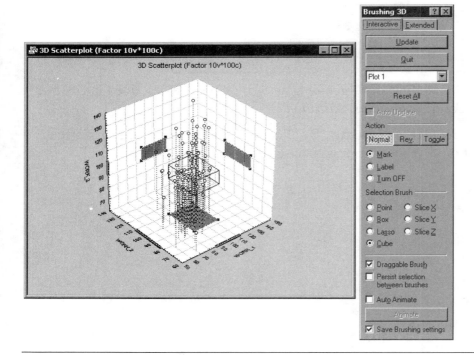

and using the **Extended** tab of the **Brushing 3D** dialog, which contains options to **Highlight** (or **De-highlight** or **Swap** the highlight attribute) points by specifying the value range coordinates of selected variables, or by reference to attributes of previously selected points.

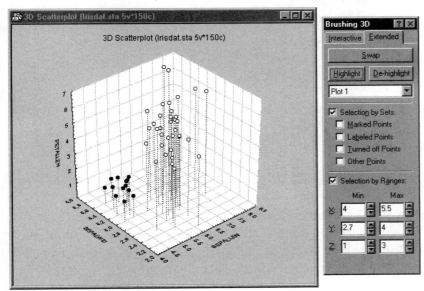

In XYZ scatterplots, the selection of ranges by values will refer to the X, Y, and Z coordinates of the data point values. For more information on brushing, see Chapter 9 – *STATISTICA Graphs: Examples and FAQ's* and the *Electronic Manual*.

Surface Plots

In the 3D Surface plot, a surface function will be fitted to a 3D scatterplot. As in the *3D Scatterplot* (see page 709), this is commonly done in order to reveal hidden patterns of data and detect the relationships among the three variables in the plot.

STATISTICA offers several types of surface functions that can be fit to the data and smoothing methods that can be applied (see Chapter 7 – *STATISTICA Graphs: Creation and Customization*). They range from the simple linear surface function to functions that will "smooth" the data and approximate a surface at a given level of stiffness (e.g., distance-weighted least squares, negative exponential smoothing, etc.). Those more complex surface-fitting methods can be particularly useful in order to uncover complex (interactive), non-linear, and non-monotonous relationships between the variables.

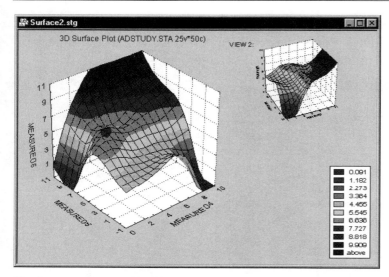

Options to customize the appearance of the wirenet, levels of shading, color palette, etc. are available in the **All Options** dialog, accessible by double-clicking on the graph background or by selecting **All Options** from the **Format** menu (see Chapter 6 – *STATISTICA Graphs: General Features* and Chapter 7 – *STATISTICA Graphs: Creation and Customization*).

The Nonlinear Estimation module. The surface fitting (and data smoothing) facilities accessible from the graphics dialogs and described below offer a selection of predefined, general-purpose exploratory procedures; you can also plot (overlay) a custom-defined function by entering a function equation (see below).

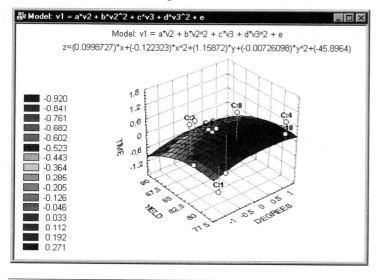

In order to fit user-defined functions to data, use **Nonlinear Estimation**, which offers comprehensive facilities to fit functions of practically unlimited complexity (including piecewise models).

A wide selection of different function minimization algorithms is provided, predefined or user-defined loss functions can be used, and the results can be explored via interactive surface plotting options (to visually examine the closeness of the fit).

Applications

Surface plots can be used in exploratory data analyses such as those illustrated in the previous section (on 3D scatterplots). Also, they are useful to visualize results from analyses, such as custom function fitting (via **Nonlinear Estimation**) or cluster analysis.

A common application of the surface plot in the context of industrial statistics is to visualize the results of central composite design experiments (see **Experimental Design** in the *Electronic Manual*). There, specific systematic values of two (or more) variables are set by the experimenter to evaluate the effect of those variables on some dependent variable of interest (e.g., strength of a synthetic fabric).

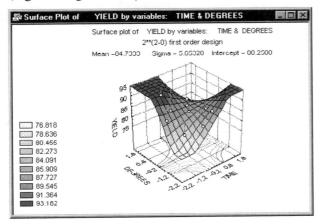

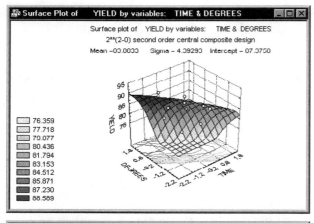

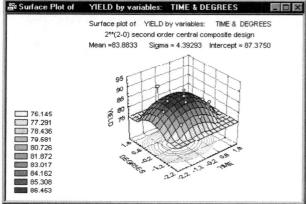

In many cases, complex non-linear relationships between variables emerge in such experiments.

Rotation

It is often useful to rotate the surface plot in order to reveal its characteristics (e.g., specific peaks and valleys) more clearly or examine the hidden sides.

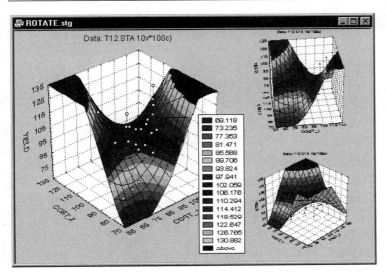

The rotation (and perspective) facilities for three-dimensional graphs are described in Chapter 7 – *STATISTICA Graphs: Creation and Customization*.

Contour Plots

Contour plots are created by fitting a 3D surface function to a 3D scatterplot; the resulting surface contours (i.e., lines of identical "altitudes") are then projected onto the *X-Y* plane. Like *Surface Plots* (see page 715), contour plots are used to examine relationships between three variables in the plot.

STATISTICA offers several types of surface functions that can be fit to the data (see Chapter 7 – *STATISTICA Graphs: Creation and Customization*) using the integrated graphics facilities. They range from the simple linear surface function to functions that will "smooth" the data and approximate a surface at a given level of stiffness (e.g., distance-weighted least squares, negative exponential smoothing, etc.).

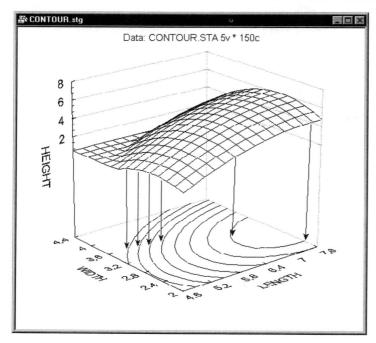

Those more complex surface-fitting methods can be particularly useful in order to uncover complex (interactive), non-linear, and non-monotonous relationships between the variables.

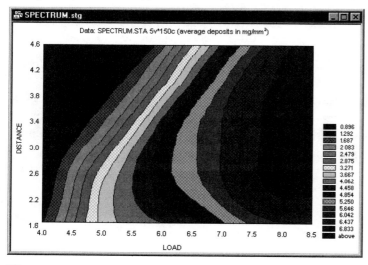

The **Nonlinear Estimation** module (see the *Electronic Manual*) offers specialized facilities to fit user-defined functions.

Applications

Contour Plots can be used in exploratory data analyses such as those mentioned in the previous two sections (on 3D scatterplots and surfaces).

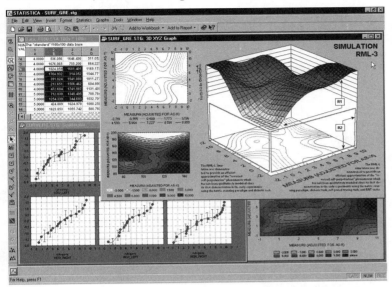

Also, they are useful to visualize results from analyses, such as custom function fitting (via **Nonlinear Estimation**) or cluster analysis. As compared to surface plots (see the previous section), they may be less effective to quickly visualize the overall shape of 3D data structures, however, their main advantage is that they allow for precise examination and analysis of the shape of the surface (contour plots display a series of undistorted horizontal "cross sections" of the surface).

A common application of the contour plot in the context of industrial statistics is to visualize the results of a central composite design experiment (see **Experimental Design** in the *Electronic Manual*).

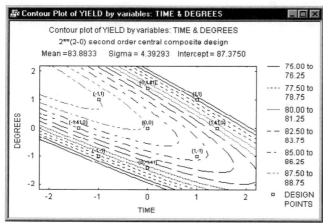

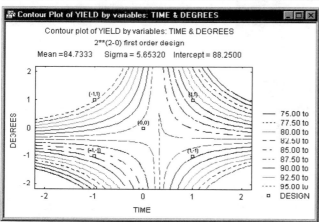

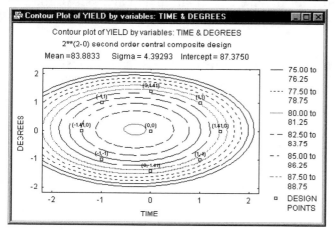

There, specific systematic values of two (or more) variables are set by the experimenter, to evaluate the effect of those variables on some dependent variable of interest (e.g., strength of a synthetic fabric). In many cases, complex non-linear relationships between variables emerge in such experiments.

Trace Plots

As in *3D Scatterplots* (see page 709), each data point in a trace plot is represented by its location in 3D space as determined by the values of the variables selected as *X*, *Y*, and *Z* (and interpreted as the *X*, *Y*, and *Z* axis coordinates). The data points are then connected sequentially (in the order encountered in the datafile) with a line to form a "trace" of a sequential process (e.g., movement, change of a phenomenon over time, etc.).

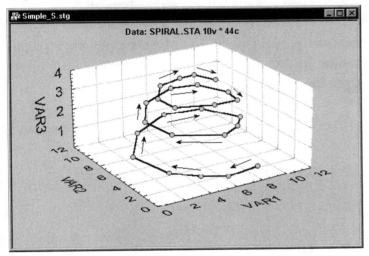

A good metaphor of the information that is best represented in a trace plot is that of the trajectory of an object in three-dimensional space.

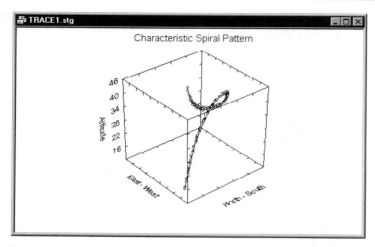

In general, any sequential process (or processes) that simultaneously produces changes in three dimensions over consecutive measurements can be studied via the trace plot.

Multiple trace plot. The multiple trace plot differs from the regular trace plot only in that it allows for a list of Z variables; thus, multiple "trajectories" can be simultaneously displayed.

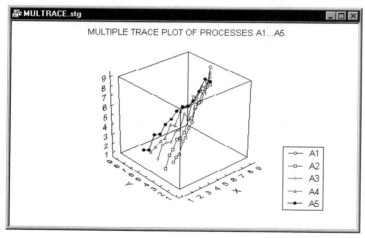

Applications

Examples of datasets that match the trajectory metaphor above could be the behavior of any multivariate time series. Suppose you measured the temperature, pollution level, and ozone content of the air in a large city over consecutive months for several years. Since those variables are likely cyclical in nature (i.e., it is cold in the winter in the northern hemisphere), a characteristic pattern would emerge; moreover, the pattern is usually not simple (e.g., linear) in nature. The prices for several commodities or multiple economic indicators could also be studied via this graph; that is, characteristic patterns or "trajectories" over time may become apparent.

3D drawings. Another application of this graph is to generate precise "3D drawings" (by providing coordinates in 3D space) of objects such as control limits, marked areas, etc. Typically, in order to customize (by overlaying) 3D data plots, 3D objects "drawn" by the trace plot facility become active parts of the 3D display, and thus they can be rotated and transformed in perspective (see below). Note that such objects cannot be drawn interactively since there is no method to control the third dimension (i.e., the "depth") in drawn objects (they can only be "drawn" by specifying 3D coordinate systems in *STATISTICA* Visual Basic).

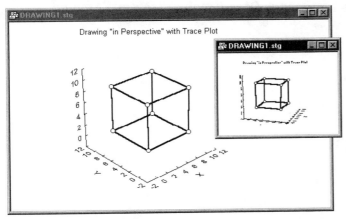

Inserting an incomplete case (i.e., a case with either one or two, but not all three values *X, Y,* and *Z*) into the dataset presented as a trace plot will break the line.

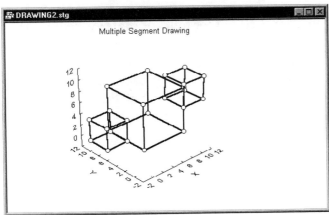

This feature can be used to create separate objects (as shown above).

Ternary Plots

A ternary (or "trilinear") plot can be used to examine relations between three or more dimensions where three of those dimensions represent components of a mixture (i.e., the relations between them are constrained such that the values of the three variables add up to the same constant for each case). One typical application of this graph is when the measured response(s) from an experiment depends on the relative proportions of three components (e.g., three different chemicals) that are varied in order to determine an optimal combination of those components (e.g., in mixture designs; see **Experimental Design** in the *Electronic Manual*).

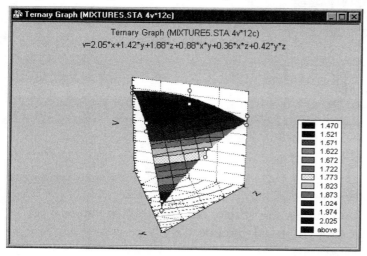

In ternary plots, the triangular coordinate systems are used to plot four (or more) variables (the components X, Y, and Z, and the responses $V1$, $V2$, etc.) in two dimensions (ternary scatterplots or contours) or three dimensions (ternary 3D scatterplots or surface plots). In order to produce ternary graphs, the relative proportions of each component (within each case) are constrained to add up to the same value (e.g., 1). By default, when the graph is produced, the proportions are rescaled so that they add to 1 in each case.

Applications

Mixture Designs. For example, suppose you have a mixture that consists of 3 components *A*, *B*, and *C*. Any mixture of the three components can be summarized by a point in the triangular coordinate system defined by the three variables.

For example, take the following 6 different mixtures of the 3 components.

A	B	C
1	0	0
0	1	0
0	0	1
0.5	0.5	0
0.5	0	0.5
0	0.5	0.5

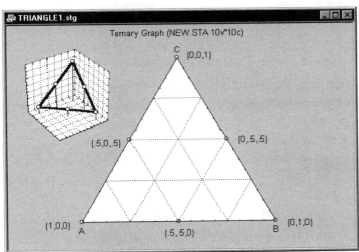

The sum for each mixture is 1.0, so the values for the components in each mixture can be interpreted as proportions. If you graph these data in a regular 3D scatterplot, it becomes apparent that the points form a triangle in the 3D space. Only the points inside the triangle where the sum of the component values is equal to 1 are valid mixtures. Therefore, you can simply plot only the triangle to summarize the component values (proportions) for each mixture.

The three components are represented by axes that run from each apex to the midpoint of the opposite baseline; the location of individual data points are determined by their values (components) on the respective three axes.

To read off the coordinates of a point in the triangular graph, you would simply "drop" a line from each respective vertex to the side of the triangle below.

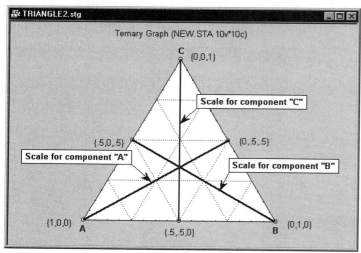

This can be further illustrated by the following example (recently discussed by Wainer, 1995).

Student performance on tests of the National Assessment of Educational Progress (NAEP) was characterized into 3 levels of proficiency (**Advanced/Proficient**, **Basic**, and **Below Basic**). Results from different regions can be compared in a ternary plot where the percentage of students at each level of proficiency is plotted on the three axes.

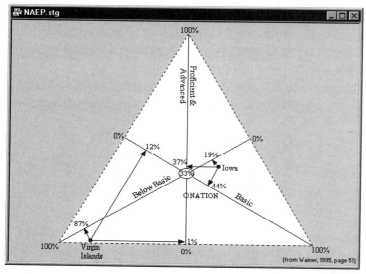

The above graph from Wainer (1995, page 51) illustrates that 37% of *Iowa* students performed at the *Advanced* or *Proficient* level, 44% at the *Basic* level, and 19% *Below Basic*. By comparison, only 1% of *Virgin Islands* students performed at the *Advanced* or *Proficient* level, 12% were at the *Basic* level, and 87% were *Below Basic*.

Wainer also discusses another interesting application of triangular coordinate plots. The plot is used to examine the percentage vote in each of Britain's three political parties in the 1987 and 1992 general elections. Interested readers should refer to Wainer, 1995 for more information.

The following types of ternary graphs are offered in *STATISTICA:*

ternary surfaces (with projected contours),

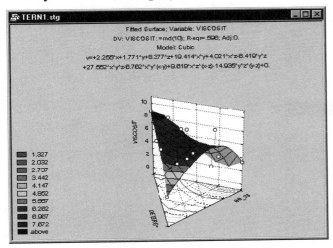

ternary 2D and 3D scatterplots (regular and multiple),

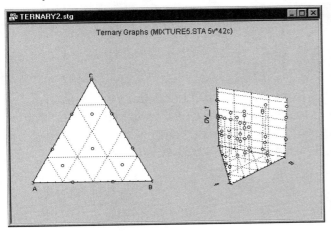

ternary 3D trace plots, space plots, and deviation plots (regular and multiple),

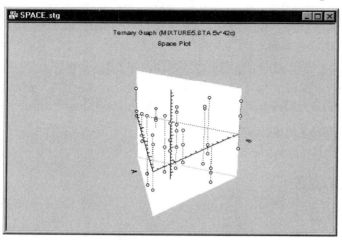

ternary 2D contour plots (area or line pattern style).

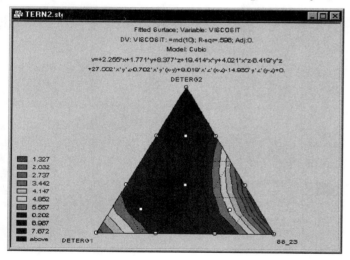

Options are provided to handle cases with invalid sums as well as out of range cases. A selection of specialized ternary fitting options (*Linear*, *Quadratic*, *Full Cubic*, and *Special Cubic* models) are available for all surface and contour plots. The surface/contour levels can be generated according to user specifications, easily redefined, customized, and/or saved as defaults (or as reusable templates).

3D Categorized Plots

This type of *Graphs* menu graph allows you to categorize 3D scatterplots (and trace plots), contour plots, and surface plots by the specified categories of a selected variable or other logical categorizations of observations.

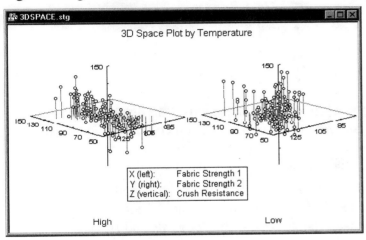

The information provided in this graph is the same as that shown in the non-categorized 3D scatterplot (page 709), surface plot (page 715), or contour plot (page 719), except that one graph is shown for each group or category specified by the user. Thus, the general purpose of this plot is to facilitate comparisons between groups or categories regarding the relationships between three or more variables

Applications

Applications for the different available 3D graph types are described in the context the of non-categorized graphs. In general, 3D XYZ graphs summarize the interactive relationships between three variables. The different ways in which data can be categorized allow you to review those relationships contingent on some other criterion (e.g., group membership).

For example, a positive relationship between age, health status, and life satisfaction may hold for females in a study, but not for the male participants.

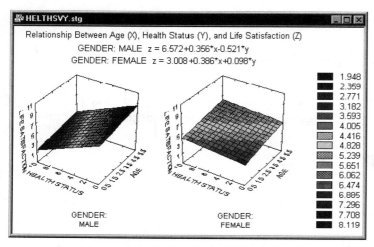

Because categories can be created via logical conditions that specify **Multiple Subsets** (see Chapter 7 – *STATISTICA Graphs: Creation and Customization*), you could produce further exploratory graphs to break down the male sample into those who are single and those who are married; you could also create one additional category for high-income males alone, etc.

From the categorized surface plot shown below (and its contour plot equivalent), you can conclude that the setting of the tolerance level in an apparatus does not affect the investigated relationship between the measurements (**Depend1**, **Depend2**, and **Height**) unless the setting is ≤**3**.

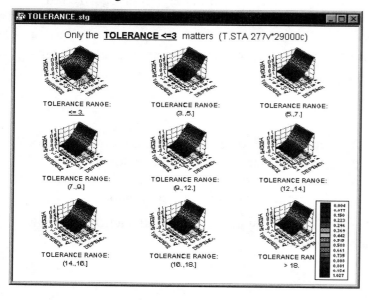

Contour plots are sometimes easier to analyze than their surface plot equivalents (as illustrated in the following example).

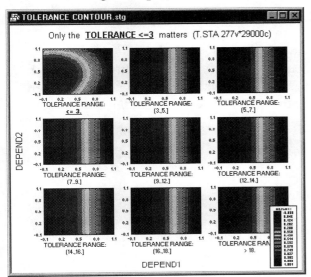

Thus, the facilities available via the *3D Categorized Plots* option provide powerful exploratory tools for studying complex relationships between variables and groups of observations.

MATRIX PLOTS

Matrix graphs summarize the relationships between several variables in a matrix of true *X-Y* plots. The most common type of matrix plot is the scatterplot matrix, which can be considered to be the graphical equivalent of the correlation matrix (e.g., see **Basic Statistics and Tables** in the *Electronic Manual*).

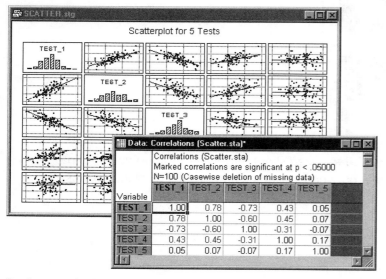

In the graph shown above, scatterplots are shown for each pair of variables; the linear regression line is also shown in each scatterplot.

Scatterplot matrices can be either square (as the one shown above) or rectangular, when two lists of variables are selected (analogous to rectangular matrices). If the matrix is square then on the diagonal of the scatterplot matrix, the distribution for each variable is represented in a histogram.

Graphs such as the one shown above provide an efficient summary of the relationships between the variables in an analysis. For example, a variable that does not correlate with any of the other variables will "stick out" and be easily identified.

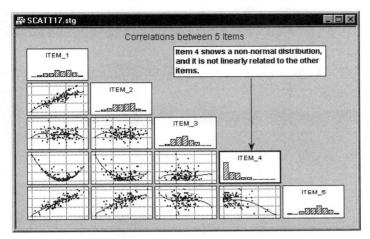

Line plot matrices. In addition to matrix scatterplots such as the ones shown above, which are typically used to summarize the relationships between some random variables, you can summarize several sequential processes in line plot matrices, such as the one shown below, which is based on a real-life dataset.

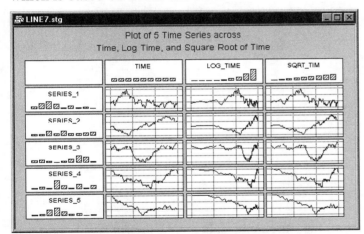

Here, several *Y* variables are plotted against a single *X* variable denoting time; thus, the "behavior" of several variables over time is summarized in this graph.

Applications

The most common application for matrix plots is to summarize the variables in an analysis (their distributions and relationships) in a single graph.

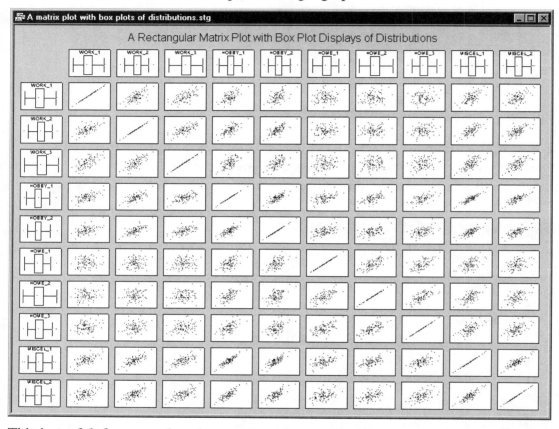

This is useful, for example, when constructing measurement scales (e.g., see **Reliability and Item Analysis** in the *Electronic Manual*) or when performing exploratory analyses of survey data (e.g., questionnaire data, economic data, process control data, etc.).

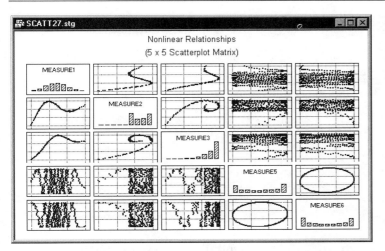

Note that the zooming (or "magnifying glass") feature described in Chapter 6 – *STATISTICA Graphs: Environment* is particularly useful for examining large matrix plots.

For example, shown below is a summary matrix scatterplot for 15 variables.

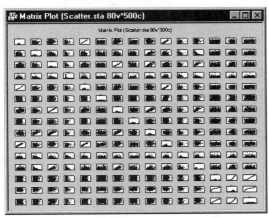

The density of this graph makes it difficult to examine individual scatterplots. However, when **Clone graph on zoom** is selected from the **View** menu, you can click the magnifying glass toolbar button ⊕, then move the magnifying glass mouse pointer to the respective scatterplot and click to expand the plot in a separate window.

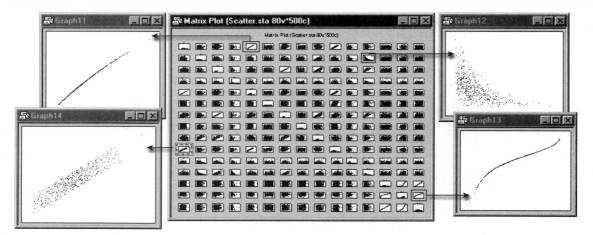

As you can see in the illustration above, you can expand any number of plots in separate windows (for more information on **Zoom** options, refer to Chapter 6 – *STATISTICA Graphs: General Features*).

In exploratory data analysis it is often useful to examine the contribution of observations that meet specific conditions to the overall shape of relations between variables. The multiple subset facility available in the matrix plot allows the user to mark subsets of cases defined by logical conditions.

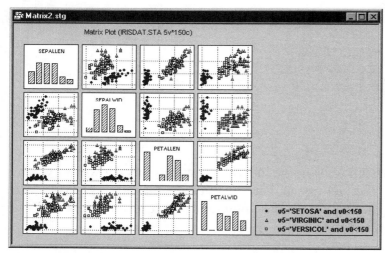

The multiple subset selection facility is described in the *Electronic Manual*

ICON PLOTS

Icon plots provide a selection of plots that represent cases or units of observation as multidimensional symbols.

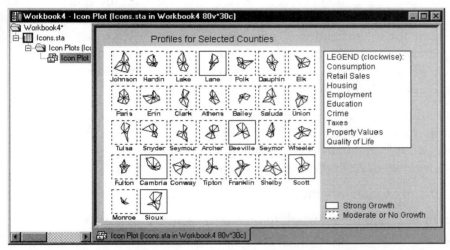

The basic idea of icon plots is to represent individual units of observation as particular graphical objects where values of variables are assigned to specific features or dimensions of the objects (usually one case = one object). The assignment is such that the overall appearance of the object changes as a function of the configuration of values.

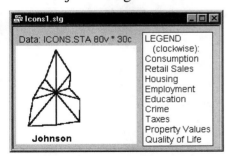

Thus, the objects are given visual "identities" that are unique for configurations of values and that can be identified by the observer. Examining such icons may help to discover specific clusters of both simple relations and interactions between variables.

Analyzing Icon Plots

The "ideal" design of the analysis of icon plots consists of five phases:

1. Select the order of variables to be analyzed. In many cases a random starting sequence is the best solution. You can also try to enter variables based on the order in a multiple regression equation, factor loadings on an interpretable factor, or a similar multivariate technique. That method can simplify and "homogenize" the general appearance of the icons, which would facilitate the identification of non-salient patterns. It can also, however, make some interactive patterns more difficult to find. No universal recommendations can be given at this point, other than to try the quicker (random order) method before getting involved in the more time-consuming method.

2. Look for any potential regularities, such as similarities between groups of icons, outliers, or specific relations between aspects of icons (e.g., "if the first two rays of the star icon are long, then one or two rays on the other side of the icon are usually short"). The circular type of icon plots (see *Taxonomy of Icon Plots*, below) is recommended for this phase.

3. If any regularities are found, try to identify them in terms of the specific variables involved.

4. Reassign variables to features of icons (or switch to one of the sequential icon plots, see *Taxonomy of Icon Plots*, below) to verify the identified structure of relations (e.g., try to move the related aspects of the icon closer together to facilitate further comparisons). In some cases, at the end of this phase it is recommended to drop the variables that appear not to contribute to the identified pattern.

5. Finally, use a quantitative method (such as a regression method, nonlinear estimation, discriminant function analysis, or cluster analysis) to test and quantify the identified pattern or at least some aspects of the pattern.

Taxonomy of Icon Plots

Most icon plots can be assigned to one of two categories: *circular* and *sequential*.

Circular icons. Circular icon plots (**Star** plots, **Sun Ray** plots, **Polygon** icons) follow a "spoked wheel" format where values of variables are represented by distances between the center ("hub") of the icon and its edges.

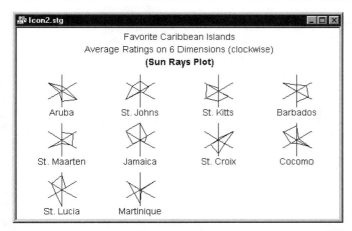

Those icons may help to identify interactive relations between variables because the overall shape of the icon may assume distinctive and identifiable overall patterns depending on multivariate configurations of values of input variables.

In order to translate such "overall patterns" into specific models (in terms of relations between variables) or verify specific observations about the pattern, it is helpful to switch to one of the sequential icon plots (see the next paragraph), which may prove more efficient when you already know what to look for.

Sequential icons. Sequential icon plots (*Column* icons, *Profile* icons, *Line* icons) follow a simpler format where individual symbols are represented by small sequence plots (of different types).

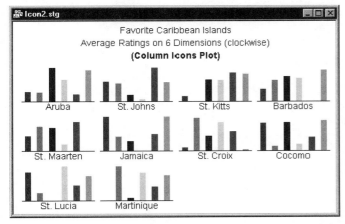

The values of consecutive variables are represented in those plots by distances between the base of the icon and the consecutive break points of the sequence (e.g., the height of the

columns shown above). Those plots may be less efficient as a tool for the initial exploratory phase of icon analysis because the icons may look alike. However, as mentioned before, they may be helpful in the phase when some hypothetical pattern has already been revealed and you need to verify it or articulate it in terms of relations between individual variables.

Pie icons. *Pie* icon plots fall somewhere in between the previous two categories; all icons have the same shape (pie) but are sequentially divided in a different way according to the values of consecutive variables.

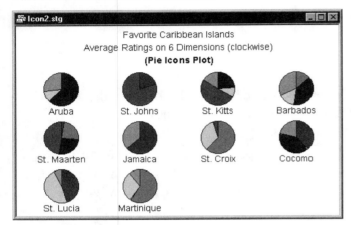

From a functional point of view, they belong rather to the sequential than circular category, although they can be used for both types of applications.

Chernoff faces. This type of icon is a category by itself. Cases are visualized by schematic faces such that relative values of variables selected for the graph are represented by variations of specific facial features.

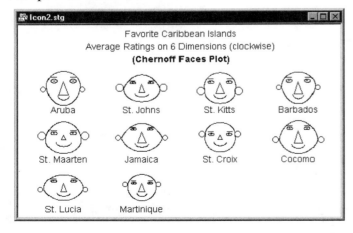

Due to its unique features, it is considered by some researchers as an ultimate exploratory multivariate technique that is capable of revealing hidden patterns of interrelations between variables that cannot be uncovered by any other technique. This statement may be an exaggeration, however. Also, it must be admitted that Chernoff Faces is a method that is difficult to use, and it requires a great deal of experimentation with the assignment of variables to facial features.

Standardization of Values

Except for unusual cases when you intend for the icons to reflect the global differences in ranges of values between the selected variables, the values of the variables should be standardized once to assure within-icon compatibility of value ranges. For example, because the largest value sets the global scaling reference point for the icons, then if there are variables that are in a range of much smaller order, they may not appear in the icon at all, e.g., in a star plot, the rays that represent them will be too short to be visible. See the *Electronic Manual* for more information on the standardization options available for this type of graph.

Applications

Icon plots are generally applicable (1) to situations where you want to find systematic patterns or clusters of observations, and (2) when you want to explore possible complex relationships between several variables. The first type of application is similar to cluster analysis; that is, it can be used to classify observations.

For example, suppose you studied the personalities of artists, and you recorded the scores for several artists on a number of personality questionnaires. The icon plot may help you determine whether there are natural clusters of artists distinguished by particular patterns of scores on different questionnaires (e.g., you may find that some artists are very creative, undisciplined, and independent, while a second group is particularly intelligent, disciplined, and concerned with publicly acknowledged success).

The second type of application (the exploration of relationships between several variables) is more similar to factor analysis; that is, it can be used to detect which variables tend to "go together." For example, suppose you were studying the structure of people's perception of cars. Several subjects completed detailed questionnaires rating different cars on numerous dimensions. In the datafile, the average ratings on each dimension (entered as the variables) for each car (entered as cases or observations) are recorded.

When you now study the Chernoff faces (each face representing the perceptions for one car), it may occur to you that smiling faces tend to have big ears; if price was assigned to the amount of smile and acceleration to the size of ears, then this "discovery" means that fast cars are more expensive. This, of course, is only a simple example; in real-life exploratory data analyses, non-obvious complex relationships between variables may become apparent.

CATEGORIZED GRAPHS

What Are Categorized Graphs?

Categorized graphs (also called casement plots, Chambers, et al., 1983) produce a series of 2D or 3D graphs (such as histograms, scatterplots, line plots, surface plots, etc.), one for each selected *category* of cases (i.e., subset of cases), for example, respondents from New York, Chicago, Dallas, etc. These "component" graphs are placed sequentially in one display, allowing for comparisons between the patterns of data shown in graphs for each of the requested groups (e.g., cities).

A variety of options is offered for selecting the subsets; the simplest of them is using a categorical variable (e.g., a variable *City*, with three values *New York*, *Chicago*, and *Dallas*). For example, the following graph shows histograms of a variable representing self-reported stress levels in each of the three cities.

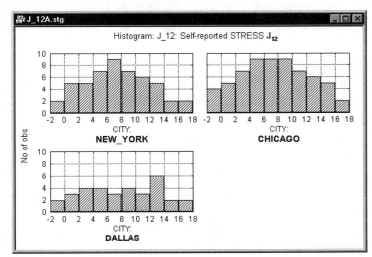

You could conclude that the data suggest that people live in *Dallas* are less likely to report being stressed, while the patterns (distributions) of stress reporting in *New York* and *Chicago* are quite similar.

Categorized graphs in *STATISTICA* also support two-way categorizations, where not one criterion (e.g., **City**) but two criteria (e.g., **City** and **Time** of the day) are used to create the subsets. Two-way categorized graphs can be thought of as "crosstabulations of graphs" where each component graph represents a cross-section of one level of one grouping variable (e.g., **City**) and one level of the other grouping variable (e.g., **Time**).

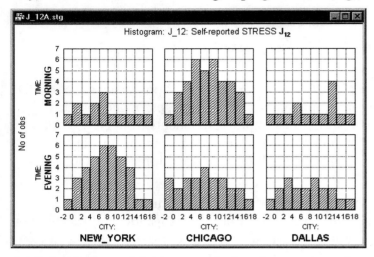

Adding this second factor reveals that the patterns of stress reporting in New York and Chicago are actually quite different when the **Time** of questioning is taken into consideration, whereas the **Time** factor makes little difference in Dallas.

Categorized graphs vs. matrix graphs. Matrix graphs (see above) also produce displays containing multiple component graphs; however, each of those component graphs are (or can be) based on the same set of cases and the graphs are generated for all combinations of variables from one or two lists. Categorized graphs require a selection of variables that normally would be selected for non-categorized graphs of the respective type (e.g., two variables for a scatterplot). However, in categorized plots, you also need to specify at least one grouping variable that contains information on group membership of each case (e.g., **Chicago**, **Dallas**). That grouping variable will not be included in the graph directly (i.e., it will not be plotted) but it will serve as a criterion for dividing all analyzed cases into separate graphs. As illustrated above, one graph will be created for each group (category) identified by the grouping variable.

Categorized Histograms

In general, histograms are used to examine frequency distributions of values of variables. For example, the frequency distribution plot shows which specific values or ranges of values of the examined variable are most frequent, how differentiated the values are, whether most observations are concentrated around the mean, whether the distribution is symmetrical or skewed, whether it is multimodal (i.e., has two or more peaks) or unimodal, etc. Histograms are also useful for evaluating the similarity of an observed distribution with theoretical or expected distributions.

The **Histogram** procedure available from the **Graphs** menu allows you to produce histograms broken down by one or two categorical variables, or by any other one or two sets of logical categorization rules (via Multiple Subsets categorization, see Chapter 7 – *STATISTICA Graphs: Creation and Customization* for details).

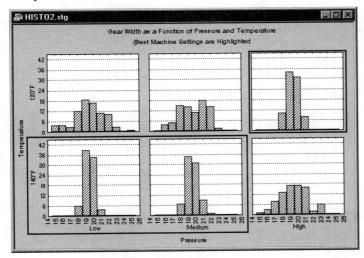

There are two major reasons why frequency distributions are of interest.

- You can learn from the shape of the distribution about the nature of the examined variable (e.g., a bimodal distribution may suggest that the sample is not homogeneous and consists of observations that belong to two populations that are more or less normally distributed).

- Many statistics are based on assumptions about the distributions of analyzed variables; histograms help you to test whether those assumptions are met.

Often, the first step in the analysis of a new dataset is to run histograms on all variables. Using categorized histograms can make the results more informative,

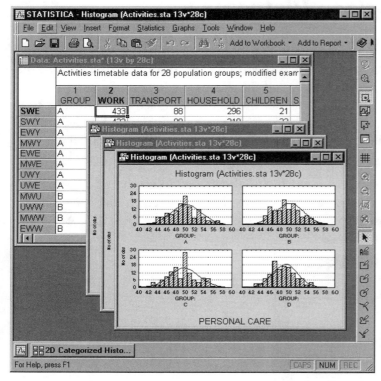

and reveal, for example, a lack of homogeneity of the sample.

Histograms vs. Breakdown

Categorized Histograms provide information similar to breakdowns (e.g., mean, median, minimum, maximum, differentiation of values, etc.; see *Basic Statistics and Tables* in the *Electronic Manual*). Although specific (numerical) descriptive statistics are easier to read in a table, the overall shape and global descriptive characteristics of a distribution are much easier to examine in a graph. Moreover, the graph provides qualitative information about the distribution that cannot be fully represented by any single index. For example, the overall skewed distribution of income may indicate that the majority of people have an income that is much closer to the minimum than maximum of the range of income. Moreover, when broken down by gender and ethnic background, this characteristic of the income distribution may be found to be more pronounced in certain subgroups. Although this information will be contained

in the index of skewness (for each sub-group), when presented in the graphical form of a histogram, the information is usually more easily recognized and remembered. The histogram may also reveal "bumps" that may represent important facts about the specific social stratification of the investigated population or anomalies in the distribution of income in a particular group caused by a recent tax reform.

Categorization of Values within Each Histogram

All histogram procedures offer the standard selection of categorization methods; see Chapter 7 – *STATISTICA Graphs: Creation and Customization* for details.

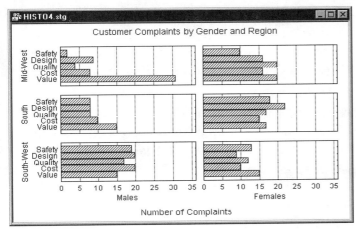

Those categorization methods divide the entire range of values of the examined variable into a number of categories or sub-ranges for which frequencies are counted and presented in the plot as individual columns or bars. See *2D Histograms - Categorization of Values* (page 671) for more details.

Categorization of Values into Component Graphs

The categorization options for assigning observations to the component graphs of the categorized histogram are equally flexible. Component graphs may be created for the levels of a categorical variable (e.g., gender), continuous variables may be categorized into a user-defined number of intervals, or user-defined logical subsetting conditions may be specified to determine each sub-group.

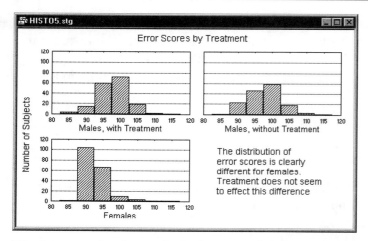

The latter option is particularly powerful, because it allows you to base the categorization on "rules" that reference more than one variable, and on the logical relationships between those variables (e.g., a subgroup might consist of all individuals who are male, 30 or older, and divorced or never married).

Categorized histograms and scatterplots. A useful application of the categorization methods for continuous variables is to represent the simultaneous relationships between three variables. Shown below is a scatterplot for two variables *Load 1* and *Load 2*.

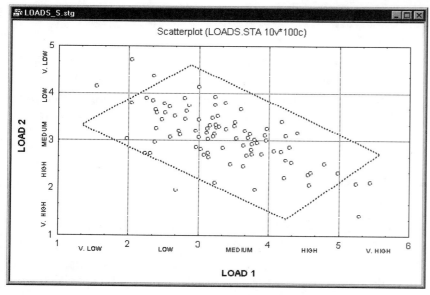

Now suppose you would like to add a third variable (**Output**) and examine how it is distributed at different levels of the joint distribution of **Load 1** and **Load 2**. The following graph could be produced:

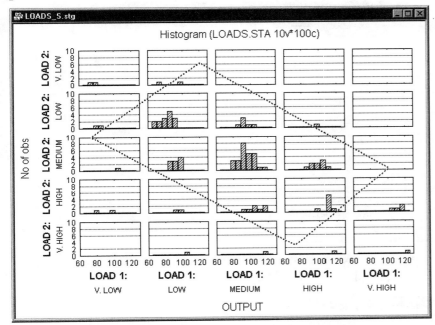

In this graph, **Load 1** and **Load 2** are both categorized into five intervals, and within each combination of intervals the distribution for variable **Output** is computed. Note that the "box" (parallelogram) encloses approximately the same observations (cases) in both graphs shown above.

Categorized Scatterplots

In general, two-dimensional scatterplots are used to visualize relations between two variables X and Y (e.g., weight and height). As mentioned before, in scatterplots, individual data points are represented by point markers in two-dimensional space, where axes represent the variables. The two coordinates (X and Y) that determine the location of each point correspond to its specific values on the two variables. If the two variables are strongly related, then the data points form a systematic shape (e.g., a straight line or a clear curve). If the variables are not related, then the points form a round "cloud."

The categorized scatterplot option allows you to produce scatterplots categorized by one or two variables. Via the **Multiple Subsets** options (see Chapter 7 – *STATISTICA Graphs: Creation and Customization*), you can also categorize the scatterplot based on logical selection conditions that define each category or group of observations.

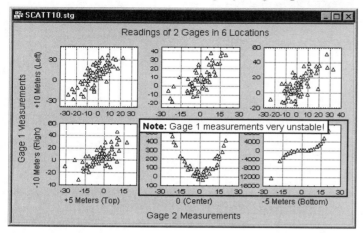

Categorized scatterplots offer a powerful exploratory and analytic technique for investigating relationships between two or more variables within different sub-groups.

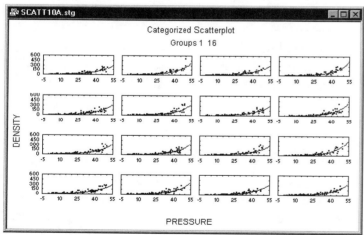

A variety of analytic options are available to enhance exploratory analyses.

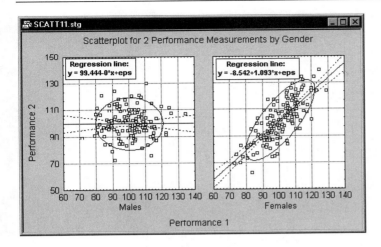

Homogeneity of Bivariate Distributions (Shapes of Relations between Variables)

Scatterplots are typically used to identify the nature of relations between two variables (e.g., blood pressure and cholesterol level), because they can provide much more information than a correlation coefficient.

For example, a lack of homogeneity in the sample from which a correlation was calculated can bias the value of the correlation. Imagine a case where a correlation coefficient is calculated from data points that came from two different experimental groups, but this fact was ignored when the correlation was calculated. Suppose the experimental manipulation in one of the groups increased the values of both correlated variables, and thus the data from each group form a distinctive "cloud" in the scatterplot (as shown in the following illustration).

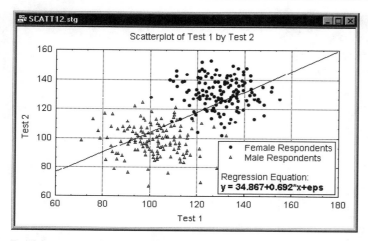

In this example, the high correlation is entirely due to the arrangement of the two groups, and it does not represent the "true" relation between the two variables, which is practically equal to *0* (as could be seen if you looked at each group separately).

If you suspect that such a pattern may exist in your data and you know how to identify the possible "subsets" of data, then producing a categorized scatterplot

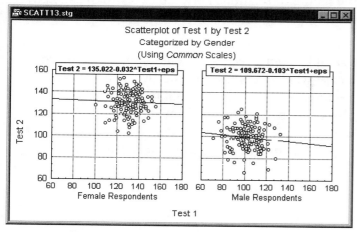

may yield a more accurate picture of the strength of the relationship between the *X* and *Y* variable, within each group (i.e., after controlling for group membership).

Fitting Functions to All Subsets
Combined or Separately to Each Subset

Categorized graphs fit separate functions to each subset of points (one function curve is drawn for each subset, as shown in the following example of an overlaid categorized scatterplot where three regression lines are plotted, one for each subset).

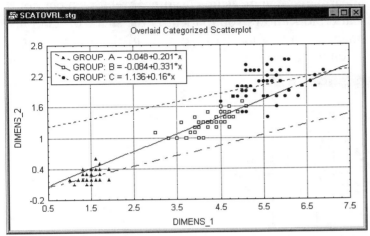

If it is desired to fit one function to all points combined (but you still need to be able to identify the members of each subset), use the **Mark Selected Subsets** option accessible from the **2D Scatterplots** dialog (see the *Electronic Manual* for more details).

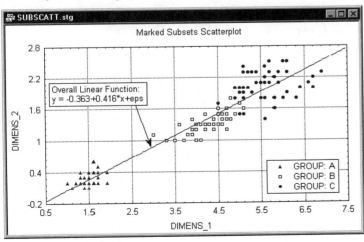

The two graphs shown above were produced from the same dataset.

Curvilinear Relations

Curvilinearity is another aspect of the relationships between variables that can be examined in scatterplots. There are no "automatic" or easy-to-use tests to measure curvilinear relationships between variables: The standard Pearson r coefficient measures only linear relations; some nonparametric correlations such as the Spearman R can measure curvilinear relations, but not non-monotonous relations. Examining scatterplots allows you to identify the shape of relations, so that later an appropriate data transformation can be chosen to "straighten" the data or choose an appropriate nonlinear estimation equation to be fit.

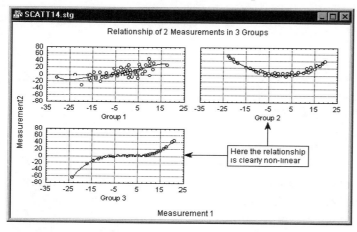

For more information, refer to the discussions on **Basic Statistics**, **Distribution Fitting**, **Multiple Regression**, and **Nonlinear Estimation** in the *Electronic Manual*.

CHAPTER

STATISTICA GRAPHS: EXAMPLES AND FAQ'S

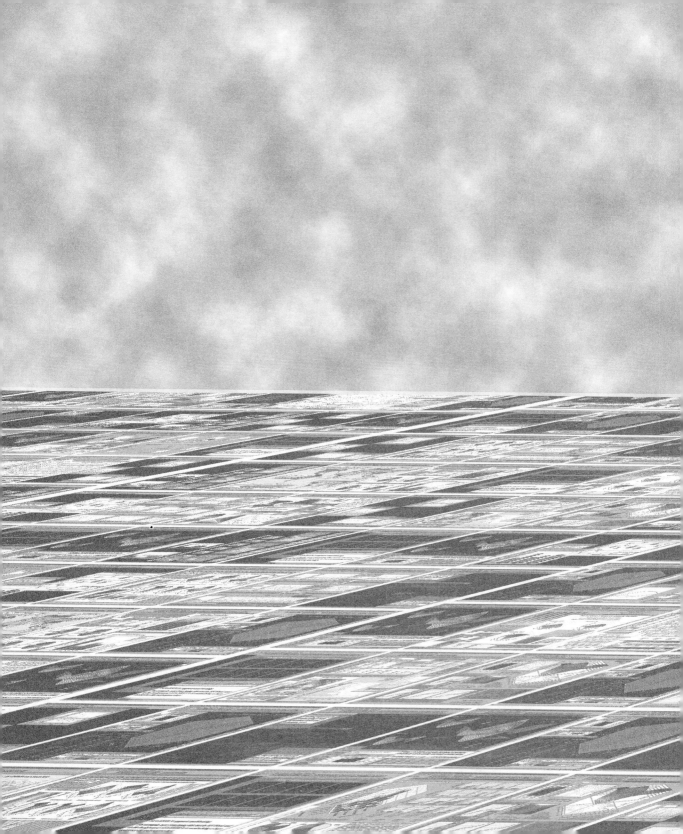

STATISTICA GRAPHS: EXAMPLES AND FAQ'S

GRAPH EXAMPLES

Several examples of the two basic categories of graphs available in *STATISTICA* (i.e., **Graphs of Input Data** (and their extended version **Graphs** menu graphs) and **Graphs of Block Data**) are illustrated in the following sections. Characteristically, **Graphs of Input Data** and **Graphs of Block Data** are quicker to define and create than **Graphs** menu graphs because they use a variety of default settings (see *General Categories of Graphs* in Chapter 6 – *STATISTICA Graphs: General Features* for details on each of the graph categories). However, **Graphs of Input Data** and **Graphs of Block Data** can be as extensively customized after creation as **Graphs** menu graphs (see Chapter 7 – *STATISTICA Graphs: Creation and Customization* for details).

Example: Creating a Graph of Input Data

Graphs of Input Data can use the context of the current selection in an analysis spreadsheet to define which variables to use, but they always use data from the input datafile to create the selected graph. For a detailed discussion of the characteristics of **Graphs of Input Data**, see Chapter 6 – *STATISTICA Graphs: General Features*.

For this example, open **Adstudy.sta** located in the Examples\Datasets subfolder of your *STATISTICA* installation folder. You can open it using the **File - Open** menu, the ⬚ Open Data button on any Startup Panel, or the ⬚ toolbar button. **Adstudy.sta** contains the results of a (hypothetical) survey given by a hotel to its customers. In addition to polling customers on

---

their preferred cola soft drinks, the surveyors asked them to rate the hotel's performance on a number of items such as room service, cleanliness, front desk service, etc.

Producing the Graph

- With **Adstudy.sta** open, select **Statistics - Basic Statistics/Tables** from either the menu bar or the *STATISTICA* Start button ▓ menu on the *STATISTICA* **Analysis** bar to display the **Basic Statistics and Tables** (Startup Panel).

- Select **Correlation matrices** from the **Basic Statistics and Tables** dialog to display the **Product-Moment and Partial Correlations** dialog.

- Click the **One variable list** button to display a variable selection dialog, and select variables **Measure01** through **Measure23** for analysis.

- Click the **OK** button on the variable selection dialog to return to the **Product-Moment and Partial Correlations** dialog.

- Click the **Summary: Correlation matrix** button on the **Quick** tab to produce the analysis spreadsheet shown below.

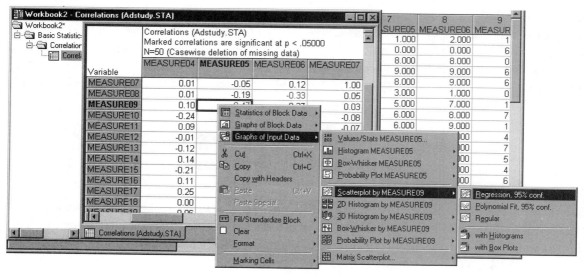

Examination of the spreadsheet of product-moment correlations reveals few correlations of significance greater than .05 (displayed in red by default). Notable among these is the correlation of **–0.47** displayed between **Measure05** and **Measure09**. We can easily create a scatterplot of these two variables by selecting their correlation value and then using the **Graphs**

of Input Data command. Right-click on that correlation and select ***Graphs of Input Data -
Scatterplot by Measure09 - Regression, 95% Conf*** from the shortcut menu. The resulting raw
data scatterplot of the two variables is shown below:

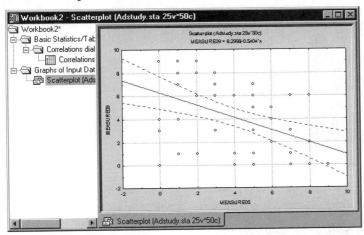

Note that by default the regression equation is displayed in the title of the graph.

Example: Creating a Graph of Block
Data Using an Input Datafile

For this example, open the ***Factor.sta*** datafile located in the Examples\Datasets subfolder of
your *STATISTICA* installation folder. You can open it using the ***File - Open*** menu, the 🖙 Open Data
button on any Startup Panel, or the 🖙 toolbar button. ***Factor.sta*** contains the (hypothetical)
results of a questionnaire used to measure people's satisfaction with their lives. The
questionnaire asked subjects questions about their work, home, hobbies, and other areas of
their lives. For a detailed discussion of the characteristics of ***Graphs of Block Data***, see Chapter
6 – *STATISTICA Graphs: General Features*.

Producing the Graph

- After opening ***Factor.sta***, select a block of data that includes the first five columns (***Work_1***,
  Work_2, ***Work_3***, ***Hobby_1***, and ***Hobby_2***) and the first ten rows (cases ***1-10***).

- Right-click in the selected block, and select ***Graphs of Block Data - Line Plot: Block Rows***
  from the shortcut menu.

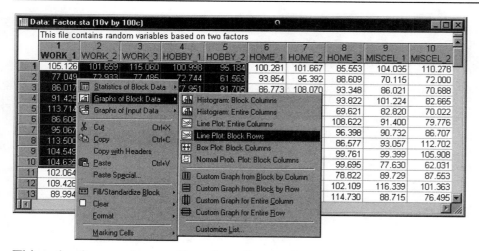

This selection produces a line plot of cases for the first ten rows in the datafile allowing you to compare the responses of the first ten respondents on five of the questions.

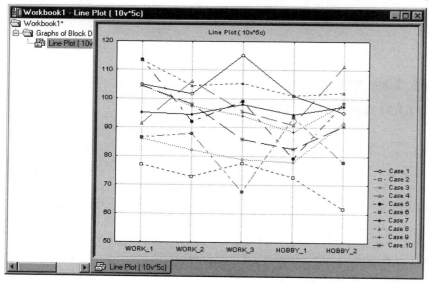

To create a line plot by rows for all rows (not just the selected rows), use the **Graphs of Block Data - Custom Graph for Entire Column** command to display the **Select Graph** dialog, and then select **2D Graphs** in the **Select Graph Category** box and **Line Plots (Case Profiles)** in the **Select Graph Type** box.

Example: Creating a Graph
of Block Data Using an Analysis
Output Spreadsheet as Input

The example above illustrates how to produce a **Graph of Block Data** from an input datafile. The following example demonstrates how to create a **Graph of Block Data** using analysis spreadsheets. For a detailed discussion of the characteristics of **Graphs of Block Data**, see Chapter 6 – *STATISTICA Graphs: General Features*.

For this example, open the **Adstudy.sta** datafile located in the Examples\Datasets subfolder of your *STATISTICA* installation folder. You can open it using the **File - Open** menu, the [🗁 Open Data] button on any Startup Panel, or the 🗁 toolbar button. **Adstudy.sta** contains the results of a (hypothetical) survey given by a hotel to its customers. In addition to polling customers on their preferred cola soft drinks, the surveyors asked them to rate the hotel's performance on a number of items such as room service, cleanliness, front desk service, etc.

Producing the Graph

- With **Adstudy.sta** open, select **Statistics - Basic Statistics/Tables** from the menu bar or the *STATISTICA* Start button ▦ menu.

- When the **Basic Statistics and Tables** (Startup Panel) is displayed, double-click **Descriptive statistics** to display the **Descriptive Statistics** dialog.

- Click the **Variables** button on this dialog to select variables for the analysis. We want to produce a summary of descriptive statistics for the 23 measures of hotel performance. To do this, select **Measure01** through **Measure23** on the **Select the variables for the analysis** dialog (or enter **3-25** in the **Select variables** box), and click the **OK** button to return to the **Descriptive Statistics** dialog.

- Click the **Summary: Descriptive statistics** button on the **Quick** tab of this dialog to produce the following spreadsheet, and click on the **Mean** column header to select the entire column of means.

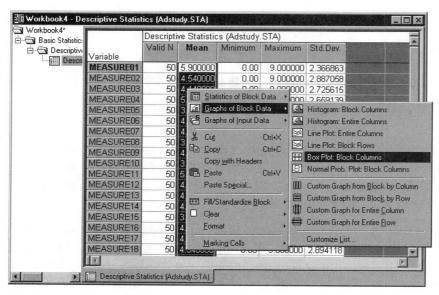

- Right-click on any highlighted cell (as shown above) and select **Graphs of Block Data - Box Plot: Block Columns** to create a box plot summarizing the distribution of mean values.

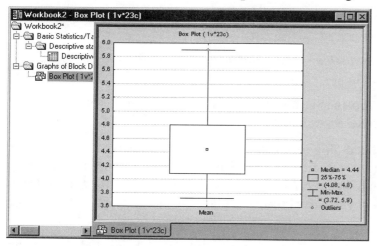

- To see the distribution of means displayed as a histogram, right-click any cell in the **Mean** column and select **Graphs of Block Data - Histogram: Entire Columns**.

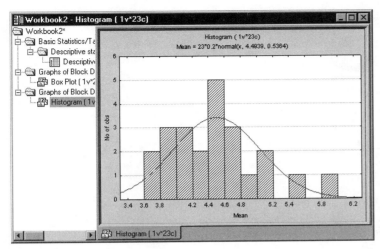

STATISTICA will create a histogram of the mean values with a normal fit applied (as shown above).

Example: Creating and Customizing a Simple 2D Graphs Menu Graph

This example illustrates methods for continuing exploratory analysis of graph data after a graph is initially created. For this example, open the *Irisdat.sta* datafile located in the Examples\Datasets subfolder of your *STATISTICA* installation folder. You can open it using the *File - Open* menu, the [Open Data] button on any Startup Panel, or the [ ] toolbar button. This datafile, based on the classic example data of Fisher (1936), contains measurements made on length and width of flower parts (sepals and petals) and the corresponding iris flower type (*Setosa*, *Versicol*, and *Virginic*). In this example, we will explore the relationship between two of the variables: sepal length (*Sepallen*) and petal length (*Petallen*).

Producing the Graph

- With the file *Irisdat.sta* open, select *Scatterplots* from the *Graphs* menu to display the *2D Scatterplots* dialog.

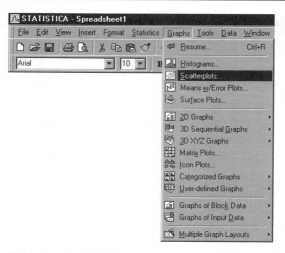

- Click the **Variables** button on the **Quick** tab of the **2D Scatterplots** dialog to display the standard variable selection dialog, specify the variables for the **X** and **Y** axes as **Petallen** and **Sepallen**, respectively, and then click the **OK** button.

- Select the **Advanced** tab of the **2D Scatterplots** dialog and note that, by default, a **Linear** fit is specified for the graph.

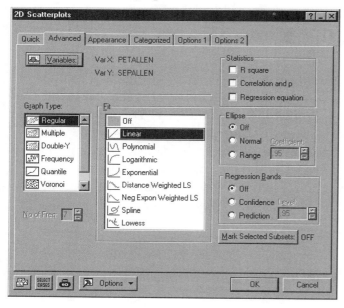

Marking selected subsets. In order to explore possible differences in the relationship between sepal length and petal length from flower type to flower type, we would like to display the flower types separately on the graph. To do this:

- Click the *Mark Selected Subsets* button on the *Advanced* tab to display the *Specify Multiple Subsets* dialog. In this dialog, you will specify three subsets to mark on the scatterplot.

- In the *Subset 1* group, select *Include if* and enter *Iristype = 'Setosa'*, and in the *Subset 2* group, select *Include if* and enter *Iristype = 'Versicol'*.

- Finally, select *Include if* and enter *Iristype = 'Virginic'* in the *Subset 3* group. The *Specify Multiple Subsets* dialog should look as shown below:

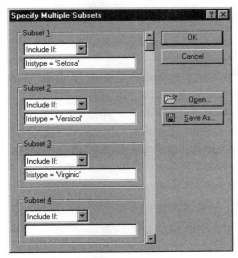

Click the *OK* button on this dialog and on the *2D Scatterplots* dialog to produce the graph shown below:

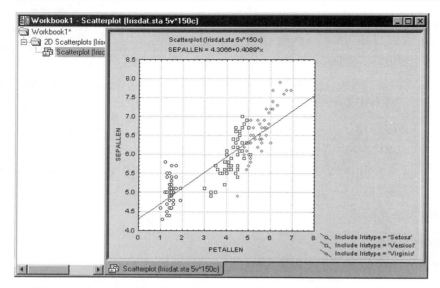

Adding a new fit line. On examination of the graph, it is obvious that the linear relationship displayed is very strongly influenced by the presence of the compact group of **Setosa** measurements that occur at lower values of both **Petallen** and **Sepallen**. It would be of some interest perhaps to explore the relationship between these variables within the **Virginic-Versicol** group, which on initial examination appears to have a slightly different trend than that calculated for all flower types together. Specifically, we would like to add another linear regression line to the plot that includes only the data in the **Versicol** and **Virginic** groups.

- Double-click on the outer background area (i.e., the area immediately outside the scatterplot) of the graph to display the **All Options** graph customization dialog.

- Next, select the **Plot: Fitting** tab. On this tab, the linear fit currently applied to the graph data is designated as number **1** in the **Fit** box.

- Click the **Add new fit** button to designate an additional fit.

Restricting the range of the fit. By default, this fit will be **Linear** as shown in the **Fit type** box. We will leave it as linear, although we could easily specify another fit type at this point. We want, however, to restrict this fit to the **Versicol** and **Virginic** data. Therefore:

- In the **Range** box in the **Fitting data** group, change the **Full Range** selection to **Custom Range**. Inspection of the graph shows we can restrict the **X** range of data to values that completely exclude the **Setosa** group.

- Enter **2.5** in the **Min** box and **7.5** in the **Max** box.

- If you would like the new fit to be displayed in a different pattern or color, you can choose from options available by clicking the **Pattern** button in the **Line** group.

- Click the **OK** button at the bottom of the **All Options** dialog to redraw the graph displaying both the original and the newly specified fit lines.

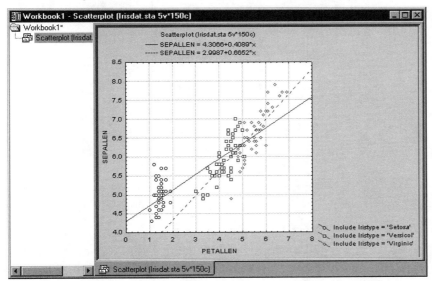

Adding custom axis labels and limit lines. Now we would like to display the lower and upper limits of the **Versicol** and **Virginic** data used in the new fit.

- Double-click the graph margin to display the **All Options** dialog.

- Select the **Axis: Custom Units** tab, and in the first line of the spreadsheet area enter **2.5** in the **Position** column, **Lower** in the **Text** column, and select the **Grid** check box. In the second line enter **7.5** and **Upper** in the **Position** and **Text** columns, respectively, and again select the **Grid** check box.

- Click the **Gridlines** button below the spreadsheet area to display the **Line Properties** dialog, where you can designate a gridline pattern and/or color that will distinguish the new gridlines to be drawn from those already present on the graph. Then click the **Close** button.

- Click the **Axis: Scale Values** tab on the **All Options** dialog and select the **Custom labels** check box in the **Display scale values** group.

- Select the **Off** option button in the **Skip values** group on the **Axis: Scale Values** tab, as well.

- Close the **All Options** dialog by clicking the **OK** button, and the graph with the limit labels and gridlines will be displayed.

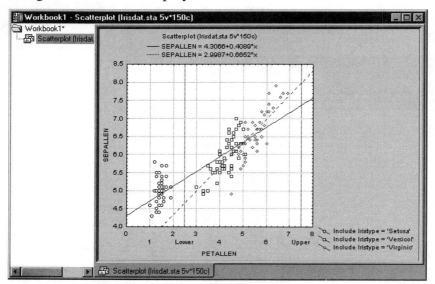

Customizing the legend. Next we will customize the legend. Double-click on the graph's outer background to once again display the **All Options** graph customization dialog. On the **Graph Titles/Text** tab, the text items associated with the graph legend can be deleted, changed, or supplemented. Note that the **Graph Titles/Text** dialog can also be displayed directly by double-clicking on the legend.

- In the **Title** drop-down box on the **Graph Titles/Text** tab, select **Right Title**. In the window above, the legend will be displayed. To modify a legend, it must first be disconnected from the graph. Select the legend text in the window, right-click on the highlighted text, and select **Disconnect Object(s) from Graph** from the resulting shortcut menu. Then, shorten the text items in the legend for better readability to **Setosa**, **Versicol**, and **Virginic**. See also *Is All Graph Text Editable?* on page 808.

- To create a legend title, place the mouse pointer on the left side of the first item (point marker symbol) on the legend, enter **Flower Type**, and press the ENTER key on your keyboard. Click the **OK** button at the bottom of the **All Options** dialog to redraw the graph.

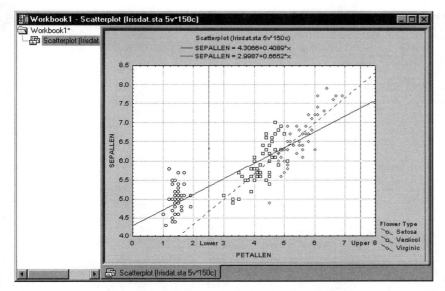

To change properties of the symbols in the legend (i.e., point markers or fit lines) double-click on the item on the graph or access the item on the appropriate tab of the **All Options** dialog (i.e., the **Plot: General** tab to change point markers or the **Plot: Fitting** tab to change the properties of the fit line). Changing the properties of these items on the graph automatically makes the changes in the legend symbols.

Changing titles and axis titles. Finally, we will change the titles and axis titles. Double-click on any graph title or axis title to display a dialog that is used to customize that title. Note that axis titles and graph titles can also be accessed from the **Axis: Title** and **Graph Titles/Text** tabs, respectively, of the **All Options** dialog.

- Double-click the x-axis title to display the **Title** dialog where **Petallen** can be changed to **Petal Length**. The y-axis label can be changed in the same fashion (i.e., change **Sepallen** to **Sepal Length**).

- Similarly, double-click the graph title to display the **Graph Titles/Text** dialog where the default **Scatterplot (Irisdat.sta 5v*150c)** can be changed to **Fisher's Iris Flower Data**.

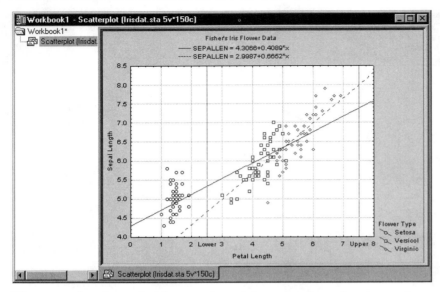

The completed graph (shown above) incorporates changes to legends and titles and the addition of labels and gridlines defining the range of data incorporated into the newly added fit.

FAQ: WORKING WITH GRAPHS

General

What Categories of Graphs Are Available in *STATISTICA*?

In addition to the specialized statistical graphs available from the output dialogs in all statistical procedures, there are two general categories or classes of graphs accessible from the *Graphs* menu, *Graphs* toolbar, shortcut (right-click) menus, and the *STATISTICA* Start button  menu:

- Input data graphs (i.e., *Graphs of Input Data* and *Graphs* menu graphs) and

- *Graphs of Block Data*.

The most important difference between these two general categories lies in the data that the graph types utilize for generating plots:

Input data graphs. *Graphs of Input Data* and their expanded version in the *Graphs* menu produce statistical summaries or other representations of the raw data in the current input data spreadsheet (typically for the entire variable(s) or its subsets if case selection conditions are used). Note that if graphs of this general category are produced using a shortcut menu from within a spreadsheet of results that does not contain the actual data (e.g., a correlation matrix), *STATISTICA* will still reach to the respective input (raw) data in order to produce the graph (e.g., a scatterplot of the variables identified by the selected cell in the correlation matrix from which the shortcut menu was opened).

Graphs of Block Data. *Graphs of Block Data*, on the other hand, are entirely independent of the concept of "input data" or "datafile." They provide a general tool to visualize numeric values in the currently selected block of any spreadsheet (which can contain values from custom defined subsets of numerical output or arbitrarily selected subsets of raw data).

Common features of the two categories of graphs. Note that these two general categories of graphs offer the same customization options and the same selection of types of graphs. For example, you can create the same, highly specialized categorized ternary graph from the input (raw) dataset and from a custom defined block of values representing results of a particular test.

Other specialized graphs. Other specialized statistical graphs that are related to a specific type of analysis [e.g., plots of means (e.g., interactions) in *ANOVA*, plots of fitted functions in *Nonlinear Estimation*] are accessible directly from results dialogs (i.e., the dialogs that contain output options from the current analysis).

Are There Different Customization Options for Each Type of Graph?

No. Once a graph is displayed on the screen, regardless of how it was requested or defined, all graph customization options available in *STATISTICA* can be used to customize it (see *When and How Can I Customize STATISTICA Graphs?*, page 788). The customization options available for all graphs include appending new plots to existing graphs and linking and embedding graphs and artwork, as well as all drawing, fitting, and graph restructuring options. Also, all these options can be used to customize graphs that were saved and later opened for additional editing.

Can Graphs Be Automatically Updated When the Datafile Changes?

Yes. All *Graphs* menu graphs can maintain automatic links to the data from which they were created as long as the graph specification dialog is active. Options for auto-updating graphs are available on the *Options 1* tab of all graph specification dialogs. (Note: Graphs that are to be dynamically updated when the datafile changes should be placed in stand-alone windows and not in a workbook.) For more information, see *Options 1 Tab* in Chapter 7 – *STATISTICA Graphs: Creation and Customization* or refer to the *Electronic Manual*.

In What Formats Can I Save *STATISTICA* Graphs?

STATISTICA graphics file format. Graphs and drawings can be saved in the *STATISTICA* system graphics file format (file name extension **.stg*). Select **Save** or **Save As** from the **File** menu. Although other graphics formats are also supported, this format is recommended whenever you intend to use the graph again in the *STATISTICA* system or link it to other application documents using Windows OLE or ActiveX conventions. Unlike other formats, the *STATISTICA* system format stores not only the graphical representation of the picture but also the **Data Editor** (see page 822) containing all data that are represented in the graph, all analytic options (fitted equations, ellipses, etc.), and other settings allowing you to

continue the graphical data analysis at a later time. This format is also most appropriate if the current graph or drawing is later to be linked to or embedded into another *STATISTICA* Graph.

Bitmap, JPEG, Metafile, and PNG (portable network graphics) format. If the graph to be saved is to be used (e.g., customized) with tools other than *STATISTICA*, placed on an Internet Web page, or used by an application that does not support OLE or ActiveX, you can choose to save the file as a **Metafile** (file name extension *.wmf*), **Bitmap File** (file name extension *.bmp*), **JPEG File** (file name extension *.jpg*), or **Portable Network Graphics File** (file name extension *.png*) by selecting the appropriate option from the **Save as type** box on the **Save As** dialog. These formats (described briefly in the next four topics) do not offer the advantages of customizability offered by the *STATISTICA* format (see above); however, they are compatible with applications that support the respective graphics file formats.

EPS format. Encapsulated PostScript (EPS) graphics files are designed for printing to a PostScript printer. To save a *STATISTICA* graph in the EPS format, you need to use a PostScript driver that is compatible with your system software. After selecting the proper printer/driver (in the **Print** dialog), specify **Encapsulated PostScript (EPS)** format as the **PostScript output format** (typically on the **PostScript** tab of that printer's **Properties** dialog). When you print the graph (to a file), it will be saved in the EPS format (with the file name extension *.eps*).

What Is the Windows Metafile Graphics Format?

The **Metafile** format, also referred to as Picture (a standard Windows graphics format used in Clipboard and disk file representations of graphs; file name extension *.wmf*) stores a picture as a set of descriptions or definitions of all components of the graph and their attributes (e.g., segments of lines, colors and patterns of those lines, specific fill patterns, text and text attributes, etc.).

Therefore, as compared to bitmaps (another standard Windows graphics format, see below), metafiles offer more flexible options for non-OLE modification and customization in other Windows applications.

For example, when you open a metafile in the Microsoft Draw program, you can "disassemble" the graph: select and modify individual lines, fill patterns, and colors; edit text and change its attributes, etc.

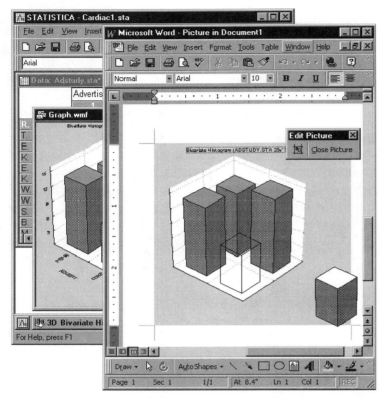

Note, however, that not all Windows applications support the complete (extended) set of metafile graphics features and attributes supported and used by *STATISTICA*, thus some aspects of *STATISTICA* Graphs (saved as metafiles) will look different when they are opened in such applications (e.g., rotated and transformed fonts can appear unrotated). For best results, copy *STATISTICA* Graphs to other applications using OLE conventions that allow you to invoke *STATISTICA* to customize or modify *STATISTICA* Graphs pasted into documents of other applications.

What Is the Bitmap Graphics Format?

The **Bitmap** format (a standard Windows graphics format used in Clipboard and disk file representations of graphs; file name extension **.bmp*) is similar to the metafile format (see above). Specifically, it stores only a representation of the picture and not the data that are plotted or any information about the analytic properties or settings used to produce the graph.

However, unlike the metafile format, the bitmap format stores only a "passive" representation of pixels that form the graph. This representation is less customizable than metafiles that store dynamic representations of all individual graph components, thus allowing selective modifications of lines, text, etc., in other Windows applications.

Bitmaps can be opened by other Windows applications, but the customization or editing options of such graph representations will be limited (typically to operations on pixels, such as stretching and shrinking, cutting and pasting, and drawing "over" the graph). As mentioned before, for best results, copy *STATISTICA* Graphs to other applications using OLE or ActiveX conventions that allow you to invoke *STATISTICA* to customize or modify *STATISTICA* Graphs pasted into documents of other applications.

What Is the JPEG Graphics Format?

JPEG is an acronym for ***Joint Photographic Experts Group***. This graphics file format is an ISO/ITU standard for storing raster (or bitmap) style (not vector style, like metafile) images in compressed form using a discrete cosine transform. Like bitmaps (see above), the customization or editing options of such graph representations will be limited (typically to operations on pixels, such as stretching and shrinking, cutting and pasting, and drawing "over" the graph). The major advantage of the JPEG format is that as compared to bitmaps, it is much more economical (uses less storage space), and it is commonly used to represent graphics components of HTML files for reports that will be viewed in an Internet or intranet setting.

What Is the PNG (Portable Network Graphics) Files Format?

PNG (Portable Network Graphics) is a graphics file format designed to store bitmap (i.e., raster, as opposed to vector/metafile) graphic images. It was introduced to be a replacement for the patented GIF format (mostly to avoid the legal restrictions associated with the patent). A PNG image can contain a variable number of colors, including a transparent color. The size of the file depends on the number of colors used in the specific image. A compression method is used (to reduce the file size) that is highly effective if a large proportion of the image is built of the pixels with the same color attributes (e.g., it is highly effective for charts and schematic line art, but somewhat less effective than the JPEG format for photographs).

As with bitmap and JPEG images, customization of the graph is limited (see above) in PNG file format. PNG files are commonly used in HTML documents. *STATISTICA* Graphs embedded into HTML documents are saved by default as PNG files (named: ***reportname_pict0001.png***, ***reportname_pict0002.png***) if the report is saved as an HTML file.

You can also save graphs in this format by selecting **JPEG Files** in the **Save as type** box in the **Save As** dialog (accessible from the graph **File** menu).

What Is the Native *STATISTICA* Graphics Format?

Unlike other graphics formats such as metafiles and bitmaps, the native *STATISTICA* graphics file format (file name extension *.stg*) not only stores the "image" of the current graph but also all information necessary to continue graph customization or graphical data analysis (including all data represented in the graph, links to data, fitted equations, embedded graphs and artwork, links to graphs and artwork, etc.). Graphs stored in this format can be opened later (in *STATISTICA*) allowing you to continue the graphical data analysis or graph customization (e.g., fit equations, add or merge new data series, etc.). Graphs saved in this format can also be dynamically linked to documents of other Windows applications using OLE conventions.

How Do I Export a *STATISTICA* Graph to Another Application?

Export via the Clipboard (and Paste or Paste Special via OLE). The quickest way to export a graph is to copy it to the Clipboard (see Chapter 6 – *STATISTICA Graphs: General Features*) and then paste it into another application. *STATISTICA* native, Windows metafile (see page 775), and bitmap (see page 776) formats are created in the Clipboard and can be used in other applications.

STATISTICA Graphs can be pasted into other application documents (e.g., word processor documents or spreadsheets) either as embedded objects or as objects linked to graph files. If *STATISTICA* Graphs are pasted to other applications via Windows OLE, they maintain their relations to *STATISTICA* and thus can be interactively edited from within the other application.

Linking *STATISTICA* Graph files via OLE. *STATISTICA* Graph files can also be inserted and linked via OLE to other applications.

Export to another file format. If the graph to be saved is to be used by an application that does not support OLE or ActiveX, you can choose to save the file as a metafile (file name extension *.wmf*), bitmap (file name extension *.bmp*), JPEG (file name extension *.jpg*), or PNG (portable network graphics file; file name extension *.png*) by selecting the appropriate option from the **Save as type** box in the **Save As** dialog.

Limitations of the standard Windows metafile format. Very large (in terms of the number of data points represented) or very complex graphs that can be produced by *STATISTICA* can exceed the capacity of the Windows metafile graphics format used in the Windows 95 and 98 systems. In those circumstances, use the JPG, PNG, or bitmap representation instead (see page 776).

What Is the Difference between a Graph and a Plot?

Each plot represents a single "series" of data. All but the simplest graphs in *STATISTICA* contain more than one plot of data.

In other words, in *STATISTICA*, the term graph applies to a complete graphical representation of (one or more) "series" of data, that is, to the entire "picture," which can be saved as a graphics document (by default, as a *STATISTICA* graphics file, file name extension *.stg*, see page 778).

There are many ways in which plots can be "put together" to form a graph, and depending on their type, some plots can require more than one sequence of values.

For example, at least three coordinated sequences of values are necessary to create a whisker plot: X-values, Y1-values (lower endpoints of whiskers), and Y2-values (upper endpoints of whiskers).

Customizing the features of a plot (the Plot: General dialog). A plot contains patterns, sizes, and all other specifications that apply to the graphical representation of only one series of data. They can be customized on the *General* dialog (accessed by selecting *General Plot Options* from the shortcut menu displayed by right-clicking on a plot), or on the *Plot: General* tab of the *All Options* dialog (accessed by double-clicking on the outer background of the graph), or on the *General* tab of the *Plot Layout* dialog (accessed by selecting *Plot Properties* from the shortcut menu displayed by right-clicking on a plot); see Chapter 7 – *STATISTICA Graphs: Creation and Customization*.

Customizing all features of a graph (the All Options dialog). The layout of a graph contains all those features and attributes that apply to the entire graph and are common to all plots as well as all features for specific plots. They include such features as titles, gridlines, global colors (backgrounds, etc.), scaling, axis labels, or categorization labels. They can be adjusted in the *All Options* dialog (see Chapter 7 – *STATISTICA Graphs: Creation and Customization*), accessible from the *Format* menu or via shortcut menus.

What Are Categorized Graphs?

Categorized graphs are created by categorizing data into subsets and then displaying each of these subsets in a separate small component graph arranged in one display. For example, one graph can represent male subjects and another one female subjects, or high blood pressure females, low blood pressure females, high blood pressure males, etc.

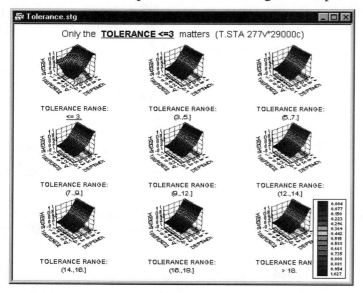

In *STATISTICA*, categorized graphs are:

- available in many output dialogs (they are automatically generated as part of output from all procedures that analyze groups or subsets of data, e.g., breakdowns, *t*-tests, ANOVA, discriminant function analysis, nonparametrics, and many others),

- accessible as part of the **Graphs of Input Data** options in the shortcut menus in all spreadsheets, and

- accessible from the **Graphs** menu where a wide variety of user-defined methods to categorize data are available.

Refer to the next topic for a review of categorization methods available in *STATISTICA*. See also, *Methods of Categorization* in Chapter 7 – *STATISTICA Graphs: Creation and Customization*.

How Do I Define Categories for Categorized Graphs?

If categorized graphs are requested from output dialogs of specific procedures that involve analyses of subsets of data, then they will automatically display the subsets that are currently analyzed (i.e., the subsets are already defined as part of the current analysis). On the other hand, the categorized graphs requested from the *Graphs* menu offer a variety of methods to specify subsets using one or two grouping variables. Also, custom-defined subset definitions can be used that can involve all variables in the current datafile.

Specifically, categories can be defined by:

- Integer values of grouping variables (*Integer Mode*),

- Dividing grouping variables into a requested number of equal-length intervals (*Categories*),

- Custom intervals (ranges) of grouping variables, defined by specific interval boundaries (*Boundaries*),

- Specific values (i.e., codes) of grouping variables (*Codes*), and

- User-defined "multiple-subset" definitions (*Multiple Subsets*) that can be entered as logical case selection conditions of virtually unlimited complexity (this categorization method can involve values of all variables in the current datafile as shown below; see *Methods of Categorization* in Chapter 7 – *STATISTICA Graphs: Creation and Customization*).

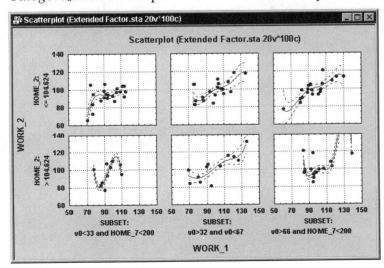

The graph shown above is a relatively complex example of a two-way categorized graph based on a mixed method of defining the subset (component) graphs. The two-way categorization means that the arrangement of the small (component) graphs in the display resembles a two-way table (crosstabulation) resulting from applying two different criteria of categorization.

For example, in the graph shown above, the two rows of graphs represent categories defined based on values of variable **Home_2** (cases where **Home_2** is less than or equal to 104.624 and cases where it is greater than 104.624). The three columns of graphs represent subsets of cases defined using specific "multiple subset" definitions based on values of variable number **0** (i.e., case numbers) and variable **Home_7**. The following is the **2D Categorized Scatterplots** dialog where the above graph was defined (select **Scatterplots** from the **Graphs - Categorized Graphs** menu).

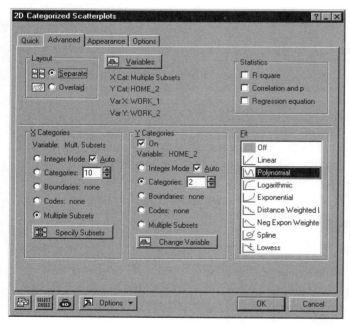

Specifically, variable **Work_1** and **Work_2** are plotted in each small graph (as variables X and Y, respectively). The first of the two categorizations (**X Categories**, or "columns" of graphs) was defined as **Multiple Subsets** in a dialog that is displayed after the **Specify Subsets** button is clicked.

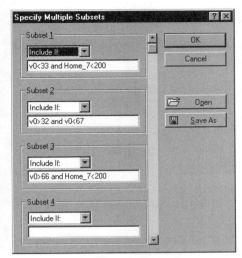

The second of the two categorizations (**Y Categories**, or "rows" of graphs) was defined by a grouping variable (**Home_2**) by dividing its range into two equal-length intervals; see the **Categories** box set to 2 in the graph specification dialog (which resulted in dividing the distribution of **Home_2** into cases below or equal to, and above 104.62, as shown in the above graph).

How Do I Produce Ternary Contour Plots and Surfaces?

Ternary plots of 3D surfaces or contours can be produced as part of the output from the analysis of mixture designs in the **Experimental Design** module. Ternary plots are also available from the **Graphs - 3D XYZ Graphs**:

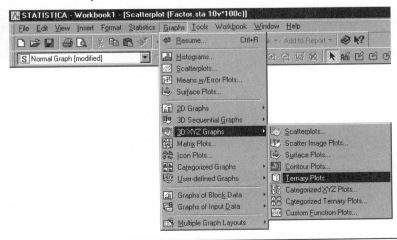

and *Graphs - Categorized Graphs* menus.

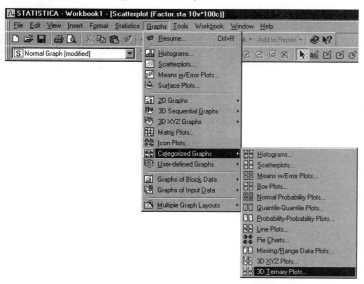

For more information, refer to the *Electronic Manual* (accessible by pressing the F1 key).

How Do I Produce Graphs with Polar Coordinates?

Polar coordinates can be selected in the graph specification dialogs for histograms, scatterplots, line plots, and stacked/sequential plots accessible from the *Graphs - 2D Graphs* menu (use the *Options 2* tab).

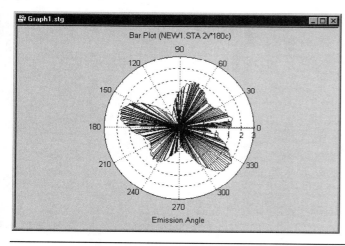

Categorized graphs with polar coordinates can also be produced.

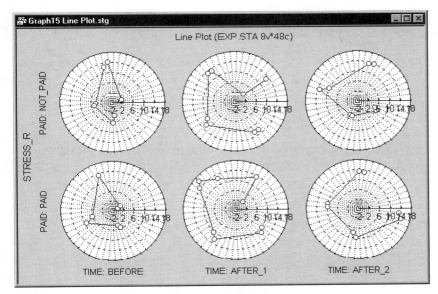

In addition, many standard (Cartesian) graphs can be converted into polar coordinates by selecting **Polar** in the **Coordinate system** box on the **Graph Layout** tab of the **All Options** dialog (see below).

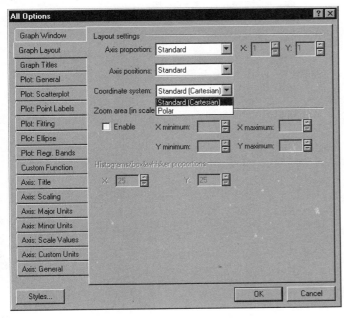

For more information, refer to the *Electronic Manual* (accessible by pressing the F1 key).

What Are Multiple Axes in Graphs?

An arrangement of axes (coordinate scales) in graphs, where two or more axes are placed parallel to each other in order to either:

- represent different units in which the variable(s) depicted in the graph can be measured (e.g., Celsius and Fahrenheit scales of temperature), or

- allow for a comparison of trends or shapes between several plots placed in one graph (e.g., one axis for each plot) which otherwise would be obscured by incompatible measurement units or ranges of values for each variable (that is an extension of the common "double-Y" type of graph).

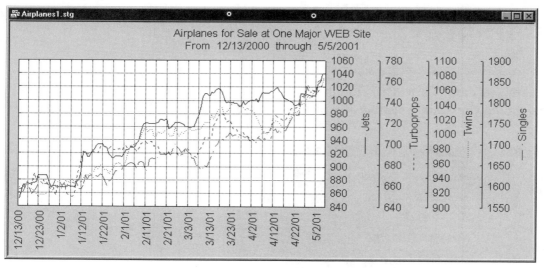

The latter instance, which requires the appropriate plot legends to be attached to each axis, is illustrated in the graph above. See also, *What Is the Axis Layout Dialog?* on page 794.

How Is the Mouse Used in Graph Applications?

In addition to the standard Windows mouse conventions for selecting objects, the mouse can be used in many more specialized applications in the graphics window in *STATISTICA*. The following is a list of representative examples:

OLE. Links or embeds foreign document files to *STATISTICA* documents by dragging them directly from the desktop or Windows *Explorer* (across application windows) and dropping them onto *STATISTICA* Graphs.

Brushing. Highlights data points from the graph by clicking on them with the brushing tool or selecting them with a *Box*, *Lasso.*, *Cube*, or a *2D* or *3D Slice*.

Zoom in and zoom out tools. Zooms in ("magnifies") or zooms out ("shrinks"), respectively, the selected area of the graph. For more information about specific mouse conventions, see the *Zoom in* ⊕ and *Zoom out* ⊖ toolbar buttons in Chapter 6 – *STATISTICA Graphs: General Features*.

Drawing tools. Adds rectangles, ovals (or circles), polylines and freehand drawings, arrows, etc. to a graph.

Resizing and moving. Resizes (drag on a "black selection square," ✛) or moves (drag the entire object, ↔) selected graph objects. For fine adjustments, see *Controlling the mouse with the keyboard in graphs*, below.

Editing polyline objects. Reshapes individual segments of the polyline drawing by dragging on either the object area black selection squares or any of the black selection squares that mark the line segments. For fine adjustments, see *Controlling the mouse with the keyboard in graphs*, below.

Rotating text. You can interactively rotate custom text by selecting it in the graph and then dragging one of the object handles (small black squares) in the desired direction. For fine adjustments (1 degree movements), see *How Do I Rotate Text?* (page 815).

Controlling the mouse with the keyboard in graphs. You can also emulate the mouse with the keyboard in order to move or resize an object by selecting the object, placing the mouse pointer over the object, and then using the keyboard cursor keys to move or resize the object.

Note that the mouse pointer will change to the appropriate tool to match the application for which it is being used. Press the ESC key to return the mouse pointer to the default mode. You can also use the mouse pointer to customize the graph (see *Common Customization Features* in Chapter 7 – *STATISTICA Graphs: Creation and Customization*).

How Do I Select an Object in a Graph?

To select an object in a graph, simply click on the object with your mouse. Once an object has been selected, press the TAB key to navigate from object to object within your graph.

How Can I Interpret a 100(1-alpha)% Confidence Interval?

We often refer to a confidence level as the probability that a specific parameter will be contained in a given interval. For example, when we fit a 95% confidence interval to a fitted line, we say there is a 95% probability that the "true" fitted line (in the population) falls between the interval. As Hahn & Meeker point out in their book on statistical intervals (*Wiley Series in Probability and Mathematical Statistics*, 1991), this definition is common, but not entirely precise:

> A 100(1-alpha)% confidence interval for an unknown quantity Theta may be characterized as follows: 'If one repeatedly calculates such intervals from many independent random samples, 100(1-alpha) % of the intervals would, in the long run, correctly bracket the true value Theta [or equivalently one would in the long run be correct 100(1-alpha)% of the time in claiming that the true value of theta is contained within the confidence interval].' More commonly, but less precisely, a two-sided confidence interval is described by a statement such as 'we are 95% confident that the interval theta-lower to theta-upper contains the unknown true parameter value of theta.' In fact the observed interval either contains theta or does not. Thus the 95% refers to the procedure for constructing a statistical interval, and not to the observed interval itself.

Graph Customization – General Features

When and How Can I Customize *STATISTICA* Graphs?

All graph customization facilities in *STATISTICA* are available when a graph window is active (thus, after a graph has been shown).

There are two major types of graph customizations:

- adding and/or editing custom graphic objects, and
- customizing the structural components of the graph.

Adding and/or editing custom graphic objects. The tools to add and edit custom graphic objects to the current graph (such as drawing, managing, and customizing objects, pasting, embedding, linking, etc.) can be accessed from the **Graph Tools** toolbar. They are described in *Graph Toolbars* in Chapter 6 – *STATISTICA Graphs: General Features*. Those options are also accessible from the graph **Edit** and **Insert** menus, and many of them can also be accessed from the shortcut menus.

Customizing the structural components of the graph. All facilities to change the structural properties of graphs (such as proportions, scales, patterns, features of individual plots, etc.) can be accessed from the **All Options** dialog (see Chapter 7 – *STATISTICA Graphs: Creation and Customization*). This dialog can be accessed from the graph **Format** menu as well as from shortcut menus described below.

Using the shortcut (right-click) menus in graphs. The easiest way to access the **All Options** dialog (as well as all other customization facilities) is to use the shortcut menus accessible by right-clicking on the graph.

Using the shortcut menus is also often faster because they provide shortcuts allowing you to directly access the nested (i.e., "second-" or "third-level") dialogs controlling the attributes of specific graph components while bypassing the **All Options** dialog.

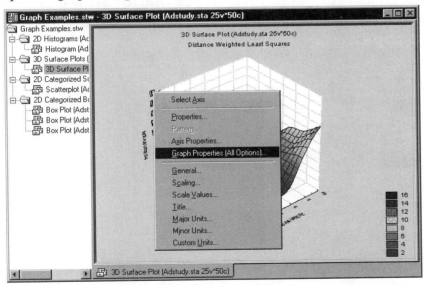

One of the graph shortcut menus has a different status than others. Specifically, unlike all other shortcut menus that are associated with specific objects, the "main" (or "background") shortcut

menu (shown below), accessible by right-clicking outside the graph axes, contains global graph customization and multi-graphics management options.

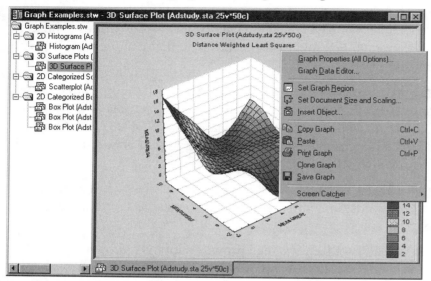

Using the left mouse button (direct access to customization dialogs).

Another mouse-based shortcut to access customization facilities for specific elements of the graph is to double-click on the specific graph object or component.

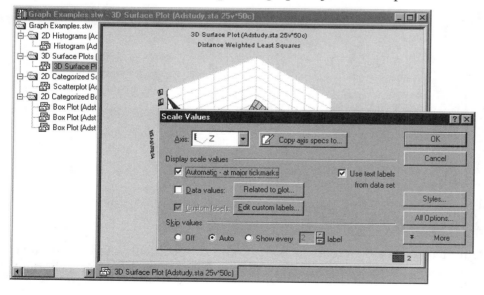

If the element you have selected this way is one of the structural components of the graph (such as a scale, point marker, or a title), then the dialog that is most commonly used to customize that type of object will automatically be displayed. For example, in the illustration shown above, double-clicking on the y-axis scale values displays the *Scale Values* dialog.

Options accessible by double-clicking on an object vs. options accessible from the shortcut menus (clicking the right mouse button).

As mentioned before, by right-clicking on an object, a list of customization dialogs and options applying to that object will be displayed in a shortcut menu (organized hierarchically with the most commonly used option at top and progressing to the dialog with the largest set of options).

On the other hand, double-clicking displays directly the most commonly used dialog, or the global dialog that applies to the object. Therefore, double-clicking saves you one step whenever you are accessing the most commonly used option.

Accessing the All Options dialog.
The same principle also applies to displaying the *All Options* dialog (see Chapter 6 – *STATISTICA Graphs: General Features*): to access this dialog, simply double-click on the empty graph space outside the graph axes. Alternatively, you can select it from the "main" shortcut menu accessible by right-clicking on the empty graph space outside the graph axes.

Customization of custom graphic objects (drawings, embedded objects).

The general principles of selecting objects and accessing their customization facilities (summarized in the two previous paragraphs) also apply to all custom graphic objects such as drawings, arrows, and embedded objects. The *Graph Tools* toolbar buttons can also be used to access most of these customization facilities. See *Graph Tools Toolbar* in Chapter 6 – *STATISTICA Graphs: General Features* for more information. For example, the buttons can be used to change the patterns, sizes, and colors of objects, and properties of predefined objects (such as arrows or error bars) or foreign objects (e.g., pasted, linked, or embedded artwork).

Customization of linked or embedded OLE objects.
To edit OLE objects in *STATISTICA* Graphs, double-click on the object; the source application will be opened in OLE server mode with the object displayed, allowing the object to be changed.

When the editing is completed, you can use any of the standard OLE conventions to exit the application's server mode and update the graph in *STATISTICA* (such as clicking on the *STATISTICA* Graph).

How Do I Add a New Plot to an Existing Graph?

Click the **Add new plot** button on the **Plot: General** tab of the **All Options** dialog (accessible from the **Format** menu). A subsequent **New plot(s)** dialog is displayed in which you can specify the plot to be added. You can also add a new plot directly to the **Graph Data Editor** (see the *Electronic Manual*).

What Is the All Options Dialog?

The **All Options** dialog contains 15 to 20 tabs that address all of the relevant customizable features for a particular graph. The tabs are grouped in clusters containing logically related items. The options on the **All Options** dialog are an all-inclusive "superset" of options accessed by double-clicking specific graph features.

Access the **All Options** dialog directly by selecting **All Options** from the graph **Format** menu, by double-clicking the background (e.g., the area outside the axes) on a graph, or by selecting **Graph Properties (All Options)** from any graph shortcut menu.

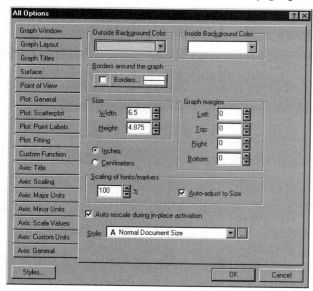

For more details, see graph customization dialogs in Chapter 7 – *STATISTICA Graphs: Creation and Customization*.

What Is the Plot Layout Dialog?

The *Plot Layout* dialog is an intermediate customization dialog that provides access to all of the relevant customizable plot features for a particular graph.

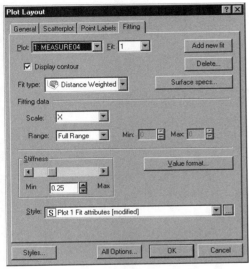

The options on the *Plot Layout* dialog are a subset of the complete set of graph options that are contained on the *All Options* dialog.

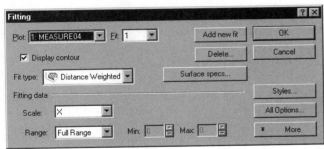

You can access the *Plot Layout* dialog by clicking the *All Options* button on any plot-specific feature customization dialog (e.g., by clicking the *All Options* button on the *Fitting* dialog, shown above) as well as from graph shortcut menus.

What Is the Axis Layout Dialog?

The *Axis Layout* dialog is an intermediate customization dialog that provides access to all of the relevant customizable axis features for a particular graph.

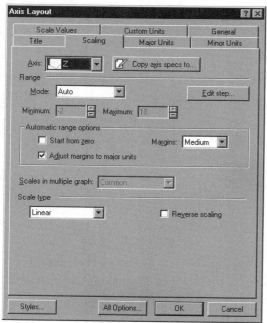

The options on the *Axis Layout* dialog are a subset of the complete set of graph options that are contained on the *All Options* dialog.

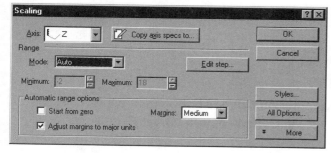

You can access the *Axis Layout* dialog by clicking the *All Options* button on any axis-specific feature customization dialog (e.g., by clicking the *All Options* button on the *Scaling* dialog, shown above) as well as from graph shortcut menus.

How Can I Adjust the Margins of a Graph?

Margins within the graph area. Click the *Set Graph Area*  button on the *Graph Tools* toolbar (see Chapter 6 – *STATISTICA Graphs: General Features*) to adjust the space between the edge of the plotting area (i.e., the borders of the graph window) and any graph components or custom graphic objects. This can be accomplished by clicking the  button and either 1) dragging the resizing squares that appear around the edges of the graph window or 2) drawing a rectangle in the place where you want the graph to be.

Note that you can also adjust margins on the *Document Size and Scaling* dialog (accessible by clicking the *Adjust the Graph Actual Size/Scaling*  button, see Chapter 6 – *STATISTICA Graphs: General Features*).

Printout margins. The printout margins (the width of the distance between the edge of the paper and the beginning of the graph area) can be adjusted in the *Print Preview* dialog (accessed from the *File* menu, see Chapter 6 – *STATISTICA Graphs: General Features*).

How Can I Change the Proportions of the Graph Window?

Use the *Document Size and Scaling* dialog (accessible by clicking the *Adjust the Graph Actual Size/Scaling*  button, see Chapter 6 – *STATISTICA Graphs: General Features*) to change the current graphs aspect ratio (i.e., the ratio of its vertical to horizontal dimensions). Note that the global default graph aspect ratio can be modified on the *Graphs 2* tab of the *Options* dialog (accessible via the *Tools* menu).

How Do I Produce Sequences of Graphs from Lists of Variables?

Specifying lists of variables for Graphs menu graphs. Most of the graph specification dialogs accessible from the *Graphs* menu allow you to select lists of variables in instances where a single variable is sufficient to define a graph. When such a list of variables is specified, *STATISTICA* cycles through the list and produces one graph for each variable (e.g., a histogram or a line plot).

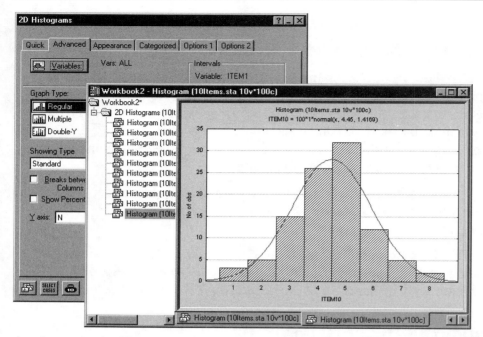

As shown in the illustration above, a separate histogram was drawn for each of the selected variables.

"Cascades" of graphs requested from output dialogs. Most of the output (results) dialogs in those statistical procedures that process lists of variables allow you to generate "cascades" of graphs for each (or each combination) of the variables in the current list. For example, such graphs can be produced from descriptive statistics, correlations, frequencies, cross-tabulations, breakdowns, and other procedures:

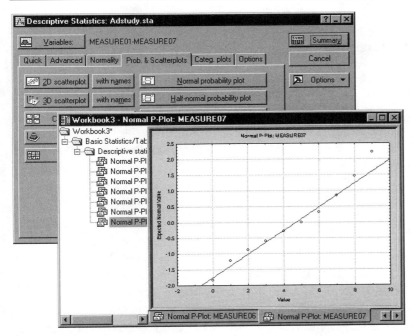

As shown in the illustration above, a probability plot is created for each variable specified in the analysis.

Automatic printouts. Note that when using either of the two methods mentioned above, you can automatically produce printouts of all graphs (in a *STATISTICA* Report) by selecting the *Also send to Report Window* check box on the *Output Manager* tab of the *Options* dialog. This report can then be edited and formatted using all the features available in the report window. For more information about reports, see Chapter 5 – *STATISTICA Reports*. Alternatively, you can send all output to a workbook (as shown above), then print the entire workbook. For more information on workbooks, see Chapter 3 – *STATISTICA Workbooks*. See also, *Printing Graphs* on page 841.

Specifying sequences of graphs in *STATISTICA* Visual Basic. Finally, multiple graphs can be generated in batch using *STATISTICA* Visual Basic. For more information on *STATISTICA* Visual Basic and how you can write macros to perform statistical analyses and create graphs, see Chapter 11 – *STATISTICA* Visual Basic.

How Do I Specify Properties for Point Markers?

Controls for modifying point markers for the various plots are located on the **General** dialog accessible by double-clicking on a point marker. You can also access these controls on the **Plot: General** tab of the **All Options** dialog (accessible via the **Format** menu). Note also that point markers (and fonts) can be increased or decreased using the **Increase Font** ⁺A or **Decrease Font** A⁺ toolbar buttons, respectively. See Chapter 6 – *STATISTICA Graphs: General Features* for more details.

How Do I Specify Area Properties?

The quickest way to modify area properties (e.g., patterns, colors, etc.) is to right-click on the area you want to modify, and select **Pattern** from the shortcut menu to display the **Area Properties** dialog. You can use this dialog to change the area color, pattern, and style in the graph. The default patterns, colors, and modes of display of consecutive plots and other components of the graphs are determined by the current selections in the **Graphs 1** tab of the **Options** dialog (accessible from the **Tools** menu, see Chapter 6 – *STATISTICA Graphs: General Features*). Note that the **Area Properties** dialog can also be accessed from the **Plot: General** tab of the **All Options** dialog (see Chapter 7 – *STATISTICA Graphs: Creation and Customization*).

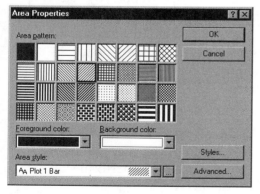

After specifying a pattern in the **Area pattern** box, you can select a different color for the area **Foreground** (i.e., the main component or the "outside" of the item) and **Background** (the "inside" of the item) with these options.

Instead of specifying area patterns and color using the options described above, you can select the style you want to use for the area from the **Area style** box. For more information about styles, see Chapter 7 – *STATISTICA Graphs: Creation and Customization*.

How Do I Specify Line Properties?

The quickest way to modify line properties (e.g., size, colors, etc.) is to right-click on the line you want to modify, and select **Pattern** from the shortcut menu to display the **Line Properties** dialog. You can use this dialog to change the line width, pattern, and color in the graph. The default patterns, colors, and modes of display of consecutive plots and other components of the graphs are determined by the current selections on the **Graphs 1** tab of the **Options** dialog (accessible from the **Tools** menu, see Chapter 6 – *STATISTICA Graphs: General Features*). Note that the **Line Properties** dialog can also be accessed by clicking the **Line** button on the **Plot: General** tab of the **All Options** dialog (see Chapter 7 – *STATISTICA Graphs: Creation and Customization*).

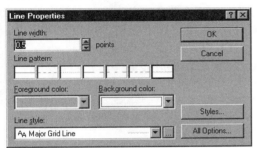

After specifying a pattern in the **Line pattern** box, you can select a different color for the line **Foreground** (i.e., the main component or the "outside" of the item) and **Background** (the "inside" of the item) with these options.

Instead of specifying line patterns and color using the options described above, you can select the style you want to use for the line from the **Line style** box. For more information about styles, see the *Electronic Manual*.

Can I Control the Resolution of Fit Lines?

Yes. Use the **Resolution** box on the **Fitting** dialog (accessible by double-clicking on the fitted line). Once a fitted function has been determined, the fit is approximated with segments on the x-axis. The **Normal** fit line is composed of 200 segments, and the number of segments increases in exponential fashion as you select **Medium**, **High**, **Very High**, or **Perfect**. Note that this option is only beneficial for fits with high curvature (i.e., a straight line fit will not be improved by

this option, but a high-level polynomial fit will be). Selecting a higher number of points will result in a smoother appearance of the fitted function in the graph; however, selecting a higher number of points will also slightly slow down the graphing procedure.

Graph Customization – Styles

What Are Graph Styles?

All of the numerous features that affect the appearance of the graph (from as elementary as the color of the font in the footnote to as general as the global features of the graph document) can be saved as individual "styles." These styles can be given custom names and later be reapplied using the **Graphics Styles** box on the left side of the **Graph Tools** toolbar or simple shortcuts (such as pressing a specific key combination or clicking a button on a custom toolbar). An intelligent system internally manages these thousands of styles and their combinations in *STATISTICA* and helps you achieve your customization objectives with a minimum amount of effort. All user-defined or modified styles will be saved automatically in the *STATISTICA* configuration file (e.g., different sets or systems of styles can be used for different projects). For further details, see the **Configuration Manager** tab of the **Options** dialog in the *Electronic Manual*.

How Can I Create a Style from a Custom Graph Title Format?

First, create a graph with custom title text. For example, open the **Adstudy.sta** datafile, and select **Scatterplots** from the **Graphs** menu to display the **2D Scatterplots** creation dialog. Select variables, and type a title (e.g., **Hotel Guest Survey**) in the **Custom title** box on the **Options 1** tab of the **2D Scatterplots** dialog. Click the **OK** button to produce the graph.

Next, double-click the custom title at the top of the graph to display the **Graph Titles/Text** dialog. Click the **Font** button and change the font type, size, color, etc. as desired. Note that the specified changes are applied immediately to the title displayed in the text box on the dialog.

Finally, click the **Styles** button on the **Graph Titles/Text** dialog to display the **Graphics Styles** dialog, and either right-click in the **Styles for Title** box or click the ellipsis button. From the resulting menu, select **Save As** to display the **Save As** dialog, and enter a name for the style you have just created (e.g., **surveytitle**). Then click the **Save** button to store your custom title style

for later use. Close the **Graphics Styles** dialog and click the **OK** button on the **Graph Titles/Text** dialog to apply the formatting features to the custom title on the current graph.

How Can I Apply a Saved Style to a Graph Title in a New Graph?

To apply the formatting captured in a saved style, highlight the title to which you want to apply the style. The current style name (most likely S **Default Top Title**) will appear in the **Graphics Styles** box on the left side of the **Graph Tools** toolbar. Click on the **Graphics Styles** box and select the desired style (e.g., **surveytitle** created in the example in the previous section). The highlighted title will instantly switch to the selected style.

Does Altering the Graph Defaults on the Options Dialog Affect Graph Styles?

Only the default graph styles will be affected by changes made on the **Graphs 1** and **Graphs 2** tabs of the **Options** dialog (accessible via the **Tools – Options** menu). All other user-defined styles will not be affected these changes. Default system graph styles change to mirror the current system graph settings (as specified on the **Graphs 1** and **Graphs 2** tabs) without any need from you to manually update them. Conversely, user-defined graph styles will retain all of their internal settings, despite any changes made to the **Graphs 1** and **Graphs 2** tabs.

If I Save a Graph with Customized Styles, How Will This Graph Appear on a Colleague's Computer Who Does Not Have Those Particular Styles?

The graph will appear exactly as it did on your computer. Although every aspect of your graph's customization is encapsulated within graph styles, all customization applied to your graphs are fully portable. In fact, your colleague can then save the styles within your graph into his/her system by doing the following: first, click on the background of the graph and then right-click on the **Graphic Styles** box on the **Graph Tools** toolbar. In the shortcut menu that is displayed, select the **Save As** command.

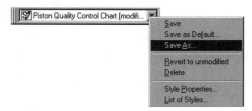

The **Save As** dialog is then displayed. Use this dialog to specify the name for the new graph style and click the **OK** button. In this manner, the graph style that was originally installed on your system has now been ported to your colleague's system in a few simple steps, using a graph as a "carrier" (see the next topic).

Can I Transfer a Graph Style from One System to Another?

Yes, you can transfer a graph style from one system to another with two different methods. The first method involves using a graph itself as a "carrier" of the graph style, while the second method makes use of the configuration manager.

The first method is to apply the desired graph style to an appropriate graph. Next, save the graph and then open it on the other system. Click the background of the graph and then right-click on the **Graphic Styles** box on the **Graph Tools** toolbar. In the shortcut menu that is displayed, select the **Save As** command to display the **Save As** dialog. Specify the name for the new graph style and click the **OK** button. Now the graph style is part of the other system's graphics library. This is a quick and recommended means of porting a graph style.

The second method is to use the **Configuration Manager** tab of the **Options** dialog (accessible via the **Tools - Options** menu). Click the **Export** button to display the **Select Configurations** dialog. Then, select the active configuration and click the **OK** button. Finally, use the **Save As** dialog to save the *.xml* file to a location where the other system can retrieve it and later import it into their system.

To import the configuration, click the **Import** button on the **Configuration Manager** tab of the **Options** dialog to display the **Open** dialog. Next, select the *.xml* file that you had previously saved and click the **Open** button to display the **Select Configurations** dialog. Here, select the configuration with the graph styles that you need to import, and click the **OK** button. After you have imported the *.xml* file, select the configuration that you have just imported on the **Configuration Manager** tab and click the **Select** button to load it. Once *STATISTICA* restarts, all of the graph styles from the other system will now be installed on the new system.

This latter method is the most efficient means of porting numerous graph styles; however, it may not be recommend because using the configuration manager will import all aspects of a system's settings (e.g., toolbar customization and output settings), not just its graph styles. If you only need to import graph styles, then the traditional method mentioned first is recommended.

What Do the Letter Icons Represent in the Graph Styles Manager?

Graphs are composed of collections of graph objects, that is, they are collections of labels, axes, fit lines, point markers, etc., as well as separate added objects such as circles, text boxes, or arrows. Graph styles address the properties of graph objects (e.g., size, color, thickness, and pattern of lines; size, shape, and color of point markers; colors and patterns used for definition of areas; size, color, and fonts for labels, titles, and scales; etc.). Graph styles are not used to add objects to graphs, but to specify the properties of objects already added to or specified for the graph. Styles can also be used to specify properties of objects that can be added to a graph later (e.g., arrows, rectangles, text, etc.).

Attributes and properties. There are two general classes of items for which a style can be created: attributes and properties. An attribute (designated with an **A**) is an object that affects the simple appearance of the graph, such as colors, line patterns, font sizes, font names, etc. A property (designated with a **P**) of a graph is an aspect of the graph that is not directly visible, such as what kind of plot to make, or what scale type to use.

Style collections. There are two additional classes of styles, collections of attributes (**AA**) and collections of properties and/or attributes (**S**). All the elements of a collection of attributes are simple attributes, **A** . All the elements of a collection of properties and/or attributes are either properties or collections of attributes, **P** and **AA**. For more details, see *Graph Styles* in Chapter 7 or the *Electronic Manual*.

User-defined styles. Any type of style that you create and save into your system will be denoted with a pencil on the icon (e.g., **A**), whether it is an attribute or property.

Graph Customization – Scales

How Do I Customize the Layout and Format of an Axis?

Double-click on the respective axis to access the *Scaling* dialog, which contains customization facilities for all features of the current axis.

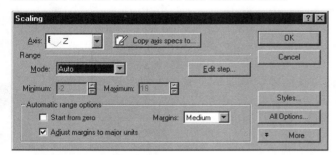

Note that the applicable features of the axis can be copied to other axes using the *Copy axis specs to* button on the *Scaling* dialog.

You can copy the features to either the corresponding (i.e., the opposite) axis or all other axes. The main scaling features of each axis can also be adjusted on the *Axis Layout* dialog (or the *All Options* dialog).

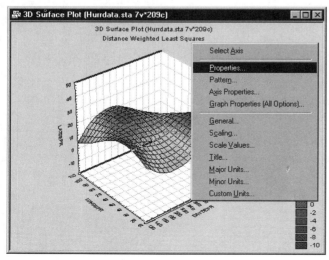

Alternatively, you can adjust individual features of an axis by selecting them from the shortcut menu accessible by right-clicking anywhere on that axis. See *Axis Options in STATISTICA* in Chapter 7 – *STATISTICA Graphs: Creation and Customization* for more details.

How Do I Replace Numeric Scale Values with Text Labels?

On the *Axis: Scale Values* tab of the *All Options* dialog (or on the *Axis Layout* dialog), select the *Use text labels from data set* check box. Note that if the variable plotted on this particular axis does not have text labels, you can create custom labels instead on the *Axis: Custom Units* tab of this dialog. Here you can create custom labels using an editable custom labels spreadsheet in which you can enter the appropriate numeric values (determining where the text labels are to be placed on the axis) and the corresponding text value labels.

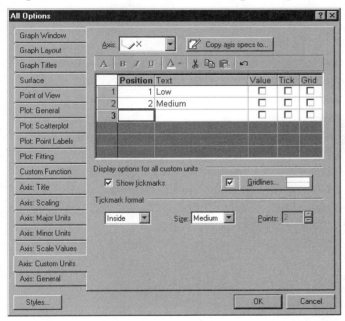

For example, if the values were entered as in the dialog shown above, then the label *Low* would be placed in the location of *1* on the axis, label *Medium* in the location of *2*, etc.

Can I Insert a Scale Break?

Yes. You can place one or more true scale "breaks" in a graph axis in order to "cut out" (i.e., "compress") certain areas of the graph space:

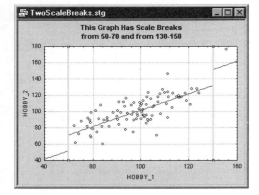

To do this, right-click on the axis in which you want the break to appear and select **Scaling** from the shortcut menu. In the resulting **Scaling** dialog, click the **More** button and add a new break by clicking the **Add new scale break** button.

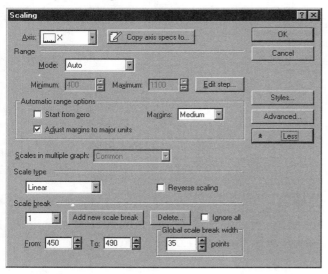

Use the **From** and **To** boxes to adjust the break location. *STATISTICA* will place the break in the specified location of the scale after you click the **OK** button on the **Scaling** dialog. These

options are also available on the **Axis: Scaling** tab of the **All Options** dialog. Note that you can add more than one scale break to an axis.

Can I Shift the Position of Axes Against the Graph?

Yes. Select the **Shift axis by** check box on the **Axis: General** tab of the **All Options** dialog to shift the position of the current axis by a specified number of points (enter a positive value to shift the axis outside the graph, enter a negative value to shift it inside). For example, when X is selected as the current scale, you can shift the x-axis upward by specifying a negative number as the shift value.

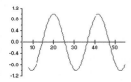

In order to imply that there is no causal relation between the X and Y variables, shift both axes by specifying a positive value in the **points** box of the respective x- and y-axes (causing a break where the x- and y-axes meet).

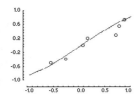

Note that the convention illustrated in the example above is not "universally" accepted, but is still used in some disciplines.

Can I Adjust the Number of Minor Units?

STATISTICA will adjust the number of minor units to the current step size value. However, you can adjust the number of minor units (as well as the default style and size of minor tickmarks) for each of the axes on the **Axis: Minor Units** tab of the **All Options** dialog (accessible from the **Format** menu). You can specify the number of tickmarks to use or have *STATISTICA* select the optimum number of minor tickmarks. The length and orientation of the tickmarks are also specified on this tab. Thickness and color of the tickmarks is addressed on the **Axis: General**

tab. For more information about the tabs of the **All Options** dialog, see Chapter 7 – *STATISTICA Graphs: Creation and Customization*.

Can I Specify Custom Locations for Tickmarks?

You can specify custom units (including their text labels, display format, and size of tickmarks) on the **Axis: Custom Units** tab of the **All Options** dialog. For more details, see *Axis: Custom Units Tab* in Chapter 7 – *STATISTICA Graphs: Creation and Customization*.

What Is the Difference between Manual and Auto Scaling?

When the axis scaling is set to **Manual** in the **Mode** box, then the minimum, maximum, and step size for the axis are determined by the current values of the **Minimum**, **Maximum**, and **Step size** (as entered in the respective boxes in the **Range** group on the **Axis: Scaling** tab of the **All Options** dialog, see Chapter 7 – *STATISTICA Graphs: Creation and Customization*). If the **Mode** is set to **Auto** (i.e., automatic), then *STATISTICA* will automatically determine the scaling based on the range of values to be plotted.

Graph Customization – Titles, Legends, Custom Text

Is All Graph Text Editable?

Yes. There are two different types of text in graphs. The first is normal editable text that you can change in the **Graph Titles/Text** dialog (e.g., by double-clicking on a title). The second is text that is automatically created and updated by *STATISTICA*, (e.g., graph legends, functions, statistics). This second type of text (and/or symbols) consists of separate "active objects" (e.g., the point marker symbol in a legend) that are automatically updated by *STATISTICA*. You can always insert new text in between active objects. Note that you can also selectively disconnect any active object from auto updating and therefore be able to edit it (but, of course, lose the auto-updating feature) by right-clicking on it and selecting **Disconnect Object(s) from Graph** from the resulting shortcut menu.

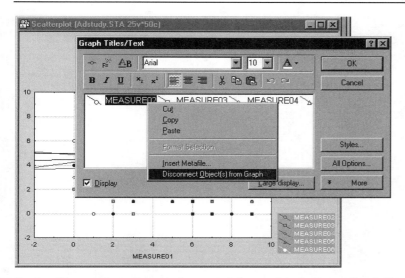

In the illustration above, the graph legend is being edited (by double-clicking on it to display the **Graph Titles/Text** dialog) and then the active object (i.e., the text **MEASURE02**) is being disconnected from the graph so it can be edited and changed to **MEASURE 02**.

Can I Customize the Location and Format of the Legend?

Fixed vs. movable legends. Legends can be treated in two ways in *STATISTICA* Graphs: either as fixed (unmovable) legends or movable legends. By default, when a graph is created, the legend is fixed (unmovable), which means that its position is automatically determined and the graph is moved to the left in the window to leave space for the legend (see the graphs below). You can make the graph legend movable so that you can reposition it in the graph and customize the text (e.g., adjust line spacing and the distance between the legend symbols and the text, etc.) and other attributes of the graph legend on the **Graph Titles/Text** dialog, (or on the **Graph Titles/Text** tab of the **All Options** dialog). To display the **Graph Titles/Text** dialog, either double-click on the legend or right-click on the legend and select **Title Properties** from the shortcut menu. For more details, see *Graph Titles/Text Tab* in Chapter 7 – *STATISTICA Graphs: Creation and Customization*.

Once you set the legend to floating text, it will become like any other added text in the graph and you can edit the text or reposition the movable legend in the graph (click on it once and then drag it to the new position on the graph).

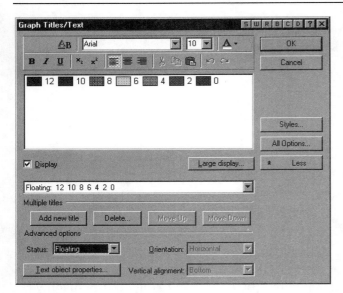

Fixed (unmovable) legends (i.e., titles) or floating (movable) legends can be temporarily removed from the graph by clearing the **Display** check box on the **Graph Titles/Text** dialog. The legend(s) can then be placed back into the graph by selecting the **Display** check box. The legends can be formatted using the mini-formatting toolbar on that dialog.

Text object properties. You can also add a custom background or frame to movable legends and/or text using the **Text Object Properties** dialog accessible by clicking the **Text object properties** button on the **Graph Titles/Text** dialog. The **Text object properties** dialog also contains options for specifying dynamic versus fixed placement of objects.

What Other Types of Legends Are Automatically Created in Graphs?

In addition to the standard fixed legend (which identifies patterns and colors used to mark individual plots in the graph), there are also other more specialized types of fixed legends. For example, there are contour legends that identify the levels in surface or contour plots:

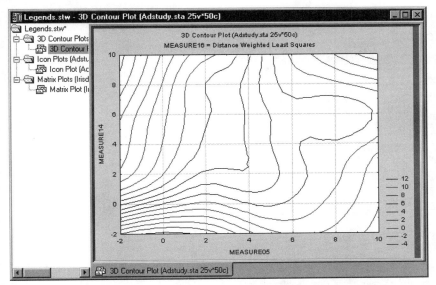

icon legends that identify the assignment of icon features to specific variables:

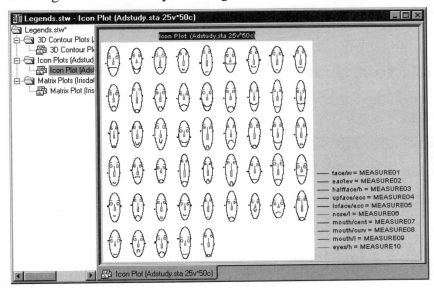

or selection legends that identify the case selection conditions used to classify cases into multiple subsets shown on the graph:

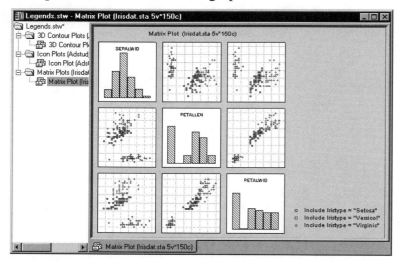

All these fixed legends can be changed to movable legends following the same conventions outlined in the previous topic.

How Can I Add a Title to a Graph?

In every graph, there are five standard graph title positions: **Title** and **Subtitle** (both at the top of graph), **Left**, **Right**, and **Footnote**.

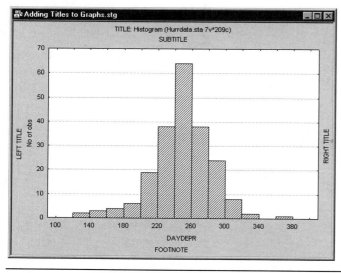

They can be edited in the *Graph Titles/Text* dialog, accessible by double-clicking on a specific title. For example, the following dialog [accessed by double-clicking on the line *TITLE: Histogram (Hurrdata.sta 7v*209c)*] shows the top title from the graph displayed above.

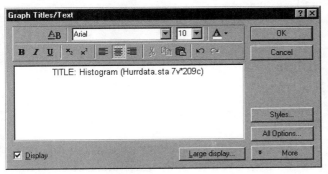

The titles can also be edited on the *Graph Titles/Text* tab of the *All Options* dialog (see Chapter 7 – *STATISTICA Graphs: Creation and Customization*).

Can I Enter a Symbol into a Graph Title?

Yes, you can easily enter symbols and special characters (such as α, β, or Σ) into a graph title. First, double-click on the title of the graph to display the *Graph Titles/Text* dialog. Change the font to *Symbol*, position the cursor at the point where you want the special character(s) to be inserted:

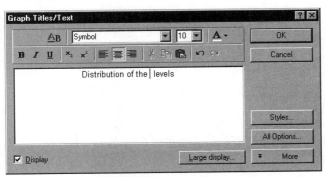

and enter the symbol(s) into the title:

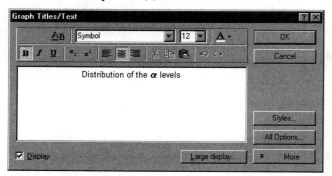

Another way to retrieve these characters is to use the Character Map program that comes installed with Microsoft Windows. This application allows you to copy these characters to the Clipboard and then paste them into your title.

Can I Convert the Standard Titles into Movable Text?

Yes. You can convert a standard title into a movable (floating) title on the **Graph Title** tab of the **All Options** dialog. In the **Advanced options** group, change the **Status** to **Floating**. See *Graph Titles/Text Tab* in Chapter 7 – *STATISTICA Graphs: Creation and Customization* for more details.

How Do I Place a Graph Title or a Footnote in a Fixed Position?

You can convert the standard title or any custom text into one of the standard graph text positions (e.g., a footnote), see the illustration on page 812. You can also convert it into moveable text (see above) and then fix it in the desired location. After the floating title is created (see *Graph Titles/Text Tab* in Chapter 7 – *STATISTICA Graphs: Creation and Customization*), click the **Text object properties** button on the **Graph Titles/Text** tab to display the **Text Object Properties** dialog. Here, you can position the title or footnote in the desired location.

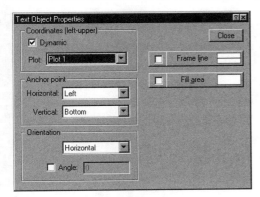

If you intend for the text to stay in a particular place in the graph area regardless of future changes to the graph scales or graph location (within the graph area), clear the **Dynamic** check box in the **Coordinates (left-upper)** group, which will keep the text in the absolute window coordinates regardless of the changes to the graph (e.g., in 5% of the window width and length from the upper-left corner). See also the previous topic.

How Do I Rotate Text?

You can select the orientation (**Horizontal**, **Vertical**, **Reversed horizontal**, or **Reversed vertical**) of floating text objects (custom text and moveable legends) in the graph in the **Orientation** group of the **Text Object Properties** dialog (accessible from the **Graphs Titles** dialog; see Chapter 7 – *STATISTICA Graphs: Creation and Customization* and the previous topic). You can also rotate the text by specifying the rotation angle (from 0 to 359° or 0 to -359°) in the **Angle** box.

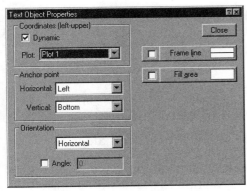

Alternatively, you can interactively rotate the text by selecting it in the graph and then dragging one of the handles (small black squares) in the desired direction,

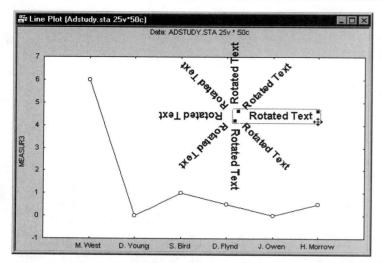

or by pressing the PAGE DOWN and PAGE UP keys to rotate text objects selected in the graph clockwise or counterclockwise, respectively, in 5° increments. To rotate in 1° increments, hold down the CTRL key while pressing PAGE DOWN or PAGE UP. The rotation of text objects takes place around the object's anchor point.

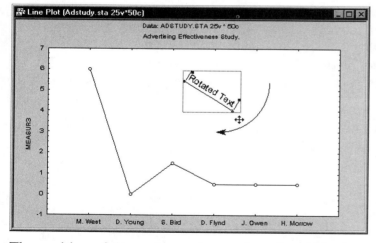

The position of the anchor point can be adjusted in the **Text Object Properties** dialog.

Technical note: Rotatability of fonts.
Some fonts cannot support rotation, and some fonts can support only limited rotation and will approximate rotation to the nearest angle to which it can be rotated. When these fonts are used, misalignment between the frame around the text and the font can occur. In this case, do not place a frame around the text. Also note that

some printer drivers cannot support rotation of some fonts (even though the text may appear properly rotated on screen).

Fitting, Plotting Functions

How Do I Fit a Function to Data?

Access the **Plot: Fitting** tab of the **All Options** dialog (see Chapter 7 – *STATISTICA Graphs: Creation and Customization*), select the appropriate plot, and click the **Add new fit** button; then, select the desired type of function or smoothing procedure in the **Fit type** box. You can adjust the fitting options (e.g., stiffness or optimization settings) and the pattern for the graphical representation of the fit on that tab as well. The pattern can also be adjusted by double-clicking on the fit line or surface in the graph. (See also, *How Do I Fit a Custom-Defined Function to the Data?* on page 819).

How Do I Display a Specific Equation for the Fitted Function?

In **Graphs** menu graphs (see Chapter 6 – *STATISTICA Graphs: General Features*), the display of the text of the fitted function equations can be requested by selecting either **In title** or **As custom text** in the **Display fit expression** box on the **Options 2** tab of the graph specification dialog. Select **Off** in the **Display fit expression** box to suppress the display of fit equations.

Note that these options can be controlled globally (i.e., for all graphs) on the **Analyses/Graphs** tab of the **Options** dialog accessible from the **Tools** menu (see Chapter 6 – *STATISTICA Graphs: General Features*).

In all single plot and non-categorized graphs where only one function is fitted, the text of the equation is displayed in the first available line of the fixed title. Depending on the number of equations to be displayed, also in categorized graphs, the equations can be displayed in the fixed titles of the graph.

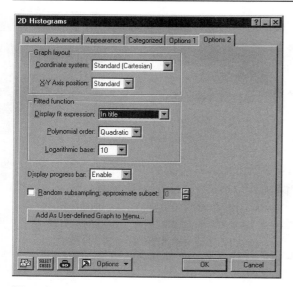

However, if more equations need to be displayed than the number of lines available in the fixed title, then *STATISTICA* will create a custom text object on the graph and place the equations there. Potentially, such lists of equations can be very long (e.g., include 256 equations), and thus the custom text object can be large and partially cover the graph. However, the location of the listing of functions can be adjusted (the list can be moved around and edited like any other custom text object, the font size reduced, etc.).

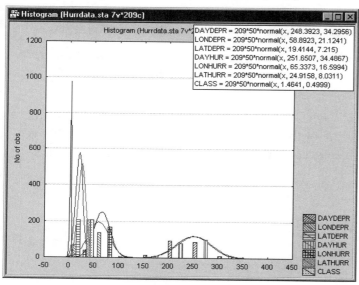

When the listing of functions is very long, it is recommended to add some space around the graph and place the text object there. You can add space around the graph using either the **Adjust the graph actual size/scaling** ⬚ toolbar button or the **Set graph area** ⬚ toolbar button (see Chapter 6 – *STATISTICA Graphs: General Features* for more details on both buttons).

How Do I Plot a Custom-Defined Function?

Select **Custom Function Plots** from either the **Graphs - 2D Graphs** or the **Graphs - 3D XYZ Graphs** menu and specify the function in the respective dialog. Also, you can add a custom function plot to any existing graph, as explained below.

Access the **Custom Function** tab of the **All Options** dialog and click the **Add new function** button. Use the options on this tab to specify the equation to be plotted in the 2D or 3D graph.

In addition to the standard math functions, a variety of functions representing distributions as well as their integrals and inverses are supported and can be plotted (including *Beta*, binomial, Cauchy, *Chi*-square, exponential, F, *Gamma*, geometric, Laplace, logistic, normal, log-normal, Pareto, Poisson, Student's *t*, and Weibull distributions).

Press the F1 key (when this tab is selected) to access the detailed syntax description and examples in the *Electronic Manual*.

How Do I Fit a Custom-Defined Function to Data?

The custom-function plotting facility (see the previous topic) accessible on the **Custom Function** tab of the **All Options** dialog (see Chapter 7 – *STATISTICA Graphs: Creation and Customization*) plots the requested (custom-defined) functions and overlays them on the existing graph, It does not fit these functions to the data. A selection of the most commonly used, predefined functions that can be fitted to the data and smoothing procedures is available from the **Plot: Fitting** tab of the **All Options** dialog (e.g., linear, logarithmic, exponential, various polynomial, distance-weighted least squares, spline, and others). See *How Do I Fit a Function to Data?* (page 817).

Comprehensive facilities to fit to data (and interactively plot in two or three dimensions) user-defined functions of practically unlimited complexity are provided in the **Nonlinear Estimation** module.

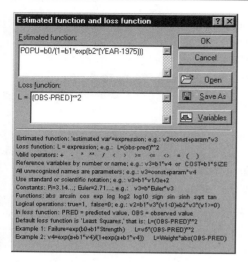

For more information on those techniques, refer to the *Electronic Manual*.

How Do I Plot a Parametric Curve?

To plot a parametric curve, select **Custom Function Plots** from the **Graphs - 2D Graphs** menu. Select the **Custom Function** tab of the **2D Custom Function Plots** dialog, and select **Parametric curve** in the **Type** box.

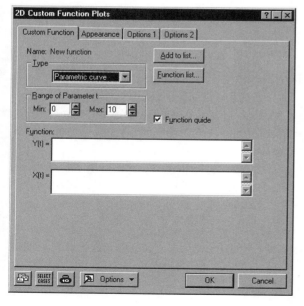

This option allows you to plot a curve in the x-y plane that is defined parametrically; i.e., a curve that is defined by two simultaneous functions of a parameter *t* that ranges over some interval (**Minimum**, **Maximum**). You can specify an equation **y = f(t)** for the y-component of the curve, and an equation **x = g(t)** for the x-component of the curve; the range of parameter *t* can be specified in the **Range of Parameter t** group (**Minimum**, **Maximum**). This option allows you to plot various curves and functions that cannot be expressed in a simple fit of **Type Function**.

Example: Plotting a circle. To plot a circle you could specify:

```
y(t)=sin(2*pi*t)
```

```
x(t)=cos(2*pi*t)
```

For **0<=t<=1** (i.e., set the **Minimum** and **Maximum** values to **0** and **1**, respectively). Note that the proportions of the plotted curves will always reflect the proportions of the X:Y scale coordinates in which the curves are plotted; therefore, for example, in order to obtain a perfect circle (as shown below), use the **Graph Layout** tab of the **All Options** dialog to set the **Axis proportion** to **Equal** (see Chapter 7 – *STATISTICA Graphs: Creation and Customization*). Alternatively, you could set the **Axis proportion** to **Custom defined** and the **X** and **Y** boxes to **1** and **1**.

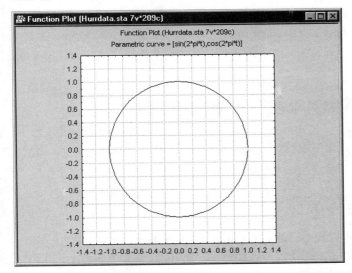

Example: Plotting a spiral. To plot a spiral, you could specify:

```
y(t)=t*cos(t)
```

```
x(t)=t*sin(t)
```

For **0<=t<=12** (i.e., set the **Minimum** and **Maximum** values to **0** and **12**, respectively).

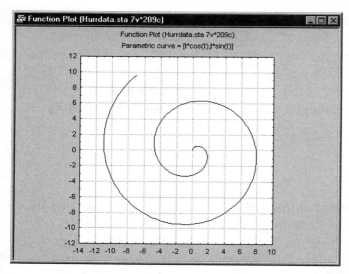

Note that the proportions of the plotted curves will always reflect the proportions of the X:Y scale coordinates in which the curves are plotted; therefore, for example, in order to obtain the spiral shown above, use the **Graph Layout** tab of the **All Options** dialog to set the **Axis proportion** to **Equal** (see Chapter 7 – *STATISTICA Graphs: Creation and Customization*). Alternatively, you could set the **Axis proportion** to **Custom defined** and the **X** and **Y** boxes to **1** and **1**.

Brushing, Interactive Analysis

The brushing facility is accessible by clicking the **Brushing** 🔍 **Graph** toolbar button or via the graph **View – Brushing** menu.

What is the Graph Data Editor?

The **Graph Data Editor** is a special version of spreadsheet that allows you to review directly the data that are plotted in the graph. This tool is useful for a variety of analytic applications, such as brushing or other forms of identifying specific data points. It also offers (sometimes) the only way to access data from those graphs, which do not plot raw values but values that have been derived, transformed, or result from specific analytic calculations, as well as values of the fitted functions. It can also be used to add additional plots of a compatible type to an existing graph.

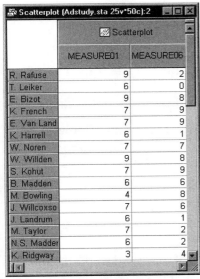

This editor is available from the graph *View* menu, the graph *Format* menu, and the general graph shortcut menu (accessible by right-clicking on the background of any graph). For more details, see *Graph Data Editor* in Chapter 6 – *STATISTICA Graphs: General Features*.

Is There a Glossary of Brushing Terms?

The following terms are used to denote specific aspects of brushing operations.

- *Highlighted:* The initial state of selected data points prior to performing some brushing action (e.g., labeling, see the term *Action*, below). Highlighting data points in the graph works like in word processors or spreadsheets where highlighting is a precursor to performing an action on the highlighted text (e.g., applying a font attribute such as boldface or italic). If no action is taken and you exit the brushing mode, the highlighted status of a point is hidden (not displayed), but preserved internally (e.g., is saved with the graph) and the points will be highlighted once again whenever you re-enter the brushing mode for the graph.

- *Action:* An operation (marking, labeling, or turning off) performed on highlighted points. Actions are executed when you click the *Update* button on the *Interactive* tab of the *Brushing* dialog (or after every operation of selecting a data point with the mouse pointer when you are in the *Auto Update* mode).

- *Selected:* A state in which points have been highlighted or further identified by execution of some action. The selected status of a point is saved with the graph. Three selection attributes are available:

- *Marked:* Identification of selected points by displaying an alternative point symbol.

- *Labeled:* Identification of selected points by displaying their respective data point labels.

- *Turned off:* Temporary removal of points from the display and fitted function calculations. The point values remain visible (on a darkened background) in the **Graph Data Editor**.

How Can Data Points Selected via Brushing Be Identified and Managed in the Graph Data Editor?

Data points that have been selected via brushing can be identified and managed in the **Graph Data Editor** (accessible by selecting **Graph Data Editor** from the **Format** menu) by selecting **Data Selected by Brushing** from the **Layouts** menu or the **Graph Data Editor** shortcut menu.

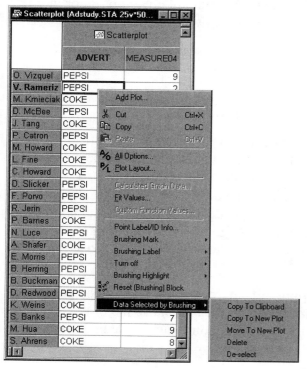

All points in the current plot of the **Graph Data Editor** that have been selected by brushing in the graph can be selectively copied to the Clipboard, copied to a new plot within the **Graph Data Editor**, moved to a new plot (i.e., deleted from their location in the original plot and pasted elsewhere), deleted, or deselected.

In addition, the attributes of points can be changed by clicking on them in the **Graph Data Editor** (or selecting them in a block) and using the options from the **Layouts** menu or the **Graph Data Editor** toolbar. Points can be **Marked**, **Labeled**, **Turned Off**, **Highlighted** (for temporary identification in the graph while in brushing mode), or **Reset**. See *Graph Data Editor Toolbar* for further details in Chapter 6 – *STATISTICA Graphs: General Features*.

How Can Brushing Be Used in Exploratory Data Analysis?

There are countless applications of brushing to explore relationships between variables and/or the contribution of specific data points or subsets to those relationships.

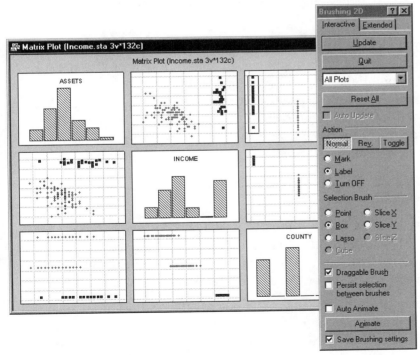

A "typical" illustration of the use of brushing in exploratory data analysis is the examination of the contribution of data points representing different ranges or different levels of one variable

to correlations between other variables, which can be inspected visually using a scatterplot matrix. For example, by including a categorical, three-level variable such as income level (**Income**) in a scatterplot matrix (see above), then by using the **Box**, **Lasso**, or **Slice X** brush, points can be selected from one income level, and the location of these points in scatterplots of all other variables (e.g., **Assets**, **Debt**) becomes immediately apparent.

How Can Brushed Data Points Be Identified and Managed in the Graph Data Editor?

All brushed points (whether marked, labeled, highlighted, or turned off) are easily identified in the **Graph Data Editor**. When data have been marked, they are displayed in bold font in the **Graph Data Editor**. Labeled cases are displayed in italics, and cases that are temporarily turned off are displayed with a darkened background. Highlighted cases are displayed in red. Cases in the **Graph Data Editor** that fall under more than one of the categories described above will have multiple overlaid formats applied, (e.g., a case that is both labeled and marked will be displayed in bold italics).

The **Graph Data Editor** toolbar provides various buttons for applying brushing actions to selected data points. That is, you can mark, label, highlight, or temporarily turn off selected data points (or unmark, unlable, dehighlight, and turn back on selected data points) in the **Graph Data Editor** using the toolbar buttons. You can also toggle the display of brushed data points using additional toolbar buttons. See *Graph Data Editor Toolbar* in Chapter 6 – *STATISTICA Graphs: General Features* for more details on using these buttons to manage brushed data points in the **Graph Data Editor**.

What Is Animated Brushing?

Typical applications of animated brushing are in exploratory data analysis using matrix plots (see the previous topic), where instead of brushing consecutive ranges of a variable (to explore the influence of various sections of its distribution), you can invoke an automatic movement of the brush (**Box**, **Lasso**, **Slice X**, **Slice Y**, **Slice Z**, or **Cube**) and watch the "results."

Specifically, a brushing region is defined in one subgraph in the matrix and is automatically moved across the subgraph (horizontally, vertically, or both). As the brushing region passes over groups of points in the subplot, corresponding points in all other plots are highlighted.

For example, in the illustration below, the rectangular region can be advanced automatically across groups of points in the **Income**, and the corresponding points will be highlighted in plots

of the other three variables. The speed and direction of the movement can be interactively controlled in the **Animation** dialog.

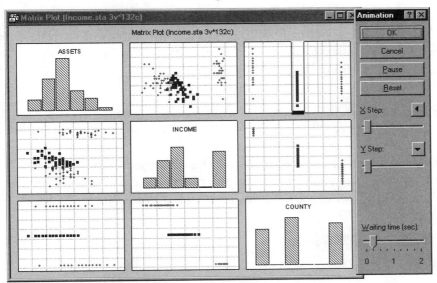

The **Animate** button is available on the **Interactive** tab of the **Brushing** dialog whenever the **Draggable Brush** check box is marked and the **Box**, **Lasso**, **Cube**, **Slice X**, **Slice Y**, or **Slice Z** brush is selected.

Draggable brush - a user controlled animation. Note that when the **Draggable Brush** check box is marked on the **Interactive** tab of the **Brushing** dialog, you can select cases via any of the six shape defining brushing tools (i.e., **Box**, **Lasso**, **Cube**, **Slice X**, **Slice Y**, or **Slice Z**), and then drag that shape to any position on the graph, thus animating the selection on your own. Note that the **Draggable Brush** check box must be selected before you make your selection in order to use the animated brushing facilities described above.

Can I Interactively Review Points That Belong to Specific Plots in Multiple Scatterplots and Other Multiple Graphs?

Identifying all points of a plot. In the (default) pointing mode (when the point tool is enabled; see the ▶ button in Chapter 6 – *STATISTICA Graphs: General Features*), click on any point that belongs to the specific plot and all points of that plot will become highlighted. They will stay highlighted for as long as you keep pressing the mouse button. If there are

many plots in the graph, and their respective point markers are small and difficult to identify, then you can click on the legend. This will also highlight all points that belong to the respective plot.

Identifying individual points of a plot. If you need to identify values of specific points in the graph, use the ***Brushing Tool*** (see Chapter 6 – *STATISTICA Graphs: General Features*; see also page 822 and the *Electronic Manual*).

How Can I Identify Specific Subsets of Data in a Graph?

Graphs menu graphs offer facilities to define subsets of cases to be identified in graphs. User-defined "multiple-subset" definitions of such subsets can be entered as logical case selection conditions of virtually unlimited complexity using facilities identical to those illustrated in producing categorized graphs (see page 780).

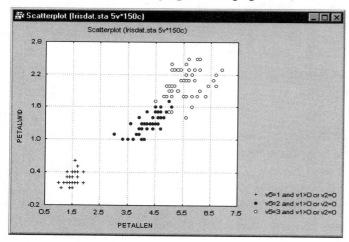

These subset identification facilities are supported in many types of **Graphs** menu graphs, including matrix plots:

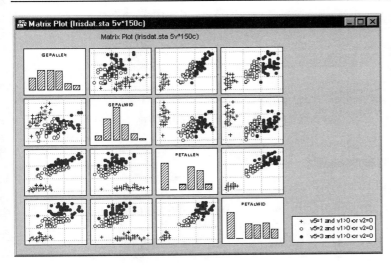

icon plots:

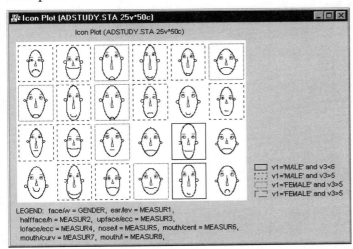

2D scatterplots, 3D scatterplots, 3D trace plots and other graphs.

Subsets can also be identified on the **Extended** tab of the **Brushing** dialog. This facility offers a wide variety of methods to identify subsets of data (which can be considered a "command driven" brushing). In this dialog, you can select ranges of values for the variables in the plot and examine their relations to other variables (for example, in a matrix plot; note that the layout of this dialog is somewhat different depending on the current type of graph).

Can I Interactively Identify Data Points in a Graph?

Yes, by marking them in the **Graph Data Editor** or labeling them selectively (see *Graph Data Editor* in Chapter 6 – *STATISTICA Graphs: General Features* or the *Electronic Manual*).

How Can I Use the Slices Brush?

2D graphs. When the brushing facility has been activated for a 2D graph such as a matrix scatterplot, available brush types in the **Selection Brush** group on the **Brushing** dialog include **Slice X** and **Slice Y**. When one of these is selected, the brushing tool can be used to define a vertical (**Slice X**) or horizontal (**Slice Y**) rectangle or "slice" of variable width on any of the individual graphs.

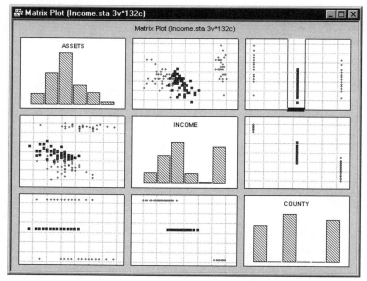

All of the point markers within the rectangular slice will become highlighted, and the corresponding point markers on the remaining matrix scatterplots will also be highlighted. Corresponding data will also be highlighted (by default colored red) in the **Graph Data Editor**.

If the **Auto Animate** checkbox on the **Interactive** tab of the **Brushing** dialog is selected, and the **Animate** button on the dialog is clicked, the slice will move back and forth on the x-axis (**Slice X**) or up and down on the y-axis (**Slice Y**) successively highlighting points that fall within the

slice on all the graphs. The movement of the slices is controlled by the options on the **Animation** dialog that appears when the animation begins.

If the **Slice X** selection brush was chosen, the **X Step** slider will control the increments at which the slice moves across the x-axis. The **Y Step** slider will control increments at which the slice moves on the y-axis if the **Slice Y** selection brush was chosen. The automatic pause between successive incremental movements of the slice is controlled by the setting of the **Waiting time** slider. The **Pause** button located near the top of the **Animation** dialog can be toggled to start and stop the animation, and a **Reset** button is available to start the animation over at the location on the graph where the slice was originally defined.

3D graphs. The slicing tools discussed above can also be applied in 3D graphs such as 3D scatterplots. In this context, the individual slice tools define a rectangular prism on the graph. All of the points within the prism will be highlighted, as well as corresponding points in the **Graph Data Editor**. In 3D graphs, slices can be defined for x-, y-, or z-axes, and the prisms can be automated in a manner analogous to that described above for slices on 2D graphs. Note that in 3D graphs, a **Cube** selection brush is also available to define a 3D prism of any size within the graph. This tool performs in a manner analogous to the **Box** tool in 2D graph brushing, and it can be animated to move incrementally within the body of the graph in the x, y, and z directions.

3D Displays – Interactive Analysis

Can I Rotate or Adjust the Perspective of a 3D Graph?

Yes. Click the **Rotate** button on the **Graph Tools** toolbar to access the interactive rotation control facility (see Chapter 6 – *STATISTICA Graphs: General Features*), or select the **Point of View** tab on the **All Options** dialog to enter specific viewpoint parameters controlling the position of the imaginary viewpoint against the 3D object. See *Viewing 3D Graphs* in Chapter 7 – *STATISTICA Graphs: Creation and Customization* for more details.

How Do I Perform Exploratory Spinning of 3D Data Displays?

Click the toolbar button to display the **Point of View Settings and Exploratory Spin** dialog, which is used for rotation, spinning (for analytic or exploratory purposes), and interactive adjustment of the point of view for three-dimensional displays.

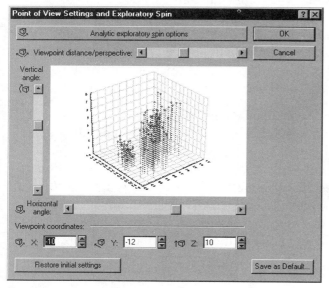

Click the **Analytic exploratory spin options** button to set the display into continuous rotation in clockwise or counter-clockwise directions.

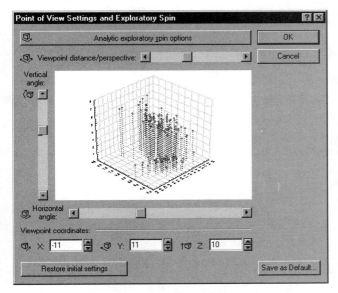

Refer to the *Electronic Manual* for more information about perspective and rotation controls in this window. Note that specific adjustments of the viewpoint and perspective (e.g., for an exact reproduction of a display) can also be made on the **Point of View** tab of the **All Options** dialog.

How Do I Change the Axis Proportions in 3D Graphs?

You can change the aspect ratio for both 2D and 3D graphs on the **Graphs 2** tab of the **Options** dialog (accessible via the **Tools** menu). For 3D graphs, use the options in the **3D graphs axis proportions** group.

See global graph options in Chapter 7 – *STATISTICA Graphs: Creation and Customization* for more details. Note that the adjustment does not modify the proportions of the graph window (only the axis proportions of the graph are modified).

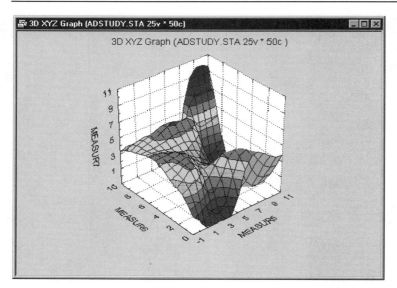

For example, the default proportions of the axes in the graph above are **1:1:1** (i.e., the proportions of the X, Y, and Z sides are of equal length and form a regular cube). By changing the respective values, you can "flatten" (e.g., increase the proportions of the x- and y-axes compared to the z-axis: **1:1:.5**)

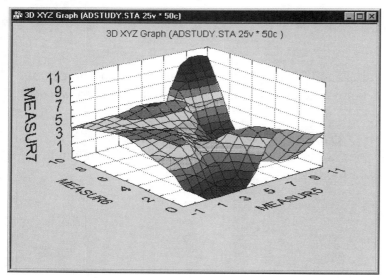

or "stretch" (e.g., increase the relative length of just the z-axis: **1:1:2**)

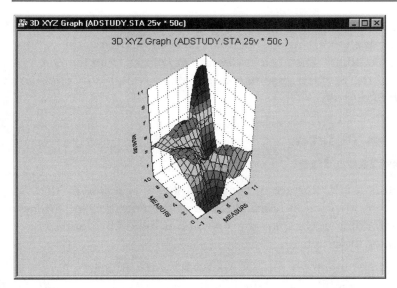

the 3D box in any direction. To change the axis proportions for a specific graph (i.e., locally), select the **Graph Layout** tab of the **All Options** dialog (see Chapter 7 – *STATISTICA Graphs: Creation and Customization*).

How Do I Define a Custom Selection of Levels for a Contour Plot or Surface Plot?

The primary controls for defining contour intervals for 3D surface or contour plots are accessed on the **Surface Specifications** dialog obtained by clicking the **Surface specs** button on the **Plot: Fitting** tab of the **All Options** plot customization dialog. (Display the **All Options** dialog by double-clicking on the outer background area of the graph.)

Number of contours/levels. The number of color levels placed on the graph is determined by default by the number of minor tickmarks spanned by the surface on the Z axis of the plot (see *Axis: Minor Units Tab* in Chapter 7 – *STATISTICA Graphs: Creation and Customization*). This default option is in the **Get shading levels from** box near the top of the **Surface Specifications** dialog (available by clicking the **Surface specs** button on the **Plot: Fitting** tab). Note that other options for determining the number of contour levels to place on the graph include major tickmarks (see *Axis: Major Units Tab* in Chapter 7 – *STATISTICA Graphs: Creation and Customization*), custom tickmarks (see *Axis: Custom Units Tab* in Chapter 7 – *STATISTICA Graphs: Creation and Customization*), and combinations of custom with major or minor tickmark options.

Specifying intervals. To control the numeric values that divide contours, use the **Custom Units** tab of the **All Options** plot customization dialog to specify as custom tickmarks the values corresponding to boundaries between contours. Then select **Custom tickmarks** from the **Get shading levels from** box near the top of the **Surface Specifications** dialog (available by clicking the **Surface specs** button on the **Plot: Fitting** tab).

How Do I Define a Custom Palette for a Contour Plot or Surface Plot?

The primary controls for defining colors for 3D surface or contour plots are accessed on the **Surface Specifications** dialog obtained by clicking the **Surface specs** button on the **Plot: Fitting** tab of the **All Options** plot customization dialog. (Display the **All Options** dialog by double-clicking on the outer background area of the graph.)

On **3D XYZ** surface plots (or contour plots, which are projections of 3D surfaces onto a 2-dimensional plane) *STATISTICA* uses a default color palette consisting of 10 colors ranging from dark green through light greens, yellows, and oranges to shades of red. This color palette can be viewed in the **Defining colors** area on the bottom half of the **Surface Specifications** dialog. The number of colors in the palette can be set anywhere in the range from 2 to 11 in the **Number of defining colors** box, and the individual color blocks can be changed by clicking the arrow at the right side of each color window. Note that you can make global changes to the color palette on the **Graphs 1** tab of the **Options** dialog (see Chapter 7 – *STATISTICA Graphs: Creation and Customization*).

Color selection by interpolation. The number of color levels required (as determined by the number of minor tickmarks or another option specified in the **Get shading levels from** box on the **Surface Specifications** dialog) is compared by *STATISTICA* against the number of color levels available in the defined palette. The colors to be applied are then picked across the range of the palette by interpolation between the specified defining colors.

Can I 3D Zoom on a Selected Cube?

Yes. To select a cube in a 3D graph, click the zoom in ⊕ button and drag the cursor ⊕₊ across the portion of the graph you want to select. A cube will be drawn, and its sides will be reflected on the three axis (as shown below). You can fine-tune the sides of the cube by dragging any of resizing handles on the side reflections (the cursor will change to a double headed arrow ↔ when the handle is selected). You can also move the cube to another location on the graph, by

dragging any of the side reflections (the cursor will change to a four headed arrow ✛ when the procedure is possible).

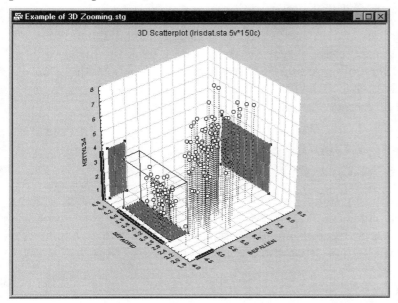

To zoom in on the selected cube, double-click on the cube. Note that if you have selected **Clone graph on zoom** from the **View** menu, the selected cube will be redrawn in a separate graph window.

Compound Graphs, Embedding, Multiple Graph Management, Clipboard

How Can I Place One *STATISTICA* Graph into Another?

Pasting one graph into another. The easiest way to place one graph into another is to copy a graph displayed in one window (press CTRL+C or click the 🖺 toolbar button), and then move to the target graph window and paste it there (press CTRL+V or click the 🖺 toolbar button). The pasted graph will be displayed on the target graph. Now you can move or resize it like every other custom graphic object.

You can also change the properties of the pasted object by selecting **Object Properties** from its respective shortcut menu (right-click on an object). You can also edit the embedded object by double-clicking on it (following the standard OLE conventions).

Linking and embedding. Graphs and artwork saved as files can also be dynamically linked or statically embedded in the current graph by using the standard OLE facility, accessible by clicking the **Graph Tools** toolbar **Insert OLE Object** button or selecting **OLE Object** from the **Insert** menu (see Chapter 6 – *STATISTICA Graphs: General Features* for a description of these operations and differences between linking and embedding).

How Can I Place a Foreign Graph or Artwork in a *STATISTICA* Graph?

The Clipboard-based operations (including linking and embedding and OLE as well as inserting) listed in the previous topic apply to all Windows-compatible graphs and artwork. Linking and embedding operations support graphs and artwork saved into bitmaps, Windows graphics metafiles, *STATISTICA* format graph files, and any OLE-compatible objects.

How Do I Place Text in a *STATISTICA* Graph?

Even large portions of text (e.g., a report several pages long) can be pasted into *STATISTICA* Graphs using the Clipboard operations mentioned in the previous two topics. Additionally, you can paste a portion of a document into the graph window using the **Paste Special** command (see Chapter 6 – *STATISTICA Graphs: General Features*). To edit and customize the text (within *STATISTICA* Graphs), double-click the text to display the **Graph Titles/Text** dialog (for custom text) or the respective OLE server application (for pasting in text via the **Paste Special** command).

Both the Clipboard-based as well as inserting operations listed in the previous topic apply to all Windows compatible graphs and artwork (linking and embedding operations support any OLE-compliant objects).

What Are Compound Graphs?

Compound graphs are those that contain other graphs. *STATISTICA* can automatically create compound graphs (e.g., in the **Quality Control** module where one display contains four different

types of graphs, or when you use the *Multiple Graph AutoLayout Wizard*, see Chapter 7 – *STATISTICA Graphs: Creation and Customization*).

Can I Represent Objects in Graphs as Expandable Icons?

Icons representing documents in Windows *Explorer* can be dragged across applications and dropped into *STATISTICA* Graphs. If the source application is OLE-compliant, the document will be displayed in the *STATISTICA* Graph.

If the source application is not OLE-compliant, then the document will be represented as an icon, either of the source application (if an association exists in Windows for the document's file extension), or of the Windows Object Packager (if no association exists).

These icons function as buttons; double-clicking on an icon will launch the application with which it is associated and open the file represented by the icon.

How Can I Copy an Entire *STATISTICA* Graph to the Clipboard?

Make sure that the window containing the graph to be copied is active, then press CTRL+C or click the ▤ toolbar button. For more information, see the *Standard Toolbar* discussion in Chapter 1 – *STATISTICA – A General Overview*.

STATISTICA Graphs can be pasted and linked or embedded in other application documents (e.g., word processor documents or spreadsheets) following standard OLE conventions. If *STATISTICA* Graphs are pasted to OLE-compatible applications, the graphs maintain their relation to *STATISTICA* and thus can be interactively edited from within the other application, or updated when the *STATISTICA* Graphs change.

If the *STATISTICA* Graph copied to the Clipboard has been saved (to a *.stg* file), you can link it in other application documents (or *STATISTICA*'s own) by selecting *Paste Link* from the *Edit* menu (or *Paste Special*).

How Can I Copy a Selected Part of a *STATISTICA* Graph to the Clipboard?

Copying an object. Select a graphic object to be copied by clicking on it (be sure that you are in default pointing mode, i.e., the *Selection Tool* ▶ button on the toolbar is clicked, see

---

Chapter 6 – *STATISTICA Graphs: General Features*). Graphic objects are all objects you have created on the screen such as a custom text, a segment of a drawing, or an embedded graph or artwork. When the object is selected, press CTRL+C. Alternatively, you can click the **Copy** toolbar button.

Copying a rectangular section of the graph. Enable the **Screen Catcher** by pressing ALT+F3 or selecting it from the **Edit** menu. The mouse pointer will change to a small circle with a cross hair; place the cross hair in the upper-left corner of the area to be copied, then drag it to the lower-right corner (a rectangle will indicate the exact area that you are selecting). When you release the mouse button, the selected area will be automatically copied to the Clipboard in the bitmap format (there is no need to click the **Copy** button). Note that the **Screen Catcher** can be used to copy any rectangular part of the screen, not only in the graph window from which it was called but any part of the screen (even including parts that belong to other applications).

Copying a specific window. The **Screen Catcher** can also capture a specific window from the screen. To copy a specific window, select **Capture Window** from the **Edit - Screen Catcher** menu, and use the mouse pointer to select the desired window. For more details on the **Screen Catcher**, see *Screen Catcher* in Chapter 6 – *STATISTICA Graphs: General Features*.

Can I Create a Blank Graph?

Yes. The quickest way to create a blank graph is to select **Blank Graph** from the **Graphs - Multiple Graph Layouts** menu. You can also select **Wizard** from the **Graphs - Multiple Graph Layouts** menu to display the **AutoLayout Wizard - Step 1** dialog. On this dialog, click the **Blank** button in the **Add Graphs** group, and then click the **OK** button to produce a "compound" graph containing one blank graph. You can then add new or existing graph objects (e.g., added text, embedded or linked objects, arrows, freehand drawings, previously saved graphs, etc.) to that blank graph.

The **Multiple Graph AutoLayout Wizard** (see below) and the **Templates** option (from the **Graphs - Multiple Graph Layouts** menu) can also be used to design and produce a custom layout. Alternatively, the **Snap to Grid** facility (see Chapter 6 – *STATISTICA Graphs: General Features*) can be used. The **Alignment Grid** (accessible from the **View** menu) and/or the dynamically updated cursor coordinates can be used to aid in the visual placement and alignment of the graph objects in the blank graph.

Can I Place Multiple Graphs on One Page?

Yes. Several graphs can be printed on one page by linking or embedding them within a blank graph (see above). Although this can be done manually using cut-and-paste (and **Snap to Grid**), the easiest method is to use either the **Multiple Graph Layouts/Templates** (see Chapter 7 – *STATISTICA Graphs: Creation and Customization*), or the **Multiple-Graph AutoLayout Wizard** (see below), which automates placement of multiple graphs on one page.

What Is the Multiple Graph AutoLayout Wizard?

The **Multiple Graph AutoLayout Wizard** can be accessed from the **Graphs** menu. The **Multiple Graph AutoLayout Wizard** assists you in selecting and arranging graphs to be placed on the same page.

Graphs can be selected from all currently open *STATISTICA* Graph windows (in all currently open *STATISTICA* modules), or from graph files previously saved to disk; blank graphs (to be filled or replaced later) can also be used. For more details, see *AutoLayout Wizard* in Chapter 6 – *STATISTICA Graphs: General Features*.

How Can I Undo Operations on Objects?

A multi-level undo option (available from the **Edit** menu, by clicking the **Undo** ↰ toolbar button, or by pressing CTRL+Z) maintains up to 32 buffers (steps), which also include operations on objects.

Printing Graphs

Do All Printer Drivers Support Rotated Fonts?

Most properly configured printers supported by Windows can handle rotated fonts; however, some printer drivers support some of the advanced printer control features used by *STATISTICA* only when they are set to a higher resolution (e.g., higher than 300 DPI) and/or when they are set to print fonts as graphics. If you encounter problems (e.g., rotated text is printed as unrotated or "uncovered" text is revealed that was supposed to be covered), consult

the documentation included with your printer for direction on printing TrueType fonts as graphics or setting your printer to a higher resolution.

Do All Printers Support the Non-Transparent Overlaying of Graphic Objects?

Most properly configured printers supported by Windows can properly handle printing of non-transparent overlays used in *STATISTICA* Graphs; see the previous topic for advice on how to configure the printer driver.

Are Fonts Set to Specific Sizes Always Printed Having the Requested Physical Sizes?

No, it depends on the current setting of the *Auto-adjust to Size* check box and the *Scaling of fonts/markers* value on the *Graphs 2* tab of the *Options* dialog (accessible via the *Tools - Options* menu). If the *Auto-adjust to Size* check box is selected, the fonts will be printed at their specified physical sizes (as set in points; 1 point = 1/72 of an inch). Specifically, the fonts will appear printed in their exact physical size, i.e.:

This is 4 point

This is 6 point

This is 8 point

This is 10 point

This is 12 point

This is 14 point

This is 16 point

This is 18 point

This is 20 point

The manner in which the font size settings as requested in the graph translate into the actual physical sizes of the fonts that are displayed or printed can be globally adjusted on the *Graphs 2* tab of the *Options* dialog, accessible from the *Tools - Options* menu. This can also be

done interactively for individual graphs using the **Decrease Font** ⧉ or **Increase Font** ⧉ buttons on the **Graph Tools** toolbar (see the next topic).

Can I Quickly Adjust Sizes of All Fonts in a Graph?

In *STATISTICA*, all graph displays and printouts can be continuously scaled. *STATISTICA* will also automatically adjust the sizes of all fonts, markers, spacing, etc., such that manual adjustments of individual font sizes are rarely necessary.

You can interactively decrease or increase the size of the selected text or point marker by clicking the **Decrease Font** ⧉ or **Increase Font** ⧉ buttons (respectively) on the **Graph Tools** toolbar (see Chapter 6 – *STATISTICA Graphs: General Features*). Each click of the toolbar button changes the font size (or point marker) by one point (i.e., one click of the **Increase Font** ⧉ button will increase the font size or point marker by one point). Note that if you have not selected any text or point markers, clicking these buttons will increase or decrease all text and point markers by one point.

What is the Difference between the Dynamic and Fixed Status of Custom Objects (e.g., Arrows, Custom Text, Drawings) in Graphs?

The difference pertains to the rules used to redraw and possibly reposition the object when the graph (and especially its scales) changes: Should it stay "fixed" where it was originally in the relation to the graph window (e.g., perfectly in the middle) but lose its relation to the specific plots, which are now drawn in a different place (fixed), or should it rather follow the plots as the scales or plot positions change (dynamic).

Specifically:

Dynamic mode. In the dynamic mode, all custom objects placed anywhere on the graph (e.g., drawings, arrows, text, embedded or linked graphs, or other artwork) will be anchored against the current *X*- and *Y*-axis scale coordinates. Thus, they will be dynamically attached to specific "logical" graph locations (and not absolute positions within the graph window). When the graph is rescaled or moved within its window, the relations between the custom (dynamic) objects and the respective graph components will not change (e.g., an arrow will still mark the

same point on a line graph even when the graph is rescaled and this point is now in a different place in the window).

Fixed mode. In the fixed mode, however, all custom objects will always remain in their absolute positions, anchored only against the proportions of the window. For example, if you add a footnote at the bottom of the graph, it will always remain on the bottom even if the graph is rescaled.

The status of objects can be adjusted in the respective **Object Properties** dialogs. Note that some objects may have two anchor points, and then each can have a different status. For example, the head of the arrow can be dynamically related to a detail in the plot, but its tail can be fixed at the very bottom of the graph window where the label or comment is located.

10

CHAPTER

STATISTICA QUERY

STATISTICA QUERY

GENERAL OVERVIEWS

General Purpose

Note: For an explanation of all technical terms used in this overview (e.g., ODBC, SQL, etc.), refer to the FAQ section at the end of this chapter (page 878).

STATISTICA Query is used to easily access data from a wide variety of databases (including many large system databases such as Oracle, MS SQL Server, Sybase, etc.) using Microsoft's OLE DB conventions. OLE DB is a powerful database technology that provides universal data integration over an enterprise's network, from mainframe to desktop, regardless of the data type. OLE DB offers a more generalized and more efficient strategy for data access than the older ODBC conventions because it allows access to more types of data and is based on the Component Object Model (COM).

STATISTICA Query supports multiple database tables; specific records (rows of tables) can be selected by entering SQL statements, which *STATISTICA* Query automatically builds for you as you select the components of the query via a simple graphical interface and/or intuitive menu options and dialogs. Therefore, an extensive knowledge of SQL is not necessary in order for you to create advanced and powerful queries of data in a quick and straightforward manner. Multiple queries based on one or many different databases can also be created to return data to an individual spreadsheet, and you can maintain connections to multiple external databases simultaneously.

Quick, Step-by-Step
Instructions

The steps necessary to retrieve external data via *STATISTICA* Query are outlined below:

1. Select **Create Query** from the **Data – Get External Data** menu (or from the **File – Get External Data** menu) to display the **Database Connection** dialog. In this dialog, select a predefined database connection (the provider, data source location, and advanced settings of the server or directory on which the data resides). Note that if you have not already created the database connection, you can do so by clicking the **New** button on the **Database Connection** dialog. The **Data Link Properties** dialog will then be displayed, which will take you through a step-by-step wizard to create a database connection. For specific documentation when you are using the **Data Link Properties** dialog, press the F1 key on your keyboard.

2. After you have selected a database connection and clicked the **OK** button on the **Data Link Properties** dialog, you will then have access to *STATISTICA* Query in which you can create your SQL statement by specifying the desired tables, fields, joins, criteria, etc. (via the **Table**, **Join,** and **Criteria** menus) to be included in your query.

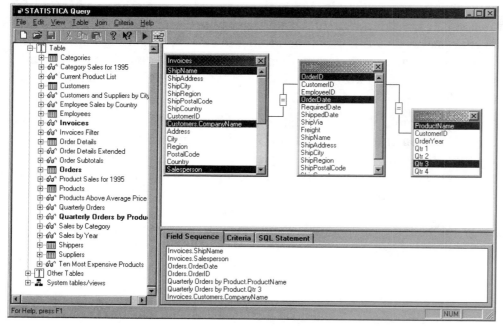

3. Once you have specified your query, select **Return Data to STATISTICA** from the **File** menu. The **Returning External Data to Spreadsheet** dialog will then be displayed in which you can specify the name of the query, where you want *STATISTICA* Query to put the data that the query returns, and additional options.

See *How Do I Query a Database Using ODBC?* (page 882) and *How Do I Query a Database Using OLE DB?* (page 882) for step-by-step examples.

"In-Place" Processing of Data on Remote Servers

The query facilities (described in the previous sections), when offered as part of the enterprise version of *STATISTICA*, are additionally enhanced by options to process data from remote servers "in-place," that is, without having to import them and create a local datafile. This technology is useful for processing extremely large datafiles where it can produce significant performance gains and allow you to process datafiles that exceed the storage capacity of the local device.

Graphic Mode Overview

In *STATISTICA* Query, the **Graphic Mode** of the user interface is the default mode of operation. In this mode the *STATISTICA* Query window is divided into three panes, the **Tree** pane on the left side, the **Graphic** pane on the upper-right side, and the **Tab** pane (containing the **Field Sequence** tab, **Criteria** tab, and **SQL Statement** tab) on the lower-right side.

The **Tree** pane displays a visual view of the tables and fields and system tables/views of the database. There are many icons within the **Tree** pane to help you easily identify parts of your Database Connection.

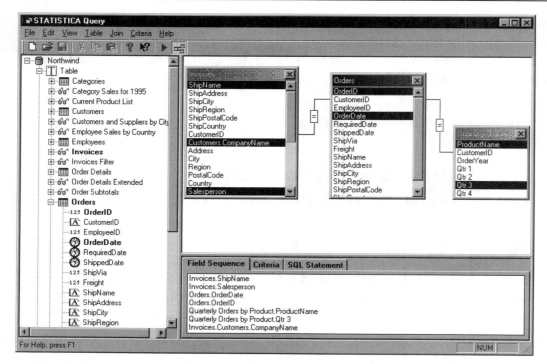

- The ⬡ icon represents the original database from which you are obtaining your external data.

- The [T] icon represents the database tables/views that have been defined by a user.

- The ⬡ icon represents the system tables/views that have been defined by the actual database.

- The ▦ icon represents tables in the database.

- The 𝓬𝓬ʳ icon represents views in the database.

- The [A̱] icon represents text fields in the database.

- The ⊕ icon represents date and time fields.

- The ¹²⁵ icon represents numerical or integer fields.

- The ⬦ icon represents object fields.

- The $ icon represents currency fields.

- The [F] icon represents all other types of fields that do not fit any of the previously defined icons.

---

850 – System Reference

The **Graphic** pane will display selected tables when you highlight and drag the table name from the **Tree** pane to the **Graphic** pane. You can also display selected tables in the **Graphic** pane by double-clicking on a field within a table in the **Tree** pane. The **Tab** pane will automatically display the field sequence, criteria, and SQL statement that you have selected via the **Tree** pane, the **Graphic** pane, and/or the menu options.

In the **Graphic Mode**, you can edit selections within the **Tree** pane and **Graphic** pane and on the **Criteria** tab (via the **Edit** menu and/or mouse and keyboard commands). You cannot, however, edit selections on the **Field Sequence** tab or the **SQL Statement** tab. If you want to edit or customize the SQL statement, then you must switch to **Text Mode**. To switch to **Text Mode**, toggle the **Graphic Mode** command on the **View** menu or click the ▣ toolbar button.

Note that when you switch from the **Graphic Mode** to the **Text Mode**, the SQL statement that is currently displayed on the **SQL Statement** tab will be retained and displayed in the identical form in the **Text Mode**. Here you can make final modifications to the SQL statement including advanced query commands that are not supported in the **Graphic Mode**. See *Additional Functionality Available in Text Mode* (page 854) for additional details.

Field Sequence Tab

The **Field Sequence** tab is only displayed when you are in the **Graphic Mode** of *STATISTICA* Query. The **Field Sequence** tab lists the table and field names that you want to include in the query in the format of **Table.Field**.

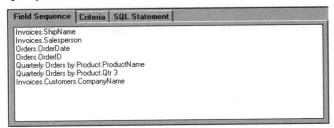

You can add a **Table.Field** to the **Field Sequence** tab (and hence to your query) in many ways. The first way in which this can be accomplished is to double-click on the field name in the **Tree** pane. Second, when a table is displayed in the **Graphic** pane, select a field within a table. Finally, in the **Tree** pane, select a field and drag it onto an empty area of the **Graphic** pane. Note that you cannot enter a **Table.Field** by using your keyboard and typing the **Table.Field** into the **Field Sequence** tab directly. To delete the selected **Table.Field** in the **Field Sequence** tab, press the DELETE key on your keyboard or select **Delete** from the **Edit** menu.

Criteria Tab

The **Criteria** tab is only displayed when you are in the **Graphic Mode** of *STATISTICA* Query. The **Criteria** tab lists the criteria that you want to include in the query.

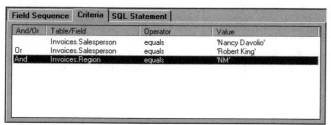

To add criteria to the **Criteria** tab (and, hence, to your query), click the **Add** button on the **Add Criteria** dialog (accessed by selecting **Add** from the **Criteria** menu) or select a field in the **Graphic** pane and drag it onto the **Criteria** tab. Note that you cannot enter criteria by using your keyboard and typing the criteria into the **Criteria** tab directly. See *How Do I Add Joins or Criteria to My Query?* (page 883) for further details.

There are several ways that you can delete the selected criteria in the **Criteria** tab. You can press the DELETE key on your keyboard, select **Delete** from the **Edit** menu, or select **Remove** from the **Criteria** menu. To remove all of the criteria listed on the **Criteria** tab, select **Remove All** from the **Criteria** menu. Note that it is possible to have multiple criteria connected with **And/Or**. If this is the case and you delete one of the criteria, then it is possible for the **And/Or** to also be deleted from another criteria (one that you have not deleted) because it is no longer applicable.

SQL Statement Tab

The **SQL Statement** tab is only displayed when you are in the **Graphic Mode** of *STATISTICA* Query.

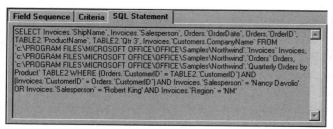

The **SQL Statement** tab lists the SQL statement that you have created using *STATISTICA* Query. Using the **Copy** option from the **Edit** menu, it is possible to copy part (or all) of the SQL statement to the Clipboard. You cannot, however, manually enter or remove text on the **SQL Statement** tab with the keyboard. To edit the SQL statement in this manner (i.e. to include advanced querying options that are not supported in the **Graphic Mode** of *STATISTICA* Query), first switch to the **Text Mode** of (via the **Graphic Mode** option from the **View** menu or the toolbar button ⚏). See *Additional Functionality Available in Text Mode* (page 854) for additional details.

Text Mode Overview

In the **Text Mode**, *STATISTICA* Query is divided into two panes: the **Tree** pane on the left side and the **Text** pane on the right side.

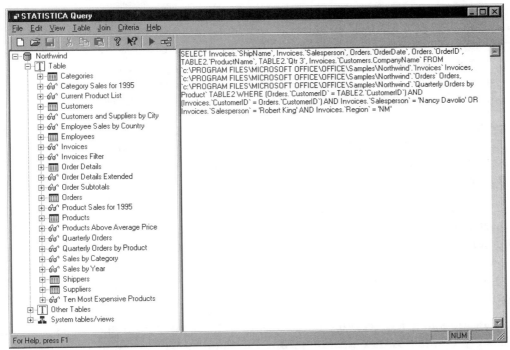

The **Tree** pane displays a visual view of the tables and fields and system tables/views of the database. The **Text** pane displays the SQL statement for you to edit. You can type directly in the **Text** pane. You can also drag names of tables and fields from the **Tree** pane to the **Text** pane. Select a table or field in the **Tree** pane and drag it onto the **Text** pane. The name of the

table or field will be inserted at the location of the mouse pointer in the **Text** pane. If you hold the SHIFT key down while you drag the name of a table onto the **Text** pane, the name and pathway of the table will be inserted at the location of the mouse pointer. If you drag the name of a field onto the **Text** pane while holding down the SHIFT key, the **Table.Field** names will be inserted at the location of the mouse pointer.

You can switch from the **Text Mode** to the **Graphic Mode** (via the **Graphic Mode** option from the **View** menu or the toolbar button ![icon]). If you choose to do this, however, you will lose your SQL statement completely. *STATISTICA* will always prompt you before you switch modes to make sure that you want to continue.

Additional Functionality Available in Text Mode

The following list is a general list of the additional SQL functionality that is not available in the **Graphic Mode** of *STATISTICA* Query, but can still be added to the query by editing the SQL statement in the **Text Mode**. This list is by no means a complete listing of the advanced SQL options available in the **Text Mode** of *STATISTICA* Query. Rather it is provided as a general overview to demonstrate the flexibility and scope of *STATISTICA* Query as an external data querying tool. For specific information on SQL syntax, please consult an SQL manual.

- Using mathematical operators to create calculations for use in criteria (e.g. +, -, *, /, MEAN, AVG, SUM, COUNT, MIN, MAX, etc.). For instance, you can add certain fields together to calculate an overall price (e.g. SELECT (SubTotal + ShippingPrice + Tax) TotalPrice FROM Orders).

- Specifying outer joins to select records that are not equivalent. This is useful if you have data from two tables that are similar but not identical. All of the records from one table will be selected whether or not there are matching records in the other table. When the records do match, they will be displayed simply as one case. When they do not match, the record will still be displayed as a case in *STATISTICA*, however, there will be missing data for cells that did not match the other table.

- Specifying subtract joins to select records that tables do not have in common. This is useful if you have data from two tables that are similar but not identical. All of the records from the first table that do not match the records from the second table will be displayed as cases in *STATISTICA*.

- Sorting rows by one or more columns. A sort can be ascending or descending. For instance, you can sort a field based on the values of another field or even multiple fields.

Graphic Mode vs. Text Mode

As mentioned before, *STATISTICA* Query offers two modes of operation, the **Graphic Mode** and the **Text Mode**. The default mode of operation is the **Graphic Mode**, and for most users is the only mode of *STATISTICA* Query ever needed to perform simple or advanced querying.

The **Graphic Mode** of *STATISTICA* Query provides you with a very intuitive, graphical means to create your query. No longer do you need to be burdened with a detailed understanding of the complex, technical terms of the SQL language. Using the simple user interface that is provided by the **Graphic Mode** of *STATISTICA* Query, you can create intricate SQL statements with only a few mouse clicks and/or menu options. Users with limited knowledge of SQL will therefore be able to create advanced and powerful queries of their data in a quick and straightforward manner. *STATISTICA* Query automatically builds the SQL statement for you as you select the components of your query via the graphical interface and/or intuitive menu commands and dialogs. See *Graphic Mode Overview* (page 849) for further details.

For users who currently have a detailed knowledge of the more complex commands of SQL language, the **Text Mode** is available for further editing and/or customization of the SQL statement that has been created in the **Graphic Mode**. To manually enter or remove text in the SQL statement with the keyboard, you need to switch to the **Text Mode**. Once you are in the **Text Mode**, you can build SQL statements including options that are not available in the **Graphic Mode** (such as outer joins; see *Additional Functionality Available in Text Mode*, page 854, for further details). Hence, by using both the **Graphic Mode** and the **Text Mode** of *STATISTICA* Query, you can create advanced queries of your data allowing you to optimize the valuable information that is contained in multiple external data sources.

Note that you can also switch from the **Text Mode** to the **Graphic Mode** (via the **Graphic Mode** command on the **View** menu or the toolbar button ▣). If you choose to do this, however, you will lose the SQL statement that has been created in the **Text Mode**. *STATISTICA* will always prompt you before you switch modes to make sure that you want to continue. See *Text Mode Overview* (page 853) for further details.

Join Overview

If the query contains more than one table, *STATISTICA* Query will automatically create a join between two tables when it detects a relationship in the original database between two fields in different tables. A join shows how data is related between two tables and determines which records *STATISTICA* Query will return as data. For example, suppose one table has the weight of objects with their associated part number and another table has part numbers and their associated product names. A join specifies that the two part number fields are equivalent and allows weights and product names to be related.

Note that when you are in the **Graphic Mode**, all joins that are created are inner joins. Inner joins select only those records that have the same value in both of the joined fields of the selected tables. *STATISTICA* Query will then combine the matching records from each table and display them as only one case in the data that is returned to the *STATISTICA* Spreadsheet. If a record does not contain a matching record in the table to which it is joined, then neither record will appear as a case in the spreadsheet.

STATISTICA Query shows that two fields are joined by displaying a black line connecting the two tables with the join's operator displayed in the center of the line.

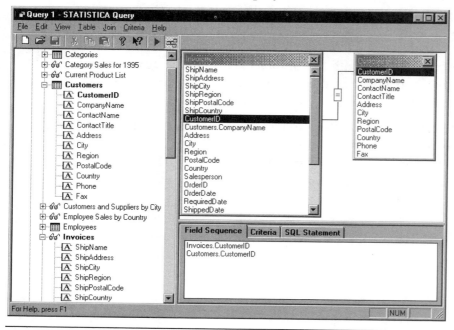

You can refine the external data that is to be returned to *STATISTICA* by specifying the operator that will be used to compare the data in the joined fields. See *Join Operators* (below) for further details.

You can manually join two fields by selecting **Add** from the **Join** menu to display the **Add Join** dialog or by selecting a field in the **Graphic** pane and dragging it to the corresponding field with which you want to create a join. By default, the join's operator will be = (equals). To edit the join's operator, double-click on the operator in the **Graphic** pane or select **Edit** from the **Join** menu to display the **Edit Join** dialog. It is also possible to specify other types of joins (i.e., outer and subtract joins), but you need to be in the **Text Mode** of *STATISTICA* Query. See *Additional Functionality Available in Text Mode* (page 854) for further details.

Join Operators

There are many different operators that you can choose from when you are specifying a join. The following is a list of the available operators in the **Add Join** dialog and their meanings.

Operator	Meaning
=	Selects records that contain equal values in the joined fields.
<>	Selects records that contain values that are not equal in the joined fields.
>	Selects records from the first specified table that contain values that are greater than values in records from the second specified table in the joined fields.
>=	Selects records from the first specified table that contain values that are greater than or equal to values in records from the second specified table in the joined fields.
<	Selects records from the first specified table that contain values that are less than values in records from the second specified table in the joined fields.
<=	Selects records from the first specified table that contain values that are less than or equal to values in records from the second specified table in the joined fields.

Criteria Overview

Criteria establishes the condition(s) that records from an external data source must meet in order to be included in the cases that are returned by the query. Therefore, criteria is part of the SQL statement that is used to limit which records are returned to the *STATISTICA* Spreadsheet. To add criteria, use the **Add Criteria** dialog by selecting **Add** from the **Criteria** menu or by selecting a field in the **Graphic** pane and then dragging it onto the **Criteria** tab.

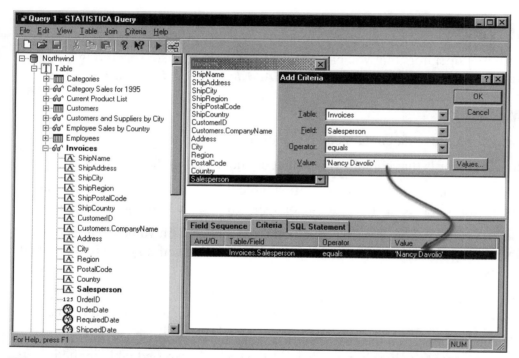

When you select a criteria field, make sure that you use the field from the correct table. The following examples are taken from fictitious databases and are provided to illustrate the various ways criteria can be specified to retrieve records.

- You can return all records that contain (or do not contain) certain value(s) you specify. For example, if you want to return only the customers who live in Louisville, specify that the field City in the Customer table **equals** Louisville.

- You can return a range of records. For example, if you want to return all orders that are between $500 and $700, specify that the field ProductAmount in the Orders table *is greater than or equal to* $500 AND the field ProductAmount *is less than or equal to* $700.

- You can return records that are one of (or are not one of) a specific group of items. For instance, if you want to return sales that were not made in the United States, Switzerland, or Germany, you will specify the field Country in the Orders table *is not one of* the values United States, Switzerland, or Germany.

- You can return records that are between (or are not between) specific values. For instance, if you want to return orders that were purchased between December 5 and December 10, you will specify that the field Date in the Order table is *between* 12-5 AND 12-10. Note that *STATISTICA* Query will automatically include AND in this criteria when you specify *between* as your operator. You simply need to select the two values.

- You can return records that begin with, end with, or contain (or do not begin with, do not end with, or do not contain) certain characters. For example, if you want to return all customers with a 330 area code, you specify that the field PhoneNumber in the Customer table *begins with* 330. Note that you can type in 330 with or without single quotes.

- You can return a record that is like (or is not like) specific criteria. For example, if you want to retrieve the personnel information on an employee named Kasmirski, but you don't remember how to spell the name, specify that the field EmployeeName in the Personnel table *is like* Ka%.

- You can return records that contain (or do not contain) missing or incomplete values. For example, if the IRS wants to identify the tax forms that have missing signatures, specify that the field Signature in the 1040 table *is null*. Note that when you choose the operator to be *is null* or *is not null*, the value field will no longer be active.

- You can return records with values that meet one criteria or another criteria. For example, you can find all customers who live in Louisville or who have a 330 area code. To select these records, you will type two sets of criteria. First, specify that the field City in the Customer table *equals* the value Louisville. Then, add another criteria, select the *Or* option button, and specify that the AreaCode field in the Customer table *equals* the value 330.

- You can return records with values that meet multiple criteria. For example, you can return all the customers who live in Louisville and who purchased items in the months of November or December. To select these records, you will type two sets of criteria. First, specify that the field City in the Customer table *equals* the value Louisville. Then, select the *And* option button and specify that the Month field in the Orders table *is one of* values November and December.

See *Criteria Operators* (below) and *Criteria Tab* (page 852) for further details.

Criteria Operators

There are many different operators that you can choose from when you are specifying criteria. The following is a list of the available operators in the **Add Criteria** dialog along with their respective meanings.

Operator	Meaning	Result	Example Table/Field	Value(s)
=	Equal to	Finds records for customers in Ohio	Customers/State	Select Ohio
<>	Not equal to	Finds customers not in Ohio	Customers/State	Select Ohio
>	Greater than	Finds orders more than $100	Orders/Price	Select 100
>=	Greater than or equal to	Finds orders placed on or later than a certain date.	Orders/Date	Select Dec-99
<	Less than	Finds orders that are less than $100	Orders/Price	Select 100
<=	Less than or equal to	Finds orders placed on or earlier than a certain date	Orders/Date	Select Nov-99
Is one of	Shows values that match a list of specified values	Finds suppliers in Tulsa, Cleveland, or Dayton	Suppliers/City	Select Tulsa, Cleveland, and Dayton
Is not one of	Shows values that do not match a list of specified values	Finds orders that were not placed in November or December	Orders/Month	Select November and December
Between	Shows a range of values between and including the two endpoints	Finds employees who were hired between 1999 and 2000	Employees/ HireDate	Select 1999 and 2000

Is not between	Shows values that are not between a range of values including the two endpoints	Finds orders that are not between $100 and $500	Orders/Price	Select 100 and 500
Begins with	Begins with specified value(s)	Finds zip codes that begin with 454	Customers/ ZipCode	Enter 454
Does not begin with	Does not begin with specified value(s)	Finds zip codes that do not begin with 454	Customers/ ZipCode	Enter 454
Ends with	Ends with specified value(s)	Finds zip codes that end in 19	Customers/ ZipCode	Enter 19
Does not end with	Does not end with specified value(s)	Finds zip codes that do not end in 19	Customers/ ZipCode	Enter 19
Contains	Contains specified value(s)	Finds products that contain sugar	Products/ Ingredients	Select sugar
Does not contain	Does not contain specified value(s)	Finds products that do not contain sugar	Products/ Ingredients	Select sugar
Is like	Uses Like and the wildcard character % to compare values	Finds names that begin with Da such as Daniel and David	Employees/ FirstName	Enter Da%
Is not like	Shows values that do not compare to the specified character(s) and wildcard character %	Finds names that do not begin with Da	Employees/ FirstName	Enter Da%
Is null	Determines whether a record has no value in the specified field	Finds records that do not have an entry for address	Customers/ Address	
Is not null	Determines whether a record has a value in the specified field	Finds records that do have an entry for address	Customers/ Address	

STATISTICA QUERY TOOLBAR

The *STATISTICA* **Query** toolbar is always available if you do not clear the **Toolbar** command on the **View** menu in *STATISTICA* Query. As with all toolbars in *STATISTICA*, you can make it a floating toolbar by holding down the mouse button when the mouse pointer is between toolbar buttons and dragging the toolbar. You can also choose to dock it to the left, right, or bottom of the workspace by dragging the floating toolbar to the area you want to place it and releasing the mouse button. Note that unlike the other *STATISTICA* toolbars, the **Query** toolbar cannot be customized.

New Query Button

Click the button to create a new query in *STATISTICA* Query. When you click this button the **Database Connection** dialog is displayed, which is used to choose the database to use in defining the query.

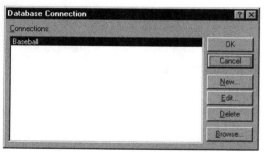

Note that you can also choose to create a new database connection (the provider, data source location, and advanced settings of the server or directory on which the data resides), by clicking **New** on the **Database Connection** dialog. The **Data Link Properties** dialog is then displayed, which takes you through a step-by-step wizard to create a database connection using OLE DB conventions.

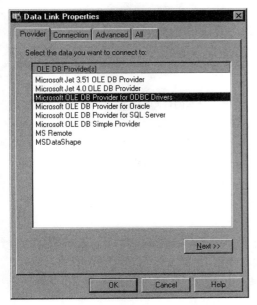

For further information, see *How Do I Set Up a Database Connection Using the OLE DB Provider for ODBC?*, (page 880).

Open Query Button

Click the button to display the standard *Open* dialog, which is used to locate and open any existing *STATISTICA* Query file (*.sqy). Note that you can open an existing file at any time to modify the query.

Save Query Button

Click the button to save the query as a *.sqy* file, replacing any previous version with the current version. This option is only available if you have already saved and named your query via the standard *Save As* dialog.

Cut Button

Click the button to cut (remove) the contents of the current selection and copy it to the Clipboard. This option is only available when you have selected text on the *Text* pane in the *Text Mode* of *STATISTICA* Query. *STATISTICA* will create in the Clipboard multiple formats

of the selected object; the specific format to be used for pasting can later be selected with the **Paste Special** options in the client application.

Copy Button

Click the button to copy the highlighted selection of the current SQL statement to the Clipboard. This option is only available when you have selected text on the **SQL Statement** tab in the **Graphic Mode** or on the **Text** pane in the **Text Mode**. *STATISTICA* will create in the Clipboard multiple formats of the selected object; the specific format to be used for pasting can later be selected with the **Paste Special** options in the client application.

Paste Button

Click the button to paste the current contents of the Clipboard into the SQL statement starting at the current cursor position in the **Text** pane. This option is only available when you are in the **Text Mode** of *STATISTICA* Query.

Help Topics Button

Click the button to access the *STATISTICA* on-line *Electronic Manual*. *STATISTICA* provides comprehensive, context-sensitive, on-line documentation for all program procedures and all options. Press the F1 key or click the help button ? on the toolbar or the caption bar of all dialog boxes to access the *Electronic Manual* (there is a total of over 100 megabytes of compressed documentation included). Due to its dynamic hypertext organization (and various facilities allowing you to customize the help system), it is usually faster to use the on-line documentation than to look for information in the printed manuals. Note that the status bar also displays short explanations of the menu commands or toolbar buttons, available when a command is highlighted or a button clicked. In addition, you can press the help key (F1) when the command is highlighted for more information on any menu command.

Context Sensitive Help Button

Click the ? button (or press SHIFT + F1) to access the on-line *Electronic Manual* in a context sensitive manner. When you click this button, the mouse pointer will change to an arrow with a question mark beside it ?. Click the toolbar button or area of *STATISTICA* Query that you are interested in learning more about. The appropriate *Electronic Manual* topic is then displayed. If an appropriate topic is not available, the **Contents** tab of the *Electronic Manual* is displayed.

▶ Return Data to *STATISTICA* Button

Click the ▶ button to run the query and return the data to *STATISTICA*. When you click this button, the **Returning External Data to Spreadsheet** dialog will be displayed, which is used to specify the name of your query and where you want *STATISTICA* Query to put the data that your query returns. This option is only available if an SQL statement has first been created.

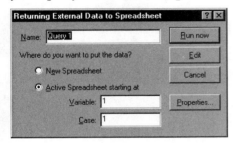

See *Return Data to STATISTICA* (page 867) for further details.

▦ Switch Mode Button

STATISTICA Query consists of two modes of operation, the **Graphic Mode** and the **Text Mode**. The default mode of operation is the **Graphic Mode**, and for most users is the only mode of *STATISTICA* Query needed to do advanced querying. When the ▦ button is selected, *STATISTICA* Query is in **Graphic Mode**. When it is deselected, *STATISTICA* Query is in **Text Mode**. See *Graphic Mode vs. Text Mode* (page 855) for further details.

Note that when switching from **Text Mode** to **Graphic Mode** the SQL statement will be lost. It is important that you save the query using the **Save** command from the **File** menu or by clicking the **Save Query** 🖫 button before switching from **Text Mode** to **Graphic Mode**. *STATISTICA* will issue a warning that the statement is about to be lost when switching modes.

If you haven't already saved the query at that time, you can click the **No** button and save the query before switching to **Graphic Mode**. However, the expectation is that you will be running the query (via the ▶ **Return Data to STATISTICA** button) in **Text Mode**.

STATISTICA QUERY WINDOW

File Menu

New

Select **New** from the **File** menu to display the **Database Connection** dialog, which is used to select a predefined database connection (the provider, data source location, and advanced settings of the server or directory on which the data resides).

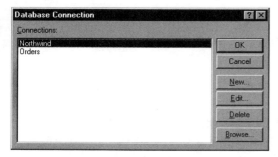

To create a new database connection, click the **New** button to display the **Data Link Properties** dialog (see *How Do I Set Up a Database Connection Using the OLE DB Provider for ODBC?*, page 880, for further details). To edit or delete the selected database connection, click the **Edit** and **Delete** buttons, respectively. To search your computer for a Microsoft Data Link file that describes the data source you want to connect to, click the **Browse** button to display the **Select Data Link File** dialog. Once you have selected the database connection in the **Connections** box, click the **OK** button to begin creating your query. You can also click the **Cancel** button to close this dialog without selecting a database connection.

Open

Select **Open** from the **File** menu to display the standard **Open** dialog, which is used to specify a *STATISTICA* Query file (**.sqy*) to open.

Save

Select **Save** from the **File** menu to save your query as a *.sqy* file, replacing any previous version with the current version. If you haven't saved the query before, then the standard **Save As** dialog is displayed. This option is only available if an SQL statement has first been created.

Save As

Select **Save As** from the **File** menu to display the standard **Save As** dialog, which is used to save a new query or save an existing query under a different name. By default, your query will be saved as a *STATISTICA* Query file (**.sqy*). This option is only available if an SQL statement has first been created. Note that the file name that you specify here may or may not be identical to the name that you specify in the **Name** field on the **Returning External Data to Spreadsheet** dialog (see below). While the name that is specified on the **Returning External Data to Spreadsheet** dialog is an internal name that only *STATISTICA* uses, the *.sqy* file name that is specified here can be used to open the SQL statement in other applications. In this manner, you will have access to all queries that you have written in *STATISTICA* Query regardless of whether or not you are actually working in *STATISTICA*.

Return Data to *STATISTICA*

Select **Return Data to STATISTICA** from the **File** menu (or press F5 or click the ▶ toolbar button) to run the query that you have created. This option is only available if an SQL statement has been created. When you select this command, the **Returning External Data to Spreadsheet** dialog is displayed.

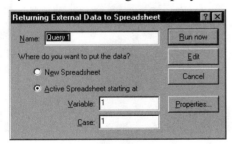

Use the **Name** field to specify the internal name that *STATISTICA* uses to refer to the current query. For instance, if multiple queries are connected to a datafile and you select **Edit Query** from the **Data – Get External Data** menu (or the **File – Get External Data** menu), this name is used in the resulting **Select Database Query** dialog that is displayed.

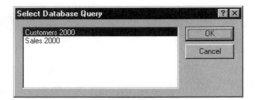

However, the *.sqy* file name that is specified via *File – Save As* is used when you select *File – Open* within *STATISTICA* Query or other applications.

Select either the *New Spreadsheet* or the *Active Spreadsheet starting at* option button to specify where you want *STATISTICA* Query to put the data that the query returns. Note that if you select the *Active Spreadsheet starting at* option button, you can then enter the exact *Variable* and *Case* number in which the returned data should begin.

Click the *Run now* button to run the query and return the external data to *STATISTICA*. Click the *Edit* button to return to *STATISTICA* Query and edit the current query. Click the *Cancel* button to return to *STATISTICA* without running the query. No external data will be returned to *STATISTICA*.

Click the *Properties* button to display the *External Data Range Properties* dialog, which is used to specify numerous options regarding the spreadsheet that contains the data that is returned by the query, both when it is first returned and also when the query is refreshed.

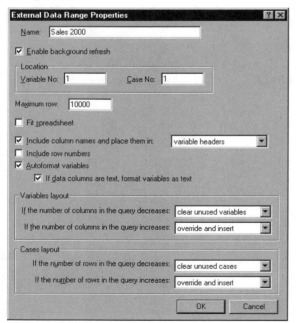

For further details on the **External Data Range Properties** dialog, see the *Electronic Manual*.

Cancel and Return to *STATISTICA*

Select **Cancel and Return to STATISTICA** from the **File** menu to cancel the query and return to *STATISTICA*. No external data will be returned to *STATISTICA*.

Edit Menu

Undo

Select **Undo** from the **Edit** menu to undo your last action. You can continue to select **Undo** to undo previous actions.

Cut

Select **Cut** from the **Edit** menu to cut the selected text or object(s) to the Clipboard.

Copy

Select **Copy** from the **Edit** menu to copy the selected text or object(s) to the Clipboard.

Paste

Select **Paste** from the **Edit** menu to paste the contents of the Clipboard into the selected location of the document.

Delete

Select **Delete** from the **Edit** menu to remove the selected item without copying it to the Clipboard. If you want to copy the item to the Clipboard while removing it, use the **Cut** option also available from the **Edit** menu. See *Criteria - Remove* (page 876) for additional details to consider when you delete a criteria.

When operating in **Graphic Mode**, this command is only available if you have selected something in either the **Graphic** pane or on the **Criteria** tab. It is only available in **Text Mode** if

you have selected something in the **Text** pane. You can also delete an item by pressing the DELETE key on your keyboard.

Select All

Select **Select All** from the **Edit** menu to select (highlight) the entire SQL statement. This option is only available when the focus is in the **Text** pane. It is not available when you are in the **Graphic Mode**.

Query Properties

Select **Query Properties** from the **Edit** menu to display the **Query Properties** dialog. Use this dialog to modify options affecting the specific query that you are currently writing, regardless of global options (see below).

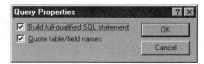

Select the **Build full-qualified SQL statement** check box to display the full-qualified SQL statement when you choose to view the SQL statement. This means that the specific database name and/or path will be included in the SQL statement. If this option is not selected, then they will not be included in the SQL statement. You can view the SQL statement by selecting the **SQL Statement** tab or by switching to **Text mode** (by toggling the 🔲 toolbar button or the **Graphic Mode** command on the **View** menu).

Select the **Quote table/field names** check box if you want to have quotation marks displayed around all table and field names in the SQL statement (this may or may not be necessary depending on the specific database format). If you do not want quotation marks around the table and field names in the SQL statement, clear this option.

Global Options

Select **Global Options** from the **Edit** menu to display the **Global Options** dialog. Use this dialog to modify the default query settings for all queries that you create.

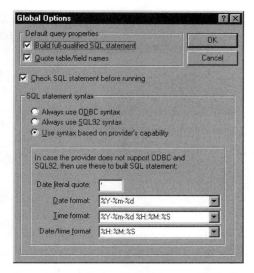

See *Query Properties* (page 870) for a description of the check boxes in the ***Default query properties*** group. Select the ***Check SQL statement before running*** check box (the default mode) to check the SQL statement for errors before the query is run.

The options in the ***SQL statement syntax*** group are used to specify the syntax for which the SQL statement is based. Select the ***Always use ODBC syntax*** option button to generate SQL statements based on the Microsoft ODBC syntax. You need to make sure that the provider supports ODBC syntax. Select the ***Always use SQL92 syntax*** option button to generate SQL statements based on the ANSI SQL-92 standard syntax. Make sure that the provider supports SQL-92 syntax. Select the ***Use syntax based on provider's capability*** option button (the default mode) to have *STATISTICA* Query make the decision between ODBC or SQL-92 syntax based on the provider's capability. Additionally, some providers have their own syntax and don't support either ODBC or SQL-92. Use the remaining fields to deal with such providers. Note that these fields require you to know how the provider generates SQL statements. For further details on providers, see *How Do I Set Up a Database Connection Using the OLE DB Provider for ODBC?* (page 880).

View Menu

Toolbar

Select **Toolbar** from the **View** menu to toggle between displaying or not displaying the **Query** toolbar at the top of the screen. If **Toolbar** is selected (the default mode) then the toolbar is displayed. Clearing this option will hide the toolbar. See *STATISTICA Query Toolbar* (page 862) for a description of the various options available on the toolbar.

Status Bar

Select **Status Bar** from the **View** menu to toggle between displaying or not displaying the text on the status bar at the bottom of the screen. If **Status Bar** is selected (the default mode) then the text is displayed. If this option is cleared, then the status bar will be hidden. Note that the status bar also conveys information about keyboard settings (e.g., whether or not the CAPS LOCK or NUM LOCK keys are depressed).

Graphic Mode

Select **Graphic Mode** from the **View** menu to toggle between the two modes of *STATISTICA* Query: the **Graphic Mode** and the **Text Mode**. If **Graphic Mode** is selected (the default mode), then the **Graphic Mode** is displayed. If this command is cleared, then the **Text Mode** is displayed. For specific details, see *Graphic Mode Overview* (page 849), *Text Mode Overview* (page 853), and the *Graphic Mode vs. Text Mode* (page 855).

Note that when you are switching from the **Graphic Mode** to the **Text Mode**, the SQL statement that is currently displayed on the SQL Statement tab will be retained and displayed in the identical form in the **Text Mode**. Here you can make final modification to the SQL statement including advanced query commands that are not supported in the **Graphic Mode** (see *Additional Functionality Available in Text Mode*, page 854, for additional details). However, if you choose to switch from **Text Mode** to **Graphic Mode**, you will lose your SQL statement completely. *STATISTICA* will always prompt you before you switch modes to make sure that you want to continue.

Table Menu

Add

Select **Add** from the **Table** menu to move the currently selected table in the **Tree** pane to the **Graphic** pane. You can add as many tables as you want. This option will not be available unless you have first selected a table in the **Tree** pane.

Note that you can also move a table to the **Graphic** pane by right-clicking on a table in the **Tree** pane and choosing **Add** in the resulting shortcut menu or by selecting the table and then dragging it to the **Graphic** pane. You can also double-click on a field within a table and then the table with the field selected will be moved to the **Graphic** pane.

Remove

Select **Remove** from the **Table** menu to remove the currently selected table from the **Graphic** pane. This option will not be available if you currently do not have any tables in the **Graphic** pane. Note that you can also remove tables in other ways. You can right-click on a table in the **Tree** pane and then choose **Remove** in the resulting shortcut menu. You can also select a table in the **Graphic** pane and then press the DELETE key on your keyboard or click the X in the upper-right corner of the table.

Remove All

Select **Remove All** from the **Table** menu to remove all of the tables from the **Graphic** pane. This option will not be available if you currently do not have any tables in the **Graphic** pane. Note that you can also remove all of the tables from the **Graphic** pane by right-clicking on any table in the **Tree** pane and choosing **Remove All** in the resulting shortcut menu.

Join Menu

Add

Select **Add** from the **Join** menu to display the **Add Join** dialog, which is used to create a join between two tables. A join shows how data is related between two tables and determines which

cases *STATISTICA* Query will return as data. If your query contains more than one table, *STATISTICA* Query will automatically create a join between two tables when it detects a relationship in the original database between two fields in different tables. Note that you can also create a join using your mouse by selecting a field in the **Graphic** pane and then dragging it to a corresponding field with which you want to create a join.

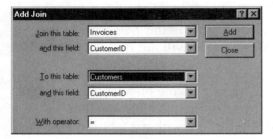

In the **Join this table, and this field, To this table, and this field** drop-down boxes, select the tables and fields that you want to join. Next, in the **With operator** drop-down box, select the operator that determines the type of comparison that *STATISTICA* Query will perform on the values in the joined fields.

Click the **Add** button to add the currently specified join to your query.

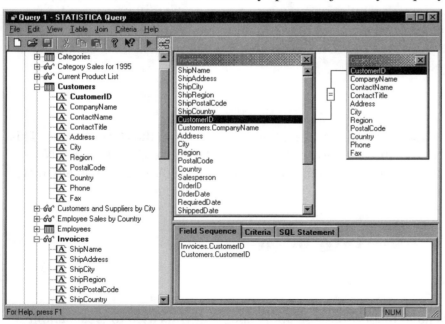

See *Join Overview* (page 856) for additional details.

Remove

Select **Remove** from the **Join** menu to remove the currently selected join (the black lines will be darker). This option will not be available unless the focus is in the **Graphic** pane and a join is selected. See *Join Overview* (page 856) for additional details.

Edit

Select **Edit** from the **Join** menu to display the **Edit Join** dialog, which is used to edit the currently selected join between two tables.

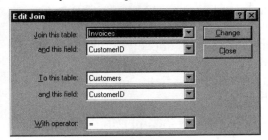

This option will not be available unless the focus is in the **Graphic** pane and a join is selected. Note that you can also edit a join by double-clicking on the operator that connects the two joined tables in the **Graphic** pane. See *Join Overview* (page 856) for additional details.

Remove All

Select **Remove All** from the **Join** menu to remove all of the joins in your query. This option will not be available if you currently do not have any joins in your query. See *Join Overview* (page 856) for additional details.

Criteria Menu

Add

Select **Add** from the **Criteria** menu to display the **Add Criteria** dialog, which is used to define criteria that records must meet in order to be included in the cases that *STATISTICA* Query will return as data. For example, you can specify criteria to select records that have a certain value (such as customers in Louisville) or a specific range of numbers (such as orders that are

between $100 and $1,000). Note that you can also add a criteria when you select a field in the **Graphic** pane and then drag it onto the **Criteria** tab.

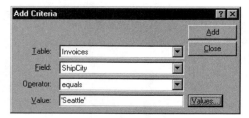

Use the **Table** drop-down box to select the appropriate table for the criteria. You can select from all of the tables that are currently in the **Graphic** pane. Use the **Field** drop-down box to select the appropriate field for the criteria. All of the fields that belong to the specified **Table** are displayed. Use the **Operator** drop-down box to select the appropriate operator for the criteria. In order for records to be returned as data, the operator that is chosen in this field must be true for the specified **Field** and **Value(s)**. Finally, enter the value(s) for *STATISTICA* Query to use when selecting records that will be returned as data in the **Value** field. For a list of all of the values found in the specified **Field**, click the **Values** button.

Click the **Add** button to add the specified criteria to your query. After you click the **Add** button, **And** and **Or** option buttons are displayed on the **Add Criteria** dialog. When the **And** option button is selected, then records must meet both the existing and the new criteria conditions (i.e. both of the criteria must be true) in order to be included in the cases that will be returned as data. When the **Or** option button is selected, then records only have to meet one of the criteria conditions (i.e. either the existing or the new criteria must be true) in order to be returned as data. See *Criteria Overview* (page 858) for additional details.

Remove

Select **Remove** from the **Criteria** menu to remove the currently selected criteria. This option will not be available unless you have selected a criteria on the **Criteria** tab. Note that pressing the DELETE key on your keyboard will also remove the selected criteria on the **Criteria** tab. Note that it is possible to have multiple criteria connected with **And/Or**. If this is the case and you delete one of the criteria, then it is possible for the **And/Or** to also be deleted from another criteria (one that you have not deleted) because it is no longer applicable. See *Criteria Overview* (page 858) for additional details.

Edit

Select *Edit* from the *Criteria* menu to display the *Edit Criteria* dialog, which is used to edit the currently selected criteria on the *Criteria* tab. This option will not be available unless the focus is in the *Tab* pane and you have selected a criteria on the *Criteria* tab.

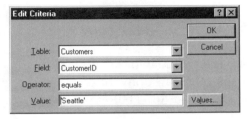

See *Criteria Overview* (page 858) for additional details.

Remove All

Select *Remove All* from the *Criteria* menu to remove all of the criteria in your query. This option will not be available if you do not have any criteria in your query. See *Criteria Overview* (page 858) for additional details.

Help Menu

Contents and Index

Select *Contents and Index* from the *Help* menu to display the index of the *STATISTICA Electronic Manual*.

About *STATISTICA* Query

Select *About STATISTICA Query* from the *Help* menu to display the *STATISTICA* Query version and copyright notice.

FAQ: WORKING WITH STATISTICA QUERY

How Do I Retrieve External Data via *STATISTICA* Query?

STATISTICA Query is used to easily access data from a wide variety of databases (including many large system databases such as Oracle, Sybase, etc.) using OLE DB conventions. The steps necessary to retrieve external data via *STATISTICA* Query are outlined below:

- Select **Create Query** from the **Data – Get External Data** menu (or the **File – Get External Data** menu) to display the **Database Connection** dialog, which is used to select a predefined database connection (the provider, data source location, and advanced settings of the server or directory on which the data resides).

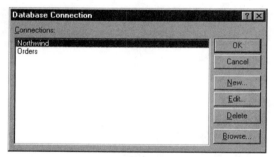

Note that if you have not already created the database connection, you can do so by clicking the **New** button on the **Database Connection** dialog. The **Data Link Properties** dialog will then be displayed, which will take you through a step-by-step wizard to create a database connection. See *How Do I Set Up a Database Connection Using the OLE DB Provider for ODBC?* (page 880) for further details.

- After you select a database connection and click the **OK** button on the **Database Connection** dialog, you now have access to *STATISTICA* Query in which you can specify the desired tables, fields, joins, criteria, etc. (via the **Table**, **Join**, and **Criteria** menus) to be included in your query.

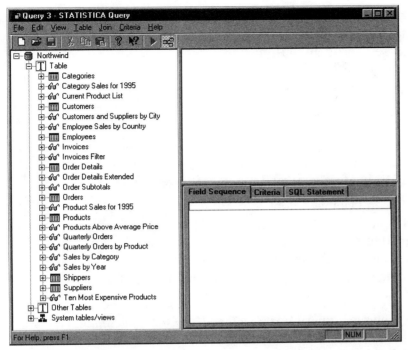

- Once you have specified your query, select **Return Data to STATISTICA** from the **File** menu or click the ▶ button.

- The **Returning External Data to Spreadsheet** dialog is then displayed, which is used to specify the name of the query, where you want *STATISTICA* Query to put the data that the query returns, and additional options.

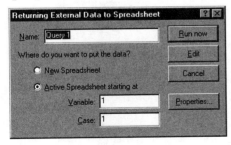

What Is SQL?

SQL (Structured Query Language) is a specialized programming language designed to specify criteria for selecting records for databases. It enables you to query an outside data source about

the data it contains and select desired subsets. Use an SQL statement to specify the desired tables, fields, rows, etc. to return as data to *STATISTICA* via *STATISTICA* Query. For detailed information on SQL syntax, please consult an SQL manual.

What Is the Difference between ODBC and OLE DB?

The differences between these two data access protocols are technical, but in the most general terms, OLE DB is newer, more advanced, and compatible with a wider selection of database technologies. In addition, OLE DB is more general, in that it includes the ODBC functionality.

Technically speaking, ODBC (Open Database Connectivity) is designed to provide access primarily to SQL data in a multi-platform environment. OLE DB (Object Linking and Embedding Database) is designed to provide access to all types of data in an OLE Component Object Model (COM) environment. OLE DB includes the SQL functionality defined in ODBC but also defines interfaces suitable for gaining access to data other than SQL data.

A query generated via ODBC is functionally similar to a query generated by OLE DB. However, be aware that if you generate a query (via *STATISTICA* Query) using an ODBC connection, *STATISTICA* uses OLE DB to connect to the ODBC connection. Because there is an intermediate connection, ODBC queries may be slower than OLE DB queries.

How Do I Set Up a Database Connection Using the OLE DB Provider for ODBC?

The necessary steps for establishing an OLE DB database connection vary depending on the type of data and database you are using. To create a database connection using the Microsoft OLE DB Provider for ODBC Drivers, select **Create Query** from the **Data – Get External Data** menu (or the **File – Get External Data** menu) to display the **Database Connection** dialog.

Next, click the **New** button on the **Database Connection** dialog to display the **Data Link Properties** dialog. Select **Microsoft OLE DB Provider for ODBC Drivers** on the **Provider** tab of the **Data Link Properties** dialog and click **Next** to display the **Connection** tab.

On the **Connection** tab of the **Data Link Properties** dialog, select the appropriate ODBC data source name (DSN) in the **Use data source name** drop-down combo box. Note that in order to use this option, the ODBC drivers (appropriate for the data format to be accessed) need to be installed on your system and a DSN must already have been created. A DSN tells the computer what ODBC driver to use, where the database is located, and any logon information (if

applicable). You create your DSN in Windows (i.e., outside of *STATISTICA*). Rather than using a previously existing DSN, you can select the **Use connection string** option button and enter or build an ODBC connection string. At this point, you can also enter information to log on to the server, enter the initial catalog to use, specify advanced network settings via the **Advanced** tab, or edit the initialization properties via the **All** tab if your specific data source requires this additional information.

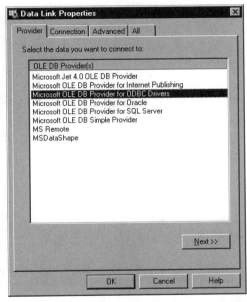

Finally, click the **Test Connection** button on the **Connection** tab to see if you successfully connected to the ODBC data. If the test connection has succeeded, click the **OK** button on the **Data Link Properties** dialog to display the **Add a Database Connection** dialog.

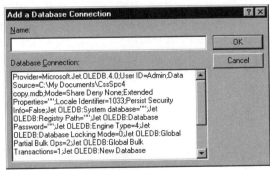

Here, type the **Name** of the database connection and click the **OK** button. Your new database connection will now be added to the **Connections** list on the **Database Connection** dialog.

Note that for detailed documentation on the additional OLE DB Providers and the other advanced options that are available on the **Data Link Properties** dialog, press the F1 key on your keyboard to display the *Microsoft Data Link Help*®.

How Do I Query a Database Using ODBC?

To use this example, you must have a database connection already set up to the **Northwind.mdb** example database. (You can choose to include this database when you install Microsoft Access on your computer.) See *How Do I Set Up a Database Connection Using the OLE DB Provider for ODBC?* (page 880) for further details.

Select **Create Query** from the **Data – Get External Data** menu (or the **File – Get External Data** menu) to display the **Database Connection** dialog. Select the database connection that you have set up to **Northwind.mdb** and click the **OK** button. At this point, *STATISTICA* Query will be available for you to actually create your query.

Use the **Tree** pane on the left side of the *STATISTICA* Query window to expand the **Customers** table and view its fields (to do this, either double-click on the word **Customers** or use the plus sign next to it). Double-click the fields **CompanyName** and **Address** to select those fields for your query (the fields are then selected and displayed in the **Graphic** pane on the right side of the dialog).

You can then use the various menu options and/or the **Text Mode** to further customize your query. When you are finished specifying your query, select **Return Data to STATISTICA** from the **File** menu or click the ▶ button to display the **Returning External Data to Spreadsheet** dialog.

On the **Returning External Data to Spreadsheet** dialog, you can name your query. You can also send your query to a new spreadsheet or to the active spreadsheet by selecting the appropriate option button. When you are finished, click the **OK** button and the data your query returns will be available on the specified spreadsheet.

How Do I Query a Database Using OLE DB?

Within *STATISTICA*, select **Create Query** from the **Data – Get External Data** menu (or the **File – Get External Data** menu) and click the **New** button on the **Database Connection** dialog to display the **Data Link Properties** dialog, which is used to create a new database connection. Note that

for detailed documentation on the **Data Link Properties** dialog, press the F1 key on your keyboard while you are on the **Data Link Properties** dialog to display the *Microsoft Data Link Help*®.

On the **Provider** tab, select the driver for your OLE DB connection. To create an OLE DB connection, select the appropriate **Microsoft Jet OLE DB Provider** for your database. Click the **Next** button to continue to the **Connection** tab, or simply click on the **Connection** tab itself.

Select the database you want to connect to in the **Select or enter a database name** field by clicking the ▤ button. Then, click the **Test Connection** button to test your connection to the database. After the test connection has succeeded, click the **OK** button on the **Data Link Properties** dialog to display the **Add a Database Connection** dialog.

On the **Add a Database Connection** dialog, enter a name for your OLE DB connection in the **Name** field. When you are finished, click the **OK** button. Your new database connection is now listed in the **Database Connection** dialog. Select the connection that you just created and click the **OK** button.

At this point, *STATISTICA* Query will be available for you to actually create your query. Use the **Tree** pane on the left side of the *STATISTICA* Query window to expand the tables and view their fields (to do this, either double-click on the table names or use the plus sign next to them). Double-click the desired fields to select those fields for your query (the fields are then selected and displayed in the **Graphic** pane on the right side of the dialog).

You can then use the various menu options and/or the **Text Mode** to further customize your query. When you are finished specifying your query, select **Return Data to STATISTICA** from the **File** menu or click the ▶ button to display the **Returning External Data to Spreadsheet** dialog.

On the **Returning External Data to Spreadsheet** dialog, you can name your query. You can also send your query to a new spreadsheet or to the active spreadsheet by selecting the appropriate option button. When you are finished, click the **OK** button and the data your query returns will be available on the specified spreadsheet.

How Do I Add Joins or Criteria to My Query?

After you have specified tables and fields to be included in your query (by selecting **Add** from the **Table** menu and then selecting fields), you can add joins and/or criteria to your query. The simplest way to do this is to select **Add** from either the **Join** or **Criteria** menu.

A join shows how data is related between two tables and determines which cases *STATISTICA* Query will return as data. On the **Add Join** dialog,

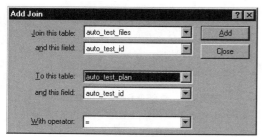

select the tables and fields that you want to join, and the operator that determines the type of comparison that *STATISTICA* Query will perform on the values in the joined fields.

Criteria establishes the condition(s) that records from an external data source must meet in order to be included in the cases that are returned by the query. For example, you can specify criteria to select only records that have a certain value (such as customers in Louisville). On the **Add Criteria** dialog,

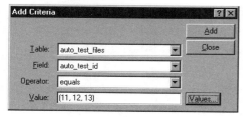

specify the **Table**, **Field**, **Operator**, and **Value** for your criteria.

Can I Save a Query and Use It Later on Another Dataset?

Yes, you can. When you are creating a query via *STATISTICA* Query, you can select **Save As** from the **File** menu to save the query as a *.sqy* file. When you open a different spreadsheet you can use the previously saved query to return data to the new spreadsheet by selecting **Open Query from File** from the **Data – Get External Data** menu (or from the **File – Get External Data** menu). In this manner, you will have access to all queries that you have written in *STATISTICA* Query even if you are working in a different application.

Can I Use More Than One Query on a Datafile?

Yes you can. Multiple queries based on one or many different databases can be created to return data to an individual spreadsheet. After you create a query via *STATISTICA* Query and select **Return Data to STATISTICA** from the **File** menu, the **Returning External Data to Spreadsheet** dialog is displayed.

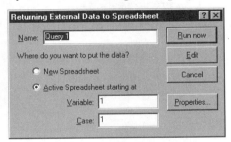

On this dialog, you can specify exactly where you want the returned data to be put on the spreadsheet. Therefore, you can specify that the first query returns data to **Variable 1** and **Case 1** of the current spreadsheet. Then you are able to create another query and specify that that data is returned to **Variable 5**, **Case 1** of the same spreadsheet. In this manner, you can maintain connections to multiple external databases simultaneously.

What Is the Difference between Graphic Mode and Text Mode in *STATISTICA* Query?

STATISTICA Query consists of two modes of operation, **Graphic Mode** and **Text Mode**. The default mode of operation is **Graphic Mode**, and for most users is the only mode of *STATISTICA* Query ever needed to do advanced querying.

The **Graphic Mode** of *STATISTICA* Query provides you with a very intuitive, graphical means to create your query. Using this simple user interface, you can create very intricate SQL statements. *STATISTICA* Query automatically builds the SQL statement for you as you select the components of your query with only a few mouse clicks and/or intuitive menu options and dialogs. Users with limited knowledge of SQL will therefore be able to create advanced and powerful queries of their data in a quick and straightforward manner.

For users who currently have a detailed knowledge of the more complex commands of SQL language, the **Text Mode** is available for further editing and/or customization of the SQL statement that has been created in the **Graphic Mode**. To manually enter or remove text in the SQL statement with the keyboard, switch to the **Text Mode** (by toggling the **Graphic Mode** command on the **View** menu or the 🔳 toolbar button). Once you are in **Text Mode**, you can build SQL statements including options that are not available in **Graphic Mode** (such as outer joins, using mathematical operators to create calculations in criteria, sorting rows by one or more columns, etc.). See *Additional Functionality Available in Text Mode* (page 854). Therefore, by using both the **Graphic Mode** and the **Text Mode** of *STATISTICA* Query, you can create advanced queries of your data allowing you to optimize the valuable information that is contained in multiple external data sources.

STATISTICA
VISUAL BASIC

continued ➡

CHAPTER

STATISTICA VISUAL BASIC

GENERAL OVERVIEWS

STATISTICA Visual Basic Language Overview

The industry standard *STATISTICA* Visual Basic language (integrated into *STATISTICA*) offers incomparably more than just a supplementary application programming language that can be used to write custom extensions. *STATISTICA* Visual Basic takes full advantage of the object model architecture of *STATISTICA* and is used to access programmatically every aspect and virtually every detail of the functionality of *STATISTICA*. Even the most complex analyses and graphs can be recorded into Visual Basic macro (SVB) programs and later be run repeatedly or edited and used as building blocks of other applications. *STATISTICA* Visual Basic adds an arsenal of more than 10,000 new functions to the standard comprehensive syntax of Microsoft Visual Basic, thus comprising one of the largest and richest development environments available.

Creating *STATISTICA* Visual Basic Programs

There are several methods in which *STATISTICA* Visual Basic programs can be created.

- *Recording a macro.* When you run an analytic procedure (from the **Statistics** menu) or create a graph (from the **Graphs** menu) the Visual Basic code corresponding to all design specifications as well as output options selected during that analysis are recorded in the

background. That code can later be executed repeatedly or edited by changing options, variables, datafiles, optionally adding a user interface, etc.

By default, the *STATISTICA* Visual Basic macros recorded in the background apply to individual analyses as selected from the **Statistics** or **Graphs** menus (e.g., **Basics Statistics**, **ANOVA**, **Categorized Scatterplot**), one separate macro per analysis. You can later use these "building blocks" to create custom applications. However, if you intend to create an SVB program file that will include a sequence of analyses that have been performed interactively (a log of analyses, including most data management operations), before you start the analyses, select **Start Recording Log of Analyses (Master Macro)** from the **Tools - Macros** menu.

For more information, see page 921; see also the FAQ section *Recording Macros* on page 942.

- *SVB development environment.* Programs can be written from scratch using the *STATISTICA* Visual Basic professional development environment featuring a convenient program editor with a powerful debugger (with breakpoints, etc.), intuitive dialog painter, and many facilities that aid in efficient code building.

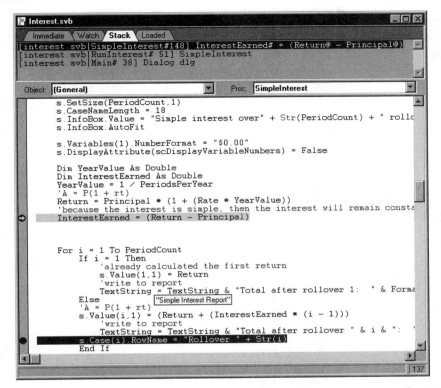

To display the SVB editor, select **New** from the **File** menu and use the **Create New Document - Macro (SVB) Program** tab.

- *Visual Basic from other applications.* SVB programs can also be developed by enhancing Visual Basic programs created in other applications (e.g., Microsoft Excel), by calling *STATISTICA* functions and procedures.

Executing *STATISTICA* Visual Basic Programs

Simple as well as very complex *STATISTICA* VB programs and macro (SVB) programs involving extensive user interface and file handling can be run from within *STATISTICA*; however, because of the industry standard compatibility of the *STATISTICA* Visual Basic and the various libraries of the *STATISTICA* system (accessible to Visual Basic), you can also access *STATISTICA* Visual Basic functions from other compatible environments (e.g., Microsoft Excel, Word, or a stand-alone Visual Basic language). Note, however, that when you run a *STATISTICA* Visual Basic program or attempt to call *STATISTICA* functions from any

other application, all calls to the *STATISTICA* specific functions (as opposed to the generic functions of Microsoft Visual Basic) will be executed only if the respective *STATISTICA* libraries are present on the computer where the execution takes place. That is, the user of the program must be a licensed user of the respective *STATISTICA* libraries of procedures.

Note that this large library of *STATISTICA* functions (more than 10,000 procedures) is transparently accessible not only to Visual Basic (either the one that is built in, or a different one), but also to calls from any other compatible programming language or environment, such as C/C++, Java, or Delphi.

Applications for *STATISTICA* Visual Basic Programs

STATISTICA Visual Basic programs can be used for a wide variety of applications, from simple macro (SVB) programs recorded to automate a specific (repeatedly used) sequence of tasks, to elaborate custom analytic systems combining the power of optimized procedures of *STATISTICA* with custom developed extensions featuring their own user interface. When properly licensed, scripts for analyses developed this way can be integrated into larger computing environments, executed from within proprietary corporate software systems, or Internet or intranet portals.

Performance of *STATISTICA* Visual Basic Programs

While the obvious advantages of Visual Basic (compared to other languages) are its ease of use and familiarity to a very large number of computer users, the possible drawback of Visual Basic programs is that they do not perform as fast as applications developed in lower level programming languages (such as C). However, that potential problem usually does not apply to *STATISTICA* Visual Basic applications, especially those that rely mostly on executing calls to *STATISTICA*'s analytic, graphics, and data management procedures. These procedures fully employ *STATISTICA*'s speed-optimized technologies to perform at a speed comparable to running the respective procedures in *STATISTICA* directly.

Structure of *STATISTICA* Visual Basic

STATISTICA Visual Basic consists of two major components: (1) The general Visual Basic Programming Environment with facilities and extensions for designing user interfaces (dialogs)

and file handling, and (2) the *STATISTICA* libraries with thousands of functions that provide access to practically all functionality of *STATISTICA*.

The Visual Basic programming environment follows the industry standard syntax conventions of the Microsoft Visual Basic Language; the few differences pertain mostly to the manner in which dialogs are created (see *Custom Dialogs, Custom User Interfaces* in the *Electronic Manual*), and they are designed to offer programmers and developers more flexibility in the way user interfaces are handled in complex programs. In the *STATISTICA* Visual Basic programming environment, dialogs can be entirely handled inside separate subroutines, which can be flexibly combined into larger multiple dialog programs; Microsoft Visual Basic is form based, where the forms or dialogs, and all events that occur on the dialogs, are handled in separate program units.

STATISTICA Visual Basic
Editor and Debugger

The *STATISTICA* Visual Basic environment includes a flexible program editor and powerful debugging tools. These facilities are described in greater detail in the section on *Editing and Customizing Recorded Macro (SVB) Programs* in the *Electronic Manual*.

When editing macro programs by typing in general Visual Basic commands or program commands specific to *STATISTICA* Visual Basic, the editor will display *type-ahead* help, to illustrate the appropriate syntax.

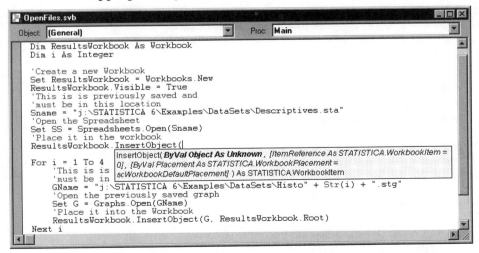

Help on the members and functions for each class (object) is also provided in-line (see also *The STATISTICA VB Object Model* in the *Electronic Manual*.

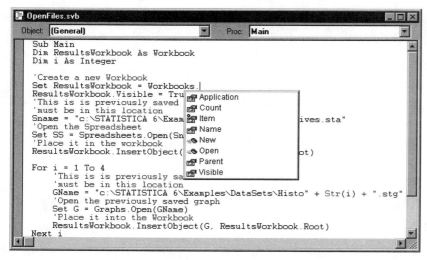

When executing a program, you can set breakpoints in the program, step through line by line, and observe and change the values of variables in the macro program, as it is running.

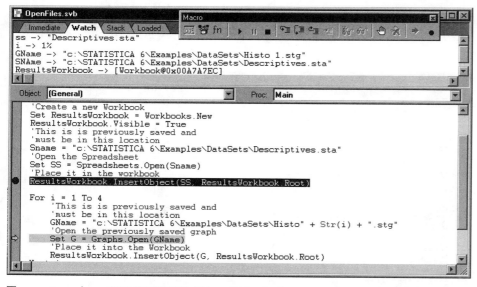

To summarize, *STATISTICA* Visual Basic is not only a powerful programming language, but it represents a very powerful professional programming environment for developing simple macros as well as complex custom applications.

Using This Documentation

The text printed in this document is also provided in the *Electronic Manual*. To test and modify the example *STATISTICA* Visual Basic (SVB) programs discussed in this documentation, copy the respective sections from the *Electronic Manual* and paste them into the *STATISTICA* Visual Basic editor window.

Macro (SVB) Programs
Window Overview

The Macro (SVB) Programs window comprises several basic components:

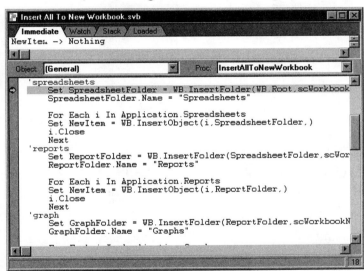

Object. The *Object* box contains a single option, *(General)*. This simply refers to the current macro that you are editing.

Proc(edure). The *Proc* box contains at least two options, *(declarations)* and *Main*. As you add functions to the macro, they will be added to this box. When you select a function from this box, the cursor will be sent directly to the function that you had chosen. If the *Proc* box is set to *(declarations)*, then you are able to enter global variables, external library inclusions, and API declarations. If the *Proc* box is set to *Main*, then the cursor will be sent to the Main

function within your macro. This is where the macro begins and ends, so it is mandatory that every macro you create must at least contain this particular function.

Visual Basic editor. The macro's code is entered and debugged within this enhanced edit field. The *STATISTICA* Visual Basic editor is a complete Visual Basic editor that incorporates the following advanced features:

- *IntelliSense® technology.* IntelliSense® can display and auto-complete all possible elements or values of objects (including their respective properties and functions).

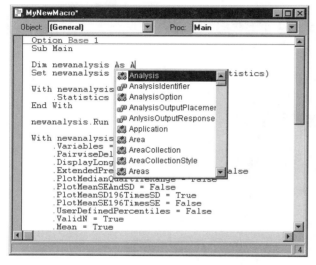

- *Syntax Highlighting.* Visual Basic keywords will be highlighted blue and comments will be highlighted green; invalid syntax will be highlighted red (these default colors can be changed; see page 897).

- *Function Page Breaking.* All functions and sub procedures will automatically have page breaks inserted between each other.

Debug Window. When the macro is running in debug mode (i.e. breakpoints have been set), a debug pane is displayed at the top of the macro document. The **Debug Window** consists of the following components:

- *Immediate.* This window is used as a *STATISTICA* Visual Basic editor while the macro itself is running. Enter or paste Visual Basic code into it, press ENTER, and it will execute; if there is a syntax error then the error message is displayed in the macro's status bar. An important use for this is to change variable values within a macro without having to reset and edit it.

- *Watch.* Enter the variables that you want to watch into the **Watch** tab's box, and it will display the variable's current value while the macro is running (depending on where the current breakpoint is set).

- *Stack.* The first line of the **Stack** tab displays the currently running library, function name, line number, and text of the current breakpoint. If the current breakpoint is within a function outside of the Main sub procedure, then a second line is displayed. This line displays the library, function name (Main), line number, and text of the location that called the current function.

- *Loaded.* Displays the currently loaded libraries that the macro is utilizing.

Global Macro (SVB)
Programs Options

You can customize the global functionality of macro (SVB) programs using the options offered on the **Macro (SVB) Programs** tab of the **Options** dialog, accessible via the **Tools** menu.

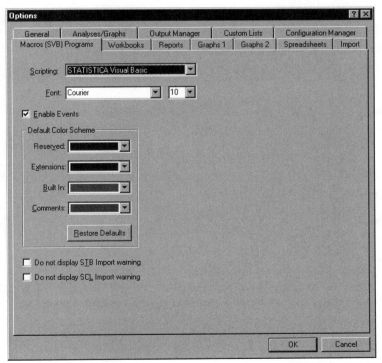

The selections made on this tab determine the defaults whenever you use a macro.

Scripting. In the *Scripting* box, select a default script to use when creating macros. Currently, *STATISTICA* supports *STATISTICA* Visual Basic only.

Font. In the *Font* boxes, specify a default font and font size to use in macro windows. Use the list box to select the font, and the smaller box (or microscrolls) to enter a font size.

Enable Events. Select the *Enable Events* check box to enable events within *STATISTICA*. By default this check box is selected. Clear this check box to disable the events running from within all open documents and from within *STATISTICA*. Even if the events for a document are set to autorun, this setting overrides the autorun specification and the events within that document are not run.

Default Color Scheme. Use the options in the *Default Color Scheme* group to specify the default colors to use for the different syntax elements in the macros. Click the *Restore Defaults* button to reset the default color scheme. You can specify colors for the following syntax components:

- *Reserved.* Reserved words are keywords within the Visual Basic language, such as `For`, `XOR`, and `Function`.

- *Extensions.* Extensions are *STATISTICA* extensions to the Visual Basic library and language. Examples of these include data types such as `Workbook`, `Spreadsheet`, and `CommandBar`, and member functions/properties such as `AnalysisOption.HideOnDeselect` and `CaseWeight.Type`.

- *Built In.* Built-in words are functions that are already built into the Visual Basic language and can be called directly. Examples of these are `Round()`, `RGB()`, and `Shell()`.

- *Comments.* Comments are any text following either an apostrophe or the keyword `REM` up until the end of the current line. Comment text is readable to you, but it is not interpreted by *STATISTICA* while the macro is running.

Do not display STB Import warning. Select the *Do not display STB Import warning* check box to not display a warning message when you import *STATISTICA* BASIC programs. It was possible to write STB programs in earlier versions of *STATISTICA*; you can import them into the current version of *STATISTICA* for backward compatibility purposes.

Do not display SCL Import warning. Select the *Do not display SCL Import warning* check box to not display a warning message when you import *STATISTICA* Command Language

programs. It was possible to write SCL programs in earlier versions of *STATISTICA*; you can import them into the current version of *STATISTICA* for backward compatibility purposes.

Controlling *STATISTICA* Events with SVB Programs

An event is an action that is typically performed by users, such as clicking a mouse button, pressing a key, changing data, or opening a spreadsheet or workbook. In *STATISTICA,* certain events are displayed, i.e., accessible to SVB, and they can be used to customize its behavior. Using programmable events, you can tailor *STATISTICA*'s behavior to your needs. Examples of events applications might include:

- building auditing systems into *STATISTICA* (by IT departments),

- building interactive demonstration programs based on workbooks,

- building customized user interfaces adhering to specific requirements of a particular application (or a specific company, e.g., to meet specific security requirements).

Events are part of a set of tools built into *STATISTICA* to make it a powerful solution building system.

Types of Events

There are three different types of events: *document-level events*, *application-level events*, and *analysis-level events*. Examples of these types of events are included in the *Electronic Manual*.

Document-level events. These events occur for open documents and in some cases, for objects within them. For example, the workbook document object can respond to the **Open**, **New**, and **Print** events and the spreadsheet document object can respond to events such as changed data values, double-clicks on a cell, etc.

Application-level events. These events occur at the level of the application itself; for example, when a new spreadsheet, report, graph, or workbook is created. You can create an event handler customizing these actions for all documents of that type. See page 1025 for a complete listing of all application-level events.

Analysis-level events. These events occur at the level of individual analyses. The term "analysis" in *STATISTICA* denotes one task selected either from the **Statistics** or **Graphs** menu,

---

which can be very small and simple (e.g., one scatterplot requested from the **Graphs** menu) or very elaborate (e.g., a complex structural equation modeling analysis selected by choosing that option from the **Statistics** menu and involving hundreds of output documents). You can create an event handler customizing actions such as activating an analysis, closing an analysis, or when output is created from an analysis (e.g., to automatically send results spreadsheets and graphs to another application).

Supported events. You can review the list of available events that can be customized by reviewing the contents of the **Proc** box in the **Document Events** window. To bring up this window for application level events, select **Tools Macro Application Events View Code**; to bring up the **Document Events** window for document-level events, first click on the respective document (e.g., spreadsheet, graph, etc), and then select **View Events View Code** (be sure to set the **Object** field to **Document**).

MACRO WINDOW

File Menu

The following commands are available from the macro (SVB) *File* menu. Many of these commands are also available from shortcut menus (accessed by right-clicking in the macro) and from the toolbar buttons.

New

Select *New* from the *File* menu (or click the ▢ toolbar button) to display the *Create New Document* dialog, and then select the *Macro (SVB) Program* tab. When creating a macro, you must specify a name for the macro in the *Name* box. A *Description* box is provided for any comments about the macro. Specify the macro type after clicking the *Macro Type* button. You can also force variables to be defined and make array indices start at one, using the check boxes at the bottom of the dialog. Note that many of the demonstration programs in this documentation do not use either option (`Option Explicit`, `Option Base 1`), and some of the example programs may not run properly if you set those options (i.e., include the statements `Option Explicit` and/or `Option Base 1` at the top of your programs).

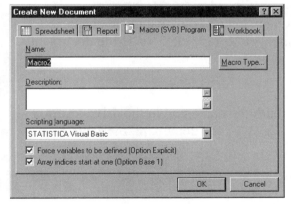

For more details on the options available when creating a new macro, see the *Electronic Manual*. See also *Creating STATISTICA Visual Basic Programs* (page 889) for more details on the various methods for creating macros.

Open

Select **Open** from the **File** menu (or click the  toolbar button) to display the **Open** dialog. From this dialog, you can open any document of a compatible type (e.g., a macro).

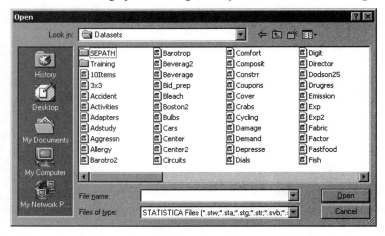

Using the **Look in** box, select the drive and directory location of the desired file. Select the file in the large box and click **Open**, or double-click the file name. You can also enter the complete path of the document in the **File name** box.

Close

Select **Close** from the **File** menu to close the current document. If you have made changes to the document since you last saved it, *STATISTICA* prompts you to save changes to the document.

Save

Select **Save** from the **File** menu to save the document with the name and in the drive location you specified when you last saved it. If you have not yet saved the document, the **Save As** dialog is displayed.

Save As

Select **Save As** from the **File** menu to display the **Save As** dialog. From this dialog, you can save the active document with the name of your choice by entering the name in the **File name**

box. Use the **Save in** box to select the appropriate drive and folder in which to save the document, or enter the complete path in the **File name** box.

By default, *STATISTICA* recommends the file type that best fits the type of document you are saving. For example, if you are saving a macro, the **Save as type** box contains the format **STATISTICA Macro Files (*.svb)**.

Save as Global Macro

Select **Save as Global Macro** from the **File** menu to display the **Save As** dialog (see above) and save the current macro as a global macro. Global macros are those that are automatically loaded when you open *STATISTICA* (i.e., you will not have to locate it and open it first). They will by default be available in the **Macro Manager** dialog (when you select **Macro - Macros** from the **Tools** menu). Global macros are the same as standard macros or SVB programs except that they are located in the same directory (by default) where the *STATISTICA* startup program file is located. When you select **Save as Global Macro** this directory will automatically be preselected. Save your frequently used macros via **Save as Global Macro** so that they are readily available when you begin *STATISTICA* the next time. Of course, you can always later and "by hand" (e.g., via the Windows operating system functions) move macros in and out of the start-up (global macro) directory.

Portability of macros containing object and class modules. SVB programs can contain references to objects and classes defined in separate files (class or object modules); these features are further explained on page 963. When an SVB program contains a reference to an external class or object module, it may use the asterisk convention in which case during execution the program will look in the global macro directory (and other standard *STATISTICA* directories) for those files. For example, if your SVB program contains the statement

`'$Include: "*OModule.svo"` then during the execution of the macro, the program will look for file **Omodule.svo** in the global macro directory. Thus, when writing programs for portability it is recommended to use the star convention, and to place object and class modules into the global macro directory.

Get External Data

The commands on this menu are for use with *STATISTICA* Query. See Chapter 10 - *STATISTICA Query* for further details about creating, editing, and using queries. Note that **Edit Query**, **Properties**, **Refresh Data**, **Delete Query**, and **Cancel Query** are only available when you have run (or are currently running) a query.

Open Query from File. Select **Open Query from File** from the **File – Get External Data** menu to display the standard open file dialog, in which you can select an SQL file to run. Note that **.sqy* files are (created and) saved in *STATISTICA* Query (by selecting **Save As** from the **File** menu).

Create Query. Select **Create Query** from the **File – Get External Data** menu to open *STATISTICA* Query and display the **Database Connection** dialog. Once you have chosen or defined a database connection, you can use *STATISTICA* Query to write your query. For details on defining a database connection, see *FAQ: Working with STATISTICA Query* in Chapter 10.

Edit Query. Select **Edit Query** from the **File – Get External Data** menu to edit the query in the active spreadsheet. Note that if you have multiple queries on the active spreadsheet, *STATISTICA* opens the query that is associated with the active cell. If you select a cell that is not connected to any query or if you select multiple cells that involve more than one query, then the **Select Database Query** dialog is displayed and you must select the query that you want to edit before *STATISTICA* Query opens.

Properties. Select **Properties** from the **File – Get External Data** menu to display the **External Data Range Properties** dialog and specify options regarding the return of external data to spreadsheets. Note that if you have multiple queries on the active spreadsheet, then *STATISTICA* displays properties for the query associated with the active cell on the spreadsheet. If you select a cell that is not connected to any query or if you select multiple cells that involve more than one query, then the **Select Database Query** dialog is displayed. On this dialog, select the query to use and click the **OK** button to display the **External Data Range Properties** dialog.

Refresh Data. Select **Refresh Data** from the **File – Get External Data** menu (or press F5) to run a query and refresh the data (retrieve the latest data from the original database). Note that if

you have multiple queries on the active spreadsheet, then *STATISTICA* refreshes data for the query associated with the active cell of the spreadsheet. If you select a cell that is not connected to any query or if you select multiple cells that involve more than one query, then the **Select Database Query** dialog is displayed so you can choose the specific query to refresh.

Delete Query. Select **Delete Query** from the **File – Get External Data** menu to delete the query from the active spreadsheet. (*STATISTICA* will always prompt you to verify that the query should be deleted.) Note that if you have multiple queries on the active spreadsheet, *STATISTICA* deletes the query associated with the active cell of the spreadsheet. If you select a cell that is not connected to any query or if you select multiple cells that involve more than one query, then the **Select Database Query** dialog is displayed so you can choose which query to delete.

Cancel Query. Select **Cancel Query** from the **File – Get External Data** menu (or press SHIFT+F5) to stop a currently running query at any time.

Workbook

Add to Analysis Workbook. Select **Add to Analysis Workbook** from the **File - Workbook** menu to add the current spreadsheet or graph to the analysis workbook. If the current spreadsheet or graph is not associated with an analysis workbook, this command is dimmed. For more information on analysis workbooks, see the *Analysis Workbook Overview* in Chapter 3 – *STATISTICA Workbooks*.

New Workbook. Select **New Workbook** from the **File - Workbook** menu to create a new workbook and add the current document to it. Note that in addition to adding the current document to a new workbook, you can select any currently open workbook from the list provided in the **File - Workbook** menu.

Active workbooks. To add the current document to an active workbook (i.e., a workbook that is open or has recently been open) select the workbook from the list of current workbooks shown on the **File - Workbook** menu. Note that if there are not any current workbooks at this time, the **File - Workbook** menu contains two options: **Add to Analysis Workbook** (which may be disabled) and **New Workbook**.

Report

Add to Analysis Report. Select **Add to Analysis Report** from the **File - Report** menu to add the current spreadsheet or graph to the analysis report. If the current spreadsheet or graph

is not associated with an analysis report, this command is dimmed. For more information on analysis reports, see the *Analysis Report Overview* in Chapter 5 – *STATISTICA Reports*.

New Report. Select *New Report* from the *File - Report* menu to create a new report and add the current document to it.

Active reports. To add the current document to an active report (i.e., a report that is open or has recently been open) select the report from the list of current reports shown on the *File - Report* menu. Note that if there are not any current reports at this time, the *File - Report* menu contains two options: *Add to Analysis Report* (which may be disabled) and *New Report*.

Output Manager

Select *Output Manager* from the *File* menu to display the *Output Manager* tab of the *Options* dialog. Use the options on this tab to specify where to direct output files. For more details, see *Three Channels for Handling Output* in Chapter 2 – *STATISTICA Output Management*.

Print Setup

Select *Print Setup* from the *File* menu to display the *Print Setup* dialog. Use this dialog to specify printing options. In the *Printer* group, you can choose a printer (using the *Name* box to view available printers). In the *Paper* group, you can choose the paper size (using the *Size* box to view a list of paper sizes), and the paper source (using the *Source* box to choose the printer's paper source). Additionally, you can choose a vertical or horizontal layout for the printed document by selecting the *Portrait* option button (for vertical) or the *Landscape* option button (for horizontal) in the *Orientation* group of the dialog.

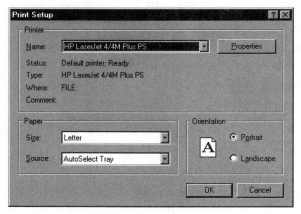

Click the **Properties** button to view options specific to the selected printer. For more information on specific printer options, consult your printer's manual.

Print

Select **Print** from the **File** menu (or press CTRL+P) to print the active macro to your default printer.

Properties

Select **Properties** from the **File** menu to display the **Macro Properties** dialog (see the **Edit** menu below).

Exit

Select **Exit** from the **File** menu to close *STATISTICA*. If you have made changes to your document since you last saved it, you will be prompted to save your changes.

Edit Menu

The following commands are available from the macro (SVB) **Edit** menu. Many of these commands are also available from shortcut menus (accessed by right-clicking in the macro) and from the toolbar buttons.

Macro Properties

Select **Macro Properties** from the **Edit** menu to display the **Macro Properties** dialog. This dialog reports the name, scripting language, and type information for the current macro.

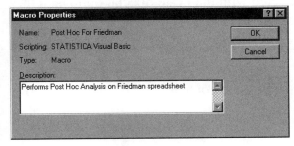

Use the **Description** box to enter or edit any comments about the macro. For more information on macro properties, press F1 or click the ?️ button in this dialog to access the *Electronic Manual*. See also, the **Macros (SVB) Programs** tab of the **Create New Document** dialog in the *Electronic Manual*.

Undo

Select **Undo** from the **Edit** menu (or press CTRL+Z) to undo your last action in the document (such as editing, moving, or copying blocks; font formatting; paragraph formatting, etc.). *STATISTICA* supports multi-level undo (with 32 buffers); therefore, you can undo multiple actions by selecting this option consecutively (up to 32 times).

Redo

Select **Redo** from the **Edit** menu (or press CTRL+Y) to counteract an **Undo** action. For example, you if your last undo action command was to undo bold formatting to the selected text, then select **Redo**, and *STATISTICA* reapplies bold formatting to the text.

Cut

Select **Cut** from the **Edit** menu (or press CTRL+X) to cut the selected text or object(s) from the document and copy them to the Clipboard. *STATISTICA* will create in the Clipboard multiple formats of the selected object; the specific format to be used for pasting can later be selected with the Paste Special options in the client application.

Copy

Select **Copy** from the **Edit** menu (or press CTRL+C) to copy the selected text or object(s) to the Clipboard. *STATISTICA* will create in the Clipboard multiple formats of the selected object; the specific format to be used for pasting can later be selected with the Paste Special options in the client application.

Paste

Select **Paste** from the **Edit** menu (or press CTRL+V) to paste the contents of the Clipboard into the selected location in the document.

Delete

Select **Delete** from the **Edit** menu (or press DEL) to delete the selected text.

Select All

Select **Select All** from the **Edit** menu (or press CTRL+A) to select the entire macro.

Find

Select **Find** from the **Edit** menu (or press CTRL+F) to display the **Find** dialog. Use this dialog to search the active document for words or phrases. Type the word or phrase that you are looking for in the **Find what** box. To start the search, click **Find Next**. *STATISTICA* highlights the first word(s) that match your search criteria. Continue clicking **Find Next** until *STATISTICA* completes the search throughout the entire document. You can stop the search at any time by clicking the **Cancel** button.

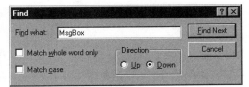

Find options. When searching a macro, select the **Match whole word only** check box if you want to restrict the search to the exact word in the **Find what** box. Select the **Match case** check box if you want to conduct a case-sensitive search for the word or phrase in the **Find what** box. Select the **Up** option button or the **Down** option button to specify the direction to search in your document.

Note that a 🔍 **Find** button is also available on the **Standard** toolbar (see Chapter 1 – *STATISTICA – A General Overview*).

Replace

Select **Replace** from the **Edit** menu (or press CTRL+H) to display the **Replace** dialog. Use this dialog to search the active document for words or phrases and replace them with different words or phrases. Specify a word or phrase for which to search (in the **Find what** box) as well as a word or phrase with which to replace it (in the **Replace with** box). To start the search, click the **Find Next** button. *STATISTICA* highlights the first word(s) that match your search

criteria. You can replace the text on an individual basis (click the **Replace** button) or globally (click the **Replace All** button). Continue clicking **Find Next** until *STATISTICA* completes the search throughout the entire document. You can stop the search at any time by clicking the **Cancel** button.

Replace options. When searching a macro, select the **Match whole word only** check box if you want to restrict the search to the exact word in the **Find what** box. Select the **Match case** check box if you want to conduct a case-sensitive search for the word or phrase in the **Find what** box.

Note that a **Replace** button is also available on the **Standard** toolbar (see Chapter 1 – *STATISTICA – A General Overview*).

Repeat Find/Replace

Select **Repeat Find/Replace** from the **Edit** menu (or press F3) to repeat the last **Find/Replace** operation you conducted.

Screen Catcher

The **Screen Catcher** utility allows you to select any rectangular area of the screen, the entire screen, or a specific window and copy it as bitmap to the Clipboard. The copied area can contain graphs, spreadsheets, or even screens from other applications (if they are currently displayed on the screen). The copied bitmap can then be pasted into *STATISTICA* documents, or any other application that supports bitmaps.

Capture Rectangle. Select **Capture Rectangle** from the **Edit - Screen Catcher** menu (or press ALT+F3) to launch the **Screen Catcher** and capture a rectangular area of the screen. After selecting this option, position the mouse pointer in one corner of the area that you want to copy. Now, click and drag the pointer over the area that you want to copy (you can drag across windows).

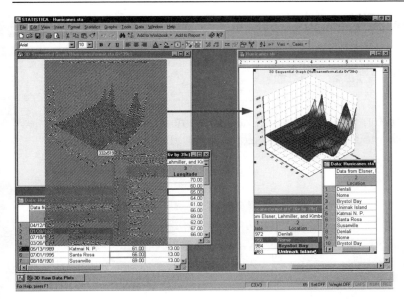

When you release the mouse button, the selected area is copied to the Clipboard.

Capture Window. Select *Capture Window* from the *Edit - Screen Catcher* menu to launch the *Screen Catcher* and capture the desired window on your desktop. After selecting this option, move the cursor over the screen until the window you want to capture is selected. Note that *STATISTICA* gives the dimensions of each window as you pass over it.

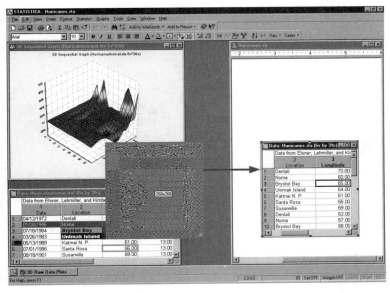

Click the window that you want to capture, and it is copied to the Clipboard (as shown above).

View Menu

The following commands are available from the macro (SVB) **View** menu. Many of these commands are also available from shortcut menus (accessed by right-clicking in the macro) and from the toolbar buttons.

Object Browser

Select **Object Browser** from the **View** menu (or press F2) to display the **Object Browser**, which lists all of *STATISTICA*'s library objects and their respective methods, properties, and events. For more information on the **Object Browser**, see the 🖳 *Object Browser Button* (page 930).

Function Browser

Select **Function Browser** from the **View** menu to display the **Function Browser**, which contains all of *STATISTICA*'s library methods. For more information on the **Function Browser**, see the fn *Function Browser Button* (page 931).

Font

Select **Font** from the **View** menu to display the **Font** dialog. Use this dialog to set the font attributes for the entire macro window.

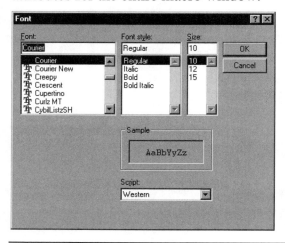

Font. Use the *Font* box and the accompanying list to select the desired font.

Font style. Select a font style using the options in the *Font style* box. Typical font styles are *Regular*, *Italic*, *Bold*, and *Bold Italic*.

Size. Use the *Size* box and the accompanying list to select the desired font size, in points. Note that 72 points equals 1 inch.

Sample. The *Sample* box displays sample text that is formatted according to the choices you have made in the *Font* dialog.

Script. Use the *Script* box to select from the available script types for the font specified in the *Font* box.

Note that the default color scheme for the *Macro (SVB Programs)* window can be adjusted on the *Macro (SVB) Programs* tab of the *Options* dialog (page 897).

Toolbars

Use the commands on the *View - Toolbars* menu to toggle the display of various toolbars. Note that toolbars can be docked to the top, left, or right side of the workspace. They can also float (e.g., be located anywhere in the workspace). For more information on a specific toolbar see the following:

Toolbar	Chapter
Standard	1
Statistics	1
Six Sigma	1
Spreadsheet	4
Tools	4
Report	5
Graphs	6
Graph Tools	6
Graph Data Editor	6
Macro	11

Customize. Select *Customize* from the *View - Toolbars* menu to display the *Customize* dialog. For details about customization, see Chapter 1 – *STATISTICA – A General Overview*.

Status Bar

Select **Status Bar** from the **View** menu to toggle the text and buttons that are displayed in the status bar in the bottom center of the screen. Note that the *STATISTICA* Start button ▨ is available regardless of whether the status bar is turned on or off.

Debug Menu

The following commands are available from the **Debug** menu. Many of these commands are also available from shortcut menus (accessed by right-clicking in the macro) and from the toolbar buttons.

Step Into

At design time select **Step Into** from the **Debug** menu (or press F8) to begin running the macro in break mode. At run time select **Step Into** to enter break mode; selecting **Step Into** again will execute the currently paused line of code and move the focus to the next line (without executing it). If an external function is called, then the steps within that function will be stepped through in the same manner. This command is also available by clicking the ⬆▤ button.

Step Over

At design time select **Step Over** from the **Debug** menu (or press SHIFT+F8) to begin running the macro in break mode. At run time select **Step Over** to enter break mode; selecting **Step Over** again will execute the currently paused line of code and move the focus to the next line (without executing it). If an external function is called, then that function will be stepped over, meaning that it will execute without being stepped through. This command is useful for skipping over external functions that have already been debugged. This command is also available by clicking the ⬇▤ button.

Step Out

At run time, select **Step Out** from the **Debug** menu (or press CTRL+F8) to execute the remaining code within the current function and return to the calling function. Once returned to the calling function, the program will enter break mode on the instruction after the original function call.

This command is useful for skipping over the steps within external functions that have already been debugged. You can also access the **Step Out** command by clicking the ⬚ button.

Run to Cursor

At run time, select **Run to Cursor** from the **Debug** menu (or press F7) to execute the program up to where the cursor is currently located (unless breakpoints are encountered). Once the line adjacent to the cursor is found, the macro will enter break mode. This is an alternative to manually entering breakpoints. You can also access the **Run to Cursor** command by clicking the ⬚ button.

Add Watch

At run time highlight a variable and select **Add Watch** from the **Debug** menu (or press CTRL+F9) to insert it into the **Watch** window, which tracks the values of the variables that are placed into it. You can also access this command by clicking the ⬚ button.

Quick Watch

At run time highlight a variable and select **Quick Watch** from the **Debug** menu (or press SHIFT+F9) to view the variable's current value in the **Immediate** window. You can also access this command by clicking the ⬚ button.

Toggle Breakpoint

Select **Toggle Breakpoint** from the **Debug** menu (or press F9) to set or unset the current line (where the cursor is located) as a breakpoint. When the macro is being run, it will pause whenever it comes across a breakpoint; the code within the breakpoint will not run. This is useful for locating program bugs by narrowing down trouble spots to specific lines of code. You can also access this command by clicking the ⬚ button.

Clear All Breakpoints

Select **Clear All Breakpoints** from the **Debug** menu (or click the ⬚ button) to turn off any breakpoints that are set in the current macro.

Set Next Statement

At run time, select **Set Next Statement** from the **Debug** menu (or click the button) to select the next line of code to execute (must be in break mode); the line that the cursor is on will be set as the next statement. The chosen statement can appear before or after the current statement, just as long as both statements are within the same function. Note that all intervening code and breakpoints will not be executed.

Show Next Statement

At run time, select **Show Next Statement** from the **Debug** menu (or click the button) to send the cursor to the next statement to be executed.

Run Menu

The following commands are available from the **Run** menu. Many of these commands are also available from shortcut menus (accessible by right-clicking in the macro) and from the toolbar buttons.

Run

Select **Run** from the **Run** menu (or press F5) to execute the macro. You can also execute the macro by clicking the ▶ **Run Macro** toolbar button (page 932).

Break

Select **Break** from the **Run** menu (or click the ‖ button) to pause the currently running macro.

Reset

Select **Reset** from the **Run** menu (or click the ■ button) to stop running (e.g., reset) the current macro.

Statistics Menu

The **Statistics** menu is available whenever any document is open. This menu provides access to all available analysis types within *STATISTICA* including **Basic Statistics/Tables**, **Multiple Regression**, **ANOVA**, **Nonparametrics**, **Distribution Fitting**, **Advanced Linear/Nonlinear Models**, **Multivariate Exploratory Techniques**, **Industrial Statistics & Six Sigma**, **Power Analysis**, and **Data-Mining**. Definitions for the various types of statistics are available in the *Statistics Toolbar* topic in Chapter 1 – *STATISTICA – A General Overview*. Also included on the **Statistics** menu are **Statistics of Block Data**, **STATISTICA Visual Basic**, and **Probability Calculator**.

Graphs Menu

The **Graphs** menu is available whenever any *STATISTICA* document is open. This menu provides access to all graph types in *STATISTICA* including **Histograms**, **Scatterplots**, **Means w/ Error Plots**, **Surface Plots**, **2D Graphs**, **3D Sequential Graphs**, **3D XYZ Graphs**, **Matrix Plots**, **Icon Plots**, **Categorized Graphs**, **User-defined Graphs**, **Graphs of Block Data**, and **Graphs of Input Data**. It also provides access to **Multiple Graph Layouts**. For more information on graphs, see Chapters 6, 7, 8, and 9 – *STATISTICA Graphs*.

Tools Menu

The following commands are available from the macro **Tools** menu. Many of these options are also available from shortcut menus (accessed by right-clicking in the macro) and from the toolbar buttons.

Analysis Bar

To take advantage of *STATISTICA*'s "multitasking" functionality, *STATISTICA*'s analyses are organized as functional units that are represented with buttons on the **Analysis** bar at the bottom of the application window (above the status bar, see the illustration below, where **Basic Statistics**, **Cluster Analysis**, and **Canonical Analysis** are running simultaneously). Normally, at least one analysis button is created, and consecutive buttons are added as you start new analyses.

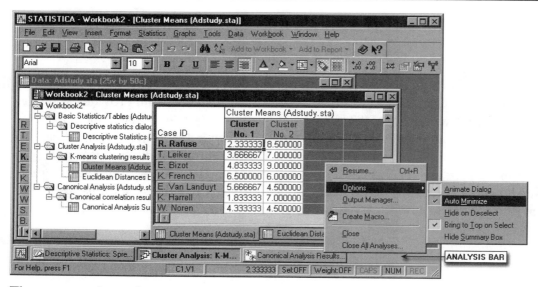

The commands on the ***Tools – Analysis Bar*** menu provide a variety of options for managing the ***Analysis*** bar.

Resume. Select ***Resume*** (or click the 🔙 toolbar button or press CTRL+R) from the ***Tools - Analysis Bar*** menu to continue the current analysis or graph. Note that you can also open the current analysis or graph by clicking on its button on the ***Analysis*** bar.

Select Analysis/Graph. Use the commands on the ***Tools - Analysis Bar - Select Analysis/Graph*** menu to select an analysis or graph from the set of active analyses and graphs. Note that you can also open an active analysis or graph by clicking on the appropriate button on the ***Analysis*** bar.

Options - Animate Dialog. Select ***Animate Dialogs*** from the ***Tools - Analysis Bar - Options*** menu to display animation when analysis dialogs are minimized or maximized. By default this command is checked.

Options - Auto Minimize. Select ***Auto Minimize*** from the ***Tools - Analysis Bar - Options*** menu to automatically minimize all analysis dialogs when you select another window in *STATISTICA* or another application. By default this command is checked. When your screen is large enough to accommodate several windows, it is recommended that you clear this option. This keeps the analysis dialogs on screen while the respective output created from these dialogs is produced, thus allowing you to use the dialogs as "toolbars" from which output can be selected.

Options - Hide on Select. Select *Hide on Select* from the *Tools - Analysis Bar - Options* menu to minimize all windows associated with a particular analysis when that analysis is deselected. By default, this command is cleared. Note that this command only applies when the results are sent to individual windows; see the discussion of the *Output Manager* tab of the *Options* dialog in Chapter 2 – *STATISTICA Output Management* for further details.

Options - Bring to Top on Select. Select *Bring to Top on Select* from the *Tools - Analysis Bar - Options* menu to activate (display at the top of *STATISTICA*) all windows associated with a particular analysis when that analysis is selected, replacing whatever dialogs were on top. This command also facilitates the organization of individual windows from various analyses. By default this option is checked. Note that this command only applies when the results are sent to individual windows; see the discussion of the *Output Manager* tab on the *Options* dialog (Chapter 2 – *STATISTICA Output Management*) for further details.

Options - Hide Summary Box. Select *Hide Summary Box* from the *Tools - Analysis Bar - Options* menu to not display the summary box, which is located at the top of certain results dialogs (such as *Multiple Regression - Results*) and contains basic summary information about the analysis. By default, this command is not checked.

Output Manager. Select *Output Manager* from the *Tools – Analysis Bar* menu to display the *Output Manager* tab of the *Options* dialog. Use the options on this tab to specify where to direct output files. For more details, see *Three Channels for Handling Output* in Chapter 2 – *STATISTICA Output Management*.

Create Macro. Select *Create Macro* from the *Tools - Analysis Bar* menu to display the *New Macro* dialog, in which you can specify a name for a new macro based on the current analysis. When you run an analytic procedure (from the *Statistics* menu) or create a graph (from the *Graphs* menu), the Visual Basic code corresponding to all design specifications as well as output options that you select are recorded in the background. Thus, when you click the *OK* button on the *New Macro* dialog, the resulting macro window displays the appropriate code to recreate the current analysis.

Minimize. Select *Minimize* from the *Tools - Analysis Bar* menu to minimize the current analysis.

Close. Select *Close* from the *Tools - Analysis Bar* menu to close the current analysis.

Close All Analyses. Select *Close All Analyses* from the *Tools - Analysis Bar* menu to close the all of the analyses/graphs on the *Analysis* bar.

Dialog Editor

Select *Dialog Editor* from the *Tools* menu (or click the 🖳 toolbar button) to display the *UserDialog Editor*, which is used to add and edit a custom dialog to your macro. For more information on this dialog, including a comprehensive description of the buttons available, see *UserDialog Editor* in the *Electronic Manual*. See also, *Can I Create Custom Dialogs and Other Interactive User Input Controls in STATISTICA Visual Basic?* (page 1007) and the *UserDialog Button* (page 929).

Object Browser

Select *Object Browser* from the *Tools* menu (or press F2 or click 🏥) to display the *Object Browser*, which lists all of *STATISTICA*'s library objects and their respective methods, properties, and events. For more information on the *Object Browser*, see *Object Browser* in the *Electronic Manual*. See also, the 🏥 *Object Browser Button* (page 930).

Function Browser

Select *Function Browser* from the *Tools* menu (or click the 📶 toolbar button) to display the *Function Browser* dialog, which contains all of *STATISTICA*'s library methods. See the 📶 *Function Browser Button* (page 931) for more information on the *Function Browser* dialog.

References

Select *References* from the *Tools* menu to display the *References* dialog

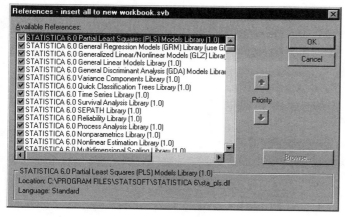

Use this dialog to view and select from all of the available object libraries on your machine.

Available References. The *Available References* box displays all of your registered object libraries. You can either use the scroll bar next to the box or the up and down arrows to navigate through the libraries. To include libraries into your *STATISTICA* project, select any of the library references in the *Available References* box and click the *OK* button.

Browse. Click the *Browse* button to search for additional reference libraries. This is useful for libraries that may not be installed or reside on another machine (in a network setting).

Import STB, Import SCL

Select *Import STB* or *Import SCL* from the *Tools* menu to import *STATISTICA* BASIC files or *STATISTICA* Command Language files, respectively. Note that because of the structural differences between the older two languages and *STATISTICA* Visual Basic, for more complex programs the process of translating the code is not automatic.

STATISTICA Visual Basic incorporates all features of the former (version 5.5) SCL, *STATISTICA* BASIC and much more. Unlike in the two older languages, in *STATISTICA* Visual Basic, all features of *STATISTICA* are accessible from the object model (including full control over event handling).

Macro

A macro is a scripted application that extends functionality to *STATISTICA* by directly accessing *STATISTICA*'s object model and manipulating it. Macros are primarily used to automate tasks done in *STATISTICA* by harnessing its power and recording it into a *STATISTICA* Visual Basic script. Additionally, macros can be written as both stand-alone scripts and library classes to extend the statistical and mathematical capabilities of *STATISTICA*.

Macros. Select *Macros* from the *Tools - Macro* menu (or click the ▶ toolbar button on the *Macro* toolbar) to display the *Macros* dialog.

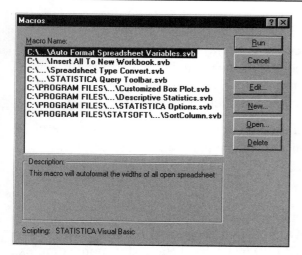

Use this dialog to run, edit, open, or delete existing macro (SVB) programs as well as to create new macros.

Start Recording Log of Analyses (Master Macro). Select *Start Recording Log of Analyses (Master Macro)* from the *Tools - Macro* menu to create an SVB program file that will include a sequence of analyses (a log of analyses) that have been performed interactively. Note that you must select this command before you start the analyses. Consult the *Electronic Manual* or the *STATISTICA Visual Basic Primer* to learn more about this powerful command and how to create recordable sequences of analyses.

Start Recording Keyboard Macro. Select *Start Recording Keyboard Macro* from the *Tools - Macro* menu to record a keyboard macro. See the *Electronic Manual* or the *STATISTICA Visual Basic Primer* for more details.

Stop Recording. Select *Stop Recording* from the *Tools - Macro* menu (or click the ■ toolbar button on the *Macro* toolbar) to stop recording a log of analyses (master macro) or a keyboard macro (see above).

Create Analysis/Graph Macro. Select *Create Analysis/Graph Macro* from the *Tools - Macro* menu to select from a list of all current analyses/graphs that are open. When you run an analytic procedure (from the *Statistics* menu) or create a graph (from the *Graphs* menu) the Visual Basic code corresponding to all design specifications as well as output options that you select are recorded. Select the appropriate analysis/graph to display that code. Note that the *New Macro* dialog will first be displayed in which you can type the macro's name and description.

Application Events - View Code. Select *View Code* from the *Tools - Macro - Application Events* menu to display the *Document Events* window, which is used to enter code to change the default behavior of application-level events. Application-level events allow you to customize the behavior of all documents of a certain type (such as all spreadsheets, all workbooks, or all reports). See *What Are Application Events and How Can They Be Controlled from STATISTICA Visual Basic?*, page 1015.

Application Events - Autorun. Select *Autorun* from the *Tools - Macro - Application Events* menu to save your application-level event code with a specific type of document so that it will be run any time that type of document is open.

Add-Ins. Select *Add-Ins* from the *Tools - Macro* menu to display the *STATISTICA Add-Ins* dialog. Add-Ins are COM server components normally written in ATL (Active Template Library) that are used to create custom user interfaces of *STATISTICA* and/or fully functional external programs.

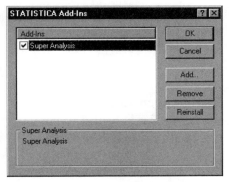

All available Add-Ins are displayed in the *Add-Ins* list. To create a new Add-In, click the *Add* button to display the *Specify Add-In to be Added* dialog, which is used to enter the program ID of the Add-In. Click the *Remove* button to delete the selected Add-In from the *Add-Ins* list. Finally, click the *Reinstall* button to register the selected Add-In to your operating system.

STATISTICA Visual Basic Editor. Select *STATISTICA Visual Basic Editor* from the *Tools - Macro - Application Events* menu (or press ALT+F11) to display a *STATISTICA* Visual Basic editor (Macro Window). Note that this chapter provides an extensive FAQ section for *STATISTICA* Visual Basic beginning on page 935.

Customize

Select *Customize* from the *Tools* menu to display the *Customize* dialog, which is used to customize toolbars, menus, and keyboard hot keys with a variety of commands. For more

information on this dialog, see Chapter 1 – *STATISTICA – A General Overview* or the *Electronic Manual*.

Options

Select **Options** from the **Tools** menu to display the **Options** dialog. In addition to general and display options, options are available for reports, file locations, custom lists, workbooks, macros, statistical analysis display, import, and edit facilities, and output management. For a discussion of the **Reports** tab of the **Options** dialog, see *Global Macro (SVB) Programs Options* (page 897).

Window Menu

The **Window** menu is available when any document is open. It provides access to commonly used commands for organizing the workspace and switching between files.

Close All

Select **Close All** (or press CTRL+L) from the **Window** menu to close all open spreadsheets, graphs, and related windows (e.g., graph data) in *STATISTICA*. This option is useful when you need to clear the screen to start a new analysis. Note that you will be prompted to save any unsaved files before they are closed.

Cascade

Select **Cascade** from the **Window** menu to arrange the open *STATISTICA* windows in an overlapping pattern so that the title bar of each window is visible.

Tile Horizontally

Select **Tile Horizontally** from the **Window** menu to arrange the open *STATISTICA* windows in a horizontal (side by side) pattern. When you select this option, *STATISTICA* automatically optimizes the display of the open windows (with the preference given to tiling horizontally).

Tile Vertically

Select *Tile Vertically* from the *Window* menu to arrange the open *STATISTICA* windows in a vertical pattern. When you select this option, *STATISTICA* automatically optimizes the display of the open windows (with the preference given to tiling vertically).

Arrange Icons

Select *Arrange Icons* from the *Window* menu to arrange all minimized windows into rows.

Windows

Select *Windows* from the *Window* menu to display the *Windows* dialog. Use this dialog to access all of the currently open windows in *STATISTICA* and manage them.

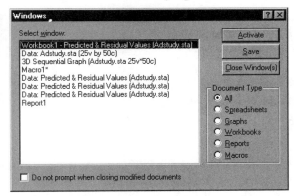

Select Window. Select either a single window or multiple windows in the *Select Window* box for manipulating its associated window in the *STATISTICA* environment. Note that you can use CTRL or SHIFT while selecting the items in the box to make multiple selections.

Activate. Click the *Activate* button to set the focus to the currently selected window in the *Select Window* box. In other words, the associated window will be brought to the front of all of the other windows. Note that because only one window can have the focus at any given time, multiple selection for this operation does not apply.

Save. Click the *Save* button to save all of the windows that are currently selected in the *Select Window* box.

Close Windows. Click the *Close Windows* button to close all of the windows that are currently selected in the *Select Window* box. Note that *STATISTICA* will prompt you to save any modified windows before closing them.

Document Type. Use the options in the *Document Type* group to filter all of the open windows within *STATISTICA* by their document type; the contents of the *Select Window* box will be updated to reflect your filter selection. You can display all open documents in the *Select Window* box (select the *All* option button), or limit the display to spreadsheets, graphs, workbooks, reports, or macros only by selecting the appropriate option button.

Help Menu

The *Help* menu is available when any document is open. It provides access to various types of help.

Contents and Index

Select *Contents and Index* from the *Help* menu to display the *STATISTICA Electronic Manual*. *STATISTICA* provides a comprehensive *Electronic Manual* for all program procedures and all options available in a context-sensitive manner by pressing the F1 key or clicking the help button **?** on the caption bar of all dialogs (there is a total of over 100 megabytes of compressed documentation included).

Due to its dynamic hypertext organization, organizational tabs (e.g., *Contents*, *Index*, *Search*, and *Favorites*), and various facilities you can use to customize the help system, it is faster to use the *Electronic Manual* than to look for information in the traditional manuals. The status bar on the bottom of the *STATISTICA* window displays short explanations of the menu options or toolbar buttons when an item is highlighted or a button is clicked.

Statistical Advisor

Select *Statistical Advisor* from the *Help* menu to launch the *Statistical Advisor*. Based on your answers to the successive questions about the nature of your research, the *Statistical Advisor* suggests the most appropriate statistical methods to use and where to find them in *STATISTICA*.

Animated Overviews

Select **Animated Overviews** from the **Help** menu to display a submenu of the available animated overviews (visual overviews and specific feature help). Select from the submenu to launch one of the animated overviews.

If you select one of these commands and you have not installed that animated overview, a *STATISTICA* message dialog will prompt you for the location of that animated overview file. For example, you can run it from your *STATISTICA* CD, or if you are running a network version, the overviews can be located on the network drive. Note that if you have the *STATISTICA* CD, you can run the overviews directly from the CD without copying them to your hard drive.

StatSoft's Home Page

Select **StatSoft's Home Page** from the **Help** menu to launch the StatSoft Home Page in your default browser. We invite you to visit the StatSoft Web page often:

- For the most recent information about *STATISTICA*, downloadable upgrades, new releases, new products, news about StatSoft, etc., and access the *What's New...* section of the Web site.

- For a comprehensive list of *Frequently Asked Questions* (including useful tips, solutions to hardware and software compatibility problems, etc.) and access the *Technical Support* section of the Web site.

- For a library of *STATISTICA* Visual Basic programs (written by users), the *Technical Support* section of the Web site (to submit your own programs to this user exchange forum, and send an e-mail to *info@statsoft.com*).

Technical Support

Select **Technical Support** from the **Help** menu to launch the *Technical Support - Getting More Help* page of the StatSoft Web site in your default browser. This page contains links to download *STATISTICA* updates and links to FAQ topics on spreadsheets, graphs, printing, reports, etc. The StatSoft Technical Support Department e-mail address, phone number, and hours are also listed. Note that most chapters of this manual also contain FAQ sections.

About *STATISTICA*

Select **About STATISTICA** from the **Help** menu to display the *STATISTICA* version, copyright notice, license information, and foreign office contact information.

MACRO TOOLBAR

The **Macro** toolbar is available when you are creating or editing a macro (SVB) program.

This toolbar provides access to commonly used buttons for formatting, writing, and running macro (SVB) programs.

▦ UserDialog Button

Click the ▦ button to display the **UserDialog Editor**. The *STATISTICA* Visual Basic environment provides all tools to program complete custom user interfaces. A powerful **UserDialog Editor** is included, which you can use to design dialogs using simple operations, such as dragging buttons to the desired locations.

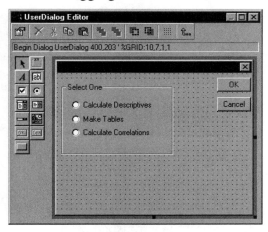

Unlike in Microsoft Visual Basic, the user-defined dialogs are stored along with the program code as data of type **UserDialog**. This method of creating dialogs allows you to create sophisticated user interfaces that can easily be edited in textual form; also, by defining the entire dialog as a variable, you can completely define dialogs inside subroutines, which can then be freely moved around the program.

However, user-defined dialogs designed in the *STATISTICA* Visual Basic environment cannot be ported directly to Microsoft Visual Basic, which uses a form-based method of creating

dialogs. This is not a serious limitation, though, but rather a design issue, in the sense that you should decide before embarking on the development of a complex program with extensive custom (user-defined) dialogs which environment you prefer. For example, if you are already familiar with the Microsoft Visual Basic environment, or would like to augment an existing program that was developed in that environment, then there would be clear advantages in staying with that language. To reiterate, virtually all *STATISTICA* Visual Basic functions are accessible from the Visual Basic environments in other applications (such as Microsoft Visual Basic, Microsoft Excel, Microsoft Word, etc.).

Object Browser Button

To review the libraries and scope of the installation of *STATISTICA* Visual Basic on your machine, display the **Object Browser** by clicking the button, selecting **Object Browser** from the **View** menu, or pressing F2.

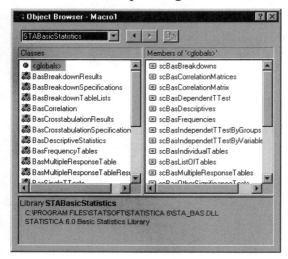

All statistical procedures and graphs and their customization are available as properties or functions to *STATISTICA* Visual Basic. To reiterate, in order to provide a transparent programming environment, the analysis objects, and properties belonging to those objects, are arranged to correspond to the respective flow of options and dialogs, as if the respective analyses were performed interactively. In addition, a large number of properties, functions, constants, and events (subroutines associated with particular user-initiated events performed on documents; e.g., if you right-click on a results spreadsheet) are available to manage documents, files, and various general analysis options.

The **Object Browser** allows you to select a *STATISTICA* library and review the classes available within it. It also includes a copy button, which allows you to copy selected functions or properties to the Clipboard for pasting into macros.

fn **Function Browser Button**

Click the fn button to display the **Function Browser**, which contains all of *STATISTICA*'s library methods.

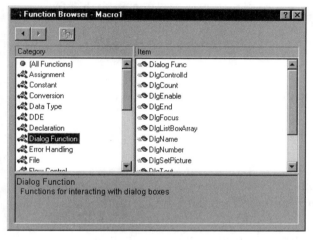

Use the arrow buttons to move backward and forward through previous selections. You can also copy your current selection to the Clipboard using the ▤ button.

Category. The **Category** box contains a list of available function categories. Select the desired category and view its specific functions within the **Item** box.

Item. The **Item** box contains a list of functions within the selected category.

Info box. The **Info box** displays the current function's name, parameters, return value, and description.

Note that the **Function Browser** dialog is modeless, meaning that you can access *STATISTICA* while the **Function Browser** dialog is displayed. This is useful because this enables you to edit your macro directly and to utilize the functionality of the **Function Browser** dialog concurrently.

▶ Run Macro Button

Click the ▶ button to execute the macro. Note this command can also be accessed from the **Run** menu or by pressing F5.

‖ Pause Button

Click the ‖ button to pause the currently running macro. This option is also available from the **Run** menu.

■ Reset Button

Click the ■ button to stop running (e.g., reset) the current macro. This option is also available from the **Run** menu.

▧ Step Into Button

At design time, click the ▧ button to begin running the macro in break mode. At run time, use this button to enter break mode; clicking **Step Into** again will execute the currently paused line of code and move the focus to the next line (without executing it). If an external function is called, then the steps within that function will be stepped through in the same manner. This command can also be accessed from the **Debug** menu or by pressing F8.

▧ Step Over Button

At design time, click the ▧ button to begin running the macro in break mode. At run time, use the **Step Over** button to enter break mode; clicking the button again will execute the currently paused line of code and move the focus to the next line (without executing it). If an external function is called, then that function will be stepped over, meaning that it will execute without being stepped through. This command is useful for skipping over external functions that have already been debugged. It can also be accessed from the **Debug** menu or by pressing SHIFT+F8.

▧ Step Out Button

At run time, click the ▧ button to execute the remaining code within the current function and return to the calling function. Once returned to the calling function, the program enters break mode on the instruction after the original function call. This command is useful for skipping

over the steps within external functions that have already been debugged. It can also be accessed from the **Debug** menu or by pressing CTRL+F8.

Run to Cursor Button

At run time, click the button to execute the program up to where the cursor is currently located (unless breakpoints are encountered). Once the line adjacent to the cursor is found, the macro will enter break mode. This is an alternative to manually entering breakpoints. This command can also be accessed from the **Debug** menu or by pressing F7.

Add Watch Button

At run time, select a variable and click the button to insert it into the **Watch** window, which tracks the values of the variables that are placed into it. This command can also be accessed from the **Debug** menu or by pressing CTRL+F9.

Quick Watch Button

At run time, select a variable and click the button to view its current value in the immediate window. This command can also be accessed from the **Debug** menu or by pressing SHIFT+F9.

Breakpoint Toggle Button

Click the button to set or unset the current line (where the cursor is located) as a breakpoint. When the macro is run, it will pause whenever it comes across a breakpoint; the code within the breakpoint will not run. This is useful for locating program bugs in specific lines of code. This command can also be accessed from the **Debug** menu or by pressing F9.

Clear All Breakpoints Button

Click the button to turn off any breakpoints that are set in the current macro.

Set Next Statement Button

At run time, click the button to select the next line of code to execute (must be in break mode); the line that the cursor is on will be set as the next statement. The chosen statement can appear before or after the current statement, just as long as both statements are within the same function. Note that intervening code and breakpoints will not be executed.

● Record Macro Button

Click the ● button to record all of your keystrokes and mouse events and save them to a new macro. This command is useful for recording extensive customization or multiple analyses.

Note that *STATISTICA* automatically records all of your keystrokes and mouse events while you are performing an analysis; however, once the dialog associated with that analysis has been closed, the recording is destroyed. Thus, this button is useful when you want to record a macro across several analyses without having to copy and paste the recordings from each analysis.

FAQ: WORKING WITH
STATISTICA VISUAL BASIC (SVB)

STATISTICA Visual Basic (integrated into *STATISTICA*) is an industry standard implementation of the Visual Basic programming language. There are on the market literally hundreds of textbooks, tutorials, courses, etc. about Visual Basic, addressed towards users with various skill levels, from novice to expert software developer. Additionally, specialized books, tutorials, etc. are available to teach users how to develop Visual Basic programs specifically to interact with particular (Microsoft Windows) applications, such as Microsoft Excel or Microsoft Word. What follows is only a very brief set of commonly asked questions to assist users who are not familiar with Visual Basic in acquiring a sufficient background to write useful programs and macro (SVB) programs to enhance the *STATISTICA* program. Additional overviews, detailed reference (linked in a context-sensitive manner to the *STATISTICA* Visual Basic editor), and numerous examples are available in the *Electronic Manual*.

General Introduction to
STATISTICA Visual Basic

What Is *STATISTICA*
Visual Basic (SVB)?

The industry standard *STATISTICA* Visual Basic language (integrated into *STATISTICA*) offers incomparably more than just a supplementary application programming language (that can be used to write custom extensions). *STATISTICA* Visual Basic takes full advantage of the object model architecture of *STATISTICA* and allows you to access programmatically every aspect and virtually every detail of the functionality of the program. Even the most complex analyses and graphs can be recorded into Visual Basic macros (see also page 942) and later be run repeatedly or edited and used as building blocks of other applications. *STATISTICA* Visual Basic adds an arsenal of more than 10,000 new functions to the standard comprehensive syntax of Microsoft Visual Basic, thus comprising one of the largest and richest development environments available.

What Are Some of the Applications of SVB?

There are countless applications for *STATISTICA* Visual Basic ranging from recording simple macros to automate routine tasks, to the development of large scale, powerful applications that take full advantage of advanced technologies offered in *STATISTICA*. The following is a selection of examples.

- You can record macro programs of your analyses to fully document all your work; by playing back the macro programs, you can recreate exactly the analyses that you performed while recording the macro (see also page 942).

- You can record macro programs of some standard analyses (reports) that you perform routinely on updated datafiles; you can then perform these routine analyses by simply running the macro program; note that practically all statistical procedures and options are accessible via *STATISTICA* Visual Basic.

- You can combine the recorded macro programs of multiple analyses into a single complex program; by running this program you can then automatically create a suite of analyses where, for example, some statistical computations are performed on the results from preceding analyses (see also page 979).

- You can modify recorded macro programs to create highly customized new statistical analyses; e.g., you could first compute the means for a set of variables, and then make customized plots of those means (see also page 979).

- You can program custom computational subroutines, using, for example, the extensive *STATISTICA* matrix library; in this manner, you can develop and prototype your own custom statistical procedures.

- You can include complex custom user interfaces in your programs; this allows you to build simple or complex custom applications, using the *STATISTICA* library of statistical and computational procedures (see also page 1007).

- You can fully customize the "behavior" of *STATISTICA* itself, by including custom procedures for various events; for example, you can call a macro program every time the user opens a datafile, requiring the user to "sign in" with a password (see also page 1015).

- You can fully customize the toolbars and options of *STATISTICA*, to produce a customized "look and feel" for specialized tasks (e.g., for data entry only); you can also create toolbars, toolbar buttons, or pop-up menus and attach macro programs to run custom procedures (see also page 1021).

- You can access from *STATISTICA* Visual Basic the libraries of other (Visual Basic) compatible programs; this allows you to include in your programs the specialized procedures and add-ins provided by other software (see also page 967).

- You can access all *STATISTICA* libraries from compatible Visual Basic programming environments; this allows you to program, for example, Microsoft Excel macro programs that automatically perform various computations in *STATISTICA*, and transfer the results back to a Microsoft Excel Spreadsheet (see also page 998).

Performance of *STATISTICA* Visual Basic programs. While the obvious advantage of Visual Basic (compared to other languages) is its ease of use and familiarity to a very large number of computer users, the possible drawback of VB programs is that they do not perform as fast as applications developed in lower-level programming languages (such as C). However, that potential problem will not apply to *STATISTICA* Visual Basic applications, especially those that rely mostly on executing calls to *STATISTICA* analytic, graphics, and data management procedures. These procedures fully employ the *STATISTICA* technology and they will perform at a speed comparable to running the respective procedures in *STATISTICA* directly.

I Know Nothing about Visual Basic or Programming; Where Do I Start?

Visual Basic is a powerful industry standard language that will allow you to develop custom programs using the features of many (compatible) standard application programs, such as Microsoft's office programs (Word, Excel, Access, etc.), and of course *STATISTICA*. Depending on your aspirations as a programmer, you may want to pursue a more systematic approach to learn the language or start by modifying ready examples, to suit your particular needs.

1. Learning Visual Basic – "becoming an expert." If you want to learn systematically about the Visual Basic language from the ground up, so to speak, we would recommend that you begin by studying a general introductory book or course on the subject. You will find in your local bookstore entire bookshelves filled with books on Visual Basic; also, your local college or university will most likely offer introductory courses on this

language. Such courses are very useful because they will allow you to familiarize yourself with basic programming concepts, and how to take advantage of many features of the language to build truly sophisticated custom applications.

2. Learning by examples and from recorded macros – "using SVB occasionally." If you have had some exposure to programming in Visual Basic or other languages, you will find many of the features of *STATISTICA* Visual Basic very familiar. You may want to read this documentation, try the numerous examples that illustrate various features of the language, and modify them to suit your needs. By running various interactive analyses, and recording them into *STATISTICA* Visual Basic program macros (see also page 942), you will quickly gain a working knowledge of the *STATISTICA* object model (see also page 939); also, much of the program code that is often needed can be generated by recording macro programs, and combining them into sophisticated custom applications.

3. Editing recorded macros – "no experience necessary..." Finally, if currently all your applications for *STATISTICA* Visual Basic are limited to occasionally rerunning analyses using different variables or different datasets, then you can simply look into the SVB scripts that are automatically recorded by *STATISTICA* when you perform any analysis (see page 942) and identify places where datafile names and variable names are entered. All you need to do is to edit them (or replace with new ones) and run the respective macros. This simple method requires very little knowledge of the *STATISTICA* Visual Basic syntax conventions.

How Do I Access SVB?

STATISTICA Visual Basic is accessible from the **Statistics** menu or by selecting **New** from the **File** menu (then select the **Macro (SVB) Program** tab on the **Create New Document** dialog). To open a previously saved *STATISTICA* Visual Basic program, select **Open** from the **File** menu. You can open multiple SVB programs at a time. For example, you can have five or more different programs opened in their separate windows and cut and paste the code between the programs or execute them in arbitrary sequences. SVB comes with an integrated environment that allows you to write, edit, verify, debug (i.e., "dry run"), and execute your programs.

What Does a Simple SVB Program Look Like?

The following program illustrates various general features of the Visual Basic (VB and SVB) language in a simple program. In *STATISTICA*, select **New** from the **File** menu, then select the **Macro** tab and create a macro called *Overview 1*.

```
Sub Main
' Bring up a Message Box, with the Ok and Cancel buttons,
' and with the following text;
' note the continuation symbol _ (underline) at the
' end of the first line, to allow the single statement
' to span two lines
    If MsgBox ("Hello! Click OK or Cancel", _
        vbOkCancel)=vbOK Then
      MsgBox "You clicked OK"
    Else
      MsgBox "You clicked Cancel"
    End If
End Sub
```

To run this program, click the ▶ icon on the toolbar, press F5, or select **Run Macro** from the **Run** menu. This program will display two message boxes: the first one shows an **OK** button and a **Cancel** button, and the next one will show a message depending on which button you clicked.

What Is Meant By the Term "Object Model?"

The power of Visual Basic in general, and of *STATISTICA* Visual Basic in particular, derives from its ability to access the features of various programs that support the Visual Basic language. For example, you can write a program that produces a results spreadsheet via *STATISTICA* and pastes that spreadsheet into a Microsoft Word document and a Microsoft Excel spreadsheet. That same program can then access individual cells in the spreadsheet, either in Microsoft Excel or *STATISTICA*, format those cells so that the contents of some cells are displayed in a different color (e.g., significant results are shown in yellow), and save all results (the document, spreadsheet, and Excel spreadsheet) to a disk. In this case you would access from a single Visual Basic program features in three different applications: *STATISTICA*, Microsoft Word, and Microsoft Excel.

The way this is generally accomplished is by organizing the respective features and functions available in an application by objects. For example, the *STATISTICA* application is an object of type **Application**; this application contains various properties and methods; see also *What Are Properties, What Are Methods?* (page 941).

For example, a property of a *STATISTICA* application is whether or not it is visible on the screen (property `Application.Visible = True` or `Application.Visible = False`). Another property of the application object is the currently active analysis (`Application.ActiveAnalysis`) or active dataset (`Application.ActiveDataSet`). There are also methods available that perform various operations on the object or inside the object. For example, the **Application** object contains a method to open a datafile (input spreadsheet).

Some of the properties are themselves different types of objects, with many properties, methods, and other objects. For example, the **ActiveDataSet** property can be assigned to an object of type **Spreadsheet**. The **Spreadsheet** object has many properties itself; among them is the **Range** property, which can be assigned to an object of type **Range**. The **Range** object has many properties, among them the **Font** property. The **Font** property can be assigned to an object of type **Font**, which has properties like **Name** (of the font), **Bold**, **Italic**, etc. So, by setting, for example, the **Bold** property of the **Font** object to **True**, you can format to bold type the contents of a cell range in a spreadsheet in a *STATISTICA* application.

In general, objects are hierarchically organized, and when you want to access a particular feature of the program (e.g., change the font in a spreadsheet cell), you need to access the proper object by "stepping" through the hierarchy of objects (e.g., from the *STATISTICA* **Application**, to the **Spreadsheet**, to the **Range** object, to the **Font** object). So part of learning to program in (*STATISTICA*) Visual Basic is to gain a good working knowledge of the hierarchical object model. Fortunately, the objects in *STATISTICA* Visual Basic are logically organized, and can be reviewed via the **Object Browser**.

The **Object Browser** is displayed by clicking the 📋 **Object Browser** button (on the **Macro** toolbar) or by selecting **Object Browser** from the **View** menu.

For an illustration of how to "navigate" the hierarchy of objects, see the example in *How Can I Change the Font (or Other Aspects) of Numbers in Spreadsheets?* (page 985).

What Are Properties, What Are Methods?

As described in the context of the description of the **Object Model** (*What Is Meant By the Term "Object Model?"*, page 939), once an object has been created, the Visual Basic program then has access to the properties and methods contained in that object. Properties can be mostly thought of as variables, methods can be mostly thought of as subroutines or functions that perform certain operations or computations inside the respective application object. For an illustration of how to "navigate" the hierarchy of objects and how to access various properties and methods, see the example in *How Can I Change the Font (or Other Aspects) of Numbers in Spreadsheets?* (page 985).

Recording Macros:
Recording an Analysis into a
STATISTICA Visual Basic Program

How Can I Record My Analysis in an SVB Program?

There are three categories of macros that can be automatically created as you run *STATISTICA*:

- Analysis Macros,

- Master Macros (logs of multiple analyses), and

- Keyboard Macros.

All three follow the identical syntax and can be modified at a later time, but because of the different ways in which each of them is created, they offer distinctive advantages and disadvantages for specific applications (for details, see also *What Exactly Is Recorded in Analysis Macros?*, page 945, and *What Exactly Is Recorded in Master Macros?*, page 948).

1. Analysis Macros. First, you can record simple Analysis Macros from an analysis to record the settings, selections, and chosen options for that specific analysis. [Note that the term "analysis" in *STATISTICA* denotes one task selected either from the **Statistics** or **Graphs** menu, which can be very small and simple (e.g., one scatterplot requested from the **Graphs** menu), or very elaborate (e.g., a complex structural equation modeling analysis selected by choosing that option from the **Statistics** menu and involving hundreds of output documents).] After selecting any of the statistical options from the **Statistics** menu or graphics options from the **Graphs** menu, all actions such as variable selections, option settings, etc. are recorded "behind the scenes"; at any time you can then transfer this recording (i.e., the Visual Basic code for that macro) to the Visual Basic program editor window. Note that the **Create Macro** command is available from every analysis via the **Options** menu or the shortcut menu (accessed by right-clicking the analysis button) when the respective analysis is minimized.

Also, your choice of datafiles, as well as case selection conditions and the weight variable, are recorded as long as those options are selected in the analysis dialog (i.e., via the  button and not from the **File** menu or the status bar). One such "stand alone" macro is created for each analysis and these macros are not "put together" automatically by *STATISTICA* (in

fact they cannot be "mechanically" combined without some editing since each of them represents a stand-alone program that starts with the appropriate declarations, etc.).

2. Master Macros (Logs). Second, you can record a Master Macro or Master Log of your entire analysis; this recording will "connect" analyses performed with various analysis options from the *Statistics* or *Graphs* menu. However, unlike simple Analysis Macros, you can turn the recording of Master Macros on or off. The Master Macro recording will begin when you turn on the recording, and it will end when you stop the recording. In between these actions, all file selections and most data management operations are recorded, as are the analyses and selections for the analyses, in the sequence in which they were chosen.

The most common application of the Master Macro would be to start *STATISTICA*, start the Master Macro recording by selecting *Start Recording Log of Analyses (Master Macro)* from the *Tools - Macro* menu, and then continue with the analyses. For example, you can compute descriptive statistics, perform some multiple regression analyses, make several histograms and scatterplots, etc. Note that during the analysis, you will see the floating *Record* toolbar to remind you that you are currently recording a Master Macro.

Finally, you stop the recording by either clicking the stop button ■ on the floating *Record* toolbar (see above) or by selecting *Stop Recording* from the *Tools - Macro* menu. At that point, the Visual Basic program that represents all actions performed or selections made during the Master Macro recording will be transferred into a *STATISTICA* Visual Basic editor window. When you run this macro "as is," the exact same analyses will be repeated (with some exceptions resulting from the logic of creating reusable programs from sequences of interactive operations performed by the user, described below).

3. Keyboard Macros. When you select *Start Recording Keyboard Macro* from the *Tools - Macro* menu, *STATISTICA* will record the actual keystrokes you are entering via the keyboard. The floating *Record* toolbar will be visible as a reminder that a recording "session" is in progress.

When you stop the recording by either clicking the stop button ■ on the floating *Record* toolbar (see above) or by selecting *Stop Recording* from the *Tools - Macro* menu, a *STATISTICA* Visual Basic editor window will open with typically a very simple program containing a single *SendKeys* command with symbols that represent all the different keystrokes you performed during the recording session.

Note that this type of macro is very simple in the sense that it will not record any context in which the recorded keystrokes are pressed and will not record their meaning (i.e., commands that these keystrokes will trigger), but this feature makes them particularly useful for some specific applications. For example, it can be very useful when it is attached to a keyboard shortcut (select **Customize** from the **Tools** menu to display the **Customize** dialog; then select the **Keyboard** tab, where you can attach the program to a shortcut key) where it could be used, for example, to quickly reenter long text, a formula, a selection of variables, or a large number of options (via keystrokes) on some complex dialog.

How Can I Record an Analysis Macro?

Here is an example of how you can record a *STATISTICA* Visual Basic Analysis Macro program; refer to *How Can I Record My Analysis in a STATISTICA Visual Basic Program?* (page 942) to learn about the differences between Analysis Macros, Master Macros (to record entire logs of several analyses and data management operations), and Keyboard Macros.

Start *STATISTICA* and open the example datafile **Exp.Sta**. Select **Basic Statistics** from the **Statistics** menu. Click the **OK** button, and on the **Descriptive Statistics** dialog, click the **Variables** button and select all variables; then click the **Summary** button to produce the results spreadsheet. Next, do one of two things: Either right-click on the minimized ("iconized") **Descriptive Statistics** dialog and select **Create macro** from the resulting shortcut menu, or maximize the **Descriptive Statistics** dialog again, and select **Create macro** from the **Options** button menu. On the **New Macro** dialog, specify a name, and click the **OK** button.

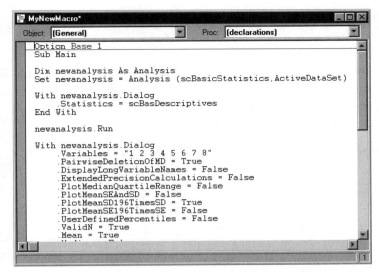

```
MyNewMacro*
Object: [General]            Proc: [declarations]

Option Base 1
Sub Main

Dim newanalysis As Analysis
Set newanalysis = Analysis (scBasicStatistics,ActiveDataSet)

With newanalysis.Dialog
    .Statistics = scBasDescriptives
End With

newanalysis.Run

With newanalysis.Dialog
    .Variables = "1 2 3 4 5 6 7 8"
    .PairwiseDeletionOfMD = True
    .DisplayLongVariableNames = False
    .ExtendedPrecisionCalculations = False
    .PlotMedianQuartileRange = False
    .PlotMeanSEAndSD = False
    .PlotMeanSD196TimesSD = True
    .PlotMeanSE196TimesSE = False
    .UserDefinedPercentiles = False
    .ValidN = True
    .Mean = True
```

If you run this program, the identical analysis is performed, i.e., the macro program will create a summary results spreadsheet with descriptive statistics for variables *1* through *8* in the currently active datafile; if the currently active datafile is **Exp.sta**, then the summary results spreadsheet will be performed for all (8) variables in that datafile.

What Exactly Is Recorded in Analysis Macros?

Analysis Macros are the ones that are created automatically and are always being recorded "behind the scenes" whenever you start an analysis from the **Statistics** or **Graphs** menus. The term "analysis" in *STATISTICA* denotes one task selected either from the **Statistics** or **Graphs** menu, which can be very small and simple (e.g., one scatterplot requested from the **Graphs** menu) or very elaborate (e.g., a complex structural equation modeling analysis selected by choosing that option from the **Statistics** menu and involving hundreds of output documents).

Whenever you choose any options from those menus, recording begins; the recording terminates when you exit the analysis, that is, when you click the final **Cancel** button to close the startup dialog. At that point, the recording is discarded and "forgotten" (unlike Master Macros; see page 948). There are several things to remember when using the Analysis Macro recording facilities. All of these are consequences of the general "rule": Only actions performed as part of and during the specific analysis being recorded will be reflected in the Analysis Macros.

Datafile selections and operations. The recording of Analysis Macros begins automatically whenever a new analysis is started from the *Statistics* or *Graphs* menu. Anything that "happened" before that is not recorded in the Analysis Macro. Thus, your specific selection of the input datafile is not recorded (the Analysis Macro always assumes that it is to be executed on the current active input datafile) unless that choice is made by clicking the [Open Data] button on the respective analysis dialog. Neither are any operations recorded that you may perform on the input datafile, such as sorting the data, subsetting of variables and/or cases, etc.

Case selection conditions, case weights. Case selection conditions and case weights are only recorded if they are specified as part of the specific analysis, by clicking on the *Select Cases* [SELECT CASES] or *Case Weights* [⚖ w] button of the respective analysis; they will not be recorded in Analysis Macros if they are specified via the respective *Tools* menu options for the input datafile (or if they have been specified prior to the current analysis for the respective input datafile). Note that in *STATISTICA*, case selection conditions and case weights can be specified either (a) relatively permanently for input datafiles and automatically stored with the files, in which case they will automatically be used by all analyses based on the input datafile, or (b) they can be specified on a per-analysis basis, in which case any new (subsequent) analyses will not use those specifications. In a sense, when you specify case weights or case selection conditions to "belong" to the datafile, those specifications become part of the datafile just like the data (numbers) themselves; when you specify case selection conditions and case weights for a particular analysis, those specifications are only applied to the respective analysis. To reiterate, case selection conditions and case weights are only recorded in Analysis Macros if they are specified via the respective analysis options (buttons), and regardless of whether or not they are specified for the current analysis only, or as a "permanent feature" of the datafile.

Changing current datafile selection conditions and weights. Because case selection conditions and case weights can be recorded into Analysis Macros, in the exact same manner as specified via the *Analysis/Graph Case Selection Conditions* and *Analysis/Graph Case Weights* dialogs (when accessed via the respective buttons of the current analysis dialog), it is possible to reset the "permanent" current datafile selection conditions and weights by running an Analysis Macro. If you run an Analysis Macro that changes the case selection conditions for the current input datafile, then all subsequent analyses will use those specifications, and the results of those analyses will be affected accordingly.

Handling output: Sending results to workbooks, reports, etc. Like case selection conditions or case weights, the selections of output options (specifications) on the *Output Manager* are only recorded if they are made from the respective analysis that is being recorded (via the *Output* options on the respective analysis *Options* [Options ▼] menu); changes

in the **Output Manager** are not recorded if they are made globally for all analyses via the **Tools** - **Options** menu. In keeping with the logic of Analysis Macros, only actions that are performed as part of a specific analysis are recorded. For example, if you recorded an Analysis Macro without making any changes (from that analysis, via the **Options - Output** option available on every dialog of the analysis), then no information about specific settings of output options will be recorded. Consequently, when you run the macro, the output will be directed to the place(s) specified by the current default settings for the output. So, for example, while during the recording of a macro, results spreadsheets might have been directed to individual spreadsheet windows; when running the macro, and while different defaults are in place, the results spreadsheets might be directed to workbooks and report windows.

How Can I Record a Master Macro (Log of Analyses)?

As described earlier (see page 942), in a Master Macro several consecutive and simultaneous analyses can be recorded (e.g., an analysis via **Multiple Regression**, followed by an analysis of the prediction residuals via **Basic Statistics**); also, most data management operations are recorded. Refer to *How Can I Record My Analysis in a STATISTICA Visual Basic Program?* (page 942) to learn about the differences between Analysis Macros, Master Macros, and Keyboard Macros; see also *What Exactly Is Recorded in a Master Macro?*, page 948, or refer to the *STATISTICA Visual Basic Primer* for additional details.

Here is an example of how you can record a *STATISTICA* Visual Basic Master Macro program – to create a complete log of all analyses from the point when you explicitly begin the Master Macro recording to the point when you terminate the recording. Remember that in Master Macros, basically all actions are recorded, including the specific datafile that is used for the analysis. Start *STATISTICA* and start the recording of the Master Macro by selecting **Recording Keyboard Macro** from the **Tools - Macro** menu; you will see the floating **Record** toolbar as a reminder that a recording "session" is in progress. Next, open the example datafile **Exp.Sta**. Select **Basic Statistics** from the **Statistics** menu. Click the **OK** button, and on the **Descriptive Statistics** dialog, click the **Variables** button and select all variables; then click the **Summary** button to produce the results spreadsheet. Next click on the stop button ■ on the floating **Record** toolbar or select **Stop Recording** from the **Tools - Macro** menu, to create the SVB program of this analysis.

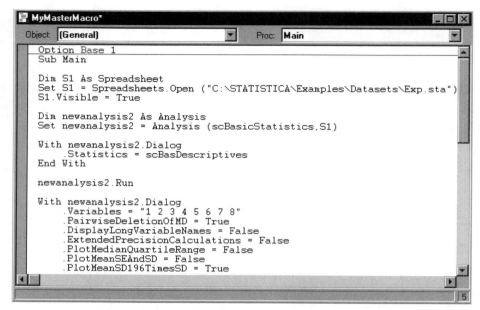

```
Option Base 1
Sub Main

Dim S1 As Spreadsheet
Set S1 = Spreadsheets.Open ("C:\STATISTICA\Examples\Datasets\Exp.sta")
S1.Visible = True

Dim newanalysis2 As Analysis
Set newanalysis2 = Analysis (scBasicStatistics,S1)

With newanalysis2.Dialog
    .Statistics = scBasDescriptives
End With

newanalysis2.Run

With newanalysis2.Dialog
    .Variables = "1 2 3 4 5 6 7 8"
    .PairwiseDeletionOfMD = True
    .DisplayLongVariableNames = False
    .ExtendedPrecisionCalculations = False
    .PlotMedianQuartileRange = False
    .PlotMeanSEAndSD = False
    .PlotMeanSD196TimesSD = True
```

Note that, unlike in the Analysis Macro shown earlier, in this case the opening of the input datafile for the analysis was explicitly recorded. If you run this program, the identical analysis is performed, i.e., the macro program will create a summary results spreadsheet with descriptive statistics for variables *1* through *8* from datafile **Exp.sta**.

To reiterate, Master Macro recordings will record entire sequences of analyses, including the selection of various global (e.g., output) options, etc. You can even record a macro of sequential analyses where some analysis is performed on a results spreadsheet created by a preceding analysis (e.g., make some custom **Scatter Icon Plots** for residuals computed by **Multiple Regression**).

What Exactly Is Recorded in Master Macros?

Datafile selections. The recording of the Master Macro begins when you select **Start Recording Log of Analyses (Master Macro)** from the **Tools** - **Macro** menu. If you select an input datafile after you start the recording, then the selection of that file will become part of the Master Macro. For example, suppose you opened a file and then started a **Multiple Regression** analysis; the recorded macro would include the following lines:

```
...
Dim S1 As Spreadsheet
Set S1 = Spreadsheets.Open ("c:\datasets\OilAnalysis.sta")
```

```
S1.Visible = True
Dim newanalysis2 As Analysis
Set newanalysis2 = Analysis (scMultipleRegression,S1)
...
```

The input datafiles selection was explicitly recorded, and the **Multiple Regression** was initialized using the datafile that was explicitly opened for this analysis. Hence, if you run this Master Macro, the analysis will be performed on the same datafile, i.e., it will be loaded from the disk prior to the **Multiple Regression** analysis.

Now suppose you started the recording of the Master Macro after you had already selected and opened an input datafile. The same analysis might be recorded like this.

```
...
Dim S1 As Spreadsheet
Set S1 = ActiveSpreadsheet
Dim newanalysis2 As Analysis
Set newanalysis2 = Analysis (scMultipleRegression,S1)
...
```

Now the recording started by defining as the input datafile the currently active input spreadsheet. If you run this Master Macro, the **Multiple Regression** analysis would be performed on the currently active input datafile, i.e., possibly a different datafile than that used while recording the Master Macro.

Data editing operations. Certain data editing operations on an input spreadsheet (datafile) are recorded as part of Master Macros. As a general rule, most editing operations that are accomplished via selections on respective dialogs (e.g., the **Sort Options** dialog) are recorded; operations that are performed via simple keyboard actions (e.g., deleting a data value by pressing the **Del** key on your keyboard) are usually not recorded. Also keep in mind that the Master Macro recording only starts when the recording is explicitly requested; so any editing operations that are performed prior to the start of the recording of course will not be reflected in the Master Macro. Here is a list of data editing operations that are recorded into Master Macros. Most operations available on the **Data** menu are recorded:

- Sorting
- Creating subsets of cases or variables
- Changes in the variable specifications
- Data transformations via formulas entered into the specifications dialogs of the respective variables
- Adding, moving, copying, and deleting variables
- Adding, moving, copying, and deleting cases

- Ranking of data
- Recoding operations
- Automatic replacement of missing data operations
- Shifting of data
- Standardizing data
- Date operations

Recordable operations performed outside the **Data** menu, include:

- Creation of new datafiles (spreadsheets)
- Opening datafiles
- All **Output Manager** options
- Filling data ranges with random values (note that this operation is recorded even though it is accomplished without going through dialogs, but by selecting it from the **Edit** menu)
- Entering case selection conditions on a global and local (one analysis) level
- Entering case weight variables on a global and local (one analysis) level

Note that recording of certain operations as part of the Master Macro logs might lead to creating ambiguous and context-dependent solutions, so they are excluded from the list of recordable tasks. This includes such operations as:

- Editing values in the spreadsheets
- Selections of cells
- Clipboard based copy, paste, and delete operations (note that this does not include the copy, move, and delete cases and variables operations listed above)
- Generally, anything not on the list of recordable data management operations (see above).

Keep these issues in mind when recording a Master Macro with data editing operations that are necessary before subsequent analyses are performed.

Recording consecutive and simultaneous analyses. To reiterate, Master Macro recordings will reproduce the exact sequence of analyses and output choices (of results spreadsheets or graphs) made during the analysis. This facility provides great flexibility and even allows you to "string together" analyses so that the first one computes certain results, while the second one analyzes those results further. For example, you could first perform a **Multiple Regression** analysis on a particular datafile, then use the option **Save** on the results dialog of the **Multiple Regression** analysis to create a stand-alone input spreadsheet of predicted and residual values from the analysis, and then compute **Basic Statistics** on the numbers in that

spreadsheet. The Master Macro recording of that sequence of analyses would look something like this (for program details see also the *STATISTICA Visual Basic Primer*):

```
...
Dim S1 As Spreadsheet
Set S1 = Spreadsheets.Open ("c:\STATISTICA\OilAnalysis.sta")
S1.Visible = True
Dim newanalysis1 As Analysis
Set newanalysis1 = Analysis (scMultipleRegression,S1)
...
Dim newanalysis2 As Analysis
Set newanalysis2 = Analysis (scBasicStatistics,ActiveDataSet)
With newanalysis2.Dialog
    .Statistics = scBasDescriptives
End With
...
```

Note how consecutive analyses (objects) are enumerated as **newanalysis1** and **newanalysis2**. The first one (**Multiple Regression**) is initialized with an explicit input datafile; the second one is initialized with the currently active dataset. When recording complex sequences of analyses like these where results of one analysis serve as the input for subsequent analyses, extra care must be taken to review and, if necessary, edit the final macro before running it to ensure that the intended sequence of **ActiveDataSet**s are chosen by the respective analyses.

Case selection conditions, case weights. Like Analysis Macros, when case selection conditions or case weights are specified during an analysis, those actions are properly recorded in the Master Macro. However, in addition, when you specify case selection conditions and case weights globally for the datafile (i.e., outside any specific analysis), those actions are recorded as well.

Handling output: Sending results to workbooks, reports, etc. Like Analysis Macros, when case output options are changed during an analysis, those selections are recorded in the macro. In addition, if global output defaults are changed via the **Tools** - **Options** menu (on the **Output Manager** tab of the **Options** dialog), then those choices are recorded into the Master Macro as well. Therefore, if you started a Master Macro, then set the global **Output Manager** option to direct all results spreadsheets and graphs to separate workbooks (for each analysis) as well as to reports, then those selections will be recorded in the Master Macro and reproduced when you execute that macro.

Master Macro recording and Analysis Macro recording. The two major modes of macro recording – Master Macros and Analysis Macros, which are always being recorded in the "background" – can be used simultaneously. In other words, you can make Analysis

Macros while recording a Master Macro. However, note that the action of creating the Analysis Macro is not itself recorded in the Master Macro.

How Do I Execute an SVB Program from within *STATISTICA*?

To run a *STATISTICA* Visual Basic program from within *STATISTICA*, in the currently active *STATISTICA* Visual Basic program editor, either click the ▶ toolbar button, press F5, or select **Run Macro** from the **Run** menu.

Components of Recorded Macros (*STATISTICA* Visual Basic Programs): Case Selection Conditions, Output Options, etc.

What Are the Major "Components" in a Typical SVB Macro Program?

When you record a macro program from an interactive analysis, as illustrated in the section on *How Can I Record My Analysis in a STATISTICA Visual Basic Program?*, page 942, you will notice several common components of the programs. *STATISTICA* Visual Basic is organized around analysis objects; for example, to run an analysis with the *STATISTICA* **Basic Statistics** module, you would first create an analysis object, with the constant `scBasicStatistics` and (optionally) with a datafile name (location of the file containing the input spreadsheet). For a complete list of modules (reference libraries) see *What Are the Names for the STATISTICA Reference Libraries (Modules)?*, page 977. To make access to the thousands of statistical functions and options available in the *STATISTICA* system as convenient as possible, *STATISTICA* Visual Basic maintains a very close correspondence between the dialogs as they are presented during interactive analyses, and the flow of the program. In a sense, once an analysis has been created, such as an analysis via the **Basic Statistics** module, you simply "program the dialogs" for the respective statistical analysis.

Here is a brief breakdown of the components of a (typical) Analysis Macro program, recorded via **Basic Statistics Descriptive Statistics**.

First, there is the statement:

```
Set newanalysis = Analysis (scBasicStatistics, ActiveDataSet)
```

Here the new analysis object is created of type **Analysis** with the constant **scBasicStatistics**, and with the current **ActiveDataSet** as the input data file; this statement will display the ***Descriptive Statistics*** (Startup Panel). Note that **scBasicStatistics** is actually a predefined (in SVB) numeric constant. If you have recorded a Master Macro, the SVB syntax would look slightly different; specifically, the input datafile would be explicitly assigned to a variable of type **Spreadsheet** (e.g., **Set S1 = ActiveSpreadsheet**), and consecutive analyses would be numbered, and initialized using that variable (e.g., **Set newanalysis1 = Analysis (scBasicStatistics,S1)**).

Next, there is the block:

```
With newanalysis.Dialog
     .Statistics = scBasDescriptives
End With
newanalysis.Run
```

The **With newanalysis.Dialog** block is a shortcut method for setting the various properties available on the startup dialog; in this case the only property is the **Statistics** property, which selects the ***Descriptive statistics*** option on the startup dialog. The **newanalysis.Run** statement "clicks" ***OK***, i.e., causes the program to proceed to the next dialog. The next dialog block sets the many properties of the ***Descriptive Statistics*** dialog.

```
With newanalysis.Dialog
     .Variables = "1 2 3 4 5 6 7 8"
        . . .
        . . .
     .CompressedStemAndLeaf = False
End With
```

Finally, at the end of the macro you have:

```
newanalysis.RouteOutput(newanalysis.Dialog.Summary).Visible = True
newanalysis.RouteOutput(newanalysis.Dialog.Histograms).Visible = True
```

These two lines select the ***Summary*** results spreadsheet and the ***Histograms*** button on the ***Normality*** tab (in a sense these commands "click" those buttons). The **RouteOutput** method takes as an argument the **Summary** or **Histograms** collection (of spreadsheets, graphs, or both) and places it into the workbook, report, etc. depending on the current selections on the ***Analysis/Graph Output Manager*** dialog. The **RouteOutput** method actually returns an object of type **AnalysisOutput**, which itself has a number of methods and properties to make it fully "programmable."

How Are Case Selection Conditions and Case Weights Handled in Recorded Macros?

It depends on whether you are recording a Master Macro or an Analysis Macro. The general rule is that case selection conditions and case weights specified via the **Select Cases** or **Case Weights** buttons in a specific analysis are recorded into the respective Analysis Macro as well as the Master Macro (if one is being recorded at the time); when case selection conditions or case weights are changed form the **Tools** menu, then those changes are only recorded in Master Macros. For details, see the topics *What Exactly Is Recorded in Analysis Macros?*, page 945, and *What Exactly Is Recorded in Master Macros?*, page 948.

How Are Output Options Handled in Recorded Macros?

It depends on whether you are recording a Master Macro or an Analysis Macro. Like with case selection conditions and case weights, if the **Analysis/Graph Output Manager** is brought up via the **Output** option for a specific analysis from the **Options** menu (of the respective analysis dialog), then the choices on that dialog will be recorded into the respective Analysis Macro as well as the Master Macro (if one is being recorded at the time); when output options are changed by choosing **Tools - Options**, and then the **Output Manager** tab, then any changes made to the output options are only recorded in Master Macros. For details, see the topics *What Exactly Is Recorded in Analysis Macros?*, page 945, and *What Exactly Is Recorded in Master Macros?*, page 948.

General *STATISTICA* Visual Basic Language Features: Data Types, Subroutines, Functions, Classes, Objects

What Data Types Are Supported in SVB?

STATISTICA Visual Basic supports several data types including *Double*, *Integer*, *Long*, *Boolean*, *String*, *Variant*, and *Object*.

The *Double* data type and the *Integer* and *Long* data types are probably the ones most commonly used in computations. Variables declared as *Double* can hold (store) real numbers, approximately in the range from +/-1.7E +/- 308 (approximately 15 digits of precision); variables declared as *Integer* can hold (store) integer numbers in the range from –32,768 to 32,767, and *Long* variables can hold (store) integer numbers in the range from –2,147,483,648 to 2,147,483,647; other common data types are *Boolean* (True [1] or False [0]) and *String* (a string variable of arbitrary length). Refer to the *Electronic Manual* for a description of all data types.

Declaring variables. Variables should be declared at the beginning of each program by using the `dim` command; for example:

```
    . . .
    Dim i As Integer, j As Long
    Dim x As Double, v As Variant
    . . .
```

The Variant data type. A data type that is particularly useful when incorporating *STATISTICA* statistical procedures into your SVB program is the *Variant* data type. A variable declared as a *Variant* data type can be empty, numeric, currency, date, string, object (see *What Is Meant By the Term "Object Model?"*, page 939), error code, null or array value. Here is an example program that demonstrates the *Variant* data type.

```
' This program illustrates the Variant data type;
' Variant's can be empty, numeric, currency, date,
' string, object, error code, null or array value;
' here, only the numeric, currency, and string
```

```
' identities are illustrated.
Sub Main
      Dim CurrentDateAndTime As Variant
' The Now function will return the current system
' time and date.
      CurrentDateAndTime = Now
      AsDouble(CurrentDateAndTime)
      AsLong(CurrentDateAndTime)
      AsCurrency(CurrentDateAndTime)
      AsString(CurrentDateAndTime)
End Sub

Sub AsDouble (x As Double)
      MsgBox "x+1="+Str(x+1)
End Sub

Sub AsLong (x As Long)
      MsgBox "x+1="+Str(x+1)
End Sub

Sub AsCurrency (x As Currency)
      MsgBox "x+1="+Str(x+1)
End Sub

Sub AsString (x As String)
      MsgBox x
End Sub
```

Note that all variables that are used in a program but are not explicitly declared, will automatically default to type *Variant*.

How Do I Assign a Value to a Variable?

Here are examples of simple declarations of variable types, and how to assign values to them:

```
' This program illustrates simple assignments of
' values to variables and elements in arrays.
Sub Main
Dim i As Long, j As Long
Dim c As String
Dim x As Double
Dim xx(10) As Double
Dim xx1(1 To 10) As Double
  i=1
```

```
        j=2
        x=i+j
' Note that the Str(x) function converts x
' to a string.
        c="1+2="+Str(x)
' This function will display a message box
' with the text "1+2=3"
        MsgBox c
        xx(0)=x
        xx1(1)=x
        xx(10)=xx(0)+xx1(1)
' Note that the _ (underline) at the end of a
' line can be used to continue a statement to the
' next line.
        c="1+2 + 1+2 =" _
            +Str(xx(10))
        MsgBox c
End Sub
```

Assigning objects to variables. When assigning objects to variables (see *What Is Meant By the Term "Object Model?"*, page 939), you need to use the syntax `Set Variable = Object`; for example:

```
Sub Main
        Dim wb As Workbook
        Dim ss As Spreadsheet
        Dim g As Graph
        Set wb=ActiveWorkbook
' Note that the _ (underline) at the end of a
' line can be used to continue a statement to the
' next line.
        Set ss=Spreadsheets.Open( _
            "j:\Statistica\Examples\Datasets\Adstudy.sta")
        ss.Visible=True
        Set g=ActiveGraph
End Sub
```

What Are Collections?

A collection is very similar to an array; however, the collection is an object (see *What Is Meant By the Term "Object Model?"*, page 939) with various methods that in many instances make dealing with a collection much more convenient than dealing with an array. In *STATISTICA* Visual Basic, all results spreadsheets and graphs from analyses are returned as collections, which makes the programmatic editing, storing, and further processing of results very

convenient. See also *How Can I Access Results Spreadsheets, Graphs, Workbooks Etc.?* (page 979).

How Do I Specify a Conditional Instruction (If ... Then)?

The following program illustrates various general features of the Visual Basic (VB, and SVB) language including a conditional instruction.

```
Sub Main
' Bring up a Message Box, with the Ok and Cancel buttons,
' and with the following text; note the continuation
' symbol _ (underline) at the end of the first line,
' to allow the single statement to span two lines.
    If MsgBox ("Hello! Click OK or Cancel", _
        vbOkCancel)=vbOK Then
      MsgBox "You clicked OK"
    Else
      MsgBox "You clicked Cancel"
    End If
End Sub
```

To run this program, click the ▶ button on the toolbar, press F5, or select **Run Macro** from the **Run** menu.

How Do I Execute an Expression in a Loop?

Here is an example of a simple (For...Next) loop. The loop is used to fill array x with values between 1 and 10.

```
' This program illustrates simple assignments of
' values to array elements.
Sub Main
Dim i As Long
Dim x(10) As Double
' Note that array x is zero-referenced by default;
' i.e., the first element in array x is x(0), and
' not x(1). Do not include the statement:
'     Option Base 1
' at the beginning of this program file, or else the
' program will not run.
 For i=0 To 10
```

```
      x(i)=i
    Next i
  End Sub
```

I Already Know about Visual Basic; Does SVB Support Advanced Language Features?

STATISTICA Visual Basic is an advanced programming environment that allows experienced programmers to build large and complex custom applications. Support is provided for subroutines and functions, as well as user-defined class modules and object modules.

What Are Subroutines and Functions?

Typically, complex applications written in *STATISTICA* (or any other) Visual Basic are structured. In other words, most programs will not consist of a single routine, but of many separate routines that will call each other. Every macro program (events, etc. are discussed in the section on *What Are Application Events and How Can They Be Controlled from STATISTICA Visual Basic?*, page 1015) consists of a *Main* program inside a `Sub Main ...` `End Sub` block. Inside the main program you can call subroutines and functions to perform repetitive tasks, or just to keep the structure of your program transparent. Here is an example program that consists of several subroutines and functions; note that the main program performs no computations, and only contains calls to the subroutines.

```
' This program will compute the sum of the
' squared values from 1 to 10. The program
' illustrates how subroutines and functions
' are declared, and how different numeric
' and string data types are declared and
' passed to subroutines and functions.
Sub Main
' Declare x as an array of double values;
' there are 10 elements in this array, which
' can be addressed as x(1), x(2), etc. (i.e.,
' the lower bound of the array is element 1,
' the upper bound is 10).
      Dim x (1 To 10) As Double
      Dim SumValue As Double, ResText As String
' The following subroutine will fill the array
' with numbers from 1 to 10.
      ComputeArray x()
```

```
' A function is called to compute the sum of
' the squared values in array x; note how
' we do not pass explicitly the lower and
' upper bounds for array x, but use functions
' (LBound, UBound) to retrieve those values.
      SumValue=ComputeSumOfSqrs ( LBound (x), _
                                 UBound(x), _
                                 x)
' The following subroutine makes the results string
' ResText.
      MakeResultsText SumValue,x(),ResText
      MsgBox ResText
End Sub


' In this subroutine, a For...Next loop is
' used to fill array x with values between
' 1 and 10.
Sub ComputeArray(x() As Double)
      Dim i As Integer
      For i =1 To 10
        x(i)=i
      Next i
End Sub


' This function computes the sum of the squared
' values in an array of double numbers. The
' function returns a double value.
Function ComputeSumOfSqrs (iFrom As Long, _
                           iTo As Long, _
                           x() As Double) As Double

      Dim i As Integer
      ComputeSumOfSqrs=0
      For i= iFrom To iTo
      ComputeSumOfSqrs=ComputeSumOfSqrs+x(i)^2
      Next i
End Function


' The following subroutine makes the results text, to be
' displayed at the end of the Main program.
Sub MakeResultsText(SumValue As Double, _
   x() As Double, _
   ResText As String)
      ResText="The sum of the square of values from " + _
                Str(LBound(x)) + _
                " to " + _
                Str(UBound(x)) + _
```

```
                       " is " + _
                    Str(SumValue)
    End Sub
```

What Is the Difference between Passing Variables to Subroutines and Functions By Value and By Reference?

If you want to pass a variable to a subroutine, run computations on it, and alter its contents, then the variable must be passed *by reference* (this is the default method how arguments are passed to subroutines or functions). Essentially, passing a variable by reference means passing the variable itself to the subroutine, and any changes made to it in the subroutine will be permanent. By contrast, if you want to pass a variable, run computations on it, and alter the temporary instance of it within the subroutine without affecting the original variable, then you must pass the variable *by value*. When a variable is passed by value to a function or subroutine, only a copy of the variable is passed, not the actual variable. If any changes are applied to the variable within the called function, then those changes will only pertain to the copy of the original variable. When passing a variable by value, its value will always be preserved, no matter what is done within the called function. To illustrate, instead of making `ComputeSumOfSqrs` in this example a function, we could make it a subroutine and pass the variable `SumValue` as an argument by reference; by using this approach, `SumValue` will be directly altered by the subroutine without needing to return a value. Note that the `ByRef` keyword in the example below would be optional, since the default method of passing arguments to subroutines or function is by reference.

```
    Sub Main
        ...
        ...
        ComputeSumOfSqrs ( LBound (x), _
                           UBound(x), _
                           x, _
                           SumValue)
        ...
        ...
    End Sub
    ' Now, ComputeSumOfSqrs is a subroutine, with arguments passed
    ' explicitly by value, or by reference (note that arrays are
    ' always passed by reference).
    Sub ComputeSumOfSqrs   ByVal iFrom As Long, _
                           ByVal iTo As Long, _
                           x() As Double, _
```

```
                  ByRef SumOfSqrs As Double
      Dim i As Integer
      SumOfSqrs=0
      For i= iFrom To iTo
      SumOfSqrs=SumOfSqrs+x(i)^2
      Next i
End Sub
      . . .
      . . .
```

What Are Local Variables vs. Global Variables?

Variables declared in a Visual Basic program are visible (i.e., they can be referenced) inside the "scope" where they are defined. The scope is the program unit inside of which the variable is defined. For example, variables declared inside a subroutine are only visible inside that subroutine. Variables declared outside any routines, before any program code, are visible to all subroutines and functions in the same file. In the programs shown below, variable *a* is either defined inside the *Main* routine, and then must be passed to the subroutine that displays its value, or it is declared as a *global* variable, outside the *Main* routine, in which case it is visible in all subroutines and functions in the same file.

Here is a simple program, where the variable must be passed, because it is defined only inside the scope of the *Main* program.

```
Sub Main
 Dim a As Long
 a=1
 DisplayVariableA a
End Sub

Sub DisplayVariableA (a As Long)
 MsgBox Str(a)
End Sub
```

Here is the functionally identical program, where the variable is declared globally.

```
Dim a As Long
Sub Main
 a=1
 DisplayVariableA
End Sub
```

```
Sub DisplayVariableA
 MsgBox Str(a)
End Sub
```

Can I Define My Own Class Modules and Object Modules?

STATISTICA Visual Basic supports class modules and objects modules. A discussion of classes and objects and the advantages (and some disadvantages) of object-oriented programming is beyond the scope of this introductory documentation; you should consult more advanced general sources on Visual Basic for details. In general, when developing large projects in Visual Basic, it is advantageous to impose as much structure on the program as possible, i.e., to break it down as much as possible into smaller functional units. This can be accomplished to a large extent by subroutines and functions; however, another way is by defining classes.

For example, you may want to use in your computations some highly customized way in which to compute some statistics to denote the central tendency for a dataset. One such statistic is, for example, the mean, but there may be additional considerations and problems that preclude you from using the simple arithmetic mean (e.g., censoring, missing data, highly skewed distributions, invalid data ranges, etc.). You may want to define a separate set of routines, options, etc., to compute the central tendency of a dataset. One way to do this is to create a class that contains the respective functions, subroutines, variables, etc. This class could for example compute the central tendency in one variable (e.g., variable ***CentralTendency***), and the method that was used as a name in another variable (e.g., variable ***MethodName***).

One of the advantages of using classes, as opposed to simple subroutines and functions, is that you can program all "things" related to the central tendency measure in a separate file. In the actual program, you would always access a variable called ***CentralTendency***; even if the method for computing the central tendency is refined or variables are added to the class, etc., you don't have to rewrite the code that uses these measures at all. In other words, the class itself can be modified or refined, or another team of programmers could develop it, while the programs that use the class do not have to be modified at all.

Classes vs. objects. The difference between classes and objects is that classes have to be explicitly declared (instantiated) in the program that uses them, while objects are automatically instantiated and available at all times. Both of these will be illustrated in a simple example below.

The asterisk (*) convention: Writing portable programs. When writing programs that include classes or objects, and that are to be deployed on other computers you

can use the asterisk (*) convention to access the include files that contain the classes and objects. Specifically, after you wrote (defined) your class or object modules, save them either in the directory where *STATISTICA* is installed (where the *STATISTICA* executable programs are located), in the directory where the current module is running from (if it has been saved), or in the global macro directory. To access modules from within an SVB program, you can then use, for example, `#Uses "*MyClassModule.svc"`, to access the previously saved class module **MyClassModule.svc**, and you do not have to use the explicit reference to the respective directory (e.g., `#Uses "J:\STATISTICA\Examples|Macros\MyClassModule.svc"`). Thus, by using the asterisk (*) convention with the `#Uses` or `$Include` directives.

A simple example. Suppose you wrote a program to retrieve some keys (key-codes) or passwords to gain access to particular datafiles or programs. The main program does nothing else but access the class (or object) that generates those keys. The method in which those keys are computed is entirely "encapsulated" inside the respective class (or object).

Create the following three files as *STATISTICA* Visual Basic macro programs. The first one defines **Class** as the **Module Type** (when creating the new Visual Basic document); call this program *KeyCodesClass*. The screen below shows the selections from the **Create New Document** and the **Module Type** dialogs to create the *KeyCodesClass* module.

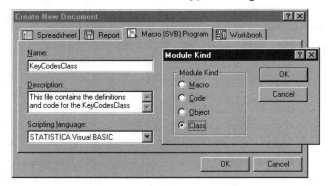

The second file defines **Object** as the **Module Type**; call this program *KeyCodesObject*. As a third file, create a standard *STATISTICA* Visual Basic macro program, and call it *RetrieveKeyCodes*; note that only this latter program will automatically have the `Sub Main ... End Sub` declarations inserted into the code.

Now enter the following code into these programs:

The Class Module: KeyCodesClass
```
     ' This file contains the code that defines the
     ' KeyCodesClass,i.e., the functions and subroutines
     ' that belong to this class. This file should be created
```

```
' (or already exists) as:
'
' J:\STATISTICA\Examples\Macros\KeyCodesClass.svc
'
' Variable KeyCode is defined globally inside this
' file; it can be accessed by the class member functions
' and subroutines.
Dim KeyCode As String

Private Sub Class_Initialize()
' This function is called when you initialize the class.
' Here you could put the code to generate the keycodes
' for various versions, etc.
' For the example, we will simply set it to a constant
' string.

KeyCode="ThisIsMyCodeFromTheClass"
End Sub

Private Sub Class_Terminate()
' This function is always called when you terminate
' the program, and this class. You could, for example,
' put some code here to update some data base, or internal
' counters that keep track of how many times the key codes
' program (class) has been used.
    MsgBox "KeyCodesClass is terminating."
End Sub

Function RetrieveCodeFromClass As String
' This is the routine that will be called to
' retrieve the code from this class.

    RetrieveCodeFromClass=KeyCode
End Function
```

The Object Module: KeyCodesObject

```
' This file contains the code that defines the
' KeyCodesObject,i.e., the functions and subroutines
' that belong to this object. This file should be created
' (or already exists) as:
'
' J:\STATISTICA\Examples\Macros\KeyCodesObject.svo
'
' Variable KeyCode is defined globally inside this
' file; it can be accessed by the member functions
' and subroutines.
```

```
    Dim KeyCode As String

    Private Sub Object_Initialize()
    ' This function is called when you initialize the object.
    ' Here you could put the code to generate the keycodes
    ' for various versions, etc.
    ' For the example, we will simply set it to a constant
    ' string.
        KeyCode="ThisIsMyCodeFromTheObject"
    End Sub

    Private Sub Object_Terminate()
    ' This function is always called when you terminate the
    ' program, and this object. You could, for example,
    ' put some code here to update some data base, or internal
    ' counters that keep track of how many times the key codes
    ' program (object) has been used.
      MsgBox "KeyCodesObject is terminating."
    End Sub

    Function RetrieveCodeFromObject As String
    ' This is the routine that will be called to
    ' retrieve the code from this object.
        RetrieveCodeFromObject=KeyCode
    End Function
```

The main program: RetrieveKeyCodes

```
    ' This program illustrates the use of classes and
    ' objects in STATISTICA Visual Basic. The following
    ' two lines include the references to definitions
    ' for a class module and an object module, respectively.
    ' (Files containing a class module have the extension
    ' .svc; files containing an object module have the
    ' extension .svo)
    ' NOTE: The directory in which these files are located may
    ' be different in your installation of STATISTICA.
    '#Uses "J:\STATISTICA\Examples\Macros\KeyCodesClass.svc"
    '#Uses "J:\STATISTICA\Examples\Macros\KeyCodesObject.svo"
    Sub Main
    ' Because KeyCodesClass is a Class, an object of this type
    ' has to be explicitly created (initialized).
        Dim Keys As New KeyCodesClass
        MsgBox Keys.RetrieveCodeFromClass
    ' KeyCodesObject is an object module, so it already exists.
    ' We can simply call it.
        MsgBox KeyCodesObject.RetrieveCodeFromObject
```

```
' The following line will explicitly destroy
' this instance of the KeyCodesClass, i.e., call
' the destructor function for that class.
        Set Keys = Nothing
End Sub
```

Before you run this program, make sure that the respective class and object module files exist in the correct locations; note that you could also save all files in the default directory for global macros (e.g., in the root directory of the *STATISTICA* installation on your machine), and then use the asterisk convention to reference the files, e.g., via `#Uses "*KeyCodesClass.svc"`. Two different "key-codes" will be retrieved. Suppose that a very elaborate (and perhaps confidential) method exists to generate the key codes. By using object (or class) modules, you could encapsulate inside the object (or class) the method for creating the keys; this method could then be modified, updated, etc., while the main program that retrieves those keys by accessing the object (class) would not have to change.

Can I Expand My SVB By Calling External DLLs?

There are two ways in which you can access external libraries from *STATISTICA* Visual Basic. The first method is the easiest, but relies on the external DLL's compliance with Visual Basic standards. In other words, if the external DLL (program) supports Visual Basic (e.g., such as Microsoft Excel), you can select **Tools - References** from the **STATISTICA Visual Basic** toolbar to load the respective libraries. The objects, functions, methods, etc. in that library are now accessible like all *STATISTICA* Visual Basic functions. So you could, for example, create from within *STATISTICA* Visual Basic an Excel Spreadsheet, perform some spreadsheet operations and computations, and then transfer the results back to a *STATISTICA* Spreadsheet (e.g., via the `.Copy` and `.Paste` methods).

Writing your own DLLs, or accessing "exported" functions in external DLLs. You can also access "exported" functions or subroutines in other DLLs, even if they were not specifically designed to work with Visual Basic. For example, if you are an experienced C++ programmer, you may want to add some highly customized computational routines to your *STATISTICA* Visual Basic program. In a sense, this method of accessing "the outside world" would allow you to add custom program options to *STATISTICA*; for example, you may have a proprietary complex algorithm, implemented in a complete program (which includes its own user interface), for evaluating stocks based on historical data. You could write a *STATISTICA* Visual Basic program to put a new option on the **Statistics** menu to execute this

program (see also, *Can I Customize the Toolbars and Menus via SVB?*, page 1021) and call your custom program when this option is chosen.

There are a number of things to consider when writing your own DLLs that are to be called by *STATISTICA* Visual Basic, or when writing *STATISTICA* Visual Basic programs to access functions in such DLLs.

Declaration of functions and subroutines in the external DLL. SVB requires that the external DLL use the *_stdcall* calling conventions. This is true of all Win32 API calls (allowing them to be used directly from SVB), but when defining your own entry points in a custom DLL, care must be taken to make sure the export is defined correctly. Here are two prototypes of sample routines that could be called from SVB:

```
extern "C" int AFX_API_EXPORT WINAPI SampleFunction(int x);
extern "C" void AFX_API_EXPORT WINAPI SampleSubroutine(int q);
```

'where:

'`extern "C"`   means this name is not "decorated" (as C programmers used to say) with C++ type information;

'`AFX_API_EXPORT` is defined to `__declspec(dllexport)`, which tells the compiler to export 'this name;

'`WINAPI`   is defined to `__stdcall`, meaning use "standard calling conventions."

Calling external functions and subroutines. To use these routines in *STATISTICA* Visual Basic, you need to "declare" them in the declarations area of your SVB program (before the first function is defined). Refer to the *Electronic Manual* for a full explanation of the `Declare` statement and its options. For example:

```
Private Declare Sub SampleSubroutine Lib _
"J:\DLLTest\Debug\DLLTest" _
        Alias "_SampleSubroutine@4" (ByVal X As Integer)
    Private Declare Function SampleFunction Lib _
"J:\DLLTest\Debug\DLLTest" _
    Alias "_SampleFunction@4" (ByVal X As Integer) _
    As Integer

Sub Main
 Dim ii As Integer
     Dim jj As Integer
     jj=1
     ii=SampleFunction(jj)
     SampleSubroutine(jj)
```

```
End Sub
```

This assumes that these routines are located in **DLLTest.dll**, and that this DLL is located in the *J:\DLLTest\Debug* directory. Note that if the DLL was located in the system search path (like the *SYSTEM32* directory), then only the DLL name is needed.

The `Alias` statement is used to define the exact name the compiler defines for this function. The `__stdcall` calling conventions dictate that the function name is preceded by an underscore ("_"), and appended by an @ and followed by the size in bytes of all the parameters. Since the example function is passing a 4-byte integer, then the name will have "@4" appended. Note that you can omit the `Alias` clause if you make an entry in the *.DEF* file for the custom DLL. If you add an entry for `SampleFunction` and `SampleSubroutine`, then the linker will create an alias without the underscore and the @*XXX* suffix.

STATISTICA Visual Basic
Development Environment

Does the SVB Environment Provide
Tools for Debugging?

Once a macro program has been specified, the *STATISTICA* Visual Basic environment provides powerful methods for testing and debugging the program. You can set individual breakpoints in the program where you want the program to stop temporarily to allow you to look at the values of variables at that point (you can set and clear breakpoints by pressing F9 or by clicking the 🖐 **Toggle breakpoint** toolbar button). You can step through the program by pressing SHIFT+F8 or clicking the ▣ **Step over** toolbar button.

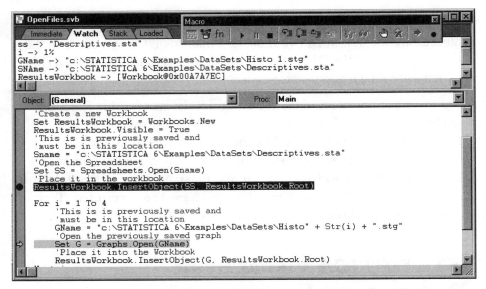

Note that the variables in the *Watch* window (tab) not only can be observed, but they can also be changed interactively as the program is running.

How Can I Get On-Line Help?

The *STATISTICA* Visual Basic language contains ("understands") all standard Visual Basic language declaration, features, and functions, as well as more than 10,000 functions, declarations, statements, etc. that provide access to the *STATISTICA* functionality. To read more about a particular Visual Basic (non-*STATISTICA* specific) statement such as *Dim*, *For...Next*, etc., highlight the respective statement and press F1.

To learn how to access the *STATISTICA* specific functions and procedures, study the *STATISTICA* object model; see also, *What Is Meant By the Term "Object Model?"* (page 939). The various features of graphs, spreadsheets, and reports are logically organized into hierarchies of objects. The objects representing the standard analyses (and graphs) are organized according to the "flow" of dialogs, as if you were performing an interactive analysis. This is best illustrated if you record an analysis in a *STATISTICA* Visual Basic macro program; see also *How Can I Record My Analysis in a STATISTICA Visual Basic Program?* (page 942). Every dialog is a separate object, and all the options available on the respective dialogs are properties. In practically all cases, after you record a macro, you will be able to match the various properties that were recorded (for the different dialogs) with the detailed descriptions of the respective options in the context of the dialog help (see also *What Are the Major "Components" in a Typical STATISTICA Visual Basic Macro Program?* on page 952).

What Is the Function Browser?

There are numerous functions specific to *STATISTICA*, and generally only available in *STATISTICA* Visual Basic. These are extensions to the general Visual Basic language, such as probability functions, matrix functions (see also page 1002), simple user input dialogs, etc. You can review these functions in the **Function Browser**, which is accessible via the SVB **View** menu or by clicking the **fn Function Browser** toolbar button.

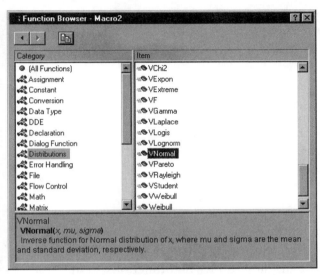

Shown below is a simple program that will "look up" the z value for $p = .975$.

```
Sub Main
        Dim p As Double
        p=VNormal(.975,0,1)
        MsgBox "For z=.975, p="+Str(p)
End Sub
```

How Are Errors Handled During the Execution of an SVB Program?

STATISTICA Visual Basic provides the same facilities for error handling as other general implementations of Visual Basic. Thus, detailed discussions on error handling can be found in the numerous general sources describing this programming language.

General Visual Basic errors, math errors, etc. A standard method for trapping general Visual Basic errors, such as math errors (e.g., division by zero), is to use the `On Error Goto ...` statement, to direct the program flow to a particular label in the program. Note that each subroutine and function should have its own `On Error Goto` statement. Here is a simple example program that will generate a division-by-zero math error.

```
Sub Main
' Set the label where to go to if an
' error occurs.
    On Error GoTo DisplayVBError
    Dim x As Double, y As Double
    y=0
' Division-by-zero error
    x=2/y
    Exit Sub
' An error occurred; Visual Basic maintains
' an object called Err with various properties;
' the Description property contains the description
' of the most recent error.
DisplayVBError:
    MsgBox Err.Description
    Exit Sub
End Sub
```

Visual Basic maintains and updates an object called **Err**. The property **Err.Description** contains the description of the most recent error. To clear the last error event, set **Err** to zero. It is generally good programming practice to write the subroutines and functions so that "clean" error exits can be accomplished, i.e., that code is provided that explicitly handles any errors.

STATISTICA analysis errors. Fatal errors (those that prevent the respective analysis from being completed) that occur as part of the analysis inside a *STATISTICA* module (e.g., incorrect selection of variables) are handled in the same way as during the interactive analysis, i.e., the program will display a message box explaining the error that has occurred, and how to remedy it.

How Are Non-Fatal Errors (Warnings) Handled During the Execution of an SVB Program?

In many instances throughout interactive analyses you may receive non-fatal error (warning) messages, for example, to inform you that some missing data were replaced by means, that some variables were dropped from the analyses, that some parameters were reset, etc. Usually, these warning will not require that the respective analyses be terminated, but the information provided in these warnings may be important for you to be aware of.

When such warning messages occur during the execution of an SVB program, the **Analysis Warnings** dialog is displayed, which lists the warnings that have occurred, when they occurred (the time and in which specific analysis), the approximate line numbers in the SVB program code where the warnings occurred, etc. If the respective non-fatal error, when it occurred during the interactive analysis, required user response (e.g., **Yes** or **No** in response to a question whether or not to continue), then that user input is also shown in the **Message** column.

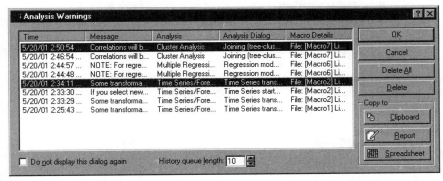

This dialog allows you to manage the warnings that occurred during your current "session" with *STATISTICA* (running *SVB* programs). Note that the warnings are collected across multiple analyses, and at any time you can create a report of these analysis, send them to a spreadsheet (so you can write SVB code to further process these messages), etc.

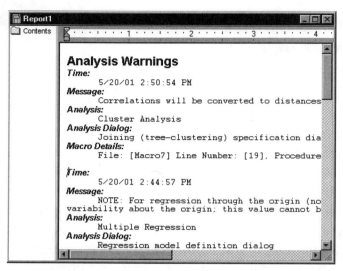

The `GetAnalysisErrorReport` **method.** When running *STATISTICA* analyses via Visual Basic from another application (e.g., from inside Microsoft Excel), it can be desirable to deal with the warnings inside the other application without displaying any *STATISTICA* specific dialogs. For example, you may want to place the warning messages into particular cells of a Microsoft Excel spreadsheet. For that purpose, the application object of *STATISTICA* (see *What Is Meant By the Term "Object Model?"*, page 939) contains a specific method:

```
Function GetAnalysisErrorReport( _
    [MakeVisible As Boolean = True]) As Spreadsheet
```

You can retrieve all information regarding non-fatal errors (warnings) that occur during a *STATISTICA* analysis into a spreadsheet object; you can then apply all the standard spreadsheet methods and properties to retrieve the desired information for display in the Microsoft Excel spreadsheet.

Does SVB include a "Dialog Painter"?

Yes, an interactive dialog editor is included that allows you to build dialogs by dragging controls from a toolbar with the mouse. For more information and examples, see the section on *Creating and Managing Custom Dialogs in SVB* (page 1007).

Using *STATISTICA* Visual Basic to Manage *STATISTICA* Analyses, Data Files, Results Spreadsheets, Graphs, etc.

How Is SVB Integrated with *STATISTICA* Data (Input), Analyses and Output?

The *STATISTICA* libraries contain all functions that allow you to fully program and customize your *STATISTICA* application; these functions can be called from other applications that support the standard Visual Basic language, such as Microsoft Visual Basic, Microsoft Excel, etc.

Organization of SVB programs. *STATISTICA* Visual Basic is organized around analysis objects (see also *What Is Meant By the Term "Object Model?"*, page 939); for example, to run an analysis with the *STATISTICA* **Basic Statistics** module, you would first create an analysis object with the constant *scBasicStatistics* and (optionally) with a datafile name (location of the file containing the input spreadsheet). For a complete list of modules (reference libraries) see *What Are the Names for the STATISTICA Reference Libraries (Modules)?*, page 977. To make access to the thousands of statistical functions and options available in the *STATISTICA* system as convenient as possible, SVB maintains a very close correspondence between the dialogs as they are presented during interactive analyses, and the flow of the SVB program. In a sense, once an analysis has been created, such as an analysis via the **Basic Statistics** module, you simply "program the dialogs" for the respective statistical analysis. Put another way, if you were to do the analysis by hand, then, after invoking the **Basic Statistics** module, you would follow through a sequence of dialogs to specify the desired analysis.

You can think of each dialog as a property (see also *What Are Properties, What Are Methods?*, page 941) of the (e.g., **Basic Statistics**) analysis, and of each option, selection, etc. on that dialog as a property of that dialog. Thus, you first invoke a module by declaring the respective analysis object, and then set the desired options etc. as properties of the analysis (and the dialogs of that analysis). Note that when designing actual programs, the automatic macro recording facilities of *STATISTICA* will do most of the programming work for you. You simply

run the desired analyses interactively, and then create the macro for those analyses (see *How Can I Record My Analysis in a STATISTICA Visual Basic Program?*, page 942); that macro will contain all of the programming code to recreate the analysis step by step, and it can easily be edited (copied, pasted) to create the desired customized application.

Moving between implied dialogs of the statistics module. To move from one dialog to the next when running *STATISTICA* interactively, you click the **OK** button (to move forward to the next dialog) and the **Cancel** button (to return to the previous dialog). SVB has two methods to accomplish this that belong to the analysis object: **Run** (to move forward to the next dialog) and **GoBack** (to return to the previous dialog).

Creating output documents. Most results from *STATISTICA* analyses are presented in results spreadsheets and graphs. When running *STATISTICA* interactively, you would create results spreadsheets and graphs by clicking on the respective buttons on the results dialogs. In SVB, each results "button" can be (implicitly) clicked by executing the respective method, that is part of the respective dialog. For example, every results dialog in *STATISTICA* has a **Summary** button (identified by the ▦ icon) to create the most "important" results from the respective analysis. In SVB, every results dialog has a **Summary** method to do the same.

Results spreadsheets and graphs. Each method that can be used to produce results spreadsheets and graphs (e.g., the **Summary** method) will return a *Collection* of spreadsheet objects, graph objects, or spreadsheet and graph objects (see also *What Are Collections*, page 957). Note that even if the respective results spreadsheet or graph consists of only a single document, it will still be returned as a collection. Thus, you can use the standard Visual Basic conventions to retrieve individual objects from the collection, etc.

RouteOutput method, and AnalysisOutput objects. When running *STATISTICA* interactively, the output spreadsheets and graphs can be sent to workbooks (by default), stand-alone windows, reports, etc., depending on the selection of options on the **Analysis/Graph Output Manager** dialog. The choices of options in that dialog are implemented by the `AnalysisOutput` object, which can be used as a "container" for the results spreadsheets and graphs. Typically, a summary results spreadsheet or graphs collection would be recorded in an Analysis Macro or Master Macro as:

```
newanalysis.RouteOutput(newanalysis.Dialog.Summary).Visible=True
```

The `RouteOutput` method takes as an argument the `Summary` collection (of spreadsheets, graphs, or both) and places it into the workbook, report, etc. depending on the current selections on the **Analysis/Graph Output Manager** dialog. The `RouteOutput` method actually returns an object of type `AnalysisOutput` which itself has a number of methods and properties to make it fully "programmable."

Note that the actual selections on the **Analysis/Graph Output Manager** dialog are typically recorded via the `OutputOption` object. Specifically, those options will be recorded either as part of Analysis Macros if they are set or changed via **Options - Output** for the specific analysis, or they will be recorded as part of Master Macros if they are set or changed in a specific analysis as well as via the **Output** tab of the **Options** dialog (accessed from the **Tools – Options** menu).

Manipulating and editing *STATISTICA* documents. Spreadsheets and graphs are only two of the document types that *STATISTICA* uses to handle input and output of statistical analyses. Other document types are (text) reports, and workbooks, which allow you to organize and manage all other documents. There are a large number of properties and methods available for each of these types of documents, that allow you to customize your results, or to access aspects of your results for further processing.

What Are the Names for the *STATISTICA* Reference Libraries (Modules)?

The following is a list of the currently available *STATISTICA* modules, and the name of the respective libraries accessible to Visual Basic.

Module (Option)	Library Name	Symbolic Constant
ANOVA*	STAMANOVA	scMANOVA
Basic Statistics	STABasicStatistics	scBasicStatistics
Canonical Analysis	STACanonical	scCanonicalAnalysis
Classification Trees	STAQuickTrees	scClassificationTrees
Cluster Analysis	STACluster	scClusterAnalysis
Correspondence Analysis	STACorrespondence	scCorrespondenceAnalysis
Discriminant Analysis	STADiscriminant	scDiscriminantAnalysis
Distribution Fitting [†]	STANonparametrics	scDistributions
Experimental Design (DOE)	STAExperimental	scDesignOfExperiments
Factor Analysis	STAFactor	scFactorAnalysis
General CHAID Models	STAGCHAID	scGCHAID
General Classification and Regression Trees	STAGTrees	scGTrees

General Discriminant Analysis Models	STAGDA	scGDA
Generalized Additive Models	STAGAM	scGAM
Generalized Linear/ Nonlinear Models	STAGLZ	scGLZ
General Linear Models	STAGLM	scGLM
General Partial Least Squares Models	STAPLS	scPLS
General Regression Models	STAGRM	scGSR
Log-Linear Analysis	STALogLinear	scLoglinearAnalysis
Multidimensional Scaling	STAMultidimensional	scMultidimensionalScaling
Multiple Regression	STARegression	scMultipleRegression
Nonlinear Estimation	STANonlinear	scNonlinearEstimation
Nonparametrics	STANonparametrics	scNonparametrics
Principal Components and Classification Analysis‡	STAFactor	scAdvancedPCA
Process Analysis Techniques	STAProcessAnalysis	scProcessAnalysis
Quality Control	STAQuality	scQualityControl
Reliability/Item Analysis	STAReliability	scReliabilityandItemAnalysis
Survival Analysis	STASurvival	scSurvivalAnalysis
Time Series	STATimeSeries	scTimeSeries
Variance Components	STAVarianceComponents	scVarianceComponents

*The **ANOVA** results dialog functions are accessible via the **General Linear Models** library.

†The **Distribution Fitting** module functions and routines are part of the **Nonparametrics** library.

‡The **Principal Components and Classification Analysis** techniques are accessible via the **Factor Analysis** library.

Note 1: The list of modules and procedures and *STATISTICA* libraries available to Visual Basic is constantly growing. Please check the StatSoft, Inc. Web site frequently (www.statsoft.com).

Note 2: The *Graph* menu graphs are all part of the *STATISTICA* main reference library; you can review the respective constants to instantiate a particular analysis (graph type) in the *Object Browser*.

How Can I Access Results Spreadsheets, Graphs, Workbooks Etc.?

The *STATISTICA* Visual Basic object model (see *What Is Meant By the Term "Object Model?"*, page 939) provides access to practically all aspects of *STATISTICA* analyses, and the results of those analyses. When you perform an interactive analysis, and record a macro program (see also *How Can I Record My Analysis in a STATISTICA Visual Basic Program?*, page 942), you will see that all of your output choices are recorded. The most important thing to remember is that all output spreadsheets and graphs are returned as a *Collection* of spreadsheet and graph objects respectively (see also *What Are Collections?*, page 957). You can retrieve and modify these objects in your program; for example, you can make the results (spreadsheet) of one analysis the input to the next analysis. Here is an example program:

```
Sub Main
Dim newanalysis As Analysis
Dim s As Object
Dim PPlot As Object
' Create the STATISTICA Basic Statistics Analysis object;
' open the input data spreadsheet (example file) 10items.sta
' NOTE: This file may reside in a different directory on
' your installation.
      Set newanalysis = Analysis (scBasicStatistics, _
            "J:\STATISTICA\Examples\Datasets\10items.sta")
' Select from the startup dialog of Basic Statistics, option
' Descriptives (the first option on the startup dialog).
      newanalysis.Dialog.Statistics = scBasDescriptives
' "Click OK", i.e., Run this dialog.
      newanalysis.Run
' On the following dialog, select all variables 1-10.
      newanalysis.Dialog.Variables = "1-10"
' Make the Summary results (descriptive statistics) spreadsheet.
      Set s=newanalysis.Dialog.Summary

' Create a 2D Probability Plots object; use the results
' spreadsheet from the previous analysis as the input spreadsheet.
      Set newanalysis = Analysis (sc2dProbabilityPlots, s.item(1))
' Request a Normal Probability plot for variable 2 (Means).
      With newanalysis.Dialog
```

```
            .Variables = 2
            .GraphType = scProbNormal
        End With
' Create the normal probability plot of means for all variables in
' file 10items.sta.
        Set PPlot=newanalysis.Dialog.Graphs
        PPlot.Visible=True
End Sub
```

However, in recorded macro programs, you will also see the so-called `RouteOutput` method, which will send the respective collection of output objects (graphs, spreadsheets) to stand-alone windows, workbooks, and/or report, depending on your current ***Analysis/Graph Output Manager*** settings. The next section briefly describes the `RouteOutput` method and `AnalysisOutput` objects.

How Can I Create a Blank Graph, and Set Data Directly into the Graph?

Practically all graphics options of *STATISTICA* are directly accessible via SVB; this makes SVB a very powerful graphics programming language. The following program shows how to create a blank 2D graph, set some data into the graph, and add an arrow.

```
Option Base 1
Sub Main
    Dim i As Integer
        Dim g As New Graph
        Dim gl As Layout2D
' Make a graph object with content.
        Set gl=g.GraphObject.CreateContent(scg2DGraph)
' Make room for 10 data points.
        gl.Plots.Add(scgSimplePlot,10)
' Make the data.
        For i=1 To 10
          gl.Plots(1).Variable(1).Value(i)=i
          gl.Plots(1).Variable(2).Value(i)=Log(i)
        Next
' Let us add an arrow to this graph.
        Dim x(2) As Double, y(2) As Double
' Point to location..
        x(1)=2
        y(1)=Log(2)
' Point from location...
        x(2)=2.5
        y(2)=1.5
```

```
      g.ExtraObjects.AddDynamicArrow(x(1),y(1),x(2),y(2))
' Finally make the graph visible.
      g.Visible=True
End Sub
```

Note that this program does not create an **Analysis** object via **Set newanalysis =...**; instead it creates a blank graph "from scratch" and then sets data directly into the graph. The graph will look like this:

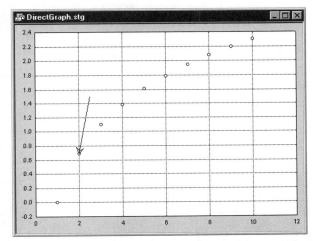

You can further customize all aspects of this graph from within the SVB program; review the different properties and methods that apply to the various graphics objects in the **Object Browser**.

What Is the RouteOutput Method,
What Is the AnalysisOutput Object?

When running *STATISTICA* interactively, the output spreadsheets and graphs can be sent to workbooks (by default), stand-alone windows, reports, etc., depending on the selection of options on the **Analysis/Graph Output Manager** dialog (accessed by selecting **Output** from the **Options ▼** button menu). The choices of options in that dialog are implemented by the **AnalysisOutput** object, which can be used as a "container" for the results spreadsheets and graphs. Typically, a summary results spreadsheet or graphs collection would be recorded as:

newanalysis.RouteOutput(newanalysis.Dialog.Summary).Visible=True

The **RouteOutput** method takes as an argument the **Summary** collection (of spreadsheets, graphs, or both) and places it into the workbook, report, etc. depending on the current

selections on the **Analysis/Graph Output Manager** dialog. The `RouteOutput` method actually returns an object of type `AnalysisOutput` which itself has a number of methods and properties to make it fully "programmable."

Note that the actual selections on the **Analysis/Graph Output Manager** dialog are recorded via the `OutputOption` object. Specifically, those options will be recorded either as part of Analysis Macros, if they are set or changed via **Options - Output** for the specific analysis, or they will be recorded as part of Master Macros, if they are set or changed in a specific analysis as well as via the **Output** tab of the **Tools - Options** dialog.

Suppose in your interactive analysis all results spreadsheets were automatically placed into a workbook; a macro recorded from such an analysis might look as shown in the following example. This example also illustrates how to access the results documents in the AnalysisOutput object (i.e., edit the recorded macro to access the results spreadsheets).

```
Sub Main
' NOTE: This file may reside in a different directory on
' your installation. Also, if you recorded this portion of
' the code via a Master Macro [option Tools - Macro - Start
' Recording of Log of Analyses (Master Macro) ], consecutive
' analyses would be enumerated as newanalysis1, newanalysis2,
' etc., and the input datafile spreadsheets would be explicitly
' assigned to variables (objects) S1, S2, etc.
        Set newanalysis = Analysis (scBasicStatistics, _
            "j:\STATISTICA\Examples\Datasets\exp.sta")
        newanalysis.Dialog.Statistics = scBasFrequencies
        newanalysis.Run
        newanalysis.Dialog.Variables = "1-8"
' Create the Analysis Output object as requested by
' the current settings in the Output Manager.
        Set r=newanalysis.RouteOutput(newanalysis.Dialog.Summary)
' Make sure that the AnalysisOutput object contains
' a Workbook.
        If (r.HasWorkbook=True) then
' We will next find the first results spreadsheet
' (frequency table) that was produced, and extract it
' from the workbook as a stand-alone spreadsheet; note
' that the objects are explicitly dimensioned in the
' following SVB code, to make it more transparent.
        Dim w as Workbook
        Set w=r.Workbook
        Dim wi As WorkbookItem
        Set wi=w.Root.Child
        While (wi.Type<>scWorkbookItemTypeSpreadsheet)
         Set wi=wi.Child
```

```
        Wend
        Dim s As Spreadsheet
        Set s=wi.Extract(scWorkbookExtractCopy)
        s.Visible=True
      End If
End Sub
```

What Kind of Output Can I Create with SVB?

There are a number of possible ways to display results, numbers, messages, etc. generated with a *STATISTICA* Visual Basic program. Most often, you will want to present results in spreadsheets, graphs, or reports. A very useful method for sending results spreadsheets and graphs to a current default "place" (workbook, stand-alone window, report) is to use the `RouteOutput` method described earlier (see page 981). However, you can also copy and paste results from *STATISTICA* into spreadsheets, text documents, etc., as long as those other types of output objects ("belonging" to other applications, such as Microsoft Excel or Word) support the Visual Basic standards.

Probably, the method you will most frequently use to display results is via spreadsheets and graphs; occasionally, you may also want to write some text or tables into a report. All of these are supported as objects, in the *STATISTICA* Visual Basic object model (see *What Is Meant By the Term "Object Model?", page 939*). For example, the following program will produce a spreadsheet, a line graph, and a report that will contain the spreadsheet and the line graph. All of these output objects will then be placed into a new workbook.

Note that the following program does not use the `RouteOutput` method (see page 981) to send the spreadsheet and graph to a default location (e.g., workbook) as currently defined via the **Analysis/Graph Output Manager**. Instead, the program illustrates how you can explicitly create spreadsheet and graph objects and move them to the location of your choosing.

```
    Sub Main
    Dim newanalysis as Analysis
    ' Create a new workbook; by default it will be
    ' empty and "invisible."
        Dim wb As New Workbook
    ' Create a new report; by default it will be
    ' empty and invisible.
        Dim r As New Report
    ' Create a new Spreadsheet; by default it will
    ' contain 10 rows and 10 columns, and it will
    ' be invisible.
```

```
        Dim ss As New Spreadsheet
' Declare an object of type Graph; we will
' create it later as a line graph.
        Dim g As Graph
        Dim NumberOfRows As Long
        Dim NumberOfColumns As Long
        Dim i As Long,j As Long
        NumberOfRows=100
        NumberOfColumns=5
' Note how we can dynamically "re-dimension" arrays.
        ReDim x(1 To NumberOfRows, _
            1 To NumberOfColumns) As Double
' Here we fill the data array x with numbers
        For i=1 To NumberOfColumns
         For j=1 To NumberOfRows
            x(j,i)=Exp(j*i/500)
         Next j
        Next i
' Next put the numbers into the spreadsheet ss. First
' we will resize the default (10 x 10) spreadsheet
' to make room for the 100 x 5 data matrix.
        ss.SetSize(NumberOfRows,NumberOfColumns)
' The following line is a shortcut method for copying
' the entire array into the spreadsheet.
        ss.Data=x
' Alternatively, the following loop could be used.
' For i=1 To NumberOfColumns
'  For j=1 To NumberOfRows
'     ss.Value(j,i)=x(j,i)
'  Next j
' Next i
'
' Create a new analysis object of type line plot; pass
' as input data the contents of spreadsheet ss.
        Set newanalysis = Analysis (sc2dLinePlots,ss)
        newanalysis.Dialog.GraphType = scLineMultiplePlot
' Select all variables in the new spreadsheet.
        newanalysis.Dialog.Variables="*"
' Create the graph, and assign it to the previously
' declared object g. Note that all results from analyses
' in STATISTICA are returned as collections, and by default
' all results objects are invisible.
        Set g=newanalysis.Dialog.Graphs.Item(1)
' Now let us "print" both results to the report.
        r.SelectionObject=ss
        r.SelectionObject=g
```

```
' Finally, move the new objects (spreadsheet ss,
' graph g, and report r) to the workbook.
      wb.InsertObject(ss,wb.Root,)
      wb.InsertObject(g,wb.Root,)
      wb.InsertObject(r,wb.Root,)
' Make the final results workbook visible.
      wb.Visible=True
  End Sub
```

After running this program, a workbook will be created that looks something like this (depending on other default systems currently active in your system).

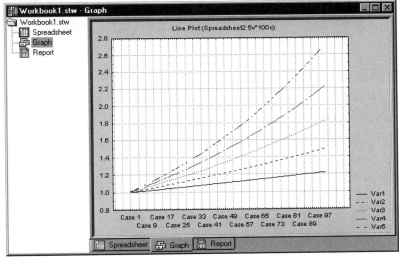

Note that further customizations, such as changing of titles, adding text, etc. can easily be achieved by further modifying various properties of the output objects, or the workbook.

How Can I Change the Font (or Other Aspects) of Numbers in Spreadsheets?

The functions in *STATISTICA* Visual Basic are organized in an intuitive hierarchical object model; see also *What Is Meant By the Term "Object Model?"* (page 939). The example program shown below illustrates the hierarchical nature of the objects, and the available respective methods and properties. Note that in order to review all available objects, methods, and properties, as well as constants that are recognized, you can always display the *Object Browser*.

The *Object Browser* can be displayed by clicking the  *Object Browser* toolbar button (on the *Macro* toolbar), or by selecting *Object Browser* from the *View* menu.

Shown below is a simple example program that will change the font of the first column of the currently active input data spreadsheet to italics and bold.

```
Sub Main
' Find the currently active input spreadsheet, and
' assign it to variable (object) InSpr; note
' that ActiveSpreadsheet is a property of the
' current default (STATISTICA) Application; if no
' input Spreadsheet is currently open, then an
' error message will be issued.
      Dim InSpr As Spreadsheet
      Set InSpr=Application.ActiveSpreadsheet
' Select all numbers in the first column of the
' spreadsheet; note that the CellsRange property
' returns an object of type Range; here is the
' complete syntax (see also the Object Browser):
' Property CellsRange(
'       FirstRow As Long, FirstColumn As Long,
'       LastRow As Long, LastColumn As Long)
'       As Range
' Note also that the underscore _ at the end of
' a line, after a blank, means that the next line
' is interpreted as a continuation of the respective
' command.
      Dim Cells  As Range
```

```
    Set Cells=InSpr.CellsRange( _
        1, 1, _
        InSpr.NumberOfCases, _
        1)
' Cells is a Range object; one of the properties
' of the Range object is the Font; now set the font;
' the font is accessible as a Font object.
        Dim CellFont As Font
        Set CellFont=Cells.Font
' Italic and Bold are properties of the Font.
        CellFont.Italic=True
        CellFont.Bold=True

End Sub
```

How Can I Modify the Appearance of Graphs (e.g., Axis Labels) from within SVB?

The customization of graphs proceeds in much the same way as the customization of spreadsheets. After a graph object is created (see also *What Is Meant By the Term "Object Model?", page 939*), various properties and methods are available to customize practically all aspects of the display. An example of how to create a blank graph, set data into the graph, and add an arrow is shown on page 980; here is an example of how to customize the font of the x-axis labels in a histogram:

```
Sub Main
' Create an analysis object; create a histogram from
' a variable in datafile Exp.sta (this example
' datafile may be located in a different directory
' on you installation).
    Dim a As Analysis
    Set a = Analysis (sc2dHistograms, _
        "j:\STATISTICA\Examples\DataSets\Exp.sta")
    a.Dialog.Variables = "5"
' Retrieve the first graph object from the collection.
' Note that all results spreadsheets and graphs
' are created in STATISTICA as collections of
' objects.
    Dim g as Graph
    Set g = a.Dialog.Graphs(1)
' Make the graph visible.
    g.Visible=True
```

```
' The Content property returns a "polymorphic"
' object, i.e., an object depending on the graph
' type that was created.
   Dim l As Object
   Set l = g.Content
' Next we want to "get to" the axes of the plot.
' Xaxis is a property of graph Content.Axes:
   Dim x As Axis2D
   Set x = l.Axes.Xaxis
' To modify the font of the x-axis labels, retrieve
' the Font property (object) of the x-axis.
   Dim xf As GraphicsFont2
   Set xf=x.Font
' Change the size of the font.
   xf.Size=20
' To change the Font itself, retrieve the FontFace
' property (object) from the GraphicsFont2 object.
   Dim xff As FontFace
   Set xff=xf.Face
' Finally, set the font to bold.
   xff.Bold=True
End Sub
```

As you can see, to set the font of the x-axis labels to bold, you retrieve the **Graph** object, then the ("polymorphic," i.e. depending on the graph) **Graph Content** object, then the **Axis2D** object, then the **GraphicsFont2** object, then the **FontFace** object, and finally set the **Bold** property. This example illustrates how learning *STATISTICA* Visual Basic means learning to navigate the *STATISTICA* object model. Fortunately, the object model is organized into a logical hierarchy, and it can be reviewed via the **Object Browser**, for additional details, see *What Is Meant By the Term "Object Model?" (*page 939).

How Can Other Aspects of Graphs Be Modified?

Here is a more complex example program that will customize various aspect of a graph:

```
' This program illustrates how create and customize
' different aspects of STATISTICA Graphs.
'
Sub Main
        Dim a As Analysis
' Create the 2D Line Plot analysis object; you
' may have to edit the path name for the input data
```

```
' file (spreadsheet), to match the installation of
' STATISTICA on your computer.
        Set a = Analysis (sc2dLinePlots, _
        "j:\STATISTICA\Examples\DataSets\Exp.sta")
        a.Dialog.GraphType = scLineMultiplePlot
        a.Dialog.Variables = "6 7 8"

'Retrieve the first graph object from the collection.
        Dim g As Graph
        Set g = a.Dialog.Graphs(1)

        Dim bFirstLevel, bSecondLevel, _
           bThirdLevel, bFourthLevel _
           As Boolean
        bFirstLevel = True
        bSecondLevel = True
        bThirdLevel = True
        bFourthLevel = True

' Customize titles, background.
        If bFirstLevel Then
           g.Titles.Item(1) = "New (better?) title"
           g.GraphWindow.Background.Color = _
           RGB (255,255,0)
        End If

' Get graph content.
        Dim l As Layout2D
        Set l = g.Content

' Customize plots: axis title and set line set to
' solid.
        If bSecondLevel Then
                'set y axis manual scale
                Dim y As Axis2D
                Set y = l.Axes.YAxis
                y.setManualRange(0,20)
                y.RangeMode = scgManualRange
                y.StepSize = 5
                y.StepMode = scgManualStep

' Set x axis title and turn off gridlines.
                Dim x As Axis2D
                Set x = l.Axes.XAxis
                x.Title = "Sequential cases"
                x.DisplayMajorGridLine = False
```

```
' Customize gridlines on y axis
                    y.MajorGridLine.ForegroundColor = RGB(0,0,0)
                    y.MajorGridLine.Type = scgSolid

        End If

        Dim pp As Plots2D
        Set pp = l.Plots

        If bThirdLevel Then
' Plot customizations

' Line patterns
                    pp.Item(1).Line.Type = scgSolid
                    pp.Item(2).Line.Type = scgSolid
                    pp.Item(3).Line.Type = scgSolid

' Make third plot stand out from the rest.
                    pp.Item(3).Marker.Size = 6
                    pp.Item(3).Line.Weight = .75
        End If

        If bFourthLevel Then
' Changing plot variable values; we will set values <= 0
' to mean of previous and next value.
                    Dim v As Variable
                    Set v = pp.Item(3).Variable(2)

                    Dim vind As Integer
                    For vind = 2 To v.ValuesCount() - 1
                            If v.Value(vind) <= 0 Then
                                    v.Value(vind) = (v.Value(vind-1) + _
                                        v.Value(vind+1))/2
                            End If
                    Next vind
        End If

        g.Visible = True
End Sub
```

This program will create a 2D line plot and customize various aspects of this plot. The resulting plot will look something like this:

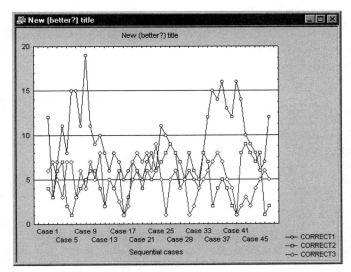

In general, in order to modify, copy, delete, etc. any aspect of a graph, locate the respective property in the object model. See also, *What Is Meant By the Term "Object Model?"* (page 939) and *How Can I Modify the Appearance of Graphs (Axis Labels) from within STATISTICA Visual Basic?* (page 987).

Can I Use SVB to Develop "Extensions" of *STATISTICA* and My Own New "Modules?"

This is one of the many useful applications for *STATISTICA* Visual Basic programs; see also *What Are Some of the Applications of STATISTICA Visual Basic?* (page 936). By recording routine analyses (see *How Can I Record My Analysis in a STATISTICA Visual Basic Program?*, page 942), you can quickly generate all the necessary program code to perform particular analyses. By combining the code from different programs, and by customizing perhaps some of the user interface (see *Can I Create Custom Dialogs and Other Interactive User Input Controls in STATISTICA Visual Basic*, page 1007), a completely new module can be created that will perform custom analyses. Shown below is an example program that will combine multiple analyses.

```
' This program will compute tests comparing the
' central tendency measures in multiple groups.
' The Basic Statistics Breakdowns options will
' be used to compute the ANOVA; the Nonparametric
' Statistics Kruskal-Wallis and Median tests will
' be computed as nonparametrics alternatives to
```

```
' ANOVA.
'
' The following variables are declared as public
' symbols, i.e., they will be "visible" to
' all routines throughout this program.
Public GroupingVariable As Integer
Public DependentVariableList() As Integer
Public ndep As Integer
Public NVars As Long
Public ResultsWorkbook As Workbook
Public Folder As WorkbookItem
Public InputSpreadsheet as Spreadsheet
Dim newanalysis As Analysis
Dim ResultsSpreadsheet As Object
' The main program will ask for variable selections
' and then perform the necessary computations by calling
' two subroutines.
Sub Main
  Set InputSpreadsheet = ActiveDataSet
' Determine the number of variables in the input datafile
      NVars = InputSpreadsheet.NumberOfVariables
' Set up an array to hold the dependent variable list
      ReDim DependentVariableList(1 To NVars)
      Dim i As Integer, ret As Integer, nindep As Integer
' Bring up the STATISTICA variable selection dialog
      ret = SelectVariables2 (InputSpreadsheet, _
            "Select dependent variables, and grouping variable", _
          1, NVars, DependentVariableList, ndep, _
          "Dependent variables:", _
          1, 1, GroupingVariable, nindep, _
          "Independent (grouping) variable: ")
      if ret=0 then GoTo Finish
' Make new workbook
    Set ResultsWorkbook = Workbooks.New
    ResultsWorkbook.Visible = True
' Call the subroutine to compute the ANOVA tables
      ANOVATable
' Call the subroutine to compute the Kruskal-Wallis
' and Median tests
      KruskalWallisAndMedianTests
Finish:
End Sub

Sub ANOVATable
      Set newanalysis = Analysis (scBasicStatistics, InputSpreadsheet)
      newanalysis.Dialog.Statistics = scBasBreakdowns
```

```
        newanalysis.Run
' NOTE: The Array function is a standard Visual Basic
' function that combines the (two in this case) arrays of integers
'into a Variant that is an array of those arrays.
        newanalysis.Dialog.Variables = _
            Array(DependentVariableList,GroupingVariable)
        newanalysis.Dialog.Codes = "*"
        newanalysis.Run
' Compute the ANOVA results spreadsheet
        Set ResultsSpreadsheet=newanalysis.Dialog.AnalysisOfVariance
' Set up a folder in the workbook, and call it "ANOVA Results"
        Set Folder=ResultsWorkbook.InsertFolder( _
            ResultsWorkbook.Root, scWorkbookLastChild)
        Folder.Name="ANOVA Results"
' Move the results spreadsheet into the workbook; remember
' that all results spreadsheets returned from analysis objects
' in STATISTICA are collections of objects (spreadsheets);
' thus we need to explicitly reference Item(1) in the collection.
        ResultsWorkbook.InsertObject( _
            ResultsSpreadsheet.Item(1), Folder, scWorkbookLastChild)
End Sub


' This subroutine uses the Nonparametric Statistics module
' to compute the Kruskal-Wallis and Median tests; the results
' will be moved into the workbook set up in the Main program.
Sub KruskalWallisAndMedianTests
        Dim i As Integer
        Set newanalysis = Analysis (scNonparametrics, InputSpreadsheet)
        newanalysis.Dialog.NonparametricStatistics = _
            scNonComparingMultipleIndependentSamples
        newanalysis.Run
' NOTE: The Array function is a standard Visual Basic
' function that turns the two arrays of integers into
' arrays of type Variant
        newanalysis.Dialog.Variables = _
            Array(DependentVariableList,GroupingVariable)
        newanalysis.Dialog.Codes = "1-2"
        Set ResultsSpreadsheet=newanalysis.Dialog.Summary
        Set Folder=ResultsWorkbook.InsertFolder( _
            ResultsWorkbook.Root, scWorkbookLastChild)
' Set up a folder in the workbook, and call it
' "Nonparametric tests"
        Folder.Name="Nonparametric tests"
' Move the results spreadsheet into the workbook; remember
' that all results spreadsheets returned from analysis objects
' in STATISTICA are collections of objects (spreadsheets);
```

```
' the .Count property will retrieve the number of objects
' (spreadsheets) in the collection.
      For i =1 To ResultsSpreadsheet.Count
            ResultsWorkbook.InsertObject( _
                ResultsSpreadsheet.Item(i), Folder, scWorkbookLastChild)
      Next i
End Sub
```

As you can see, the SVB environment not only allows you to automate routine analyses, but also in a sense to "program your own statistical package" with various nonstandard options tailored to your specific needs.

How Can I Attach a Macro Program to a Toolbar Button (Keyboard Command, or Menu Option)?

One common application of *STATISTICA* Visual Basic macro programs is to "enhance" the *STATISTICA* functionality by creating toolbar buttons or keyboard commands (e.g., CTRL+key sequences) that allow you to quickly perform a series of operations or routine analyses. Shown below is a simple program that will sort a spreadsheet in a workbook, by the (first) selected column.

```
' This program will sort the spreadsheet data by
' the currently selected column (the first column
' in a range of selected columns). The program
' assumes that the spreadsheet is inside the
' currently active workbook.

Sub Main
' The following line will cause the program to
' resume at label ErrOut, if any error occurs (e.g.,
' if a workbook or spreadsheet can't be found).
      On Error GoTo ErrOut
      Dim ss As Spreadsheet
      Dim wb As Workbook
      Dim r As Range
      Dim ifault As Integer
      ifault=1
      On Error GoTo ErrOut
      Set wb=ActiveWorkbook
      Set ss=wb.ActiveItem.Object
      ifault=2
      Set r=ss.ActiveCell
```

```
            ifault=0
            ss.SortData(r.Column)
            Exit Sub
    ErrOut:
            If ifault=1 Then
               MsgBox "Nothing to sort; the program assumes that you" + _
                 "have a Spreadsheet with numbers highlighted in a Workbook."
            ElseIf ifault=2 Then
               MsgBox "No column is currently highlighted; nothing to sort."
            Else
               MsgBox "Cannot sort data."
            End If
    End Sub
```

First, create this macro by selecting **File New**, and then on the **Macro** tab call this new macro program **SortColumn**; then click the **OK** button. Next type (or paste) in this program, and save it (e.g., as **SortColumn.svb**).

Select **Customize** from the **Tools** menu, and select **Macros** in the **Categories** list.

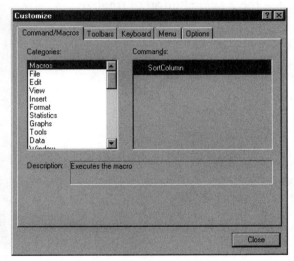

Now drag the **SortColumn** macro in the **Command** list onto the toolbar. A new toolbar button will be created, identified by the name of the macro.

In the illustration above, a "smiley face" was added to the **SortColumn** button. To do this, select **Tools - Customize**; then, while the **Customize** dialog remains displayed, right-click on the **SortColumn** button and select **Button Appearance**.

StatSoft
Copyright © StatSoft, 2001

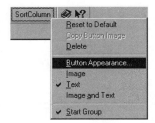

The **Button Appearance** dialog is displayed, where you can select a predefined icon or make a new one for the new button.

Attaching the macro to a shortcut key. To "connect" this macro program to a shortcut key, launch the **Customize** dialog again (**View - Toolbars - Customize**), and select the **Keyboard** tab. Select **Macros** in the **Category** box, and select **Macro** in the **Set shortcut key for** box. In the **Commands** list select **SortColumn**, and then choose CTRL+ALT+S as the key sequence for executing the **SortColumn** macro program. To do this, click in the **Press new shortcut key** box, and then press CTRL+ALT+S; this key sequence will now be displayed in this box, and the **Assign** button will become available. Click the **Assign** button to finalize your choice.

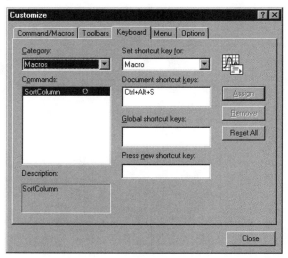

After you **Close** this dialog, the **SortColumn** macro program can now be executed from the keyboard by pressing CTR+ALT+S.

Attaching the macro to a menu. To attach the macro program to a menu item, select the **Commands/Macros** tab on the **Customize** dialog. Select **Macros** in the **Categories** box, and the **SortColumn** macro will be displayed in the **Commands** box.

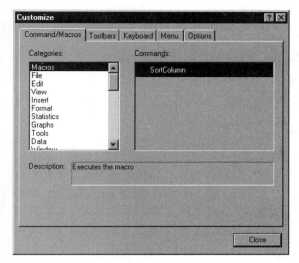

Click on the **SortColumn** macro, and drag it to the desired menu and menu location (e.g., the end of the **Edit** menu).

As you can see, the new macro was appended now as a menu item at the end of the **Edit** menu.

Running *STATISTICA* Visual Basic Programs from Other Applications

Can I Execute an SVB Program from within Other Visual Basic Compatible Applications?

Yes; in general, because of the industry standard compatibility of *STATISTICA* Visual Basic, you can execute *STATISTICA* Visual Basic programs from any other Visual Basic compatible environment (e.g., Microsoft Excel, Microsoft Word, or a stand alone Visual Basic language; in practice, you would typically call *STATISTICA* functions from VB in another application). However, there are a few considerations and limitations that you need to be aware of:

STATISTICA libraries must be properly licensed and installed. When you run a *STATISTICA* Visual Basic program or attempt to call *STATISTICA* functions from any other application, all calls to the *STATISTICA* specific functions (as opposed to the generic functions of Microsoft Visual Basic) will be executed only if the respective *STATISTICA* libraries are present on the computer where the execution takes place – that is, *the user of the program must be a licensed user of the respective STATISTICA libraries of procedures.* Note that this large library of *STATISTICA* functions (more than 10,000 procedures) is transparently accessible not only to Visual Basic (either the one that is built in, or a different one), but also to calls from any other compatible programming language or environment, such as C/C++, Java, or Delphi.

The proper STATISTICA libraries must be loaded. In order to access the thousands of library functions available in *STATISTICA*, the respective object reference libraries must be loaded. For example, in Microsoft Excel, use option **Tools - References** and then select the **STATISTICA Object** library, and any other libraries that you would want to access.

The STATISTICA Application object must be explicitly created. When you run *STATISTICA* Visual Basic programs from with the *STATISTICA* program editor, then a few shortcuts are available. In particular, the *STATISTICA* program editor assumes that the current application is *STATISTICA*, so it doesn't have to be initialized. However, when running *STATISTICA* from within another application (e.g., from within the Microsoft Excel Visual

Basic editor), you have to explicitly declare the *STATISTICA Application* object; see also *What Is Meant By the Term "Object Model?"* (page 939).

Some other limitations and differences between SVB and Visual Basic in other applications. While the objects of the *STATISTICA* libraries are fully exposed and accessible to all compatible Visual Basic programming environments in other applications, there are a few functions that are only available in SVB, i.e., when the Visual Basic program is run from within *STATISTICA*. (Likewise, there are some functions in other applications that are not accessible in the SVB environment.) Specifically, all user interface functions, such as user-defined dialogs or functions for retrieving variable lists or value lists (via dialogs from the user) are closely tied to the *STATISTICA* application itself. Therefore, when designing a Visual Basic program to run from a "foreign" application (e.g., from within Microsoft Excel), it is best to design the user interface (dialogs) using the tools available in that application; those tools are usually designed such that they allow you to program interfaces with an overall look and feel that makes them compatible with all other interfaces (dialogs) used in the respective application.

How Can I Run a *STATISTICA* Analysis from Microsoft Excel?

First, if you want to run *STATISTICA* from within the Visual Basic environment in Microsoft Excel, be aware of the issues (and some limitations) described in *Can I Execute a STATISTICA Visual Basic Program from within Other Visual Basic Compatible Applications?* (page 998).

To illustrate, we will create an example program that can be run from Visual Basic within Excel. After starting Excel, create a new Worksheet. Then select **Macro - Visual Basic Editor** from the **Tools** menu. From the Visual Basic **Tools** menu, select **References**.

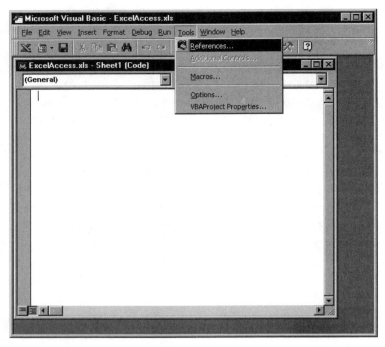

The **References** dialog allows you to select the libraries (objects) that you would like to be visible inside the Visual Basic program. To make *STATISTICA* visible, select the **STATISTICA Object Library**, and the **STATISTICA Basic Statistics Library** (for the current version of *STATISTICA*); then click the **OK** button.

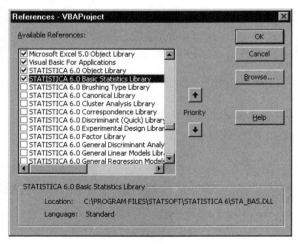

Now type or paste the following program into the program editor.

```
Sub ExcelTest()
' Run the STATISTICA application; create the STATISTICA
' Application object and assign it to variable (object) x.
    Set x = CreateObject("Statistica.Application")
' Create a STATISTICA Basic Statistics object (i.e., run the
' Basic Statistics module; start it with datafile exp
' (note: the actual location of that datafile may be
' different on your installation).
    Set a = x.Analysis(scBasicStatistics, _
        "j:\STATISTICA\Examples\Datasets\exp.sta")
' the following 7 lines of code will produce a summary results
' spreadsheet from the STATISTICA Basic Statistics module.
    a.Dialog.Statistics = scBasDescriptives
    a.Run
    a.Dialog.Variables = "5-8"
    Set out = a.Dialog.Summary
 ' Select all rows and columns in the STATISTICA results spreadsheet.
    out.Item(1).SelectAll
' Copy the highlight selection (all rows and columns in the
' Summary results spreadsheet.
    out.Item(1).Copy

' Set the cursor to cell A1 in the currently active Excel Spreadsheet.
    Range("A1").Select
' Paste in the summary statistics.
    ActiveSheet.PasteSpecial Format:="Biff4"
End Sub
```

When you run this Visual Basic program from inside Microsoft Excel (Visual Basic Editor), it will paste the results from the **Summary** results spreadsheet of the **Basic Statistics - Descriptive Statistics** analysis into the current Excel Spreadsheet. Note that this is accomplished without the user ever seeing or having to interact with the *STATISTICA* application; the program runs entirely invisibly, and the results are displayed inside the Excel spreadsheet. This simple example illustrates the power and versatility of the *STATISTICA* Visual Basic object model: All analysis and graphics options and methods available in *STATISTICA* are fully exposed in the respective object libraries, and even advanced and complex analyses can be automated and performed routinely "behind the scenes" from within any other Visual Basic compatible application.

Handling Large Computational Problems, Matrix Operations

What Is the Capacity of SVB, and Can It Handle Large Computational Problems?

STATISTICA Visual Basic is a powerful programming language which combines the general Visual Basic Programming Environment [with facilities and extensions for designing user interfaces (dialogs) and file handling], with the *STATISTICA* libraries with thousands of functions that provide access to practically all functionality of *STATISTICA*. It is well suited to handle large computational problems.

SVB supports local data arrays with up to 8 dimensions, and there is no limit to the size of the arrays (all memory available in your system can be used, including virtual memory, e.g., correlation matrices 1,000 x 1,000 and larger can be computed [with a single function call] on most systems), so custom procedures involving operations on large multidimensional matrices can be developed (a comprehensive selection of matrix operations are also supported, see *Does SVB Support Matrix Operations?*, below). Matrices can be dynamically allocated or redimensioned in run time.

Performance of *STATISTICA* Visual Basic programs. While the obvious advantage of Visual Basic (compared to other languages) is its ease of use and being familiar to a very large number of computer users, the possible drawback of VB programs is that they do not perform as fast as applications developed in lower-level programming languages (such as C). However, that potential problem will not apply to *STATISTICA* Visual Basic applications, especially those that rely mostly on executing calls to *STATISTICA* analytic, graphics, and data management procedures. These procedures fully employ the *STATISTICA* technology and they will perform at a speed comparable to running the respective procedures in *STATISTICA* directly.

Does SVB Support Matrix Operations?

STATISTICA Visual Basic contains a large number of designated matrix and statistical functions, which make the SVB environment ideal for prototyping algorithms or for developing custom statistical procedures. The matrix and statistical functions are documented

in detail in the section on the *STATISTICA Matrix Function Library* (see the *Electronic Manual*).

One major advantage of using the *STATISTICA* library of matrix functions, instead of writing these functions "by hand" in Visual Basic, is that the former will evaluate much faster. For example, when you want to invert large matrices, the ***MatrixInverse*** function will perform the actual matrix inversion using the highly optimized (compiled) algorithms of *STATISTICA*.

Include file: *STB.svx*. To provide convenient access to the matrix functions without requiring you to pass explicitly the dimensions of the arrays that are being passed, *STATISTICA* includes a file with function interfaces: ***STB.svx***. You may want to open this file to review the functions and see how they provide simplified access to the actual matrix library. It is recommended that you routinely include that file at the beginning of your program when you intend to use functions from the *STATISTICA* libraries of matrix and statistical functions.

In order to insure the portability of all *STATISTICA* Visual Basic macros, you can specify an asterisk in front of the root file name for the standard include files. When *STATISTICA* detects the *, it will automatically search for the ***STB.svx*** file in standard places (e.g., the directory where the macro has been saved, the global macro directory, and the directory where the *STATISTICA* application is located). Thus it is not necessary to state the complete path for the ***STB.svx*** file when including this file in your *STATISTICA* Visual Basic programs via the $Include statement.

Reviewing the matrix functions in the function browser. You can review the available matrix functions either in the *Electronic Manual* or in the ***Function Browser***, which also lists various other functions available in *STATISTICA* Visual Basic. The ***Function Browser*** is accessible via the SVB ***View*** menu or by clicking on the 🔢 ***Function Browser*** toolbar button.

Example. The following example illustrates how to use the functions in the *STATISTICA* library of matrix functions, to perform some basic matrix operations: Namely, matrix inversion and matrix multiplication:

```
' The following line is very important!
' It contains the "interfaces" to the STATISTICA
' Matrix and Statistical Function libraries.
'
'$include: "*STB.svx"
' The next statement will cause all arrays to be
' declared by default as 1-referenced, i.e., the
' first element in an array will be element 1, not 0.
' Note that the matrix and statistical functions
' will work fine with 0-referenced arrays; however,
```

```
' it can be confusing to write programs with 0-
' referenced arrays, because many of the matrix
' functions expect arguments based on 1-referenced
' arrays.
Option Base 1
Sub Main
' Set up two 4 x 4 matrices.
  Dim a(4,4) As Double, ainv(4,4) As Double
  Dim i As Long, j As Long
' Fill the matrix with random numbers
  For i=1 To 4
   For j=1 To 4
    a(i,j)=Rnd(10)
   Next j
  Next i
' Display the original matrix.
  MatrixDisplay (a, "Original Matrix A")
' Invert the matrix.
  MatrixInverse(a,ainv)
' Display the inverted matrix.
  MatrixDisplay (ainv, "A, Inverse")
' Multiply a*a-inverse
  MatrixMultiply(a,ainv,a)
  MatrixDisplay(a,"A * A-inverse")
End Sub
```

How Can I Use the Data in a Spreadsheet with Matrix Functions? (How Do I Copy Data Efficiently from Spreadsheets to Matrices, and Back?)

STATISTICA Visual Basic contains several functions to quickly move columns (variables), rows (cases), or the entire data matrix from spreadsheets to dimensioned arrays, and back. The following example shows how you can use the .**VData(variable)**, **CData(case)**, and .**Data** properties of the spreadsheet object to retrieve particular columns (variables) or rows (cases) of data from the respective spreadsheet, or to copy the entire data matrix.

```
' This program will fill a dimensioned array
' x(100,10) with random numbers, and then
' transfer the contents of that array to
' a spreadsheet.
'
' To use any of the matrix functions, make sure to
```

```
' include the following line, to include the file
' with the the matrix function interfaces.
'$include: "*STB.svx"

Sub Main
      Dim i As Integer
        Dim j As Integer
' Create the new Spreadsheet; by default it will have
' 10 rows and 10 columns.
      Dim ss As New Spreadsheet
' The next statement will resize the new Spreadsheet to
' 100 cases and 10 variables.
      ss.SetSize(100,10)
' Dimension array x.
      ReDim x(1 To ss.NumberOfCases, _
              1 To ss.NumberOfVariables) As Double
' Fill array x with random numbers.
      For i = 1 To ss.NumberOfCases
       For j=1 To ss.NumberOfVariables
        x(i,j)=Rnd(1)
       Next j
      Next i
' Copy the contents of array x to the new spreadsheet.
      ss.Data=x
' Make the resulting spreadsheet visible.
      ss.Visible=True
' Dimension another array y.
      ReDim y(1 To ss.NumberOfCases, _
              1 To ss.NumberOfVariables) As Double
' With the next line, the entire contents of spreadsheet
' ss is transferred to the array y.
      y=ss.Data
' Alternatively, you could also have used the .Vdata(Variable)
' property, to copy the contents column-by-column, or the
' Cdata(Case) property, to copy the contents row by row.
'
' The following code will add first 10, and then 100 to
' each random number; so the final spreadsheet that will
' be displayed will contain numbers in the range from
' 110 to 111.
      ReDim yVector(1 To ss.NumberOfCases) As Double
      For i=1 To ss.NumberOfVariables
' Copy column i from the spreadsheet to the vector.
        yVector=ss.VData(i)
' Use matrix function MatrixElemAdd to add 10 to each
' element in yVector.
```

```
        MatrixElemAdd (yVector, 10, yVector)
' Copy the contents of the vector to column i of the
' spreadsheet.
        ss.VData(i)=yVector
        Next i

        ReDim yVector(1 To ss.NumberOfVariables)
        For i=1 To ss.NumberOfCases
' Copy row i from the spreadsheet to the vector.
        yVector=ss.CData(i)
' Use matrix function MatrixElemAdd to add 100 to each
' element in yVector.
        MatrixElemAdd (yVector, 100, yVector)
' Copy the contents of the vector to row i of the
' spreadsheet.
        ss.CData(i)=yVector
      Next i
End Sub
```

Note that internally, the `.VData(variable)`, `CData(case)`, and `.Data` properties perform copying operations either from or to a dimensioned array. You can only use these properties as arguments in matrix function calls if those functions do not write values back to the respective input matrix (so you could not, for example, use them directly in the call to the `MatrixElemAdd` function as shown above). Also, the dimensions of the arrays used in conjunction with these properties must match the respective dimensions (row, column, or both) of the spreadsheet.

Changing individual values in a spreadsheet. When using the `Vdata`, `CData`, and `Data` properties it is also important to remember that they will actually "move" data; for example, the `Data` property will move the data from the respective spreadsheet to the specified array. Therefore, when performing operations in a loop, for example in order to update particular cells in the spreadsheet only, it is much more efficient to use the `Value(case,variable)` property.

Creating and Managing Custom Dialogs in *STATISTICA* Visual Basic

Can I Create Custom Dialogs and Other Interactive User Input Controls in SVB?

The *STATISTICA* Visual Basic environment provides all tools to program complete custom user interfaces. A powerful **User-Dialog Editor** is included to design dialogs using simple operations like dragging buttons to the desired locations. Unlike in Microsoft Visual Basic, the user-defined dialogs will be stored along with the program code as data of type **UserDialog**. This method of creating dialogs allows you to create sophisticated user interfaces that can easily be edited in textual form; also, by defining the entire dialog as a variable, you can completely define dialogs inside subroutines, which can be freely moved around the program.

However, user-defined dialogs designed in the *STATISTICA* Visual Basic environment cannot be ported directly to Microsoft Visual Basic, which uses a form-based method of creating dialogs. This is not a serious limitation, though, but rather a design issue, in the sense that you should decide before embarking on the development of a complex program with extensive custom (user-defined) dialogs which environment you would prefer. For example, if you are already familiar with the Microsoft Visual Basic environment, or would like to augment an existing program that was developed in that environment, then there would be clear advantages in staying with that language. To reiterate, virtually all *STATISTICA* Visual Basic functions are accessible from the Visual Basic environments in other applications (such as Microsoft Visual Basic, Microsoft Excel, Microsoft Word, etc.).

The following example illustrates how to create a simple dialog and "service" the user choices on this dialog. Start by creating a new macro: Select **New** from the **File** menu, select the **Macro** tab, and create a new macro called *SimpleDialog*.

Creating the dialog. Next select **Dialog Editor** from the **Tools** menu or click the 📠 **Dialog Editor** toolbar button to display the **UserDialog Editor**.

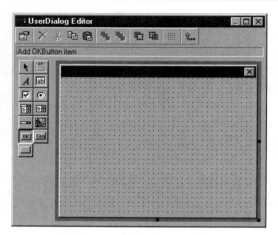

Click the **OK** button on the left side of this dialog, and then click in the upper-right corner of the window; an **OK** button will be inserted in that location.

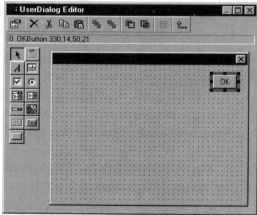

Note that you can further edit the size and location of the **OK** button by clicking on it and dragging it to the desired location, or resizing it. Repeat these steps for the **Cancel** button and for a user-defined button. The latter can be produced by clicking the blank button toolbar button on the lower-left of the toolbar on the **UserDialog**, and then clicking on the desired location in the dialog editor.

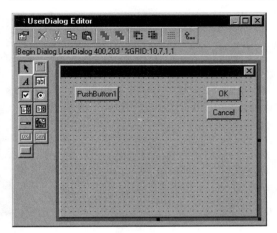

By default, the new button is labeled **Pushbutton1**; to change that, double-click on it to display the **Edit PushButton Properties** dialog. Then edit the fields as follows:

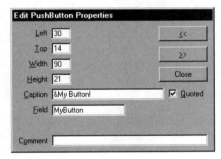

Note that the **Caption** for the new button is **&My Button!** When you click **Close**, the actual caption of the button will change to **My Button!**, i.e., with the **M** underlined. Thus, **M** is the keyboard accelerator that will "click" this button. The other important information in the **Edit PushButton Properties** dialog is the ID of the button, entered into the **Field** box. The ID for this button is **MyButton**; this ID will be referenced throughout the SVB program to service the button, i.e., to identify when the button was pushed, and to respond inside the program.

Now close the **Edit PushButton Properties** dialog, and close the **UserDialog Editor** dialog by clicking on the 🔧 toolbar button in the **UserDialog** editor. The SVB program will now contain the following code.

```
Sub Main
        Begin Dialog UserDialog 400,203 ' %GRID:10,7,1,1
                OKButton 300,14,70,21
                CancelButton 300,42,70,21
                PushButton 30,14,90,21,"&My Button!",.MyButton
```

```
            End Dialog
            Dim dlg As UserDialog
            Dialog dlg
    End Sub
```

Remember that to learn more about the different keywords and statements used in this program so far, you can always select the respective text, and then press F1 to display the *Electronic Manual*, which explains the syntax for the respective keyword or statement and provides simple examples on how to use them.

Servicing the new dialog, simple dialogs. If you run the program created thus far, the dialog we designed will be displayed on the screen; when you click on any button the program will terminate. The next task is to "connect" specific programming instructions to different user actions on the dialog. For example, you could display message boxes to indicate which button the user clicked.

The simplest way to do this is to use the codes returned by the `Dialog` method; this method will return a **0** when **Cancel** is clicked, **-1** for **OK**, and different integers greater than 0 to enumerate the other controls on the dialog. So in this example, clicking the **My Button** button would return a **1**. Here is the program that would service all buttons.

```
    Sub Main
        Begin Dialog UserDialog 400,203 ' %GRID:10,7,1,1
            OKButton 300,14,70,21
            CancelButton 300,42,70,21
            PushButton 30,14,90,21,"&My Button!",.MyButton
        End Dialog
        Dim dlg As UserDialog
      Dim ReturnId As Integer
' NOTE: here we added parenthesis around the argument
' (dlg) for the Dialog method; this allows us to
' retrieve the return code from that method.
      ReturnId = Dialog (dlg)
      Select Case ReturnId
      Case -1
        MsgBox "The OK button was pressed"
      Case 0        '  Cancel Button
        MsgBox "The CANCEL button was pressed"
      Case 1        '  The 'first' button on the Dialog,
        MsgBox "The MY BUTTON was pressed"
    End Select
End Sub
```

This program uses the standard Visual Basic `Select Case` statement to execute the code for the desired message box, based on the ID number returned by the `Dialog` method; you could of course also accomplish the same effect by using `If ... Then ... Else` statements. (All general Visual Basic statements are documented in the general SVB syntax help.)

How Are the Different Controls in Custom Dialogs "Serviced?"

Here is a fairly elaborate example program to illustrate how to service the various elements commonly used in dialogs.

```
' This program demonstrate how one can service various
' controls in a complex dialog.
'
' The following arrays will be used to handle the text
' in the list box and combo box.
Dim ListArray(4) As String
Dim ComboArray(5) As String


' In this program, the Main routine calls a subroutine
' that displays the dialog; note that you can "string together"
' many such subroutines to construct complex programs.
Sub Main
    If DisplayDialog = False Then
        GoTo Finish
    End If
    Exit Sub
Finish:
End Sub


' This function brings up the dialog, and services the various
' controls in the dialog using the DialogFunc (dialog function)
' method. See also the general STATISTICA Visual Basic syntax
' help for additional details.
Function DisplayDialog As Boolean
    DisplayDialog=True
        Begin Dialog UserDialog 390,336,"Demonstating Dialog Controls",

        .MyDialogFunction ' %GRID:10,7,1,1
            OKButton 300,14,70,21,.OkButton
            CancelButton 300,42,70,21,.CancelButton
            PushButton 30,14,160,21,"&Reset to defaults",.MyButton
            Text 30,49,90,14,"&Combo box:",.TextForMyComboBox
```

```
                    ComboBox 130,49,160,70,ComboArray(),.MyComboBox
          Text 30,126,100,28,"&Text box (multiple lines):", _
.TextForTextBox
                    TextBox 130,126,160,35,.MyTextBox,1
                    Text 30,175,90,14,"Value:",.TextForMyTextBox2
                    TextBox 130,168,160,21,.MyTextBox2
                    Text 30,196,60,14,"List box:",.TextForMyListBox
                    DropListBox 130,196,160,91,ListArray(),.MyListBox
                    CheckBox 30,231,130,14,"My Checkbox",.MyCheckBox
                    GroupBox 30,252,180,63,"&Group Box",.MyGroupBox1
                    OptionGroup .MyOptionButtons
                          OptionButton 40,273,150,14,"My Option Button &1"
                          OptionButton 40,294,150,14,"My Option Button &2"
          End Dialog
          Dim dlg As UserDialog
' The following routine will initialize various controls in the
' dialog (fill the combo box, list box, check buttons, etc.)
          InitializeUserDialogFields (dlg)
TryAgain:
' The following line will display the dialog.
          DisplayDialog=Dialog(dlg)
' If the user exited the dialog via OK, then DisplayDialog will
' be set to True; in that case, retrieve the chosen settings for
' the controls, and check for their correctness; if an error
' occurred, display the dialog again.
          If DisplayDialog=True Then
            If RetrieveUserDialogFields (dlg)=0 Then
              MsgBox "Error in the dialog specs; try again.", vbCritical
              GoTo TryAgain
            End If
          End If
End Function


' This is the (private) dialog function that services the UserDialog.
' The name of this function must match the name given to the Dialog
' (see the .MyDialogFunction above).
' Three parameters are passed to the dialog function. Refer to the
' general syntax help for additional details about each parameter.
Private Function MyDialogFunction(DlgItem$, Action%, SuppValue&) As
Boolean

          Select Case Action%

          Case 1 ' Dialog box initialization
          Case 2 ' Value changing or button pressed
' Set the default return code for MyDialogFunction to True; this will
```

```
' mean that the dialog will not close; only on OK and Cancel set this
' value to False, i.e., close the dialog.
                MyDialogFunction = True

                Select Case DlgItem

                Case "MyButton"
' Reset all controls to their initial defaults
                DlgValue "MyListBox", 0
                DlgText "MyComboBox", ComboArray(1)
                DlgText "MyTextBox", _
                        "Initial text for multiple line edit field"
                DlgText "MyTextBox2",".5"
                DlgValue "MyCheckBox", True
                DlgValue "MyOptionButtons", 1
                MyDialogFunction=True

                Case "MyListBox"
                  MsgBox "New combo box selection:" _
                       + ListArray(SuppValue)

                Case "OkButton"
                  MyDialogFunction=False

                Case "CancelButton"
                  MsgBox "The CANCEL button was pressed"
                  MyDialogFunction=False

                End Select

        End Select
End Function

' This subroutine initializes the list box, combo box, text
' fields, etc.
' Note that the entries in list boxes and combo boxes are
' zero-referenced, i.e., the first item in the list is
' referenced as element 0.
Sub InitializeUserDialogFields (dlg)
    ListArray(0)="List entry 0"
    ListArray(1)="List entry 1"
    ListArray(2)="List entry 2"
    ListArray(3)="List entry 3"
    ListArray(4)="List entry 4"
    dlg.MyListBox=0
```

```
        ComboArray(0)="Combobox entry 0"
        ComboArray(1)="Combobox entry 1"
        ComboArray(2)="Combobox entry 2"
        ComboArray(3)="Combobox entry 3"
        ComboArray(4)="Combobox entry 4"
        ComboArray(5)="Combobox entry 5"
        dlg.MyComboBox=ComboArray(1)

        dlg.MyTextBox="Initial text for multiple line edit field"
        dlg.MyTextBox2=".5"

          dlg.MyCheckBox=True

' Note that the option buttons in a group of option buttons are
' zero-referenced, i.e., the first button in the group is referenced
' as element 0; thus here we set the second option button
          dlg.MyOptionButtons=1

End Sub


Function RetrieveUserDialogFields (dlg) As Boolean
On Error GoTo InvalidInput
Dim xval As Double
              RetrieveUserDialogFields=True
        MsgBox "My combo box is set to: " + dlg.MyComboBox
        MsgBox "My multi-line text box is set to: " + dlg.MyTextBox
        xval=CDbl(dlg.MyTextBox2)
        MsgBox "My value was set to: " + Str(xval)
        MsgBox "My (0-referenced) element chosen in My Listbox is: " _
            + Str(dlg.MyListBox)
        MsgBox "My check box is set to: " + Str(dlg.MyCheckBox)
        MsgBox "My options buttons are set to (0-referenced): " _
            + Str(dlg.MyOptionButtons)
        Exit Function
InvalidInput:
        RetrieveUserDialogFields=False
End Function
```

This example program is fairly complex and illustrates how to interact with the different standard Windows dialog controls. Remember that the *Electronic Manual* contains detailed documentation for the statements, declarations, and data types used in this program.

Defining dialogs in subroutines; defining sequences of dialogs. In this example program, the dialog is defined in and displayed by a subroutine (function), rather than the main program. Thus, by defining different dialogs in different subroutines, you can build

---

elaborate programs with complex flow control, i.e., sequences of dialogs that depend on prior user choices or results of computations.

Zero referencing in ListBox, DropListBox, and OptionGroup controls.
Remember that the elements in `ListBox`, `DropListBox`, and `OptionGroup` controls are zero referenced, i.e., the first element is referenced as number 0 (zero), the second as number 1, etc. Note that this is the default way in which arrays are referenced unless `Option Base 1` is set; if you have problems running this program, make sure that `Option Base 1` is not set at the beginning of your program.

Retrieving numeric values. The program also illustrates how standard text controls can be used to return numeric values (see the `.MyTextBox2` control). Specifically, the program retrieves the user entered value as text, and later checks whether the text can be converted into a valid value of type `Double`; this is accomplished by defining an `On Error Goto` label in the `RetrieveUserDialogFields` function, where the program control will resume when an error occurs in the `CDbl` conversion function.

Changing controls from inside the dialog function. When you run the program and click the **Reset to defaults** button (with the dialog ID *"MyButton"*), all fields will be reset to their defaults. This is done via the `DlgText` and `DlgValue` functions, as is illustrated in case *"MyButton"* in the dialog function.

Customizing the Appearance and "Behavior" of *STATISTICA* via *STATISTICA* Visual Basic

What Are Application Events and How Can They Be Controlled from SVB?

An event is an action that is typically performed by a user, such as clicking a mouse button, pressing a key, changing data, or opening a spreadsheet or workbook. In *STATISTICA* certain events are visible to the outside world, i.e., accessible to *STATISTICA* Visual Basic, and they can be used to customize its behavior. Using programmable events you can tailor the program's behavior to the users' needs. Examples of events applications might include:

- building auditing systems into *STATISTICA* (by IT departments),
- building interactive demonstration programs based on workbooks,

---

- building customized user interfaces adhering to specific requirements of a particular application (or a specific company, e.g., to meet specific security requirements).

Events are an important part of the set of tools built into *STATISTICA* to make it a powerful solution building system.

There are three basic types of events: those pertaining to *STATISTICA* documents (e.g., spreadsheets), those pertaining to the *STATISTICA* application as a whole (e.g., opening a spreadsheet), and those pertaining to specific analyses (e.g., **Basic Statistics**, or **2D Scatterplots**).

Document-level events. These events occur for open documents and in some cases, for objects within them. For example, the workbook document object can respond to the **Open**, **New**, and **Print** events; the spreadsheet document can respond to events such as changed data values, or double-clicking on a cell.

Application-level events. These events occur at the level of the application itself, for example, when a new report, graph, or workbook is created. You can create an event handler customizing these actions for all documents.

Analysis-level events. These events occur at the level of a specific analysis, for example, when you close an analysis using the **Basic Statistics** methods, or before an analysis (e.g., **Basic Statistics**) is sending results to a spreadsheet or graph. Note that the term "analysis" in *STATISTICA* denotes one task selected either from the **Statistics** or **Graphs** menu, which can be very small and simple (e.g., one scatterplot requested from the **Graphs** menu) or very elaborate (e.g., a complex structural equation modeling analysis selected by choosing that option from the **Statistics** menu and involving hundreds of output documents). A typical application using analysis-level events would be to implement custom handling of output, for example, to send all graphs and results spreadsheets to another application.

A simple example. Here is a very simple example of how a *STATISTICA* Visual Basic program can be connected to the application-level event of starting a new analysis. Start by selecting **View code** from the **Tools - Macro - Application events** menu. A Visual Basic editor will open; in the **Object** box of that editor, select **Document**, and then select **AnalysisNew** in the **Proc** box. Enter the simple **MsgBox** instruction as shown below.

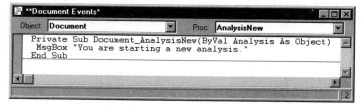

Next select **Autorun** from the **Tools - Macro - Application events** menu and close the program editor. Every time that you start a new analysis (e.g., select **Basic Statistics** from the **Statistics** menu), the message **You are starting a new analysis** will be displayed. In fact, you can now exit the program, respond **Yes** when prompted to save the documents events, and the next time that you start *STATISTICA* and select an analysis, the message will be displayed. To turn off this program, deselect **Autorun** from the **Tools - Macro - Application events** menu.

How Can I Create a Cell-Function Spreadsheet Using Spreadsheet Events?

One of the most basic functions of the designated spreadsheet software (such as Microsoft Excel) is to automatically recompute cells in the datafile when any of the input data (cells) are changed. For example, you can set up a complex budget for a project so that when you change the values for particular budget items, the entire budget will be recomputed based on the newly supplied values.

The same functionality can be programmed into *STATISTICA* Spreadsheets by attaching an SVB macro to certain spreadsheet events as demonstrated in this example. Note that practically all spreadsheet (and other) events can be customized, thus providing the tools to build very sophisticated and highly customized automated data operations right "into" the spreadsheet.

How is the macro created? First, create a datafile and set up the necessary cells. Datafile **CellFunctionDemo.sta** (available in the Examples\Macros directory which can be found in the directory in which you installed *STATISTICA*) contains prices for various items on a holiday shopping list.

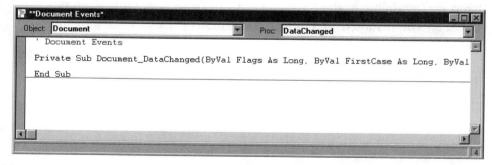

Note that in the spreadsheet shown, the *Item Cost* and *Coupon* variables are data input variables and the *Final Item Cost* and *Total Cost of All Items* variables are derived or computed variables.

Entering the computations for derived cells (programming the DataChanged event). After entering the basic information, using the *View* menu, select *Events - View Code*. This displays the SVB program editor for document level events (i.e., events that apply to the newly created spreadsheet document). In the *Object* box of the SVB editor (*Document Events* dialog), select *Document*; in the *Proc* box, select the *DataChanged* event. Shown below is the appearance of the SVB editor after the *DataChanged* subroutine has been created in the editor.

```
** Document Events*
Object: Document                              Proc: DataChanged
    ' Document Events
    Private Sub Document_DataChanged(ByVal Flags As Long, ByVal FirstCase As Long, ByVal
    End Sub
```

Now type in the following program into the SVB editor (you can, of course, omit the comments).

```
Private Sub Document_DataChanged(ByVal Flags As Long, _
    ByVal FirstCase  As Long, ByVal FirstVar As Long, _
    ByVal LastCase As Long, ByVal LastVar  As Long, _
    ByVal bLast As Boolean)
```

```
' Only process the data if there was a change in
' the data area of the spreadsheet.
    If (Flags And scNotifyDCData) Then
    If FirstVar = 3 Or FirstVar = 4 Or _
     LastVar = 4 Or LastVar  = 4 Then
     MsgBox "These are derived fields And they cannot be edited."
    End If

    Const V1 As Long = 1
    Const V2 As Long = 2
    Const VDest As Long = 3
    Const VResult As Long = 4
    Dim s As Spreadsheet

    Set s = ActiveSpreadsheet
' We need to recalculate first to update the final cost of each item;
' we only need to do so for the range of cases that have changed.
    Dim j As Long
    For j = FirstCase To LastCase
     Dim x1 As Double
     Dim x2 As Double
' If the source data is missing data, then substitute 0.
     If (s.MissingData(j, V1)) Then
      x1 = 0
     Else
      x1 = s.Value(j,V1)
     End If
' If the source data is missing data, then substitute 0.
     If (s.MissingData(j,V2)) Then
      x2 = 0
     Else
      x2 = s.Value(j,V2)
     End If
' Calculate new destination variable
      s.Value(j, VDest) = x1 - x2
     Next j
    Dim i As Long
    Dim TotalVal As Double
' Iterate through each cell in variable 3 and add it to
' TotalVal.
    For i = 1 To s.NumberOfCases
       TotalVal = TotalVal + s.Value(i, VDest)
    Next i
' Update the cell's value to reflect total cost of all items.
     s.Value(s.NumberOfCases, VResult) = TotalVal
    End If
End Sub
```

This macro defines the computations for the cells in the spreadsheet that will be performed every time the data in the input variables are changed.

Write protecting the derived cells. Also, we want to make sure that certain cells are "protected," i.e., users should not be able to type values into the cells that are derived (by computation from other cells). Some of that protection is already implemented in the macro shown above, which checks whether or not the user attempted to type a value into a derived cell. For this example, let us also "catch" the *BeforeDoubleClick* event for the cells in the third and fourth variables of our example spreadsheet. Select in the **Proc** field of the SVB editor the **BeforeDoubleClick** event, and then enter the code as shown below:

```
Private Sub Document_BeforeDoubleClick(ByVal Flags As Long, ByVal
CaseNo As Long, ByVal VarNo As Long, Cancel As Boolean)
   If VarNo = 3 Or VarNo = 4 Then
MsgBox  "These are derived fields and they cannot be edited."
      Cancel = True
   End If
End Sub
```

Saving the spreadsheet and *AutoRun*. Finally, before saving the macro and the datafile, click on the data spreadsheet once more, and using the **View** menu select **Events - Autorun**.

This will cause the new macro to run automatically every time you open the data spreadsheet. Next, save the spreadsheet and run the macro. You are now ready to compute your holiday shopping budget using your customized spreadsheet. If you try to "cheat" by double-clicking on one of the computed fields to enter a (lower) total value, a message will be displayed.

This simple example illustrates how you could build very sophisticated "cost models" that can also include dialogs, automatic analyses with the *STATISTICA* statistical or graphics functions, or any of the more than 10,000 automation functions available in the *STATISTICA* system, thus vastly expanding the functionality of ordinary spreadsheets.

Can I Customize the Toolbars and Menus via SVB?

STATISTICA provides full access to all of its customization options via *STATISTICA* Visual Basic functions. A detailed discussion of the techniques available for customizing toolbars, menus, etc., and how to use these methods to fully customize or expand *STATISTICA* is beyond the scope of this introduction. There are many excellent books available that discuss in detail the `CommandBars` object and how to use it in (Microsoft) Windows applications (e.g., to customize Microsoft Office applications). In general, customizing toolbars, menus, pop-up menus, etc. is useful in order to:

- Provide quick access to frequently used options or methods;

- Program highly customized events; e.g., to customize the shortcut menu when the user clicks on certain objects by adding macro programs or links to other applications to that menu;

- Program fully customized versions of *STATISTICA* dedicated to perform only few predefined tasks (e.g., a version for data entry only);

- Add new options to *STATISTICA* developed by other vendors; those options could be invoked from a *STATISTICA* Visual Basic macro program, if they are accessible as reference libraries.

Example. The following example illustrates how to produce a custom toolbar, and how to attach to them particular events. After running this program, a new toolbar will be added to your installation of *STATISTICA*.

```
' This program illustrates how to add a custom toolbar
' to your STATISTICA program. Use the Tools
' menu option Customize and the Toolbar tab to remove
' the custom toolbar from your application, after you
' added it via this program (or write another program
' that will remove the toolbar via the CommandBar.Delete
' method).
'
' The following identifiers are specific to STATISTICA
```

```
' and its commands; you can find a complete list of these
' identifiers in file Menus.h. They define the command ID's
' that will launch Basic Statistics, Regression,
' Nonparametrics, and ANOVA.
Const ID_ANALYSIS_BASICSTATISTICS As Long = 10301
Const ID_ANALYSIS_REGRESSION As Long        = 10302
Const ID_ANALYSIS_NONPARAMETRICS As Long    = 10304
Const ID_ANALYSIS_ANOVA As Long             = 10323
' This constant is only locally defined for this macro
' program.
Const ID_RUNOTHERS As Long = 30000

Sub Main
' Retrieve the CommandBars object for the current
' application; note that CommandBars is a property
' of the Application object, so when running a VB
' program from another application, you need to explicitly
' declare the STATISTICA application object, and CommandBars
' as a property of that object.
    Dim bars As CommandBars
    Set bars = CommandBarOptions.CommandBars(scBarTypeToolBar)
    Dim newBar As CommandBar

' The new toolbar will be called "Custom"
    Set newBar = bars.Add("Custom")
' Insert the following toolbar options in the toolbar
    newBar.InsertButton 1, ID_ANALYSIS_BASICSTATISTICS , , _
        "Basic Stats"
    newBar.InsertButton 2, ID_ANALYSIS_NONPARAMETRICS, , _
        "Nonparametrics"
' Insert a vertical separator line.
    newBar.InsertSeparator 3
' Make this a floating toolbar.
    newBar.Position = scBarFloating
' Set the display mode.
    newBar.Item(1).DisplayMode=scCommandDisplayTextAndImage
    newBar.Item(2).DisplayMode=scCommandDisplayTextAndImage

' Make a pop-up menu item on the new toolbar, labeled "Others"
    Dim menu As CommandBarItem
    Set menu = newBar.InsertPopupMenu(4, "Others")
' Insert the following menu option under the pop-up menu "Others"
    menu.Items.InsertButton 1, ID_ANALYSIS_REGRESSION, , "Regression"
    menu.Items.InsertButton 2, ID_ANALYSIS_ANOVA , , "ANOVA"
    menu.Items.Item(1).DisplayMode=scCommandDisplayTextAndImage
    menu.Items.Item(2).DisplayMode=scCommandDisplayTextAndImage
```

```
' Insert as a third option in "Others" a button that will run a
' macro program when selected. That macro program could, for example,
' run another application, or perform other customizations.
    menu.Items.InsertMacroButton 3, "j:\erase\svb\1.svb"
    menu.Items.Item(3).Caption="My Macro"
End Sub
```

This program will create the following toolbar.

To remove the new toolbar, select **Customize** from the **Tools** menu, and then select the **Toolbars** tab; select the new toolbar, and **Delete** it. (You could also write an SVB program to remove the new toolbar, using the `CommandBar.Delete` method.)

TECHNICAL NOTE: CHANGING APPLICATION-LEVEL EVENTS (A FLEXIBLE WAY TO CUSTOMIZE *STATISTICA*)

An event is an action that is typically performed by users, such as clicking a mouse button, pressing a key, changing data, or opening a spreadsheet or workbook. In *STATISTICA*, certain events are directly accessible to customization, and you can change the default behavior of the program when the specific events are triggered (or when the events are triggered in specific parts of the documents). Using programmable events offers a particularly flexible and "far reaching" method to tailor the program behavior to your specific needs.

For example, you can prevent another user from saving changes made to a file, (see *Can I Write Protect a Workbook File?* in Chapter 3 – *STATISTICA Workbooks*) or make specific sections of a spreadsheet dependent solely on other (e.g., automatically recalculated) values and protect them from being changed directly by editing the specific cells. You could also create a backup file of a particular workbook to a specific location whenever you save that workbook (so you wouldn't have to save it twice). Those are just a few examples of countless customizations that can easily be accomplished by customizing events.

Event Types

Events are part of a set of tools built into *STATISTICA* to make it a powerful solution building system. There are three different types of events: Document-level events, Application-level events, and Analysis-level events.

Document-level events. Document-level events occur only for open documents and in some cases, for objects within them. They allow you to customize the behavior of open documents.

Application-level events. Application-level events occur at the level of the application itself. They allow you to customize the behavior of all documents of a certain type (such as all spreadsheets, or all workbooks, or all reports, etc.). For example, you could customize *STATISTICA* so that all reports are automatically saved to a backup folder or all spreadsheets are automatically saved (without prompting) before they are closed.

Analysis-level events. These events occur at the level of a specific analysis, for example, when you close an analysis using the *Basic Statistics* methods, or before an analysis (e.g., *Basic Statistics*) is sending results to spreadsheet or graph. A typical application using analysis-level events would be to implement custom-handling of output, for example, to send all graphs and results spreadsheets to another application.

Application-Level Event Commands

The following table gives the application-level events that are available in *STATISTICA*. Document-level and analysis-level event commands are discussed in the respective document chapters (workbooks in Chapter 3 – *STATISTICA Workbooks*, etc.). For more information on events, see *What Are Application Events and How Can They Be Controlled from STATISTICA Visual Basic?* (page 1015) and the examples in the *Electronic Manual*.

Command	Action
AnalysisActivate	Executes when a *STATISTICA* analysis within the *STATISTICA* environment has received the focus.
AnalysisBeforeClose	Executes when a *STATISTICA* analysis within *STATISTICA* is about to be closed.
AnalysisBeforeOutput	Executes when an analysis within *STATISTICA* is about to be processed.
AnalysisDeactivate	Executes when an analysis has lost the focus to another window within *STATISTICA*.
AnalysisNew	Executes when *STATISTICA* creates a new analysis.
AnalysisOnClose	Executes when *STATISTICA* closes an analysis.
GraphActivate	Executes when a *STATISTICA* graph within *STATISTICA* has received the focus.
GraphBeforeClose	Executes when a *STATISTICA* graph within *STATISTICA* is about to be closed.
GraphBeforeDoubleClick	Executes when you double-click anywhere on a graph within *STATISTICA*.
GraphBeforePrint	Executes when you print a graph within *STATISTICA*.

GraphBeforeRightClick	Executes when you right-click anywhere on a graph within *STATISTICA*.
GraphBeforeSave	Executes when you save changes made to the graph within *STATISTICA*.
GraphDeactivate	Executes when a graph has lost the focus to another window within *STATISTICA*.
GraphOnClose	Executes when *STATISTICA* closes a graph.
GraphOpen	Executes when *STATISTICA* opens a graph.
GraphSelectionChanged	Executes when the focus within a Graph in *STATISTICA* has moved.
OnActiveDataSet	Executes whenever the program makes reference to the `ActiveDataSet`; this event handler would, for example, be useful to automatically replace in an SVB program references to an `ActiveDataSet` with a custom routine.
OnClose	Executes when *STATISTICA* is being closed.
OnInit	Executes when *STATISTICA* is opening (initialized).
ReportActivate	Executes when a *STATISTICA* Report within the *STATISTICA* environment has received the focus.
ReportBeforePrint	Executes when you print a report within *STATISTICA*.
ReportBeforeRightClick	Executes when you right-click anywhere on a report within *STATISTICA*.
ReportBeforeSave	Executes when you save changes made to the report within *STATISTICA*.
ReportDeactivate	Executes when a report has lost the focus to another window within *STATISTICA*.
ReportNew	Executes when *STATISTICA* creates a new report.
ReportOnClose	Executes when *STATISTICA* closes a report.
ReportOpen	Executes when *STATISTICA* opens a report.
ReportSelectionChanged	Executes when the focus within a report in *STATISTICA* has moved.

SpreadsheetActivate	Executes when a *STATISTICA* Spreadsheet within *STATISTICA* has received the focus.
SpreadsheetBeforeClose	Executes when a *STATISTICA* Spreadsheet within *STATISTICA* is about to be closed.
SpreadsheetBeforeDoubleClick	Executes when you double-click anywhere on a spreadsheet within *STATISTICA*.
SpreadsheetBeforePrint	Executes when you print a spreadsheet within *STATISTICA*.
SpreadsheetBeforeRightClick	Executes when you right-click anywhere on the spreadsheet within *STATISTICA*.
SpreadsheetBeforeSave	Executes when you save changes made to the spreadsheet within *STATISTICA*.
SpreadsheetDataChanged	Executes when data within a cell or block of cells of a spreadsheet within *STATISTICA* has been altered. This will also execute when a variable's Data type or Display format has changed.
SpreadsheetDeactivate	Executes when a spreadsheet has lost the focus to another window within *STATISTICA*.
SpreadsheetNew	Executes when *STATISTICA* creates a new spreadsheet.
SpreadsheetOnClose	Executes when *STATISTICA* closes a spreadsheet.
SpreadsheetOpen	Executes when *STATISTICA* opens a spreadsheet.
SpreadsheetSelectionChange	Executes when the focus within a spreadsheet in *STATISTICA* has moved.
SpreadsheetStructureChanged	Executes when the size of a spreadsheet within *STATISTICA* is changed.
WorkbookActivate	Executes when a *STATISTICA* Workbook within the *STATISTICA* environment has received the focus.
WorkbookBeforeClose	Executes when you close a workbook in *STATISTICA*.
WorkbookBeforePrint	Executes when you print a workbook in *STATISTICA*.
WorkbookBeforeRightClick	Executes when you right-click anywhere on a workbook in *STATISTICA*.

WorkbookBeforeSave	Executes when you save changes made to a workbook in *STATISTICA*.
WorkbookDeactivate	Executes when a workbook in *STATISTICA* has lost the focus to another window.
WorkbookNew	Executes when *STATISTICA* creates a new workbook.
WorkbookOnClose	Executes when a workbook in *STATISTICA* is being closed.
WorkbookOpen	Executes when *STATISTICA* opens a workbook.
WorkbookSelectionChanged	Executes when the focus within a workbook in *STATISTICA* has moved.

A
APPENDIX

REFERENCES

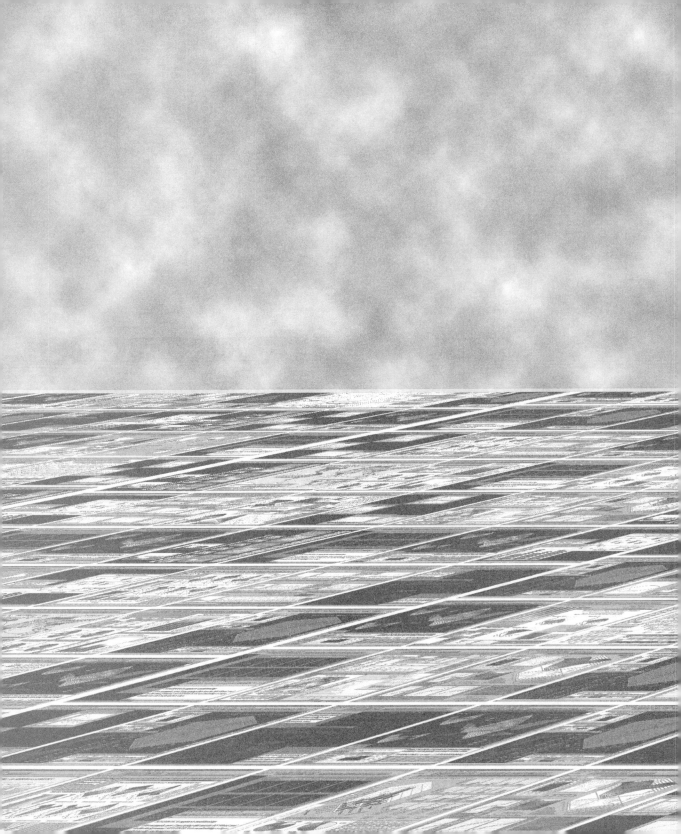

REFERENCES

A

Abraham, B., & Ledolter, J. (1983). *Statistical methods for forecasting*. New York: Wiley.

Adorno, T. W., Frenkel-Brunswik, E., Levinson, D. J., & Sanford, R. N. (1950). *The authoritarian personality*. New York: Harper.

Agresti, Alan (1996). *An Introduction to Categorical Data Analysis*. New York: Wilely.

Akaike, H. (1973). Information theory and an extension of the maximum likelihood principle. In B. N. Petrov and F. Csaki (Eds.), *Second International Symposium on Information Theory*. Budapest: Akademiai Kiado.

Akaike, H. (1983). Information measures and model selection. *Bulletin of the International Statistical Institute: Proceedings of the 44th Session, Volume 1*. Pages 277-290.

Aldrich, J. H., & Nelson, F. D. (1984). *Linear probability, logit, and probit models*. Beverly Hills, CA: Sage Publications.

Almon, S. (1965). The distributed lag between capital appropriations and expenditures. *Econometrica, 33*, 178-196.

American Supplier Institute (1984-1988). *Proceedings of Supplier Symposia on Taguchi Methods*. (April, 1984; November, 1984; October, 1985; October, 1986; October, 1987; October, 1988), Dearborn, MI: American Supplier Institute.

Anderson, O. D. (1976). *Time series analysis and forecasting*. London: Butterworths.

Anderson, S. B., & Maier, M. H. (1963). 34,000 pupils and how they grew. *Journal of Teacher Education, 14*, 212-216.

Anderson, T. W. (1958). *An introduction to multivariate statistical analysis*. New York: Wiley.

Anderson, T. W. (1984). *An introduction to multivariate statistical analysis* (2nd ed.). New York: Wiley.

Anderson, T. W., & Rubin, H. (1956). Statistical inference in factor analysis. *Proceedings of the Third Berkeley Symposium on Mathematical Statistics and Probability*. Berkeley: The University of California Press.

Andrews, D. F. (1972). Plots of high-dimensional data. *Biometrics, 28*, 125-136.

ASQC/AIAG (1990). *Measurement systems analysis reference manual*. Troy, MI: AIAG.

ASQC/AIAG (1991). *Fundamental statistical process control reference manual*. Troy, MI: AIAG.

AT&T (1956). *Statistical quality control handbook, Select code 700-444*. Indianapolis, AT&T Technologies.

Auble, D. (1953). Extended tables for the Mann-Whitney statistic. *Bulletin of the Institute of Educational Research, Indiana University, 1*, No. 2.

B

Bagozzi, R. P., & Yi, Y. (1989). On the use of structural equation models in experimental design. *Journal of Marketing Research, 26,* 271-284.

Bagozzi, R. P., Yi, Y., & Singh, S. (1991). On the use of structural equation models in experimental designs: Two extensions. *International Journal of Research in Marketing, 8,* 125-140.

Bailey, A. L. (1931). The analysis of covariance. *Journal of the American Statistical Association, 26,* 424-435.

Bails, D. G., & Peppers, L. C. (1982). *Business fluctuations: Forecasting techniques and applications.* Englewood Cliffs, NJ: Prentice-Hall.

Bain, L. J. (1978). *Statistical analysis of reliability and life-testing models.* New York: Decker.

Bain, L. J. and Engelhart, M. (1989) *Introduction to Probability and Mathematical Statistics.* Kent, MA: PWS.

Baird, J. C. (1970). *Psychophysical analysis of visual space.* New York: Pergamon Press.

Baird, J. C., & Noma, E. (1978). *Fundamentals of scaling and psychophysics.* New York: Wiley.

Barcikowski, R., & Stevens, J. P. (1975). A Monte Carlo study of the stability of canonical correlations, canonical weights, and canonical variate-variable correlations. *Multivariate Behavioral Research, 10,* 353-364.

Barker, T. B. (1986). Quality engineering by design: Taguchi's philosophy. *Quality Progress, 19,* 32-42.

Barlow, R. E., & Proschan, F. (1975). *Statistical theory of reliability and life testing.* New York: Holt, Rinehart, & Winston.

Barlow, R. E., Marshall, A. W., & Proschan, F. (1963). Properties of probability distributions with monotone hazard rate. *Annals of Mathematical Statistics, 34,* 375-389.

Barnard, G. A. (1959). Control charts and stochastic processes. *Journal of the Royal Statistical Society,* Ser. B, *21,* 239.

Bartholomew, D. J. (1984). The foundations of factor analysis. *Biometrika, 71,* 221-232.

Bates, D. M., & Watts, D. G. (1988). *Nonlinear regression analysis and its applications.* New York: Wiley.

Bayne, C. K., & Rubin, I. B. (1986). *Practical experimental designs and optimization methods for chemists.* Deerfield Beach, FL: VCH Publishers.

Becker, R. A., Denby, L., McGill, R., & Wilks, A. R. (1986). Datacryptanalysis: A case study. *Proceedings of the Section on Statistical Graphics, American Statistical Association,* 92-97.

Belsley, D. A., Kuh, E., and Welsch, R. E. (1980). *Regression Diagnostics.* New York: Wiley.

Bendat, J. S. (1990). *Nonlinear system analysis and identification from random data.* New York: Wiley.

Bentler, P. M, & Bonett, D. G. (1980). Significance tests and goodness of fit in the analysis of covariance structures. *Psychological Bulletin, 88,* 588-606.

Bentler, P. M. (1986). Structural modeling and Psychometrika: A historical perspective on growth and achievements. *Psychometrika, 51,* 35-51.

Bentler, P. M. (1989). *EQS Structural equations program manual.* Los Angeles, CA: BMDP Statistical Software.

Bentler, P. M., & Weeks, D. G. (1979). Interrelations among models for the analysis of moment structures. *Multivariate Behavioral Research, 14,* 169-185.

Bentler, P. M., & Weeks, D. G. (1980). Linear structural equations with latent variables. *Psychometrika, 45,* 289-308.

Benzécri, J. P. (1973). *L'Analyse des Données: T. 2, I' Analyse des correspondances.* Paris: Dunod.

Berkson, J. (1944). Application of the Logistic Function to Bio-Assay. *Journal of the American Statistical Association, 39,* 357-365.

Berkson, J., & Gage, R. R. (1950). The calculation of survival rates for cancer. *Proceedings of Staff Meetings, Mayo Clinic, 25,* 250.

Bhote, K. R. (1988). *World class quality.* New York: AMA Membership Publications.

Binns, B., & Clark, N. (1986). The graphic designer's use of visual syntax. *Proceedings of the Section on Statistical Graphics, American Statistical Association*, 36-41.

Birnbaum, Z. W. (1952). Numerical tabulation of the distribution of Kolmogorov's statistic for finite sample values. *Journal of the American Statistical Association*, 47, 425-441.

Birnbaum, Z. W. (1953). Distribution-free tests of fit for continuous distribution functions. *Annals of Mathematical Statistics*, 24, 1-8.

Bishop, C. (1995). *Neural Networks for Pattern Recognition*. Oxford: University Press.

Bishop, Y. M. M., Fienberg, S. E., & Holland, P. W. (1975). *Discrete multivariate analysis*. Cambridge, MA: MIT Press.

Bjorck, A. (1967). Solving linear least squares problems by Gram-Schmidt orthonormalization. *Bit*, 7, 1-21.

Blackman, R. B., & Tukey, J. (1958). *The measurement of power spectral from the point of view of communication engineering*. New York: Dover.

Blackwelder, R. A. (1966). *Taxonomy: A text and reference book*. New York: Wiley.

Blalock, H. M. (1972). *Social statistics* (2nd ed.). New York:McGraw-Hill

Bliss, C. I. (1934). The method of probits. *Science*, 79, 38-39.

Bloomfield, P. (1976). *Fourier analysis of time series: An introduction*. New York: Wiley.

Bock, R. D. (1963). Programming univariate and multivariate analysis of variance. *Technometrics*, 5, 95-117.

Bock, R. D. (1975). *Multivariate statistical methods in behavioral research*. New York: McGraw-Hill.

Bolch, B.W., & Huang, C. J. (1974). *Multivariate statistical methods for business and economics*. Englewood Cliffs, NJ: Prentice-Hall.

Bollen, K. A. (1989). *Structural equations with latent variables*. New York: John Wiley & Sons.

Borg, I., & Lingoes, J. (1987). *Multidimensional similarity structure analysis*. New York: Springer.

Borg, I., & Shye, S. (in press). *Facet Theory*. Newbury Park: Sage.

Bouland, H. and Kamp, Y. (1988). Auto-association by multilayer perceptrons and singular value decomposition. *Biological Cybernetics 59*, 291-294.

Bowker, A. G. (1948). A test for symmetry in contingency tables. *Journal of the American Statistical Association*, 43, 572-574.

Bowley, A. L. (1897). Relations between the accuracy of an average and that of its constituent parts. *Journal of the Royal Statistical Society*, 60, 855-866.

Bowley, A. L. (1907). *Elements of Statistics*. London: P. S. King and Son.

Box, G. E. P. (1953). Non-normality and tests on variances. *Biometrika*, 40, 318-335.

Box, G. E. P. (1954a). Some theorems on quadratic forms applied in the study of analysis of variance problems: I. Effect of inequality of variances in the one-way classification. *Annals of Mathematical Statistics*, 25, 290-302.

Box, G. E. P. (1954b). Some theorems on quadratic forms applied in the study of analysis of variance problems: II. Effect of inequality of variances and of correlation of errors in the two-way classification. *Annals of Mathematical Statistics*, 25, 484-498.

Box, G. E. P., & Anderson, S. L. (1955). Permutation theory in the derivation of robust criteria and the study of departures from assumptions. *Journal of the Royal Statistical Society*, 17, 1-34.

Box, G. E. P., & Behnken, D. W. (1960). Some new three level designs for the study of quantitative variables. *Technometrics*, 2, 455-475.

Box, G. E. P., & Cox, D. R. (1964). An analysis of transformations. *Journal of the Royal Statistical Society*, 26, 211-253.

Box, G. E. P., & Cox, D. R. (1964). An analysis of transformations. *Journal of the Royal Statistical Society, B26*, 211-234.

Box, G. E. P., & Draper, N. R. (1987). *Empirical model-building and response surfaces*. New York: Wiley.

Box, G. E. P., & Jenkins, G. M. (1970). *Time series analysis*. San Francisco: Holden Day.

Box, G. E. P., & Jenkins, G. M. (1976). *Time series analysis: Forecasting and control*. San Francisco: Holden-Day.

Box, G. E. P., & Muller, M. E. (1958). A note on the generation of random normal deviates. *Annals of Mathematical Statistics*, 29, 610=611. Inversion method (Muller, 1959)

Box, G. E. P., & Tidwell, P. W. (1962). Transformation of the independent variables. *Technometrics, 4*, 531-550.

Box, G. E. P., & Wilson, K. B. (1951). On the experimental attainment of optimum conditions. *Journal of the Royal Statistical Society*, Ser. B, *13*, 1-45.

Box, G. E. P., Hunter, W. G., & Hunter, S. J. (1978). *Statistics for experimenters: An introduction to design, data analysis, and model building*. New York: Wiley.

Breiman, L., Friedman, J. H., Olshen, R. A., & Stone, C. J. (1984). *Classification and regression trees*. Monterey, CA: Wadsworth & Brooks/Cole Advanced Books & Software.

Brenner, J. L., et al. (1968). Difference equations in forecasting formulas. *Management Science, 14*, 141-159.

Brent, R. F. (1973). *Algorithms for minimization without derivatives*. Englewood Cliffs, NJ: Prentice-Hall.

Breslow, N. E. (1970). A generalized Kruskal-Wallis test for comparing *K* samples subject to unequal pattern of censorship. *Biometrika, 57*, 579-594.

Breslow, N. E. (1974). Covariance analysis of censored survival data. *Biometrics, 30*, 89-99.

Bridle, J.S. (1990). Probabilistic interpretation of feedforward classification network outputs, with relationships to statistical pattern recognition. In F. Fogelman Soulie and J. Herault (Eds.), *Neurocomputing: Algorithms, Architectures and Applications*, 227-236. New York: Springer-Verlag.

Brigham, E. O. (1974). *The fast Fourier transform*. Englewood Cliffs, NJ: Prentice-Hall.

Brillinger, D. R. (1975). *Time series: Data analysis and theory*. New York: Holt, Rinehart. & Winston.

Broomhead, D.S. and Lowe, D. (1988). Multivariable functional interpolation and adaptive networks. *Complex Systems 2*, 321-355.

Brown, D. T. (1959). A note on approximations to discrete probability distributions. *Information and Control, 2*, 386-392.

Brown, M. B., & Forsythe, A. B. (1974). Robust tests for the equality of variances. *Journal of the American Statistical Association, 69*, 264-267.

Brown, R. G. (1959). *Statistical forecasting for inventory control*. New York: McGraw-Hill.

Browne, M. W. (1968). A comparison of factor analytic techniques. *Psychometrika, 33*, 267-334.

Browne, M. W. (1974). Generalized least squares estimators in the analysis of covariance structures. *South African Statistical Journal, 8*, 1-24.

Browne, M. W. (1982). Covariance Structures. In D. M. Hawkins (Ed.) *Topics in Applied Multivariate Analysis*. Cambridge, MA: Cambridge University Press.

Browne, M. W. (1984). Asymptotically distribution free methods for the analysis of covariance structures. *British Journal of Mathematical and Statistical Psychology, 37*, 62-83.

Browne, M. W., & Cudeck, R. (1990). Single sample cross-validation indices for covariance structures. *Multivariate Behavioral Research, 24*, 445-455.

Browne, M. W., & Cudeck, R. (1992). Alternative ways of assessing model fit. In K. A. Bollen and J. S. Long (Eds.), *Testing structural equation models*. Beverly Hills, CA: Sage.

Browne, M. W., & DuToit, S. H. C. (1982). AUFIT (Version 1). A computer programme for the automated fitting of nonstandard models for means and covariances. Research Finding WS-27. Pretoria, South Africa: Human Sciences Research Council.

Browne, M. W., & DuToit, S. H. C. (1987). Automated fitting of nonstandard models. Report WS-39. Pretoria, South Africa: Human Sciences Research Council.

Browne, M. W., & DuToit, S. H. C. (1992). Automated fitting of nonstandard models. *Multivariate Behavioral Research, 27*, 269-300.

Browne, M. W., & Mels, G. (1992). *RAMONA User's Guide.* The Ohio State University: Department of Psychology.

Browne, M. W., & Shapiro, A. (1989). Invariance of covariance structures under groups of transformations. Research Report 89/4. Pretoria, South Africa: University of South Africa Department of Statistics.

Browne, M. W., & Shapiro, A. (1991). Invariance of covariance structures under groups of transformations. *Metrika*, *38*, 335-345.

Browne, M.W., & Cudeck, R. (1993). Alternative ways of assessing model fit. In K. A. Bollen & J. S. Long, (Eds.), *Testing structural equation models.* Beverly Hills, CA: Sage.

Brownlee, K. A. (1960). *Statistical Theory and Methodology in Science and Engineering.* New York: John Wiley.

Buffa, E. S. (1972). *Operations management: Problems and models* (3rd. ed.). New York: Wiley.

Buja, A., & Tukey, P. A. (Eds.) (1991). *Computing and Graphics in Statistics.* New York: Springer-Verlag.

Buja, A., Fowlkes, E. B., Keramidas, E. M., Kettenring, J. R., Lee, J. C., Swayne, D. F., & Tukey, P. A. (1986). Discovering features of multivariate data through statistical graphics. *Proceedings of the Section on Statistical Graphics, American Statistical Association*, 98-103.

Burman, J. P. (1979). Seasonal adjustment - a survey. *Forecasting, Studies in Management Science, 12*, 45-57.

Burns, L. S., & Harman, A. J. (1966). *The complex metropolis, Part V of profile of the Los Angeles metropolis: Its people and its homes.* Los Angeles: University of Chicago Press.

Burt, C. (1950). The factorial analysis of qualitative data. *British Journal of Psychology, 3*, 166-185.

C

Campbell D. T., & Fiske, D. W. (1959). Convergent and discriminant validation by the multitrait-multimethod matrix. *Psychological Bulletin*, *56*, 81-105

Carmines, E. G., & Zeller, R. A. (1980). *Reliability and validity assessment.* Beverly Hills, CA: Sage Publications.

Carrol, J. D., Green, P. E., and Schaffer, C. M. (1986). Interpoint distance comparisons in correspondence analysis. *Journal of Marketing Research, 23*, 271-280.

Carroll, J. D., & Wish, M. (1974). Multidimensional perceptual models and measurement methods. In E. C. Carterette and M. P. Friedman (Eds.), *Handbook of perception.* (Vol. 2, pp. 391-447). New York: Academic Press.

Cattell, R. B. (1966). The scree test for the number of factors. *Multivariate Behavioral Research, 1*, 245-276.

Cattell, R. B., & Jaspers, J. A. (1967). A general plasmode for factor analytic exercises and research. *Multivariate Behavioral Research Monographs.*

Chambers, J. M., Cleveland, W. S., Kleiner, B., & Tukey, P. A. (1983). *Graphical methods for data analysis.* Bellmont, CA: Wadsworth.

Chan, L. K., Cheng, S. W., & Spiring, F. (1988). A new measure of process capability: Cpm. *Journal of Quality Technology, 20*, 162-175.

Chen, J. (1992). Some results on 2(nk) fractional factorial designs and search for minimum aberration designs. *Annals of Statistics, 20*, 2124-2141.

Chen, J., & Wu, C. F. J. (1991). Some results on s(nk) fractional factorial designs with minimum aberration or optimal moments. *Annals of Statistics, 19*, 1028-1041.

Chen, J., Sun, D. X., & Wu, C. F. J. (1993). A catalog of two-level and three-level fractional factorial designs with small runs. *International Statistical Review, 61*, 131-145.

Chernoff, H. (1973). The use of faces to represent points in k-dimensional space graphically. *Journal of American Statistical Association, 68*, 361-368.

Christ, C. (1966). *Econometric models and methods.* New York: Wiley.

Clarke, G. M., & Cooke, D. (1978). *A basic course in statistics*. London: Edward Arnold.

Clements, J. A. (1989). Process capability calculations for non-normal distributions. *Quality Progress*. September, 95-100.

Cleveland, W. S. (1979). Robust locally weighted regression and smoothing scatterplots. *Journal of the American Statistical Association*, 74, 829-836.

Cleveland, W. S. (1984). Graphs in scientific publications. *The American Statistician*, 38, 270-280.

Cleveland, W. S. (1985). *The elements of graphing data*. Monterey, CA: Wadsworth.

Cleveland, W. S. (1993). *Visualizing data*. Murray Hill, NJ: AT&T.

Cleveland, W. S., Harris, C. S., & McGill, R. (1982). Judgements of circle sizes on statistical maps. *Journal of the American Statistical Association*, 77, 541-547.

Cliff, N. (1983). Some cautions concerning the application of causal modeling methods. *Multivariate Behavioral Research*, 18, 115-126.

Cochran, W. G. (1950). The comparison of percentages in matched samples. *Biometrika*, 37, 256-266.

Cohen, J. (1977). *Statistical power analysis for the behavioral sciences. (*Rev. ed.). New York: Academic Press.

Cohen, J. (1983). *Statistical power analysis for the behavioral sciences. (2nd Ed.)*. Mahwah, NJ: Lawrence Erlbaum Associates.

Cohen, J. (1994). The earth is round ($p < .05$). *American Psychologist, 49,* 9971003.

Cole, D. A., Maxwell, S. E., Arvey, R., & Salas, E. (1993). Multivariate group comparisons of variable systems: MANOVA and structural equation modeling. *Psychological Bulletin, 114*, 174-184.

Connor, W. S., & Young, S. (1984). Fractional factorial experiment designs for experiments with factors at two and three levels. In R. A. McLean & V. L. Anderson (Eds.), *Applied factorial and fractional designs*. New York: Marcel Dekker.

Connor, W. S., & Zelen, M. (1984). Fractional factorial experiment designs for factors at three levels. In R. A. McLean & V. L. Anderson (Eds.), *Applied factorial and fractional designs*. New York: Marcel Dekker.

Conover, W. J. (1974). Some reasons for not using the Yates continuity correction on 2 x 2 contingency tables. *Journal of the American Statistical Association*, 69, 374-376.

Conover, W. J., Johnson, M. E., & Johnson, M. M. (1981). A comparative study of tests for homogeneity of variances with applications to the outer continental shelf bidding data. *Technometrics, 23*, 357-361.

Cook, R. D. (1977). Detection of influential observations in linear regression. *Technometrics, 19*, 15-18.

Cook, R. D., & Nachtsheim, C. J. (1980). A comparison of algorithms for constructing exact D-optimal designs. *Technometrics, 22*, 315-324.

Cook, R. D., & Weisberg, S. (1982). *Residuals and Influence in Regression*. (Monographs on statistics and applied probability). New York: Chapman and Hall.

Cooke, D., Craven, A. H., & Clarke, G. M. (1982). *Basic statistical computing*. London: Edward Arnold.

Cooley, J. W., & Tukey, J. W. (1965). An algorithm for the machine computation of complex Fourier series. *Mathematics of Computation, 19*, 297-301.

Cooley, W. W., & Lohnes, P. R. (1971). *Multivariate data analysis*. New York: Wiley.

Cooley, W. W., & Lohnes, P. R. (1976). *Evaluation research in education*. New York: Wiley.

Coombs, C. H. (1950). Psychological scaling without a unit of measurement. *Psychological Review, 57*, 145-158.

Coombs, C. H. (1964). *A theory of data*. New York: Wiley.

Corballis, M. C., & Traub, R. E. (1970). Longitudinal factor analysis. *Psychometrika, 35*, 79-98.

Corbeil, R. R., & Searle, S. R. (1976). Restricted maximum likelihood (REML) estimation of variance components in the mixed model. *Technometrics, 18*, 31-38.

Cormack, R. M. (1971). A review of classification. *Journal of the Royal Statistical Society, 134*, 321-367.

Cornell, J. A. (1990a). *How to run mixture experiments for product quality.* Milwaukee, Wisconsin: ASQC.

Cornell, J. A. (1990b). *Experiments with mixtures: designs, models, and the analysis of mixture data* (2nd ed.). New York: Wiley.

Cox, D. R. (1957). Note on grouping. *Journal of the American Statistical Association, 52*, 543-547.

Cox, D. R. (1959). The analysis of exponentially distributed life-times with two types of failures. *Journal of the Royal Statistical Society, 21*, 411-421.

Cox, D. R. (1964). Some applications of exponential ordered scores. *Journal of the Royal Statistical Society, 26*, 103-110.

Cox, D. R. (1970). *The analysis of binary data.* New York: Halsted Press.

Cox, D. R. (1972). Regression models and life tables. *Journal of the Royal Statistical Society, 34*, 187-220.

Cox, D. R., & Oakes, D. (1984). *Analysis of survival data.* New York: Chapman & Hall.

Cramer, H. (1946). *Mathematical methods in statistics.* Princeton, NJ: Princeton University Press.

Crowley, J., & Hu, M. (1977). Covariance analysis of heart transplant survival data. *Journal of the American Statistical Association, 72*, 27-36.

Cudeck, R. (1989). Analysis of correlation matrices using covariance structure models. *Psychological Bulletin, 105*, 317-327.

Cudeck, R., & Browne, M. W. (1983). Cross-validation of covariance structures. *Multivariate Behavioral Research, 18*, 147-167.

Cutler, S. J., & Ederer, F. (1958). Maximum utilization of the life table method in analyzing survival. *Journal of Chronic Diseases, 8*, 699-712.

D

Dahlquist, G., & Bjorck, A. (1974). *Numerical Methods.* Englewood Cliffs, NJ: Prentice-Hall.

Daniel, C. (1976). *Applications of statistics to industrial experimentation.* New York: Wiley.

Daniell, P. J. (1946). Discussion on symposium on autocorrelation in time series. *Journal of the Royal Statistical Society*, Suppl. *8*, 88-90.

Daniels, H. E. (1939). The estimation of components of variance. *Supplement to the Journal of the Royal Statistical Society, 6*, 186-197.

Darlington, R. B. (1990). *Regression and linear models.* New York: McGraw-Hill.

Darlington, R. B., Weinberg, S., & Walberg, H. (1973). Canonical variate analysis and related techniques. *Review of Educational Research, 43*, 433-454.

DataMyte (1992). *DataMyte handbook.* Minnetonka, MN.

David, H. A. (1995). First (?) occurrence of common terms in mathematical statistics. *The American Statistician, 49*, 121-133.

Davies, P. M., & Coxon, A. P. M. (1982). *Key texts in multidimensional scaling.* Exeter, NH: Heinemann Educational Books.

Davis, C. S., & Stephens, M. A. Approximate percentage points using Pearson curves. *Applied Statistics, 32*, 322-327.

De Boor, C. (1978). *A practical guide to splines.* New York: Springer-Verlag.

DeCarlo, L. T. (1998). Signal detection theory and generalized linear models, *Psychological Methods*, 186-200.

De Gruitjer, P. N. M., & Van Der Kamp, L. J. T. (Eds.). (1976). *Advances in psychological and educational measurement.* New York: Wiley.

de Jong, S (1993) SIMPLS: An Alternative Approach to Partial Least Squares Regression, *Chemometrics and Intelligent Laboratory Systems*, 18, 251-263

de Jong, S and Kiers, H. (1992) Principal Covariates regression, *Chemometrics and Intelligent Laboratory Systems*, 14, 155-164

Deming, S. N., & Morgan, S. L. (1993). *Experimental design: A chemometric approach.* (2nd ed.). Amsterdam, The Netherlands: Elsevier Science Publishers B.V.

Deming, W. E., & Stephan, F. F. (1940). The sampling procedure of the 1940 population census. *Journal of the American Statistical Association*, *35*, 615-630.

Dempster, A. P. (1969). *Elements of Continuous Multivariate Analysis*. San Francisco: Addison-Wesley.

Dempster, A. P., Laird, N. M., & Rubin, D. B. (1977). Maximum likelihood from incomplete data via the EM algorithm. *Journal of the Royal Statistical Society*, *39*, 1-38.

Dennis, J. E., & Schnabel, R. B. (1983). *Numerical methods for unconstrained optimization and nonlinear equations*. Englewood Cliffs, NJ: Prentice Hall.

Derringer, G., & Suich, R. (1980). Simultaneous optimization of several response variables. *Journal of Quality Technology*, *12*, 214-219.

Devroye, L. (1986). *Non-uniform random variate generation*. New York: Springer.

Diamond, W. J. (1981). *Practical experimental design*. Belmont, CA: Wadsworth.

Dijkstra, T. K. (1990). Some properties of estimated scale invariant covariance structures. *Psychometrika*, *55*, 327-336.

Dinneen, L. C., & Blakesley, B. C. (1973). A generator for the sampling distribution of the Mann Whitney *U* statistic. *Applied Statistics*, *22*, 269-273.

Dixon, W. J. (1954). Power under normality of several non-parametric tests. *Annals of Mathematical Statistics*, *25*, 610-614.

Dixon, W. J., & Massey, F. J. (1983). *Introduction to statistical analysis* (4th ed.). New York: McGraw-Hill.

Dobson, A. J. (1990). *An introduction to generalized linear models*. New York: Chapman & Hall.

Dodd, B. (1979). Lip reading in infants: Attention to speech presented in- and out-of-synchrony. *Cognitive Psychology*, *11*, 478-484.

Dodge, Y. (1985). *Analysis of experiments with missing data*. New York: Wiley.

Dodge, Y., Fedorov, V. V., & Wynn, H. P. (1988). *Optimal design and analysis of experiments*. New York: North-Holland.

Dodson, B. (1994). *Weibull analysis*. Milwaukee, Wisconsin: ASQC.

Doyle, P. (1973). The use of Automatic Interaction Detection and similar search procedures. *Operational Research Quarterly*, *24*, 465-467.

Duncan, A. J. (1974). *Quality control and industrial statistics*. Homewood, IL: Richard D. Irwin.

Duncan, O. D., Haller, A. O., & Portes, A. (1968). Peer influence on aspiration: a reinterpretation. *American Journal of Sociology*, *74*, 119-137.

Dunnett, C. W. (1955). A multiple comparison procedure for comparing several treatments with a control. *Journal of the American Statistical Association*, *50*, 1096-1121.

Durbin, J. (1970). Testing for serial correlation in least-squares regression when some of the regressors are lagged dependent variables. *Econometrica*, *38*, 410-421.

Durbin, J., & Watson, G. S. (1951). Testing for serial correlations in least squares regression. II. *Biometrika*, *38*, 159-178.

Dykstra, O. Jr. (1971). The augmentation of experimental data to maximize |X'X|. *Technometrics*, *13*, 682-688.

E

Eason, E. D., & Fenton, R. G. (1974). A comparison of numerical optimization methods for engineering design. *ASME Paper 73-DET-17*.

Edgeworth, F. Y. (1885). Methods of statistics. In *Jubilee Volume, Royal Statistical Society*, 181-217.

Efron, B. (1982). *The jackknife, the bootstrap, and other resampling plans*. Philadelphia, Pa. Society for Industrial and Applied Mathematics.

Eisenhart, C. (1947). The assumptions underlying the analysis of variance. *Biometrics*, *3*, 1-21.

Elandt-Johnson, R. C., & Johnson, N. L. (1980). *Survival models and data analysis*. New York: Wiley.

Elliott, D. F., & Rao, K. R. (1982). *Fast transforms: Algorithms, analyses, applications*. New York: Academic Press.

Elsner, J. B., Lehmiller, G. S., & Kimberlain, T. B. (1996). Objective classification of Atlantic hurricanes. *Journal of Climate, 9,* 2880-2889.

Enslein, K., Ralston, A., & Wilf, H. S. (1977). *Statistical methods for digital computers.* New York: Wiley.

Euler, L. (1782). Recherches sur une nouvelle espece de quarres magiques. *Verhandelingen uitgegeven door het zeeuwsch Genootschap der Wetenschappen te Vlissingen, 9,* 85-239. (Reproduced in *Leonhardi Euleri Opera Omnia.* Sub auspiciis societatis scientiarium naturalium helveticae, 1st series, 7, 291-392.)

Evans, M., Hastings, N., & Peacock, B. (1993). *Statistical Distributions.* New York: Wiley.

Everitt, B. S. (1977). *The analysis of contingency tables.* London: Chapman & Hall.

Everitt, B. S. (1984). *An introduction to latent variable models.* London: Chapman and Hall.

Ewan, W. D. (1963). When and how to use Cu-sum charts. *Technometrics, 5,* 1-32.

F

Fahlman, S.E. (1988). Faster-learning variations on back-propagation: an empirical study. In D. Touretzky, G.E. Hinton and T.J. Sejnowski (Eds.), *Proceedings of the 1988 Connectionist Models Summer School*, 38-51. San Mateo, CA: Morgan Kaufmann.

Fausett, L. (1994). *Fundamentals of Neural Networks.* New York: Prentice Hall.

Fayyad, U. M., Piatetsky-Shapiro, G., Smyth, P., & Uthurusamy, R. (Eds.). 1996. *Advances in Knowledge Discovery and Data Mining.* Cambridge, MA: The MIT Press.

Fayyad, U. S., & Uthurusamy, R. (Eds.) (1994). *Knowledge Discovery in Databases; Papers from the 1994 AAAI Workshop.* Menlo Park, CA: AAAI Press.

Feigl, P., & Zelen, M. (1965). Estimation of exponential survival probabilities with concomitant information. *Biometrics, 21,* 826-838.

Feller, W. (1948). On the Kolmogorov-Smirnov limit theorems for empirical distributions. *Annals of Mathematical Statistics, 19,* 177-189.

Fetter, R. B. (1967). *The quality control system.* Homewood, IL: Richard D. Irwin.

Fienberg, S. E. (1977). *The analysis of cross-classified categorical data.* Cambridge, MA: MIT Press.

Finn, J. D. (1974). *A general model for multivariate analysis.* New York: Holt, Rinehart & Winston.

Finn, J. D. (1977). Multivariate analysis of variance and covariance. In K. Enslein, A. Ralston, and H. S. Wilf (Eds.), *Statistical methods for digital computers. Vol. III.* (pp. 203-264). New York: Wiley.

Finney, D. J. (1944). The application of probit analysis to the results of mental tests. *Psychometrika, 9,* 31-39.

Finney, D. J. (1971). *Probit analysis.* Cambridge, MA: Cambridge University Press.

Fisher, R. A. (1918). The correlation between relatives on the supposition of Mendelian inheritance. *Transactions of the Royal Society of Edinbrugh, 52,* 399-433.

Fisher, R. A. (1922). On the interpretation of *Chi-square* from contingency tables, and the calculation of *p. Journal of the Royal Statistical Society, 85,* 87-94.

Fisher, R. A. (1922). On the mathematical foundations of theoretical statistics. *Philosophical Transactions of the Royal Society of London*, Ser. A, 222, 309-368.

Fisher, R. A. (1926). The arrangement of field experiments. *Journal of the Ministry of Agriculture of Great Britain, 33,* 503-513.

Fisher, R. A. (1928). The general sampling distribution of the multiple correlation coefficient. *Proceedings of the Royal Society of London*, Ser. A, *121*, 654-673.

Fisher, R. A. (1935). *The Design of Experiments.* Edinburgh: Oliver and Boyd.

Fisher, R. A. (1936). *Statistical Methods for Research Workers (6th ed.).* Edinburgh: Oliver and Boyd.

Fisher, R. A. (1936). The use of multiple measurements in taxonomic problems. *Annals of Eugenics, 7,* 179-188.

Fisher, R. A. (1938). The mathematics of experimentation. *Nature, 142,* 442-443.

Fisher, R. A., & Yates, F. (1934). The 6 x 6 Latin squares. *Proceedings of the Cambridge Philosophical Society, 30,* 492-507.

Fisher, R. A., & Yates, F. (1938). *Statistical Tables for Biological, Agricultural and Medical Research.* London: Oliver and Boyd.

Fleishman, A. E. (1980). Confidence intervals for correlation ratios. *Educational and Psychological Measurement, 40,* 659670.

Fletcher, R. (1969). *Optimization.* New York: Academic Press.

Fletcher, R., & Powell, M. J. D. (1963). A rapidly convergent descent method for minimization. *Computer Journal, 6,* 163-168.

Fletcher, R., & Reeves, C. M. (1964). Function minimization by conjugate gradients. *Computer Journal, 7,* 149-154.

Fomby, T.B., Hill, R.C., & Johnson, S.R. (1984). *Advanced econometric methods.* New York: Springer-Verlag.

Ford Motor Company, Ltd. & GEDAS (1991). *Test examples for SPC software.*

Fouladi, R. T. (1991). *A comprehensive examination of procedures for testing the significance of a correlation matrix and its elements.* Unpublished master's thesis, University of British Columbia, Vancouver, British Columbia, Canada.

Franklin, M. F. (1984). Constructing tables of minimum aberration p(nm) designs. *Technometrics, 26,* 225-232.

Fraser, C., & McDonald, R. P. (1988). COSAN: Covariance structure analysis. *Multivariate Behavioral Research, 23,* 263-265.

Freedman, L. S. (1982). Tables of the number of patients required in clinical trials using the logrank test. *Statistics in Medicine, 1,* 121129.

Friedman, M. (1937). The use of ranks to avoid the assumption of normality implicit in the analysis of variance. *Journal of the American Statistical Association, 32,* 675-701.

Friedman, M. (1940). A comparison of alternative tests of significance for the problem of *m* rankings. *Annals of Mathematical Statistics, 11,* 86-92.

Fries, A., & Hunter, W. G. (1980). Minimum aberration 2 (kp) designs. *Technometrics, 22,* 601-608.

Frost, P. A. (1975). Some properties of the Almon lag technique when one searches for degree of polynomial and lag. *Journal of the American Statistical Association, 70,* 606-612.

Fuller, W. A. (1976). *Introduction to statistical time series.* New York: Wiley.

G

Gaddum, J. H. (1945). Lognormal distributions. *Nature, 156,* 463-466.

Gale, N., & Halperin, W. C. (1982). A case for better graphics: The unclassed choropleth map. *The American Statistician, 36,* 330-336.

Galil, Z., & Kiefer, J. (1980). Time- and space-saving computer methods, related to Mitchell's DETMAX, for finding D-optimum designs. *Technometrics, 22,* 301-313.

Galton, F. (1882). Report of the anthropometric committee. In *Report of the 51st Meeting of the British Association for the Advancement of Science, 1881,* 245-260.

Galton, F. (1885). Section H. Anthropology. Opening address by Francis Galton. *Nature, 32,* 507- 510.

Galton, F. (1885). Some results of the anthropometric laboratory. *Journal of the Anthropological Institute, 14,* 275-287.

Galton, F. (1888). Co-relations and their measurement. *Proceedings of the Royal Society of London, 45,* 135-145.

Galton, F. (1889). *Natural Inheritance.* London: Macmillan.

Galton, F. (1889). *Natural Inheritance.* London: Macmillan.

Ganguli, M. (1941). A note on nested sampling. *Sankhya, 5,* 449-452.

Gara, M. A., & Rosenberg, S. (1979). The identification of persons as supersets and subsets in free-response personality descriptions. *Journal* of Personality and Social Psychology, 37, 2161-2170.

Gara, M. A., & Rosenberg, S. (1981). Linguistic factors in implicit personality theory. Journal of Personality and Social Psychology, 41, 450-457.

Gardner, E. S., Jr. (1985). Exponential smoothing: The state of the art. Journal of Forecasting, 4, 1-28.

Garthwaite, P. H. (1994) An Interpretation of Partial Least Squares, Journal of the American Statistical Association, 89 NO. 425, 122-127.

Garvin, D. A. (1987). Competing on the eight dimensions of quality. Harvard Business Review, November/December, 101-109.

Gatsonis, C., & Sampson, A. R. (1989). Multiple correlation: exact power and sample size calculations. Psychological Bulletin, 106, 516524.

Gbur, E., Lynch, M., & Weidman, L. (1986). An analysis of nine rating criteria on 329 U. S. metropolitan areas. Proceedings of the Section on Statistical Graphics, American Statistical Association, 104-109.

Gedye, R. (1968). A manager's guide to quality and reliability. New York: Wiley.

Gehan, E. A. (1965a). A generalized Wilcoxon test for comparing arbitrarily singly-censored samples. Biometrika, 52, 203-223.

Gehan, E. A. (1965b). A generalized two-sample Wilcoxon test for doubly-censored data. Biometrika, 52, 650-653.

Gehan, E. A., & Siddiqui, M. M. (1973). Simple regression methods for survival time studies. Journal of the American Statistical Association, 68, 848-856.

Gehan, E. A., & Thomas, D. G. (1969). The performance of some two sample tests in small samples with and without censoring. Biometrika, 56, 127-132.

Geladi, P. and Kowalski, B. R. (1986) Partial Least Squares Regression: A Tutorial, Analytica Chimica Acta, 185, 1-17.

Gerald, C. F., & Wheatley, P. O. (1989). Applied numerical analysis (4th ed.). Reading, MA: Addison Wesley.

Gibbons, J. D. (1976). Nonparametric methods for quantitative analysis. New York: Holt, Rinehart, & Winston.

Gibbons, J. D. (1985). Nonparametric statistical inference (2nd ed.). New York: Marcel Dekker.

Gifi, A. (1981). Nonlinear multivariate analysis. Department of Data Theory, The University of Leiden. The Netherlands.

Gifi, A. (1990). Nonlinear multivariate analysis. New York: Wiley.

Gill, P. E., & Murray, W. (1972). Quasi-Newton methods for unconstrained optimization. Journal of the Institute of Mathematics and its Applications, 9, 91-108.

Gill, P. E., & Murray, W. (1974). Numerical methods for constrained optimization. New York: Academic Press.

Gini, C. (1911). Considerazioni sulle probabilita a posteriori e applicazioni al rapporto dei sessi nelle nascite umane. Studi Economico-Giuridici della Universita de Cagliari, Anno III, 133-171.

Glass, G V., & Hopkins, K. D. (1996). Statistical methods in education and psychology. Needham Heights, MA: Allyn & Bacon.

Glass, G. V., & Stanley, J. (1970). Statistical methods in education and Psychology. Englewood Cliffs, NJ: Prentice-Hall.

Glasser, M. (1967). Exponential survival with covariance. Journal of the American Statistical Association, 62, 561-568.

Gnanadesikan, R., Roy, S., & Srivastava, J. (1971). Analysis and design of certain quantitative multiresponse experiments. Oxford: Pergamon Press, Ltd.

Goldberg, D. E. (1989). Genetic Algorithms. Reading, MA: Addison Wesley.

Golub, G. and Kahan, W. (1965). Calculating the singular values and pseudo-inverse of a matrix. SIAM Numerical Analysis, B 2 (2), 205-224.

Golub, G. H. and van Load, C. F. (1996) Matrix Computations, The Johns Hopkins University Press

Golub, G. H., & Van Loan, C. F. (1983). Matrix computations. Baltimore: Johns Hopkins University Press.

Gompertz, B. (1825). On the nature of the function expressive of the law of human mortality. *Philosophical Transactions of the Royal Society of London*, Series A, 115, 513-580.

Goodman, L .A., & Kruskal, W. H. (1972). Measures of association for cross-classifications IV: Simplification of asymptotic variances. *Journal of the American Statistical Association*, 67, 415-421.

Goodman, L. A. (1954). Kolmogorov-Smirnov tests for psychological research. *Psychological Bulletin*, 51, 160-168.

Goodman, L. A. (1971). The analysis of multidimensional contingency tables: Stepwise procedures and direct estimation methods for models building for multiple classification. *Technometrics*, 13, 33-61.

Goodman, L. A., & Kruskal, W. H. (1954). Measures of association for cross-classifications. *Journal of the American Statistical Association*, 49, 732-764.

Goodman, L. A., & Kruskal, W. H. (1959). Measures of association for cross-classifications II: Further discussion and references. *Journal of the American Statistical Association*, 54, 123-163.

Goodman, L. A., & Kruskal, W. H. (1963). Measures of association for cross-classifications III: Approximate sampling theory. *Journal of the American Statistical Association*, 58, 310-364.

Goodnight, J. H. (1980). Tests of hypotheses in fixed effects linear models. *Communications in Statistics*, A9, 167-180.

Grant, E. L., & Leavenworth, R. S. (1980). *Statistical quality control* (5th ed.). New York: McGraw-Hill.

Green, P. E., & Carmone, F. J. (1970). *Multidimensional scaling and related techniques in marketing analysis*. Boston: Allyn & Bacon.

Green, P. J. & Silverman, B. W. (1994) *Nonparametric regression and generalized linear models: A roughness penalty approach*. New York: Chapman & Hall.

Greenacre, M. J. & Hastie, T. (1987). The geometric interpretation of correspondence analysis. *Journal of the American Statistical Association*, 82, 437-447.

Greenacre, M. J. (1984). *Theory and applications of correspondence analysis*. New York: Academic Press.

Greenacre, M. J. (1988). Correspondence analysis of multivariate categorical data by weighted least-squares. *Biometrica*, 75, 457-467.

Greenhouse, S. W., & Geisser, S. (1958). Extension of Box's results on the use of the *F* distribution in multivariate analysis. *Annals of Mathematical Statistics*, 29, 95-112.

Greenhouse, S. W., & Geisser, S. (1959). On methods in the analysis of profile data. *Psychometrika*, 24, 95-112.

Grizzle, J. E. (1965). The two-period change-over design and its use in clinical trials. *Biometrics*, 21, 467-480.

Gross, A. J., & Clark, V. A. (1975). *Survival distributions: Reliability applications in the medical sciences*. New York: Wiley.

Gruska, G. F., Mirkhani, K., & Lamberson, L. R. (1989). *Non-Normal data Analysis*. Garden City, MI: Multiface Publishing.

Gruvaeus, G., & Wainer, H. (1972). Two additions to hierarchical cluster analysis. *The British Journal of Mathematical and Statistical Psychology*, 25, 200-206.

Guttman, L. (1954). A new approach to factor analysis: the radex. In P. F. Lazarsfeld (Ed.), *Mathematical thinking in the social sciences*. New York: Columbia University Press.

Guttman, L. (1968). A general nonmetric technique for finding the smallest coordinate space for a configuration of points. *Pyrometrical*, 33, 469-506.

Guttman, L. B. (1977). What is not what in statistics. *The Statistician*, 26, 81107.

H

Haberman, S. J. (1972). Loglinear fit for contingency tables. *Applied Statistics*, 21, 218-225.

Haberman, S. J. (1974). *The analysis of frequency data*. Chicago: University of Chicago Press.

Hahn, G. J., & Shapiro, S. S. (1967). *Statistical models in engineering*. New York: Wiley.

Hakstian, A. R., Rogers, W. D., & Cattell, R. B. (1982). The behavior of numbers of factors rules with simulated data. *Multivariate Behavioral Research, 17*, 193-219.

Hald, A. (1949). Maximum likelihood estimation of the parameters of a normal distribution which is truncated at a known point. *Skandinavisk Aktuarietidskrift, 1949*, 119-134.

Hald, A. (1952). *Statistical theory with engineering applications.* New York: Wiley.

Harlow, L. L., Mulaik, S. A., & Steiger, J. H. (Eds.) (1997). *What if there were no significance tests.* Mahwah, NJ: Lawrence Erlbaum Associates.

Harman, H. H. (1967). *Modern factor analysis.* Chicago: University of Chicago Press.

Harris, R. J. (1976). The invalidity of partitioned U tests in canonical correlation and multivariate analysis of variance. *Multivariate Behavioral Research, 11*, 353-365.

Harrison, D. & Rubinfield, D. L. (1978). Hedonic prices and the demand for clean air. *Journal of Environmental Economics and Management, 5*, 81-102.

Harry, M. J. & Schroeder, R. (2000). *Six Sigma: The breakthrough management strategy revolutionizing the world's top corporations.* New York: Doubleday.

Hart, K. M., & Hart, R. F. (1989). *Quantitative methods for quality improvement.* Milwaukee, WI: ASQC Quality Press.

Hartigan, J. A. & Wong, M. A. (1978). Algorithm 136. A *k*-means clustering algorithm. *Applied Statistics, 28*, 100.

Hartigan, J. A. (1975). *Clustering algorithms.* New York: Wiley.

Hartley, H. O. (1959). Smallest composite designs for quadratic response surfaces. *Biometrics, 15*, 611-624.

Harville, D. A. (1977). Maximum likelihood approaches to variance component estimation and to related problems. *Journal of the American Statistical Association, 72*, 320-340.

Haskell, A. C. (1922). *Graphic Charts in Business.* New York: Codex.

Haviland, R. P. (1964). *Engineering reliability and long life design.* Princeton, NJ: Van Nostrand.

Hayduk, L. A. (1987). *Structural equation modeling with LISREL: Essentials and advances.* Baltimore: The Johns Hopkins University Press.

Haykin, S. (1994). *Neural Networks: A Comprehensive Foundation.* New York: Macmillan Publishing.

Haykin, S. (1994). *Neural Networks: A Comprehensive Foundation.* New York: Macmillan College Publishing.

Hays, W. L. (1981). *Statistics* (3rd ed.). New York: CBS College Publishing.

Hays, W. L. (1988). *Statistics* (4th ed.). New York: CBS College Publishing.

Heiberger, R. M. (1989). *Computation for the analysis of designed experiments.* New York: Wiley.

Hemmerle, W. J., & Hartley, H., O. (1973). Computing maximum likelihood estimates for the mixed A.O.V. model using the W transformation. *Technometrics, 15*, 819-831.

Henley, E. J., & Kumamoto, H. (1980). *Reliability engineering and risk assessment.* New York: Prentice-Hall.

Hettmansperger, T. P. (1984). *Statistical inference based on ranks.* New York: Wiley.

Hibbs, D. (1974). Problems of statistical estimation and causal inference in dynamic time series models. In H. Costner (Ed.), *Sociological Methodology 1973/1974* (pp. 252-308). San Francisco: Jossey-Bass.

Hill, I. D., Hill, R., & Holder, R. L. (1976). Fitting Johnson curves by moments. *Applied Statistics. 25*, 190-192.

Hilton, T. L. (1969). *Growth study annotated bibliography. Princeton, NJ: Educational Testing Service Progress Report 69-11.*

Hochberg, J., & Krantz, D. H. (1986). Perceptual properties of statistical graphs. *Proceedings of the Section on Statistical Graphics, American Statistical Association*, 29-35.

Hocking, R. R. (1985). *The analysis of linear models.* Monterey, CA: Brooks/Cole.

Hocking, R. R. (1996). *Methods and Applications of Linear Models. Regression and the Analysis of Variance*. New York: Wiley.

Hocking, R. R., & Speed, F. M. (1975). A full rank analysis of some linear model problems. *Journal of the American Statistical Association, 70*, 707-712.

Hoerl, A. E. (1962). Application of ridge analysis to regression problems. *Chemical Engineering Progress, 58*, 54-59.

Hoerl, A. E., & Kennard, R. W. (1970). Ridge regression: Applications to nonorthogonal problems. *Technometrics, 12*, 69-82.

Hoff, J. C. (1983). *A practical guide to Box-Jenkins forecasting*. London: Lifetime Learning Publications.

Hoffman, D. L. & Franke, G. R. (1986). Correspondence analysis: Graphical representation of categorical data in marketing research. *Journal of Marketing Research, 13*, 213-227.

Hogg, R. V., & Craig, A. T. (1970). *Introduction to mathematical statistics*. New York: Macmillan.

Holzinger, K. J., & Swineford, F. (1939). *A study in factor analysis: The stability of a bi-factor solution*. University of Chicago: Supplementary Educational Monographs, No. 48.

Hooke, R., & Jeeves, T. A. (1961). Direct search solution of numerical and statistical problems. *Journal of the Association for Computing Machinery, 8*, 212-229.

Hosmer, D. W and Lemeshow, S. (1989), *Applied Logistic Regression*, John Wiley & Sons, Inc.

Hotelling, H. (1947). Multivariate quality control. In Eisenhart, Hastay, and Wallis (Eds.), *Techniques of Statistical Analysis*. New York: McGraw-Hill.

Hotelling, H., & Pabst, M. R. (1936). Rank correlation and tests of significance involving no assumption of normality. *Annals of Mathematical Statistics, 7*, 29-43.

Hoyer, W., & Ellis, W. C. (1996). A graphical exploration of SPC. *Quality Progress, 29*, 65-73.

Hsu, P. L. (1938). Contributions to the theory of Student's *t* test as applied to the problem of two samples. *Statistical Research Memoirs, 2*, 1-24.

Huba, G. J., & Harlow, L. L. (1987). Robust structural equation models: implications for developmental psychology. *Child Development, 58*, 147-166.

Huberty, C. J. (1975). Discriminant analysis. *Review of Educational Research, 45*, 543-598.

Huynh, H., & Feldt, L. S. (1970). Conditions under which mean square ratios in repeated measures designs have exact *F*-distributions. *Journal of the American Statistical Association, 65*, 1582-1589.

I

Ireland, C. T., & Kullback, S. (1968). Contingency tables with given marginals. *Biometrika, 55*, 179-188.

J

Jaccard, J., Weber, J., & Lundmark, J. (1975). A multitrait-multimethod factor analysis of four attitude assessment procedures. *Journal of Experimental Social Psychology, 11*, 149-154.

Jacobs, D. A. H. (Ed.). (1977). *The state of the art in numerical analysis*. London: Academic Press.

Jacobs, R.A. (1988). Increased Rates of Convergence Through Learning Rate Adaptation. *Neural Networks 1 (4)*, 295-307.

Jacoby, S. L. S., Kowalik, J. S., & Pizzo, J. T. (1972). *Iterative methods for nonlinear optimization problems*. Englewood Cliffs, NJ: Prentice-Hall.

James, L. R., Mulaik, S. A., & Brett, J. M. (1982). *Causal analysis. Assumptions, models, and data*. Beverly Hills, CA: Sage Publications.

Jardine, N., & Sibson, R. (1971). *Mathematical taxonomy*. New York: Wiley.

Jastrow, J. (1892). On the judgment of angles and position of lines. *American Journal of Psychology, 5*, 214-248.

Jenkins, G. M., & Watts, D. G. (1968). *Spectral analysis and its applications*. San Francisco: Holden-Day.

Jennrich, R. I. (1970). An asymptotic test for the equality of two correlation matrices. *Journal of the American Statistical Association, 65*, 904-912.

Jennrich, R. I. (1977). Stepwise regression. In K. Enslein, A. Ralston, & H.S. Wilf (Eds.),

Statistical methods for digital computers. New York: Wiley.

Jennrich, R. I., & Moore, R. H. (1975). Maximum likelihood estimation by means of nonlinear least squares. *Proceedings of the Statistical Computing Section*, American Statistical Association, 57-65.

Jennrich, R. I., & Sampson, P. F. (1968). Application of stepwise regression to non-linear estimation. *Technometrics, 10*, 63-72.

Jennrich, R. I., & Sampson, P. F. (1976). Newton-Raphson and related algorithms for maximum likelihood variance component estimation. *Technometrics, 18*, 11-17.

Jennrich, R. I., & Schuchter, M. D. (1986). Unbalanced repeated-measures models with structured covariance matrices. *Biometrics, 42*, 805-820.

Jennrich. R. I. (1977). Stepwise discriminant analysis. In K. Enslein, A. Ralston, & H.S. Wilf (Eds.), *Statistical methods for digital computers*. New York: Wiley.

Johnson, L. W., & Ries, R. D. (1982). *Numerical Analysis* (2nd ed.). Reading, MA: Addison Wesley.

Johnson, N. L. (1961). A simple theoretical approach to cumulative sum control charts. *Journal of the American Statistical Association, 56*, 83-92.

Johnson, N. L. (1965). Tables to facilitate fitting *SU* frequency curves. *Biometrika, 52*, 547.

Johnson, N. L., & Kotz, S. (1970). *Continuous univariate distributions, Vol I and II*. New York: Wiley.

Johnson, N. L., Kotz, S., Blakrishnan, N. (1995). *Continuous univariate distributions: Volume II.* (2nd Ed). NY: Wiley.

Johnson, N. L., & Leone, F. C. (1962). Cumulative sum control charts - mathematical principles applied to their construction and use. *Industrial Quality Control, 18*, 15-21.

Johnson, N. L., Nixon, E., & Amos, D. E. (1963). Table of percentage points of pearson curves. *Biometrika, 50*, 459.

Johnson, N. L., Nixon, E., Amos, D. E., & Pearson, E. S. (1963). Table of percentage points of Pearson curves for given 1 and 2, expressed in standard measure. *Biometrika*, 50, 459-498.

Johnson, P. (1987). *SPC for short runs: A programmed instruction workbook.* Southfield, MI: Perry Johnson.

Johnson, S. C. (1967). Hierarchical clustering schemes. *Psychometrika, 32*, 241-254.

Johnston, J. (1972). *Econometric methods*. New York: McGraw-Hill.

Jöreskog, K. G. (1973). A general model for estimating a linear structural equation system. In A. S. Goldberger and O. D. Duncan (Eds.), *Structural Equation Models in the Social Sciences*. New York: Seminar Press.

Jöreskog, K. G. (1974). Analyzing psychological data by structural analysis of covariance matrices. In D. H. Krantz, R. C. Atkinson, R. D. Luce, and P. Suppes (Eds.), *Contemporary Developments in Mathematical Psychology, Vol. II*. New York: W. H. Freeman and Company.

Jöreskog, K. G. (1978). Structural analysis of covariance and correlation matrices. *Psychometrika, 43*, 443-477.

Jöreskog, K. G., & Lawley, D. N. (1968). New methods in maximum likelihood factor analysis. *British Journal of Mathematical and Statistical Psychology, 21*, 85-96.

Jöreskog, K. G., & Sörbom, D. (1979). *Advances in factor analysis and structural equation models.* Cambridge, MA: Abt Books.

Jöreskog, K. G., & Sörbom, D. (1982). Recent developments in structural equation modeling. *Journal of Marketing Research, 19*, 404-416.

Jöreskog, K. G., & Sörbom, D. (1984). *Lisrel VI. Analysis of linear structural relationships by maximum likelihood, instrumental variables, and least squares methods.* Mooresville, Indiana: Scientific Software.

Jöreskog, K. G., & Sörbom, D. (1989). *Lisrel 7. A guide to the program and applications.* Chicago, Illinois: SPSS Inc.

Judge, G. G., Griffith, W. E., Hill, R. C., Luetkepohl, H., & Lee, T. S. (1985). *The theory and practice of econometrics*. New York: Wiley.

Juran, J. M. (1960). Pareto, Lorenz, Cournot, Bernnouli, Juran and others. *Industrial Quality Control, 17*, 25.

Juran, J. M. (1962). *Quality control handbook.* New York: McGraw-Hill.

Juran, J. M., & Gryna, F. M. (1970). *Quality planning and analysis.* New York: McGraw-Hill.

Juran, J. M., & Gryna, F. M. (1980). *Quality planning and analysis* (2nd ed.). New York: McGraw-Hill.

Juran, J. M., & Gryna, F. M. (1988). *Juran's quality control handbook* (4th ed.). New York: McGraw-Hill.

K

Kachigan, S. K. (1986). *Statistical analysis: An interdisciplinary introduction to univariate & multivariate methods.* New York: Redius Press.

Kackar, R. M. (1985). Off-line quality control, parameter design, and the Taguchi method. *Journal of Quality Technology, 17*, 176-188.

Kackar, R. M. (1986). Taguchi's quality philosophy: Analysis and commentary. *Quality Progress, 19*, 21-29.

Kahneman, D., Slovic, P., & Tversky, A. (1982). *Judgment under uncertainty: Heuristics and biases.* New York: Cambridge University Press.

Kaiser, H. F. (1958). The varimax criterion for analytic rotation in factor analysis. *Pyrometrical, 23*, 187-200.

Kaiser, H. F. (1960). The application of electronic computers to factor analysis. *Educational and Psychological Measurement, 20*, 141-151.

Kalbfleisch, J. D., & Prentice, R. L. (1980). *The statistical analysis of failure time data.* New York: Wiley.

Kane, V. E. (1986). Process capability indices. *Journal of Quality Technology, 18*, 41-52.

Kaplan, E. L., & Meier, P. (1958). Nonparametric estimation from incomplete observations. *Journal of the American Statistical Association, 53*, 457-481.

Karsten, K. G., (1925). *Charts and graphs.* New York: Prentice-Hall.

Kass, G. V. (1980). An exploratory technique for investigating large quantities of categorical data. *Applied Statistics, 29*, 119-127.

Keats, J. B., & Lawrence, F. P. (1997). Weibull maximum likelihood parameter estimates with censored data. *Journal of Quality Technology, 29*, 105-110.

Keeves, J. P. (1972). *Educational environment and student achievement.* Melbourne: Australian Council for Educational Research.

Kendall, M. G. (1940). Note on the distribution of quantiles for large samples. *Supplement of the Journal of the Royal Statistical Society, 7*, 83-85.

Kendall, M. G. (1948). *Rank correlation methods.* (1st ed.). London: Griffin.

Kendall, M. G. (1975). *Rank correlation methods* (4th ed.). London: Griffin.

Kendall, M. G. (1984). *Time Series.* New York: Oxford University Press.

Kendall, M., & Ord, J. K. (1990). *Time series* (3rd ed.). London: Griffin.

Kendall, M., & Stuart, A. (1977). *The advanced theory of statistics. (Vol. 1).* New York: MacMillan.

Kendall, M., & Stuart, A. (1979). *The advanced theory of statistics* (Vol. 2). New York: Hafner.

Kennedy, A. D., & Gehan, E. A. (1971). Computerized simple regression methods for survival time studies. *Computer Programs in Biomedicine, 1*, 235-244.

Kennedy, W. J., & Gentle, J. E. (1980). *Statistical computing.* New York: Marcel Dekker, Inc.

Kenny, D. A. (1979). *Correlation and causality.* New York: Wiley.

Keppel, G. (1973). *Design and analysis: A researcher's handbook.* Engelwood Cliffs, NJ: Prentice-Hall.

Keppel, G. (1982). *Design and analysis: A researcher's handbook* (2nd ed.). Engelwood Cliffs, NJ: Prentice-Hall.

Keselman, H. J., Rogan, J. C., Mendoza, J. L., & Breen, L. L. (1980). Testing the validity conditions for repeated measures *F* tests. *Psychological Bulletin, 87*, 479-481.

Khuri, A. I., & Cornell, J. A. (1987). *Response surfaces: Designs and analyses*. New York: Marcel Dekker, Inc.

Kiefer, J., & Wolfowitz, J. (1960). The equivalence of two extremum problems. *Canadian Journal of Mathematics, 12*, 363-366.

Kim, J. O., & Mueller, C. W. (1978a). *Factor analysis: Statistical methods and practical issues*. Beverly Hills, CA: Sage Publications.

Kim, J. O., & Mueller, C. W. (1978b). *Introduction to factor analysis: What it is and how to do it*. Beverly Hills, CA: Sage Publications.

Kirk, D. B. (1973). On the numerical approximation of the bivariate normal (tetrachoric) correlation coefficient. *Psychometrika, 38*, 259-268.

Kirk, R. E. (1968). *Experimental design: Procedures for the behavioral sciences*. (1st ed.). Monterey, CA: Brooks/Cole.

Kirk, R. E. (1982). *Experimental design: Procedures for the behavioral sciences*. (2nd ed.). Monterey, CA: Brooks/Cole.

Kirk, R. E. (1995). *Experimental design: Procedures for the behavioral sciences*. Pacific Grove, CA: Brooks-Cole.

Kirkpatrick, S., Gelatt, C.D. and Vecchi, M.P. (1983). Optimization by simulated annealing. *Science 220 (4598)*, 671-680.

Kivenson, G. (1971). *Durability and reliability in engineering design*. New York: Hayden.

Klecka, W. R. (1980). *Discriminant analysis*. Beverly Hills, CA: Sage.

Klein, L. R. (1974). *A textbook of econometrics*. Englewood Cliffs, NJ: Prentice-Hall.

Kleinbaum, D. G. (1996). *Survival analysis: A self-learning text*. New York: Springer-Verlag.

Kline, P. (1979). *Psychometrics and psychology*. London: Academic Press.

Kline, P. (1986). *A handbook of test construction*. New York: Methuen.

Kmenta, J. (1971). *Elements of econometrics*. New York: Macmillan.

Knuth, Donald E. (1981). *Seminumerical algorithms*. 2nd ed., Vol 2 of: *The art of computer programming*. Reading, Mass.: Addison-Wesley.

Kohonen, T. (1982). Self-organized formation of topologically correct feature maps. *Biological Cybernetics, 43*, 59-69.

Kolata, G. (1984). The proper display of data. *Science, 226*, 156-157.

Kolmogorov, A. (1941). Confidence limits for an unknown distribution function. *Annals of Mathematical Statistics, 12*, 461-463.

Korin, B. P. (1969). On testing the equality of k covariance matrices. *Biometrika, 56*, 216-218.

Kramer, M.A. (1991). Nonlinear principal components analysis using autoassociative neural networks. *AIChe Journal 37 (2)*, 233-243.

Kruskal, J. B. (1964). Nonmetric multidimensional scaling: A numerical method. *Pyrometrical, 29*, 1-27, 115-129.

Kruskal, J. B., & Wish, M. (1978). *Multidimensional scaling*. Beverly Hills, CA: Sage Publications.

Kruskal, W. H. (1952). A nonparametric test for the several sample problem. *Annals of Mathematical Statistics, 23*, 525-540.

Kruskal, W. H. (1975). Visions of maps and graphs. In J. Kavaliunas (Ed.), *Auto-carto II, proceedings of the international symposium on computer assisted cartography*. Washington, DC: U. S. Bureau of the Census and American Congress on Survey and Mapping.

Kruskal, W. H., & Wallis, W. A. (1952). Use of ranks in one-criterion variance analysis. *Journal of the American Statistical Association, 47*, 583-621.

Ku, H. H., & Kullback, S. (1968). Interaction in multidimensional contingency tables: An information theoretic approach. *J. Res. Nat. Bur. Standards Sect. B, 72*, 159-199.

Ku, H. H., Varner, R. N., & Kullback, S. (1971). Analysis of multidimensional contingency tables. *Journal of the American Statistical Association, 66*, 55-64.

Kullback, S. (1959). *Information theory and statistics*. New York: Wiley.

Kvålseth, T. O. (1985). Cautionary note about R2. *The American Statistician, 39*, 279-285.

L

Lagakos, S. W., & Kuhns, M. H. (1978). Maximum likelihood estimation for censored exponential survival data with covariates. *Applied Statistics, 27,* 190-197.

Lakatos, E., & Lan, K. K. G. (1992). A comparison of sample size methods for the logrank statistic. *Statistics in Medicine, 11,* 179191.

Lance, G. N., & Williams, W. T. (1966). A general theory of classificatory sorting strategies. *Computer Journal, 9,* 373.

Lance, G. N., & Williams, W. T. (1966). Computer programs for hierarchical polythetic classification ("symmetry analysis"). *Computer Journal, 9,* 60.

Larsen, W. A., & McCleary, S. J. (1972). The use of partial residual plots in regression analysis. *Technometrics, 14,* 781-790.

Lawless, J. F. (1982). *Statistical models and methods for lifetime data.* New York: Wiley.

Lawley, D. N., & Maxwell, A. E. (1971). *Factor analysis as a statistical method.* New York: American Elsevier.

Lawley, D. N., & Maxwell, A. E. (1971). *Factor analysis as a statistical method* (2nd. ed.). London: Butterworth & Company.

Lebart, L., Morineau, A., and Tabard, N. (1977). *Techniques de la description statistique.* Paris: Dunod.

Lebart, L., Morineau, A., and Warwick, K., M. (1984). Multivariate descriptive statistical analysis: Correspondence analysis and related techniques for large matrices. New York: Wiley.

Lee, E. T. (1980). *Statistical methods for survival data analysis.* Belmont, CA: Lifetime Learning.

Lee, E. T., & Desu, M. M. (1972). A computer program for comparing *K* samples with right-censored data. *Computer Programs in Biomedicine, 2,* 315-321.

Lee, E. T., Desu, M. M., & Gehan, E. A. (1975). A Monte-Carlo study of the power of some two-sample tests. *Biometrika, 62,* 425-532.

Lee, S., & Hershberger, S. (1990). A simple rule for generating equivalent models in covariance structure modeling. *Multivariate Behavioral Research, 25,* 313-334.

Lee, Y. S. (1972). Tables of upper percentage points of the multiple correlation coefficient. *Biometrika, 59,* 175189.

Legendre, A. M. (1805). *Nouvelles Methodes pour la Determination des Orbites des Cometes.* Paris: F. Didot.

Lehmann, E. L. (1975). *Nonparametrics: Statistical methods based on ranks.* San Francisco: Holden-Day.

Levenberg, K. (1944). A method for the solution of certain non-linear problems in least squares. *Quarterly Journal of Applied Mathematics II (2),* 164-168.

Lewicki, P., Hill, T., & Czyzewska, M. (1992). Nonconscious acquisition of information. *American Psychologist, 47,* 796-801.

Lieblein, J. (1953). On the exact evaluation of the variances and covariances of order statistics in samples form the extreme-value distribution. *Annals of Mathematical Statistics, 24,* 282-287.

Lieblein, J. (1955). On moments of order statistics from the Weibull distribution. *Annals of Mathematical Statistics, 26,* 330-333.

Lilliefors, H. W. (1967). On the Kolmogorov-Smirnov test for normality with mean and variance unknown. *Journal of the American Statistical Association, 64,* 399-402.

Lim, T.-S., Loh, W.-Y., & Shih, Y.-S. (1997). An emprical comparison of decision trees and other classification methods. *Technical Report 979,* Department of Statistics, University of Winconsin, Madison.

Lindeman, R. H., Merenda, P. F., & Gold, R. (1980). *Introduction to bivariate and multivariate analysis.* New York: Scott, Foresman, & Co.

Lindman, H. R. (1974). *Analysis of variance in complex experimental designs.* San Francisco: W. H. Freeman & Co.

Linfoot, E. H. (1957). An informational measure of correlation. *Information and Control, 1,* 50-55.

Linn, R. L. (1968). A Monte Carlo approach to the number of factors problem. *Psychometrika, 33,* 37-71.

Lipson, C., & Sheth, N. C. (1973). *Statistical design and analysis of engineering experiments*. New York: McGraw-Hill.

Lloyd, D. K., & Lipow, M. (1977). *Reliability: Management, methods, and mathematics*. New York: McGraw-Hill.

Loehlin, J. C. (1987). *Latent variable models: An introduction to latent, path, and structural analysis*. Hillsdale, NJ: Erlbaum.

Loh, W.-Y, & Shih, Y.-S. (1997). Split selection methods for classification trees. *Statistica Sinica, 7*, 815-840.

Loh, W.-Y., & Vanichestakul, N. (1988). Tree-structured classification via generalized discriminant analysis (with discussion). *Journal of the American Statistical Association, 83*, 715-728.

Long, J. S. (1983a). *Confirmatory factor analysis*. Beverly Hills: Sage.

Long, J. S. (1983b). *Covariance structure models: An introduction to LISREL*. Beverly Hills: Sage.

Longley, J. W. (1967). An appraisal of least squares programs for the electronic computer from the point of view of the user. *Journal of the American Statistical Association, 62*, 819-831.

Longley, J. W. (1984). *Least squares computations using orthogonalization methods*. New York: Marcel Dekker.

Lord, F. M. (1957). A significance test for the hypothesis that two variables measure the same trait except for errors of measurement. *Psychometrika, 22*, 207-220.

Lorenz, M. O. (1904). Methods of measuring the concentration of wealth. *American Statistical Association Publication, 9*, 209-219.

Lowe, D. (1989). Adaptive radial basis function non-linearities, and the problem of generalisation. *First IEEE International Conference on Artificial Neural Networks*, 171-175, London, UK.

Lucas, J. M. (1976). The design and use of cumulative sum quality control schemes. *Journal of Quality Technology, 8*, 45-70.

Lucas, J. M. (1982). Combined Shewhart-CUSUM quality control schemes. *Journal of Quality Technology, 14*, 89-93.

M

MacCallum, R. C., Browne, M. W., & Sugawara, H. M. (1996). Power analysis and determination of sample size for covariance structur modeling. *Psychological Methods, 1,* 130149.

Maddala, G. S. (1977) *Econometrics*. New York: McGraw-Hill.

Maiti, S. S., & Mukherjee, B. N. (1990). A note on the distributional properties of the Jöreskog-Sörbom fit indices. *Psychometrika, 55*, 721-726.

Makridakis, S. G. (1983). Empirical evidence versus personal experience. *Journal of Forecasting, 2*, 295-306.

Makridakis, S. G. (1990). *Forecasting, planning, and strategy for the 21st century*. London: Free Press.

Makridakis, S. G., & Wheelwright, S. C. (1978). *Interactive forecasting: Univariate and multivariate methods* (2nd ed.). San Francisco, CA: Holden-Day.

Makridakis, S. G., & Wheelwright, S. C. (1989). *Forecasting methods for management* (5th ed.). New York: Wiley.

Makridakis, S. G., Wheelwright, S. C., & McGee, V. E. (1983). *Forecasting: Methods and applications* (2nd ed.). New York: Wiley.

Makridakis, S., Andersen, A., Carbone, R., Fildes, R., Hibon, M., Lewandowski, R., Newton, J., Parzen, R., & Winkler, R. (1982). The accuracy of extrapolation (time series) methods: Results of a forecasting competition. *Journal of Forecasting, 1*, 11-153.

Malinvaud, E. (1970). *Statistical methods of econometrics*. Amsterdam: North-Holland Publishing Co.

Mandel, B. J. (1969). The regression control chart. *Journal of Quality Technology, 1*, 3-10.

Mann, H. B., & Whitney, D. R. (1947). On a test of whether one of two random variables is stochastically larger than the other. *Annals of Mathematical Statistics, 18*, 50-60.

Mann, N. R., Schafer, R. E., & Singpurwalla, N. D. (1974). *Methods for statistical analysis of reliability and life data*. New York: Wiley.

Mann, N. R., Scheuer, R. M, & Fertig, K. W. (1973). A new goodness of fit test for the two-parameter Weibull or exteme value distribution. *Communications in Statistics*, 2, 383-400.

Mantel, N. (1966). Evaluation of survival data and two new rank order statistics arising in its consideration. *Cancer Chemotherapy Reports*, 50, 163-170.

Mantel, N. (1967). Ranking procedures for arbitrarily restricted observations. *Biometrics*, 23, 65-78.

Mantel, N. (1974). Comment and suggestion on the Yates continuity correction. *Journal of the American Statistical Association*, 69, 378-380.

Mantel, N., & Haenszel, W. (1959). Statistical aspects of the analysis of data from retrospective studies of disease. *Journal of the National Cancer Institute*, 22, 719-748.

Marascuilo, L. A., & McSweeney, M. (1977). *Nonparametric and distribution free methods for the social sciences*. Monterey, CA: Brooks/Cole.

Marple, S. L., Jr. (1987). *Digital spectral analysis*. Englewood Cliffs, NJ: Prentice-Hall.

Marquardt, D.W. (1963). An algorithm for least-squares estimation of non-linear parameters. *Journal of the Society of Industrial and Applied Mathematics 11 (2)*, 431-441.

Marsaglia, G. (1962). Random variables and computers. In J. Kozenik (Ed.), *Information theory, statistical decision functions, random processes: Transactions of the third Prague Conference*. Prague: Czechoslovak Academy of Sciences.

Mason, R. L., Gunst, R. F., & Hess, J. L. (1989). *Statistical design and analysis of experiments with applications to engineering and science*. New York: Wiley.

Massey, F. J., Jr. (1951). The Kolmogorov-Smirnov test for goodness of fit. *Journal of the American Statistical Association*, 46, 68-78.

Masters (1995). *Neural, Novel, and Hybrid Algorithms for Time Series Predictions*. New York: Wiley.

Matsueda, R. L., & Bielby, W. T. (1986). Statistical power in covariance structure models. In N. B. Tuma (Ed.), *Sociological methodology.*

Washington, DC: American Sociological Association.

McArdle, J. J. (1978). A structural view of structural models. Paper presented at the *Winter Workshop on Latent Structure Models Applied to Developmental Data, University of Denver, December, 1978.*

McArdle, J. J., & McDonald, R. P. (1984). Some algebraic properties of the Reticular Action Model for moment structures. *British Journal of Mathematical and Statistical Psychology, 37*, 234-251.

McCleary, R., & Hay, R. A. (1980). *Applied time series analysis for the social sciences*. Beverly Hills, CA: Sage Publications.

McCullagh, P. & Nelder, J. A. (1989). *Generalized linear models* (2nd Ed.). New York: Chapman & Hall.

McDonald, R. P. (1980). A simple comprehensive model for the analysis of covariance structures. *British Journal of Mathematical and Statistical Psychology, 31*, 59-72.

McDonald, R. P. (1989). An index of goodness-of-fit based on noncentrality. *Journal of Classification, 6*, 97-103.

McDonald, R. P., & Hartmann, W. M. (1992). A procedure for obtaining initial value estimates in the RAM model. *Multivariate Behavioral Research, 27*, 57-76.

McDonald, R. P., & Mulaik, S. A. (1979). Determinacy of common factors: A nontechnical review. *Psychological Bulletin, 86*, 297-306.

McDowall, D., McCleary, R., Meidinger, E. E., & Hay, R. A. (1980). *Interrupted time series analysis*. Beverly Hills, CA: Sage Publications.

McKenzie, E. (1984). General exponential smoothing and the equivalent ARMA process. *Journal of Forecasting, 3*, 333-344.

McKenzie, E. (1985). Comments on 'Exponential smoothing: The state of the art' by E. S. Gardner, Jr. *Journal of Forecasting, 4*, 32-36.

McLachlan, G. J. (1992). *Discriminant analysis and statistical pattern recognition*. New York: Wiley.

McLain, D. H. (1974). Drawing contours from arbitrary data points. *The Computer Journal, 17,* 318-324.

McLean, R. A., & Anderson, V. L. (1984). *Applied factorial and fractional designs.* New York: Marcel Dekker.

McLeod, A. I., & Sales, P. R. H. (1983). An algorithm for approximate likelihood calculation of ARMA and seasonal ARMA models. *Applied Statistics,* 211-223 (Algorithm AS).

McNemar, Q. (1947). Note on the sampling error of the difference between correlated proportions or percentages. *Psychometrika, 12,* 153-157.

McNemar, Q. (1969). *Psychological statistics* (4th ed.). New York: Wiley.

Melard, G. (1984). A fast algorithm for the exact likelihood of autoregressive-moving average models. *Applied Statistics, 33,* 104-119.

Mels, G. (1989). A general system for path analysis with latent variables. M. S. Thesis: Department of Statistics, University of South Africa.

Mendoza, J. L., Markos, V. H., & Gonter, R. (1978). A new perspective on sequential testing procedures in canonical analysis: A Monte Carlo evaluation. *Multivariate Behavioral Research, 13,* 371-382.

Meredith, W. (1964). Canonical correlation with fallible data. *Psychometrika, 29,* 55-65.

Miettinnen, O. S. (1968). The matched pairs design in the case of all-or-none responses. *Biometrics, 24,* 339352.

Miller, R. (1981). *Survival analysis.* New York: Wiley.

Milligan, G. W. (1980). An examination of the effect of six types of error perturbation on fifteen clustering algorithms. *Psychometrika, 45,* 325-342.

Milliken, G. A., & Johnson, D. E. (1984). *Analysis of messy data: Vol. I. Designed experiments.* New York: Van Nostrand Reinhold, Co.

Milliken, G. A., & Johnson, D. E. (1992). *Analysis of messy data: Vol. I. Designed experiments.* New York: Chapman & Hall.

Minsky, M.L. and Papert, S.A. (1969). *Perceptrons.* Cambridge, MA: MIT Press.

Mitchell, T. J. (1974a). Computer construction of "D-optimal" first-order designs. *Technometrics, 16,* 211-220.

Mitchell, T. J. (1974b). An algorithm for the construction of "D-optimal" experimental designs. *Technometrics, 16,* 203-210.

Mittag, H. J. (1993). *Qualitätsregelkarten.* München/Wien: Hanser Verlag.

Mittag, H. J., & Rinne, H. (1993). *Statistical methods of quality assurance.* London/New York: Chapman & Hall.

Monro, D. M. (1975). Complex discrete fast Fourier transform. *Applied Statistics, 24,* 153-160.

Monro, D. M., & Branch, J. L. (1976). The chirp discrete Fourier transform of general length. *Applied Statistics, 26,* 351-361.

Montgomery, D. C. (1976). *Design and analysis of experiments.* New York: Wiley.

Montgomery, D. C. (1985). *Statistical quality control.* New York: Wiley.

Montgomery, D. C. (1991) *Design and analysis of experiments* (3rd ed.). New York: Wiley.

Montgomery, D. C. (1996). *Introduction to Statistical Quality Control* (3rd Edition). New York:Wiley.

Montgomery, D. C., & Wadsworth, H. M. (1972). Some techniques for multivariate quality control applications. *Technical Conference Transactions.* Washington, DC: American Society for Quality Control.

Montgomery, D. C., Johnson, L. A., & Gardiner, J. S. (1990). *Forecasting and time series analysis* (2nd ed.). New York: McGraw-Hill.

Mood, A. M. (1954). *Introduction to the theory of statistics.* New York: McGraw Hill.

Moody, J. and Darkin, C.J. (1989). Fast learning in networks of locally-tuned processing units. *Neural Computation 1 (2),* 281-294.

Morgan, J. N., & Messenger, R. C. (1973). THAID: A sequential analysis program for the analysis of nominal scale dependent variables. *Technical report,* Institute of Social Research, University of Michigan, Ann Arbor.

Morgan, J. N., & Sonquist, J. A. (1973). Problems in the analysis of survey data, and a proposal.

*Journal of the American Statistical Association,
58*, 415-434.

Morris, M., & Thisted, R. A. (1986). Sources of error
in graphical perception: A critique and an
experiment. *Proceedings of the Section on
Statistical Graphics, American Statistical
Association*, 43-48.

Morrison, A. S., Black, M. M., Lowe, C. R.,
MacMahon, B., & Yuasa, S. (1973). Some
international differences in histology and survival
in breast cancer. *International Journal of Cancer*,
11, 261-267.

Morrison, D. (1967). *Multivariate statistical methods*.
New York: McGraw-Hill.

Morrison, D. F. (1990). *Multivariate statistical
methods*. (3rd Ed.). New York: McGraw-Hill.

Moses, L. E. (1952). Non-parametric statistics for
psychological research. *Psychological Bulletin*,
49, 122-143.

Mulaik, S. A. (1972). *The foundations of factor
analysis*. New York: McGraw Hill.

Muller, M.E., (1959). A comparison of methods for
generating normal deviates on digital computers.
Journal of the ACM, 6, 376-383.

Murphy, K. R., & Myors, B. (1998). *Statistical power
analysis: A simple general model for traditional
and modern hypothesis tests*. Mahwah, NJ:
Lawrence Erlbaum Associates.

Muth, J. F. (1960). Optimal properties of
exponentially weighted forecasts. *Journal of the
American Statistical Association*, *55*, 299-306.

N

Nachtsheim, C. J. (1979). *Contributions to optimal
experimental design*. Ph.D. thesis, Department of
Applied Statistics, University of Minnesota.

Nachtsheim, C. J. (1987). Tools for computer-aided
design of experiments. *Journal of Quality
Technology*, *19*, 132-160.

Nelder, J. A., & Mead, R. (1965). A Simplex method
for function minimization. *Computer Journal*, *7*,
308-313.

Nelson, L. (1984). The Shewhart control chart - tests
for special causes. *Journal of Quality Technology,
15*, 237-239.

Nelson, L. (1985). Interpreting Shewhart X-bar control
charts. *Journal of Quality Technology, 17*, 114-
116.

Nelson, W. (1982). *Applied life data analysis*. New
York: Wiley.

Nelson, W. (1990). *Accelerated testing: Statistical
models, test plans, and data analysis*. New York:
Wiley.

Neter, J., Wasserman, W., & Kutner, M. H. (1985).
*Applied linear statistical models: Regression,
analysis of variance, and experimental designs*.
Homewood, IL: Irwin.

Neter, J., Wasserman, W., & Kutner, M. H. (1989).
Applied linear regression models (2nd ed.).
Homewood, IL: Irwin.

Newcombe, Robert G. (1998). Two-sided confidence
intervals for the single proportion: comparison of
seven methods. *Statistics in Medicine, 17*,
857872.

Neyman, J., & Pearson, E. S. (1931). On the problem
of *k* samples. *Bulletin de l'Academie Polonaise
des Sciences et Lettres*, Ser. A, 460-481.

Neyman, J., & Pearson, E. S. (1933). On the problem
of the most efficient tests of statistical hypothesis.
*Philosophical Transactions of the Royal Society
of London*, Ser. A, *231*, 289-337.

Nisbett, R. E., Fong, G. F., Lehman, D. R., & Cheng,
P. W. (1987). Teaching reasoning. *Science, 238*,
625-631.

Noori, H. (1989). The Taguchi methods: Achieving
design and output quality. *The Academy of
Management Executive, 3*, 322-326.

Nunally, J. C. (1970). *Introduction to psychological
measurement*. New York: McGraw-Hill.

Nunnally, J. C. (1978). *Psychometric theory*. New
York: McGraw-Hill.

Nussbaumer, H. J. (1982). *Fast Fourier transforms
and convolution algorithms* (2nd ed.). New York:
Springer-Verlag.

StatSoft
Copyright © StatSoft, 2001

O

O'Brien, R. G., & Kaiser, M. K. (1985). MANOVA method for analyzing repeated measures designs: An extensive primer. *Psychological Bulletin, 97,* 316-333.

Okunade, A. A., Chang, C. F., & Evans, R. D. (1993). Comparative analysis of regression output summary statistics in common statistical packages. *The American Statistician, 47,* 298-303.

Olds, E. G. (1949). The 5% significance levels for sums of squares of rank differences and a correction. *Annals of Mathematical Statistics, 20,* 117-118.

Olejnik, S. F., & Algina, J. (1987). Type I error rates and power estimates of selected parametric and nonparametric tests of scale. *Journal of Educational Statistics, 12,* 45-61.

Olson, C. L. (1976). On choosing a test statistic in multivariate analysis of variance. *Psychological Bulletin, 83,* 579-586.

O'Neill, R. (1971). Function minimization using a Simplex procedure. *Applied Statistics, 3,* 79-88.

Ostle, B., & Malone, L. C. (1988). *Statistics in research: Basic concepts and techniques for research workers* (4th ed.). Ames, IA: Iowa State Press.

Ostrom, C. W. (1978). *Time series analysis: Regression techniques.* Beverly Hills, CA: Sage Publications.

O'Sullivan, F. (1985). Comments on "Some aspects of the spline smoothing approach to non-parametric regression curve fitting." *Journal of the Royal Statistical Society,* B, 47, 39-40.

Overall, J. E., & Speigel, D. K. (1969). Concerning least squares analysis of experimental data. *Psychological Bulletin, 83,* 579-586.

P

Page, E. S. (1954). Continuous inspection schemes. *Biometrics, 41,* 100-114.

Page, E. S. (1961). Cumulative sum charts. *Technometrics, 3,* 1-9.

Palumbo, F. A., & Strugala, E. S. (1945). Fraction defective of battery adapter used in handie-talkie. *Industrial Quality Control, November,* 68.

Pande, P. S., Neuman, R. P., Cavanagh, R. R. (2000). *The Six Sigma way: How GE, Motorola, and other top companies are honing their performance.* New York: McGraw.

Pankratz, A. (1983). *Forecasting with univariate Box-Jenkins models: Concepts and cases.* New York: Wiley.

Parker, D.B. (1985). *Learning logic. Technical Report TR-47,* Cambridge, MA: MIT Center for Research in Computational Economics and Management Science.

Parzen, E. (1961). Mathematical considerations in the estimation of spectra: Comments on the discussion of Messers, Tukey, and Goodman. *Technometrics, 3,* 167-190; 232-234.

Parzen, E. (1962). On estimation of a probability density function and mode. *Annals of Mathematical Statistics 33,* 1065-1076.

Patil, K. D. (1975). Cochran's Q test: Exact distribution. *Journal of the American Statistical Association, 70,* 186-189.

Patterson, D. (1996). *Artificial Neural Networks.* Singapore: Prentice Hall.

Peace, G. S. (1993). *Taguchi methods: A hands-on approach.* Milwaukee, Wisconsin: ASQC.

Pearson, E. S., and Hartley, H. O. (1972). *Biometrika tables for statisticians, Vol II.* Cambridge: Cambridge University Press.

Pearson, K. (1894). Contributions to the mathematical theory of evolution. *Philosophical Transactions of the Royal Society of London,* Ser. A, *185,* 71-110.

Pearson, K. (1895). Skew variation in homogeneous material. *Philosophical Transactions of the Royal Society of London,* Ser. A, *186,* 343-414.

Pearson, K. (1896). Regression, heredity, and panmixia. *Philosophical Transactions of the Royal Society of London,* Ser. A, *187,* 253-318.

Pearson, K. (1900). On the criterion that a given system of deviations from the probable in the case of a correlated system of variables is such that it can be reasonably supposed to have arisen from

random sampling. *Philosophical Magazine*, 5th Ser., *50*, 157-175.

Pearson, K. (1904). On the theory of contingency and its relation to association and normal correlation. *Drapers' Company Research Memoirs*, Biometric Ser. I.

Pearson, K. (1905). Das Fehlergesetz und seine Verallgemeinerungen durch Fechner und Pearson. A Rejoinder. *Biometrika*, *4*, 169-212.

Pearson, K. (1908). On the generalized probable error in multiple normal correlation. *Biometrika*, *6*, 59-68.

Pearson, K., (Ed.). (1968). *Tables of incomplete beta functions* (2nd ed.). Cambridge, MA: Cambridge University Press.

Pedhazur, E. J. (1973). *Multiple regression in behavioral research*. New York: Holt, Rinehart, & Winston.

Pedhazur, E. J. (1982). *Multiple regression in behavioral research* (2nd ed.). New York: Holt, Rinehart, & Winston.

Peressini, A. L., Sullivan, F. E., & Uhl, J. J., Jr. (1988). *The mathematics of nonlinear programming*. New York: Springer.

Peto, R., & Peto, J. (1972). Asymptotically efficient rank invariant procedures. *Journal of the Royal Statistical Society*, *135*, 185-207.

Phadke, M. S. (1989). *Quality engineering using robust design*. Englewood Cliffs, NJ: Prentice-Hall.

Phatak, A., Reilly, P. M., and Penlidis, A. (1993) An Approach to Interval Estimation in Partial Least Squares Regression, *Analytica Chimica Acta*, 277, 495-501

Piatetsky-Shapiro, G. (Ed.) (1993). *Proceedings of AAAI-93 Workshop on Knowledge Discovery in Databases*. Menlo Park, CA: AAAI Press.

Piepel, G. F. (1988). Programs for generating extreme vertices and centroids of linearly constrained experimental regions. *Journal of Quality Technology, 20*, 125-139.

Piepel, G. F., & Cornell, J. A. (1994). Mixture experiment approaches: Examples, discussion, and recommendations. *Journal of Quality Technology, 26*, 177-196.

Pigou, A. C. (1920). *Economics of Welfare*. London: Macmillan.

Pike, M. C. (1966). A method of analysis of certain class of experiments in carcinogenesis. *Biometrics, 22*, 142-161.

Pillai, K. C. S. (1965). On the distribution of the largest characteristic root of a matrix in multivariate analysis. *Biometrika, 52*, 405-414.

Plackett, R. L., & Burman, J. P. (1946). The design of optimum multifactorial experiments. *Biometrika, 34*, 255-272.

Polya, G. (1920). Uber den zentralen Grenzwertsatz der Wahrscheinlichkeitsrechnung und das Momentenproblem. *Mathematische Zeitschrift, 8*, 171-181.

Porebski, O. R. (1966). Discriminatory and canonical analysis of technical college data. *British Journal of Mathematical and Statistical Psychology, 19*, 215-236.

Powell, M. J. D. (1964). An efficient method for finding the minimum of a function of several variables without calculating derivatives. *Computer Journal, 7*, 155-162.

Pregibon, D. (1997). Data Mining. *Statistical Computing and Graphics, 7*, 8.

Prentice, R. (1973). Exponential survivals with censoring and explanatory variables. *Biometrika, 60*, 279-288.

Press, W. H., Flannery, B. P., Teukolsky, S. A., Vetterling, W. T. (1992). *Numerical recipies (2nd Edition)*. New York: Cambridge University Press.

Press, W.H., Teukolsky, S.A., Vetterling, W.T. and Flannery, B.P. (1992). *Numerical Recipes in C: The Art of Scientific Computing (Second ed.)*. Cambridge University Press.

Press, William, H., Flannery, B. P., Teukolsky, S. A., Vetterling, W. T. (1986). *Numerical recipies*. New York: Cambridge University Press.

Priestley, M. B. (1981). *Spectral analysis and time series*. New York: Academic Press.

Pyzdek, T. (1989). *What every engineer should know about quality control*. New York: Marcel Dekker.

Pyzdek, T. (1999). *Quality Engineering Handbook (Quality and Reliability, 57)*. New York: Marcel Dekker.

Pyzdek, T. (2000). *The Handbook for Quality Management*. New York: Quality Publishing.

Pyzdek, T. (2001). *The Six Sigma handbook: A complete guide for greenbelts, blackbelts, and managers at all levels*. New York: McGraw.

Q

Quinlan, J.R., & Cameron-Jones, R.M. (1995). Oversearching and layered search in empirical learning. *Proceedings of the 14th International Joint Conference on Artificial Intelligence, Montreal* (Vol. 2). Morgan Kaufman, 1019-10244.

R

Ralston, A., & Wilf, H.S. (Eds.). (1960). *Mathematical methods for digital computers*. New York: Wiley.

Ralston, A., & Wilf, H.S. (Eds.). (1967). *Mathematical methods for digital computers* (Vol. II). New York: Wiley.

Randles, R. H., & Wolfe, D. A. (1979). *Introduction to the theory of nonparametric statistics*. New York: Wiley.

Rannar, S., Lindgren, F., Geladi, P, and Wold, S. (1994) A PLS Kernel Algorithm for Data Sets with Many Variables and Fewer Objects. Part 1: Theory and Algorithm, *Journal of Chemometrics*, 8, 111-125.

Rao, C. R. (1951). An asymptotic expansion of the distribution of Wilks' criterion. *Bulletin of the International Statistical Institute*, *33*, 177-181.

Rao, C. R. (1952). *Advanced statistical methods in biometric research*. New York: Wiley.

Rao, C. R. (1965). *Linear statistical inference and its applications*. New York: Wiley.

Rath & Strong (2000). *Rath & Strong's Six Sigma pocket guide*. Lexington, MA: Rath & Strong, Inc.

Rhoades, H. M., & Overall, J. E. (1982). A sample size correction for Pearson chi-square in 2 x 2 contingency tables. *Psychological Bulletin*, *91*, 418-423.

Rinne, H., & Mittag, H. J. (1995). *Statistische Methoden der Qualitätssicherung (3rd. edition)*. München/Wien: Hanser Verlag.

Ripley, B. D. (1981). *Spacial statistics*. New York: Wiley.

Ripley, B. D. (1996). *Pattern recognition and neural networks*. Cambridge: Cambridge University Press.

Ripley, B. D., (1996) *Pattern Recognition and Neural Networks*, Cambridge University Press

Rodriguez, R. N. (1992). Recent developments in process capability analysis. *Journal of Quality Technology*, *24*, 176-187.

Rogan, J. C., Keselman, J. J., & Mendoza, J. L. (1979). Analysis of repeated measurements. *British Journal of Mathematical and Statistical Psychology*, *32*, 269-286.

Rosenberg, S. (1977). New approaches to the analysis of personal constructs in person perception. In A. Landfield (Ed.), *Nebraska symposium on motivation* (Vol. 24). Lincoln, NE: University of Nebraska Press.

Rosenberg, S., & Sedlak, A. (1972). Structural representations of implicit personality theory. In L. Berkowitz (Ed.). *Advances in experimental social psychology* (Vol. 6). New York: Academic Press.

Rosenblatt, F. (1958). The Perceptron: A probabilistic model for information storage and organization in the brain. *Psychological Review 65*, 386-408.

Roskam, E. E., & Lingoes, J. C. (1970). MINISSA-I: A Fortran IV program for the smallest space analysis of square symmetric matrices. *Behavioral Science*, *15*, 204-205.

Ross, P. J. (1988). *Taguchi techniques for quality engineering: Loss function, orthogonal experiments, parameter, and tolerance design*. Milwaukee, Wisconsin: ASQC.

Roy, J. (1958). Step-down procedure in multivariate analysis. *Annals of Mathematical Statistics*, *29*, 1177-1187.

Roy, J. (1967). *Some aspects of multivariate analysis*. New York: Wiley.

Roy, R. (1990). *A primer on the Taguchi method*. Milwaukee, Wisconsin: ASQC.

Royston, J. P. (1982). An extension of Shapiro and Wilk's W test for normality to large samples. *Applied Statistics, 31*, 115-124.

Rozeboom, W. W. (1979). Ridge regression: Bonanza or beguilement? *Psychological Bulletin, 86,* 242-249.

Rozeboom, W. W. (1988). Factor indeterminacy: the saga continues. *British Journal of Mathematical and Statistical Psychology, 41,* 209-226.

Rubinstein, L.V., Gail, M. H., & Santner, T. J. (1981). Planning the duration of a comparative clinical trial with loss to follow-up and a period of continued observation. *Journal of Chronic Diseases, 34,* 469479.

Rumelhart, D.E. and McClelland, J. (eds.) (1986). *Parallel Distributed Processing, Vol 1.* Cambridge, MA: MIT Press.

Rumelhart, D.E., Hinton, G.E. and Williams, R.J. (1986). Learning internal representations by error propagation. In D.E. Rumelhart, J.L. McClelland (Eds.), *Parallel Distributed Processing, Vol 1.* Cambridge, MA: MIT Press.

Runyon, R. P., & Haber, A. (1976). *Fundamentals of behavioral statistics.* Reading, MA: Addison-Wesley.

Ryan, T. P. (1989). *Statistical methods for quality improvement.* New York: Wiley.

Ryan, T. P. (1997). *Modern Regression Methods.* New York: Wiley.

S

Sandler, G. H. (1963). *System reliability engineering.* Englewood Cliffs, NJ: Prentice-Hall.

Satorra, A., & Saris, W. E. (1985). Power of the likelihood ratio test in covariance structure analysis. *Psychometrika, 50,* 83-90.

Saxena, K. M. L., & Alam, K. (1982). Estimation of the noncentrality parameter of a chi squared distribution. *Annals of Statistics, 10,* 1012-1016.

Scheffé, H. (1953). A method for judging all possible contrasts in the analysis of variance. *Biometrica, 40,* 87-104.

Scheffé, H. (1959). *The analysis of variance.* New York: Wiley.

Scheffé, H. (1963). The simplex-centroid design for experiments with mixtures. *Journal of the Royal Statistical Society, B25,* 235-263.

Scheffé, H., & Tukey, J. W. (1944). A formula for sample sizes for population tolerance limits. *Annals of Mathematical Statistics, 15,* 217.

Scheines, R. (1994). Causation, indistinguishability, and regression. In F. Faulbaum, (Ed.), *SoftStat '93. Advances in statistical software 4.* Stuttgart: Gustav Fischer Verlag.

Schiffman, S. S., Reynolds, M. L., & Young, F. W. (1981). *Introduction to multidimensional scaling: Theory, methods, and applications.* New York: Academic Press.

Schmidt, F. L., & Hunter, J. E. (1997). Eight common but false objections to the discontinuation of significance testing in the analysis of research data. In Harlow, L. L., Mulaik, S. A., & Steiger, J. H. (Eds.), *What if there were no significance tests.* Mahwah, NJ: Lawrence Erlbaum Associates.

Schmidt, P., & Muller, E. N. (1978). The problem of multicollinearity in a multistage causal alienation model: A comparison of ordinary least squares, maximum-likelihood and ridge estimators. *Quality and Quantity, 12,* 267-297.

Schmidt, P., & Sickles, R. (1975). On the efficiency of the Almon lag technique. *International Economic Review, 16,* 792-795.

Schmidt, P., & Waud, R. N. (1973). The Almon lag technique and the monetary versus fiscal policy debate. *Journal of the American Statistical Association, 68,* 11-19.

Schnabel, R. B., Koontz, J. E., and Weiss, B. E. (1985). A modular system of algorithms for unconstrained minimization. *ACM Transactions on Mathematical Software, 11,* 419-440.

Schneider, H. (1986). *Truncated and censored samples from normal distributions.* New York: Marcel Dekker.

Schneider, H., & Barker, G.P. (1973). *Matrices and linear algebra* (2nd ed.). New York: Dover Publications.

Schönemann, P. H., & Steiger, J. H. (1976). Regression component analysis. *British Journal of Mathematical and Statistical Psychology, 29,* 175-189.

Schrock, E. M. (1957). *Quality control and statistical methods.* New York: Reinhold Publishing.

Schwarz, G. (1978). Estimating the dimension of a model. *Annals of Statistics, 6*, 461-464.

Scott, D. W. (1979). On optimal and data-based histograms. *Biometrika, 66*, 605-610.

Searle, S. R. (1987). *Linear models for unbalanced data.* New York: Wiley.

Searle, S. R., Casella, G., & McCullock, C. E. (1992). *Variance components.* New York: Wiley.

Searle, S., R., Speed., F., M., & Milliken, G. A. (1980). The population marginal means in the linear model: An alternative to least squares means. *The American Statistician, 34*, 216-221.

Seber, G. A. F., & Wild, C. J. (1989). *Nonlinear regression.* New York: Wiley.

Sebestyen, G. S. (1962). *Decision making processes in pattern recognition.* New York: Macmillan.

Seder, L. A. (1962). *Quality improvement.* In J. M. Juran. *Quality control handbook.* New York: McGraw-Hill.

Sen, P. K., & Puri, M. L. (1968). On a class of multivariate multisample rank order tests, II: Test for homogeneity of dispersion matrices. *Sankhya, 30*, 1-22.

Serlin, R. A., & Lapsley, D. K. (1993). Rational appraisal of psychological research and the good-enough principle. In G. Keren & C. Lewis (Eds.), *A handbook for data analysis in the behavioral sciences: Methodological issues* (pp. 199-228). Hillsdale, NJ: Lawrence Erlbaum Associates.

Serlin. R. A., & Lapsley, D. K. (1985). Rationality in psychological research: The good-enough principle. *American Psychologist, 40*, 7383.

Shapiro, A., & Browne, M. W. (1983). On the investigation of local identifiability: A counter example. *Psychometrika, 48*, 303-304.

Shapiro, S. S., Wilk, M. B., & Chen, H. J. (1968). A comparative study of various tests of normality. *Journal of the American Statistical Association, 63*, 1343-1372.

Shepherd, A. J. (1997). *Second-Order Methods for Neural Networks.* New York: Springer.

Shewhart, W. A. (1931). *Economic control of quality of manufactured product.* New York: D. Van Nostrand.

Shewhart, W. A. (1939). *Statistical method from the viewpoint of quality.* Washington, DC: The Graduate School Department of Agriculture.

Shirland, L. E. (1993). *Statistical quality control with microcomputer applications.* New York: Wiley.

Shiskin, J., Young, A. H., & Musgrave, J. C. (1967). *The X-11 variant of the census method II seasonal adjustment program.* (Technical paper no. 15). Bureau of the Census.

Shumway, R. H. (1988). *Applied statistical time series analysis.* Englewood Cliffs, NJ: Prentice Hall.

Siegel, A. E. (1956). Film-mediated fantasy aggression and strength of aggressive drive. *Child Development, 27*, 365-378.

Siegel, S. (1956). *Nonparametric statistics for the behavioral sciences.* New York: McGraw-Hill.

Siegel, S., & Castellan, N. J. (1988). *Nonparametric statistics for the behavioral sciences* (2nd ed.) New York: McGraw-Hill.

Simkin, D., & Hastie, R. (1986). Towards an information processing view of graph perception. *Proceedings of the Section on Statistical Graphics, American Statistical Association*, 11-20.

Sinha, S. K., & Kale, B. K. (1980). *Life testing and reliability estimation.* New York: Halstead.

Smirnov, N. V. (1948). Table for estimating the goodness of fit of empirical distributions. *Annals of Mathematical Statistics, 19*, 279-281.

Smith, D. J. (1972). *Reliability engineering.* New York: Barnes & Noble.

Smith, K. (1953). Distribution-free statistical methods and the concept of power efficiency. In L. Festinger and D. Katz (Eds.), *Research methods in the behavioral sciences* (pp. 536-577). New York: Dryden.

Sneath, P. H. A., & Sokal, R. R. (1973). *Numerical taxonomy.* San Francisco: W. H. Freeman & Co.

Snee, R. D. (1975). Experimental designs for quadratic models in constrained mixture spaces. *Technometrics, 17*, 149-159.

Snee, R. D. (1979). Experimental designs for mixture systems with multi-component constraints. *Communications in Statistics - Theory and Methods, A8(4)*, 303-326.

Snee, R. D. (1985). Computer-aided design of experiments - some practical experiences. *Journal of Quality Technology, 17,* 222-236.

Snee, R. D. (1986). An alternative approach to fitting models when re-expression of the response is useful. *Journal of Quality Technology, 18,* 211-225.

Sokal, R. R., & Mitchener, C. D. (1958). A statistical method for evaluating systematic relationships. *University of Kansas Science Bulletin, 38,* 1409.

Sokal, R. R., & Sneath, P. H. A. (1963). *Principles of numerical taxonomy.* San Francisco: W. H. Freeman & Co.

Soper, H. E. (1914). Tables of Poisson's exponential binomial limit. *Biometrika, 10,* 25-35.

Speckt, D.F. (1990). Probabilistic Neural Networks. *Neural Networks 3 (1),* 109-118.

Speckt, D.F. (1991). A Generalized Regression Neural Network. *IEEE Transactions on Neural Networks 2 (6),* 568-576.

Spirtes, P., Glymour, C., & Scheines, R. (1993). *Causation, prediction, and search.* Lecture Notes in Statistics, V. 81. New York: Springer-Verlag.

Spjotvoll, E., & Stoline, M. R. (1973). An extension of the *T*-method of multiple comparison to include the cases with unequal sample sizes. *Journal of the American Statistical Association, 68,* 976-978.

Springer, M. D. (1979). *The algebra of random variables.* New York: Wiley.

Spruill, M. C. (1986). Computation of the maximum likelihood estimate of a noncentrality parameter. *Journal of Multivariate Analysis, 18,* 216-224.

Steiger, J. H. (1979). Factor indeterminacy in the 1930's and in the 1970's; some interesting parallels. *Psychometrika, 44,* 157-167.

Steiger, J. H. (1980a). Tests for comparing elements of a correlation matrix. *Psychological Bulletin, 87,* 245-251.

Steiger, J. H. (1980b). Testing pattern hypotheses on correlation matrices: Alternative statistics and some empirical results. *Multivariate Behavioral Research, 15,* 335-352.

Steiger, J. H. (1988). Aspects of person-machine communication in structural modeling of correlations and covariances. *Multivariate Behavioral Research, 23,* 281-290.

Steiger, J. H. (1989). *EzPATH: A supplementary module for SYSTAT and SYGRAPH.* Evanston, IL: SYSTAT, Inc.

Steiger, J. H. (1990). Some additional thoughts on components and factors. *Multivariate Behavioral Research, 25,* 41-45.

Steiger, J. H., & Browne, M. W. (1984). The comparison of interdependent correlations between optimal linear composites. *Psychometrika, 49,* 11-24.

Steiger, J. H., & Fouladi, R. T. (1992). *R2: A computer program for interval estimation, power calculation, and hypothesis testing for the squared multiple correlation. Behavior Research Methods, Instruments, and Computers, 4,* 581582.

Steiger, J. H., & Fouladi, R. T. (1997). Noncentrality interval estimation and the evaluation of statistical models. In Harlow, L. L., Mulaik, S. A., & Steiger, J. H. (Eds.), *What if there were no significance tests.* Mahwah, NJ: Lawrence Erlbaum Associates.

Steiger, J. H., & Hakstian, A. R. (1982). The asymptotic distribution of elements of a correlation matrix: Theory and application. *British Journal of Mathematical and Statistical Psychology, 35,* 208-215.

Steiger, J. H., & Lind, J. C. (1980). Statistically-based tests for the number of common factors. Paper presented at the annual Spring Meeting of the Psychometric Society in Iowa City. May 30, 1980.

Steiger, J. H., & Schönemann, P. H. (1978). A history of factor indeterminacy. In S. Shye, (Ed.), *Theory Construction and Data Analysis in the Social Sciences.* San Francisco: Jossey-Bass.

Steiger, J. H., Shapiro, A., & Browne, M. W. (1985). On the multivariate asymptotic distribution of sequential chi-square statistics. *Psychometrika, 50,* 253-264.

Stelzl, I. (1986). Changing causal relationships without changing the fit: Some rules for generating equivalent LISREL models. *Multivariate Behavioral Research, 21,* 309-331.

Stenger, F. (1973). Integration formula based on the trapezoid formula. *Journal of the Institute of Mathematics and Applications, 12,* 103-114.

Stevens, J. (1986). *Applied multivariate statistics for the social sciences.* Hillsdale, NJ: Erlbaum.

Stevens, W. L. (1939). Distribution of groups in a sequence of alternatives. *Annals of Eugenics, 9,* 10-17.

Stewart, D. K., & Love, W. A. (1968). A general canonical correlation index. *Psychological Bulletin, 70,* 160-163.

Steyer, R. (1992). *Theorie causale regressionsmodelle* [Theory of causal regression models]. Stuttgart: Gustav Fischer Verlag.

Steyer, R. (1994). Principles of causal modeling: a summary of its mathematical foundations and practical steps. In F. Faulbaum, (Ed.), *SoftStat '93. Advances in statistical software 4.* Stuttgart: Gustav Fischer Verlag.

Stone, M. and Brooks, R. J. (1990) Continuum Regression: Cross-validated Sequentially Constructed Prediction Embracing Ordinary Least Squares, Partial Least Squares, and Principal Components Regression, *Journal of Royal Statistical Society,* 52, No. 2, 237-269.

Student (1908). The probable error of a mean. *Biometrika, 6,* 1-25.

Swallow, W. H., & Monahan, J. F. (1984). Monte Carlo comparison of ANOVA, MIVQUE, REML, and ML estimators of variance components. *Technometrics, 26,* 47-57.

T

Taguchi, G. (1987). *Jikken keikakuho* (3rd ed., Vol I & II). Tokyo: Maruzen. English translation edited by D. Clausing. *System of experimental design.* New York: UNIPUB/Kraus International

Tanaka, J. S., & Huba, G. J. (1985). A fit index for covariance structure models under arbitrary GLS estimation. *British Journal of Mathematical and Statistical Psychology, 38,* 197-201.

Tanaka, J. S., & Huba, G. J. (1989). A general coefficient of determination for covariance structure models under arbitrary GLS estimation.

British Journal of Mathematical and Statistical Psychology, 42, 233-239.

Tatsuoka, M. M. (1970). *Discriminant analysis.* Champaign, IL: Institute for Personality and Ability Testing.

Tatsuoka, M. M. (1971). *Multivariate analysis.* New York: Wiley.

Tatsuoka, M. M. (1976). Discriminant analysis. In P. M. Bentler, D. J. Lettieri, and G. A. Austin (Eds.), *Data analysis strategies and designs for substance abuse research.* Washington, DC: U.S. Government Printing Office.

Taylor, D. J., & Muller, K. E. (1995). Computing confidence bounds for power and sample size of the general linear univariate model. *The American Statistician, 49,* 4347.

Thorndike, R. L., & Hagen, E. P. (1977). *Measurement and evaluation in psychology and education.* New York: Wiley.

Thurstone, L. L. (1931). Multiple factor analysis. *Psychological Review, 38,* 406-427.

Thurstone, L. L. (1947). *Multiple factor analysis.* Chicago: University of Chicago Press.

Timm, N. H. (1975). *Multivariate analysis with applications in education and psychology.* Monterey, CA: Brooks/Cole.

Timm, N. H., & Carlson, J. (1973). *Multivariate analysis of non-orthogonal experimental designs using a multivariate full rank model.* Paper presented at the American Statistical Association Meeting, New York.

Timm, N. H., & Carlson, J. (1975). Analysis of variance through full rank models. *Multivariate behavioral research monographs,* No. 75-1.

Tracey, N. D., Young, J., C., & Mason, R. L. (1992). Multivariate control charts for individual observations. *Journal of Quality Technology, 2,* 88-95.

Tribus, M., & Sconyi, G. (1989). An alternative view of the Taguchi approach. *Quality Progress, 22,* 46-48.

Trivedi, P. K., & Pagan, A. R. (1979). Polynomial distributed lags: A unified treatment. *Economic Studies Quarterly, 30,* 37-49.

Tryon, R. C. (1939). *Cluster Analysis*. Ann Arbor, MI: Edwards Brothers.

Tucker, L. R., Koopman, R. F., & Linn, R. L. (1969). Evaluation of factor analytic research procedures by means of simulated correlation matrices. *Psychometrika, 34*, 421-459.

Tufte, E. R. (1983). *The visual display of quantitative information*. Cheshire, CT: Graphics Press.

Tufte, E. R. (1990). *Envisioning information*. Cheshire, CT: Graphics Press.

Tukey, J. W. (1953). *The problem of multiple comparisons*. Unpublished manuscript, Princeton University.

Tukey, J. W. (1962). The future of data analysis. *Annals of Mathematical Statistics, 33*, 1-67.

Tukey, J. W. (1967). An introduction to the calculations of numerical spectrum analysis. In B. Harris (Ed.), *Spectral analysis of time series*. New York: Wiley.

Tukey, J. W. (1972). Some graphic and semigraphic displays. In *Statistical Papers in Honor of George W. Snedecor*; ed. T. A. Bancroft, Arnes, IA: Iowa State University Press, 293-316.

Tukey, J. W. (1977). *Exploratory data analysis*. Reading, MA: Addison-Wesley.

Tukey, J. W. (1984). *The collected works of John W. Tukey*. Monterey, CA: Wadsworth.

Tukey, P. A. (1986). A data analyst's view of statistical plots. *Proceedings of the Section on Statistical Graphics, American Statistical Association*, 21-28.

Tukey, P. A., & Tukey, J. W. (1981). Graphical display of data sets in 3 or more dimensions. In V. Barnett (Ed.), *Interpreting multivariate data*. Chichester, U.K.: Wiley.

U

Upsensky, J. V. (1937). *Introduction to Mathematical Probability*. New York: McGraw-Hill.

V

Vale, C. D., & Maurelli, V. A. (1983). Simulating multivariate non-normal distributions. *Psychometrika, 48*, 465-471.

Vandaele, W. (1983). *Applied time series and Box-Jenkins models*. New York: Academic Press.

Vaughn, R. C. (1974). *Quality control*. Ames, IA: Iowa State Press.

Velicer, W. F., & Jackson, D. N. (1990). Component analysis vs. factor analysis: some issues in selecting an appropriate procedure. *Multivariate Behavioral Research, 25*, 1-28.

Velleman, P. F., & Hoaglin, D. C. (1981). *Applications, basics, and computing of exploratory data analysis*. Belmont, CA: Duxbury Press.

Von Mises, R. (1941). Grundlagen der Wahrscheinlichkeitsrechnung. *Mathematische Zeitschrift, 5*, 52-99.

W

Wainer, H. (1995). Visual revelations. *Chance, 8*, 48-54.

Wald, A. (1939). Contributions to the theory of statistical estimation and testing hypotheses. *Annals of Mathematical Statistics, 10*, 299-326.

Wald, A. (1945). Sequential tests of statistical hypotheses. *Annals of Mathematical Statistics, 16*, 117-186.

Wald, A. (1947). *Sequential analysis*. New York: Wiley.

Walker, J. S. (1991). *Fast Fourier transforms*. Boca Raton, FL: CRC Press.

Wallis, K. F. (1974). Seasonal adjustment and relations between variables. *Journal of the American Statistical Association, 69*, 18-31.

Wang, C. M., & Gugel, H. W. (1986). High-performance graphics for exploring multivariate data. *Proceedings of the Section on Statistical Graphics, American Statistical Association*, 60-65.

Ward, J. H. (1963). Hierarchical grouping to optimize an objective function. *Journal of the American Statistical Association, 58*, 236.

Warner B. & Misra, M. (1996). Understanding Neural Networks as Statistical Tools. *The American Statistician, 50*, 284-293.

Weatherburn, C. E. (1946). *A First Course in Mathematical Statistics*. Cambridge: Cambridge University Press.

Wei, W. W. (1989). *Time series analysis: Univariate and multivariate methods*. New York: Addison-Wesley.

Weibull, W. (1951). A statistical distribution function of wide applicability. *Journal of Applied Mechanics, September*.

Weibull, W., (1939). A statistical theory of the strength of materials. *Ing. Velenskaps Akad. Handl., 151*, 1-45.

Weigend, A.S., Rumelhart, D.E. and Huberman, B.A. (1991). Generalization by weight-elimination with application to forecasting. In R.P. Lippmann, J.E. Moody and D.S. Touretzky (Eds.) *Advances in Neural Information Processing Systems 3*, 875-882. San Mateo, CA: Morgan Kaufmann.

Welch, B. L. (1938). The significance of the differences between two means when the population variances are unequal. *Biometrika, 29*, 350-362.

Welstead, S. T. (1994). *Neural network and fuzzy logic applications in C/C++*. New York: Wiley.

Werbos, P.J. (1974). *Beyond regression: new tools for prediction and analysis in the behavioural sciences*. Ph.D. thesis, Harvard University, Boston, MA.

Wescott, M. E. (1947). Attribute charts in quality control. *Conference Papers, First Annual Convention of the American Society for Quality Control*. Chicago: John S. Swift Co.

Wheaton, B., Múthen, B., Alwin, D., & Summers G. (1977). Assessing reliability and stability in panel models. In D. R. Heise (Ed.), *Sociological Methodology*. New York: Wiley.

Wheeler, D. J., & Chambers, D.S. (1986). *Understanding statistical process control*. Knoxville, TN: Statistical Process Controls, Inc.

Wherry, R. J. (1984). *Contributions to correlational analysis*. New York: Academic Press.

Whitney, D. R. (1948). *A comparison of the power of non-parametric tests and tests based on the normal distribution under non-normal alternatives*. Unpublished doctoral dissertation, Ohio State University.

Whitney, D. R. (1951). A bivariate extension of the U statistic. *Annals of Mathematical Statistics, 22*, 274-282.

Widrow, B., and Hoff Jr., M.E. (1960). Adaptive switching circuits. *IRE WESCON Convention Record*, 96-104.

Wiggins, J. S., Steiger, J. H., and Gaelick, L. (1981). Evaluating circumplexity in models of personality. *Multivariate Behavioral Research, 16*, 263-289.

Wilcoxon, F. (1945). Individual comparisons by ranking methods. *Biometrica Bulletin, 1*, 80-83.

Wilcoxon, F. (1947). Probability tables for individual comparisons by ranking methods. *Biometrics, 3*, 119-122.

Wilcoxon, F. (1949). *Some rapid approximate statistical procedures*. Stamford, CT: American Cyanamid Co.

Wilde, D. J., & Beightler, C. S. (1967). *Foundations of optimization*. Englewood Cliffs, NJ: Prentice-Hall.

Wilks, S. S. (1943). *Mathematical Statistics*. Princeton, NJ: Princeton University Press.

Wilks, S. S. (1946). *Mathematical statistics*. Princeton, NJ: Princeton University Press.

Williams, W. T., Lance, G. N., Dale, M. B., & Clifford, H. T. (1971). Controversy concerning the criteria for taxonometric strategies. *Computer Journal, 14*, 162.

Wilson, G. A., & Martin, S. A. (1983). An empirical comparison of two methods of testing the significance of a correlation matrix. *Educational and Psychological Measurement, 43*, 11-14.

Winer, B. J. (1962). *Statistical principles in experimental design*. New York: McGraw-Hill.

Winer, B. J. (1971). *Statistical principles in experimental design* (2nd ed.). New York: McGraw Hill.

Winer, B. J., Brown, D. R., Michels, K. M. (1991). *Statistical principals in experimental design. (3rd ed.)*. New York: McGraw-Hill.

Wolfowitz, J. (1942). Additive partition functions and a class of statistical hypotheses. *Annals of Mathematical Statistics, 13,* 247-279.

Wolynetz, M. S. (1979a). Maximum likelihood estimation from confined and censored normal data. *Applied Statistics, 28,* 185-195.

Wolynetz, M. S. (1979b). Maximum likelihood estimation in a linear model from confined and censored normal data. *Applied Statistics, 28,* 195-206.

Wonnacott, R. J., & Wonnacot, T. H. (1970). *Econometrics.* New York: Wiley.

Woodward, J. A., & Overall, J. E. (1975). Multivariate analysis of variance by multiple regression methods. *Psychological Bulletin, 82,* 21-32.

Woodward, J. A., & Overall, J. E. (1976). Calculation of power of the *F* test. *Educational and Psychological Measurement, 36,* 165-168.

Woodward, J. A., Bonett, D. G., & Brecht, M. L. (1990). *Introduction to linear models and experimental design.* New York: Harcourt, Brace, Jovanovich.

Woodward, J. A., Douglas, G. B., & Brecht, M. L. (1990). *Introduction to linear models and experimental design.* New York: Academic Press.

Y

Yates, F. (1933). The principles of orthogonality and confounding in replicated experiments. *Journal of Agricultural Science, 23,* 108-145.

Yates, F. (1937). *The Design and Analysis of Factorial Experiments.* Imperial Bureau of Soil Science, Technical Communication No. 35, Harpenden.

Yokoyama, Y., & Taguchi, G. (1975). *Business data analysis: Experimental regression analysis.* Tokyo: Maruzen.

Youden, W. J., & Zimmerman, P. W. (1936). Field trials with fiber pots. *Contributions from Boyce Thompson Institute, 8,* 317-331.

Young, F. W, & Hamer, R. M. (1987). *Multidimensional scaling: History, theory, and applications.* Hillsdale, NJ: Erlbaum

Young, F. W., Kent, D. P., & Kuhfeld, W. F. (1986). Visuals: Software for dynamic hyper-dimensional graphics. *Proceedings of the Section on Statistical Graphics, American Statistical Association,* 69-74.

Younger, M. S. (1985). *A first course in linear regression* (2nd ed.). Boston: Duxbury Press.

Yuen, C. K., & Fraser, D. (1979). *Digital spectral analysis.* Melbourne: CSIRO/Pitman.

Yule, G. U. (1897). On the theory of correlation. *Journal of the Royal Statistical Society, 60,* 812-854.

Yule, G. U. (1907). On the theory of correlation for any number of variables treated by a new system of notation. *Proceedings of the Royal Society,* Ser. A, *79,* 182-193.

Yule, G. U. (1911). *An Introduction to the Theory of Statistics.* London: Griffin.

Z

Zippin, C., & Armitage, P. (1966). Use of concomitant variables and incomplete survival information in the estimation of an exponential survival parameter. *Biometrics, 22,* 665-672.

Zupan, J. (1982). *Clustering of large data sets.* New York: Research Studies Press.

Zwick, W. R., & Velicer, W. F. (1986). Comparison of five rules for determining the number of components to retain. *Psychological Bulletin, 99,* 432-442.

B
APPENDIX

STATISTICA FAMILY
OF PRODUCTS

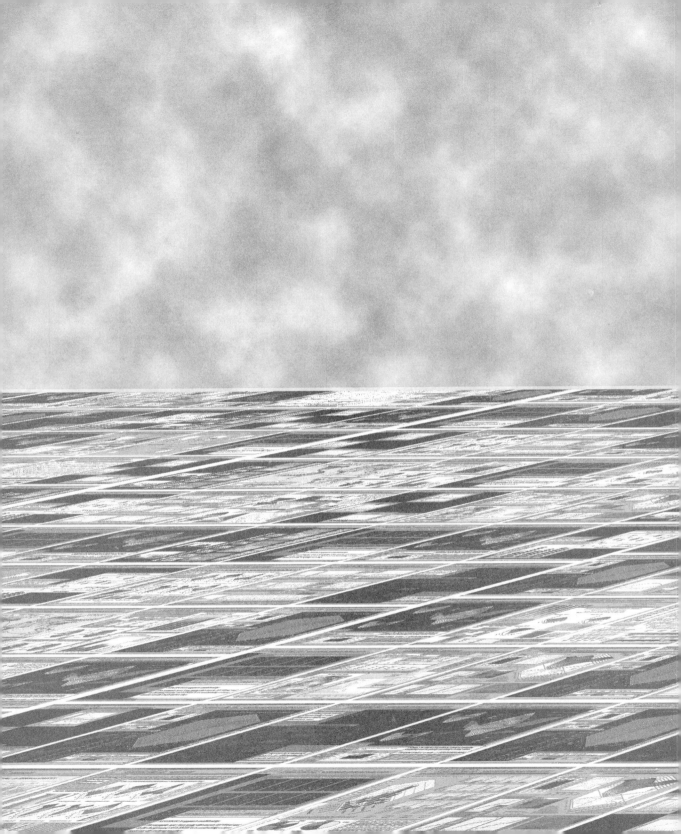

STATISTICA FAMILY OF PRODUCTS

Common System Features. All *STATISTICA* products offer fully customizable user interfaces (with simplified shortcut templates for novices), flexible output management, presentation-quality reporting, full OLE and ActiveX support, and Web enablement. Also, all products include data management optimized to handle large datasets, interactive database query tools, and a wide set of import and export facilities. All products handle datasets of unlimited size (with "quadruple" precision calculations), multiple input files, and multitasking. A broad selection of interactive visualization and graphics and drawing tools of the highest quality is fully integrated into each product, and each includes a complete set of automation options and a professional Visual Basic development environment.

 STATISTICA **Base (a stand-alone product).** Offers a comprehensive set of essential statistics in a user-friendly package and offers all the performance, power, and ease of use of the *STATISTICA* technology (see above).

- All *STATISTICA* Graphics Tools
- Basic Statistics & Tables
- Distribution Fitting
- Multiple Linear Regression
- Analysis of Variance
- Nonparametrics, and more.

 STATISTICA **Advanced Linear/Non-Linear Models (add-on).** Offers a wide array of the most advanced modeling and forecasting tools on the market, including automatic model selection facilities and extensive interactive visualization tools.

- General Linear Models
- Generalized Linear/Nonlinear Models
- General Regression Models
- General Partial Least Squares Models
- Variance Components
- Survival Analysis
- Nonlinear Estimation
- Fixed Nonlinear Regression
- Log-linear Analysis of Frequency Tables
- Time Series/Forecasting
- Structural Equational Modeling, and more.

 ***STATISTICA* Multivariate Exploratory Techniques (add-on).** Offers a broad selection of exploratory techniques for various types of data, with extensive, interactive visualization tools.

- Cluster Analysis
- Factor Analysis
- Principal Components & Classification Analysis
- Canonical Analysis
- Reliability/Item Analysis
- Classification Trees
- Correspondence Analysis
- Multidimensional Scaling
- Discriminant Analysis
- General Discriminant Analysis Models, and more.

 ***STATISTICA* Neural Networks (stand-alone or add-on).** Contains the most comprehensive selection of neural network methods with intelligent problem solvers and automatic wizards; a C-code generator add-on available.

- Intelligent Problem Solver Wizard
- Automatic Search for Best Architecture

- Multilayer Perceptrons
- Radial Basis Function Networks, and many others
- Self-Organizing Feature Map
- Back Propagation
- Conjugate Gradient Descent
- Numerous Analytical Graphs
- Resampling (Cross Validation, Bootstrap)
- Sensitivity Analysis, ROC curves
- Network Ensembles
- API interface, and more.

 STATISTICA Power Analysis (add-on). An extremely precise and user-friendly, specialized tool for analyzing all aspects of statistical power and sample size calculation.

- Sample size calculation
- Confidence Interval estimation
- Statistical distribution calculators, and more.

Industrial Solutions, Six Sigma Tools

 STATISTICA Quality Control Charts (stand-alone or add-on). Offers fully customizable (e.g., callable from other environments), easy and quick to use, versatile charts with a selection of automation options, and user-interface shortcuts to simplify routine work (a comprehensive tool for Six Sigma methods).

- Multiple Chart (Six Sigma Style) Reports and Displays
- X-bar and/or R Charts; S^2, Np, P, U, C Charts
- Pareto Charts
- Process Capability and Performance Indices
- Moving Average/Range Charts, EWMA Charts
- Nominal Charts
- Target Charts
- Short Run Charts

- CuSum (Cumulative Sum)
- Runs Tests
- Multiple Process Streams, and more.

 ***STATISTICA* Process Analysis (add-on).** A comprehensive package for process capability, Gage R&R, and other quality control/improvement applications (a comprehensive tool for Six Sigma methods).
- Process/Capability Analysis Charts
- Ishikawa (Cause and Effect) Diagrams
- Gage Repeatability & Reproducibility
- Variance Components for Random Effects
- Weibull Analysis
- Sampling Plans, and more.

 ***STATISTICA* Design of Experiments (add-on).** Features the largest selection of DOE and related visualization techniques including interactive desirability profilers (a comprehensive tool for Six Sigma methods).
- Fractional Factorial Designs
- Mixture Designs
- Latin Squares
- Search for Optimal 2^{k-p} Designs
- Residual Analysis & Transformations
- Optimization of Single/Multiple Response Variables
- Central Composite Designs
- Taguchi Designs
- Minimum Aberration & Maximum Unconfounding
- 2^{k-p} Fractional Factorial Designs with Blocks
- Constrained Surfaces
- D- and A-Optimal Designs
- Desirability Profilers, and more

STATISTICA Enterprise Systems

In addition to the common features listed above, *STATISTICA* Enterprise Systems offer a wide selection of tools for collaborative work, Web browser-based user interface (using the optional *STATISTICA Web Server*), specialized databases, and a highly optimized interface to enterprise-wide data repositories, including options to rapidly process large datasets from remote servers in-place, without creating local copies.

 STATISTICA Enterprise-Wide Data Mining System (Data Miner). The most comprehensive selection of data mining solutions on the market, with an icon-based, extremely easy to use user interface. Features a selection of completely integrated and automated ready to deploy "as is" (but also easily customizable) systems of specific data mining solutions for a wide variety of business applications. Offered optionally with a choice of deployment and on-site training services. The data mining solutions are driven by powerful procedures from five modules that can also be used interactively and/or used to build, test, and deploy new solutions:

- General Slicer/Dicer Explorer with OLAP
- General Classifier
- General Modeler/Multivariate Explorer
- General Forecaster
- General Neural Networks Explorer

 STATISTICA Enterprise-Wide Data Analysis System (SEDAS). An integrated multi-user software system designed for general-purpose data analysis and business intelligence applications in research, marketing, finance, and other industries. *SEDAS* can optionally offer the statistical functionality available in any or all of the *STATISTICA* products (listed above). In addition, it features:

- Integration with data warehouses
- Intuitive query and filtering tools
- Easy-to use administration tools
- Automatic report distribution
- Alarm notification, and more.

 ***STATISTICA* Enterprise-Wide SPC System (SEWSS).** Based on state-of-the-art connectivity technologies, *SEWSS* is designed for local and global enterprise quality control and improvement applications, including Six Sigma; it offers real-time monitoring and alarm notification for the production floor, a comprehensive set of analytical tools for engineers, sophisticated reporting features for management, Six Sigma reporting options. It also features:

- Web-enabled user interface and reporting tools; interactive querying tools
- User-specific interfaces for operators, engineers, managers, analysts, etc.
- Groupware functionality for sharing queries, special applications, etc.
- Open-ended alarm notification including cause/action prompts
- Scalable, customizable, integratable into existing database/ERP systems, and more.

modifying
  area properties, 798
  line properties, 799
  point markers, 798
modules
  accessing in SVB, 977
mouse conventions, 122
  decrease column width, 131
  drag-and-drop, 131
  graph applications, 786
  increase column width, 131
  left mouse button, 122
  microscrolls, 132
  multiple selections, 130
  other operations, 131
  reordering items in list, 133
  right mouse button, 123
  selecting items in workbook, 131
  split scrolling, 132
  toolbar configuration, 133
  variable speed scrolling, 132
movable
  graph title, 814
  legend, 809
move back command (format
  menu), 492
move command (data menu), 297,
  299
move commands (edit menu), 257
  cases, 257
  variables, 257
move forward command (format
  menu), 492
move to back button (graph tools
  toolbar), 566
move to back command (format
  menu), 492
move to front button (graph tools
  toolbar), 565
move to front command (format
  menu), 492
moving between implied dialogs, in
  SVB programs, 976
Multidimensional Scaling, 84
multimedia tables, 9
multiple analysis support, 37
multiple graph autolayout wizard,
  838, 841

multiple graph layouts
  templates, 608
  wizard, 606
multiple graph layouts commands
  (graphs menu), 535
  blank graph, 535
  templates, 535
  wizard, 535
multiple graphs, 827, 838, 841
  points, identifying, 827
  via embedding, 838
  via linking, 838
  via pasting, 837
multiple input spreadsheets, 324
multiple queries, 885
Multiple Regression, 68
multiple selections, 130
multiple subsets, 603
multitasking, 117
multivariate exploratory techniques
  buttons (statistics toolbar), 114
multivariate exploratory techniques
  command (statistics menu), 79
multivariate exploratory techniques
  commands (statistics menu)
  canonical analysis, 81
  classification trees, 82
  cluster analysis, 79
  correspondence analysis, 83
  discriminant analysis, 84
  factor analysis, 80
  general discriminant analysis
    models, 85
  general multidimensional scaling,
    84
  principal components and
    classification analysis, 81
  reliability/item analysis, 82

N

navigation tree (reports), 402
negative exponential
  2D fits, 598
  3D surfaces, 599
new command (file menu), 179, 241,
  408, 469, 866, 901

new file button (standard toolbar),
  101
new from selection command
  (format menu), 274
new mark button (tools toolbar), 323
new query button (query toolbar),
  862
new report command (file menu),
  246, 412, 474, 906
new selection conditions include
  button (tools toolbar), 323
new workbook command (file
  menu), 246, 411, 473, 905
non-fatal errors during SVB
  execution, 973
nonlinear estimation, 756
Nonlinear Estimation, 76
non-monotonous relations, 676, 756
non-outlier range, in 2D box plot,
  690
nonparametric correlations, 676, 756
Nonparametric Statistics, 70
nonparametric tests command
  (statistics menu), 91
non-transparent overlay of graphic
  objects, 842
normal command (format menu),
  274
normal ellipse, 631
normal fit - histogram (graphs of
  input data), 529
normal probability plot: block
  columns (graphs of block data),
  526
normal probability - probability plot
  (graphs of input data), 530
normal probability - probability plot
  by (graphs of input data), 533
normal probability plot, in graphs
  menu 2D graphs, 679
normal probability plots, 586
  2D graphs, 501
  categorized graphs, 520
notes and comments in workbooks,
  138